Biology of Industrial Microorganisms

BIOTECHNOLOGY

JULIAN E. DAVIES, *Editor*
Biogen, S.A.
Geneva, Switzerland

BIOTECHNOLOGY SERIES

Other volumes in preparation

Biology of Industrial Microorganisms

Edited by

Arnold L. Demain
Massachusetts Institute of Technology
Cambridge, Massachusetts

Nadine A. Solomon
Massachusetts Institute of Technology
Cambridge, Massachusetts

1985

THE BENJAMIN/CUMMINGS PUBLISHING COMPANY, INC.
Advanced Book Program
Menlo Park, California

LONDON • AMSTERDAM • DON MILLS, ONTARIO • SYDNEY • TOKYO

7269 - 1918

CHEMISTRY

Library of Congress Cataloging in Publication Data

Main entry under title:

Biology of industrial microorganisms.

    Bibliography: p.
    Includes index.
    1. Industrial microbiology. I. Demain, A. L.
(Arnold L.), 1927–   . II. Solomon, N. A. (Nadine A),
1950–
QR53.B53 1985   660′.62   84-21621
ISBN 0-8053-2451-8

abcdefghij-MA-8987654

CONTRIBUTORS

Arthur L. Koch
Dept. of Microbiology
Indiana University
Bloomington, Indiana 47405

Moselio Schaecter
Dept. of Molecular Biology
School of Medicine
Tufts University
Boston, Massachusetts 02111

Norberto J. Palleroni
Hoffmann La Roche
Microbiology Dept.
Nutley, New Jersey 07110

Lee A. Bulla
Dean of Agriculture's Office
University of Wyoming
Laramie, Wyoming 82070

James A. Hoch
Division of Cell Biology
Scripps Institute for Biomedical Research
La Jolla, California 92037

J. Gregory Zeikus
Michigan Biotechnology Institute
Michigan State University
East Lansing, Michigan 48824

Shukuo Kinoshita
Kyowa Hakko Kogyo Co., Ltd.
Ohtemachi, Chiyoda-ku
Tokyo, Japan

Albert G. Moat
Marshall University School of Medicine
Huntington, West Virginia 25701

Meyer J. Wolin
Division of Laboratories and Research
NY State Department of Health
Empire State Plaza
Albany, New York 12201

Terry L. Miller
Division of Laboratories and Research
NY State Department of Health
Empire State Plaza
Albany, New York 12201

Israel Goldberg
The Hebrew University-Hadassah Medical School
Jerusalem, Israel

Bland S. Montenecourt
Dept. of Biology
Lehigh University
Bethelem, PA 18015

Charles F. Hirsch
Merck, Sharp, & Dobne Research Laboratories
Rahway, New Jersey 07065

Pamela A. McCann-McCormick
Merck, Sharp, & Dobne Research Laboratories
Rahway, New Jersey 07065

Mary P. Lechevalier
Walksmann Institute of Microbiology
Rutgers University
New Brunswick, New Jersey 08903

Hubert Lechevalier
Walksmann Institute of Microbiology
Rutgers University
New Brunswick, New Jersey 08903

Graham Stewart
Prod-Res. Department
Labatt Brewing Co., Ltd.
London, Ontario
Canada N6A 4M3

Inge Russell
Prod-Res. Department
Labatt Brewing Co., Ltd.
London, Ontario
Canada N6A 4M3

Herman J. Phaff
Dept. of Food Science and Technology
University of Calif., Davis
Davis, CA 95616

Joan W. Bennett
Dept. of Biology
Tulane University
New Orleans, LA 70118

John F. Peberdy
Fungal Genetics and Biochemistry Lab
Department of Botany
The University of Nottingham/School of Biological Sciences
University Park
Nottingham NG7 2RD
United Kingdom

Douglas Eveleigh
Dept. of Microbiology & Biochemistry
Cook College
Rutgers University
New Brunswick, New Jersey 08903

Claude H. Nash III
Schering-Plough
Bloomfield, New Jersey 07003

Rajanikant J. Mehta
Smith Kline & French Laboratories
Philadelphia, Pennsylvania 19101

Christopher Ball
Panlabs Genetics, Inc.
Bellevue, Washington 98004

Willard A. Taber
Dept. of Biology
Texas A & M University
College Station, Texas 77843

CONTENTS

PREFACE

Both the newer literature on biotechnology and the older literature on industrial microbiology are filled with descriptions of important commercial processes. Thus one can easily obtain media recipes and descriptions of conditions for fermentations conducted in vessels ranging from 50 ml Erlenmeyer flasks to 100,000 gallon agitated fermentors. However, what do you do when you want to know something about the organism itself — for example, its nutrition, physiology, ecology, taxonomy, or genetics? Such information, if it exists at all, is scattered throughout the literature, often in obscure journals. How can one rationally develop a commercial process without having some appreciation for the biology of the organism involved in the process? This volume is an attempt to solve this problem by assembling, for the first time, information on the biology of industrial microorganisms. We have attempted to cover a large assemblage of industrially important genera of unicellular bacteria, actinomycetes, yeasts, and molds.

We are indebted to our authors, each of whom is an expert on a particular group of organisms. Though they are extremely busy individuals, they agreed to the task because they realized its importance and necessity.

Unavoidably, the chapters are somewhat uneven in treatment. Our authors could review only what was already published about each group, and the basic literature is very spotty; in some genera, genetics is heavily stressed but nutrition is not, whereas the opposite is true in other genera. Indeed, the detection of such unevenness might have the beneficial effect of stimulating researchers to "fill in the gaps" by working in areas ignored in the past.

Some of our readers may be surprised when they see that we include *Escherichia coli* as an industrial microorganism and, indeed, devote the first chapter to this organism. We did this in recognition of the impact that recombinant DNA technology has had on our field in the last decade. Without the information provided by years of basic research on this bacterium, the current revolution in applied biology would not have taken place. Our main wish is that this book will be useful for all investigators, both young and old, who suddenly find themselves working on a new biotechnological project involving a microorganism they have never encountered before.

Cambridge, Massachusetts   **Arnold L. Demain**
                                                         Nadine A. Solomon

The World and Ways of
E. coli

Arthur L. Koch
Moselio Schaechter

I. PROLOGUE

Until recently, *Escherichia coli* was not an organism of industrial importance. This has changed with the advent of recombinant DNA technology, which has developed around *E. coli* and its associated viruses and plasmids because much of the basic work had been done with them. *E. coli* is the best studied and exploited of cellular organisms. This prior knowledge of the physiology and genetic organization of *E. coli* led to the development of genetic manipulative techniques. It still is unmatched for gene cloning. Foreign genes can be introduced into its genome or that of its plasmids and viruses with relative ease and predictability.

The other side of the coin is that *E. coli* does not do all the things that a genetic engineer may want. It does not always express foreign genes at industrially useful levels and does not secrete proteins into the medium in significant amounts. Still, it occupies a central role in this sort of work even if it may not necessarily find itself in a large-scale fermentor. It will continue to be an intermediate host for genes to be exploited in eukaryotes such as yeast.

In this chapter, we consider aspects of *E. coli* biology, physiology, molecular biology, and genetics. We shall try to combine them with evolutionary and ecological generalizations to piece together a picture of this organism's role in the world. We shall use teleology to try to guess why individual mechanisms are at work but shall depend on experiments to test the quality of our guesses. Many of the guesses and speculations depend on studies with strain K-12 but probably apply generally to enterics and to other gram-negative rods.

II. WHY *ESCHERICHIA COLI?*

After Robert Koch developed methods to culture pure clones of bacteria, many organisms were quickly isolated. Among these was *Bacillus coli*. First isolated by a pediatrician, Escherich, in 1884 from the feces of a cholera patient, it was subsequently found to be a normal inhabitant of the intestinal tract of humans and many vertebrates. It occurs almost universally in their excreta but is not normally found in soil and water. On this account, it became of interest to sanitarians, since its presence in water supplies could be regarded as direct evidence of sewage pollution (Pipes, 1982). It is ubiquitous in sewage and feces; therefore it is a better measuring stick than any particular organism that might cause disease.

Questions soon arose concerning how to define it, how to count it, and how to distinguish it from other organisms. *Escherichia coli*, as it has come to be known, is a short, usually motile rod with a diameter less than one micron, incapable of making spores. It is gram-negative and not putrefactive. It ferments glucose and lactose with the formation of gas containing approximately one mole of hydrogen to one mole of carbon dioxide. It is oxidase-negative and capable of growing aerobically in the presence of bile salts, detergents, and dyes. A number of other criteria distinguish it from various other related species. Today, for ecological purposes, they are usually called "coliform" or "fecal coliform" organisms, in an attempt to lump together many related organisms that possess a similar niche in the intestine. These include such genera as *Escherichia*, *Citrobacter*, and *Klebsiella* (*Enterobacter*). For microbiologists examining water supplies, there are standardized procedures that differ from country to country but depend on the characteristics mentioned above (see Cooke, 1974; Rand et al., 1976).

Over the years, organisms identified as *E. coli* have been subdivided on the basis of differences in biochemical details of energy production by fermentation, hemolysis, and other metabolic characters (see Edwards and Ewing, 1972; Ørskov, 1981). Beyond this, subclassification has been by (1) serological methods, (2) sensitivity to phages, and (3) sensitivity to colicins. This has led to the conclusion that natural isolates of *E. coli* can differ a great deal, more in fact than separate species within other groups of bacteria or higher organisms.

Isozyme studies show abundant protein polymorphism. This should not distract us from the major idea that there is a collection of fairly similar microorganisms living in a niche provided by the intestine of animals for which the rubric "fecal coliform" is possibly the best.

In order to colonize a sterile newborn mammal or adult mammal, *E. coli* must be able to survive at least temporarily in the external environment. It does not remain long in bodies of water or soils. Only when there is fecal contamination can *E. coli* be found in large number: water and soil are not its long-range habitats; it is just passing through. On the other hand, *E. coli* is well suited for growth and survival in the relatively rich environment of the mammalian gut. Of course, in the gut it is in direct competition with a large number of other organisms and must adapt to intermittent nutritional deprivation. It uses resources available in the intestine very effectively. Thus its uptake systems for small peptides and sugars like glucose and lactose are very avid; that is, the K_m for uptake is very small. Living on "hand-me-downs" and "scraps," as it were, it cannot afford an absolute nutritional requirement for sophisticated biochemicals such as vitamins, amino acids, and so on. It can use many low-molecular-weight compounds, but it can also do without them. The basis for this concept is that most strains can be grown with a few simple ingredients: a source of phosphate, nitrogen, sulfur, magnesium, a few trace metals, and just one of many simple compounds for use as a carbon and energy source. Unlike many other prokaryotes and most eukaryotes, *E. coli* generally requires no complex preformed organic molecules. Occasionally, however, strains are found that need vitamin B_l or other growth factors.

This simple nutrition makes it very easy to isolate mutants, to conduct radioisotope-labeling experiments, and to grow the organisms at defined and reproducible rates. Therefore we can see why *E. coli* was picked as the premier organism in the development of microbial physiology and molecular biology. In retrospect, we can understand why Delbrück adopted *E. coli* and its T bacteriophages as the system to study "important" life processes. Particularly fortunate was the choice of strain K-12 by Lederberg and Tatum to start their studies of genetic processes in bacteria. Strain K-12 had originally been isolated from a patient but had been carried in Clifton's laboratory at Stanford for many years. It still contained an F plasmid, which made history in our understanding of sexual systems in bacteria. We now know that plasmids, such as the original F factor, are cellular parasites or symbionts, capable of being transmitted to other microorganisms of a similar type. A great leap forward by Cavalli and Hayes was the production of Hfr strains, capable of transmitting their chromosome to recipient F⁻ cells. This led to the concept of episomes, which states that plasmids and certain lysogenic viruses can exist independently or be integrated in the host chromosome (see Judson, 1979, for an historical treatment).

With these genetic tools in hand and the discovery of plasmids bearing antibiotic resistance factors, molecular biologists were off and running, making families of derivative organisms and producing genetic vehicles that allow DNA to be inserted and removed from microorganisms at will. Seen from *E. coli*'s point of view, the consequences are that it has evolved and diversified much more in the last 30 years than it did in the previous two hundred million years since the first mammalian hosts arose. (For the range of this diversification, see any molecular genetics text; e.g., Lewin, 1983.)

At the time (middle Jurassic) when mammals became abundant on this planet, a niche was created in the gut ecosystem that allowed some organisms to evolve and to become today's *E. coli*. The important evolutionary point is that living as a commensal in the intestine of a very early mammal probably posed problems similar to those of living in the intestine of a mammal today (Koch, 1971). There would have been the same problems of overpopulation and of short supply or temporary absence of particular nutrients. Thus these organisms are subject to chronic starvation as well as to temporarily favorable conditions when an abundance of nutrients is present. Under these circumstances, the best competitor is what ecologists call an r-strategist, the kind that grows as fast as possible during feast. However, it must be able to change its style of living quickly at the end of brief periods of plenty and be able to compete for scarce resources. It must become what is called a K-strategist. Therefore it is an organism that must play each strategy to the hilt (Koch, 1979).

For a dozen or so compounds, we know of elaborate genetic regulatory systems that permit the organism to adapt to fluctuations of their availability in a most efficient, even prescient way (Savageau, 1976; Lewin, 1983, pp. 219–235). The study of these control processes serves as the basis for much of the speculations presented here. At this point, we should caution the reader that current science has explored only a small fraction of the physiological and genetic mechanisms of the *E. coli* repertoire. The organism must have many extremely sophisticated control systems to accommodate to changes in its environment.

III. HOW GROWTH KINETICS CAN BE USED TO STUDY CELLULAR PROCESSES

Because life begets more life of the same kind, the kinetics of growth has positive feedback and is autocatalytic; that is, the rate at which new life is produced is proportional to the amount that there is. This can be expressed diagrammatically:

$$\curvearrowleft N \curvearrowright \tag{1}$$

or mathematically:

$$dN/dt = \lambda N \tag{2}$$

In this equation, λ is the specific growth rate, N is the number of bacteria, and t is time. This differential equation can be solved, yielding

$$N = N_0 e^{\lambda t} \tag{3}$$

or

$$N = N_0 2^{t/T_2} \quad \text{and} \quad T_2 = \ln(2)/\lambda \tag{4}$$

where N_0 is the number at the time taken as zero time and T_2 is the doubling time. Equations (3) and (4) apply when growth is in a steady state, that is, when all environmental conditions are such that the growth rate, λ, does not change with time. Under these conditions, growth takes place in an environment where the limiting factor is the living organism itself. These equations apply when neither the level of nutrients nor the level of toxic compounds in the environment changes appreciably while growth is taking place.

Hinshelwood (1946) noted that there are other kinetic systems that also behave autocatalytically in that the growth data fit Eq. (3). In addition to the simplest case, cyclic series of reactions can follow as in this scheme:

$$A \rightarrow B \rightarrow C \rightarrow D \rightarrow \ \rightarrow X \tag{5}$$

In such a system, any or all combinations of components would increase exponentially as in Eq. (3).

The state of affairs for a growing bacterium is a complex interweaving of many series of cyclical reactions—all essential for the production of new bacteria. Details of how these many cycles interact with each other are given in metabolic charts and in microbiology, biochemistry, and bacterial physiology texts (see, for example, Mandelstam et al., 1983). Yet no matter how complicated the interconnections, no matter how complex the kinetics of each individual reaction or the production of the enzyme needed for each step, it can be shown mathematically that a steady state is eventually established (Hinshelwood, 1952). Before the steady state of growth is achieved, the kinetics of separate cellular kinetics differ from one another and are very complicated indeed. Afterward, growth will be represented by Eq. (3) as the simple autocatalytic system of Scheme (1). This is the reason why the steady state condition, also known as "balanced growth" has become so important in microbial physiology. Under this condition, the differential rates of all processes going on within the bacteria are the same and equal to λ, the specific growth rate. It is satisfying that within our experimental ability, a constant growth rate is achieved when the environment is kept constant. This has required attention to details; for example, frequent dilution of batch cultures for many cycles is needed before composition and rate of processes become constant and every component increases in proportion to $e^{\lambda t}$ and satisfies Eq. (3).

Imagine that someplace in the reaction maze X is converted to Y; then, from ordinary chemical kinetics, $dY/dt = kX$, and from the balanced growth condition $dY/dt = \lambda Y$. Consequently, the first-order or pseudo-first-order rate constant can be calculated from

$$k = \lambda Y/X \qquad (6)$$

In this way, mainly with *E. coli*, the rates of many cellular processes have been studied, and important generalizations have emerged that will be considered throughout the remainder of this chapter.

One of the major consequences of this type of study is the realization that in spite of the large number of reactions, the essentials of the biochemical growth processes of microorganisms in balanced growth can be summarized quite well by a simple two-component model (Koch, 1971). It was shown that the large number of components in the cell do not need to be taken into account explicitly, but that two suffice. This simplification is based on an economic point: the constituents needed in the largest amount in bacterial cells are proteins. Although there are many kinds of proteins serving a variety of functions for the cell, the same ribosomal system makes them all. Therefore the metabolic Scheme (7):

$$R \underset{k}{\overset{c}{\rightleftarrows}} P \qquad (7)$$

summarizes the kinetic fact that the cell uses its protein-synthetic machinery (*R*) in order to make all denominations of protein (*P*), and it needs a consortium of those proteins to make more protein-synthetic machinery. In Scheme (7), *k* is the reaction rate constant for the rate at which cell protein is created per unit amount of ribosomal system, and *c* is the rate constant at which proteins of the cell act to synthesize the entire set of components needed for protein synthesis.

Although the steady state assumption justifies this formulation, we may need to defend the argument that this simple scheme is appropriate, since it lumps all proteins together no matter what their function and does not explicitly mention other vital constituents such as DNA. The omission of DNA is justified because *E. coli* maintains a constant proportion of DNA to protein. This is done by coupling the frequency of the initiations of chromosome replication to the rate of growth. It follows from Scheme (7) that

$$\lambda = \sqrt{kc} \qquad (8)$$

What this expression tells us is that if an environmental change alters the growth rate by changing one rate constant, it must change that constant a great deal. For example, if only *c* changes, it must change 100-fold for each tenfold change in the specific growth rate. As will be discussed below, to the first approximation, the cell does make large changes in *c*, the rate of manufacture of ribosomal and associated proteins.

This type of approach has been exploited by Maaløe and his school (see Ingraham et al., 1983) and by others. Bremer (1975) has measured many constituent processes affecting the growth rate of *E. coli* in different growth media. He considered six growth parameters. They are Ψ_r, the fraction of functioning RNA polymerase engaged in the synthesis of rRNA; c_r, the rRNA chain growth rate; β_p, the fraction of total RNA polymerase functioning; α_p, the differential rate of RNA polymerase protein synthesis; c_p, the peptide chain growth rate; and β_r, the fraction of active ribosomes. In Eqs. (7) and (8), he replaced c with $\psi_r c_r \beta_p$ and k with $\alpha_p c_p \beta_r$. His analysis, based on actual measurements, showed that these quantities change somewhat with growth rate. Thus regulation is taking place at several sites, consistent with the findings of Harvey (1973) and Gausing (1982) that β_r, the fraction of polysomes that function, increases with growth rate. Only β_p, the fraction of the total RNA polymerase that is functional, remains invariant. These changes are the fine print. They will be mentioned below, together with other details important at the extremes of growth rate.

IV. GROWTH KINETICS AT LOW SUBSTRATE CONCENTRATION

In many natural, industrial, and laboratory circumstances, bacterial growth is limited not by the biomass but by the delivery of some limiting nutrient. In particular, the chemostat (the name given to a type of continuous culture apparatus) allows balanced growth at different rates by adjusting the rate of addition of a medium containing a limiting growth factor. The basic idea is to add the medium at a given rate to a culture vessel whose volume is maintained constant by siphoning off a corresponding volume of culture fluid. Eventually, this leads to a steady state in which the concentration and growth rate of bacteria in the growth chamber are constant. One required growth factor is present in a limiting concentration, so its consumption effectively decreases its concentration. This becomes the chemical regulator of the "chemostat." Obviously, different chemicals can be the limiting substance, depending on the design of the medium.

When required nutrients become limiting, selection will be for bacteria with uptake systems that are highly efficient. The original conception of the kinetics of the bacterial permease system is based on the work of Jacques Monod, who likened growing bacteria to an enzyme system and symbolized the process as follows:

$$B + S \rightarrow BS \rightarrow 2B \tag{9}$$

In this reaction scheme, bacteria react with a substrate to form a complex that, in due time, decomposes into two bacteria, each of which serves as a new autocatalytic unit. This equation is formally equivalent to the Michaelis-Menten treatment of enzyme kinetics. The assumption of such a mechanism is that the growth rate, λ, is analogous to the basic law of enzymology, setting λ proportional to v and λ_{max} to V:

$$v = VS/(K + S) \quad \text{or} \quad \lambda = \lambda_{max}S/(K + S) \tag{10}$$

This equation is widely accepted and serves as a basis for microbial ecological thought. It is a good but not an accurate formulation. An older and sometimes better approximation than Monod's assumes that when the substrate concentration is low, the growth rate is directly proportional to the concentration, but that when the concentration is above a threshold value, the growth rate is independent of concentration (Koch, 1982, and see Fig. 1.1). These two conditions define Blackman's "Law of the Minimum":

$$\lambda = a_1S; \quad S < \lambda_{max}/a_1 \tag{11}$$

$$\lambda = \lambda_{max}; \quad S \geq \lambda_{max}/a_1 \tag{12}$$

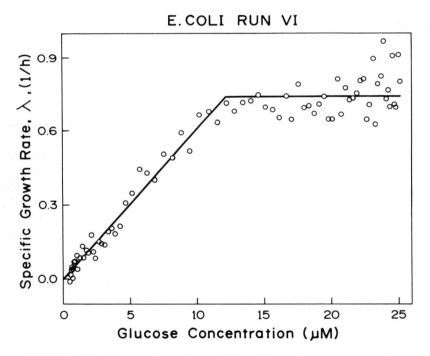

E. COLI RUN VI

FIGURE 1.1 Growth dependence of *E. coli* on glucose concentration. Data from Koch (1982). Growth was measured turbidimetrically with a computer-linked spectrophotometer. A more accurate fit than the "Law of the Minimum" was found to be consistent with the effects of the outer membrane in impeding glucose uptake. The fit shown here exhibits the important point that growth is accurately proportional to substrate concentration when the concentration is low.

The original equations of growth in a chemostat were based on the assumption of the Monod growth law. They have been presented in many places (e.g., Kubitschek, 1970), and we shall not repeat them here. Instead, we shall reformulate them as follows: When a chemostat is functioning at a steady state and organisms are sufficiently numerous to lower the concentration of limiting nutrients from the value in the input reservoir to a value at which Eq. (11) applies, then one of the chemostat equations is

$$D = \lambda = a_1 S \tag{13}$$

where D is the dilution or washout rate. D is equal to the ratio of the inflow rate to the chemostat divided by its volume. The importance of this reformulation is that it removes the emphasis from the affinity constant concept and replaces it with an emphasis on the apparent second-order rate constant, a_1, for the "bimolecular" interaction of substrate with the organism. This quantity is dependent on physical processes of diffusion into the cell. This change in emphasis is important for certain industrial applications in which the removal of low concentrations of substance from large bodies of water is the problem at hand.

The second relationship needed to define the system is the same for the Monod or Blackman formulations and is

$$W = (S_0 - S)/Y \tag{14}$$

where Y is the yield constant of the number of organisms created by the conversion of the limiting nutrient into biomass, W, and S_0 is the concentration of the limiting nutrient in the inflow.

This discussion of chemostat operation is appropriate in a chapter about *E. coli* because the chemostat was developed to grow *E. coli*; it also turns out that *E. coli* approximates the ideals that the equations put to mind. In an ideal chemostat, there should be no interaction between the organisms other than the competition for a single substrate. The organisms should be homogenous enough and the physical conditions (stirring, aeration, temperature control) constant enough that it is a good approximation that every organism "sees" the same environment. The regular division of many strains of this bacterium and its reduced adherence to surfaces make *E. coli* the prototype organism for chemostat culture. On the other hand, it is just the breakdown of these characters (particularly the last) that limit the chemostat as a tool for physiology and genetics.

V. THE BIOLOGY OF SIZE AND SHAPE

We usually think that microbes are small in order to decrease problems of import of nutrients and export of waste materials. On this basis, they should be as small as possible, contingent on all other biological constraints. A spherical shape may be the easiest to achieve, but it is the worst possible one for an organism to have because it minimizes the surface-to-volume ratio. Except for budding yeasts, there are no microbes that have an approximately spherical or ovoid shape *throughout* their life cycle. Cocci must achieve shapes other than spherical during cell division, although in the diplococcal stage, certain cocci appear as two equal, nearly spherical segments.

For the gram-negative organism, the rigid wall that gives it its shape is a very thin single-layered network of peptidoglycan. Peptidoglycan (or murein) is a two-dimensional polymer in which repeating saccharide chains of N-acetyl glucosamine alternating with N-acetyl muramic acid extend in one direction and short peptide chains cross-link the glycan chains. This sacculus is so thin and the pressure inside the living cell so great that the fabric must be in a highly stretched, extended conformation.

A formulation of how such a structure grows, achieves a nonspherical rod shape, and is even able to divide has recently been presented based on mathematics formally equivalent to those used in the calculation of the shapes of soap bubbles (Koch, 1983). For gram-positive cocci and rods, the external wall is created by splitting septa to form poles or by elongation of the side walls. The latter is an inside-to-outside growth process whereby a new layer is added inside the older wall and later expands when it comes to bear stress. The mechanism used by the thick-walled gram-positive organisms do not apply to gram-negative bacteria. The gram-negative organisms have only one apparent alternative, that is, to involve different enzymatic machinery or mechanisms for wall enlargement in different regions of the cell envelope. Because of the thinness of the peptidoglycan surface, growth must take place by random (diffuse) attachment of oligopeptidoglycan units secreted through the cytoplasmic membrane. The surface area is enlarged after multiple covalent links have been formed by transpeptidation between the terminal alanine of the pentapeptide chains and unlinked peptide chains of nearby stress-bearing peptidoglycan. Then highly specific cleavages are made, but only at stress-bearing peptidoglycan bonds that are already bridged by the new peptidoglycan. As this happens, the new material comes to bear the stress, and the surface becomes enlarged. The physics of this situation leads to elongation of the cylindrical portion of the cell as long as the poles are relatively rigid. By the trick of making peptide bridges before breaking others, the bacterium solves its structural problem the way any engineer would.

Gram-negative cells also use a particular mechanism in order to divide because they cannot form a thick septum and then split it, as gram-positive organisms do. Rather, the biochemical energetics of wall formation must be modified in some way that is equivalent to lowering the surface tension of a soap bubble. How this actually happens is not yet known, but there are a number of conceivable mechanisms. One will be described for illustrative purposes: At sites where constriction is to take place, several new oligopeptidoglycan units are linked to each other. Such a group or "raft" is then inserted into the membrane in the same way that single oligopeptidoglycan units are added during side wall extension. Thus a larger surface area is created at the cleavage site than where elongation of the side walls is taking place. Such a change alters the biophysics such that constriction and division take place by the same process that causes a flowing stream of water from a faucet to break spontaneously into a series of drops. The theoretical and actual shapes of a gram-negative bacterium are shown in Figs. 1.2a and 1.2b, respectively.

Thus the surface stress theory is able to explain gram-negative cell division without postulating the need for special proteins, such as tubulin, which in eukaryotic cells allows the transduction of metabolic energy into the physical work that constricts the cell. Consequently, the dimensions of a typical bacterium are determined by the biophysical properties related to the internal pressure and the energetics of biosynthesis. The length depends on triggering events that initiate the change in mechanism that starts the constriction.

VI. HOW DOES *E. COLI* KEEP CERTAIN CHEMICALS OUT AND LET OTHERS IN?

In order to digest fats, vertebrates pour bile into their intestines. Bile salts, a major constituent of bile, are detergents that damage unprotected biological membranes. Consequently, all bacteria in the vertebrate intestine must find a way of protecting their membranes from the deleterious effects of these substances. The gram-negative answer to this hydrophobic pollution is to make a very special second membrane, external to the peptidoglycan sacculus

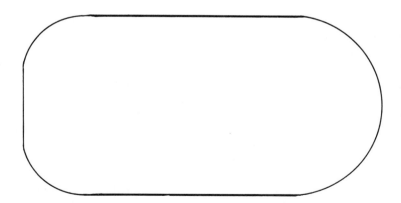

FIGURE 1.2a   Theoretical shape of *E. coli*. The figure shows the shape of an organism that has just separated from its sister on the right-hand side and has not had sufficient time for the initially flattened pole to become hemispherical. On the assumption that the parameter of wall growth analogous to surface tension has a value such that growth of the cylindrical region leads to extension, the right-hand pole shape has been calculated as discussed in Koch (1983) for a local decrease of 2.5-fold in this parameter in the region of the cell where division ensues.

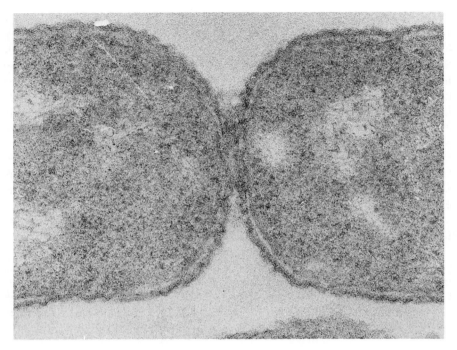

FIGURE 1.2b   An actual electron micrograph of a median section of *E. coli.* (*Courtesy of Dr. I. D. J. Burdett.*)

and the cytoplasmic membrane, whose role is to effect such protection. To serve this purpose, the external membrane must be different in structure from the cytoplasmic membrane.

The so-called outer membrane of gram-negative bacteria, *E. coli* included, has an external fringe of long polysaccharide molecules. In the enterics, this is known as the O-antigen. This substance is hydrophilic and effectively excludes hydrophobic substances. Mutants defective in O-antigen biosynthesis are quite permeable to hydrophobic compounds. On the other hand, the lipid leaflets of the outer membrane are hydrophobic and prevent the passage of hydrophilic molecules. Seemingly nothing could cross the outer membrane, but an additional kind of structure exists for the entry of small hydrophilic molecules. The outer membrane contains nonspecific channels for the passage of small hydrophilic nutrients; the channels are made up of proteins called porins. This provides an entry mechanism that is not as selective as the individual specific uptake systems of the cytoplasmic membrane; that is, porins are biochemical doughnuts extending across the membrane. Such channels cannot be too wide, or else hydrophobic compounds may also get through. In reality, they allow the passage only of hydrophilic substances smaller than 600–800 daltons (Nikaido, 1979).

However, enterics are able to use some polar substances larger than this exclusion limit, such as vitamin B_{12}, iron chelates, and amylomaltose chains with more than four glucose units. For example, under iron limitation, bacteria must and do develop special mechanisms for the utilization of iron. They do this by excreting avid iron chelators and then reabsorbing the iron in the chelated form. The solution to the problem of permeation of compounds larger than 800 daltons is to fabricate specific systems for transport across the outer membrane. These mechanisms are highly specialized, and so far only a few kinds have been found in the outer membrane of *E. coli*. In this location, they cannot be coupled to the usual cellular supplies of metabolic energy, and they do not carry out active transport. These special transport mechanisms constitute an Achilles' heel, since various viruses, plasmids, and colicins use them as attachment sites to enter the bacteria. Thus the bacterium is always in a potential "double bind"; it may lose, whether it keeps or genetically deletes these outer membrane transport mechanisms.

For comparison, the gram-positive bacteria protect themselves by wearing a thick hydrophilic coat of peptidoglycan, polysaccharides, or polypeptides outside their cytoplasmic membrane. Thus they do not have an outer membrane. In general, this solution does not seem to be as felicitous as that of the gram-negatives because gram-positives tend to be more sensitive to deleterious hydrophobic compounds such as detergents, dyes, and certain antibiotics.

VII. TRANSPORT THROUGH THE CYTOPLASMIC MEMBRANE

To scavenge successfully for low-molecular-weight compounds, *E. coli* has developed sophisticated, active transport systems with very avid uptake capability (Harold, 1978; Rosen, 1978). They are located on the cytoplasmic membrane, where they have access to cellular energy supplies. Typically, the apparent Michaelis-Menten constants for uptake of glucose, lactose, etc. are in the micromolar range for gram-negative bacteria. This is to be contrasted with the millimolar range for the facilitated diffusion mechanism used by certain yeasts that grow profusely in a carbohydrate-rich environment like grape juice.

One class of active transport mechanisms is powered by the chemiosmotic coupling. Protons are pumped out of the cell, driven by respiration or by a special ATPase. Proton flux back into the cell via special symporters or antiporters leads to active transport and accumulation against a concentration gradient. For this kind of mechanism to work, the total amount of return paths must be limited. A good example of transport is typified by the *lac* permease. This intrinsic membrane protein is able to transport galactosides in either direction across the membrane, but only if there is an accompanying proton. It returns to its orginal conformation for a next cycle of transport, carrying *neither* the proton *nor* the galactoside. It can accumulate against a concentration gradient or speed transport compared with a simple carrier mechanism.

A second major system is the phosphotransferase system (PTS). This system uses the very high-energy bond of phosphoenolpyruvate to simultaneously transport certain sugars and phosphorylate them. The third class of active transport mechanism somehow uses the energy of ATP to force transport. Sometimes two types of energy-coupling processes may be used for the transport of a single molecule. We do not yet understand in detail how any of these mechanisms work, though these are active areas of research employing all available techniques. We can appreciate that they do work effectively to pump large quantities of certain organic chemicals needed as carbon and energy sources or trace amounts of particular compounds and metals that may be present in only extremely low concentrations. Ordinarily, trace metals need not be added to chemically defined media because the essential ones are present as contaminants in sufficient amounts even in reagent-grade chemicals. Thus it is very difficult to prepare media that are, say, sufficiently molybdenum-poor to demonstrate a molybdenum deficiency.

VIII. ENERGY TRANSDUCTION

The heavy industry of the bacterial cell is involved in trapping energy to transduce it into useful forms (e.g., PEP, ATP, and proton gradients). This energy is then used for transport and biosynthesis. _E. coli_ can use about 30 known organic compounds for both biosynthesis and catabolism for energy production. For some of the more oxidized compounds, the organism extracts energy fermentatively. With glucose, _E. coli_ carries out a mixed acid fermentation that yields lactic, acetic, and succinic acids, the equivalent of formic acid (CO_2 plus H_2), and ethanol. For a wider range of compounds, energy may be obtained oxidatively. The oxidant can be nitrate if oxygen is absent, but oxygen is the preferred terminal electron acceptor.

The electron transport chain is not a static arrangement as it is in, say, mammalian mitochondria. Rather, the components are altered by the cell, depending on the terminal oxidant and even in response to its concentration. Thus at low partial pressures of oxygen, a path is elaborated that extrudes fewer protons than when the pressure is higher.

Coupling and Reversibility of Biosynthetic Processes

An analogy with a car comes to mind. Imagine driving a car up a mountain and running out of gasoline partway to the top. A car with the simplest imaginable engine not only would stop climbing but would coast back down the hill, making the engine turn backwards. Like a real car, the living state is remote from equilibrium and demands the continuing expenditure of energy to remain so.

Thus most organisms do not autolyze when a source of energy is removed. This is not surprising because of two factors: The first is the analog of the fact that the car would roll down faster out of gear than when the cylinders are forced to move. Biochemically, biosynthetic systems are tightly coupled and, by and large, transfer bonds only in the higher-energy state. With some exceptions, they do not act as hydrolases. The second fact is that, like the car, organisms have brakes. It seems likely that there are biochemical and physiological brakes that prevent the biosynthetic systems under de-energized conditions from functioning inappropriately and perversely. Clearly, some brakes are set when the cell enters the stationary phase. Other organisms are capable of specialized brake-setting processes such as sporulation that remove the biomass from the sphere of active biochemistry and allow the spore to persist for very long periods of time.

The *E. coli* version of setting the brakes is more subtle. When resources gradually come into short supply, cell division outruns biomass production until smaller cells result with only a single genome and with few ribosomes. These cells are capable of suspended animation for very long periods (see Koch, 1976). After being starved for a few days, *E. coli* cells are ready to go and start at a moment's notice. This has clear advantages over the sporulation strategy of certain bacteria, since it takes them valuable time to revert to the active growing state when conditions become favorable. Possibly, this is a partial reason why enteric bacteria have not developed a spore state. Instead, they have developed quick ways to enter active metabolism when nutrients become available.

IX. SUBSTRATE UTILIZATION AND ITS REGULATION

Species of *Pseudomonas* can metabolize a tremendous range of carbon sources. *E. coli* has a much more limited repertoire, which reflects the resources that *may* be available in the intestine. It can use short polypeptides, amino acids, some short-chained carboxylic acids, and a handful of carbohydrates (including nucleosides).

The regulation of substrate utilization is very sophisticated. Consider the best-studied case, that of lactose utilization. The major details of the regulation are so well known that we need not go into them here (see, for example, Miller and Reznikoff, 1980; Lewin, 1983). Rather, we point out a single aspect that is beyond those usually presented. It follows from the observation that while lactose is the natural β-galactoside that is episodically available in sufficient quantities to merit metabolism, this substance is not an inducer of the *lac* operon. Rather, the actual natural inducer is its transformation product, allolactose; this is made inside the cell from lactose by β-galactosidase. The implication of this finding is that induction depends on the presence of both lactose *and* previously induced β-galactosidase. This deviously simple strategy means that the regulatory system responds to the availability of lactose in a way that is dependent on past lactose availability. If lactose has been available at frequent intervals, induction will be much more responsive. The organism has a short life compared with the life of its hosts or even the eating habits of some of its hosts. Imagine the case in which a host had been provided with lactose but there had been several bacterial generations of lactose abstinence before this compound was reintroduced. If the previous exposure had been recent enough, β-galactosidase and permease, although partially diluted out by fresh growth, would still be present in sufficient amounts to quickly make inducing concentrations of allolactose. After long growth periods with no β-galactosides, induction by lactose would depend on the basal level of *lac* operon products. The basal level results from occasional dissociation of the repressor from the operator region, leading to new synthesis. This level can be affected by mutation and by catabolite repression. In addition, the function of both β-galactosidase and permease is affected by other substances such as glucose and other galactosides. This implies that the control mechanism of the cell for this one function is sophisticated and not only responds to the immediate circumstances, but modifies that response in a reasonable way, by weighting the odds on the basis of *experience of the recent past*.

X. OPTIMUM STRATEGIES IN A FLUCTUATING ENVIRONMENT

E. coli can grow in a variety of culture media, ranging from those that are nutritionally quite meager and contain a single organic compound to those that are rich in kinds and amounts of utilizable nutrients. It makes a considerable difference to the cells, and they respond in a rather dramatic fashion. In poorer media, the organisms grow more slowly and have a chemical composition different from that in rich medium. The extremes of balanced growth rates at 37° vary from one doubling every 13 hours to three to four doublings per hour, resulting in cells that are very different in many characteristics. At still slower rates, the cells alternate between a growing and a resting state. The fastest doubling times are achieved only with strains that have not gone through many cycles of mutagenesis in the laboratory.

The difference in the organisms is most readily apparent in the volume of cells, which can vary over tenfold from the slowest- to the fastest-growing cultures. The most striking chemical difference is the increase in the protein-synthesizing machinery, in the form of ribosomes and other factors. Over a range of growth rates, the content of ribosomes per unit DNA is roughly proportional to the growth rate, and the amount of protein per unit DNA is nearly constant. Therefore if in Eq. (6) we call X the ribosomes and Y the proteins, the rate of protein synthesis is directly proportional to the ribosomes. Expressed in another way, the production rate of proteins is directly proportional to the amount of ribosomes present. Consequently, each ribosome functions at the same rate of biosynthesis no matter what the nutritional environment.

If we could follow a single ribosome, it would spend the same time waiting for a message and the same time adding an amino acid to a growing chain no matter whether the cell it was in doubles once every 13 hours or every 15 minutes. When growing *very* slowly, cells carry an extra load of nonfunctioning or slow-functioning ribosomes. However, these ribosomes synthesize proteins immediately when such cultures are fed rich nutrients (Koch and Deppe, 1971). This reserve may be a small insurance price for a destitute *E. coli* to pay because cells with too few ribosomes would be unable to start growing quickly when conditions become favorable.

Chain growth rates of various macromolecules have been measured in various ways and seem to be quite constant for proteins, RNA, and DNA at moderate and fast growth rates. These experimental findings permit the generalization that the machinery involved in macromolecular synthesis works at a constant efficiency over a wide range of environmental conditions (Maaløe and Kjeldgaard, 1966; Ingraham et al., 1983). All this implies that macromolecular syntheses are regulated mainly at the level of their initiation and not by altering the speed of chain elongation. Each synthetic apparatus can be likened to an automobile assembly line whose output is regulated by how often the assembly of a car is started rather than by changing the speed of the assembly line itself.

Efficiency of operation is also seen in the manner whereby *E. coli* adjusts to sudden shifts from one growth condition to another. For example, a slow-growing culture can be transferred abruptly to a rich medium by the addition of extra nutrients. Under these "shift-up" conditions, cells accelerate their growth and macromolecular synthesis with a very short lag (Kjeldgaard et al., 1958). They start making ribosomal RNA and assembling ribosomes within a minute or less. Thus they sense their new environment very rapidly and take advantage of the new circumstances right away. Conversely, when cells that are growing fast are placed in a poorer medium ("shift-down") by centrifugation or filtration and resuspension, accumulation of RNA stops immediately.

To understand how these rapid adjustments of macromolecular synthesis take place, it is helpful to focus on the synthesis of protein and on the protein-synthesizing machinery. The problem can be reduced to asking how the cell knows to make a given amount of ribosomal RNA per unit time. Does the cell regulate RNA polymerase so that it recognizes different classes of promoters and consequently makes different proportions of different kinds of RNAs? Probably not. Maaløe (see Ingraham et al., 1983) conjectured that the behavior of RNA polymerase may be simply explained from general regulatory principles for the synthesis of enzymes for anabolism and catabolism. He called it the "passive control" model because it does not postulate any special new regulatory element. It works like this: At constant temperature, the nutritional conditions determine the growth rate of the organism. A medium in which the organisms can grow fast contains more *kinds* of utilizable nutrients. These will inhibit the synthesis of the enzymes involved in their biosynthetic pathways. Thus cells growing in a medium rich in amino acids will not make the enzymes involved in the biosynthesis of the amino acids. A similar effect is seen for purines, pyrimidines, and a large number of other low-molecular-weight compounds.

Now cells that are not making this vast array of enzymes are spared from the need of engaging RNA polymerase in forming the corresponding messages. The model simply states that under these conditions, more RNA polymerase molecules or other limiting replication factors are available to synthesize RNA from genes that are not repressed (see Fig. 1.3). In other words, proteins that are made constitutively (including ribosomal proteins) will be favored when

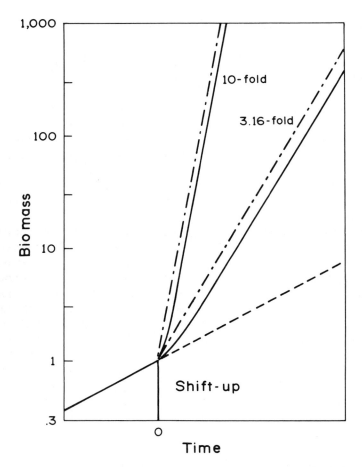

FIGURE 1.3 Idealized shift-up experiment. On a semilogarithmic plot, bacterial growth is shown. Curves are shown for hypothetical growth of biomass (turbidity) for an enrichment increasing the steady-state growth rate by $\sqrt{10} = 3.16$ and by tenfold. The solid curves show the calculated time course if this is achieved by instantaneously changing only the rate constant c of Scheme (6). The dash-dotted curves show the consequence if the bacteria could instantaneously change the specific growth rate. Actual cultures of *E. coli* have an intermediate response due to slight alterations as discussed in the text and by making more efficient use of ribosomes.

the synthesis of the inducible systems is shut off. The better the new medium, the larger the content of ribosomal RNA, etc. The passive model accounts for much of the qualitative changes that *E. coli* makes to an improved nutritional status; there are, as mentioned, a number of minor changes in rates of the different subprocesses in protein synthesis and in the synthesis of the machinery for protein synthesis.

XI. A CASE OF STARVATION: AMINO ACID DEPRIVATION

There are several ways in which *E. coli* can be starved for required amino acids. Wild-type strains can make their own and do not need preformed amino acids, but starvation can be imposed by omitting from the medium amino acids required by auxotrophic mutants, by placing mutants with temperature-sensitive activating enzymes at nonpermissive conditions, or by adding inhibitory amino acid analogs. In all cases, the result is the same; when the organism is starved for a single amino acid, all protein synthesis is stopped. The reason for this complete blockage is that a ribosome traveling down a messenger RNA molecule will not go past the site for which a charged tRNA is missing. The lack of any given amino acid will therefore have the same overall consequence as the lack of any other.

The organism reacts to these circumstances in a rather more complex way than might be imagined at first glance. The lack of amino acids is translated into a series of metabolic events that make considerable teleological sense. Thus most strains of *E. coli* do not accumulate RNA under these circumstances. The advantage of this behavior is seen when we compare wild-type strains with mutants that do accumulate RNA when faced with the same problem. These mutants are called "relaxed," in contrast with the wild types, which are called "stringent." This type of mutant, originally found in 1956, has been extensively studied ever since (see Maaløe, 1979). The most telling physiological differences are seen when cultures of these two kinds of strains are re-fed after a couple of hours of starvation. The stringent cells begin to grow at a near-normal rate with an exceedingly short lag, less than a minute. The relaxed cells, on the other hand, require several hours to achieve the growth rate characteristic of that medium; moreover, they have difficulty forming inducible enzymes. It follows that under competitive conditions, the stringent strains will outpace the relaxed ones in a very short time. Not surprisingly, natural isolates are stringent.

A good deal of available information explains how this comes about. Stringent strains under starvation conditions do not completely shut off RNA synthesis. Rather, they are selective in what they make. They continue to produce certain messenger RNAs that are involved in cell maintenance, such as those coding for enzymes necessary for the preservation of cell integrity related to membrane and cell wall metabolism. On the other hand, the synthesis of ribosomal and transfer RNA is blocked in starved stringent cells. Relaxed strains make these kinds of RNAs, which, in the absence of synthesis of ribosomal proteins, become excess baggage and interfere with the assembly of ribosomes once the metabolic block is lifted. Thus stringent cells find themselves at an advantage when the missing amino acid is supplied: they can start to grow right away.

To some extent, the biochemical mechanism that dictates the stringent response is known (see Ingraham et al., 1983). Selective inhibition of the synthesis of ribosomal RNA and some classes of messenger RNA seems to be due to the production of a factor that tells RNA polymerase which classes of genes to transcribe and which ones not to transcribe. This factor is guanosine tetraphosphate, G4P, also known as "magic spot" because its novel occurrence on paper chromatograms raised an appropriate eyebrow. There are other related compounds, such as guanosine pentaphosphate, G5P, that also play a role in this phenomenon. It appears that G4P has the ability to fine tune the specificity of RNA polymerase and that in its presence those RNAs characteristic of the stringent response are made. Under normal growth conditions, very little G4P is made, but its level shoots up upon amino acid starvation. The interesting thing is that this compound serves to link two biochemically unrelated pathways, RNA synthesis and protein synthesis.

How is this done? When a ribosome travels down the messenger RNA and encounters a site where there is no corresponding charged tRNA, it carries out what has been called the "idling reaction," a biochemical series of events that results in the synthesis of G4P as an alternative to a peptide bond. This synthesis requires the action of a so-called "stringent factor," an enzyme that is missing in some of the relaxed strains. Thus the difference between stringent and relaxed cells is that the former can make a large amount of G4P, whereas the latter cannot. A moment's reflection will tell us that under normal growth conditions, there must be some mechanism to keep the synthesis of G4P in check, else the cells may think they are starved for amino acids. In fact, little is known about the participation of G4P in regulation during normal growth.

Obviously, the important lesson is that *E. coli* cells in their habitat respond to the presence or absence of amino acids and other building blocks. The paired phenomena of enzyme repression and stringency allow for greatest efficiency in their response. When amino acids, etc. are present, cells do not waste their resources in making enzymes involved in their biosynthesis. When these resources are no longer available, the bacteria carry out just those biosynthetic reactions that will allow them to grow rapidly when these compounds become available again.

It should not be thought that the shift-up and shift-down are simply opposite cases of the same phenomenon. For nonextreme shift-up situations, it suffices to assume that the bulk of the adaptation to the improved conditions is that one parameter, c, in Scheme (7) is increased very rapidly (largely owing to the passive model effect), while the other stays nearly constant (constant efficiency model). On the other hand, on the shift-down, ribosomal efficiency is very small because of lack of amino acids, etc., and ribosomal synthesis

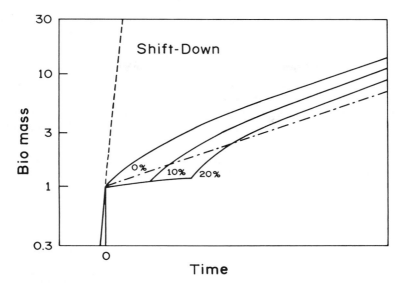

FIGURE 1.4 Idealized shift-down experiment. The curves are calculated on the assumption that at the moment of shift-down, c and k defined by Scheme (7) decrease 100-fold. Subsequently, after enough new synthesis of protein occurs with derepression of biosynthesis of enzymes involved in the limiting biosynthetic pathway, k increases tenfold. The several curves were calculated on the basis of the indicated level of new protein synthesis needed to achieve enough biosynthetic capability to relieve the growth limitations.

becomes very small because of the stringent response. Only after a delay (see Fig. 1.4) does biosynthetic ability for the missing growth factors become instituted, and c increases, while k remains at a small value. Then a long period of growth is needed before the cells shift their constituents to achieve slower, but balanced, growth with a lower content of protein-synthetic machinery.

XII. SEGREGATING THE GENOME

Early in the beginning of life, nucleic acid fragments must have existed inside of cells as discrete pieces coding for different functions. There must have been a very strong selection for mechanisms that ensured the precise segregation of complete sets of genome. In fact, today we find that an extraordinarily complex process, mitosis, has evolved in the eukaryotes to do just that. The prokaryotes have evolved a different solution to the problem. They have lined up nearly all their genes in a single molecule, a long covalently linked DNA in the form of a circle. As bacteria have grown more and more complicated, the number of genes for essential functions has increased until today we find that

E. coli contains some 1027 identified genes and probably a total of 3000 genes (Bachman, 1983).

In exchange for the solution to the problem of segregation, *E. coli* must solve a very demanding topological problem. Its DNA is extraordinarily long, about 1000 times the length of the cell. Moreover, this great length of DNA is wadded up in a relatively small volume. The space that contains the DNA lies more or less in the center of the bacteria and is called the *nucleoid* to distinguish it from a true nucleus. Unlike the nucleus, it is not bound by a limiting membrane. What holds it together is not known, but some evidence indicates that it regards the cytoplasm as if it were a different, nonmiscible aqueous phase. Within the nucleoid, DNA takes up about 10% of the volume — an extraordinary concentration if one takes into account that in the test tube a solution even ten times more dilute is a gel!

It is fair to say that we understand little about the mechanical and physical aspects of DNA metabolism in bacteria, although we know a good deal about its biochemistry (Kornberg, 1980). We know that in a short period of time, 40 minutes at 37°, all the DNA is replicated. It helps a little to add that replication is bidirectional; that is, DNA synthesis starts at a precise place and proceeds semiconservatively away from it in both directions. The two replication machines run into each other at a terminus, some 180° from the origin of replication. How this immense plate of linked-together spaghetti can carry out this process with precision and without obvious entanglement is not known. In an interesting confirmation of the Maaløe-Kjeldgaard notion that chain growth rate of every macromolecule is the same regardless of the growth rate of the cells, DNA replication takes 40 minutes in *E. coli* growing fast or growing fairly slowly. How can this take place in cells whose generation time is half that long, 20 minutes? The answer is another example of regulation at the level of initiation of synthesis. That is, in balanced growth, bacteria initiate DNA replication just as often as they divide. Although the process takes 40 minutes, in a bacterium growing at a 20-minute generation time, a new round starts every 20 minutes. This results in more than one replication event going on at the same time. Such multifork replication can be demonstrated experimentally, and yet the nucleoid is still not tangled up!

There is no real understanding of how the cell knows to start DNA replication at a precise point in the cell cycle, but it does. At least the volume of newly made cells in a balanced population has a coefficient of variation usually as small as 10% (Koppes et al., 1978). The most popular notion is that somehow the cell recognizes that it has reached a certain critical size and that this triggers a new round of replication (Helmstetter et al., 1979). It has been proposed by Pritchard and his colleagues (see Mandelstam et al., 1983) that what is sensed is the concentration of an inhibitor of the initiation process. If this substance is made only at the start of the cell cycle, it will be diluted out by cell growth, and when its concentration drops below a critical point, initiation will begin.

The site on the chromosome where initiation is controlled is the origin of replication. An example of this control is illustrated by the following experiment: It is possible to substitute the normal origin of replication by that of a plasmid integrated into the chromosome. Under a certain experimental condition (temperature shift), one can shut off the *E. coli* origin and have chromosome replication driven from the plasmid origin. However, under normal conditions when the host origin functions, the plasmid origin is totally shut off. This suggests that there are separate regulatory processes, each specific for a given origin of replication. Thus we may not know the nature of the biochemical signal that sets off initiation of DNA replication, but at least we are aware of its existence.

XIII. GENETIC PARASITES

"Fleas have little fleas on their backs to bite them . . . ." This ditty is not true *ad infinitum*, but it is true to the extent that all cellular forms of life, no matter how small, have parasites of one sort or another. The more interesting ones can alter the genetic system of their host and either add genetic functions or modify existing ones. Bacterial viruses, the bacteriophages, can do these things in one of two ways. They can take over the host in a virulent fashion and convert it into a virus-producing machine; or they can integrate their genome into the genome of the bacteria. This condition, called lysogeny, is not detectable unless it manifests itself in one of two ways. Either the integrated viral genes express a function that is not otherwise characteristic of the bacteria, or they "pop out" from the host DNA and make a new virus.

Peaceful coexistence can also take place with plasmids, genetic parasites that are not typical viruses but are just pieces of DNA. Like bacteriophages, some can also integrate into the chromosome or lead an independent existence in the cytoplasm of the host cell. In order to survive as independent entities, they must replicate precisely as often as the host chromosome. Many plasmids (nontransmissible ones) cannot inject themselves into host cells like the viruses and depend for transfer on mechanisms programmed by the hosts or its viruses for exchanging DNA, such as conjugation, transduction, or transformation. Transmissible plasmids code for their own conjugation system, which also can serve to transmit host genes. From the point of view of the bacterium, such transfer cannot be indiscriminating or else the host genome would be subject to frequent and destabilizing changes. To avoid promiscuous exchanges, bacteria have evolved systems of enzymatic restrictions whereby they are capable of hydrolyzing foreign DNA while protecting their own DNA by specific chemical modifications. Both processes involve recognition of specific DNA sequences by specific restriction and modification enzymes. In turn, these enzymes become very useful in the practice of genetic engineering because they can cause DNA to be cleaved at precisely determined sites.

Since plasmids can be seen as extensions of the bacterial genome, one can properly ask why they exist in the first place. Are they small "carryalls" that supplement the "trunks" of genetic information carried by the chromosome?

We can assume that it is to the advantage of bacteria and viruses to store genetic information in a dense fashion. The world of little genomes has several examples of high-density packing. For instance, small RNA phages have stretches of nucleic acid that code for several proteins by having overlapping coding sequences. Must *E. coli* also practice genetic austerity? In fact, only a few of the genes of *E. coli* are duplicated, and almost all of its DNA probably can be accounted for as necessary genetic material, although some of it may be used only infrequently (see Koch, 1976). Of course, some traits, while dispensable under some circumstances, become essential for survival under others. We count among these the ability of many plasmid-borne genes to neutralize antibiotics. In many cases, resistance mechanisms are carried by plasmids because, in the long run, a chromosome location may not be "cost-effective" (Koch, 1981). As an example of a chromosomally located protective mechanism, *E. coli* has a photo-repair system to rectify its DNA after it has been damaged by ultraviolet irradiation. As far as we know, it is useful only to remove dimers of thymine made when DNA is exposed to ultraviolet light and requires near-ultraviolet light for the repair. Since ultraviolet irradiation is not one of the perils that *E. coli* encounters in the vertebrate gut, nor is near-ultraviolet light available there, this function must be related to some other physiological phenomenon that is essential for survival. No doubt this is in the essential part of the propagation process of getting from host to host.

XIV. EPILOGUE

Because *E. coli* has been so well studied, our considerations of the holistic biology of microoganisms center around this species. Physiological and bio-chemical studies have led us to recognize the problems that this organism has faced and solved. Studies of this organism point to intricacies of molecular biology and biochemistry that we now understand or are beginning to under-stand. These will have their counterparts in the other organisms discussed in this volume. It can be anticipated that the take-home lessons from the study of this organism will have application to all the others.

REFERENCES

Bachman, B. J. (1983) *Microbiol. Rev. 47*, 180–230.
Bremer, H. (1975) *J. Theoret. Biol. 53*, 115–124.
Cooke, E. M. (1974) Escherichia coli *and Man.* Churchill Livingstone, Edinburgh.
Edwards, R. R., and Ewing, W. H. (1972) *Identification of Enterobacteriaceae*, 3rd edition, Burgess Publications Co., Minneapolis.
Gausing, K. (1982) *Trends Biochem. Sci. 7*, 65–67.
Harold, F. M. (1978) in *The Bacteria*, Vol. VI (Gunsalus, I. C., Ornston, L. N., and Sokatch, J. R., eds.), pp. 473–521, Academic Press, New York.

Harvey, R. J. (1973) *J. Bacteriol. 114*, 309–322.

Helmstetter, C. E., Pierucci, O., Weinberger, M., Holmes, M., and Tang, M-S. (1979) in *The Bacteria*, Vol. VII (Gunsalus, I. C., Sokatch, J. R., and Ornston, L. N., eds.), pp. 517–579, Academic Press, New York.

Hinshelwood, C. N. (1946) *The Chemical Kinetics of the Bacterial Cell*, Clarenden Press, Oxford.

Hinshelwood, C. (1952) *J. Chem. Soc.* 745–755.

Ingraham, J. L., Maaløe, O., and Neidhardt, F. C. (1983) *Growth of the Bacterial Cell*, Sinauer, Sunderland, Mass.

Judson, H. F. (1979) *The Eight Days of Creation – The Makers of the Revolution in Biology*, Simon and Schuster, New York.

Kjeldgaard, N. O., Maaløe, O., and Schaechter, M. (1958) *J. Gen. Microbiol. 19*, 607–616.

Koch, A. L. (1971) *Adv. Microbiol. Physiol. 6*, 147–217.

Koch, A. L. (1976) *Prosp. Biol. and Med. 20*, 44–63.

Koch, A. L. (1979) in *Strategies of Microbial Life in Extreme Environments* (Shilo, M., ed.), pp. 261–279, Dahlem Conference, Berlin.

Koch, A. L. (1981) *Microbiol. Rev. 45*, 355–378.

Koch, A. L., and Deppe, C. S. (1971) *J. Mol. Biol. 55*, 549–562.

Koch, A. L. (1982) in *Overproduction of Microbial Products* (Krumphanzl, V., Sikyta, B., and Vanek, Z., eds.), pp. 571–580, Academic Press, London.

Koch, A. L. (1983) *Adv. Microbiol. Physiol. 24*, 301–367.

Koppes, L. J. H., Woldringh, C. L., and Nanninga, N. (1978) *J. Bacteriol. 134*, 423–433.

Kornberg, A. (1980) *DNA Replication*, Freeman and Co., San Francisco.

Kubitschek, H. I. (1970) *Introduction to Research in Continuous Cultures*, Prentice-Hall, Englewood Cliffs, N. J.

Lewin, B. (1983) *Genes*, John Wiley and Sons, New York.

Maaløe, O. (1979) in *Biological Regulation and Development*, Vol. I (Goldberger, R. F., ed.), pp. 482–542, Plenum Press, New York.

Maaløe, O., and Kjeldgaard, N. O. (1966) *Control of Macromolecular Synthesis*, W. A. Benjamin, Inc., New York.

Mandelstam, J., McQuillen, K., and Dawes, I. (1983) *Biochemistry of Bacterial Growth*, John Wiley and Sons, New York.

Miller, J. H., and Reznikoff, W. S. (1980) *The Operon*, 2nd edition, Cold Spring Harbor Laboratory, Cold Spring Harbor, New York.

Ørskov, F. (1981) in *The Prokaryotes*, Vol. II (Starr, M. P., Stolp, H., Trüper, H. G., Balows, A., and Schlegel, H. G., eds.), pp. 1128–1134, Springer, Berlin.

Nikaido, H. (1979) in *Bacterial Outer Membranes: Biogenesis and Function* (Inouye, M., ed.), pp. 361–407, John Wiley and Sons, New York.

Pipes, W. D. (1982) *Bacterial Indicators of Pollution*, CRC Press, Boca Raton, Fla.

Rand, M. C., Greenberg, A. E., and Tares, M. J. (1976) *Standard Methods for the Examination of Water and Waste Water*, 14th edition, American Public Health Association, Washington, D.C.

Rosen, B. P. (1978) *Bacterial Transport*, Marcel Dekker, New York.

Savageau, M. A. (1976) *Biochemical Systems Analysis*, Addison-Wesley, Reading, Mass.

2

Biology of *Pseudomonas* and *Xanthomonas*

Norberto J. Palleroni

I. INTRODUCTION

The term "pseudomonad" has a precise morphological meaning. It applies to any rod-shaped, gram-negative, polarly flagellated, nonsporulating bacterium. From a phylogenetic point of view, however, the term fails to define the properties of a single natural group of prokaryotes, since the above characteristics can be found in members of about 20 currently accepted genera of bacteria.

In this chapter we examine the biological properties of two groups of pseudomonads assigned to two important genera, *Pseudomonas* and *Xanthomonas*. It is appropriate to treat these two genera together, but the reasons for adopting such a temperament became apparent only recently, since there is nothing in the superficial resemblance of the organisms suggesting a closer relationship of *Pseudomonas* and *Xanthomonas* than that of either genus to other groups of pseudomonads.

The current definition of *Pseudomonas* (Doudoroff and Palleroni, 1974) is as follows. They are gram-negative rods, usually less than 1 μm in diameter and 1.5–5 μm in length. The cells do not have prosthecae and are not surrounded by sheaths. They are immotile or motile by one to several polar flagella; lateral flagella of shorter wavelength may be produced. No resting stages are known.

They are aerobic; but in some instances, nitrate can be used as an alternative electron acceptor under anaerobic conditions. Many species accumulate poly-β-hydroxybutyrate (PHB) as a carbon reserve material. A variety of pigments, but not xanthomonadins, may be produced. They do not grow at pH 4.5 or lower, and organic growth factors usually are not required. They are oxidase-positive or -negative and catalase-positive. Though generally chemoorganotrophic, some species can also live autotrophically using CO and/or H_2 as energy sources.

We could easily fit *Xanthomonas* in this definition by appropriate adjustments of the variable characteristics: *Xanthomonas* cells do not accumulate PHB, are monoflagellated, do not use nitrate as an alternate electron acceptor, usually require growth factors, and are chemoorganotrophic and pathogenic for plants. The G + C content of the DNA goes from 63 to 69 moles %. Strictly speaking, the only *Xanthomonas* characteristic never found in *Pseudomonas* species is the production of xanthomonadins. To this we can add the fact that *Xanthomonas* species cannot use asparagine as both carbon and nitrogen source, a characteristic proposed as diagnostic for the genus by Dye (1962), and that growth is inhibited by low concentrations of triphenyltetrazolium chloride (Dye and Lelliott, 1974), but we do not know the behavior of all *Pseudomonas* species for these two properties. In addition, most *Xanthomonas* species hydrolyze starch, a property that is not very common among *Pseudomonas* species.

These two definitions separate *Pseudomonas* and *Xanthomonas* from most of the other genera of pseudomonads, but in truth an absolute phenotypic segregation is practically impossible. Fortunately, as we shall discuss in the next section, some methods of modern taxonomic analysis can define rather precisely natural groups at the level of genera or higher and give a measure of their relatedness. In the particular instance of some traditional genera like *Pseudomonas*, these new methods have revealed an internal heterogeneity to such a degree that a further subdivision of the genus into other genera appears highly desirable. The new methodology has also shown that some groups within *Pseudomonas* are more closely related to *Xanthomonas* than any species of these two genera are related to other taxa of aerobic pseudomonads.

II. TAXONOMY

The taxonomy of the two genera will be treated here only in general terms. Since we shall frequently refer to "natural relationships" and not simply to artificial classifications of practical determinative value, our taxonomic considerations should come at the end of the biology discussion. Modern approaches to the natural relationships among organisms are not simply based on the results of a few arbitrarily chosen practical tests. True phylogenetic classifications should emerge from a comprehensive mass of information on the organisms involved with respect to their morphology, nutrition, metabolism, biochemistry, genetics, and ecology. For practical reasons this natural order will not be followed here, and the taxonomy section will serve as an introduction to the main scheme of classification and to the names used in subsequent discussion.

A. *Pseudomonas*

In the course of almost one century since the creation of this genus, a large number of species names have been assigned to it. Aerobic pseudomonads are widespread in nature, performing an important role in the degradation of organic matter and causing food spoilage and various diseases of plants and animals. Many saprophytic strains with a bewildering variety of properties were isolated very easily from numerous natural sources, and the consequence, in taxonomic terms, was that the number of species names grew to alarming proportions. Many other strains were isolated from diseased plants and assigned to *Pseudomonas*. As for the saprophytes, the taxonomic problems of the phytopathogenic species became acute, since many of the species were classified and named on the basis of the plant from which they were isolated without much consideration for the host specificity or for properties other than pathogenicity.

Changes in the definition of the genus throughout the years contributed to formidable nomenclatural hypertrophy. However, recently the definition has gained in precision, although in some instances the changes represent only artificial limitations of doubtful phylogenetic value.

Work started at the Department of Bacteriology of the University of California at Berkeley in the early 1960's with the purpose of creating some order in the confused state of *Pseudomonas* taxonomy through an extensive phenotypic study of a large number of strains. At the center of the phenotypic characterization was a convenient methodological modernization of the nutritional approach introduced by den Dooren de Jong in 1926, by means of which many strains could be analyzed for growth on a large number of organic compounds used as the sole source of carbon and energy. These extensive studies of metabolic pathways in *Pseudomonas* were a valuable guide for the selection of additional independent characters and resulted in a minimum of redundancy in the taxonomic conclusions.

These studies permitted a more precise definition of known species, the creation of new ones, and the identification of groups of species related by many characters, including similar DNA base composition, nutritional properties, metabolic pathways, as well as metabolic regulatory mechanisms (Stanier et al., 1966). The phenotypic characterization was followed by studies of homology of DNA sequences by DNA–DNA hybridization, with results that supported the phenotypic analysis (Palleroni et al., 1972). At this point, much of the collected information was strongly suggestive of a marked heterogeneity in *Pseudomonas*, but the methodology used until then lacked appropriate resolving power for an experimental evaluation of distant relationships.

Studies performed in other laboratories (Doi and Igarashi, 1965; Dubnau et al., 1965) demonstrated that ribosomal RNA (rRNA) is the transcription product of a very conservative part of the genome and that rRNA/DNA hybridization could be used as a probe for the estimation of distant relationships among organisms. Since *Pseudomonas* included many species among which there was no detectable DNA homology, application of such methodology to members of various species groups appeared attractive indeed. These studies represented the first instance of application of this approach to the solution of taxonomic problems of a large and complex group of organisms. The results were most rewarding and now serve as the basis for the current classification of *Pseudomonas* species into groups (Palleroni et al., 1973) (Fig. 2.1).

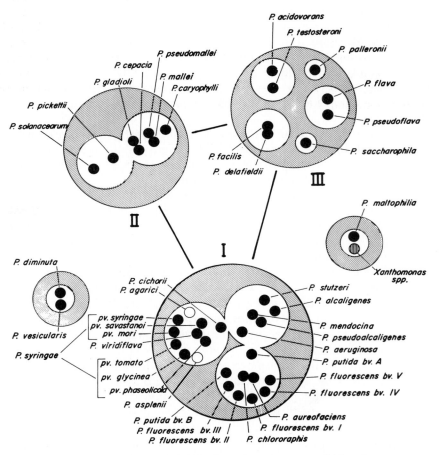

FIGURE 2.1 Ribosomal RNA homology groups within the genus *Pseudomonas*. The RNA homology groups are indicated by large shaded circles, within which the white circles represent DNA homology groups. Distances among the RNA groups are not drawn to scale. (Reprinted, with permission, from Palleroni, 1984.)

   Pseudomonas can be subdivided into five RNA homology groups. These groups are related to each other at different levels; that is, the phylogenetic distances among groups are unequal. Groups I and V are closer to one another than to other groups, and groups II and III are members of a second cluster. The affinities of group IV are not clear at present, but it is certain that it is only distantly related to the other four *Pseudomonas* groups.

   These relative distances among the groups were suggested by the hybridization data (Palleroni et al., 1973) and were confirmed more recently by the work of Woese and collaborators on the basis of a finer analysis of the 16S component of ribosomal RNA (Stackebrandt and Woese, 1981) (Fig. 2.2). The method they used is based on comparisons among the sequences of 16S oligonucleotides of different organisms obtained by T-1 ribonuclease treatment. Such comparisons are finally expressed in similarity coefficients, which can be represented in dendrogram form. These authors have examined members of many genera, and Fig. 2.2 clearly shows that the natural affinities of some *Pseudomonas* groups to other bacterial genera of different morphology and physiological properties are higher than the natural relationships among the true pseudomonads as defined by basic phenotypic criteria.

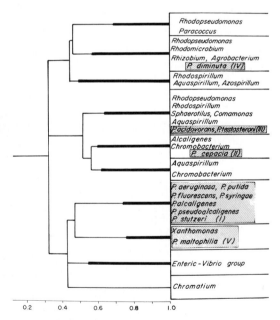

FIGURE 2.2   Relationships among some gram-negative bacterial genera, adapted from Stackebrandt and Woese (1981). Representatives of the five RNA homology groups of *Pseudomonas* are enclosed in shaded areas. The scale represents similarity coefficients among 16S oligoribonucleotide catalogs. Thick bars represent variation of similarity coefficients within each group. (Reprinted, with permission, from Palleroni, 1983.)

For a number of reasons, Group I deserves to retain the genus designation. It is a large group constituted of pigmented and nonpigmented pseudomonads. The pigmented species are the so-called fluorescent pseudomonads, one of which is the type species of the genus, *Pseudomonas aeruginosa*. This species can be easily isolated from soil where it lives freely as a saprophyte, but under certain conditions it can be pathogenic for animals and, more rarely, for plants. The species is very homogeneous; this may be related to properties restricting genetic exchange with other species. The base composition of the DNA is different from that of other fluorescent pseudomonads, the G + C content being the highest of the group (67 moles %), and vectors such as bacteriophages are very host-specific. Strains of *P. aeruginosa* isolated from various sources around the world share many distinctive phenotypic properties (Stanier et al., 1966) and a high degree of DNA homology (Palleroni et al., 1972).

Pseudomonas putida and *Pseudomonas fluorescens* are two important saprophytic species of group I, very common in soil and water. Both are very complex, particularly *P. fluorescens*, which is perhaps the most heterogeneous species of the genus (Jessen, 1965; Stanier et al., 1966). They can be subdivided into a number of biovars, some of which may deserve independent species status. Two of the biovars of *P. fluorescens* defined by Stanier et al. (1966) are now named *Pseudomonas chlororaphis* and *Pseudomonas aureofaciens*, as originally described.

A separate subgroup of fluorescent pseudomonads includes the plant pathogenic species *Pseudomonas cichorii* and *Pseudomonas syringae*. The latter is a collective species comprising a large number of pathovars (Young et al., 1978; Palleroni, 1984).

The nonpigmented members of group I include two denitrifying species as members of a small subgroup, *Pseudomonas stutzeri* and *Pseudomonas mendocina* (Palleroni et al., 1970), and another small cluster, *Pseudomonas alcaligenes* and *Pseudomonas pseudoalcaligenes* (Ralston-Barrett et al., 1976).

RNA homology group II is constituted mainly of pathogenic species. One of them, *Pseudomonas cepacia*, is particularly interesting from a biochemical point of view because it is the most nutritionally versatile of all aerobic pseudomonads (Stanier et al., 1966; Ballard et al., 1970). The group includes important animal pathogens (*Pseudomonas mallei* and *Pseudomonas pseudomallei*) and an important and complex plant pathogenic species, *Pseudomonas solanacearum*, of very wide host range.

Pseudomonas pickettii is the only species of group II for which pathogenic propensities have not yet been clearly defined, although one of its biovars may be significant in human disease (Pickett and Greenwood, 1980).

Group III is dominated by hydrogen pseudomonads: *Pseudomonas saccharophila*, *Pseudomonas facilis*, *Pseudomonas flava*, *Pseudomonas pseudoflava*, and *Pseudomonas palleronii*. A purely chemoorganotrophic species of the group, *Pseudomonas delafieldii*, is so similar to *Pseudomonas facilis* (Davis et al., 1970; Ralston et al., 1972) that the two names should be considered synonyms.

Two other chemoorganotrophic species are included in group III as a separate subgroup, *Pseudomonas acidovorans* and *Pseudomonas testosteroni*. Sharply different from the fluorescent organisms of group I, these two species were examined taxonomically with special interest by the Berkeley group (Stanier et al., 1966).

As was mentioned before, group IV is quite distant from other groups of the genus. Its two species, *Pseudomonas diminuta* and *Pseudomonas vesicularis*, have unique properties (Ballard et al., 1968) not found in any of the other *Pseudomonas* species, and they will be useful in defining a new genus in the future.

Finally, group V includes the species *Pseudomonas maltophilia* and species of *Xanthomonas*. Palleroni et al. (1973) presented the first pieces of convincing evidence of the phylogenetic relationship of these two taxa and of the fact that they are closer to group I than to any other group within the genus. A recent proposal for assigning *P. maltophilia* to the genus *Xanthomonas* (Swings et al., 1983) is rather unfortunate in our opinion. These two groups of bacteria share little DNA homology and can be differentiated by a group of striking properties (flagellar number, pigment composition, pathogenicity, etc.) on the basis of which the creation of a new genus for *P. maltophilia* seems justified.

In any case, the similarities between *P. maltophilia* and *Xanthomonas* species are indeed remarkable, being expressed not only in the high RNA homology values and a concomitant high similarity of 16S oligoribonucleotides, but also in unique details of the composition of the outer membrane and lipids (Palleroni, 1978).

The number of *Pseudomonas* species that have been left out of the current scheme of classification is very large. Of them, about 60 species are included in a separate section of the first volume of *Bergey's Manual of Systematic Bacteriology* (Palleroni, 1984). This miscellaneous group of unclassified species is composed of (a) species that have been temporarily preserved in the approved list of bacterial names (Skerman et al., 1980) and represented in culture collections and (b) other species not included in the approved list but represented by available strains on which a substantial amount of research has been done. Among the latter, interesting examples are seven species of autotrophic pseudomonads, five of which are able to live on CO as sole energy and carbon source, and a well-characterized group of marine pseudomonads. Many of the species of this miscellaneous group eventually will be assigned to RNA homology groups, while others may be excluded altogether from the genus. In a recent paper by Byng et al. (1983) the following species of the miscellaneous section have been assigned to each of the groups indicated in parentheses: *Pseudomonas synxantha* and *Pseudomonas fragi* (I), *Pseudomonas pyrrocinia* and *Pseudomonas andropogonis* (II), *Pseudomonas alboprecipitans* (a synonym of *Pseudomonas avenae*) (III).

The scheme of internal subdivision of the genus *Pseudomonas* into five homology groups first proposed by Palleroni et al. (1973) has received confirmation in many laboratories, not only by a more detailed study of RNA sequences, but also by a study of the regulatory mechanisms of biosynthetic pathways leading to aromatic amino acids (Byng et al., 1980, 1983; Whitaker et al., 1981a, b).

B. *Xanthomonas*

The genus was created by Dowson in 1939 for a unique cluster of strains within the old genus *Phytomonas*, and its taxonomic career in many ways was similar to that of the fluorescent phytopathogens. A large number of species were assigned to it, whose names derived from the host plant of origin. In the eighth edition of *Bergey's Manual* (1974), both the fluorescent phytopathogens and the genus *Xanthomonas* appeared strongly reduced. Over 100 nomenspecies of *Xanthomonas* were lumped into a single species, *Xanthomonas campestris* (Dye and Lelliott, 1974), while the fluorescent phytopathogens giving a negative oxidase reaction were reduced to the single species *P. syringae* (Doudoroff and Palleroni, 1974). As was expected, many plant pathologists reacted strongly to these temperaments and, without objecting to the soundness of such radical change on taxonomic grounds, insisted on the convenience of maintaining the names in the categories of pathovars for practical reasons. Aside from *X. campestris* with its numerous pathovars, the species of *Xanthomonas* presently recognized are *Xanthomonas fragariae*, *Xanthomonas albilineans*, *Xanthomonas axonopodis*, and *Xanthomonas ampelina*.

As we have mentioned before, the position of *Xanthomonas* in a natural classification of pseudomonads is as part of a group together with *P. maltophilia*, and there are a number of important differences for the differentiation of these two taxons.

III. MORPHOLOGY AND CYTOLOGY

The cells of strains of *Pseudomonas* and *Xanthomonas* are relatively short rods less than 1 μm in diameter. Long cells are predominant in some strains. The cells are usually arranged either singly or in pairs, and less frequently in chains. Few details may be seen within the small cells of pseudomonads under the phase microscope; but under certain conditions, many *Pseudomonas* species are capable of accumulating large amounts of very refractile granules of poly-β-hydroxybutyrate. This accumulation is favored in media with low nitrogen and high carbon content.

In thin sections observed under the electron microscope the cell envelopes of pseudomonads are typical of gram-negative organisms. In the microscopic analysis of the cell envelopes of one species, *P. aeruginosa*, freeze-etching studies have been particularly informative (Lickfield et al., 1972; Gilleland et al., 1973). The chemical composition and the physicochemical properties of the outer membrane of pseudomonads have been examined in detail in the laboratories of P. Meadow and S. G. Wilkinson in England and of H. Nikaido at

Berkeley. The outer membrane and lipid composition are characteristically different in members of the various RNA homology groups. Species other than *P. aeruginosa* are now attracting the attention of biochemists. Analysis of the composition of the membranes of *P. cepacia* has been reported by Manniello et al. (1979) and by Anwar et al. (1983). For a discussion of taxonomic implications of membrane composition, see Palleroni (1978, 1981).

Many strains of *Xanthomonas* have cells surrounded by abundant capsular material, both in their natural plant habitats and in artificial media containing sugars. The mucous material has been named xanthan gum, and in some species it consists of an anionic heteropolysaccharide composed of glucose, mannose, glucuronate, and small amounts of pyruvate and acetate. Some nonmucoid mutants of these strains have exopolysaccharides built of glucose, rhamnose, and galactose units (Whitfield et al., 1981). These polysaccharides may function in the colonization of the plant host (Morris et al., 1977).

Pseudomonas and *Xanthomonas* flagella are polar, but some species of *Pseudomonas* may also produce lateral flagella of shorter wavelength (Palleroni et al., 1970). One species, *P. mallei*, is permanently nonflagellated (Redfearn et al., 1966). The number of flagella per pole is taxonomically significant. *Xanthomonas* cells rarely have more than one polar flagellum per cell. In contrast, its relative *P. maltophilia* is multiflagellated.

Pili or fimbriae can also be observed in some species of *Pseudomonas*. Several roles have been attributed to fimbriae in pseudomonads: attachment of pathogens to host cells and interference with phagocytosis (Buchanan and Pearce, 1979), formation of star-shaped cell aggregates (Heumann, 1962), twitching motility (Bradley, 1980), attachment of phages (Bradley, 1972), and participation in conjugation (Bradley, 1983). To our knowledge, no pili have been reported for *Xanthomonas* strains.

Pseudomonas species can be divided into two groups according to their sensitivity to ethylenediaminetetraacetate (EDTA). At low concentrations this chelator rapidly alters the cell integrity of *P. aeruginosa*, and lysis ensues (Wilkinson, 1970). Other *Pseudomonas* species (e.g., *P. maltophilia*) are less sensitive to EDTA.

Pigments are produced by several *Pseudomonas* species and by *Xanthomonas*. The character of pigment production is not universal within *Pseudomonas*, but it is of considerable value in the differentiation of species.

Fluorescent pigments are among the most notorious and characterize one large subgroup within group I, as mentioned before. They are produced most abundantly under conditions of iron starvation (Meyer, 1977; Meyer and Abdallah, 1978). Fluorescent pigments have been known for a long time, and the structure of pyoverdine *Pa* of *P. aeruginosa* has been elucidated only recently (Wendenbaum et al., 1983). Other important pigments are derivatives of phenazine; pyocyanine, the well-known blue pigment of *P. aeruginosa*; and various pigments of *P. fluorescens* biovars and of *P. cepacia*. All these pigments usually diffuse into the medium, while others, such as the xanthomonadins of *Xanthomonas*, the carotenoids of several *Pseudomonas* species (*P. mendocina*, *P. flava*, *P. palleronii*), the blue pigment of *P. fluorescens* biovar IV (*P. lemonnieri*), and the purple pigment indigoidine of *P. indigofera* are much less soluble in water and remain associated with cell structures.

Xanthomonadins are produced by most *Xanthomonas* strains. They are yellow water-insoluble pigments having the structure of brominated aryl polyenes (Andrewes et al., 1973; Starr et al., 1977).

IV. NUTRITION

Strains of *Pseudomonas* have very simple nutritional requirements. The genus still includes members requiring one or more organic growth factors, but in all likelihood these species will be assigned to other genera in the future. Simple mineral media (Stanier et al., 1966; Palleroni and Doudoroff, 1972), supplemented with a single organic compound of low molecular weight as the sole carbon and energy source, allow good growth. The number of compounds utilizable by strains of *Pseudomonas* is very large, and this is one of the most striking physiological features of the group, namely, a remarkable nutritional versatility, which is responsible for the active participation of *Pseudomonas* in the carbon cycle in nature.

The list of organic compounds utilizable for growth by *Pseudomonas* typically excludes most of the C-1 compounds (CO_2 being an exception). These compounds are used for growth by various types of microorganisms, some of which have some superficial resemblances to *Pseudomonas* and still carry (although incorrectly) this generic designation.

Species of *Xanthomonas* are more demanding. They usually require methionine, glutamate, and nicotinate in various combinations as growth factors (Dye and Lelliott, 1974). Tryptophan can relieve the need for nicotinate, of which it is the metabolic precursor (Wilson and Henderson, 1963). The main carbon source of the medium is usually a monosaccharide, glucose being the most commonly used. Extensive systematic studies of nutrition have not been performed for *Xanthomonas*; thus we do not have information comparable to the enormous mass of nutritional data available for many species of *Pseudomonas*. General media for the cultivation of *Xanthomonas* usually contain yeast extract, glucose, and calcium carbonate.

No species of *Pseudomonas* or *Xanthomonas* can fix dinitrogen. Ammonium salts, nitrate, or amino acids can be used by most species of *Pseudomonas* as nitrogen sources (Stanier et al., 1966). The aerobic assimilation of nitrate involves reduction to ammonia, which eventually is incorporated in all the nitrogen compounds of the cell. The formation of nitrite from nitrate is the only step in common with the anaerobic dissimilation of nitrate (denitrification), but some gene products are also shared (Goldflam and Rowe, 1983).

As for *Xanthomonas*, ammonium salts are better nitrogen sources than nitrate. To this day, we have no explanation for the curious report by Patel and Kulkarni (1949) that *Xanthomonas malvacearum* uses KNO_3 but not $NaNO_3$ as a nitrogen source, and yet this observation is cited uncritically in many reviews. Use of various amino acids as carbon and nitrogen sources by *Xanthomonas* was studied by Kotasthane et al. (1965). Use of asparagine as a nitrogen source but not as a carbon source was described by Starr (1946) and proposed by Dye (1962) as a diagnostic property of the genus.

V. METABOLISM

Members of both genera have a metabolism typically respiratory, with oxygen as the final electron acceptor. They are absolute aerobes, but some species of *Pseudomonas* can use nitrate instead of oxygen (nitrate respiration or denitrification). This system is derepressed under anaerobic conditions. The capacity for denitrification is absent in *Xanthomonas*.

The cytochrome composition of a number of *Pseudomonas* species was reported by Stanier et al. (1966), Sands et al. (1967), Davis (1967), Auling et al. (1978), and Ballard et al. (1968). Most *Pseudomonas* species give a positive oxidase reaction, which is diagnostic for the presence of cytochrome *c* in gram-negative bacteria (Stanier et al., 1966; P. Baumann et al., 1968). Other species give a weak or negative reaction, and this is also the case with *Xanthomonas*, which has a very low cytochrome *c* content (Hochster and Nozzolillo, 1960).

All *Pseudomonas* species that have been examined have a functional tricarboxylic acid cycle typical of absolute aerobes (Weitzman and Jones, 1968), and for biosynthetic purposes the glyoxylate cycle operates as an anaplerotic system (Kornberg and Madsen, 1958). The Krebs cycle has also been identified in *Xanthomonas* (Madsen and Hochster, 1959).

Much work has been done on the biochemistry and metabolism of several *Pseudomonas* species, but in general our knowledge of the metabolism of many other species and of *Xanthomonas* species is considerably more limited. The catabolism of some organic compounds by *Pseudomonas* strains involves the participation of oxygenases. Some of them are responsible for the initial attack of many aliphatic and aromatic hydrocarbons and therefore are of great importance in the degradation of oil components and in the treatment of oil spills. Other oxygenases are specific for terpenoid compounds, such as camphor, in degradative pathways of great biochemical complexity (Gunsalus et al., 1971).

The degradative pathways of aromatic compounds such as mandelate, benzoate, naphthalene, and their derivatives have been studied in great detail, and both the pathways and their regulatory mechanisms have phylogenetic implications of great interest. The degradation of many of these compounds by *Pseudomonas* species of group I proceeds through the so-called beta-ketoadipate pathway. This intermediate originates after "dearomatization" of dihydroxylated intermediates (e.g., catechol) by the action of a specific 1,2-dioxygenase, which catalyzes a so-called "ortho" cleavage. This mechanism of ring fission is also used by members of group II. In contrast, in members of group III the metabolism of aromatic compounds does not converge to beta-ketoadipate, and cleavage of the aromatic ring occurs by the action of a 2,3-dioxygenase ("meta" cleavage). This division is not absolute, since organisms of group I may carry plasmids with genes encoding enzymes specific for a meta cleavage of the aromatic intermediates (Austen and Dunn, 1980).

Many studies on the metabolism of aromatic compounds have been reported in the literature, and there is also much valuable information on the regulatory mechanisms and on the immunology and amino acid sequences of selected enzymes. Valuable reviews and those of Ornston (1971) and Clarke and Ornston (1975).

About 30 years ago, one of the most popular species of *Pseudomonas* in the study of carbohydrate metabolism was *P. saccharophila*. This is a hydrogen pseudomonad, now included in group III (Fig. 2.1), which was discovered by M. Doudoroff in 1940. An unusual form of glucose dissimilation was discovered by Entner and Doudoroff (1952) in this organism. One of the intermediates, 6-phosphogluconate, undergoes dehydration, and 2-keto-3-deoxy-6-phosphogluconate is formed. This last compound has in its structure the direct precursors of pyruvate and triosephosphate, to which it gives rise by the action of a specific aldolase. This pathway of glucose degradation is now known to occur in other species and in catabolic pathways of other monosaccharides among the pseudomonads.

Fluorescent pseudomonads (group I) oxidize glucose by peripheral pathways converging to 6-phosphogluconate (Eisenberg et al., 1974), with interesting species variations (Vicente and Cánovas, 1973) or variations induced by the oxygen tension (Hunt and Phibbs, 1981). The phosphoenolpyruvate–phosphotransferase system has been shown to operate in the metabolism of fructose by various *Pseudomonas* species (L. Baumann and P. Baumann, 1975, P. Baumann and L. Baumann, 1975; Sawyer et al., 1977a, b).

The Entner-Doudoroff pathway is functional in *Xanthomonas* (Katznelson, 1955, 1958), but gluconate is not respired, although Dye and Lelliott (1974) report that this compound may be utilized after a delay.

Many hydrolases have been identified in *Pseudomonas* species: proteases (Morihara, 1964; Morihara et al., 1965), amylases (Markowitz et al., 1956; Robyt and Ackerman, 1971), PHB depolymerases (Delafield et al., 1965; Lusty and Doudoroff, 1966), pectin hydrolases (Hildebrand, 1971; Wilkie et al., 1973; Ohuchi and Tominaga, 1973, 1975), xylanases (Maino et al., 1974), various glycosidases (Hayward, 1977), cellulases (Gehring, 1962; Lange and Knösel, 1970). Glycosidases are also found in *Xanthomonas* (Hayward, 1977; Dekker and Candy, 1979), and pectolytic activity was also detected, with significant differences among the various pathovars, some of which have a very low activity (Dye, 1960). The various hydrolases of *Xanthomonas* and of phytopathogenic species of *Pseudomonas* probably play an important role in pathogenesis.

The metabolism of mono- and dicarboxylic acids by *P. aeruginosa* and the relationship with the glyoxylate cycle have been examined by Chapman and Duggleby (1967). The acids enter the tricarboxylic acid cycle after degradation to the stage of acetyl-CoA. Metabolism of various acids has been reported by Shilo and Stanier (1957), Hurlbert and Jakoby (1965), and Cooper and Kornberg (1964).

The degradation of amides by *P. aeruginosa* has been described in great detail by Clarke and her collaborators in many papers covering both the biochemistry and the genetics. Certain aliphatic amides are hydrolyzed by the same amidase, but Clarke and collaborators succeeded in a number of cases in altering the specificity of the enzyme, suggesting mechanisms that may occur during evolution in the process of differentiation of species (Clarke, 1974).

Most amino acids are excellent carbon sources for *Pseudomonas* strains, and the nutritional patterns have interesting taxonomic implications (Stanier et al., 1966). Permeases with various degrees of specificity have been reported (Rosenfeld and Feigelson, 1969; Hechtman and Scriver, 1970). Different catabolic pathways may be found in different species for a given amino acid. This occurs, for instance, in the degradation of tryptophan, which can proceed by a pathway that includes the intermediate beta-ketoadipate, in common with other pathways of metabolism of aromatic compounds (group I), or through quinoline derivatives as intermediates (group III) (Behrman, 1962; Stanier et al., 1966). Coordinate metabolite induction has been demonstrated in the aromatic pathway of tryptophan degradation by fluorescent pseudomonads (Palleroni and Stanier, 1964).

Interesting variations in the catabolism of lysine have been described by Miller and Rodwell (1971), Chang and Adams (1971), and Fothergill and Guest (1977). Three different pathways converge to produce glutarate, which is converted to acetyl-CoA.

In some instances, arginine degradation by pseudomonads includes the so-called arginine dihydrolase system, which provides energy for motility under anaerobic conditions (Sherris et al., 1959). Other pathways of arginine catabolism can also be identified (Miller and Rodwell, 1971; Stalon and Mercenier, 1984).

Pseudomonas species offer a great wealth of biochemical variations in the amino acid biosynthetic pathways and their regulatory mechanism. Because amino acids may be readily utilized for growth, elucidation of the biosynthetic pathways and their control can be difficult. The genetic aspects of these pathways are also rich in information. Often the degree of clustering of the genes involved is radically different from that found in the enteric bacteria (Mee and Lee, 1967).

In the biosynthesis of amino acids of the aspartate family, high concentrations of either lysine or threonine can cause feedback inhibition of aspartokinase in species of group I. In contrast, in group III organisms, sensitivity to individual amino acids is lower (Cohen et al., 1969).

Participation of acetylated intermediates in arginine biosynthesis presents interesting variations in pseudomonads, where de novo acetylation is replaced by transacetylation among intermediates, thus becoming energetically more efficient (Udaka, 1966; Leisinger et al., 1972). The catabolic arginine dihydrolase system mentioned above has steps in common with the biosynthetic pathway, but the overlapping reactions can be distinguished by their biochemical and regulatory properties (Stalon et al., 1972).

The work of Jensen and his collaborators on the regulatory mechanisms of the pathways of aromatic amino acid biosynthesis has demonstrated that a clear differentiation of members of the various RNA groups can be achieved. The work introduces a methodology useful in assigning new strains to any of the groups. The biosynthetic pathways of two of the aromatic amino acids, phenylalanine and tyrosine, have been particularly informative. Comparative allosteric studies of 3-deoxy-D-arabinoheptulosonate-7-phosphate (DAHP) synthetase permits the certain identification of organisms of groups IV and V, but not a clear-cut differentiation of strains of groups I, II, and III (Whitaker et al., 1981a). Such a goal can be reached on the basis of the regulatory pattern of the reactions specific for the tyrosine branch of the pathway (Byng et al., 1980). In addition, the phenylalanine branch provides a fine-tuning probe to further differentiate DNA homology groups within the large RNA groups (Whitaker et al., 1981b). General considerations on the evolutionary significance of the various patterns of aromatic amino acid biosynthesis for the pseudomonads (including *Xanthomonas*) have been summarized in a recent paper by Byng et al. (1983).

VI. PRODUCTS OF PSEUDOMONADS

A wide variety of *Pseudomonas* and *Xanthomonas* products is recorded in literature, but we shall consider only a few examples here. A bibliographic source of products of pseudomonads may be found in the work of Laskin and Lechevalier (1973). Compounds of interest include phenazines, pyrrole and quinoline derivatives, keto acids, vitamin B_{12}, enzymes (proteases, amylases, cellulases, lipases), enzyme inhibitors, and antibiotics.

Among the species described in the new version of *Bergey's Manual* (Palleroni, 1984), three have been isolated in antibiotic screening programs, *Pseudomonas acidophila*, *Pseudomonas mesoacidophila*, and *P. pyrrocinia*. *P. pyrrocinia* has been assigned by Byng et al. (1980) to group II and produces pyrrolnitrin (Imanaka et al., 1965). *P. acidophila* and *P. mesoacidophila* were described by Imada et al. (1980a, b), but no assignment to any of the RNA homology groups has been suggested so far. In view of the acidophilic character of these species, which is absent from most *Pseudomonas* species, further studies may be necessary to decide whether these organisms are in fact species of *Pseudomonas*.

One *Xanthomonas* product, the extracellular heteropolysaccharide xanthan, has considerable commercial importance because of its interesting physical properties. One application is in the enhancement of oil recovery, for which there is increasing demand (Sandvick and Maerker, 1977). Chemically defined media for high yields of xanthan gum have been formulated (Souw and Demain, 1979). The possibility of catastrophic consequences for plant health by accidents or sabotage in the handling of enormous masses of *Xanthomonas* cells has not received all the attention that it deserves (Starr, 1981).

A novel group of pseudomonad products may be found in the field of restriction endonucleases for use in analytical studies on DNA and in recombinant DNA work. As can be seen in Table 2.1, *Xanthomonas* occupies a much more important position than *Pseudomonas* in the literature of restriction enzymes, which is reflected in a more prominent representation in catalogs of commercially available endonucleases. This is surprising when one considers the well-deserved reputation of *Pseudomonas* species for versatility and biochemical ingenuity.

VII. GENETICS

Members of the genus *Pseudomonas* have considerable biological importance and are very easily handled under laboratory conditions. Some species are agents of serious nosocomial infections, and many species display marked nutritional and biochemical versatility. Therefore it is not surprising that the genus should have attracted the attention not only of biochemists, but also of geneticists. At present, *P. aeruginosa* is one of the bacterial species better known from the genetic standpoint. The principal merit for the development of genetic studies of the pseudomonads goes to Professor B. W. Holloway and his colleagues at Monash University in Australia (Holloway et al., 1979).

Most genetic studies of *P. aeruginosa* have been carried out on a strain of Australian origin, strain PAO. A second strain, PAT, isolated in South Africa, has also received much attention. At present we have a genetic map of PAO with about 100 markers and a less detailed map for PAT resembling that of PAO, as expected, in view of the marked homogeneity of the species and the high level of DNA homology shared by different strains.

The three basic mechanisms of genetic exchange in prokaryotes — conjugation, transduction, and transformation — have been found to occur in members of the genus *Pseudomonas*. The first two processes have been the most useful in mapping work. Plasmid FP2, which acts as a fertility factor in the mobilization of the chromosome, was used in the initial stages of gene mapping of PAO, and later other plasmids were added to the arsenal. Several of these plasmids had the same transfer origin as FP2, but one of them, FP110, was found to have a different attachment site, which made it more effective in the mobilization of late chromosome markers.

Plasmid R68 of the incompatibility group I (see below) and R91 of group 10 have good chromosome mobilizing ability (*cma*) in strain PAT and are

TABLE 2.1   Restriction Enzymes Produced by Pseudomonads

Enzyme	Organism*	Specifcity	Reference
Xam I	Xanthomonas amaranthicola	GTCGAC	Arrand et al. (1978)
Xba I	Xanthomonas badrii	T′CTAGA	Zain and Roberts (1977)
Xho I	Xanthomonas holcicola	C′TCGAG	Gingeras et al. (1978)
Xho II	Xanthomonas holcicola	Pu′GATCPy	New England Biolabs Catalog (1983)
Xma I	Xanthomonas malvacearum	C′CCGGG	Endow and Roberts (1977)
Xma II	Xanthomonas malvacearum	CTGCAG	Endow and Roberts (1977)
Xma III	Xanthomonas malvacearum	C′GGCCG	Kunkel et al. (1979)
Xmn I	Xanthomonas manihotis	GAANN′NNTTC	Lin et al. (1980)
Xni I	Xanthomonas nigromaculans	CGATCG	BRL Catalog (1982)
Xor I	Xanthomonas oryzae	CTGCAG	Wang et al. (1980)
Xor II	Xanthomonas oryzae	CGAT′CG	Wang et al. (1980)
Xpa I	Xanthomonas papavericola	C′TCGAG	Gingeras et al. (1978)
Pae R7 I	Pseudomonas aeruginosa	C′TCGAG ⎫	
Pfa I	Pseudomonas facilis	GATC ⎬	New England Biolabs Catalog (1983)
Pma I	Pseudomonas maltophilia	CTGCAG ⎭	

* The nomenspecies of *Xanthomonas* in this list are now considered pathovars of the species *Xanthomonas campestris.*

capable of mobilizing the chromosome from different origins, a property that has facilitated the demonstration of chromosome circularity of PAT (Watson and Holloway, 1978a, b). When used in strain PAO, R68 shows very poor *cma*, but one variant with enhanced chromosome mobilizing (*ecm*) capacity (R68.45) could be isolated, and circularity of the PAO chromosome was eventually demonstrated (Royle et al., 1981) (Fig. 2.3).

The *ecm* ability of R68.45 is related to the presence of an insertion sequence (IS21), whose origin is a duplication of a segment already present in R68 (Willetts et al., 1981). Somehow, this characteristic allows R68.45 to interact with the bacterial chromosome and promote its mobilization. R68.45 can be transferred to many gram-negative bacteria. It is extensively used at present for the construction of "primes," which are plasmids carrying a piece of bacterial chromosome. The survival of these plasmids in a recipient cell depends on preventing loss of the extra segment by recombination. This is achieved by using as a recipient either a strain of a different species or a recombination-deficient mutant of the same species as the donor.

Transduction is a common occurrence in *P. aeruginosa* but rare in other species. Numerous transducing phages can be isolated, and transduction has been used for fine-mapping purposes (Holloway and Krishnapillai, 1975). In contrast, the impact of transformation on *Pseudomonas* genetic research has been much less significant. However, the calcium chloride-induced transformation may be effective in *Pseudomonas* (Sano and Kageyama, 1977; Hara et al., 1981), and this method of genetic exchange probably will be exploited more frequently in future studies.

P. putida, an important soil pseudomonad, recently has been introduced to active genetic research, and the circularity of its genetic map has been demonstrated at Monash University (Dean and Morgan, 1983). The map has significant differences from that of *P. aeruginosa*. Although the two organisms are classified in the same RNA homology group, they have important phenotypic differences (Stanier et al., 1966). The genetic properties of other species of the genus are only imperfectly known at present.

It is fair to assume that *Xanthomonas* will soon be explored for its genetic properties. As expected, promiscuous plasmids of incompatibility group I can be transferred to *Xanthomonas* species, and transmission of these plasmids from *Xanthomonas* to other gram-negative bacteria is possible (Lai et al., 1977).

VIII. PLASMIDS, PHAGES, AND BACTERIOCINS

Many bacteria, including pseudomonads, carry plasmids, which are extrachromosomal elements of double-stranded DNA capable of autonomous replication. In fact, the variety of plasmids found in the few species of *Pseudomonas* that have been examined is as large as that of any other group of bacteria.

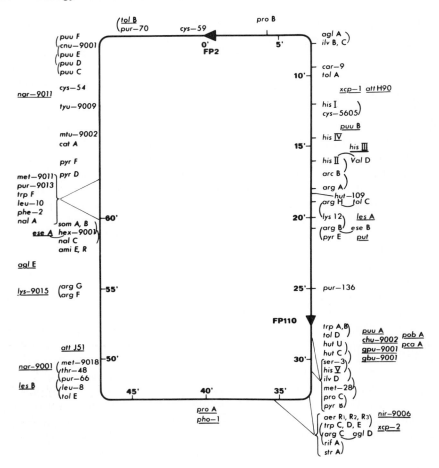

FIGURE 2.3 Chromosome map of *P. aeruginosa* PAO. Symbols: lines joining locus abbreviation to map were located by interrupted mating with FP2 donors; round brackets join markers contransducible with phages F116, F116L, G101, and E79tv-1. Abbreviations are underlined for markers not yet precisely located. Abbreviations are as follows. Anabolic markers: *arg*, arginine; *aro*, aromatic acids; *car*, carbamoyl; *cys*, cysteine; *his*, histidine; *ilv*, isoleucine-valine; *lys*, lysine; *met*, methionine; *phe*, phenylalanine; *pro*, proline; *pur*, purine, *pyr*, pyrimidine; *ser*, serine; *thr*, threonine; and *trp*, tryptophan. Catabolic markers: *ami*, amidase; *arc*, arginine; *cat*, catechol; *chu*, choline; *cnu*, carnosine; *cbu*, guanidinobutyrate; *gpu*, guanidinopropionate; *hex*, hexose; *hut*, histidine; *mtv*, mannitol; *pca*, protocatechuate; *pob*, *p*-hydroxybenzoate; *put*, proline; *puu*, purine; *tyu*, tryosine; and *val*, D-valine. Resistance markers: *agl*, aminoglycoside; *chl*, chloramphenicol resistance; *ese*, phage E79; *fpa*, *p*-fluorophenylalanine; *fus*, fusidic acid hypersensitivity; *nal*, nalidixic acid; *rif*, rifampin; *spc*, spectinomycin; *str*, streptomycin. Other markers:

Plasmids carry genes conferring properties such as resistance to antibiotics, phages, bacteriocins, and various toxic chemicals, or they can provide the capacity to degrade many different organic compounds. Thus they contribute to the marked nutritional versatility of this group. As we have seen, plasmids can also have chromosomal-mobilizing ability and act as fertility factors, or they can interact in various ways with bacteriocins, phages, or other plasmids. An excellent review on *Pseudomonas* plasmids with particular emphasis on *P. aeruginosa* has been written by Jacoby (1979). Valuable information is to be found also in the reviews by Chakrabarty (1976) and Jacoby and Shapiro (1977). A recent review by Foster (1983) updates the aspect of plasmid-determined resistance to antibiotics, with many examples referring to *Pseudomonas* species.

Different plasmids may or may not be able to coexist in the same cell, and on this basis they can be classified into different incompatibility groups. At least ten such groups can be defined in *Pseudomonas* (Jacoby and Shapiro, 1977). Of these, group 1 (IncP-1) includes plasmids of wide host range that can be transferred to members of many gram-negative genera. Group 2 (IncP-2) comprises some very large plasmids, some of which carry genes specifying enzymes for the degradation of unusual organic compounds.

Because many plasmids transmit the property of antibiotic resistance, they have considerable importance in medicine, and the spread of *Pseudomonas* infections in hospitals after the introduction of antibiotic therapy is of great medical concern (Doggett, 1979). In contrast, there are few recorded instances of participation of plasmids in plant pathogenesis by pseudomonads. In one instance the conversion of tryptophan to the plant hormone indoleacetic acid by enzymes coded for by plasmid genes in the phytopathogen *P. syringae* pv. *savastanoi* explains the formation of galls by this organism in the diseased plant (Comai and Kosuge, 1980). Production of the toxin syringomycin correlates with the presence of a plasmid in the pathogen (González and Vidaver, 1977).

Few reports on the detection of plasmids in *Xanthomonas* are found in the literature. Lin et al. (1979) discovered plasmids in the species *Xanthomonas manihotis*, and Kado and Liu (1981) developed a rapid technique for the detection of plasmids of a wide range of sizes with which they detected plasmids in *Xanthomonas pruni* and *Xanthomonas vitians* (these three species are now considered to be pathovars of *X. campestris*).

Wide-range plasmids of the IncP-1 group, such as RP4 and RK2, have been transferred to *X. campestris* pv. *vesicatoria* by Lai et al. (1977). It is likely that these plasmids may promote the transfer of other nonconjugative

---

aer, aeruginocin production; *att*, prophage attachment site; *les*, lysogenic establishment; *nar*, nitrate reductase; *nir*, nitrite reductase; *pho*, alkaline phosphatase; *som*, somatic antigen; *sup*, suppressor activity; *tol*, aeruginocin tolerance; *xcp*, extracellular protease. (Reprinted, with permission, from Royle et al., 1981; courtesy of B.W. Holloway.)

plasmids in *Xanthomonas*. Investigations exploring the possibility of formation of primes of plasmids like R68.45 may open new routes for genetic research in strains of this genus.

Many *Pseudomonas* phages that can be classified into various morphological types have been identified (Bradley, 1967; Holloway and Krishnapillai, 1975). Most of the phages contain DNA. Host specificity is predominant, and useful applications in the identification can be derived, but lack of correlation with other criteria of classification is common, particularly for some groups (M. P. Starr, personal communication). Phage sensitivity tests, either by themselves or in combination with other tests, have been used for identification of phytopathogenic species (or pathovars) (Billing, 1963, 1970; Crosse and Garrett, 1963). Identification of the origin of nosocomial infections by *P. aeruginosa* can be done by phage sensitivity tests (phage typing), but the information is particularly valuable when taken in conjunction with other typing procedures (Bergan, 1978; Brokopp and Farmer, 1979).

As was mentioned before, lysogeny is very common in *P. aeruginosa*, and numerous transducing phages have been described (Holloway and Krishnapillai, 1975). A general discussion of *Pseudomonas* phages is to be found in a review by Bradley (1967).

Xanthomonas phages have been reported for some pathovars of *X. campestris* (Okabe and Goto, 1963; Vidaver, 1976). As for the *Pseudomonas* phages, the host range of phages of this species may not always correlate with other properties of the hosts; but even so, in particular instances the practical value of the phage sensitivity tests in identification may be considerable (Dye et al., 1964; Hayward, 1964).

Pseudomonads are able to produce bacteriocins, that is, proteins that have a lethal effect on other strains of the same species. The bacteriocins of *P. aeruginosa* have been named pyocins and aeruginosins, but strictly speaking, both names are inappropriate (Palleroni, 1984). They can be classified into two types, one (S type) amorphous and readily hydrolyzed by proteases and a second type (R) with the appearance of phage components that are resistant to proteolysis (Bradley, 1967). Bacteriocins are specific and can be used in intraspecific classifications ("typing") (Govan, 1978; Brokopp and Farmer, 1979; Vidaver et al., 1972; Vidaver and Buckner, 1978). The practical use of phages and bacteriocins for the control of plant diseases has been reviewed by Vidaver (1976).

IX. SEROLOGY

Because of its medical importance, *P. aeruginosa* is the species of the genus that has been studied in greatest detail for its antigenic properties. Two main practical applications of these studies on animal pathogens are the protection of humans and animals against infections by means of vaccines and sera and the possibility of identification of the source of infections in epidemiological studies. In the case of plant pathogens, the main goal is identification.

Early investigations on the antigenic structure of *P. aeruginosa* could recognize two main types of antigens, one thermolabile (H-antigen), located in the cell appendages (mainly flagella), and a second type that is thermostable (O-antigen), corresponding to the lipopolysaccharide, located in the outermost layer of the outer membrane. Animals treated with whole cells produce antibodies to both types of antigens, and the respective sera give agglutination reactions when in contact with the same type of cells. Immune sera obtained early during immunization are the best, since the IgM antibodies involved in cell agglutination reactions are produced more abundantly at this stage.

A number of O- and H-antigens have been recognized in *P. aeruginosa* and comprise the bases of several typing systems. This is particularly helpful to medical microbiologists.

Aside from the main surface antigenic determinants, various other components of the *P. aeruginosa* cells can elicit antibodies in laboratory animals. These include the extracellular slime, various exoenzymes, toxins, some antigens common to many other bacteria (Høiby, 1975), and a heat-labile component present in the cytoplasm, which has been characterized in some detail (Sompolinsky et al., 1980a, b). The serological characterization and typing methods have been reviewed by Véron and Berche (1976), Lányi and Bergan (1978), and Brokopp and Farmer (1979). Papers covering various aspects of identification of plant pathogenic species of *Pseudomonas* by serological methods include those by Otta and English (1971), Guthrie (1968), Lucas and Grogan (1969a, b), and Otta (1977).

From the taxonomic point of view, *Xanthomonas* has not received much help from immunological approaches. Sometimes pathovar (or species) differentiation has been possible (Elrod and Braun, 1947; Lovrekovich and Klement, 1965), but correlations with pathogenicity may be low or absent (Charudattan et al., 1973; Schaad, 1976). Construction of serogroups may be possible, but cross-reactions are quite frequent (Yano et al., 1979).

As was mentioned before, the immunological properties of various cell components, the ability to produce bacteriocins and the sensitivity to those produced by other strains, and the sensitivity to bacteriophages are all very important criteria, not only for the taxonomy of pseudomonads but also for practical epidemiological applications. It is generally accepted that phage, bacteriocin, and serotyping, as well as sensitivity to antibiotics and selected biochemical properties, are not absolutely reliable when taken individually, but they become very useful when considered in relation to one another.

X. PATHOGENICITY

A. Plant Pathogenicity

Three of the five *Pseudomonas* RNA homology groups (I, II, and V) include plant-pathogenic species. A wide variety of symptoms are caused in plants by phytopathogenic *Pseudomonas*. They include cankers, rots, wilts, leaf spots, twig and blossom blights, and galls. Some strains cause blights in cultivated mushrooms. One or more of the pathogen's products, such as hydrolytic enzymes, toxins, or plant hormones, may be involved in the production of the symptoms. References to the production of hydrolytic enzymes that may have a participation in pathogenesis have been given before in our discussion on metabolism of pseudomonads.

Aside from the specific role as pathogen by means of invasion of plant tissues, the species *P. syringae* was also found to participate in frost damage to plants by ice nucleation (Maki et al., 1974).

Necrosis, wilts, and rots are common symptoms produced by *Xanthomonas*. Characteristically, the lesions are frequently rich in a yellow and sticky bacterial slime (Starr, 1981).

B. Animal Pathogenicity

The most important *Pseudomonas* species pathogenic for man and animals are *P. mallei* and *P. pseudomallei*. They are closely related (Redfearn et al., 1966; Rogul et al., 1970), and the diseases produced by them have marked similarities (Redfearn and Palleroni, 1975).

In the last 20 years the widespread use of antibiotics has resulted in increasing frequency of infections by antibiotic-resistant pseudomonads, which are normally opportunistic pathogens, particularly in patients in which the normal defenses are depressed. Therefore it is not surprising to find recent references to isolation and identification procedures for *Pseudomonas* species that, not too long ago, passed largely unnoticed in clinical practice (Gilardi, 1971; Hugh and Gilardi, 1980).

In hospital infections, *P. aeruginosa* is the most important etiologic agent of the genus. The factors responsible for its pathogenic potentialities are only partly defined, and they include not only the endotoxin (LPS) characteristic of gram-negatives, but also extracellular products, toxins, and enzymes. The most serious toxin is exotoxin A, a protein that inhibits protein synthesis at the translation level. The book edited by Doggett (1979) presents a general view of *P. aeruginosa* infections, with interesting chapters on cell envelopes, plasmids, toxins, typing methods, genetics, various immunological aspects, antibiotic resistance, and modern trends of therapy.

XI. ECOLOGY AND ISOLATION

A large variety of natural materials are good sources for the isolation of saprophytic pseudomonads as long as the materials have a pH close to neutrality, have some soluble organic matter and oxygen tension that is not too low, and are not normally exposed to continuous high temperature. Soil, freshwater and seawater, foods, clinical materials, plants, and animals are sources that almost invariably yield pseudomonads by appropriate methods of isolation.

Considering the versatility of the group, it is difficult in most cases to draw precise ecological conclusions from isolation data. Marine pseudomonads have an absolute sodium requirement (Baumann et al., 1972), which can easily be demonstrated; but in other examples the factors affecting the distribution of pseudomonads in nature are imperfectly known, particularly in the cases when the pseudomonads are found consistently in association with higher organisms. Thus the presence of fluorescent pseudomonads in the rhizosphere of wheat plants (Sands and Rovira, 1971), of *P. rhodos* in the rhizosphere of alder trees (Heumann, 1962), of *P. maltophilia* in the rhizosphere of cultivated plants of the family Cruciferae (Debette and Blondeau, 1980), of some *Pseudomonas* species on *Lolium* leaves, and the fact that *P. fluorescens* strains are generally associated with leaf surfaces (Austin and Goodfellow, 1979) are observations still lacking satisfactory explanations.

Soils of tropical regions are sources of *P. pseudomallei* (Redfearn et al., 1966), and mud, but not dry soils, is a source of carboxydobacteria (Zavarzin and Nozhevnikova, 1977).

The example of the plant pathogenic pseudomonads (*Xanthomonas* included) is more straightforward. Repeated isolations of certain pathovars from lesions of certain plants clearly suggest unique interactions between the pathogen and the host as ecological niche. However, selective pressure in the various niches may affect speciation in different ways, and the botanical or horticultural region may have a definite influence in differentiation of ecotypes (Garrett et al., 1966).

The diseased plant seems to be the only source for the isolation of many of these pseudomonads, and the pathogen's survival outside of the plant seems poor, in spite of the fact that all the pathogenic pseudomonads are facultative and should live well in common artificial media. Constant association with living plants is important for survival of *P. syringae* pathovars (Schroth et al., 1981), even if the association involves a temporary epiphytic state (Ercolani et al., 1974).

Direct isolation from soil easily gives saprophytic species of pseudomonads, but few solid media are sufficiently specific for the isolation of a given species. Good examples are the solid mineral medium with an overlayer of PHB for the direct isolation of organisms producing enzymes active on this polymer (Delafield et al., 1965) and the medium proposed by Sands and Rovira (1970) with antibiotics, which facilitates the isolation of fluorescent pseudomonads.

During this century, bacteria have been a rich source of biochemical information because of their unparalleled variety of metabolic activities. Many metabolic pathways have been elucidated in pseudomonads isolated by enrichment methods from soil or water. In several instances the carbon source and the conditions for the enrichment proved effective for the specific isolation of certain species or groups of pseudomonads. For instance, strains of the subgroup acidovorans within group III (*P. acidovorans* and *P. testosteroni*) can be enriched by using higher dicarboxylic acids (Stanier et al., 1966). In particular, *cis,cis-* and *cis,trans*-muconic acids are effective in enrichment for *P. acidovorans* (Robert-Gero et al., 1969). Imidazole derivatives facilitate the isolation of *P. testosteroni* (Coote and Hassal, 1973; Hassal, 1966; Hassal and Rabie, 1966).

Denitrification conditions can be used for the enrichment of *P. aeruginosa*, *P. stutzeri*, *P. mendocina*, and other denitrifiers. Enrichments under an atmosphere of H_2, O_2, and CO_2 or CO and O_2 give chemolithotrophic hydrogen pseudomonads or carboxydobacteria, respectively.

Lesions on higher plants are natural enrichment cultures from which it is feasible to obtain pathogenic pseudomonads. From the yellow slime abundant in some of the lesions produced by *Xanthomonas* in plants, strains of this group can be isolated in media containing yeast extract and glucose, where they give characteristic mucous yellow colonies. This is not invariably so, particularly when the materials are low in cell number. Some *Xanthomonas* strains do not give slime, and some yellow isolates are not *Xanthomonas*.

The review articles by Palleroni (1981), Bergan (1981), Schroth et al. (1981), Stolp and Gadkari (1981), Starr (1981), and Hugh and Gilardi (1980) and the laboratory guide edited by Schaad (1980) are recommended for further discussions on habitats, methods of isolation, cultivation, and maintenance of cultures of pseudomonads.

REFERENCES

Andrewes, A. G., Hertzberg, S., Liaaen-Jensen, S., and Starr, M. P. (1973) *Acta Chem. Scand. 27*, 2383-2395.

Anwar, H., Brown, M. R. W., Cozens, R. M., and Lambert, P. A. (1983) *J. Gen. Microbiol. 129*, 499-507.

Arrand, J. R., Myers, P. A., and Roberts, R. J. (1978) *J. Mol. Biol. 118*, 127-135.

Auling, G., Reh, M., Lee, C. M., and Schlegel, H. G. (1978) *Int. J. Syst. Bacteriol. 28*, 82-95.

Austen, R. A., and Dunn, N. W. (1980) *J. Gen. Microbiol. 117*, 521-528.

Austin, B., and Goodfellow, M. (1979) *Int. J. Syst. Bacteriol. 29*, 373-378.

Ballard, R. W., Doudoroff, M., Stanier, R. Y., and Mandel, M. (1968) *J. Gen. Microbiol. 53*, 349-361.

Ballard, R. W., Palleroni, N. J., Doudoroff, M., Stanier, R. Y., and Mandel, M. (1970) *J. Gen. Microbiol. 60*, 199-214.

Baumann, L., and Baumann, P. (1975) *Arch. Microbiol. 105*, 241-248.

Baumann, L., Baumann, P., Mandel, M., and Allen, R. D. (1972) *J. Bacteriol. 110*, 402-429.

Baumann, P., and Baumann, L. (1975) *Arch. Microbiol. 105*, 225-240.

Baumann, P., Doudoroff, M., and Stanier, R. Y. (1968) *J. Bacteriol. 95*, 58-73.

Behrman, E. J. (1962) *Nature 196*, 150-152.

Bergan, T. (1978) in *Methods in Microbiology*, Vol. X (Bergan, T., and Norris, J. R., eds.), pp. 169-199, Academic Press, New York.

Bergan, T. (1981) in *The Prokaryotes*, Vol. I (Starr, M. P., Stolp, H., Trüper, H. G., Balows, A., and Schlegel, H. G., eds.), pp. 666-700, Springer, New York.

Bethesda Research Laboratories Catalog (1981–1982) Gaithersburg, MD 20877.

Billing, E. (1963) *J. Appl. Bacteriol. 26*, 193–210.

Billing, E. (1970) *J. Appl. Bacteriol. 33*, 478–491.

Bradley, D. E. (1967) *Bacteriol. Rev. 31*, 230–314.

Bradley, D. E. (1972) *J. Gen. Microbiol. 72*, 303–319.

Bradley, D. E. (1980) *Can. J. Microbiol. 26*, 146–154.

Bradley, D. E. (1983) *J. Gen. Microbiol. 129*, 2545–2556.

Brokopp, C. D., and Farmer, J. J. (1979) in *Pseudomonas aeruginosa: Clinical Manifestations of Infection and Current Therapy* (Doggett, R. G., ed.), pp. 89–133, Academic Press, New York.

Buchanan, T. M., and Pearce, W. A. (1979) in *Bacterial Outer Membranes: Biogenesis and Function* (Inouye, M., ed.), pp. 475–514, John Wiley and Sons, New York.

Byng, G. S., Whitaker, R. J., Gherna, R. L., and Jensen, R. A. (1980) *J. Bacteriol. 144*, 247–257.

Byng, G. S., Johnson, J. L., Whitaker, R. J., Gherna, R. L., and Jensen, R. A. (1983) *J. Mol. Evol. 19*, 272–282.

Chakrabarty, A. M. (1976) *Ann. Rev. Genet. 10*, 7–30.

Chang, Y.-F., and Adams, F. (1971) *Biochem. Biophys. Res. Comm. 45*, 570–577.

Chapman, P. J., and Duggleby, R. G. (1967) *Biochem. J. 103*, 7C–9C.

Charudattan, R., Stall, R. E., and Batchelor, D. L., (1973) *Phytopathol. 63*, 1260–1265.

Clarke, P. H. (1974) *Symp. Soc. Gen. Microbiol. 24*, 183–217.

Clarke, P. H., and Ornston, L. N. (1975) in *Genetics and Biochemistry of Pseudomonas* (Clarke, P. H., and Richmond, M. H., eds.), pp. 191–340, John Wiley and Sons, New York.

Cohen, G. N., Stanier, R. Y., and LeBras, G. (1969) *J. Bacteriol. 99*, 791–801.

Comai, L., and Kosuge, T. (1980) *J. Bacteriol. 143*, 950–957.

Cooper, R. A., and Kornberg, H. L. (1964) *Biochem. J. 91*, 82–91.

Coote, J. G., and Hassal, H. (1973) *Biochem. J. 132*, 409–422.

Crosse, J. E., and Garrett, C. M. E. (1963) *J. Appl. Bacteriol. 26*, 159–177.

Davis, D. H. (1967) "Studies on the Gram Negative Hydrogen Bacteria and Related Organisms," Ph.D. thesis, University of California, Berkeley.

Davis, D. H., Stanier, R. Y., Doudoroff, M., and Mandel, M. (1970) *Arch. Microbiol. 70*, 1–13.

Dean, H. F., and Morgan, A. F. (1983) *J. Bacteriol. 153*, 485–497.

Debette, J., and Blondeau, R. (1980) *Can. J. Microbiol. 26*, 460–463.

Dekker, R. F. H., and Candy, G. P., (1979) *Arch. Microbiol. 122*, 297–299.

Delafield, F. P., Doudoroff, M., Palleroni, N. J., Lusty, C. J., and Contopoulou, R. (1965) *J. Bacteriol. 90*, 1455–1466.

den Dooren de Jong, L. E. (1926) *Bijdrage tot de kennis van het mineralisatieproces*, Nijgh and van Ditmar Uitgevers-Mij, Rotterdam.

Doggett, R. G. (ed.) (1979) *Pseudomonas aeruginosa: Clinical Manifestations of Infection and Current Therapy*, Academic Press, New York.

Doi, R. H., and Igarashi, R. T. (1965) *J. Bacteriol. 90*, 384–390.

Doudoroff, M. (1940) *Enzymologia 9*, 59–72.

Doudoroff, M., and Palleroni, N. J. (1974) in *Bergey's Manual of Determinative Bacteriology*, 8th ed. (Buchanan, R. E., and Gibbons, N. E., eds.), pp. 217–243, Williams and Wilkins, Baltimore.

Dowson, W. J. (1939) *Zantralbl. Bakteriol. Parasitenk. Infektionskr. Hyg. Abt. II, 100*, 177–193.

Dubnau, D., Smith, I. Morell, P., and Marmur, J. (1965) *Proc. Nat. Acad. Sci. U.S.A. 54*, 491–498.

Dye, D. W. (1960) *N. Zeal. J. Sci. 3*, 61–69.

Dye, D. W. (1962) *N. Zeal. J. Sci. 5*, 393–416.

Dye, D. W., and Lelliott, R. A. (1974) in *Bergey's Manual of Determinative Bacteriology*, 8th ed. (Buchanan, R. E., and Gibbons, N. E., eds.), pp. 243–249, Williams and Wilkins, Baltimore.

Dye, D. W., Starr, M. P., and Stolp, H. (1964) *Phytopathol. Z. 51*, 394–407.

Eisenberg, R. C., Butters, S. J., Quay, S. C., and Friedman, S. B. (1974) *J. Bacteriol. 120*, 147–153.

Elrod, R. P., and Braun, A. C. (1947) *J. Bacteriol. 53*, 500–518.

Endow, S. A., and Roberts, R. J. (1977) *J. Mol. Biol. 112*, 521–529.

Entner, N., and Doudoroff, M. (1952) *J. Biol. Chem. 196*, 853–862.

Ercolani, G. L., Hagedorn, D. J., Kelman, A., and Rand, R. E. (1974) *Phytopathology 64*, 1330–1339.

Foster, T. J. (1983) *Microbiol. Rev. 47*, 361–409.

Fothergill, J. C., and Guest, J. R. (1977) *J. Gen. Microbiol. 99*, 139–155.

Garrett, C. M. E., Panagopoulos, C. G., and Crosse, J. E. (1966) *J. Appl. Bacteriol. 29*, 342–356.

Gehring, F. (1962) *Phytopathol. Z. 43*, 383–407.

Gilardi, G. L. (1971) *Appl. Microbiol. 21*, 414–419.

Gilleland, H. E., Stinnett, J. D., Roth, I. L., and Eagon, R. G. (1973) *J. Bacteriol. 113*, 417–432.

Gingeras, T. R., Myers, P. A., Olson, J. A., Hanberg, F. A., and Roberts, R. J. (1978) *J. Mol. Biol. 118*, 113–122.

Goldflam, M., and Rowe, J. J. (1983) *J. Bacteriol. 155*, 1446–1449.

González, C. F., and Vidaver, A. K. (1977) *Proc. Amer. Phytopathol. Soc. 107* (4).

Govan, J. R. W. (1978) in *Methods in Microbiology*, Vol. X (Bergan, T., and Norris, J. R., eds.), pp. 61–91, Academic Press, New York.

Gunsalus, I. C., Tyson, C. A., Tsai, R. L., and Lipscomb, J. D. (1971) *Chemico-Biological Interact. 4*, 75–78.

Guthrie, J. W. (1968) *Phytopathology 58*, 716–717.

Hara, T., Aumayr, A., and Veda, S. (1981) *J. Gen. Appl. Microbiol. 27*, 109–114.

Hassal, H. (1966) *Biochem. J. 101*, 22P.

Hassal, H., and Rabie, F. (1966) *Biochim. Biophys. Acta 115*, 521–523.

Hayward, A. C. (1964) *J. Appl. Bacteriol. 27*, 265–277.

Hayward, A. C. (1977) *J. Appl. Bacteriol. 43*, 407–411.

Hechtman, P., and Scriver, C. R. (1970) *J. Bacteriol. 104*, 857–863.

Heumann, W. (1962) *Biologische Zentralbl. 81*, 341–354.

Hildebrand, D. C. (1971) *Phytopathology 61*, 1430–1436.

Hochster, R. M., and Nozzolillo, C. G. (1960) *Can. J. Biochem. Physiol. 38*, 79–93.

Høiby, N. (1975) *Scand. J. Immunol. 4* (suppl. 2), 187–196.

Holloway, B. W., and Krishnapillai, V. (1975) in *Genetics and Biochemistry of Pseudomonas* (Clarke, P. H., and Richmond, M. H., eds.), pp. 99–132, John Wiley and Sons, New York.

Holloway, B. W., Krishnapillai, V., and Morgan, A. F. (1979) *Microbiol. Rev. 43*, 73–102.

Hugh, R., and Gilardi, G. L. (1980) in *Manual of Clinical Microbiology*, 3d. ed. (Lenette, E. H., Balows, A., Hausler, W. J., and Truant, J. P., eds.), pp. 289–317, American Society for Microbiology, Washington, D.C.

Hunt, J. C., and Phibbs, P. V. (1981) *Biochem. Biophys. Res. Comm. 102*, 1393–1399.

Hurlbert, R. E., and Jacoby, W. B. (1965) *J. Biol. Chem. 240*, 2772–2777.

Imada, A., Kintaka, K., and Haibara, K. (1980a) U.S. Patent 4,225,586.

Imada, A., Kitano, K., and Asai, M. (1980b) U.S. Patent 4,229,436.

Imanaka, H., Kohsaka, M., Tamura, G., and Arima, K. (1965) *J. Antibiot. 18*, 205–206.

Jacoby, G. A. (1979) in *Pseudomonas aeruginosa: Clinical Manifestations of Infection and Current Therapy*, (Doggett, R. G., ed.), pp. 271–309, Academic Press, New York.

Jacoby, G. A., and Shapiro, J. A. (1977) in *DNA Insertion Elements, Plasmids and Episomes* (Bukhari, A. E., Shapiro, J. A., and Adhya, S. L., eds.), pp. 639–656, Cold Spring Harbor Laboratory, Cold Spring Harbor, New York.

Jessen, O. (1965) *Pseudomonas aeruginosa and Other Fluorescent Pseudomonads*, Munksgaard, Copenhagen.

Kado, C. I., and Liu, S.-T. (1981) *J. Bacteriol. 145*, 1365–1373.

Katznelson, H. (1955) *J. Bacteriol. 70*, 469–475.

Katznelson, H. (1958) *J. Bacteriol. 75*, 540–543.

Kornberg, H. L., and Madsen, N. B. (1958) *Biochem. J. 68*, 549–557.

Kotasthane, W. V., Padhya, A. C., and Patel, M. K. (1965) *Indian Phytopathol. 18*, 154–159.

Kunkel, L. M., Silberklang, M., and McCarthy, B. J. (1979) *J. Mol. Biol. 132*, 133–139.

Lai, M., Panapoulos, N. J., and Shaffer, S. (1977) *Phytopathology 67*, 1044–1050.

Lange, E., and Knösel, D. (1970) *Phytopathol. Z. 69*, 315–329.

Lányi, B., and Bergan, T. (1978) in *Methods in Microbiology*, Vol. X (Bergan, T., and Norris, J. R., eds.), pp. 93–168, Academic Press, New York.

Laskin, A. I., and Lechevalier, H. A. (1973) *Handbook of Microbiology*, Vol. III, *Microbial Products*, CRC Press, Cleveland.

Leisinger, T., Hass, D., and Hegarty, M. B. (1972) *Biochim. Biophys. Acta 262*, 214–219.

Lickfield, K. G., Achterrath, H., Hentrich, F., Kolehmainen-Sevens, L., and Persson, A. (1972) *J. Ultrastruct. Res. 38*, 27–45.

Lin, B.-C., Day, H.-J., Chen, S.-J., and Chien, M.-C. (1979) *Bot. Bull. Acad. Sinica 20*, 157–171.

Lin, B.-C., Chien, M.-C., and Lou, S.-Y. (1980) *Nucl. Acids Res. 8*, 6189–6198.

Lovrekovich, L., and Klement, Z. (1965) *Phytopathol. Z. 52*, 222–228.

Lucas, L. T., and Grogan, R. G. (1969a) *Pathopathology 59*, 1908–1912.

Lucas, L. T., and Grogan, R. G. (1969b) *Phytopathology 59*, 1913–1917.

Lusty, C. J., and Doudoroff, M. (1966) *Proc. Nat. Acad. Sci U.S.A. 56*, 960–965.

Madsen, N. B., and Hochster, R. M. (1959) *Can. J. Microbiol. 5*, 1–8.

Maino, A. L., Schroth, M. N., and Palleroni, N. J. (1974) *Phytopathology 64*, 881–885.

Maki, L. R., Galyan, E. L., Chien, M. C., and Caldwell, D. R. (1974) *Appl. Microbiol. 28*, 456–459.

Manniello, J. M., Heymann, H., and Adair, F. N. (1979) *J. Gen. Microbiol. 112*, 397–400.

Markowitz, A., Klein, H. P., and Fischer, E. H. (1956) *Biochim. Biophys. Acta 19*, 267–273.

Mee, B. J., and Lee, B. T. O. (1967) *Genetics 55*, 709–722.

Meyer, J. M. (1977) "Pigment fluorescent et metabolisme du fer chez *Pseudomonas fluorescens*," Ph.D. thesis, Strasbourg, France.

Meyer, J. M., and Abdallah, M. A. (1978) *J. Gen. Microbiol. 107*, 319–328.

Miller, D. L., and Rodwell, V. W. (1971) *J. Biol. Chem. 246*, 2758–2764.

Morihara, K. (1964) *J. Bacteriol. 88*, 745–757.

Morihara, K., Tsuzuki, H., Oka, T., Inoue, H., and Ebata, M. (1965) *J. Biol. Chem. 240*, 3295–3304.

Morris, E. R., Rees, D. A., Young, G., Walkinshaw, M. D., and Darke, A. (1977) *J. Mol. Biol. 110*, 1–16.

New England Biolabs Catalog (1982–1983) Beverly, MA 01915.

Ohuchi, A., and Tominaga, T. (1973) *Ann. Phytopathol. Soc. Japan 39*, 417–424.

Ohuchi, A., and Tominaga, T. (1975) *Bull. Nat. Inst. Agric. Sci., Ser. C*, No. 29, 45–63.

Okabe, N., and Goto, M. (1963) *Ann. Rev. Phytopathol. 1*, 397–418.

Ornston, L. N. (1971) *Bacteriol. Rev. 35*, 87–116.

Otta, J. D. (1977) *Phytopathology 67*, 22–26.

Otta, J. D., and English, H. (1971) *Phytopathology 61*, 443–452.

Palleroni, N. J. (1978) *The Pseudomonas Group*, Meadowfield Press, Shildon, England.

Palleroni, N. J. (1981) in *The Prokaryotes*, Vol. I (Starr, M. P., Stolp, H., Trüper, H. G., Balows, A., and Schlegel, H. G., eds.), pp. 655–665, Springer, New York.

Palleroni, N. J. (1983) *BioScience 33*, 370–377.

Palleroni, N. J. (1984) in *Bergey's Manual of Systematic Bacteriology*, Vol. I (Kreig, N. R., ed.), pp. 141–199, Williams and Wilkins, Baltimore.

Palleroni, N. J., and Doudoroff, M. (1972) *Ann. Rev. Phytopathol. 10*, 73–100.

Palleroni, N. J., and Stanier, R. Y. (1964) *J. Gen. Microbiol. 35*, 319–334.

Palleroni, N. J., Doudoroff, M., Stanier, R. Y., Solanes, R. E., and Mandel, M. (1970) *J. Gen. Microbiol. 60*, 215–231.

Palleroni, N. J., Ballard, R. W., Ralston, E., and Doudoroff, M. (1972) *J. Bacteriol. 110*, 1–11.

Palleroni, N. J., Kunisawa, R., Contopoulou, R., and Doudoroff, M. (1973) *Int. J. Syst. Bacteriol. 23*, 333–339.

Patel, M. K., and Kulkarni, Y. S. (1949) *Indian Phytopathol. 2*, 62–64.

Pickett, M. J., and Greenwood, J. R. (1980) *J. Gen. Microbiol. 120*, 439–446.

Ralston, E., Palleroni, N. J., and Doudoroff, M. (1972) *J. Bacteriol. 109*, 465–466.

Ralston-Barrett, E., Palleroni, N. J., and Doudoroff, M. (1976) *Int. J. Syst. Bacteriol. 26*, 421–426.

Redfearn, M. S., and Palleroni, N. J. (1975) in *Diseases Transmitted from Animals to Man*, 6th ed. (Hubbert, W. T., McCulloch, W. F., and Schnurrenberger, P. R., eds.), pp. 110–128, C. C. Thomas, Springfield, Ill.

Redfearn, M. S., Palleroni, N. J., and Stanier, R. Y. (1966) *J. Gen. Microbiol. 43*, 293–313.

Robert-Gero, M., Poiret, M., and Stanier, R. Y. (1969) *J. Gen. Microbiol. 57*, 207–214.

Robyt, J. F., and Ackerman, R. J. (1971) *Arch. Biochem. Biophys. 145*, 105–114.

Rogul, M., Brendle, J. J., Haapala, D. K., and Alexander, A. D. (1970) *J. Bacteriol. 101*, 827–835.

Rosenfeld, H., and Feigelson, P. (1969) *J. Bacteriol. 97*, 697–704.

Royle, P. L., Matsumoto, H., and Holloway, B. W. (1981) *J. Bacteriol. 145*, 145–155.

Sands, D. C., and Rovira, A. D. (1970) *Appl. Microbiol. 20*, 513–514.

Sands, D. C., and Rovira, A. D. (1971) *J. Appl. Bacteriol. 34*, 261–275.

Sands, D. C., Gleason, F. H., and Hildebrand, D. C. (1967) *J. Bacteriol. 94*, 1785–1786.

Sandvik, E. I., and Maerker, J. M. (1977) in *Extracellular Microbial Polysaccharides* (Sanford, P. A., and Laskin, A., eds.), pp. 242–264, American Chemical Society, Washington, D.C.

Sano, Y., and Kageyama, M. (1977) *J. Gen. Appl. Microbiol. 23*, 183–186.

Sawyer, M. H., Baumann, P., and Baumann, L. (1977a) *Arch. Microbiol. 112*, 49–55.

Sawyer, M. H., Baumann, P., and Baumann, L. (1977b) *Arch. Microbiol. 112*, 169–172.

Schaad, N. W. (1976) *Phytopathology 66*, 770–776.

Schaad, N. W. (ed.) (1980) *Laboratory Guide for Identification of Plant Pathogenic Bacteria*, Bacteriology Committee of American Phytopathological Society, St. Paul, Minn.

Schroth, M. N., Hildebrand, D. C., and Starr, M. P. (1981) in *The Prokaryotes*, Vol. I (Starr, M. P., Stolp, H., Trüper, H. G., Balows, A., and Schlegel, H. G., eds.), pp. 701–718, Springer, New York.

Sherris, J. C., Shoesmith, J. G., Parker, M. T., and Breckon, D. (1959) *J. Gen. Microbiol. 21*, 389–396.

Shilo, M., and Stanier, R. Y. (1957) *J. Gen. Microbiol. 16*, 482–490.

Skerman, V. B. D., McGowan, V., and Sneath, P. H. A. (1980) *Int. J. Syst. Bacteriol. 30*, 225–420.

Sompolinsky, D., Hertz, J. B., Høiby, N., Jensen, K., Mansa, B., and Samra, Z. (1980a) *Acta Path. Microbiol. Scand. 88*, 143–149.

Sompolinsky, D., Hertz, J. B., Høiby, N., Jensen, K., Mansa, B., Pedersen, V. B., and Samra, Z. (1980b) *Acta Path. Microbiol. Scand. 88*, 253–260.

Souw, P., and Demain, A. L. (1979) *Appl. Environ. Microbiol. 37*, 1186–1192.

Stackebrandt, E., and Woese, C. R. (1981) in *Molecular and Cellular Aspects of Microbial Evolution* (Carlile, M. J., Collins, J. F., and Moseley, B. E. B., eds.), pp. 1–32, Cambridge University Press, Cambridge, England.

Stalon, V., and Mercenier, A. (1984) *J. Gen. Microbiol. 130*, 69–76.

Stalon, V., Ramos, F., Pierard, A., and Wiame, J.-M. (1972) *Eur. J. Biochem. 29*, 25–35.

Stanier, R. Y., Palleroni, N. J., and Doudoroff, M. (1966) *J. Gen. Microbiol. 43*, 159–271.

Starr, M. P. (1946) *J. Bacteriol. 51*, 131–143.

Starr, M. P. (1981) in *The Prokaryotes*, Vol. I (Starr, M. P., Stolp, H., Trüper, H. G., Balows, A., and Schlegel, H. G., eds.), pp. 742–763, Springer, New York.

Starr, M. P., Jenkins, C. L., Bussey, L. B., and Andrewes, A. G. (1977) *Arch. Microbiol. 113*, 1–9.

Stolp, H., and Gadkari, D. (1981) in *The Prokaryotes*, Vol. I (Starr, M. P., Stolp, H., Trüper, H. G., Balows, A., and Schlegel, H. G., eds.), pp. 719–741, Springer, New York.

Swings, J., DeVos, P., van den Mooter, M., and De Ley, J. (1983) *Int. J. Syst. Bacteriol. 33*, 409–413.

Udaka, S. (1966). *J Bacteriol. 91*, 617–621.

Véron, M., and Berche, P. (1976) *Bull. Inst. Pasteur 74*, 295–337.

Vicente, M., and Cánovas, J. L. (1973) *J. Bacteriol. 116*, 908–914.

Vidaver, A. K. (1976) *Ann. Rev. Phytopathol. 14*, 451–465.

Vidaver, A. K., and Buckner, S. (1978) *Can. J. Microbiol. 24*, 14–18.

Vidaver, A. K., Mathys, M. L., Thomas, M. E., and Schuster, M. L. (1972) *Can. J. Microbiol. 18*, 705–713.

Wang, R. Y.-H., Shedlarski, J. G., Farber, M. B., Kuebling, D., and Ehrlich, M. (1980) *Biochim. Biophys. Acta 606*, 371–385.

Watson, J. M., and Holloway, B. W. (1978a) *J. Bacteriol. 133*, 1113–1125.

Watson, J. M., and Holloway, B. W. (1978b) *J. Bacteriol. 136*, 507–521.

Weitzman, P. D. J., and Jones, D. (1968) *Nature 219*, 270–272.

Wendenbaum, S., Demange, P., Dell, A., Meyer, J. M., and Abdallah, M. A. (1983) *Tetrahedr. Lett. 24*, 4877–4880.

Whitaker, R. J., Byng, G. S., Gherna, R. L., and Jensen, R. A. (1981a) *J. Bacteriol. 145*, 752–759.

Whitaker, R. J., Byng, G. S., Gherna, R. L., and Jensen, R. A. (1981b) *J. Bacteriol. 147*, 526–534.

Whitfield, C., Sutherland, I. W., and Cripps, R. E. (1981) *J. Gen. Microbiol. 124*, 385–392.

Wilkie, P. J., Dye, D. W., and Watson, D. R. W. (1973) *N. Z. J. Agr. Res. 16*, 315–323.

Wilkinson, S. G. (1970) *J. Bacteriol. 104*, 1035–1044.

Willetts, N. S., Crowther, C., and Holloway, B. W. (1981) *Plasmid 6*, 30–52.

Wilson, R. G., and Henderson, L. M. (1963) *J. Bacteriol. 85*, 221–229.

Yano, T., Pestana de Castro, A. F., Lauritis, J. A., and Namekata, T. (1979) *Ann. Phytopathol. Soc. Jpn. 45*, 1–8.

Young, J. M., Dye, D. W., and Wilkie, P. J. (1978) *N. Z. J. Agr. Res. 21*, 153–177.

Zain, B. S., and Roberts, R. J. (1977) *J. Mol. Biol. 115*, 249–255.

Zavarzin, G. A., and Nozhevnikova, A. N. (1977) *Microb. Ecol. 3*, 305–326.

Biology of the Bacilli

Lee A. Bulla, Jr.
James A. Hoch

I. INTRODUCTION

The genus *Bacillus* represents an extremely diverse group of rod-shaped bacteria that form refractile endospores. The spores are more resistant than vegetative cells to heat, drying, irradiation, and other adverse conditions. Most of the bacilli are gram-positive and are motile by lateral or peritrichous flagella; some are nonmotile. They are unicellular and reproduce by binary fission; mycelia are not formed. The spore-forming bacilli characteristically are chemoheterotrophs whose metabolism allows dissimilation of organic substances by strict respiration, strict fermentation, or both respiration and fermentation. A variety of substrates can be utilized, and some species, but not all, require growth factors. Molecular oxygen serves as the terminal electron acceptor in respiratory metabolism, although nitrate-oxygen can serve as a replacement. Typically, the spore-forming bacilli inhabit the soil and live as saprophytes. However, some species are pathogenic and cause disease by producing toxins; a few actually can invade animal tissues. A wide range of peptide antibiotics are produced by the bacilli; most of them are active against gram-positive bacteria, although a few affect gram-negatives.

II. TAXONOMY AND CLASSIFICATION

The bacilli are classified according to the eighth edition of *Bergey's Manual of Determinative Bacteriology* (Buchanan and Gibbons, 1974) in the family Bacillaceae. The type genus is *Bacillus*. The most reliable and distinguishing characteristic used for diagnostic and identification purposes is endospore formation. There is some serological information on the bacilli, but it is somewhat limited. In those species examined, flagellar H-antigens serve to recognize and delineate serotypes. The H-antigens most extensively used for this purpose are those of the subspecies of *Bacillus thuringiensis*.

Taxonomic relationships of species have been characterized and compared by DNA–DNA hybridization and genetic recombination methods such as transformation and transduction. These relationships are discussed later in this chapter. The bacilli studied most extensively are those that can be readily and easily cultured in artificial media and those that are medically and industrially important.

Because spore formation is the most prominent trait for classifying *Bacillus* species, mention should be made of the genus *Clostridium* (discussed in a later chapter), whose members also produce endospores. The primary difference between these two genera of spore formers is their contrasting oxygen tolerances. The clostridia generally are strict anaerobes. Any observable growth by some species in the presence of oxygen is very slight and does not result in sporulation. For the most part the bacilli are catalase-positive, whereas the clostridia are catalase-negative. Spore-forming bacteria other than the bacilli and clostridia have been found in the alimentary tracts of animals. Such spore formers are much larger in size than *Bacillus* and *Clostridium* species; and sometimes, two spores are formed per cell, whereas only one is generally produced per bacillus or clostridial cell. Little is known about the taxonomic relationships of these intestinal dwellers either to each other or to the bacilli and clostridia.

The eighth edition of *Bergey's Manual of Determinative Bacteriology* (Buchanan and Gibbons, 1974) subdivides the genus *Bacillus* into two groups. Group I contains 22 species with rather well-accepted and well-defined characteristics. Group II contains 26 species whose delineation is not as well-accepted and recognized. The reader is referred to *Bergey's Manual* (Buchanan and Gibbons, 1974) for taxonomic descriptions and differential properties of the species in Groups I and II. For a more extensive and thorough coverage, see *The Genus Bacillus* (Gordon et al., 1973).

III. SPORULATION

The life cycle of the bacilli is characterized by two distinct steps: vegetative growth and sporulation. Vegetative cell division is typified by the formation of division septa that are initiated midway along the plasma membrane and often at a site where membranelike vesicles occur. These vesicles, referred to as mesosomes by Bechtel and Bulla (1976), may appear randomly along the periphery of vegetative cells and are associated with the plasma membrane. The division septum appears to be an extension of the cell wall and cytoplasmic membrane and to be continuous with, or in close association with, these membranelike vesicles. Once completely formed, the septum serves as the common boundary between two new daughter cells, which can further undergo new cell division.

The sequence of spore development is diagrammed in Fig. 3.1. The example provided is that of *B. thuringiensis* sporulation, one that generally represents this phenomenon in all bacilli. Sporulation is summarized here according to the conventional sporulation stages: stage I — axial filament formation; stage II — forespore septum formation; stage III — engulfment, change in stainability of membranes and cytoplasm, and formation of forespore; stages IV to VI — formation of exosporium, primordial cell wall, cortex, and spore coats accompanied by transformation of the spore nucleoid; and stage VII — spore maturation and sporangial lysis. For *B. thuringiensis* an additional event that does not normally occur in other bacilli is the formation of a proteinaceous parasporal crystal that is first visible at stage III of sporulation and is sometimes accompanied by an ovoid inclusion of unknown function. The parasporal crystal is insecticidal; some of its properties are discussed later in this chapter.

Axial Filament and Forespore Septum Formation—Stages I and II

The transition from vegetative growth to sporulation is accompanied by condensation of the nucleoid (N) into a compact and elongated axial filament (AF, stage I, Fig. 3.1). Forespore septa (FS) are initiated immediately following stage I and are recognizable as invaginations of the plasma membrane (stage II; see arrows). Mesosomes are not associated with the invaginations initially but are prevalent slightly later and throughout forespore development. Upon completion of the septum the area of cytoplasm destined to be incorporated into the forespore is termed the "incipient forespore." The forespore septum is easily distinguished from the vegetative division septum. The former lacks visible cell wall material and is similar in appearance to properly fixed membrane.

Engulfment—Stage III

Once the forespore septum is completed, engulfment of the forespore commences by the movement of the junction of the forespore septum and plasma membrane toward one pole of the cell. Mesosomes are sometimes present at the junction of the septum and plasma membrane during engulfment. As the process nears completion, mesosomes are not always readily apparent at this junction. Completion of engulfment occurs when the septum becomes detached from the plasma membrane, isolating the incipient forespore from the mother cell cytoplasm.

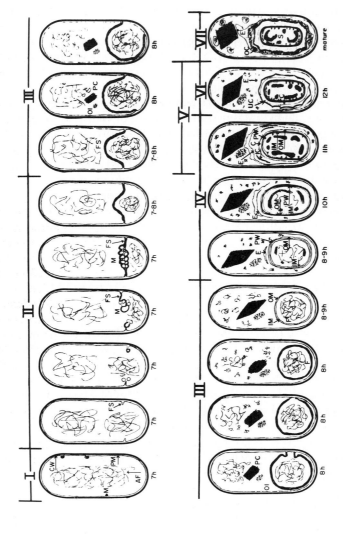

FIGURE 3.1 Diagrammatic scheme of sporulation in *B. thuringiensis*. M = mesosome, CW = cell wall, PM = plasma membrane, AF = axial filament, FS = forespore septum, IF = incipient forespore, OI = ovoid inclusion, PC = parasporal crystal, F = forespore, IM = inner membrane, OM = outer membrane, PW = primordial cell wall, E = exosporium, LC = lamellar spore, OC = outer spore coat, C = cortex, UC = undercoat, S = mature spore in an unlysed sporangium.

Immediately following septum detachment, several changes occur that lead to development of the forespore (F). One of the most noticeable changes is the decrease in electron density of the incipient forespore membrane. The incipient forespore membrane also loses its mesosomelike appearance and is transformed into the inner (IM) and outer (OM) forespore membranes. Highton (1972) observed a similar reduction in electron density of the forespore membrane of *Bacillus cereus* 569H/24. Bechtel and Bulla (1982) have observed that *B. thuringiensis* cells that have reached this particular stage of sporulation are incapable of returning to vegetative growth, and therefore they have concluded that the decrease in electron density of the forespore membrane reflects commitment to sporulation.

A second significant change is the staining quality of the mother cell (sporangial) and forespore cytoplasms. The sporangial cytoplasm of cells that have completed engulfment stain more heavily and appear more granular than either the cytoplasm of forespores or cells not having completed the process. Such a change may be characteristic of senescence and also may reflect a change in plasma membrane permeability.

A third, and dramatic, change occurs in the appearance of the plasma membrane. Prior to engulfment the plasma membrane (PM) is irregular and not distinctly trilaminar. After engulfment, however, the membrane is distinctly trilaminar with the outer electron-dense portion being thicker than the inner one. The same phenomenon occurs with the plasma membrane of *B. cereus* sporulating cells (Highton, 1972) and has been interpreted as reflecting a change in the molecular structure of the plasma membrane, which consequently alters the permeability of that membrane. Freese et al. (1970) demonstrated diminution of the membrane-associated glucose phosphoenolpyruvate transferase system after vegetative growth of *Bacillus subtilis*. Lang and Lundgren (1970) showed quantitative changes in membrane lipid composition during sporulation of *B. cereus*. Also, Bulla et al. (1971) observed derepression of membrane-bound enzyme activities of *B. thuringiensis* involved with energy production by the electron transport system and oxidation of acetyl coenzyme A via terminal respiratory pathways, that is, tricarboxylic acid and glyoxylic acid cycles. Another possible implication of the change in electron density of the plasma membrane is the active transport of cations (Eisenstadt and Silver, 1972; Scribner et al., 1975). Scribner et al. (1975) described active transport of potassium, magnesium, calcium, and manganese and demonstrated that each cation transport system is independently regulated during sporulation, which results in various rates of cation uptake. Coincidentally, calcium and potassium uptake is high during forespore formation (Scribner et al., 1975; Vinter, 1969).

Spore Wall Development—Stages IV to VI

Following engulfment (stage III) the sporulation stages are not readily dissociable from one another (stages IV to VI). However, numerous processes occur during these stages, including primordial cell wall, cortex, and peptidoglycan synthesis; exosporium formation; development of the spore coats; and transformation of the spore nucleoid. As was reported for *Bacillus sphaericus* (Holt et al., 1975), the spore nucleoid changes from a dispersed state characteristic of the vegetative nucleoid to a fibrous circular region to an electron-translucent matrix. Little information is available on the physical and chemical properties of nucleic acids contained in either the developing or the mature spore, although deoxyribonucleic acid (DNA) of mature spores appears to be different physically from that of vegetative cells (Doi, 1969).

The primordial cell wall (PW; also called cortical membrane (Walker, 1970) and germ cell wall (Murrell et al., 1971)) is the first part of the spore wall to form (stage IV). It is observed between the inner and outer forespore membranes. Concomitant with primordial cell wall appearance, electron-dense, fibrous masses (Fr) appear with the spore nucleoid (SN); also the mother cell nucleoid (N) begins to fragment, and the exosporium (E) is initiated at this time. The lamellar spore coat (LC) is first seen on the lateral sides of the spore between the exosporium and the outer forespore membrane, as also described for *B. sphaericus* (Holt et al., 1975) and *B. cereus* T (Murrell et al., 1971). Eventually, it surrounds the entire spore. While the lamellar coat is being formed, three other modifications also occur: (1) the electron-dense, fibrous spore nucleoid becomes an electron-transparent homogeneous structure; (2) the primordial cell wall thickens; and (3) the exosporium engulfs the spore. Once the primordial cell wall has attained maximum thickness, the cortex (C) develops. The cortex is located between the primordial cell wall and the outer forespore membrane and appears less electron-dense than the primordial cell wall. Between the developing lamellar spore coat and outer membrane is the undercoat (UC). The undercoat otherwise has been called the inner spore coat in *B. sphaericus* (Holt et al., 1975) and *B. cereus* T (Murrell et al., 1971). After cortex and lamellar spore coat development, the outer fibrous spore coat (OC) forms. Unlike some bacilli, *B. thuringiensis* does not produce a discrete peptidoglycan layer.

Mature Spore—Stage VII

The mature, dormant spore is the ultimate stage of sporulation and represents a cryptobiotic state in which metabolic and developmental activity is at a minimum or is nonexistent. Once appropriate environmental conditions are encountered, however, the spore can reverse itself metabolically and physiologically and proceed toward vegetative growth.

IV. SPORE GERMINATION

The sequence of events that leads from the dormant spore to the vegetative cell has been divided into at least three events—activation, germination, and outgrowth (Keynan and Halvorson, 1965). Activation can be accomplished by sublethal heat treatment (heat shock) in an appropriate nutritional milieu and by exposure to low pH, thiol compounds, or strong oxidizing agents.

The cytological sequence of spore germination and outgrowth is diagrammed in Fig. 3.2 and is summarized here according to the following stages: stage I — spore; stage II — germination, conversion of spore to an active cell marked by vegetative appearance of cytoplasm; stage III — outgrowth, swelling of germinated cell; stage IV — outgrowth, elongation of cell; stage V — outgrowth, emergence of elongated cell from spore coats and exosporium; and stage VI — cell division, resumption of vegetative growth.

For a detailed explanation of the cytological and biochemical events that accompany spore activation, germination, and outgrowth, please refer to the review articles by Bulla et al. (1980) and Setlow (1981).

V. METABOLISM AND PHYSIOLOGY

As was stated above, the bacilli are chemoheterotrophs capable of utilizing a rather wide range of organic compounds as respiratory substrates as well as carbohydrates for fermentative purposes. Most bacilli are mesophilic, growing quite well at temperatures between 30°C and 45°C. Some species, however, are thermophilic and thrive at temperatures as high as 65°C but not at temperatures below 45°C.

The species of bacilli that can grow anaerobically do so primarily because of their capacity to ferment sugars. *B. cereus, Bacillus licheniformis,* and *Bacillus subtilis*, for example, ferment simple sugars by a distinctive fermentation that produces the major end-products 2,3-butanediol, glycerol, and CO_2 as well as small amounts of lactate and ethanol. *Bacillus polymyxa* and *Bacillus macerans*, on the other hand, are capable of catabolizing polysaccharides such as starch and pectins in addition to monosaccharides. *B. polymyxa* converts sugar substrates into 2,3-butanediol, ethanol, CO_2, and H_2, whereas the products of sugar fermentation by *B. macerans* are ethanol, acetone, acetate, formate, CO_2, and H_2. Formation of H_2 as a major end-product and the incapacity to produce glycerol distinguish these two organisms from *B. cereus, B. licheniformis*, and *B. subtilis* and therefore can be used as a diagnostic tool for characterizing and identifying these species and others that appear closely related on the basis of factors such as optimal growth temperatures, spore size and shape, and cell growth. Another variation of sugar fermentation includes homolactic fermentation as exemplified by *Bacillus coagulans*.

Aerobic metabolism generally consists of glucose dissimilation via the Embden-Meyerhof-Parnas pathway and subsequent oxidation of pyruvate through the tricarboxylic acid (TCA) cycle. However, terminal oxidation is usually not associated with vegetative cell growth but with the onset of sporogenesis. During vegetative proliferation there is an accumulation of organic acids, such as pyruvate, in the culture medium with a resultant pH value as low as 4–5. At the end of exponential growth and during the transition of vegeta-

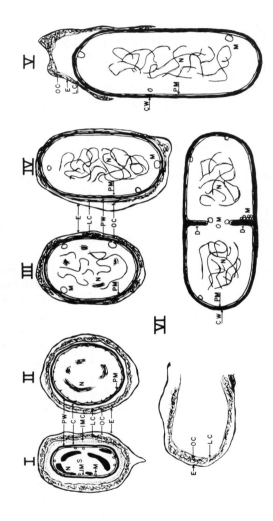

FIGURE 3.2 Diagrammatic scheme of germination and outgrowth of the *B. thuringiensis* spore. M = mesosome, IM = inner membrane, N = nucleoid, Fr = fibrous nucleoid, PW = primordial cell wall, C = cortex, UC = undercoat, LC = lamellar spore coat, OC = outer fibrous spore coat, E = exosporium, S = mature spore cytoplasm, PM = plasma membrane, CW = cell wall, D = division septum.

tive cells to sporulating cells, the organic acids in the medium are oxidized by terminal respiratory reactions to high-energy compounds, and the pH of the culture medium rises to above neutrality. Sporulation consists of a series of energy-utilizing biosynthetic activities; apparently, this surge in production of high-energy compounds and metabolites is a prerequisite to proper spore development and maturation.

Because the energy requirement is great for sporulation and related biosynthetic events, it is not surprising to find that early forespores are laden with adenosine triphosphate (ATP) and large amounts of reduced nicotinamide adenine dinucleotide (NADH) (Singh et al., 1977). Presumably, these energy compounds are supplied to the forespore by the mother cell, whose terminal respiratory machinery is intact and functioning at a high rate. Forespores apparently do not possess significant levels of TCA cycle enzymes; consequently, they probably depend on the mother cell for energy production.

A metabolic event that is associated with sporulation is the accumulation of dipicolinic acid (DPA). Young forespores somehow accumulate DPA, as well as calcium by facilitated diffusion from the mother cell cytoplasm, and a chelate of calcium dipicolinate is formed as the forespore matures. In fully developed spores the calcium dipicolinate complex represents 10–15% of the spore dry weight, and the compound may be implicated in spore thermostability.

Another metabolic feature of sporulation is the synthesis of a unique peptidoglycan (localized in the cortex; see Fig. 3.1) that consists of three repeating subunits: (1) a muramic lactam with no amino acids attached, (2) an alanine subunit with only an L-alanyl residue, and (3) a tetrapeptide subunit having the sequence L-ala-D-glu-*meso*-diaminopimelic acid-D-ala. These components occur in a general ratio of 3.5:1:2, respectively. Very little cross-linking occurs between the tetrapeptide chains. The cortical peptidoglycan is similar chemically in both *B. subtilis* and *B. sphaericus*, for example, whereas the vegetative cell wall peptidoglycans of these two organisms are quite different.

The protein content of spores is different from that of vegetative cells. During sporulation the compliment of proteins changes in both the mother cell and the developing spore. Setlow (1981) has delineated three groups or classes of proteins associated with sporulation: (1) group I proteins, which occur in both the mother cell and forespore at similar levels, (2) group II proteins, which are present in the mother cell but not in the forespore, and (3) group III proteins, which are associated with the forespore but not with the mother cell.

Group I proteins include glycolytic enzymes, a few enzymes involved in terminal electron transport, enzymes for vegetative cell wall biosynthesis, and some enzymes responsible for amino acid catabolism. Group II proteins include most of the TCA cycle enzymes, alanine dehydrogenase, several enzymes specific for biosynthesis of cortex, DPA biosynthetic pathway enzymes, and some nonspecific endoproteases. Among the group III proteins are glucose dehydrogenase, enzymes for cortex processing and salvaging of by-products of cortex synthesis, aspartase, low-molecular-weight proteins associated with the spore interior, and a variety of proteases.

The outer spore coat is composed mainly of protein and accounts for about 80% of the total spore protein. The outer spore coat itself represents 30–60% of the dry weight of the spore. Characteristic of spore coat proteins is an unusually high content of cysteine and of hydrophobic amino acids. They are extremely resistant to treatments and reagents that normally solubilize most proteins.

VI. GENETICS

Most of the available knowledge about the structure of chromosomes of bacilli is dominated by studies of one species, *B. subtilis*. This organism is the most widely studied and thoroughly mapped gram-positive microorganism. Initially, genetic studies were begun in this species as a result of the discovery of DNA-mediated transformation (Spizizen, 1958). At that time, *B. subtilis* was the only bacterium capable of DNA-mediated transformation and the only one capable of growing on simple, chemically defined media. These attributes were advantageous to many genetic and biochemical studies of metabolic pathways, cell wall synthesis, cell growth and division, recombination, and a host of other processes. In addition, *B. subtilis* has become the focal point of studies on sporulation and germination because of the ease of genetic analysis. However, genetic studies in other species of the genus *Bacillus* have been carried out for a number of other reasons. These studies have provided insight into the comparative genetic structure of the chromosomes of members of the genus.

Methods of Genetic Analyses

The key to the generation of a complete, uninterrupted genetic map of *B. subtilis* has been the combination of genetic analyses afforded by DNA-mediated transformation and phage PBS1 transduction. DNA-mediated transformation allows fine structure analyses of relatively short segments of the chromosome, whereas PBS1-mediated transduction is most useful as a means to study large areas of the chromosome. Analysis of the available DNA-mediated transformation data (Henner and Hoch, 1980) indicates that the apparent size of the transforming DNA fragment is approximately 30 kilobase pairs of DNA. However, the practical limit of DNA-mediated transformation is about 15 kilobase pairs of DNA. That is, studies of genetic markers (or loci) separated by less than 15 kilobase pairs of DNA are feasible. Genetic markers farther apart than 15 kilobase pairs of DNA are less amenable to study because of the low frequency of cotransformation and ambiguities arising from double transformations.

Linkage of widely separated genetic loci can be accomplished by PBS1-mediated transduction. It is estimated that the DNA fragment transferred by this means is at least ten times longer than that of DNA-mediated transformation (Henner and Hoch, 1980). Because the probability of cotransfer of genetic markers seems to be a direct function of the size of the DNA fragment being transferred, the linkage values for any two markers are at least tenfold higher in PBS1 transduction than DNA-mediated transformation. This feature allows the joining of isolated genetic regions that cannot be linked by DNA-mediated transformation. The result in *B. subtilis* is a contiguous circular genetic map.

Other transduction systems have been reported for *B. subtilis*. Bacteriophages SP10 and SPP1 are capable of mediating transduction, although there are some associated procedural difficulties. Bacteriophage SP10 is unable to plaque on the common 168 strain of *B. subtilis*. Therefore its usefulness is limited. Rigorous methods for SPP1-mediated transduction have not been determined. Recombination values between genetic markers using either bacteriophage are similar to those observed in DNA-mediated transformation. Bacteriophage SPβ, a resident lysogen in *B. subtilis* 168–derived strains, is capable of mediating specialized transduction of markers linked to its attachment site on the chromosome (Zahler and Korman, 1981). Recently, S. A. Zahler and colleagues (personal communication) have developed a system to insert SPβ at different sites on the chromosome. This method has potential for isolating specialized transducing bacteriophages for any chromosomal region.

Protoplast fusion of different genetically marked *B. subtilis* strains has been shown to give rise to metastable diploids with interesting properties (Schaeffer et al., 1976). Although such diploids have not been exploited for genetic analysis, they will be potentially useful for complementation and dominance studies.

Bacteriophage-mediated transduction systems have been used for genetic analysis of *B. pumilus*, *B. cereus*, *B. thuringiensis*, *B. anthracis. B. amyloliquefacieus*, *B. megaterium*, and *B. licheniformis*. In addition, DNA-mediated transformation has been an important method for the study of *B. lichenifor-mis*. It probably is not too unreasonable to presume that transducing bacteriophages can be found for any of the members of this genus if sufficient time and effort are expended. A more enlightened strategy might be to transfer desirable traits from *Bacillus* species to *B. subtilis*, where they can be more easily manipulated.

Genetic Maps of Bacillus Species

B. subtilis. A recent version (Henner and Hoch, 1980) of the compiled genetic map of *B. subtilis* is shown in Fig. 3.3. If one compares this map to the genetic map of *Escherichia coli*, some similarities can be noted (Bachmann and Low, 1980). One feature of both maps is the relative clustering of genes for important biosynthetic functions and the presence of genetically silent regions of the chromosome. It has been suggested that clustering is a consequence of the structure of the folded nucleoid. Genes required for essential functions may be located at the surface of the nucleoid, whereas relatively nonessential genes could be buried within the structure. A determination of whether the silent

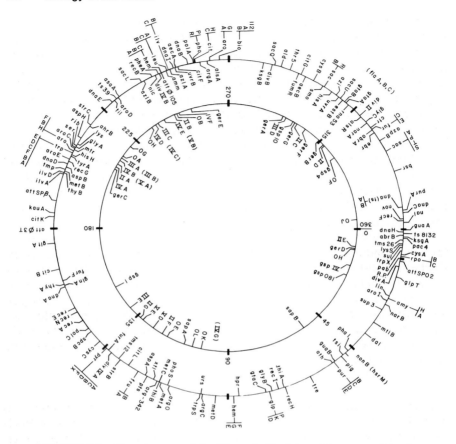

FIGURE 3.3   Genetic map of *Bacillus subtilis*. (From Henner and Hoch, 1980.)

regions are transcriptionally inactive as well as genetically inactive might shed some light on this question.

***Bacillus pumilus* and *B. licheniformis*.**   On the basis of physiological and morphological studies, *Bacillus pumilus* and *B. licheniformis* are thought to be closely related to *B. subtilis* (Smith et al., 1952). DNA homology as assessed by DNA–DNA hybridization studies revealed that only 16–18% homology exists between *B. pumilus* and *B. subtilis* (Lovett and Young, 1969). *B. licheniformis* showed 24% homology to *B. subtilis* and 9% homology to *B. pumilus*. This lack of homology is reflected in the inability to obtain transformation of auxotrophic *B. subtilis* strains with heterologous DNA. However, the recent discovery that restriction enzyme–deficient *B. subtilis* strains are transformed at higher levels by heterologous DNA indicates that heterologous transformation is dependent upon factors other than strict DNA–DNA homology (Hoshino et al., 1980). The nucleotide sequences conserved in the DNA of these organisms most likely corresponds to those genes that evolved early and were passed on relatively unchanged from the progenitor of the various *Bacillus* species. Thus, heterologous transformation studies have revealed that significantly higher homology exists between species for ribosomal protein genes (Dubnau et al., 1965) than for enzymes of amino acid biosynthetic pathways.

Genetic studies of *B. pumilus* using PBS1 transduction have revealed that several linkage groups in *B. pumilus* are identical to those observed in *B. subtilis* (Lovett and Young, 1970, 1971). The similarity includes the *lys-ser-trp-ilv-ile* and *argA-leu-phe* linkage groups that have been studied in other members of the species. Furthermore, the *argO-met-aro-cys* linkage group seems to be identical to that of *B. subtilis*.

Mapping studies in *B. licheniformis* have been carried out by using transformation and transduction mediated by bacteriophage SP-15, a bacteriophage similar to PBS1 (Perlack and Thorne, 1981). The chromosomal map obtained is virtually identical to the *B. subtilis* map. Moreover, the location of genes coding for the glutamyl peptide (*pep* mutations) characteristic of *B. licheniformis* have been located on the chromosome. These results indicate that there is a strong conservation of gene order among these related *Bacillus* species. As the species evolved from a common progenitor, the amount of translocation or inversion of segments of the chromosome has been minimal.

Bacillus megaterium.   Recently, the *Bacillus megaterium* chromosome has been the subject of extensive genetic analysis through the use of MP-13-mediated transduction (Garbe and Vary, 1982). Although *B. megaterium* has been reported to have only 8% DNA homology with *B. subtilis* (Sharp et al., 1980), some conservation of gene order has been observed between the two species. Transductional analyses of the leucine genes has shown that the linkage group *leu-ilv-hem-phe* is probably identical to that of *B. subtilis* (J. C. Garbe, M. A. Frazen, and P. S. Vary, unpublished). In addition, the order of the leucine biosynthetic genes is identical to the order in *B. subtilis*. The order of loci within the tryptophan operon seems to be the same as that for *B. subtilis*, and the tryptophan operon is linked to a histidine locus as in *B. subtilis*. However, linkage to *trp* or *rib*, *ser*, or *tyr* mutations was not found, a result suggesting that the order of genes around the tryptophan operon differs significantly in the two species (J. P. Callahan and P. S. Vary, unpublished).

B. thuringiensis.   A renewed interest in the insect pathogen *B. thuringiensis* has led to some revealing studies of its genetic map. The linkage group *met-(ArgC)-met-pyr* is identical to that observed in *B. subtilis* (G. D. Garsomian, N. J. Robillard, and C. B. Thorne, unpublished). The other linkage groups *pur-nal-str-pur* and *phe-argA* also appear to be similar to their counterparts in *B. subtilis*. A major divergence from this apparent conservation of gene order is in the *trp* region, where the leucine genes, along with additional DNA, seem to be inserted between *trp* and *hisH*. In *B. subtilis*, *B. licheniformis*, and *B. megaterium* the *hisH* locus is immediately adjacent to the *trp* operon, whereas in *B. thuringiensis* the two loci are separated by about half the transducing fragment, which may be equivalent to 50–100 genes. The order of loci in this linkage group is *trp-leu-his-lys-cys*.

The genetic maps of the various species of the genus *Bacillus* have revealed that there is a conservation of gene order among closely related species. More distantly related species show some conservation of gene order in certain regions of the chromosome and less in others. There is some evidence to suggest that loss of conservation results from the movement of large fragments of the chromosome from one location to another, i.e., translocations.

Gene Cloning

In the past few years there has been an extensive effort on the part of many groups of investigators to develop gene-cloning systems for members of the genus *Bacillus*. Because an exhaustive survey of all the published work in this area is beyond the scope of this chapter (Gryczan, 1982), we shall attempt to outline the general methods used for gene cloning as well as point out some of the problems that have been encountered.

Cloning via *Escherichia coli*

B. subtilis DNA inserted into *E. coli* phage and plasmid vectors can be assayed by transformation as readily as free *B. subtilis* DNA. This feature makes cloning in *E. coli* an attractive alternative to direct cloning in *B. subtilis*, especially because *E. coli* cloning systems have been well worked out and are easy to use. Bacteriophage lambda systems, such as the Charon phage (Blattner et al., 1977), are suitable for cloning fragments of DNA up to 20 kilobases in size. *B. subtilis* chromosomal libraries prepared in these vectors have yielded many different regions of the chromosome in cloned form (Ferrari et al., 1981). Although the strategy for preparing these libraries was devised to maximize the completeness of the library, it has been the experience of several investigators that certain regions of the chromosome are not represented in the library. There are several possible explanations for this observation. One obvious explanation is that the region in question resides on a chromosomal fragment that is too large for the capacity of the vector. The presence of strong transcriptional promoters on a chromosomal fragment also leads to the apparent absence of the fragment from a library. K. Bott and associates (unpublished) have discovered that Charon phage containing the strong ribosomal RNA promoters on an insert form much smaller plaques than normally expected. Apparently, strong promoters on an insert interfere with phage replication. With the exception of certain unclonable regions, the Charon and λgtWESλB libraries have proven extremely useful for the isolation of many areas of the chromosome.

The recently developed bifunctional cosmid cloning vehicle pQB79-1 may prove to be useful in the isolation of large fragments of chromosomes (Aubert et al., 1982). This cosmid should have a capacity of approximately 40 kilobase pairs of DNA. Several regions of the *B. subtilis* chromosome have been isolated from a library prepared in pQB79-1. As more use is made of this vector, it will be of interest to see whether structures such as strong promoters on the insert will be maintained stably in *E. coli* and *B. subtilis*.

Chromosomal libraries of *B. subtilis* DNA have been prepared in *E. coli* plasmids like pMB9 (Hutchison and Halvorson, 1980) and in shuttle plasmids such as pHV33 (Rapoport et al., 1979). Shuttle plasmids are generally constructed by preparing a hybrid plasmid between an *E. coli* plasmid such as pBR322 and a *Staphylococcus aureus* plasmid such as pC194 in order to obtain a gram-positive replicon and an expressible antibiotic resistance gene. Such plasmids have the advantage of replicating both in *B. subtilis* and *E. coli* and allowing the original cloning experiments to be carried out in *E. coli*. Shuttle plasmids of this type may find wide use among the genus *Bacillus*.

Direct Cloning in *B. subtilis*

Initial attempts to clone *B. subtilis* genes made use of *S. aureus* plasmids because no naturally occurring *B. subtilis* plasmids were available. The first of such cloning experiments reported was the insertion of DNA fragments into the plasmid pUB110, resulting in complementation of the *trpC2* mutation (Keggins et al., 1978). Since that time, many investigators have tried to construct libraries of *B. subtilis* fragments in a variety of *S. aureus* plasmids with little success. One possible explanation for this problem is the demonstrated dependence of plasmid transformation on oligomeric forms of the plasmid (Canosi et al., 1978). It appears that the transformation process requires a recombination event (Michel et al., 1982). A unique cloning system for heterologous DNA was designed to take advantage of the requirement for recombination. *B. licheniformis* DNA was successfully cloned into *S. aureus* plasmids by using a recipient strain carrying an *S. aureus* plasmid as the source of DNA homology (Gryczan et al., 1980). This system is probably applicable to the cloning of a variety of heterologous DNAs in *Bacillus* species that can be transformed. Because of the difficulties in obtaining recombinants, the direct cloning systems have been superseded by the shuttle vector systems where the primary cloning event is carried out in *E. coli*.

An interesting method of bacteriophage cloning has been used to clone the *spoOB* and *spoOF* genes (Kawamura et al., 1981), wherein chromosomal DNA fragments were ligated to fragments of bacteriophage p11 DNA. The ligated DNA was used to transform a p11 lysogen to Spo$^+$. The lysogen within these transformants was then induced, and those phage with an integrated Spo gene were isolated by virtue of their ability to affect specialized transduction of the Spo$^+$ character. The success of this method depends upon complementation, and the technique should be applicable to all recessive selectable genes.

Plasmids for the Analysis of Gene Regulatory Signals

Several plasmids have recently become available for the analysis of gene regulatory signals. Successful expression of a gene requires three basic elements: (1) a promoter to allow RNA polymerase to bind and initiate transcription of the gene, (2) a ribosome binding site to allow the messenger RNA to bind to ribosome so that the messenger RNA can be translated, and (3) a translation start codon. Plasmids carrying a gene lacking one or more of these elements along with an easily assayable protein product to determine whether the gene is on or off have proven useful for the study of regulatory signals.

Donnelly and Sonenshein (1982) have constructed a plasmid for the study of promoters. This shuttle plasmid contains the β-galactosidase (*lacZ*) gene of *E. coli* preceded by a portion of the *trp* operon of *E. coli* that provides a ribosome binding site recognizable by *B. subtilis*. The *trp* portion of the plasmid is not preceded by a promoter. Therefore there is no expression of β-galactosidase in either *B. subtilis* or *E. coli*. A convenient, unique *Hind*III restriction endonuclease site preceding the *trp* portion allows the cloning of DNA fragments. The cloned fragments with promoter activity directed toward β-galactosidase will result in the expression of β-galactosidase. Such clones can be easily isolated on X-gal plates where Lac$^+$ colonies are blue and Lac$^-$ colonies are white. This system is convenient for probing shotgun collections of fragments for the promoter activity. Furthermore, assay of β-galactosidase allows an assessment of promoter strength.

A second "promoter probe" type of plasmid has been exploited by Goldfarb et al. (1982). A shuttle plasmid was constructed containing the TN9-derived chloramphenicol resistance gene that lacks a promoter but is preceded by the TN9 ribosome binding site that is not recognized by *B. subtilis*. Immediately preceding this gene is a unique *Hind*III restriction endonuclease cloning site. Cloning restriction fragments into the *Hind*III site results in chloramphenicol resistance if a promoter, ribosome binding site, and initiation codon are placed in the correct direction and in frame with the chloramphenicol resistance gene. Because the insert must be in frame, only one third of the possible promoters will be detected by this system. However, the chloramphenicol resistance does provide a direct selection for promoter-containing fragments.

A third system to study regulatory elements is derived from the β-galactosidase fusion vector of Casadaban et al. (1980). A shuttle plasmid has been constructed with a *lacZ* gene missing all the regulatory signals and the first eight codons of *lacZ* (F. Ferrari and J. Hoch, unpublished). Immediately preceding the gene is a small synthetic oligonucleotide linker containing sites for the restriction endonucleases *Eco*RI, *Sma*I, and *Bam*HI. DNA fragments cloned into these sites require a promoter, ribosome binding site, and initiation codon for expression of β-galactosidase. In addition, the fragment must be in frame with *lacZ*. The advantage of this system is the variety of restriction endonucleases that can be used to generate fragments compatible with the cloning sites.

These plasmids and others being constructed in numerous laboratories will allow further dissection of regulatory elements required for gene expression in *Bacillus* species. The complexity of *B. subtilis* RNA polymerase and its regulatory subunits has already been established (Losick and Pero, 1981; Doi, 1982). The apparent higher stringency for interaction of ribosome binding sites and ribosomes in this organism is another one of its surprising features (Murray and Rabinowitz, 1982).

The enormous amount of research that has developed *B. subtilis* into the preeminent gram-positive organism for studies in molecular biology forms a solid basis for exploitation of this organism in industrial and health-related areas.

VII. INDUSTRIAL APPLICATIONS

Not only do the bacilli provide a variety of fascinating systems and features to investigate at the fundamental level, they also offer powerful applications for human use. Three general areas of industrial applications involving the bacilli are presented here, although there are others as well. The three include enzymes, antibiotics, and insecticides.

Enzymes

The bacilli elaborate an expansive array and an abundant supply of enzymes. Because of the extensive turnover of proteins and the complex biosynthetic activities that accompany cell growth and spore development, it is no wonder that this group of organisms has such a vigorous enzymatic capacity. Kornberg et al. (1968) have listed approximately 60 different enzymes contained in the spore alone. Of all the enzymes characterized for the bacilli, those that hydrolyze starch have had the broadest use and application commercially. Particular enzymes of this group include α-amylase, β-amylase, and amyloglucosidases used currently in the food and brewing industries. Extracellular enzymes such as cellulase and xylanase are potentially useful for conversion of various waste products and natural commodities into more utilizable substrates and components such as animal feeds, fermentation starter compounds, single-cell protein, and various fuels, to name a few. A large number of *Bacillus* species produce enzymes with the above properties and capacities.

Surely, there are more and different enzymes among the many species. They await only further discovery and exploration.

Another class of enzymes are the proteases that include esterases, metallo-proteases, and alkaline serine proteases. These proteases are useful not only for industrial purposes but also for laboratory experimentation and diagnostic procedures. Generally, they are produced at the end of vegetative growth and during the onset of sporulation. A full complement of specific proteolytic enzymes resides with the developing spore and lasts through sporangial lysis. Presumably, these proteases are sporulation-related because they are found only immediately before and during the sporulation cycle.

Antibiotics

The bacilli produce a large variety of antibiotic substances, most of which are peptides. The chief producers are *B. subtilis* strains (approximately 65–70 different peptide antibiotics synthesized) and *Bacillus brevis* (responsible for as many as 20–25 peptide antibiotics). The peptide antibiotics produced by species of the genus *Bacillus* are effective primarily against gram-positive bacteria, although some are active against gram-negative bacteria, and some against molds and yeasts as well. Several have application in medicine, plant disease control, food preservation, and as research tools.

There are three main classes of peptide antibiotics produced by the bacilli: (1) edeines, linear basic peptides that inhibit DNA synthesis; (2) bacitracins, cyclic peptides that inhibit cell wall synthesis; and (3) the gramicidin–poly-myxin–tyrocidin group, linear or cyclic peptides that modify membrane structure or function. Many of the antibiotics contain D-amino acids and amino acids that do not occur in proteins. Furthermore, the edeines possess certain constituents that are not amino acids, e.g., spermidine. It should be noted that antibiotics other than peptides are synthesized by the bacilli such as butirosin from *Bacillus circulans* and proticin from *B. licheniformis* subsp. *mesentericus*. A comprehensive review of peptide antibiotics of *Bacillus* has been written by Katz and Demain (1977).

As is true for protease synthesis, most of the antibiotics are formed near the end of exponential growth and just before and during sporulation. Such a temporal relationship is not always true because changes in cultural conditions and nutrient availability, as well as genetic modifications, can affect synthesis of both the antibiotics and the proteases. One interesting observation regarding the peptide antibiotics is that they apparently are not susceptible to the proteolytic enzymes with which they are in immediate contact as well as to a number of peptidases and proteases of animal and plant origin. Another intriguing feature of these peptides is that transfer RNA, messenger RNA, and ribosomes evidently are not involved in their synthesis. The amino acid sequence is determined by an enzyme.

Insecticides

Several bacilli have insecticidal properties and, as such, are called entomo-pathogens. Foremost among the entomopathogenic bacilli is *B. thuringiensis* because it shares the status of many chemical compounds designed for use in controlling economically and biomedically important insects. Other bacilli, in-cluding *Bacillus popilliae*, *Bacillus lentimorbus*, and *B. sphaericus*, are prime candidates for use as insecticides, but *B. thuringiensis* is the only one that has been successfully produced on a large scale and formulated as an active insec-ticidal material in many countries. A number of subspecies of *B. thuringiensis* synthesize insecticidal toxins with some variation in host specificity (Bulla et al., 1975; Miller et al., 1983). Most strains of *B. thuringiensis* are toxic to lep-idopteran insects such as moths, but one subspecies called *israelensis* kills dipteran insects — e.g., mosquitoes and black flies — instead.

The primary insecticidal activity of *B. thuringiensis* lies within a large, proteinaceous crystal that is synthesized and crystallized within the mother cell during sporulation. Once the parasporal crystal has been ingested by a susceptible insect, it is solubilized and activated. Subsequent interaction of the toxic moiety with midgut epithelial cell membrane upsets transport and passage of nutrients through the midgut cells and into the hemolymph. Thus feeding and nutrient uptake are impaired. The result is dehydration and ultimate death.

Biochemical and biophysical evidence indicates that the toxins of both the moth- and mosquito-killing strains of *B. thuringiensis* have an apparent molecular weight of 6.8×10^4 and that they are derived from larger protoxic monomers (native crystal subunit) whose apparent molecular weight is 1.34×10^5. Interestingly, the protoxic protein is contained within the parasporal crystal and in the coats of the dormant spore, a characteristic indicating that toxin synthesis is a sporulation specific event.

Both *B. popilliae* and *B. lentimorbus* are pathogenic to various scarabaeid beetles (Bulla et al., 1978). The bacteria, when ingested by beetle larvae, invade the hemocoel, where they undergo vegetative proliferation and sporula-tion, causing death of the larvae. The large number of spores that accumulates within the insect is released to the surrounding environment upon death and decay of the insect. These spores, eaten by other beetle larvae, germinate in the insect gut and begin the infectious process again. The infection caused by *B. popilliae* and *B. lentimorbus* is called "milky disease" because of the milky ap-pearance of the hemolymph containing spores of the bacteria.

Although *B. popilliae* and *B. lentimorbus* are closely related, they are different in one striking way. *B. popilliae*, like *B. thuringiensis*, forms a parasporal crystal within the parent cell during sporulation, whereas *B. lenti-morbus* does not. Evidently, the crystals of *B. popilliae* are not toxic to beetle larvae in the same fashion that *B. thuringiensis* crystals are to moth and mos-quito larvae. Attempts to kill beetle larvae by feeding *B. popilliae* crystals have not been successful. It is believed that the cause of beetle death by *B. popilliae* and *B. lentimorbus* is bacteremia rather than any toxic compounds produced by either organism. The best means for commercially producing *B. popilliae* and *B. lentimorbus* is by artificially infecting individual larvae. Consequently, widespread use and application of these bacteria is precluded by insufficient supplies.

B. sphaericus is another bacillus that exhibits good insecticidal activity against mosquitoes. Unlike *B. thuringiensis*, however, no parasporal crystal is formed. Apparently, the toxin is associated with the vegetative cell wall and not the spore, although spores of *B. sphaericus* do kill mosquito larvae. Characterization of the toxin is not complete; therefore elucidation of its biochemical and biological properties awaits further investigation.

VIII. CONCLUSION

The bacilli provide a wealth of biological material that can be used for fundamental research in molecular biology, genetics, physiology, biochemistry, and cellular development. They also can be exploited for a number of industrial applications. Innovative uses of some bacilli are found throughout the world, but more widespread commercial utilization and production of bacilli await further study. Genetic engineering techniques may be advanced to expand and broaden our uses of the bacilli in many different ways. Use of their end products could decrease our dependence on chemicals produced by other than biological means.

REFERENCES

Aubert, E., Fargette, F., Fouet, A., Klier, A., and Rapoport, G. (1982) in *Molecular Cloning and Gene Regulation in Bacilli* (Ganesan, A. T., Chang, S., and Hoch, J. A., eds.), pp. 11–24, Academic Press, New York.

Bachman, B. S., and Low, K. B. (1980) *Microbiol. Rev. 44*, 1–56.

Bechtel, D. B., and Bulla, L. A., Jr. (1976) *J. Crit. Rev. Microbiol. Bacteriol. 127*, 1477–1481.

Bechtel, D. B., and Bulla, L. A., Jr. (1982) *J. Ultrastruct. Res. 79*, 121–132.

Blattner, F. R., Williams, B. G., Blechl, A. E., Thompson, K. P., Faber, H. E., Furlong, L. A., Grunwald, D. J., Kiefer, D. O., Moore, D. D., Scham, J. W., Sheldon, E. L., and Smithies, O. (1977) *Science 196*, 161–169.

Buchanan, R. E., and Gibbons, N. E. (eds.) (1974) *Bergey's Manual of Determinative Bacteriology*, 8th ed., pp. 529–575, Williams and Wilkins, Baltimore.

Bulla, L. A., Jr., St. Julian, G., and Rhodes, R. A. (1971) *Can. J. Microbiol. 17*, 1073–1079.

Bulla, L. A., Jr., Rhodes, R. A., and St. Julian, G. (1975) *Ann. Rev. Microbiol. 29*, 163.

Bulla, L. A., Jr., Costilow, R. N., and Sharpe, E. S. (1978) *Adv. Appl. Microbiol. 23*, 1–18.

Bulla, L. A., Jr., Bechtel, D. B., Kramer, K. J., Shethna, Y. I., Aronson, A. I. and Fitz-James, P. C. (1980) *CRC Crit. Rev. Microbiol. 8*, 147–204.

Canosi, U., Morelli, G., and Trautner, T. A. (1978) *Mol. Gen. Genet. 166*, 259–267.

Casadaban, M. J., Chou, J., and Cohen, S. N. (1980) *J. Bacteriol. 143*, 971–980.

Doi, R. H. (1969) in *The Bacterial Spore* (Gould, G., and Hurst, A., eds.), p. 125, Academic Press, New York.

Doi, R. H. (1982) in *The Molecular Biology of the Bacilli* (Dubnau, D., ed.), pp. 71–110, Academic Press, New York.

Donnelly, C. E., and Sonenshein, A. L. (1982) in *Molecular Cloning and Gene Regulation in Bacilli* (Ganesan, A. T., Chang, S., and Hoch, J. A., eds.), pp. 63–72, Academic Press, New York.

Dubnau, D., Smith, I., Morell, P., and Marmur, J. (1965) *Proc. Nat. Acad. Sci. U.S.A. 54*, 491–498.

Eisenstadt, E., and Silver, S. (1972) in *Spores V* (Halvorson, H. O., Hanson, R., and Campbell, L. L., eds.), p. 180, American Society for Microbiology, Washington, D.C.

Ferrari, E., Henner, D. J., and Hoch, J. A. (1981) *J. Bacteriol. 146*, 430–432.

Freese, E., Klofat, W., and Gallius, E. (1970) *Biochem. Biophys. Acta 222*, 265.

Garbe, J. C., and Vary, P. S. (1982) in *Sporulation and Germination* (Levinson, H. S., Sonenshein, A. L., and Tipper, D. J., eds.), pp. 83–87, American Society for Microbiology, Washington, D.C.

Goldfarb, D. S., Rodriguez, R. L., and Doi, R. H. (1982) *Proc. Nat. Acad. Sci. U.S.A. 79*, 5886–5890.

Gordon, R. E., Haynes, W. C., and Pang, C. H.-N. (1973) in *The Genus Bacillus,* Agriculture Handbook No. 427, 283 pp., U. S. Department of Agriculture, Washington, D.C.

Gryczan, T. J. (1982) in *The Molecular Biology of the Bacilli* (Dubnau, D. E., ed.), pp. 307–329, Academic Press, New York.

Gryczan, T. J., Contente, S., and Dubnau, D. (1980) *Mol. Gen. Genet. 177*, 456–467.

Henner, D. J., and Hoch, J. A. (1980) *Microbiol. Rev. 44*, 57–82.

Highton, P. J. (1972) in *Spores V* (Halvorson, H. O., Hanson, R., and Campbell, L. L., eds.), p. 13, American Society for Microbiology, Washington, D.C.

Holt, S. C., Gauthier, J. J., and Tipper, D. J. (1975) *J. Bacteriol. 122*, 1322.

Hoshino, T., Uozumi, T., Beppu, T., and Arima, K. (1980) *Agr. Biol. Chem. 44*, 621–623.

Hutchison, K. W., and Halvorson, H. O. (1980) *Gene 8*, 267–278.

Katz, E., and Demain, A. L. (1977) *Bacteriol. Rev. 41*, 449–474.

Kawamura, F., Shimotsu, H., Saito, H., Hirochika, H., and Kobayashi, Y. (1981) in *Sporulation and Germination* (Levinson, H. S., Sonenshein, A. L., and Tipper, D. J., eds.), pp. 109–113, American Society for Microbiology, Washington, D.C.

Keggins, K. M., Lovett, P. S., and Duvall, E. J. (1978) *Proc. Nat. Acad. Sci. U.S.A. 75*, 1423–1427.

Keynan, A., and Halvorson, H. (1965) in *Spores III* (Campbell, L. L., Halvorson, H. O., eds.), p. 74, American Society for Microbiology, Ann Arbor, Michigan.

Kornberg, A., Spudich, J. A., Nelson, D. L., and Deutscher, M. P. (1968) *Ann. Rev. Biochem. 37*, 51–78.

Lang, D. R., and Lundgren, D. G. (1970) *J. Bacteriol. 101*, 483.

Losick, R., and Pero, J. (1981) *Cell 25*, 582–584.

Lovett, P. S., and Young, F. E. (1969) *J. Bacteriol. 100*, 658–661.

Lovett, P. S., and Young, F. E. (1970) *J. Bacteriol. 101*, 603–608.

Lovett, P. S., and Young, F. E. (1971) *J. Bacteriol. 106*, 697–699.

Michel, B., Niaudet, B., and Ehrlich, S. D. (1982) in *Molecular Cloning and Gene Regulation in Bacilli* (Ganesan, A. T., Chang, S., and Hoch, J. A., eds.), pp. 73–81, Academic Press, New York.

Miller, L. K., Lingg, A. J., and Bulla, L. A., Jr. (1983) *Science 219*, 715–721.

Murray, C. L., and Rabinowitz, J. C. (1982) in *Molecular Cloning and Gene Regulation in Bacilli* (Ganesan, A. T., Chang, S., and Hoch, J. A., eds.), pp. 271–285, Academic Press, New York.

Murrell, W. G., Ohye, D. F., and Gordon, R. A. (1971) in *Spores*, Vol. IV (Campbell, L. L., ed.), p. 1, American Society for Microbiology, Bethesda, Md.

Perlack, F. J., and Thorne, C. B. (1981) in *Sporulation and Germination* (Levinson, H. S., Sonenshein, A. L., and Tipper, D. J., eds.), pp. 78–82, American Society for Microbiology, Washington, D.C.

Rapoport, G., Klier, A., Billault, A., Fargette, F., and Dedonder, R. (1979) *Mol. Gen. Genet. 176*, 239–246.

Schaeffer, P., Cami, B., and Hotchkiss, R. D. (1976) *Proc. Nat Acad. Sci. U.S.A. 73*, 2151–2155.

Scribner, H. E., Mogelson, J., Eisenstadt, E., and Silver, S. (1975) in *Spores*, Vol. VI (Gerhardt, P., Costilow, R. N., and Sadojj, H. L., eds.), pp. 346–355, American Society for Microbiology, Washington, D.C.

Setlow, P. (1981) in *Sporulation and Germination* (Levinson, H. S., Sonenshein, A. L., and Tipper, D. J., eds.), pp. 13–28, American Society for Microbiology, Washington, D.C.

Sharp, R. J., Bown, K. J., and Atkinson, A. (1980) *J. Gen. Microbiol. 117*, 201–210.

Singh, R. P., Setlow, B., and Setlow, P. (1977) *J. Bacteriol. 130*, 1130–1138.

Smith, N. R., Gorden, R. E., and Clark, F. E. (1952) *Aerobic Sporeforming Bacteria*, Monograph No. 16, U. S. Department of Agriculture, Washington, D.C.

Spizizen, J. (1958) *Proc. Nat. Acad. Sci. U.S.A. 44*, 1072–1078.

Vinter, V. (1969) in *The Bacterial Spore* (Gould, G., and Hurst, A., eds.), p. 73, Academic Press, New York.

Walker, P. D. (1970) *J. Appl. Bacteriol. 33*, 1.

Zahler, S. A., and Korman, R. Z. (1981) in *Sporulation and Germination* (Levinson, H. S., Sonenshein, A. L., and Tipper, D. J., eds.), pp. 101–103, American Society for Microbiology, Washington, D.C.

Biology of Spore-Forming Anaerobes

J. Gregory Zeikus

I. INTRODUCTION

This chapter focuses on the general biological properties of obligately anaerobic spore-forming bacteria, which are of current interest in the development of biotechnology for chemical and fuel production from renewable biomass substrate (i.e., biopolymers or soluble saccharides) fermentations (Ng et al., 1983; Zeikus, 1979, 1980a; D. Wang et al., 1978; Zeikus and Ng, 1982; Sinskey, 1983; Wise, 1983; Tsao et al., 1982) and from simple organic pyrolysis substrate (i.e., CO, H_2–CO_2, CH_3OH) fermentations (Ng et al., 1983; Zeikus, 1980a, 1983c). Anaerobic bacteria offer potential for large-scale production of chemicals and fuels because their catabolism involves incomplete degradative oxidations of multicarbon compounds or synthesis of larger molecules from simple carbon substrates (Zeikus, 1980a; Wise, 1983).

The spore-forming anaerobes are recognized as being distinct because they make a resting cell, the spore, which is morphologically and metabolically differentiated from growing or vegetative cells. Spores display enhanced tolerance to environmental agents that are inhibitory to species growth (e.g., solvents, acids, oxygen). Spore-forming anaerobes are of special interest to industrial microbiology because sporulating anaerobes are more easily maintained and manipulated. Traditionally, spore-forming anaerobes were all considered to be members of the genus *Clostridium*. However, current biological understanding recognizes several genera of obligately anaerobic spore-forming bacteria based on species with substantial differences in spore structure–function and in catabolic enzyme function. Furthermore, at the forefront of taxanomic understanding are the new facts that several species of *Clostridium* are more phylogenetically related to species that have not yet been shown to sporulate than to other *Clostridium* species themselves.

Historically, *Clostridium* species have been important in applied microbiology largely because of their role in human and animal disease (Smith, 1973; Smith and Dowell, 1974), the retting of flax (Rombouts and Pilnik, 1980), and their limited use in the industrial production of acetone-butanol (Moreira, 1983; Spivey, 1978). More recently, *Clostridium* species have been of scientific interest because of their importance in tree disease (Schink et al., 1981), potato crop destruction (Lund, 1972), anaerobic waste treatment (Zeikus, 1980b), tertiary oil recovery (Moses and Springham, 1982), and large-scale production of chemicals and fuels (Zeikus, 1980a).

Table 4.1 summarizes several potential applications of spore-forming anaerobes or their enzymes in the industrial production of chemicals and fuels. Anaerobes make a limited number of gases, organic acids, or solvents as fermentation products (Zeikus, 1980a). Thus increasing the range of different chemicals produced by anaerobic transformations requires either that the normal fermentation products themselves be further modified by chemical processing to yield new products, such as a conversion of fermentation acids to organic acid esters (Datta, 1981), or that chemical substrates be transformed to useful end products (e.g., racemic ketones to chiral alcohols) by immobilization technology using whole cells or enzymes from anaerobes (Lamed et al., 1981).

Table 4.1 lists some of the anaerobic spore-forming bacteria that may provide transformation systems for chemical and fuel production from renewable resources. The industrial uses for these products have been well reviewed elsewhere (Palsson et al., 1981; Sinskey, 1983; Wise, 1983). Interest in acetone and butanol production by *Clostridium acetobutylicum* fermentation has received considerable attention (Moreira, 1983; Spivey, 1978; Barber et al., 1979; Allcock et al., 1981; Haggstrom and Molin, 1980; Gibbs, 1983; Walton and Martin, 1979). In pure culture, *C. acetobutylicum* ferments a variety of low-cost waste substrates including starch, whey, hexose, and pentose-derived wood sugars (Maddox, 1980, 1982; Mes-Hartree and Saddler, 1982; Beesch, 1982). The major goals for improving the acetone-butanol fermentation are to utilize cheaper substrates, increase the final solvent concentration, and most importantly, to enhance the amount of fermentation substrate transformed to solvents in lieu of forming waste products (i.e., H_2, acetate, butyrate). *Clostridium biejerinckii* fermentations differ from those of *C. acetobutylicum* because the former species produces isopropanol as a further reduction product of acetone metabolism (Krouwel et al., 1983; George et al., 1983).

TABLE 4.1 Potential Utility of Selected Spore-forming Anaerobes in Microbial and Enzyme Technology

Application	Transformation System Organisms-Catalyst	Substrate	Products
I. Chemical- and fuel-producing microbial fermentations			
	A. Solvent Production		
	1. *Clostridium acetobutylicum*	starch, hexose, pentose	acetone–butanol
	2. *Clostridium beijerinckii*	starch, hexose, pentose	isopropanol-butanol
	3. *Clostridium thermocellum*	cellulose	ethanol
	4. *Clostridium thermohydrosulfuricum*	starch, hexose, pentose	ethanol
	5. *Clostridium thermosaccharolyticum*	hexose, pentose	ethanol
	6. *C. thermocellum* and *C. thermohydro-sulfuricum*	cellulose and hemicellulose	ethanol
	7. *C. thermocellum* and *C. thermosaccha-rolyticum*	cellulose and hemicellulose	ethanol
	B. Acid Production		
	1. *Clostridium thermoaceticum*	hexose, pentose, CO_2–H_2, CO	acetic acid
	2. *Clostridium thermoautotrophicum*	saccharides, CH_3OH, CO, H_2–CO_2	acetic acid
	3. *Butyribacterium methylotrophicum*	H_2–CO_2, CO–CH_3OH	acetic acid
	4. *B. methylotrophicum*	hexose, CH_3OH–CO_2	butyric acid
	5. *Clostridium kluyveri*	ethanol-acetate	caproic acid
	6. *Sporolactobacillus inulinus*	saccharides	lactic acid
	7. *Clostridium propionicum*	saccharides	propionic acid
	8. *C. propionicum*	lactic acid	acrylic acid
	C. Methane Production		
	1. *C. butyricum* and *Methanosarcina barkeri*	pectin	methane

81

TABLE 4.1 (continued)

Application	Transformation System Organisms-Catalyst	Substrate	Products
	D. Vitamins		
	1. *B. methylotrophicum*	methanol	vitamin B_{12}
	2. *C. acetobutylicum*	starch wastes	riboflavins
II. Biocatalytic enzyme transformations	A. Depolymerization		
	1. *C. thermocellum* cellulase	cellulose-hemicellulose	xylose, xylobiose, cellobiose, glucose
	2. *C. thermosulfurogenes* pectinase	pectin	soluble sugars
	3. *C. histolyticum* protease	collagen	amino acids and peptides
	4. *C. acetobutylicum* amylase	starch	glucose, dextrins
	5. *C. thermohydrosulfuricum* glucoamylase	starch	glucose
	pullulanase	starch	dextrins
	6. *C. thermosulfarogenes* B-amylase	starch	maltose
	B. Hydrogenation-Dehydrogenation		
	1. *C. thermohydrosulfuricum* reversible alcohol dehydrogenase	alcohols, ketones	chiral ketones and alcohols
	2. *C. butylicum* hydrogenase	saccharides	hydrogen
	3. *Clostridium* species hydrogenase	enoates, aldehydes	alcohols

82

Thermophilic ($\geq 60°C$) ethanol production by direct biopolymer fermentations of *Clostridium* species is of interest because of low substrate costs, potentially enhanced productivity rates at high temperatures, and less energy-intensive product-recovery methods associated with continuous solvent removal at $60°C$ by reduced pressure distillation during the fermentation (D. Wang et al., 1978; Zeikus and Ng, 1982; Zeikus, 1979; Wiegel, 1982; Avgerinos and Wang, 1980; Zeikus et al., 1981; Saddler and Chan, 1982). *Clostridium thermocellum* ferments cellulose to ethanol as the major reduced end product (Weimer and Zeikus, 1977); or in co-culture with either *Clostridium thermohydrosulfuricum* (Ng et al., 1981) or *Clostridium thermosaccharolyticum* (Avgerinos and Wang, 1980), it can ferment both hemicellulose and cellulose to ethanol. In monoculture, strains of *C. thermohydrosulfuricum* or *C. thermosaccharolyticum* ferment a wide variety of saccharides, including starch, lactose, various pentoses, and hexose sugars (Zeikus and Ng, 1982; Avgerinos and Wang, 1980). Industrial ethanol is not produced by thermophilic fermentations because the ethanol concentrations are too low for economical recovery. More improvements are needed to enhance the concentration and yield of ethanol made by thermoanaerobes, and/or novel technology must be established to make ethanol recovery at low concentrations ($\leq 4\%$) economically feasible.

A larger variety of organic acids can be produced by anaerobic fermentations than solvents. The acetic acid fermentation of *Clostridium thermoaceticum* has received the greatest attention because this species makes nearly 3 moles of acetic per mole of glucose (G. Y. Wang et al., 1983; Ljungdahl, 1983). This means that aside from cells, no waste carbon is produced. This species also makes acetic acid from H_2–CO_2 or CO (Kerby and Zeikus, 1983). *Clostridium thermoautotrophicum* has a wide substrate range and can also synthesize acetic acid from methanol (Weigel et al., 1981). *Butyribacterium methylotrophicum* fermentations can be modulated by mixtures of one-carbon substrates to make either acetate or butyrate as the major end product during fermentation of inexpensive pyrolysis feed stocks (Lynd et al., 1982; Lynd and Zeikus, 1983; Kerby et al., 1983). *Clostridium kluyveri* fermentations are novel for synthesis of four- and six-carbon acids from two-carbon precursors (Thauer et al., 1977). *Sporolactobacillus* species form lactic acid as virtually the sole end product of saccharide fermentations (Norris et al., 1981). Recently, *Clostridium propionicum* fermentations have been investigated for synthesis of acrylic acid in lieu of propionate, the normal reduced end product of this species (Sinskey et al., 1981). At present the acid fermentations of these anaerobic spore-forming bacteria are of limited practical significance because of the inability of species to form high concentrations of free organic acids and the economics of current product-recovery methods at low free acid concentrations.

Spore-forming bacteria are of tremendous importance in anaerobic digestors for conversion of agricultural, municipal, and industrial wastes into methane (Zeikus, 1980b). Spore-forming species do not themselves form methane as a catabolic end product; but as a result of fermenting high-molecular-weight biopolymers, organic acids, and alcohols, they produce the immediate precursors (i.e., H_2–CO_2, CH_3OH, acetate, formate) used by methanogenic bacteria. Methanogenic species in turn metabolically communicate (Zeikus, 1983b) with spore-forming species to enhance the rate of total organic destruction directly by removing toxic end products (i.e., H_2, acetic acid) or indirectly by not allowing certain metabolites to form (i.e., long-chain fatty acids and alcohol. Research is urgently needed to develop defined, stable starter cultures of spore-forming anaerobes and methanogens that can be used as inocula for waste biomethanation processes. In this regard, a defined co-culture comprised of *Clostridium butyricum* and *Methanosarcina barkeri* has been described for the complete mineralization of pectin into methane and CO_2 (Schink and Zeikus, 1982).

Anaerobic spore formers may have potential in the industrial production of certain vitamins in spite of displaying lower cell yields (i.e., g cell/g substrate) than aerobic microbes grown on the same substrates. This can be explained by the need for very high levels of certain types of vitamins to perform the unique multi- and single-carbon transformation reactions inherent to catabolism of some spore-forming anaerobes. Thus methanol fermentations of *B. methylotrophicum* and other acetogen fermentations can yield high levels of vitamin B_{12} (Zeikus et al., 1980a; Hollriegl et al., 1982; Perlman and Seman, 1963). Previously, the spent cells from *C. acetobutylicum* fermentations were sold as animal feed supplements because of high riboflavin levels (Gibbs, 1983).

Anaerobes originated very early on earth and have had a long time to evolve highly efficient enzymes to perform catabolic chemical transformations under energy stressed conditions. Anaerobic spore formers produce very active depolymerases and unique oxidoreductases that may be of industrial importance in chemicals production. This is even more exciting now because of the possibility of using recombinant DNA technology to produce even higher levels of active enzymes from anaerobes by cloning and expressing their genes in aerobic microbes (Snedecor and Gomez, 1983). Thermophilic saccharidases from spore-forming anaerobes include an active extracellular cellulase from *C. thermocellum* (Ng and Zeikus, 1981a; Johnson et al., 1982a, b) and pectinase from *Clostridium thermosulfurogenes* (Schink and Zeikus, 1983a). Active exo-enzymes produced by mesophilic *Clostridium* species include the collagenase of *Clostridium histolyticum* (Tsuru, 1973) and α-amylase of *C. acetobutylicum* (Ensley et al., 1975). Immobilized whole cells or enzymes of *Clostridium* species have been used to produce chiral alcohol or ketones with a novel alcohol dehydrogenase present in thermoanaerobic species (Lamed et al., 1981), to produce hydrogen from waste saccharide in mesophilic species with active hydrogenase (Karube et al., 1982) or to reduce exotic enoates and aldehydes to alcohols by hydrogen in mixtures of *Clostridium* species that contain hydrogenase and alcohol dehydrogenase (Egerer and Simon, 1982).

The purpose of this chapter is to provide both a general description of the unifying biological properties of spore-forming anaerobes and a specific analysis of the physiological features of several species that are of current interest in biotechnology. This chapter is not intended to be encyclopedic; its aim is to draw together the pertinent reference material that forms the foundation for this topic.

II. GENERAL BIOLOGY

A. Cultivation

Obligate anaerobes lack the enzymes required to use oxygen as a catabolic substrate. To isolate and grow these bacteria, special conditions are often required. Most spore-forming anaerobes tolerate oxygen exposure to a certain extent but grow only when the dissolved O_2 partial pressure is low. Species with high tolerance to oxygen, for example, *Clostridium perfringens*, can be readily grown in a complex medium with sodium thioglycolate (0.05%) as the medium reductant and with an air headspace provided that the medium is heated at 75°C for 20 min to lower the O_2 solubility prior to inoculation with an active starter culture (Duncan and Strong, 1968). Fastidious species with low tolerance to oxygen, for example, *C. thermoaceticum*, can be grown unicarbonotrophically in a minimal medium with sodium sulfide (0.04%) as the reductant and with an H_2-CO_2 (80%–20%) headspace, provided that the medium is made with special precautions to remove oxygen (Kerby and Zeikus, 1983).

A variety of methods have been described for growing fastidious anaerobes in the absence of oxygen (Gerhardt, 1981). Species are isolated on agar roll tubes or on agar plates in anaerobic glove box chambers and are then transferred to a liquid medium. These procedures employ cultivation vessels (i.e., test tubes, serum vials, fermentors) sealed from air and gassed with an O_2-free gas. Culture media are generally made anoxic by their preparation under an N_2-CO_2 (95%–5%) gas headspace and by the syringe addition of a medium reductant after dispensing autoclaved media into sealed containers. Cysteine and sodium sulfide are the most common medium reductants, but they are often metabolized by anaerobes. Nonmetabolizable titanium III salts have been described as sulfur-free medium reductants for growth of fastidious anaerobes at low redox values (Zehnder and Wuhrman, 1976; Moench and Zeikus, 1983a). Growth of anaerobes that consume or produce copious amounts of gas necessitates the use of specialized containers that tolerate high gas pressure. Balch and Wolfe (1976) described a simple glass tube system that is useful for the routine growth of anaerobes at less than 4 atm gas pressure. Inverted incubation of these tubes is useful to prevent gas leakage through the rubber bungs that seal the tubes. Media and culture conditions for the enrichment, isolation, and growth of the described *Clostridium* species were excellently reviewed by Gottschalk et al. (1981).

B. Habitat

Although spore-forming anaerobes have traditionally been isolated from spoiled foods, they abound in a variety of natural and synthetic (e.g., anaerobic biogas digestor) environments. The primary habitats for spore-forming anaerobes are the anoxic muds and sediments of freshwater and marine ecosystems where organic matter is being vigorously decomposed (Gottschalk et al., 1981). In freshwater lake sediments the population of C. butyricum, a prevalent hydrolytic pectin degrading bacterium, displays a seasonal response in relation to nutrient availability; the highest numbers of pectin degraders are present in the fall, when surface algal sedimentation rates are highest (Schink and Zeikus, 1982). Soil environments that constantly change in relation to water activity (from waterlogged to dry), O_2 content (from aerobic to anoxic), and temperature (from freezing to greater than 60°C from thermal absorption or microbial enhanced self-heating) also contain anaerobes that produce dessication-, oxygen-, and temperature-resistant spores. Thus even extreme thermophilic species like C. thermohydrosulfuricum are readily isolated from surface soils (Weigel et al., 1979).

Secondary habitats for the spore-forming anaerobes include animal gastrointestinal tracts and manures (Gottschalk et al., 1981) and the wet woods of living trees (Schink et al., 1981). The anaerobes in these habitats appear to have been acquired from contact with the sedimentary–mud–soil environment. Only recently have extreme environments been recognized as habitats of spore-forming anaerobes (Zeikus, 1979). Thermal, volcanic–hot spring ecosystems (i.e., thermal waters and sediments) can contain diverse and a high number ($> 10^4$) of Clostridium species (Zeikus et al., 1980b; Weigel et al., 1981). The hypersaline (i.e., 20% total salts) sediments of the Dead Sea (Oren, 1983) and Great Salt Lake (Zeikus, 1983b) contain diverse populations of obligate anaerobes, including spore formers.

C. Species Diversity and Description

The described species of spore-forming anaerobes are quite diverse and can be categorized into three broad physiological groups: mesophiles, thermophiles, and halophiles (see Table 4.2). Mesophilic species generally grow well above 20°C but below 50°C. Thermophilic species grow well above 55–60°C but not below 35–40°C. The halophilic extreme hypersalinophiles are unique because they require high salt for growth (i.e., 5–15% NaCl). Clostridium lortetii is the only described halophilic spore-forming anaerobe, and it grows optimally at about 40°C (Oren, 1983).

Species diversity is greatest among mesophilic spore formers, and other references should be examined for a more complete list of described Clostridium (Gottschalk et al., 1981), Desulfotomaculum (Pfennig et al., 1981), and Sporolactobacillus (Norris et al., 1981) species than those described in Table 4.2. Species diversity among mesophilic and thermophilic spore formers is based on different criteria, including spore ultrastructure (e.g., presence or absence

TABLE 4.2 Examples of Species Diversity Among Spore-Forming Anaerobes

Broad Physiological Category	Species Name	Habitat	Source	Major Reduced Catabolic End Products
I. Mesophiles	Clostridium acetobutylicum	soil	starch, carbohydrates	H₂, butanol, acetone, ethanol
	C. beijerinckii	soil	starch, carbohydrates	H₂, butanol, +/− isopropanol, acetone
	C. barkeri	sediments	carbohydrates, nicotinic acid	lactate, butyrate, propionate
	C. butyricum	sediments, wetwood	carbohydrates, pectin	butyrate, H₂
	C. cellobioparum	rumen	cellulose, carbohydrates	ethanol, formate
	C. formicoaceticum	muds	carbohydrates, lactate	acetate, +/− formate
	C. aceticum	sediments	carbohydrates	acetate
	Butyribacterium methylotrophicum	sediments	carbohydrates, lactate, CH₃OH, CO, H₂/CO₂	acetate, butyrate
	Desulfotomaculum acetooxidans	cow manure	ethanol, acetate + SO₄⁼	H₂S, CO₂
	Sporolactobacillus inulinus	soil	carbohydrates	lactate
II. Thermophiles	Clostridium thermoautotrophicum	thermal springs, mud	carbohydrates, H₂–CO₂, methanol, formate	acetate
	C. thermoaceticum	soil, manure	carbohydrates, H₂–CO₂, CO	acetate
	C. thermocellum	soil, manure	cellulose, carbohydrates	ethanol, H₂
	C. thermohydrosulfuricum	soil, hot springs	starch, carbohydrates	ethanol, lactate
	C. thermosaccharolyticum	soil, manure	starch, carbohydrates	butyrate, lactate, +/− ethanol
	C. thermosulfurogenes	thermal springs	pectin, starch, carbohydrates	ethanol, lactate, H₂, +/− methanol
	Desulfotomaculum nigrificans	soil, thermal springs	lactate, ethanol + SO₄⁼	acetate, H₂S
III. Halophiles	C. lortetii	Dead Sea sediments	carbohydrates, amino acids	butyrate, acetate, H₂

87

of a cortex or coat layers); cell and sporangium morphology; major reduced end product formed (e.g., solvents or acids); types and amounts of acids (e.g., lactic, formic, acetic, or butyric) or solvents (e.g., butanol, ethanol, acetone, or isopropanol) formed; ability to perform dissimilatory sulfate reduction; and energy sources metabolized (e.g., H_2–CO_2, polymers, nitrogenous compounds, sugars, organic acids, or methanol).

It is clear that the understanding of the diversity of spore-forming anaerobes in nature is limited by the amount of research devoted to understanding the general biology of these bacteria, especially in extreme environments such as acid, alkaline, hypersaline, or thermal habitats, where little is known about the anoxic microbial world.

Thus far, four genera of obligately anaerobic (i.e., non-O_2-catabolizing) spore-forming bacteria have been described: *Clostridium, Desulfotomaculum, Sporolactobacillus,* and *Butyribacterium.* The latter three genera are distinguished from *Clostridium* by the following criteria. *Desulfotomaculum* species are able to utilize sulfate as a catabolic electron acceptor and produce sulfide as the major reduced end product (Pfennig et al., 1981). *Desulfotomaculum* species are further distinguished by their ability to completely or incompletely oxidize fatty acids. For example, *Desulfotomaculum nigrificans* cannot oxidize acetate plus sulfate to CO_2 and H_2S, whereas *Desulfotomaculum acetooxidans* has this metabolic function. *Sporolactobacillus* species make lactic acid as the sole product of carbohydrate fermentation. They are similar to *Clostridium* species in lacking catalase and being obligate anaerobes but are considered closely related to *Bacillus* species that catabolize oxygen because of a similar cell wall chemistry (Norris et al., 1981). *Butyribacterium* species form branched cells and endospores that are not brightly refractile and do not have a spore cortex. *Butyribacterium methylotrophicum* (Zeikus et al., 1980a) is the only described species, and it is distinct from *Eubacterium* species, and *Eubacterium limosum* in particular, because of three main taxonomic criteria (Lewis and Sutter, 1981): it forms branched cells, it forms spores, and it does not form the copious slime that is so characteristic of *E. limosum.* The spore structure and cell branching character also distinguish *B. methylotrophicum* from *Clostridium* species. It is similar to some acetogenic *Clostridium* and non-spore-forming *Acetobacterium* species in terms of sharing a common mechanism of acetate synthesis from one carbon metabolism (Kerby et al., 1983). *Clostridium* species vary enormously and are identified and distinguished in terms of their energy and metabolism (Thauer et al., 1977; Gottschalk et al., 1981).

C. acetobutylicum grows at 37°C on a variety of carbohydrates, including starch. The main fermentation end products include acetate, butyrate, acetone, ethanol, butanol, and copious amounts of hydrogen and CO_2 (McCoy et al., 1930). This species differs from *C. beijerinkii* (syn. *C. butylicum*), which forms isopropanol as an additional end product often in lieu of acetone (George et al., 1983; George and Chen, 1983). Both species are of industrial interest because they produce high concentrations of solvents (especially butanol). Other *Clostridium* species form butanol including *Clostridium pasteurianium*, *Clostridium aurantibutyricum*, *Clostridium felsinum*, *Clostridium sporogenes*, and *Clostridium cadaveris* but in lower yields (Gottschalk et al., 1981; George et al., 1983).

C. butyricum and *Clostridium barkeri* are examples of very diverse mesophilic *Clostridium* species; both produce butyrate as the major end product of fermentation but not in association with butanol formation. *C. butyricum* stains ferment a variety of carbohydrates including pectin into acetate, butyrate, lactate, ethanol, CO_2, and copious amounts of hydrogen (Schink and Zeikus, 1982; Schink et al., 1981). *C. barkeri* is ultrastructurally distinct from *C. butyricum* (Stadtman et al., 1972) but similar to *B. methylotrophicum* (Zeikus et al., 1980a). Lactic and butyric acid are the main end products of *C. barkeri* glucose metabolism, whereas nicotinic acid is fermented to propionic acid, acetic acid, CO_2, and NH_2 (Stadtman et al., 1972).

Clostridium cellobioparum and *C. thermocellum* are the only well-documented cellulose decomposing *Clostridium* species in pure culture. *C. cellobioparum* optimally ferments a wide variety of carbohydrates at 30–37°C. The main end products are formate, lactate, acetate, butyrate, ethanol, CO_2, and low amounts of hydrogen (Chung, 1976). *C. thermocellum* grows optimally at 60–65°C by fermenting cellulose, cellobiose, and glucose but not starch, pentose sugars, or hemicellulose (Ng et al., 1977; Ng and Zeikus, 1982). The main fermentation products are ethanol, acetate, lactate, CO_2, and high levels of hydrogen.

Several *Clostridium* species are recognized as acetogens because they form acetate as the main product of carbohydrate fermentation by a corrinoid-dependent one-carbon metabolism (Ljungdahl and Wood, 1982). *C. thermoaceticum* optimally ferments hexoses and pentoses at 60°C (Fontaine et al., 1942), or it can metabolize H_2–CO_2 or CO as energy source (Kerby and Zeikus, 1983). *C. thermoautotrophicum* differs from *C. thermoaceticum* mainly in the ability to metabolize methanol as energy source (Weigel et al., 1981). *Clostridium aceticum* and *Clostridium formicoaceticum* are both mesophiles that do not utilize glucose or xylose but ferment fructose, glutamate, and ethanol (Andreesen et al., 1970; Braun et al., 1981). *C. formicoaceticum* does not utilize formate, H_2–CO_2, serine, ethylene, or glycol as energy sources, whereas these substrates are fermented by *C. aceticum* (Braun et al., 1981).

Three different thermophilic *Clostridium* species have been described that ferment starch and other saccharides but not cellulose to ethanol. *C. thermosaccharolyticum* is recognized as distinct from *C. thermohydrosulfuricum* on the basis of wall ultrastructure, DNA homology, fermentation pattern, and the capacity to form sulfide from thiosulfate reduction (Matteuzzi et al., 1978; Sleytr and Glauert, 1976; Weigel et al., 1979; Hsu and Ordal, 1970; Zeikus et al., 1980b). *C. thermosaccharolyticum* forms acetate, butyrate, lactate, and high levels of H_2 and CO_2 during growth but forms ethanol as a major product during sporulation (Hsu and Ordal, 1970). *C. thermohydrosulfuricum* produces ethanol, acetate, CO_2, lactate, and low levels of hydrogen during growth (Weigel et al., 1979; Ziekus et al., 1980b). Acetate and H_2S are the main end products of carbohydrate fermentation when *C. thermohydrosulfuricum* is grown in the presence of thiosulfate. *C. thermosulfurogenes* actively ferments pectin to methanol, ethanol, acetate, lactate, CO_2, and high levels of H_2 (Schink and Zeikus, 1983b). This species is recognized by its ability to transform thiosulfate into elemental sulfur in lieu of H_2S. *C. thermosulfurogenes* produces an active, thermal-stable, extracellular pectate hydrolase (Schink and Zeikus, 1983a).

D. Taxonomy

The extreme amount of biological diversity exhibited by the described spore-forming anaerobes suggests that these bacteria do not form a common taxon. Rather, they are a heterogenous assemblage with several apparent major genera. The taxonomy of anaerobes (i.e., understanding the evolutionary relatedness and similarity of different species) is in a constant state of flux; but with the use of molecular techniques (i.e., analysis of nucleic acid homologies, cell wall chemistry, protein homology, etc.) as well as detailed biochemical–metabolic analysis, some future trends in recognizing differences in spore-forming genera and species are apparent.

At present it could be argued by several lines of evidence that the genus *Clostridium* should be further subdivided into several genera based on extreme physiological differences including temperature and salt range for growth and the mechanism of acetate synthesis. Extreme thermophiles such as *C. thermohydrosulfuricum* are not considered closely related to mesophiles like *C. cellobioparum* because extreme thermophiles require multigene differences to account for thermal stable enzymes and membrane lipids (Zeikus, 1979). DNA guanosine plus cytosine and homology studies indicate that some of the thermophilic *Clostridium* species are more closely related to themselves than to mesophilic species or to other thermophilic spore formers. Matteuzzi et al. (1978) reported the following DNA G + C values (in moles %) for anaerobic spore formers: *C. thermohydrosulfuricum*, 32; *C. thermosaccharolyticum*, 31; *D. nigrificans*, 45; *C. thermoaceticum*, 54; *C. formicoaceticum*, 34; *C. butyricum*, 27.5; *C. acetobutylicum*, 29; and *Clostridium felsineum*, 25. Furthermore, these authors reported that the DNA homology values between *C. thermohydrosulfuricum* and other spore-forming anaerobes were as high as 49% for *C. thermosaccharolyticum* strains, but no (0%) homology was observed for *D. nigrificans*, *C. thermoaceticum*, *C. butyricum*, and *C. acetobutylicum*. On the other hand, the homoacetogenic mesophilic species *C. formicoaceticum* displayed less DNA homology toward *C. thermoaceticum* (3%) than toward *C. butyricum* (8%). These results suggest that different *Clostridium* species (e.g., *C. formicoaceticum*, *C. thermohydrosulfuricum*, *C. butyricum*, and *C. thermoaceticum*) could be subdivided into four different genera by using taxonomic splitter logic and not using spore formation as the prime generic criterion.

The taxonomic disposition of *C. barkeri*, *E. limosum* and *B. methylotrophicum* is of particular interest. All three species share a common wall and cytoplasmic architecture (see Fig. 4.1) with unusually thick wall end caps and internal disc structures (Zeikus et al., 1980a; Stadtman et al., 1972; Thompson and Zeikus, unpublished data). The metabolic properties of *B. methylotrophicum* are very similar to those of *E. limosum* (Genthner et al., 1981; Genthner and Bryant, 1982). However, spores have not been detected in *E. limosum*, and this is why it is currently considered a *Eubacterium* (Lewis and Sutter, 1981). Both *E. limosum* and *B. methylotrophicum* have the same B2-type peptidoglycan with D-*lys* and L-*orn* as interpeptide bridges (Guinand et al., 1969; O. Kandler, personal communication). Both *C. barkeri* and *B. methylotrophicum* form spores but not readily, and this is difficult to document with cells in thin section (Stadtman et al., 1972; Zeikus et al., 1980a; Thompson and Zeikus, unpublished data; T. C. Stadtman, personal communication). There are sufficient physiological differences among the three organisms to confirm species distinction (Stadtman et al, 1972; Zeikus et al., 1980a; Moench and Zeikus, 1983b; Genthner and Bryant, 1982), but it is suggested here that they should all belong to the genus *Butyribacterium* originally described by Barker and Hass (1944). Analysis of 16S ribosomal RNA oligonucleotide catalogs of anaerobes indicated that *C. barkeri* and *E. limosum* were identical and that these species were more closely related to *Acetobacterium woodii* than other *Eubacterium* or *Clostridium* species (Tanner et al., 1981, 1982).

The recent study by Tanner et al. (1982) is of particular interest because it showed the great phylogenic diversity that exists in the genus *Clostridium* and the close relatedness of some non-spore-forming species with spore-forming species. One ought not to forget that 16S RNA oligonucleotide analysis itself is only accurate in identification of supra generic taxa (i.e., families, order, kingdoms), but it is now evident that the ability to form or not to form

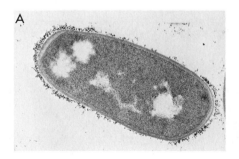 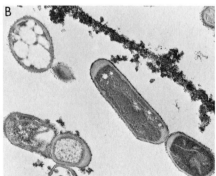

FIGURE 4.1   Electron photomicrographs of thin sections of *B. methylotrophicum*. (a) Vegetative cells. (b) Sporulating cells with mature spores in lower left corner (from T. Thompson and J. Zeikus, unpublished results).

(which is actually the absence of evidence) a spore alone is not a valid criterion for genus assignment of anaerobic bacteria.

E. Spore Properties

Spores or clostridial stages classically refer to swollen, spindle-club-shaped, or refractive cells that are derived from rod-shaped vegetative cells. Spore formation is often difficult to document because this process requires different combinations of exacting environmental conditions (i.e., dissolved oxygen, pH, temperature, nutrient levels, mineral components, activation-inhibitor substances) in different species. Spores of *C. thermoaceticum* are not readily refractile in liquid media but are seen when the organism is grown on an agar medium (see Fig. 4.2a, data of R. Kerby). Spores of some anaerobes like *B. methylotrophicum* are most difficult to detect, and optimized sporulation media have not been developed. A complex medium can be formulated that enhances both a given species' ability to sporulate and the heat resistance of its spores (Duncan and Strong, 1968), but explanations for why this is so are not well understood. Recently, defined media have been described for growth and sporulation of *C. thermocellum* (Johnson et al., 1981; Hyun et al., 1983), *C. acetobutylicum* (Long et al., 1983), and *C. thermohydrosulfuricum* (Hyun et al., 1983).

Sporulation is a survival mechanism in response to factors that uncouple growth and metabolism. Apparently, sporulation is regulated in *Clostridium botulinum* by classic mechanisms of glucose catabolite repression, since the addition of 1.0 M^{-5} cyclic 3′, 5′ adenosine monophosphate reversed glucose repression of sporulation from 30% to 80% (Emeruwa and Hawirko, 1975). Likewise, exacting environmental conditions including specific organic acids, amino acids, and cations influence the rate spores germinate into vegetative cells of *Clostridium bifermentans* (Waites and Wyatt, 1971).

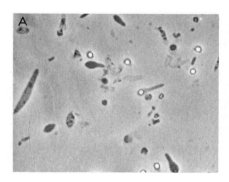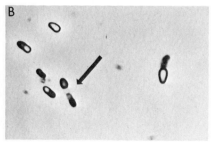

FIGURE 4.2 Phase-contrast photomicrographs of spore-forming anaerobes. (a) Sporulating cells of *C. thermoaceticum* (photograph courtesy of R. L. Kerby). (b) Sporulating (left) cells and a mature spore (right) of *B. methylotrophicum* (from Zeikus et al., 1980a).

The morphology and ultrastructure of spores vary enormously in anaerobes. Some *Clostridium* (Oren, 1983) and *Desulfotomaculum* (Pfennig, et al., 1981) species actually form gas vacuoles within the sporgangium. *Clostridium* species form bright white-refractile endospores that vary enormously in shape, size, and ultrastructure (Fitz-James, 1962; Hoeniger et al., 1968; Oren, 1983; Hyun et al., 1983). Figs. 4.2a and 4.3 illustrate these spore features for *C. thermoaceticum*, *C. thermocellum*, and *C. thermohydrosulfuricum*. *B. methylotrophicum* forms less-bright and blue refractile endospores that lack complex spore envelope layers and are notably devoid of a cortex layer (see Figs. 4.2b and 4.1).

The heat resistance of *D. nigrificans* (Donnelly and Busta, 1980) and *C. thermohydrosulfuricum* (Hyun et al., 1983) spores is much greater (see Table 4.3) than those of the aerobic species *Bacillus stearothermophilus*, which is

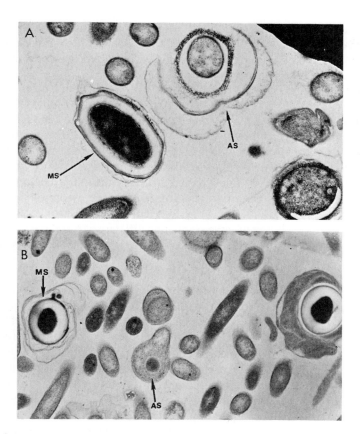

FIGURE 4.3 Electron photomicrographs of thin sections of sporulating cells showing mature spores (MS) and abortive spores (AS) of thermophilic *Clostridium* species. (a) *C. thermocellum*. (b) *C. thermohydrosulfuricum* (from Hyun et al., 1983).

TABLE 4.3   **Heat Resistance of Spores as a Function of Decimal Reduction Times***

Species	100°C	110°C	121°C
Clostridium thermocellum LQRI	200	8	0.5
C. thermosulfurogenes 4B	520	33	2.5
C. thermohydrosulfuricum 39E	770	123	11
Desulfotomaculum nigrificans[a]			5.6
Bacillus subtilis[b]			0.9
B. stearothermophilus FS1518[b]			3.0
B. stearothermophilus FS617[b]			0.8
C. sporogenes PA3679[b]			1.3

Header spanning columns 100°C, 110°C, 121°C: *D-Values (min.)*

* The time required to reduce the viable spore number by one $\log_{10}$ cycle.
[a,b] Data obtained from Hyun et al. (1983) and Donnelly and Busta (1980).

used in industry for sterility tests of autoclaving practice. The ability of *C. thermohydrosulfuricum* spores to withstand normal autoclaving (20 min at 121°C) points to some potential problems with maintaining pure cultures of thermophiles in industry or in research labs.

The heat resistance of spores from thermoanaerobic species is proportional to the ratio of spore cortex volume to cytoplasmic volume (Hyun et al., 1983). This value was 1.4 for *C. thermocellum* spores (see Fig. 4.3a) and 5.5 for *C. thermohydrosulfuricum* spores (see Fig. 4.3b). The cortex of *C. thermohydrosulfuricum* is notably thicker in relation to cytoplasmic diameter than that of *C. thermocellum* or *C. thermosulfurogenes*.

The presence of a very thick-walled structure around a resting cell membrane, which functions in osmoregulation to keep the cytoplasm osmotically dehydrated, is generally responsible for spore heat resistance (Gould and Dring, 1975). Thus spores of anaerobes can vary enormously in heat resistance depending upon the thickness of the cortex or the expanded peptidoglycan polymer layer iteself and the positively charged ions associated with it. In this regard, spores of *B. methylotrophicum* lack a cortex but have a very thick peptidoglycan layer, and this probably accounts for their ability to withstand 10 min of heat treatment at 75–80°C (Zeikus et al., 1980a).

Spores of anaerobic bacteria are also resistant to many other extreme conditions. It is common practice to maintain preserved stocks of spore-forming anaerobes in the presence of oxygen but in a desiccated state on sterile soil particles. The viability of spore soil stocks is quite remarkable. I obtained a spore soil stock of *Clostridium pasteurianum* from Professor McCoy at the University of Wisconsin that was originally prepared by Winogradsky. These 80-year-old spores are still viable and readily germinate! Bacterial spores are generally more tolerant to organic solvents and acids, but interestingly, some solvents are actually produced in response to sporulation. Sporulating cells of *C. thermosaccharolyticum* induce a NADP-linked alcohol dehydrogenase and form ethanol at high concentrations (Hsu and Ordal, 1970). Sporulation mutants of

C. acetobutylicum that are unable to form clostridial stages did not produce solvents (Jones et al., 1982).

F. Growth and Metabolism

Obligate anaerobes usually grow in media with low redox potential (< 150 mv), and as a result of their catabolism, anaerobes can establish a redox potential lower than -400 mv, depending upon the fermentation substrates and end products. The inhibition of anaerobic species growth by oxygen is a multifaceted phenomenon related directly to oxygen's raising the redox potential and altering catabolic carbon and electron flow rates and indirectly to the generation of chemically active hydroxyl radicals, peroxides, and super oxide anions that are toxic to anaerobic species that lack super oxide dismutase or catalase as detoxifying agents (Morris, 1975). O'Brien and Morris (1971) performed a detailed study on the effects of oxygen on that growth and metabolism of *C. acetobutylicum*, which lacks catalase. Altering medium redox potential to above $+100$ mv by high levels of oxygen or H_2O_2 was not lethal; it just inhibited growth and metabolism, but cells readily recovered after the removal of these oxidizing agents. Notably, the addition of reduced O_2 partial pressures to lower the medium redox to -50 to -100 mv did not alter the rate of glucose consumption or the molar growth yield but increased acetate and decreased butyrate production.

C. acetobutylicum elicits an interesting behavioral response by developing fruiting bodies when colonies on agar plates are exposed to oxygen (Jones et al., 1980). Vegetative and sporulating cells are restricted to the basal region of the structure, while the trunk region contains mainly spores.

Our understanding of energy conservation mechanisms (i.e., ATP synthesis and consumption) during growth and metabolism in anaerobes has changed dramatically in the past ten years. The excellent review by Thauer et al. (1977) explains two major mechanisms for energy conservation in spore-forming species. Substrate level phosphorylation (SLP) occurs in all described species, and electron transport–mediated phosphorylation (ETP) was postulated to occur in acetogenic, sulfate-reducing, and propionate-producing species (Thauer et al., 1977). In SLP the dehydrogenating partial reactions and the non-redox processes of catabolism are associated with production and consumption of an energy-rich compound (e.g., pyruvate dehydrogenase coupled to acetate kinase). In ETP the electrochemical potential between redox partners of different potential is used to drive the phosphorylation of ADP. This process is dependent on the location of hydrogen and electron carriers in the cell membrane and on generation of a proton motive force to activate membrane-bound ATPase. Unfortunately, much of the evidence used to support the presence of either mechanism in anaerobes has been based on growth yield data, which are now quite subject to question because of the effect of pH changes on growth. A third mechanism of energy conservation, transport-coupled phosphorylation (TCP), which is not present in obligate aerobes, has recently been discovered in anaerobes (see Konings and Veldkamp, 1983). Here anaerobes gain ATP via the generation of a proton motive associated with end product efflux. These authors reported that *Streptococcus cremoris* gained more ATP and enhanced growth yields when cultured at lower than higher external lactate and H^+ concentrations because lactic acid export was thermodynamically favored and resulted in a higher proton motive force. This is a very exciting and important finding, which was previously overlooked in anaerobes. Thus it is now possible to postulate that solvent production by spore-forming anaerobes may also be associated with a membrane-linked proton-consuming and ATP-generating process. For example, a membrane-bound butyrate-to-butanol reduction process in *C. acetobutylicum* may be coupled to ATP synthesis.

Growth and metabolism of anaerobic spore formers is altered not only by high O_2, acid, or solvent concentrations. Various exogenous electron donors and acceptors can dynamically alter the direction and rate of intraspecies carbon and electron flow pathways. For example, hydrogen addition can inhibit species growth and increase reduced fermentation products (Chung, 1976; Lamed and Zeikus, 1980). Nitrate addition to *C. perfringens* glucose fermentations caused increased acetate formation and decreased hydrogen, ethanol, and butyrate formation, but without a decrease in growth rate or yield during nitrate reduction (Ishimoto et al., 1974). This feature will be covered more completely in Part III of this chapter.

G. Genetics

Genetic understanding of spore-forming anaerobes is quite primitive, although it holds great promise and studies on industrial species have begun (Snedecor and Gomez, 1983). Procedures for isolation of auxotrophic mutants have been reported for *C. perfringens* (Sebald and Costilow, 1975) and *C. thermocellum* (Mendez and Gomez, 1982). Solvent-producing and sporulation mutants have been described for *C. acetobutylicum* (Jones et al., 1982). Also, a transformation system has been developed for this organism (Reid et al., 1983). The most exciting results to date have been the cloning and expression of two endoglucanases from *C. thermocellum* into *E. coli* (Cornet et al., 1983a), which has enabled further genetic characterization of the cel or cellulose degradation genes (Cornet et al., 1983b).

III. COMPARATIVE CATABOLISM OF SOLVENTOGENESIS AND ACETOGENESIS

A. *C. acetobutylicum*

Figure 4.4 depicts a plausible pathway to account for the directed flow of carbon and electrons during saccharide fermentation by *C. acetobutylicum*. Most of the enzyme activities required to account for this pathway have not been well-characterized in this organism. Saccharides are transformed to a common

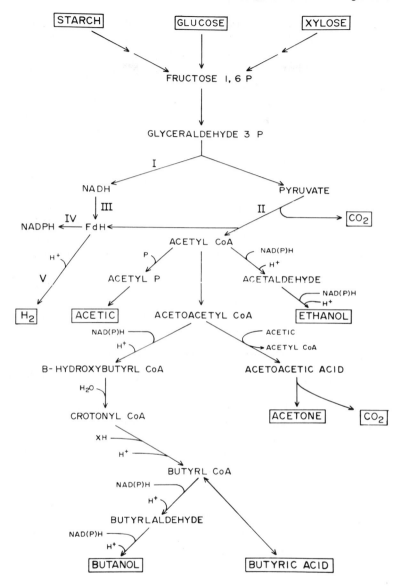

FIGURE 4.4   Proposed carbon and electron flow pathway for *C. acetobutylicum* saccharide fermentation. Enzyme-catabolizing reactions I–IV have been characterized by Petitdemange et al. (1976, 1977). I, glyceraldehyde 3 phosphate dehydrogenase; II, pyruvate–ferredoxin oxidoreductase; III, ferredoxin–NAD oxidoreductase; IV, ferredoxin–NADP oxidoreductase; and V, hydrogenase. Each arrow represents one or more enzyme-catalyzed reactions. The flow of electrons from NADH to reduced ferredoxin to H_2 explains why the organism produces high partial pressures of H_2 and does not consume H_2.

intermediary metabolite, fructose 1-6 diphosphate, which is cleaved into glyceraldehyde-3-phosphate, and NADH and pyruvate are formed by normal glycolysis mechanisms. The further transformation of the carbon and electrons in pyruvate and the oxidation of NADH are performed by a complicated interconnected series of electrochemical transformations. NADH is oxidized either by ferredoxin–NAD oxidoreductase or by the formation of reduced end products directly (e.g., alcohol dehydrogenase). Pyruvate is transformed via pyruvate ferredoxin oxidoreductase into acetyl-CoA, reduced ferredoxin, and CO_2. Either reduced ferredoxin is oxidized via hydrogenase and H_2 is produced, or NADPH is formed by ferredoxin–NADP oxidoreductase activity. Acetyl-CoA is transformed in either of three directions, to yield acetic acid via phosphotransacetylase and acetate kinase, to yield acetoacetyl-CoA by condensation, or to yield ethanol via acetyl-CoA reductase and alcohol dehydrogenase. Acetoacetyl-CoA is further transformed by one of two routes: to yield acetone via an exchange reaction with acetic acid and a further decarboylation or to yield butyryl-CoA via a series of reductions. Finally, butyryl-CoA is transformed either to butyric acid or to butanol by butyryl-CoA reductase and alcohol dehydrogenase.

Little is known about the enzymes responsible for saccharide transformation or uptake in *C. acetobutylicum*. Ensley et al. (1975) reported that starch was transformed to glucose via α-amylase and glucoamylase. Recently, a carboxymethyl cellulase and cellobiose activity were also demonstrated (Allcock and Woods, 1981).

The key enzymes that control the intracellular direction of electron flow were studied by Petitdemange et al. (1976, 1977). These authors demonstrated significant levels of the following enzymes in crude cell extracts: NAD-linked glyceraldehyde-3-phosphate dehydrogenase; NAD-linked alcohol dehydrogenase; NAD-linked aldehyde dehydrogenase; pyruvate–ferredoxin oxidoreductase, and ferredoxin-NAD(P) oxidoreductases. The very high levels of H_2 produced during saccharide fermentations were accounted for by the activity of NAD-ferredoxin reductase (Petitdemange et al., 1976). The function of ferredoxin-NADP reductase activity was suggested to be only anabolic (Petitdemange et al., 1977). An NADH-rubredoxin oxidoreductase was purified from *C. acetobutylicum*, but the physiological function of this diaphorase activity is not known (Petitdemange et al., 1979).

Regulation of solvent production in *C. acetobutylicum* fermentations is an interesting subject. Figure 4.5 represents a glucose fermentation time course for *C. acetobutylicum*; it depicts the classic association (i.e., secondary metabolic) of solvent production with the stationary phase of growth. Jones et al. (1982) have shown that cell morphology dramatically changes during the fermentation and swollen clostridial forms were involved in the conversion of acids to solvents. These same authors (Jones et al., 1982) showed that sporulation mutants unable to form clostridial stages did not produce solvents. It is worth noting that butanol production is associated with butyrate consumption. Wood et al. (1945) very eleoquently showed by use of C-13 tracer studies

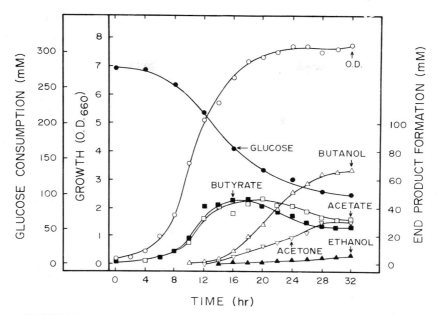

FIGURE 4.5 Association of solvent production with stationary phase (i.e., secondary metabolism) during glucose fermentation by *C. acetobutylicum* at pH 4.5 (unpublished data of Hong, Bellows, and Zeikus).

that butyrate was the precursor for butanol production. This fact is overlooked in the current literature.

Recently, a lot of interest has been directed toward manipulating environmental conditions in order to improve and understand solvent production. Monot et al. (1982) reported on the effect of various metal ions and acetate on solvent production from glucose in a defined medium. The presence of a high fatty acid concentration and high proton activity was shown to enhance or induce solvent production (Andersch et al., 1982; Bahl et al., 1982b; Nishio et al., 1983; Gottschal and Morris, 1981a). Continuous solvent production was achieved in chemostats, and the influence of various nutrient and cultural conditions on glucose fermentation was examined (Bahl et al., 1982a; Gottschal and Morris, 1981b).

The physiological basis that accounts for butanol tolerance in *C. acetobutylicum* strains is not apparent. Butanol tolerant mutants that grow above 15 g of butanol per liter have been selected by serial transfers (Lin and Blaschek, 1983). Van den Westhuizen et al. (1982) demonstrated that autolysin-deficient mutants were more tolerant of solvents than the wild-type strain.

B. *C. thermocellum*

The catabolism by *C. thermocellum* is of interest because the organism produces ethanol during rapid growth (doubling time of 6 h) on crystalline

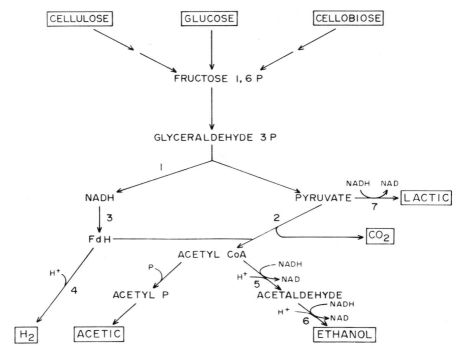

FIGURE 4.6 Carbon and electron flow pathway for saccharide catabolism of *C. thermocellum*. The organism contains the following oxidoreductases: 1, glyceraldehyde 3-P dehydrogenase; 2, pyruvate ferredoxin oxidoreductase; 3, NAD–ferredoxin oxidoreductase; 4, hydrogenase; 5, acetyl-CoA reductase (NAD); 6, acetaldehyde reductase (NAD); and 7, pyruvate reductase (NAD). Modified from Lamed and Zeikus (1980).

cellulose (Weimer and Zeikus, 1977) as a consequence of producing an active extracellular cellulase (Ng et al., 1977). Figure 4.6 provides an overall catabolic scheme that depicts the flow of carbon and electrons during saccharide fermentation by *C. thermocellum*.

   C. thermocellum adheres very tightly to cellulose during the initiation of cellulose hydrolysis because cells produce a multicomponent cellulose-binding factor (Bayer et al., 1983) that contains the cellulase complex and is released into the medium during growth (Lamed et al., 1983). The extracellular cellulase complex contains xylanase activity and high endoglucanase activity that is not inhibited by oxygen or high concentrations of cellobiose (Ait et al., 1979; Garcia-Martinez et al., 1980; Ng and Zeikus, 1981a). In addition, the extracellular cellulase complex uniquely contains high exoglucanase activity that is inhibited by oxygen and cellobiose (Johnson et al., 1982a, b; Johnson and Demain, 1984). The extracellular cellulase complex of *C. thermocellum* may have industrial utility. *C. thermocellum* cellulase displays far greater heat stability (Ng and Zeikus, 1981b) and activity (Johnson et al., 1982a) on a crude protein basis than the cellulase prepared from best *Trichoderma reesei* strains. Several different endoglucanase activities have been purified and biochemically characterized from *C. thermocellum* (Petre et al., 1981; Ng and Zeikus, 1981b), including two cloned into *E. coli* (Bequin et al., 1983; Cornet et al., 1983b).

   C. thermocellum ferments glucose or cellobiose and produces ethanol, lactic acid, acetic acid, CO_2, and high H_2 partial pressures during growth. The enzymatic activities responsible for the catabolism of glucose and cellobiose in *C. thermocellum* have been reviewed elsewhere (Zeikus and Ng, 1982). Glucose is transported and phosphorylated by an inducible glucose permease and hexokinase, whereas cellobiose is transported by an inducible permease and phosphorylated by cellobiose phosphorylase (Ng and Zeikus, 1982). Glycolysis results in the generation of NADH and pyruvate, which are transformed according to the enzymatic activities and routes indicated in Fig. 4.6. *C. thermocellum* contains only NAD-linked catabolic reductases, and the function of NADP-ferredoxin oxidoreductase is anabolic (Lamed and Zeikus, 1980). The flow of electrons from NADH to reduced ferredoxin explains why *C. thermocellum* forms high H_2 partial pressures during growth. It is of interest to note that the NAD-linked acetaldehyde reductase does not display detectable ethanol oxidation activity (i.e., is not reversible) and H_2 does not inhibit species growth.

   Metabolic associations comprised of *C. thermocellum* and noncellulolytic species including either *Thermoanaerobacter ethanolicus* (Weigel, 1982), *C. thermosaccharolyticum* (D. Wang et al., 1978), and *C. thermohydrosulfuricum* (Ng et al., 1981) have been reported to significantly enhance the ethanol yield of cellulose fermentations. Figure 4.7 illustrates a time course for Solka Floc fermentation by *C. thermocellum* in the absence or presence of *C. thermohydrosulfuricum*. The metabolic basis for increasing ethanol yields by greater than twofold when *C. thermocellum* is grown in co-culture with *C. thermohydrosulfuricum* is complex and not completely understood. The cellulase of *C. thermocellum* forms mainly cellodextrins, cellobiose, xylobiose, and some glucose and xylose during fermentation of cellulose and hemicellulose (Ng and Zeikus, 1981a). The kinetic features of glucose and cellobiose consumption by the two thermophiles suggests that *C. thermohydrosulfuricum* can compete with *C. thermocellum* for cellobiose and glucose and that only the former species consumes the pentoses (Ng et al., 1981; Ng and Zeikus, 1982). In co-culture, cellulose fermentations reducing sugars do not accumulate in the medium, and a significant portion of the cellulose hydrolysis products generated by *C. thermocellum*'s cellulase must be fermented by *C. thermohydrosulfuricum* to account for nearly equal numbers of both species in co-culture. The higher ethanol yield in co-culture appears as a consequence of the saccharide fermentation of *C. thermohydrosulfuricum*, which has a higher ethanol yield than *C. thermocellum*. The final ethanol concentration reported in monoculture or co-culture cellulose fermentations of *C. thermocellum* is <2%.

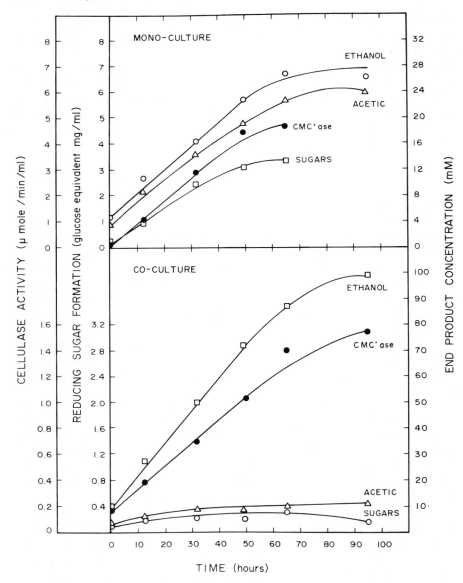

FIGURE 4.7   Time course of cellulose fermentation in monoculture of *C. thermocellum* and in co-culture with *C. thermohydrosulfuricum*. Experiments were performed in serum vials that contained 1% Solka Floc. Cultures were incubated at 60°C without shaking and pH control (from Ng et al., 1981).

Recently, the mechanism of ethanol tolerance in *C. thermocellum* was studied. Herrero and Gomez (1980) selected an ethanol-resistant strain that, unlike the parent strain, grew at ethanol concentrations greater than 20 g/l but less than 35 g/l. However, these low concentrations of ethanol inhibited the growth rate of the *C. thermocellum* mutant. The low ethanol tolerance of *C. thermocellum* was ascribed to the general effect of a solvent on increasing membrane fluidity (Herrero and Gomez, 1982) and to specific inhibition of some glycolytic enzyme(s) involved in transformation of hexose into glyceraldehyde-3-phosphate (Herrero, 1983).

C. C. thermohydrosulfuricum

Ethanol yields can often be as high as 1.6–1.8 mol ethanol per mole of hexose consumed during glucose fermentation by *C. thermohydrosulfuricum* strain 39E (Zeikus et al., 1980b; Ng et al., 1981; Lovitt et al., 1984). *C. thermohydrosulfuricum* forms the same end products as *C. thermocellum* but in drastically different ratios. Both species ferment saccharides by the Embden-Meyerhof-Parnas path, but hydrogen inhibits growth of *C. thermohydrosulfuricum* (Zeikus et al., 1980b, 1981). *C. thermohydrosulfuricum* contains a reversible NADP-linked alcohol–ketone/aldehyde oxidoreductase (Lamed and Zeikus, 1981). It also differs from *C. thermocellum* in containing the following saccharidases: cellobiase, xylose isomerase, glucoamylase, and pullulanase (Ng and Zeikus, 1982; Zeikus et al., 1981; Hyun, Ben-Bassat, and Zeikus, unpublished data).

Figure 4.8 illustrates a proposed saccharide catabolism pathway that can explain the significant differences between electron flow in *C. thermohydrosulfuricum* and *C. thermocellum* (see Fig. 4.6). Cell extracts of *C. thermohydrosulfuricum* contain the following enzyme activities: glyceraldehyde-3-phosphate dehydrogenase, pyruvate ferredoxin oxidoreductase, fructose 1-6 bis phosphate–activated lactate dehydrogenase, ferredoxin NAD(P) oxidoreductases, phosphotransacetylase, acetate kinase, and both NADP- and NAD-linked acetaldehyde dehydrogenase and ethanol dehydrogenase (Lovitt and Zeikus, unpublished data). Electron flow is notably different in this organism because reduced ferredoxin is the donor for both NADPH and NADH, which are both oxidized when coupled to ethanol production. These enzymatic features also explain why the organism produces low levels of hydrogen.

End products dramatically inhibit metabolism and growth of *C. thermohydrosulfuricum* because high hydrogen partial pressure or ethanol concentration readily reverses electron flow. This results in overreduction of internal electron carriers and a cessation of carbon flow. For example, high hydrogen partial pressure inhibits glucose metabolism, as a consequence of NAD depletion caused by the reverse function of hydrogenase increasing internal pools of reduced ferredoxin, which in turn enhance NADH levels because of ferredoxin-NAD reductase activity (Ben-Bassat et al., 1981).

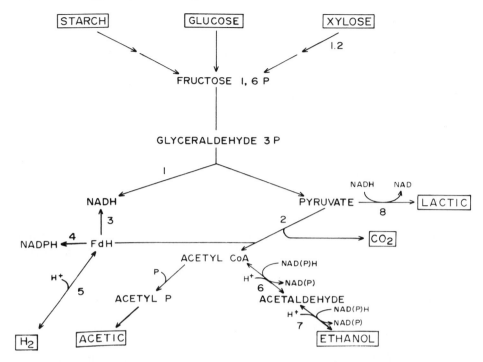

FIGURE 4.8 Proposed carbon and electron flow pathways for saccharide catabolism of *C. thermohydrosulfuricum* strain 39E. Key enzymes include: 1, glyceraldehyde 3-P dehydrogenase; 2, pyruvate dehydrogenase; 3, ferredoxin–NAD oxidoreductase; 4, ferredoxin–NADP oxidoreductase; 5, hydrogenase; 6, NAD(P)-linked acetyl-CoA reductases; 7, NAD(P)-linked alcohol dehydrogenases. The in vivo direction of reactions 3 and 4 explains why this species produces so much ethanol and so little H_2 and why H_2 at high concentration inhibits growth by flowing backwards and overreducing NADP and NAD pools.

The mechanism of ethanol tolerance for *C. thermohydrosulfuricum* is also quite different from that observed for *C. thermocellum* (Lovitt et al., 1984). These authors compared solvent tolerance of the wild-type strain 39E, which would not grow at 60°C in 2% ethanol, with an ethanol-adapted mutant strain 39EA that grows at 56 g/l ethanol. Notably, both strains displayed high solvent tolerance and grew at 60°C in >80 g/l methanol, a result indicating unique membrane stability. Table 4.4 shows that the basis for low ethanol tolerance in the wild-type strain was due to reversed electron flow caused by alcohol dehydrogenase at elevated ethanol concentrations. Hence high ethanol tolerance is related to alterations in the electron flow pathway of the mutant. The mutant produces ethanol at greater than 40 g/l ethanol; unfortunately, the yield (i.e., mol ethanol/mol saccharide fermented) is lower than that of the

TABLE 4.4 **Influence of Ethanol Concentration on Glucose Fermentation by the Parent and Mutant Strains of** *Clostridium thermohydrosulfuricum*[a]

Ethanol (% w/v)	Growth (O.D._660_)	pH	Glucose Consumed (μmol) tube	End Product (mM/100 mM glucose)				Carbon Recovery (%)	Ethanol Yield (mol ethanol/ mol glucose)
				Lactate	Acetate	Ethanol	CO_2		
Parent Strain 39E									
0	1.15	6.3	220	23	8	155	163	93	1.5
0.4	1.1	6.4	212	19	8	147	153	86	1.5
0.8	0.75	6.5	63	14	22	112	133	91	1.1
1.6	0.3	6.6	27	30	37	63	100	93	0.6
Mutant Strain 39EA									
0	0.8	5.0	164	80	10	92	102	92	0.9
0.8	0.7	5.1	155	85	9	87	96	94	0.9
2.0	0.8	5.0	211	66	7	103	110	90	1.0
4.0	0.6	5.0	157	82	8	74	83	87	0.7

[a] Experimental conditions: Anaerobic pressure tubes contained 10 ml of TYE medium with 0.4% glucose, 2 μCi·^{14}C-glucose, and the amount of solvent indicated. Tubes were assayed after growth was completed at 60°C. Data obtained from Lovitt et al. (1984).

wild-type in the absence of high ethanol concentration. Thus further genetic modifications are needed to alter the carbon and electron flow path of the tolerant mutant to achieve a high ethanol yield.

The NADP-linked alcohol/aldehyde ketone oxidoreductase present in *C. thermohydrosulfuricum* has a very broad substrate range and is detected in other non-spore-forming thermoanaerobes but not in *C. thermocellum* (Lamed and Zeikus, 1981). This enzyme is active and stable at either high temperature ($> 80°C$) and low solvent concentration or high solvent concentration (50% propanol) and moderate (40°C) temperature (Lamed and Zeikus, 1981; Lamed et al., 1981). Practical utility of the enzyme was shown by immobilized cell or enzyme production of chiral alcohols and NADPH, its use as an enzyme electrode for detection of alcohols or carbonyls and as a nucleotide cofactor regenerating system (Lamed et al., 1981).

D. *C. thermoaceticum*

The catabolic pathway of this species has received the greatest attention because of its novelty in producing 3 moles of acetate per mole of glucose. This is achieved by a novel carbon and electron flow pathway involving the synthesis of one acetate molecule by a series of one-carbon transformation reactions (for reviews of this subject, see Wood et al. (1982), Ljungdahl and Wood (1982)).

Figure 4.9 illustrates the glucose–homoacetate fermentation of *C. thermoaceticum*. Glucose is transformed into 2 pyruvate and 2 NADH by normal glycolysis. One pyruvate (on the left side of Fig. 4.9) is transformed into acetate, CO_2, and reduced ferredoxin. This CO_2 is then reduced to formate via an NADPH-linked CO_2 reductase (Thauer, 1972). Formate is then activated and reduced on folate carriers to CH_3–THF. The electron donors involved in these reactions include NADH, and cytochromes and menaquinones may also be involved (Gottwald et al., 1975). Methyl THF is then carboxylated by the other pyruvate (on the right side of Fig. 4.9) to yield 2 acetates via a transcarboxylation (Schulman et al., 1973).

The transcarboxylase reaction mechanism shown is now questioned because cell extracts contain high levels of carbon monoxide dehydrogenase activity (Diekert and Thauer, 1978) and the physiological role of the enzyme in synthesis of acetate from glucose (Drake et al., 1981). CO dehydrogenase of *C. thermoaceticum* is a nickel-containing enzyme (Diekert and Thauer, 1980; Drake et al., 1980) that functions in the corrinoid-dependent carbonylation of a methyl group into CH_3CO-CoA (Hu et al., 1982). A lot of work remains to be done to explain how a carbonyl group is formed during glucose fermentation and to understand what enzymes are responsible for electron flow during the entire process of acetate synthesis.

Little is known about the physiological mechanisms of end product inhibition in *C. thermoaceticum*, although wild-type strains do not grow below pH 6.5. Both high-proton and acetate-tolerant mutants have been obtained from

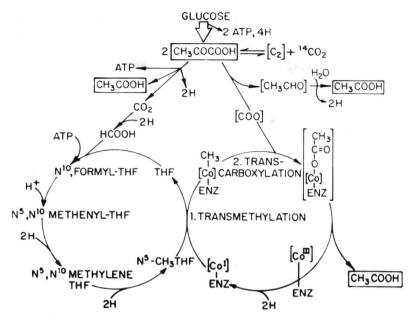

FIGURE 4.9 Homoacetogenic pathway proposed for *C. thermoaceticum* grown on glucose. From Ljungdahl and Wood (1982). Abbreviations: THF, tetrahydrofolate; Co-ENZ, corrinoid enzyme.

the parent strain (G. Y. Wang et al., 1978; Schwartz and Keller, 1982). Schwartz and Keller (1982) have reported on the influence of pH on acetic acid production of mutants that can grow between 7 and 4.5. The highest produced acid concentrations were about 15 and 20 g/l at pH 6 and 7, respectively. It was discovered that the redox potential influenced growth such that minimal values of -360, -300, and -240 mv were needed to initiate growth at pH 7, 6, and 5. *Clostridium thermoaceticum* was also shown to be more sensitive to free acetic acid than to either acetate or pH (Schwartz and Keller, 1982). It would appear that these mutants are just able to grow at low pH values but are similar to the parent strain because free acid production stops near the pK_a of acetic acid. Perhaps this is due to a chemical equilibrium between the external medium and the internal cellular acetic acid concentrations.

E. *B. methylotrophicum*

The ability of anaerobes to grow on single-carbon substrates and synthesize larger, more complex molecules is novel and has some exciting potential industrial applications (Datta, 1981; Zeikus, 1983c). The metabolism of one-carbon compounds by chemotrophic anaerobes has been reviewed (Zeikus, 1983a). *B. methylotrophicum* is an interesting model anaerobe because it can perform nearly homo-butyrate or homo-acetate fermentations from mixtures of single-carbon substrates or from saccharide fermentations (Zeikus et al., 1980a; Lynd and Zeikus, 1983). Growth on high CO partial pressures (>2.0 atm) requires adaptation and selection of a CO-tolerant strain (Lynd et al., 1982). This *B. methylotrophicum* strain displayed a growth rate of 0.05 h^{-1} and a K_s for CO of 28–56 μm.

Table 4.5 compares the stoichiometries of single-carbon transformations achieved at the end of *B. methylotrophicum* fermentations. The concentration and fraction of acetate and butyrate formed depend on the substrate composition. Homo-acetate fermentations are nearly achieved with H_2–CO_2, CO, or CH_3OH–CO as substrates, whereas homo-butyrate fermentations are nearly achieved with CH_3OH–CO_2 as the substrate. Fermentation of CH_3OH–HCOOH yields nearly a 60/40 blend of butyrate and acetate. *B. methylotrophicum* also ferments formate alone to acetate and CO_2 as final end products (Kerby and Zeikus, unpublished results).

The flow of single carbons (see Fig. 4.10) during catabolism of one-carbon compounds by acetogenic bacteria was recently deciphered by ^{13}C nuclear magnetic resonance analysis of ^{13}C-substrate transformations (Kerby et al., 1983). This model is useful in explaining the fermentation stoichiometries obtained on single-carbon mixtures and understanding the unique metabolism of acetogenic bacteria, including *B. methylotrophicum*. The key intermediary metabolite of this pathway is acetyl-CoA, which is formed by the carbonylation of a methyl intermediate. CO dehydrogenase catalyzes this reaction in acetogens (Hu et al., 1982), and it is present in high activity in *B. methylotrophicum* cell extracts (Lynd et al., 1982). Acetyl-CoA is the direct precursor to either catabolic end products (i.e., acetate or butyrate) or cells, and this explains why these organisms are such efficient synthesizers, since they gain energy by formation of a common two-carbon intermediary metabolite. During acetyl-CoA synthesis from one-carbon substrates, carbon flows via two different routes (i.e., to a methyl (left side of Fig. 4.10) and a carbonyl intermediate (right side of Fig. 4.10), which are linked to each other by a carboxyl in-

TABLE 4.5   Comparison of Catabolic Single-Carbon Transformations of *Butyribacterium methylotrophicum*[a]

Substrates (μmol consumed)	Products (μmol produced)
1. 3260 H_2 + 1620 CO_2	693 acetate + 18 butyrate
2. 4431 CO	816 acetate + 2403 CO_2
3. 2742 CH_3OH + 543 CO_2	735 butyrate + 196 acetate
4. 4200 CH_3OH + 2056 HCOOH	854 butyrate + 643 acetate + 16 CO_2
5. 1573 CH_3OH + 1417 CO	153 butyrate + 747 acetate + 164 CO_2

[a] Values obtained after fermentation of single-carbon substrates were completed. Note that the balance of substrate carbon and electrons was accounted for in cell synthesis, which is quite significant. For example, in transformation 3, 25% of the methanol was transformed to cells. Data from Lynd and Zeikus (1983), Lynd et al. (1982), and Kerby et al. (1983).

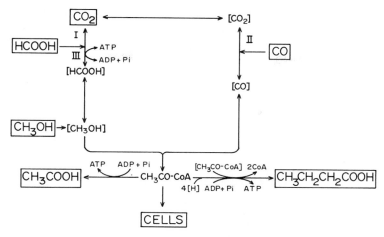

FIGURE 4.10   Single-carbon catabolism flow model proposed for acetogenic bacteria that synthesize acetate or butyrate from single-carbon compounds. Acetyl-CoA is the direct precursor for acetic or butyric acids and
cells. [CH$_3$] and [CO] represent the immediate methyl and carbonyl precursors for acetyl-CoA synthesis and are also precursors for the synthesis of
CO$_2$. This scheme predicts that two distinct formyl-level intermediates,
[HCOOH] and [CO], are linked by formate, CO$_2$, and a carboxyl intermediate [CO$_2$]. The Roman numerals indicate the following enzymatic
activities: I, formate dehydrogenase; II, CO dehydrogenase; and III,
formyl-THF synthetase (from Kerby et al., 1983).

termediate). When acetogens grow on methanol and CO, they conserve energy
by a non-redox reaction associated with the direct condensation of the two
substrates to yield acetate, and this accounts for the observed 1:1 substrate
consumption stoichiometry. A stoichiometry of ten CH$_3$OH and two CO$_2$
transformed into three butyrates can also be explained by the model. Six
methanols are transformed to the methyl position in six acetyl-CoA, while
four methanols are oxidized to CO$_2$; this results in six carboxyl intermediates
being reduced to the carbonyl position of six acetyl-CoA. The six acetyl-CoA
are then condensed and reduced to three butyrates, and this achieves the required redox balance.

The enzymes that perform these proposed one-carbon transformations
have not been well characterized in *B. methylotrophicum*. This organism has
hydrogenase, formate dehydrogenase, and CO dehydrogenase activity when
grown on one-carbon substrates (Lynd et al., 1982); but at this point it is
assumed that one-carbon transformations by acetogens utilize the same enzymes (i.e., folate- and corrinoid-dependent) present in *C. thermoaceticum*
grown on glucose (Wood et al., 1982). The mechanism of energy conservation
during growth on one-carbon compounds is not known except for SLP via
acetate kinase, butyrate kinase, and possible formyl tetrahydrofolate lyase

during methanol oxidation. Growth yield data (Lynd et al., 1982; Lynd and Zeikus, 1983) on certain substrates (e.g., CO, H_2–CO_2) suggest ETP, and TCP is also possible on all growth substrates.

REFERENCES

Ait, N., Creuzet, N., and Forget, P. (1979) *J. Gen. Microbiol. 113*, 399–402.

Allcock, E. R., and Woods, D. R. (1981) *Appl. Environ. Microbiol. 41*, 539–541.

Allcock, E. R., Reid, S. J., Jones, D. T., and Woods, D. R. (1981) *Appl. Environ. Microbiol. 42*, 929–935.

Andersch, W., Bahl, H., and Gottschalk, G. (1982) *Biotechnol. Lett. 4*, 29–32.

Andreesen, J. R., Gottschalk, G., and Schlegel, H. G. (1970) *Arch. Mikrob. 72*, 154–174.

Avgerinos, G. C., and Wang, D. I. C. (1980) in *Annual Reports on Fermentation Processes* (Tsao, G., ed.), Vol. IV, pp. 165–191, Academic Press, New York.

Bahl, H., Andersch, W., and Gottschalk, G. (1982a) *Eur. J. Appl. Microbiol. Biotechnol. 15*, 201–205.

Bahl, H., Andersch, W., Braun, K., and Gottschalk, G. (1982b) *Eur. J. Appl. Microbiol. Biotechnol. 14*, 17–20.

Balch, W. E., and Wolfe, R. S. (1976) *Appl. Environ. Microbiol. 32*, 781–791.

Barber, J. M., Robb, F. T., Webster, J. R., and Woods, D. R. (1979) *Appl. Environ. Microbiol. 37*, 433–437.

Barker, H. A., and Hass, G. (1944) *J. Bacteriol. 47*, 301–305.

Bayer, H. A., Kening, R., and Lamed, R. (1983) *J. Bacteriol. 156*, 818–827.

Beesch, S. C. (1982) *Appl. Microbiol. 1*, 85–95.

Ben-Bassat, A., Lamed, R., and Zeikus, J. G. (1981) *J. Bacteriol. 146*, 192–199.

Bequin, P., Cornet, P., and Millet, J. (1983) *Biochimie 65*, 495.

Braun, M., Mayer, F., and Gottschalk, G. (1981) *Arch. Microbiol. 128*, 288–293.

Chung, K.-T. (1976) *Appl. Environ. Microbiol. 31*, 342–348.

Cornet, P., Tronik, D., Millet, J., and Aubert, J.-P. (1983a) *FEMS Microbiol. Lett. 16*, 137–141.

Cornet, P., Millet, J., Bequin, P., and Aubert, J.-P. (1983b) *Bio/Technol. 1*, 589–594.

Datta, R. (1981) *Biotechnol. Bioeng. Symp. Ser. 11*, 521–532.

Diekert, G. B., and Thauer, R. K. (1978) *J. Bacteriol. 136*, 597–606.

Diekert, G., and Thauer, R. K. (1980) *FEMS Microbio. Lett. 7*, 187–189.

Donnelly, L. S., and Busta, F. F. (1980) *Appl. Environ. Microbiol. 40*, 721–725.

Drake, H. L., Hu, S.-I., and Wood, H. G. (1980) *J. Biol. Chem. 255*, 7174–7180.

Drake, H. L., Hu, S.-I., and Wood, H. G. (1981) *J. Biol. Chem. 256*, 11137–11144.

Duncan, C. L., and Strong, D. H. (1968) *Appl. Microbiol. 16*, 82–89.

Egerer, P., and Simon, H. (1982) *Biotechnol. Lett. 4*, 501–506.

Emeruwa, A. C., and Hawirko, R. Z. (1975) *Arch. Microbiol. 105*, 67–71.

Ensley, B., McHugh, J. J., and Barton, L. L. (1975) *J. Gen. Appl. Microbiol. 21*, 51–59.

Fitz-James, P. C. (1962) *J. Bacteriol. 84*, 104–114.

Fontaine, F. E., Peterson, W. H., McCoy, E., Johnson, M. J., and Ritter, G. T. (1942) *J. Bacteriol. 43*, 701–705.

Garcia-Martinez, D. V., Shinmyo, A., Madia, A., and Demain, A. L. (1980) *Eur. J. Appl. Microbiol. 9*, 189–197.

Genthner, B. R. S., and Bryant, M. P. (1982) *Appl. Environ. Microbiol. 43*, 70–74.

Genthner, B. R. S., Davis, C. L., and Bryant, M. P. (1981) *Appl. Environ. Microbiol. 42*, 12–19.

George, H. A., Johnson, J. L., Moore, W. E. C., Holdeman, L. V., and Chen, J. S. (1983) *Appl. Environ. Microbiol. 45*, 1160–1163.

George, H. A., and Chen, J.-S. (1983) *Appl. Environ. Microbiol. 46*, 321–327.

Gerhardt, P. (1981) *Manual of Methods for General Bacteriology,* American Society for Microbiology, Washington, D.C.

Gibbs, D. F. (1983) *Trends Biotechnol. 1*, 12–15.

Gottschal, J. C., and Morris, J. G. (1981a) *FEMS Microbiol. Lett. 12*, 385–389.

Gottschal, J. C., and Morris, J. G. (1981b) *Biotechnol. Lett. 3*, 525–530.

Gottschalk, G., Andreesen, J. R., and Hippe, H. (1981) in *The Prokaryotes* (Starr, M. P., ed.), pp. 1768–1803, Springer, New York.

Gottwald, M., Andreesen, J. R., LeGall, J., and Ljungdahl, L. G. (1975) *J. Bacteriol. 122*, 325–328.

Gould, G. W., and Dring, G. J. (1975) *Nature 258*, 402–405.

Guinand, M., Ghuysen, J.-M., Schleifer, K. H., and Kandler, O. (1969) *Biochem. 8*, 200–207.

Haggstrom, L., and Molin, N. (1980) *Biotechnol. Lett. 2*, 241–246.

Herrero, A. A. (1983) *Trends Biotechnol. 1*, 49–53.

Herrero, A. A., and Gomez, R. F. (1980) *Appl. Environ. Microbiol. 40*, 571–577.

Herrero, A. A., and Gomez, R. F. (1982) *Biochem. Biophys. Acta 693*, 195–204.

Hoeniger, J. F. M., Stuart, P. F., and Holt, S. C. (1968) *J. Bacteriol. 96*, 1818–1834.

Hollriegl, V., Lamm, L., Rowod, J., Horig, J., and Renz, P. (1982) *Arch. Microbiol. 132*, 155–158.

Hsu, E. G., and Ordal, Z. J. (1970) *J. Bacteriol. 102*, 369–376.

Hu, S.-I., Drake, H. L., and Wood, H. G. (1982) *J. Bacteriol. 149*, 440–448.

Hyun, H. H., Zeikus, J. G., Longin, R., Millet, J., and Ryter, A. J. (1983) *J. Bacteriol. 156*, 1332–1337.

Ishimoto, M., Umeyama, M., and Chiba, S. (1974) *Z. Allg. Mikrob. 14*, 115–121.

Johnson, E. A., and Demain, A. L. (1984) *Arch. Microbiol. 137*, 135–138.

Johnson, E. A., Madia, A., and Demain, A. L. (1981) *Appl. Environ. Microbiol. 41*, 1060–1062.

Johnson, E. A., Reese, E. T., and Demain, A. L. (1982a) *J. Appl. Biochem. 4*, 64–71.

Johnson, E. A., Sakajoh, M., Halliwell, G., Madia, A., and Demain, A. L. (1982b) *Appl. Environ. Microbiol. 43*, 1125–1132.

Jones, D. T., Webster, J. R., and Woods, D. R. (1980) *J. Gen. Microbiol. 116*, 195–200.

Jones, D. T., Van den Westhuizen, A., Long, S., Allcock, E. R., Reid, S. J., and Woods, D. R. (1982) *Appl. Environ. Microbiol. 43*, 1434–1439.

Karube, I., Urano, N., Matsunaga, T., and Suzuki, S. (1982) *Eur. J. Appl. Microbiol. Biotechnol. 16*, 5–9.

Kerby, R., and Zeikus, J. G. (1983) *Curr. Microbiol. 8*, 27–30.

Kerby, R., Niemczura, W., and Zeikus, J. G. (1983) *J. Bacteriol. 155*, 1208–1218.

Konings, W. N., and Veldkamp, H. (1983) in *Microbes in Their Natural Environments* (Slater, J. H., Whittenbury, R., and Wimpenny, J. W. T., eds.) pp. 153–198, Society for General Microbiology Symposium, Cambridge, England.

Krouwel, P. G., Groot, W. J., Koosen, N. W. F., and Van der Laan, W. F. M. (1983) *Enzyme Microbiol. Technol. 5*, 46–54.

Lamed, R., and Zeikus, J. G. (1980) *J. Bacteriol. 144*, 569–578.

Lamed, R. J., and Zeikus, J. G. (1981) *Biochem. J. 195*, 183–190.

Lamed, R. J., Keinan, E., and Zeikus, J. G. (1981) *Enzyme Microbiol. Technol. 3*, 144–148.

Lamed, R., Setter, E., and Bayer, E. A. (1983) *J. Bacteriol. 156*, 828–836.

Lewis, R. P., and Sutter, V. L. (1981) in *The Prokaryotes* (Starr, M. P., ed.), pp. 1903–1911, Springer, New York.

Lin, Y.-L., and Blaschek, H. P. (1983) *Appl. Environ. Microbiol. 45*, 966–974.

Ljungdahl, L. G. (1983) in *Organic Chemicals from Biomass* (Wise, D. L., ed.) pp. 219–248, Benjamin Cummings, Menlo Park, Calif.

Ljungdahl, L. G., and Wood, H. G. (1982) in *B_{12}*, Vol. II (Dolphin, D., ed.), John Wiley and Sons, New York.

Long, S., Jones, D. T., and Wood, D. R. (1983) *Appl. Environ. Microbiol. 45*, 1389–1393.

Lovitt, R. W., Longin, R., and Zeikus, J. G. (1984) *J. Appl. Environ. Microbiol. 48*, 171–177.

Lund, B. M. (1972) *J. Appl. Bacteriol. 35*, 609–614.

Lynd, L., and Zeikus, J. G. (1983) *J. Bacteriol. 153*, 1415–1423.

Lynd, L., Kerby, R., and Zeikus, J. G. (1982) *J. Bacteriol. 149*, 255–263.

Maddox, I. S. (1980) *Biotechnol. Lett. 1*, 493–498.

Maddox, I. S. (1982) *Biotechnol. Lett. 4*, 23–28.

Matteuzzi, D., Hollaus, F., and Biaruti, B. (1978) *Int. J. Systematic Bacteriol. 28*, 528–531.

McCoy, E., Fred, E. B., Peterson, W. H., and Hastings, E. G. (1930) *J. Infect. Diseases 46*, 118–137.

Mendez, B., and Gomez, R. F. (1982) *Appl. Environ. Microbiol. 43*, 495–496.

Mes-Hartree, M., and Saddler, J. N. (1982) *Biotechnol. Lett. 4*, 247–252.

Moench, T. T., and Zeikus, J. G. (1983a) *J. Microbiol. Meth. 1*, 199–202.

Moench, T. T., and Zeikus, J. G. (1983b) *Curr. Microbiol. 9*, 151–154.

Monot, F., Martin, J.-R., Petitdemange, H., and Gay, R. (1982) *Appl. Environ. Microbiol. 44*, 1318–1324.

Moreira, A. R. (1983) in *Organic Chemicals from Biomass* (Wise, D. L., ed.), pp. 385–406, Benjamin Cummings, Menlo Park, Calif.

Morris, J. G. (1975) *Adv. Microbiol. Physiol. 12*, 169–245.

Moses, V., and Springham, D. G. (1982) in *Bacteria and the Enhancement of Oil Recovery*, Applied Science Publishing, Essex, England.

Ng, T. K., and Zeikus, J. G. (1981a) *Appl. Environ. Microbiol. 42*, 231–240.

Ng, T. K., and Zeikus, J. G. (1981b) *Biochem. J. 199*, 341–350.

Ng, T. K., and Zeikus, J. G. (1982) *J. Bacteriol. 150*, 1391–1399.

Ng, T. K., Weimer, P. J., and Zeikus, J. G. (1977) *Arch. Microbiol. 114*, 1–7.

Ng, T. K., Ben-Bassat, A., and Zeikus, J. G. (1981) *Appl. Environ. Microbiol. 41*, 1337–1343.

Ng, T. K., Busche, R. M., McDonald, C. C., and Hardy, R. W. F. (1983) *Science 219*, 733–740.

Nishio, N., Biebl, H., and Meiners, M. (1983) *J. Ferment. Technol. 61*, 101–104.

Norris, J. R., Berkeley, R. C. W., Logan, N. A., and O'Donnell, A. G. (1981) in *The Prokaryotes* (Starr, M. P., ed.), pp. 1711–1742, Springer, New York.

O'Brien, R. W., and Morris, J. G. (1971) *J. Gen. Microbiol. 68*, 307–318.

Oren, A. (1983) *Arch. Microbiol. 136*, 42–48.

Palsson, B. O., Fthi-Afshar, S., Rudd, D. F., and Lighfoot, E. N. (1981) *Science 213*, 513–517.

Perlman, D., and Seman, J. B. (1963) *Biotechnol. Bioeng. 5*, 21–25.

Petitdemange, H., Cherrier, C., Raval, G., and Gay, R. (1976) *Biochim. Biophys. Acta 421*, 334–347.

Petitdemange, H., Cherrier, C., Bengone, J. M., and Bray, R. (1977) *Can. J. Microbiol. 23*, 152–160.

Petitdemange, H., Marczak, R., Blusson, H., and Gay, R. (1979) *Biochem. Biophys. Res. Comm. 91*, 1258–1265.

Petre, J., Longin, R., and Millet, J. (1981) *Biochimie 63*, 629–639.

Pfennig, N., Widdel, F., and Truper, H. G. (1981) in *The Prokaryotes* (Starr, M. P., ed.), pp. 925–940, Springer, New York.

Reid, S. J., Allcock, E. R., Jones, D. T., and Wood, D. R. (1983) *Appl. Environ. Microbiol. 45*, 305–307.

Rombouts, F. M., and Pilnik, W. (1980) in *Economic Microbiology Microbial Enzymes and Bioconversions* (Rose, A. H., ed.), Vol. 5, pp. 227–282, Academic Press, New York.

Saddler, J. N., and Chan, M. K.-H. (1982) *Eur. J. Appl. Microbiol. and Biotechnol. 16*, 99–104.

Schink, B., and Zeikus, J. G. (1982) *J. Gen. Microbiol. 128*, 393–404.

Schink, B., and Zeikus, J. G. (1983a) *FEMS Microbiol. Lett. 17*, 295–298.

Schink, B., and Zeikus, J. G. (1983b) *J. Gen. Microbiol. 129*, 1149–1158.

Schink, B., Ward, J. C., and Zeikus, J. G. (1981) *Appl. Environ. Microbiol. 42*, 526–532.

Schulman, M., Ghambeer, R. K., Ljungdahl, L. G., and Wood, H. G. (1973) *J. Biol. Chem. 248*, 6255–6261.

Schwartz, R. D., and Keller, F. A., Jr. (1982) *Appl. Environ. Microbiol. 43*, 1385–1392.

Sebald, M., and Costilow, R. N. (1975) *Appl. Microbiol. 29*, 1–16.

Sinskey, A. J. (1983) in *Organic Chemicals from Biomass* (Wise, D. L., ed.), pp. 1–67, Benjamin Cummings, Menlo Park, Calif.

Sinskey, A. J., Akedo, M., and Cooney, C. L. (1981) in *Trends in the Biology of Fermentations for Fuels and Chemicals* (Hollander, A., ed.), pp. 473–492, Plenum Press, New York.

Sleytr, U. B., and Glauert, A. M. (1976) *J. Bacteriol. 126*, 869–882.

Smith, L. D. S. (1973) in *Handbook of Microbiology* (Laskin, A. I., and Lechevalier, H. A., eds.), pp. 89–96, CRC Press, Cleveland, Ohio.

Smith, L. D. S., and Dowell, V. R. (1974) in *Manual of Clinical Microbiology* (Lennette, E. H., Spaulding, E. H., and Truant, J. P., eds.), pp. 376–380, American Society for Microbiology, Washington, D.C.

Snedecor, B. R., and Gomez, R. F. (1983) in *Organic Chemicals from Biomass* (Wise, D. L., ed.), pp. 93–108, Benjamin Cummings, Menlo Park, Calif.

Spivey, M. J. (1978) *Proc. Biochem. 13*, 2–4.

Stadtman, E. R., Stadtman, T. C., Pastan, I., and Smith, L. D. S. (1972) *J. Bacteriol. 110*, 758–760.

Tanner, R. S., Stackebrandt, E., Fox, G. E., and Woese, C. R. (1981) *Curr. Microbiol. 5*, 35–38.

Tanner, R. S., Stackebrandt, E., Fox, G. E., Gupta, R., Magrum, L. J., and Woese, C. R. (1982) *Curr. Microbiol. 7*, 127–132.

Thauer, R. K. (1972) *FEBS Lett. 27*, 111–115.

Thauer, R. K., Jungermann, K., and Decker, K. (1977) *Bacteriol. Rev. 41*, 100–180.

Tsao, G. T., Ladisch, M. R., Voloch, M., and Bienkowski, P. (1982) *Proc. Biochem. 17*, 34–38.

Tsuru, D. (1973) in *Handbook of Microbiology* (Laskin, A. I., and Lechevalier, H. A., eds.), pp. 593–624, CRC Press, Cleveland, Ohio.

Van den Westhuizen, A., Jones, D. T., and Woods, D. R. (1982) *Appl. Environ. Microbiol. 44*, 1277–1281.

Waites, W. M., and Wyatt, L. R. (1971) *J. Gen. Microbiol. 67*, 215–222.

Walton, M. T., and Martin, J. L. (1979) in *Microbiol Technology*, Vol. I, 2nd ed. (H. J. Peppler and D. Perlman, eds.), pp. 187–209, Academic Press, New York.

Wang, D. I. C., Cooney, C. L., Wang, S.-D., Gordon, J., and Wang, G. Y. (1978) in *Proceedings of the Second Annual Symposium on Fuels from Biomass*, pp. 537–570, Department of Energy, Washington, D.C.

Wang, G. Y., Wang, D. I. C., and Fleischaker, H. (1978) *AICHE Symp. Ser. 182,74*, 105–110.

Weimer, P. J., and Zeikus, J. G. (1977) *Appl. Environ. Microbiol. 33*, 289–297.

Wiegel, J. (1982) *Experientia 38*, 151–156.

Wiegel, J., Ljungdahl, L. G., and Rawson, J. R. (1979) *J. Bacteriol. 139*, 800–810.

Wiegel, J., Braun, M., and Gottschalk, G. (1981) *Curr. Microbiol. 5*, 255–260.

Wise, D. L. (ed.) (1983) *Organic Chemicals from Biomass*, 465 pp., Benjamin Cummings, Menlo Park, Calif.

Wood, H. G., Brown, R. W., and Werkman, C. H. (1945) *Arch. Biochem. 6*, 243–260.

Wood, H. G., Drake, H. L., and Hu, S.-I. (1982) in *Proceedings of the Biochemistry Symposium* (E. E. Snell, ed.), pp. 29–56, Annual Reviews, Palo Alto, California.

Zehnder, A. J. B., and Wuhrman, K. (1976) *Science 194*, 1165–1166.

Zeikus, J. G. (1979) *Enzyme Microb. Technol. 1*, 243–259.

Zeikus, J. G. (1980a) *Ann. Rev. Microbiol. 34*, 423–456.

Zeikus, J. G. (1980b) in *Anaerobic Digestion* (Stafford, D. A., Wheatley, B. I., and Hughes, D. E., eds.), pp. 61–87, Applied Science Publishers, London.

Zeikus, J. G. (1983a) in *Adv. Microb. Phys.*, Vol. XXIV, pp. 215–299.

Zeikus, J. G. (1983b) in *Microbes in Their Natural Environments* (Slater, J. H., Whittenbury, R., and Wimpenny, J. W. T., eds.), Society for General Microbiology Symposium 34, pp. 423–462, Cambridge University Press, Cambridge, England.

Zeikus, J. G. (1983c) in *Organic Chemicals from Biomass* (Wise, D. L., ed.), pp. 359–384, Benjamin Cummings, Menlo Park, Calif.

Zeikus, J. G., and Ng, T. K. (1982) in *Annual Reports on Fermentation Processes* (Tsao, G., ed.), Vol. V, Chapter 7, pp. 253–289, Academic Press, New York.

Zeikus, J. G., Lynd, L. H., Thompson, T. E., Krzycki, J. A., Weimer, P. J., and Hegge, P. W. (1980a) *Curr. Microbiol. 3*, 381–386.

Zeikus, J. G., Ben-Bassat, A., and Hegge, P. W. (1980b) *J. Bacteriol. 143*, 432–440.

Zeikus, J. G., Ben-Bassat, B., Ng, T. K., and Lamed, R. J. (1981) in *Trends in the Biology of Fermentations for Fuels and Chemicals* (Hollander, A., ed.), pp. 442–461, Plenum Press, New York.

5

Glutamic Acid Bacteria

Shukuo Kinoshita

I. INTRODUCTION

Up until the 1950's, no appropriate commercial processes for production of natural (L) amino acids existed except by isolation from natural proteins. Thus strenuous efforts were made to establish chemical synthetic methods, but these were useful only for experimental purposes or, at best, for small-scale production.

Our discovery of a potent glutamic acid–producing microorganism, *Micrococcus glutamicus* (later named *Corynebacterium glutamicum*) (Kinoshita et al., 1957a), provided a novel approach to establish commercial methods for producing natural amino acids. Within a decade a flood of papers appeared reporting similar findings in other microorganisms. Although most of these microorganisms seemed quite similar to *C. glutamicum*, each strain was named differently, which caused confusion.

Comparative taxonomic studies of these strains were later made by several investigators, and the various coryneform bacteria proved to be similar to the microorganism originally described. Thus this chapter deals with the group as "coryneform glutamic acid–producing bacteria" or, more simply, "glutamic acid bacteria." On the other hand, additional glutamic acid bacteria were found later that could be distinguished from the above group because of their ability to utilize hydrocarbons as a sole carbon source for growth.

The basic biology of both types of glutamic acid bacteria is stressed in this chapter; discussion of the fermentation processes and their related mechanisms will be relegated to minimum description. The reader may find useful other reviews dealing with the mechanisms of amino acid production either by direct fermentation processes (Kinoshita, 1959; Kinoshita and Nakayama, 1978) or by bioconversions (H. Yamada and Kumagai, 1978).

II. CLASSIFICATION OF GLUTAMIC ACID BACTERIA

There has been considerable confusion concerning the classification of the glutamic acid bacteria because of the multiplicity of names given to the individual strains: *Micrococcus glutamicus* (later named *Corynebacterium glutamicum*), *Brevibacterium flavum*, *Brevibacterium lactofermentum*, *Brevibacterium divaricatum*, *Brevibacterium ammoniagenes*, *Brevibacterium thiogenitalis*, *Microbacterium ammoniaphilum*, *Corynebacterium lilium*, *Corynebacterium callunae*, *Corynebacterium herculis*, and so on (Abe and Takayama, 1972). All are potent glutamic acid producers yielding glutamic acid at 30–50% of glucose consumed. *M. glutamicus* has the following characteristics: gram-positive, nonsporulating, nonmotile, pleomorphic short rods, generally yellowish colonies, having a G + C content of 53–55% in DNA, requiring biotin and growing at 30°C. Most strains can utilize acetic acid, ethanol, glucose, or sucrose for glutamic acid production.

To clarify the differences and similarities among the various strains of glutamic acid producers, several extensive studies have been made. Abe et al. (1967) examined 212 strains of glutamic acid producers. They grouped these strains into 12 types, 60% of which belonged to one type, which possesses urease, reduces nitrate, and produces acid from glucose. It is noteworthy that almost all of the representative strains employed for commercial production of glutamic acid were of this type. Details of the classification of the glutamic acid bacteria may be found in the work of Abe and Takayama (1972).

Chemotaxonomy has also been an important tool for the classification and identification of these microorganisms. *M. glutamicus*, having *meso*-diaminopimelic acid in its cell wall, was shifted into *C. glutamicum* (Kinoshita et al., 1960a). Keddie and co-workers (Keddie and Cure, 1978; Keddie and Bousfield, 1980) classified coryneform bacteria into 13 groups and defined *Corynebacterium sensu stricto* as bacteria that contain *meso*-diaminopimelic acid, arabinose, galactose, mycolic acid having 22–38 carbon chains in the cell wall and MK-8 or MK-9 as the major menaquinone, and 51–59% G + C in DNA. This definition is quite comparable to the taxon proposed by Yamada and Komagata (1972) for bacteria belonging to group 1 of *Corynebacterium*. Almost all of the glutamic acid bacteria can be grouped into the taxon of *Corynebacterium sensu stricto* proposed by Keddie and co-workers. For other information on the chemotaxonomy of glutamic acid bacteria the reader is referred to K. Suzuki et al. (1981), Uchida and Aida (1979), and Minnikin et al. (1978).

TABLE 5.1   DNA Homology in Glutamic Acid Bacteria

Strains	DNA Homology*
C. glutamicum ATCC 13032	100.00
C. acetoacidophilum ATCC 13870	76.2
B. lactofermentum ATCC 13655	102.1
B. divaricatum ATCC 14020	93.4
B. saccharolyticum ATCC 14066	77.3
B. roseum ATCC 13825	73.5
B. immariophilium ATCC 14068	76.2
B. ammoniagenes ATCC 13745	75.6
M. ammoniaphilum ATCC 15354	78.4

From Komatsu and Kaneko, (1980).
C.: Corynebacterium, B: Brevibacterium, M.: Microbacterium.
* Homology to C. glutamicum ATCC 13032.

Komatsu and Kaneko (1980) reported on DNA homology among the glutamic acid bacteria. As shown in Table 5.1, considerable homology was found among nine of the glutamic acid bacteria. Another study demonstrated that C. glutamicum ATCC 13032, C. lilium ATCC 15990, B. flavum ATCC 14067, and B. divaricatum ATCC 14020 can be grouped in the same DNA homology cluster, which is distinct from that of B. ammoniagenes ATCC 6872, Microbacterium flavum, and C. diphtheriae ATCC 11913 (K. Suzuki et al., 1981).

Considering all the findings of conventional taxonomy, chemotaxonomy, and DNA homology, it is reasonable to regard the glutamic acid bacteria as a single species in the genus Corynebacterium.

III. MORPHOLOGY OF GLUTAMIC ACID BACTERIA

Glutamic acid bacteria are usually rod-shaped and tend to form a V-shape when snapping division occurs. The alteration in the cell cycle between rods and cocci is shown in Fig. 5.1. When the bacteria are cultivated in rich media, remarkable pleomorphic alterations are not observed. However, elongation, swelling, and branching of cells occur frequently when the bacteria are cultivated in media with low biotin concentrations or with an antibiotic. Electron micrographs are shown in Figs. 5.2 and 5.3. The snapping cell division presumably results from sudden fission of the outer layer after the inner layer is invaginated to form a septum (Fig. 5.2(1,2)). Electron micrographs of thin sections of plasmolysed cells reveal that the cell wall is composed of three layers, but another layer with high electron density can be observed inside the peptidoglycan layer (Fig. 5.3(1)). Freeze-etch preparations of cells show both concave and convex faces caused by fracturing of the cell wall (Fig. 5.3(2–4)). This observation suggests the existence of a layer having hydrophobic bonding in the cell wall. In cells cultivated in biotin-deficient medium, bleblike projec-

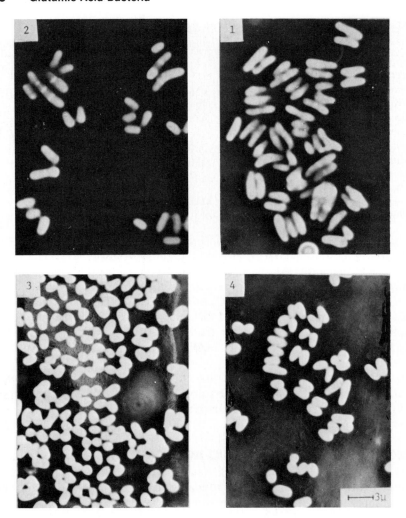

FIGURE 5.1 Morphological change in *Corynebacterium glutamicum* ATCC 13032 during growth. Phase contrast micrographs are shown with the cells after 2 h cultivation (lag phase, 1), 5 h (logarithmic phase, 2), 10 h (late logarithmic phase, 3), and 24 h (stationary phase, 4). (From Abe et al., 1967.)

tions on the cell surface are sometimes observed with morphological changes such as elongation and swelling (Fig. 5.3(3)). The alteration of fine structure in these cells is also observed in the freeze-etch electron micrograph (Fig. 5.3(4)). Similar structural changes are observed in cells cultivated with penicillin added to the biotin-sufficient medium. The correlation between the fine structure of cell walls and the excretion of glutamic acid is of great interest.

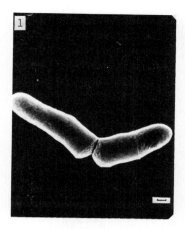
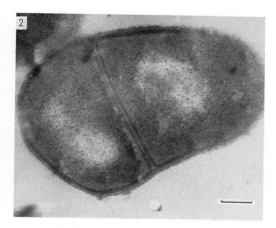

FIGURE 5.2   Scanning electron micrograph of snapping division (1) and thin section micrograph of septum formation in *Corynebacterium glutamicum* ATCC 13032. The bars inserted in the photographs correspond to 0.2 μm in length. (From Ochiai and Takayama, 1982.)

IV. FACTORS INFLUENCING GROWTH AND VIABILITY

Glutamic acid bacteria without exception require biotin for their growth, but some strains also require thiamine (Kinoshita and Tanaka, 1972). Studies on *B. divaricatum* NRRL B-2311 indicated that the biotin requirement is caused by the deficiency of two enzymes mediating early steps of biotin biosynthesis from pimelic acid; one is pimelyl-CoA synthetase, responsible for the first step, and the other is 7-keto-8-aminopelargonic acid synthetase catalyzing the second step (Izumi et al., 1981). This was confirmed with *B. lactofermentum*, in which it was demonstrated that intermediates of the third or later steps of the biotin biosynthetic pathway are able to replace biotin (Okumura et al., 1962). Biotin is known to be an essential factor for the biosynthesis of fatty acids as the coenzyme of acetyl-CoA carboxylase. Accordingly, control of the biotin concentration in the medium to a low level depresses fatty acid synthesis. In this case, oleic acid, which is the principal constituent of fatty acids in glutamic acid bacteria, is able to replace biotin (Takinami et al., 1968).

Under conditions of nutrient sufficiency, the cells grow rapidly even in a simple, chemically defined medium as shown in Table 5.2 (Araki et al., 1973). When cells are cultured under starvation conditions, they require a thiol compound such as cysteine. To achieve massive growth in media containing relatively high concentrations of sugar, the addition of iron salts and chelating or chelated substances (e.g., ferrichrome, 8-hydroxyquinoline, or rutin) is indispensable. In a natural medium containing 1% meat extract, 1% peptone, 0.5% yeast extract, and 0.3% NaCl the optimum temperature and pH for

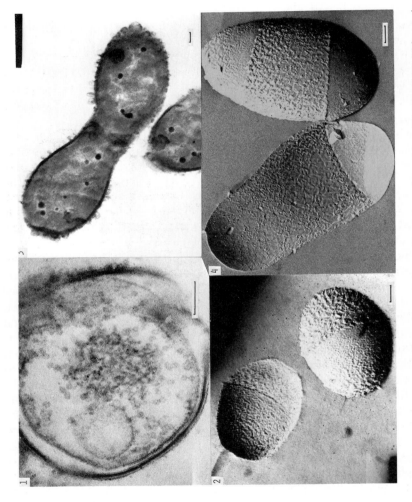

FIGURE 5.3 Electron micrographs of cell envelope of *Corynebacterium glutamicum* ATCC 13032. Thin section of plasmolysed cell (1), freeze-etch preparation of normally grown cells (2), thin section of cells cultivated in biotin-deficient medium (3), and freeze-etch preparation of the above cells (4). Bar is 0.2 μm. (From Ochiai and Takayama, 1982.)

TABLE 5.2   Composition of a Minimal Medium

Glucose 0.5%, $(NH_4)_2SO_4$ 0.15%, KH_2PO_4 0.1%,
K_2HPO_4 0.3%, $MgSO_4 \cdot 7H_2O$ 0.01%, $CaCl_2 \cdot 2H_2O$ 0.0001%,
Biotin 30 $\mu g/l$, Thiamine 100 $\mu g/l$, $MnCl_2 \cdot 4H_2O$ 0.0007%,
$FeSO_4 \cdot 7H_2O$ 0.001%, NaCl 100 $\mu g/ml$ (pH 7.8 adjusted with KOH)

From Araki et al. (1973).

growth are at 30°C and at 7.0, respectively; under such conditions the doubling time is approximately 60 min.

Glutamic acid bacteria are sensitive to various kinds of antibiotics. For example, minimum inhibitory concentrations for *B. lactofermentum* are 0.1–0.5 IU/ml for penicillin G; 0.5–1 $\mu g/ml$ for tetracycline, oxytetracycline, polyoxin B, and chloramphenicol; and 0.05–1 $\mu g/ml$ for streptomycin (Shiio et al., 1970).

When cultivated with sufficient amounts of biotin, the extracellular production of glutamic acid occurs only when sublethal concentrations of penicillin or a detergent is added in early exponential growth (Somerson and Phillips, 1962). In contrast, addition of such inhibitors at late exponential or at stationary growth phase results in neither glutamic acid production nor marked growth inhibition even at high concentrations. This phenomenon suggests a significant rigidity of the cell wall, and the cells do not easily undergo lysis even with sonication or lysozyme treatment. Complete cell lysis occurs only by first cultivating cells with a high concentration of penicillin added at late exponential phase and then sonicating or adding lysozyme (Komatsu, 1979).

V. MUTATION AND GENETIC RECOMBINATION

A. Biochemical Mutants

The success in industrial production of glutamic acid stimulated further interest in finding producing strains for other amino acids. Wild strains of glutamic acid bacteria are able to produce only a few amino acids extracellularly, such as glutamic acid, valine, proline, glutamine, and alanine. For the extracellular production of a desired amino acid, changes in cellular metabolism and/or regulatory controls are required. Thus attempts were made to induce auxotrophic mutants and regulatory mutants. The purpose of utilizing auxotrophic mutants is to negate feedback control mechanisms by limiting the intracellular accumulation of feedback inhibitors or repressors. For example, an ornithine fermentation was achieved by application of an arginine auxotrophic mutant (Kinoshita et al., 1957b), and a year later a successful lysine fermentation was devised with homoserine auxotrophic mutants (Kinoshita et al., 1958; Nakayama et al., 1961). The purpose of regulatory mutants is to bypass feedback control mechanisms by using feedback-insensitive mutants that are insensitive to end product inhibition or to end product repression. A variety of such mutants that are resistant to amino acid antimetabolites have been induced; for instance, an S-(α-aminoethyl)-L-cysteine-resistant mutant can be used for lysine production. Most amino acids are produced today by use of mutants that contain combinations of auxotrophic and regulatory mutations. Even more potent producing strains have been obtained by eliminating the ability of the organism to degrade the product and by improving cell permeability in favor of excretion of the product.

Mutation of glutamic acid bacteria was originally conducted with ultraviolet light, X-rays or γ-rays. In recent years, however, chemical mutagens such as N-methyl-N'-nitro-N-nitrosoguanidine and ethylmethane sulfonate have been employed instead of the above physical mutagens because of the higher frequencies of mutation and simpler treatments. Up to now, a wide variety of mutants that are resistant or sensitive to analogs of amino acids, purines, pyrimidines, vitamins, organic acids, and antibiotics have been isolated. Changes in ability to utilize carbon sources or nitrogen sources and in sensitivity to temperature have also become important. Relationships between various mutants and their products are summarized in Table 5.3. More detailed information has been published in recent reviews (Kinoshita and Nakayama, 1978; Nakayama, 1982).

B. Temperate Phage and Transduction

Among the methods for genetic recombination in bacteria, cell fusion mediated by F-factor, transduction by phage, and transformation by DNA are known. There are, however, only a few cases of application of these methods for industrial purposes. Success was reported recently in the construction of *Serratia marcescens* strains overproducing arginine, histidine, isoleucine, or threonine; temperate-phage PS-20 was used for transduction of several characters involving nutrient requirements and metabolic regulation (Komatsubara et al., 1980). On the other hand, in one study on glutamic acid bacteria belonging to *Corynebacterium* or *Brevibacterium*, the transduction of auxotrophic characters has not been successful employing temperate phages (Shapiro, 1976). In another study, however, temperate-phage CP-119 of *Brevibacterium* was able to transduce auxotrophic characters at a frequency of 10^{-7} (Momose et al., 1976). Recently, four kinds of temperate phages were isolated from *C. glutamicum* having a polyhedral head and a tail with a contractile sheath. They were classified into two molecular weight groups, i.e., $10–11 \times 10^6$ daltons and $25–26 \times 10^6$ daltons (Katsumata and Furuya, 1983). Their use for transduction is under investigation.

C. Cell Fusion

In addition to mutation, cell fusion is becoming a promising method for the breeding of amino acid producers. The method depends primarily upon the preparation of protoplasts from bacterial cells. Protoplasts of *B. flavum* are

TABLE 5.3 Amino Acid–Producing Mutants of Glutamic Acid Bacteria

Produced Amino Acid	Phenotypes	Some Identified Enzymatic Modifications
Glutamic acid	Auxotrophic for oleate; auxotrophic for glycerol; sensitive to lysozyme; temperature-sensitive; esculetin-resistant	Transport altered; deficient in pyruvate dehydrogenase; deficient in isocitrate lyase
Glutamine	Sulfaguanidine-resistant	
Arginine	2-Thiazole-3-alanine-resistant; D-serine-sensitive; sulfaguanidine-resistant	Derepressed arginine enzymes; desensitized acetylglutamate kinase
Citrulline	Auxotrophic for arginine	Deficient in argininosuccinate synthetase
Ornithine	Auxotrophic for arginine	Deficient in ornithine transcarbamylase
Proline	Sulfaguanidine-resistant; 3,4-dehydroproline-resistant; auxotrophic for isoleucine	
Tryptophan	Auxotrophic for phenylalanine and tyrosine; 5-methyltryptophan-resistant	Densensitized anthranilate synthetase
Phenylalanine	Auxotrophic for tyrosine; p-fluorophenylalanine-resistant	Desensitized prephenate dehydratase and chorismate mutase
Tyrosine	Auxotrophic for phenylalanine; p-fluorophenylalanine-resistant	Deficient in prephenate dehydratase
Lysine	Auxotrophic for homoserine; auxotrophic for leucine; thialysine-resistant; sensitive to fluoropyruvate	Deficient in homoserine dehydrogenase, desensitized aspartokinase; deficient in pyruvate dehydrogenase
Isoleucine	α-Amino-β-hydroxyvalerate-resistant; thiaisoleucine-resistant	Desensitized homoserine dehydrogenase
Homoserine	Auxotrophic for threonine	Deficient in homoserine kinase
Aspartic acid	Bradytrophic for glutamate	Deficient in citrate synthetase
Threonine	α-Amino-β-hydroxyvalerate-resistant	Desensitized homoserine dehydrogenase
Valine	Auxotrophic for leucine; auxotrophic for isoleucine; 2-thiazole-3-alanine-resistant; thialysine-resistant	Desensitized acetohydroxy acid synthetase
Leucine	Bradytrophic for phenylalanine; 2-thiazole-3-alanine-resistant; thialysine-resistant	Desensitized isopropylmalate synthetase; desensitized acetohydroxy acid synthetase
Alanine	Arginine hydroxyamate–resistant	
Histidine	2-Thiazole-3-alanine-resistant; triazolealanine-resistant; sulfaguanidine-resistant	Desensitized phosphoribosyl-ATP synthetase

prepared by treating cells with lysozyme under isotonic conditions, after cultivation with a low concentration of penicillin (Kaneko and Sakaguchi, 1979). Fusion of these protoplasts is achieved in the presence of polyethylene glycol, yielding a variety of recombinant strains. For example, protoplasts from two strains that had different auxotrophic characters and were resistant to either streptomycin or rifampicin were fused, yielding recombinant strains at a frequency of 10^{-6} to 10^{-7} per protoplast used that were resistant to both antibiotics. Auxotrophic characters of some of the fusants were different from those of parent strains, a result proving that these were recombinant strains. Similarly, with *C. glutamicum*, recombinant strains were obtained by fusing protoplasts, one of which was strain KY9182, resistant to streptomycin and requiring phenylalanine (Str^R, Phe^-), and the other of which was strain KY9684, resistant to rifampicin and requiring homoserine and leucine (Rif^R, Hom^-, Leu^-) (Katsumata et al., 1980). Strains doubly resistant to streptomycin and rifampicin were obtained as recombinants with the frequency shown in Table 5.4. Among them the phenotypes of unselected chromosomal markers are summarized in Table 5.5; all of possible phenotypic characters were realized that were stable enough that dissociated characters were not generated. These results suggest that the recombinants were derived by haploid recombination via protoplast fusion. By further development of this recombination system it is expected that the protoplast fusion method will yield many recombinants that will be useful for amino acid production.

D. Gene Manipulation

In recent years, recombinant DNA technology has been extensively applied for the production of scarce and medically valuable polypeptides. A number of genes have been cloned and expressed in *E. coli* by the use of a suitable plasmid vector and by insertion alongside a powerful promoter. It is hoped that such gene manipulation will be applied for improvement in amino acid production. Two examples of success in threonine overproduction have been reported (Debabov et al., 1982; Miwa et al., 1981), in which a threonine operon was introduced into *E. coli* K-12 by transformation using a hybrid plasmid constructed from pBR322. Under optimum conditions, threonine was produced at a higher potency than that of existing strains. Also a tryptophan operon has been introduced into *Escherichia coli* W3110 that contained a deleted *trp* operon and was deficient in tryptophan repressor and tryptophanase. The transformant thus obtained produced tryptophan when anthranilic acid was fed, and some improvement in production was seen (Aiba et al., 1982).

Recombinant DNA technology could also be useful for improving glutamic acid bacteria. For this purpose, basic knowledge and technology concerning the genes of these bacteria should be developed. There are, however, various problems that hinder rapid progress in this field. These include the limited knowledge of the molecular genetics of these bacteria, the diversity of

TABLE 5.4 Formation of Rifampicin- and Streptomycin-Resistant Recombinants by Polyethylene Glycol(PEG)-mediated Protoplast Fusion in *Corynebacterium glutamicum*

Crossing	Treatment	Colony Formation per ml	Frequency per Minority Parent
KY9182 (Str^R Phe$^-$) × KY9684 (Rif^R Hom$^-$ Leu$^-$)	None	90	1.32×10^{-7}
	PEG 4000 (40%)	1323	1.95×10^{-6}
	PEG 6000 (33%)	1267	1.86×10^{-6}

From Katsumata et al. (1980).

TABLE 5.5 **Phenotypic Variation of Unselected Chromosomal Markers in Rifampicin- and Streptomycin-Resistant Fusants Obtained from KY9182(Str^R Phe$^-$) × KY9684(Rif^R Hom$^-$ Leu$^-$)**

	Recombinants per No. of Colonies Tested (%)		
Phenotype	No Treatment	PEG 4000 Treatment	PEG 6000 Treatment
Phe$^-$	86.7	40.1	37.2
Hom$^-$ Leu$^-$	11.1	8.9	12.4
Prototroph	2.2	20.0	22.9
Hom$^-$	0	2.3	1.3
Leu$^-$	0	2.9	2.1
Hom$^-$ Phe$^-$	0	2.0	3.7
Leu$^-$ Phe$^-$	0	4.0	4.0
Hom$^-$ Leu$^-$ Phe$^-$	0	19.8	16.4

From Katsumata et al. (1980).

the producer strains used, and the complexity of the amino acid biosynthetic pathways. Recent efforts have made it possible to isolate some plasmids of glutamic acid bacteria (Sakaguchi and Tanaka, 1979; Katsumata and Furuya, 1982). Thus success is likely in the near future with respect to development of host–vector systems in glutamic acid bacteria and cloning of genes for amplifying particular enzyme functions that are limiting steps in the biosynthetic pathways of desired amino acids.

VI. CARBOHYDRATE METABOLISM

C. glutamicum is capable of assimilating glucose, fructose, sucrose, maltose, and ribose, but mannose, galactose, and sorbose are used only weakly. Recently, a sucrase was purified from *B. divaricatum* NRRL B-2311, which is specific for sucrose and requires phosphate for its activity. Two key enzymes of the hexose monophosphate (HMP) shunt, G-6-P dehydrogenase and 6-phosphogluconate dehydrogenase (both requiring NADP), were demonstrated in *C. glutamicum* ATCC 13058 (Kinoshita and Tanaka, 1972). The fact that the strain produces glutamic acid in high yield from gluconate or D-ribose suggests the operation of the HMP shunt for assimilation of these compounds. Enzymes of the Embden-Meyerhof pathway (EMP) such as hexokinase, glyceroaldehyde-3-phosphate dehydrogenase, triosephosphate isomerase, phosphoglycerate kinase, enolase, phosphoglyceromutase, and pyruvate kinase have been demonstrated in *B. flavum* 2247 (Shiio et al., 1961).

Studies using 1-^{14}C- and 6-^{14}C-glucose demonstrated that 10–38% of accumulated glutamic acid is formed via the HMP shunt from glucose (Oishi and Aida, 1965). Recently, involvement of the HMP shunt in glutamate formation from glucose was calculated to be 13% in *M. ammoniaphilum*, using results obtained with ^{13}C-NMR spectrometry (Walker et al., 1982). As for the

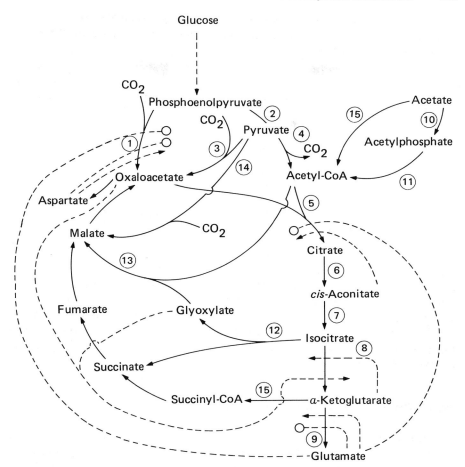

FIGURE 5.4 Central and anaplerotic pathways and regulation patterns in glutamic acid bacteria. 1, phosphoenolpyruvate carboxylase; 2, pyruvate kinase; 3, pyruvate carboxylase; 4, pyruvate dehydrogenase; 5, citrate synthase; 6,7, aconitase; 8, isocitrate dehydrogenase; 9, glutamate dehydrogenase; 10, acetate kinase; 11, phosphate acetyltransferase; 12, isocitrate lyase; 13, malate synthetase; 14, malic enzyme; 15, α-ketoglutarate dehydrogenase. $\text{---}\!\!\!>$, inhibition; $\text{----}\!\bigcirc$, repression. (From Shiio and Sugimoto, 1981.)

metabolic route from C-3 compounds to the TCA cycle, $^{13}CO_2$ was fixed into the C-1 position of glutamate in the presence of pyruvate by *C. glutamicum,* a species in which the NADP-specific malic enzyme is thought to participate in CO_2 fixation (Kinoshita and Tanaka, 1972). The existence of phosphoenolpyruvate (PEP) carboxylase (1 in Fig. 5.4) was established later in *B. flavum* (Shiio and Ujigawa, 1978) and shown to play a key role in entry of C-3 compounds into the TCA cycle; mutants defective in this enzyme require glutamate for growth. PEP-carboxylase is subject to severe feedback inhibition by aspartic acid but is activated by fructose-1,6-diphosphate or acetyl-CoA; the inhibition is reversed by acetyl-CoA. The affinity of PEP-carboxylase to PEP as substrate is ten times lower than that of pyruvate kinase, but the activities of both enzymes seem to be balanced by regulatory mechanisms involving activators and inhibitors as described above.

Recently, a biotin-dependent pyruvate carboxylase was recognized in *B. lactofermentum* 2256 (Tosaka et al., 1979). A lysine-producing mutant derived from the above strain fixed $^{13}CO_2$ in the γ-position of lysine in the presence of pyruvate. The CO_2-fixation activity of the cells was stimulated by addition of biotin to the medium, and the requirement of glutamate by a mutant defective in PEP carboxylase was replaced by a high concentration of biotin. These results indicate the importance of pyruvate carboxylase in the metabolism of C-3 compounds through the TCA cycle. The presence of pyruvate dehydrogenase was also proved in *B. flavum* (Shiio and Ujigawa, 1978) and *B. lactofermentum* (Tosaka et al., 1979).

It is reasonable to speculate that the activity of α-ketoglutarate dehydrogenase would exert a considerable effect on glutamate production. Early work showed that this enzyme is extremely limited in comparison with other enzymes in the TCA cycle (Kinoshita and Tanaka, 1972). However, recent work provides evidence for the presence of this enzyme in *B. flavum*, but it is so unstable that it must be extracted only after stabilization with glycerol, Mg^{++}, and thiamine pyrophosphate (Shiio and Ujigawa-Takeda, 1980). In addition, the activity of α-ketoglutarate dehydrogenase is only one hundredth that of glutamate dehydrogenase. Such differences in the activities of the two enzymes could be responsible for glutamic acid overproduction in preference to complete oxidation via the TCA cycle.

All other enzymes involved in the TCA cycle have been found in glutamic acid bacteria. Isocitrate dehydrogenase requires NADP as a cofactor, and the NADPH that is formed reacts with glutamate dehydrogenase to yield glutamate. This coupling plays a key role in the production of glutamate. Mutants devoid of citrate synthase or aconitase require glutamate (Kinoshita and Tanaka, 1972). The regulatory mechanisms controlling these enzymes in *B. flavum* 2247 are summarized in Fig. 5.4 (Shiio and Sugimoto, 1981). Under biotin-excessive growth conditions, synthesized glutamic acid does not permeate out through the cell membranes; it accumulates intracellularly. The high concentration of glutamate in the cells inhibits citrate synthase directly or indirectly, resulting in a high intracellular concentration of aspartate. This in turn leads to inhibition of PEP carboxylase and to accumulation of pyruvate, which is then converted to lactate or succinate. Transient accumulation of these organic acids is often observed in fermentations; they are finally degraded to carbon dioxide and water through the TCA or glyoxylate cycle.

In *B. flavum*, acetate is incorporated into acetyl-CoA by acetate kinase (10 in Fig. 5.4) and phosphoacetyltransferase (11 in Fig. 5.4), but direct formation of acetyl-CoA from acetate by acetyl-CoA synthetase (15 in Fig. 5.4) is not observed. Acetate-nonassimilating mutants were found to be devoid of one of the above two enzymes or of isocitrate lyase (12 in Fig. 5.4) (Shiio et al., 1969). Acetyl-CoA formed from acetate is metabolized through the glyoxylate cycle. Isocitrate lyase and malate synthetase (13 in Fig. 5.4), which are the key enzymes in glyoxylate cycle, are present in *C. glutamicum* 541 and *B. flavum*. The formation of these two enzymes is stimulated in acetate-containing media. Isocitrate lyase activity in *C. glutamicum* considerably decreases under biotin limitation (Kinoshita and Tanaka, 1972). Inhibition of isocitrate dehydrogenase occurs by oxaloacetate in concert with glyoxylate; this plays an important role in the metabolic switch of the TCA cycle to the glyoxylate cycle (Shiio and Ozaki, 1968).

NMR-analysis of *M. ammoniaphilum* cultured under biotin-deficient conditions indicates that 58% of the glutamate formed from glucose is produced via the pathway involving phosphoenolpyruvate, citrate, and α-ketoglutarate; the remainder is produced via the glyoxylate or TCA cycle (Walker et al., 1982).

VII. BIOSYNTHESIS AND METABOLISM OF AMINO ACIDS

A. Biosynthesis of Amino Acids and Its Regulation

General features of the biosynthetic pathways of amino acids in glutamic acid bacteria closely resemble those in *E. coli* (see Kinoshita, 1959; Kinoshita and Nakayama, 1978; Nakayama, 1982).

As for ammonia-assimilating pathways, high activity of NADPH-specific glutamic acid dehydrogenase is present in glutamic acid bacteria, indicating that the enzyme plays a key role in ammonia assimilation. Alanine dehydrogenase and leucine dehydrogenase have not been found, and only a weak activity of aspartase was detected (Kinoshita and Tanaka, 1972).

Recently, glutamate synthase activity (2 in Fig. 5.5) was detected in *Corynebacterium* sp. C91 (hydrocarbon assimilating) (Vandecasteele et al., 1975), *C. glutamicum* ATCC 13032, and *B. flavum* 2247 (Sung et al., 1982). In the latter two strains, formation of the enzyme is repressed by a relatively high concentration of ammonium ion, a result suggesting that the enzyme functions only under low ammonium conditions. This was supported by the fact that *B. flavum* mutants that require glutamic acid owing to a defect in glutamate dehydrogenase could be induced and their glutamate synthase activities were the same as that of the parent strain (Shiio and Ujigawa, 1978). In *Corynebacterium* sp.C91, glutamate synthase formation was not completely repressed by ammonium ion, and feedback inhibition by glutamate affected only glutamate dehydrogenase. It is thus reasonable to speculate that glutamate synthase is actually involved in glutamate formation in strain C91.

The amino group thus introduced into glutamate is further utilized to form other amino acids by transaminase action. Recently, the transaminases in

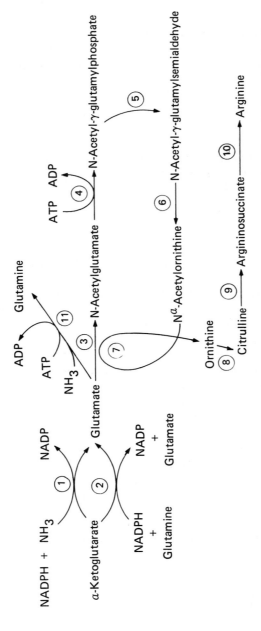

FIGURE 5.5 Biosynthetic pathways of glutamic acid, glutamine and arginine in glutamic acid bacteria. 1, glutamate dehydrogenase; 2, glutamate synthase; 3, N-acetylglutamate synthetase; 4, N-acetyl-γ-glutamatekinase; 5, N-acetylglutamate-γ-semialdehyde dehydrogenase; 6, N-acetylornithine-δ-aminotransferase; 7, N-α-acetylornithine glutamate acetyltransferase; 8, ornithine carbamyltransferase; 9, argininosuccinate synthetase; 10, argininosuccinate lyase.

130

B. flavum 2247 were isolated and separated into fractions specific for prephenate, aspartate, phenylalanine, or valine; most had high substrate specificity except for valine transaminase (Mori et al., 1981). An aspartate-requiring mutant of *B. flavum* 2247 was found to be devoid of aspartate transaminase.

Concerning lysine biosynthesis, all the enzymes on the diaminopimelate pathway are present in *C. glutamicum*, *B. flavum*, and *B. lactofermentum*. Glutamate dehydrogenase from *C. glutamicum* catalyzes the following additional reaction (Oshima et al., 1965):

$$\alpha\text{-amino-}\epsilon\text{-ketopimelic acid} + NH_3 + NADPH + H^+ \rightleftharpoons$$
$$\alpha,\epsilon\text{-diaminopimelic acid} + NADP$$

However, since a *B. flavum* mutant defective in glutamate dehydrogenase does not require lysine for growth, the above reaction does not seem to participate in lysine biosynthesis (Shiio and Ujigawa, 1978).

Aspartokinase in *C. glutamicum* ATCC 13032 or *B. flavum* 2247, the first enzyme in lysine biosynthetic pathway, is subject to concerted feedback inhibition by lysine and threonine (Nakayama et al., 1966). In *B. lactofermentum* 2256, however, at least two isoenzymes of aspartokinase appear to exist, one of which is sensitive only to lysine and the other to lysine plus threonine (Tosaka et al., 1978a). Unlike *E. coli*, dihydrodipicolinate synthetase in the glutamic acid bacteria (the first enzyme in the branch to lysine) is not regulated by lysine, but its formation is repressed by leucine (Tosaka et al., 1978b). While homoserine dehydrogenase in the glutamic acid bacteria is inhibited by threonine, its formation is controlled by methionine (Kinoshita et al., 1960b).

The methionine biosynthetic pathway has been elucidated in *C. glutamicum* ATCC 13032 (Kase and Nakayama, 1974) and *B. flavum* 2247 (Ozaki and Shiio, 1982) and is shown in Fig. 5.6. All the enzymes (1 to 6) have been found in the above two strains. The pathway involving enzymes 1, 2, and 3 appears to be the main route for methionine synthesis.

As for arginine, its biosynthetic pathway is shown in Fig. 5.5, which includes an acetyl-transfer reaction (7 in Fig. 5.5) from N-acetyl-L-orinithine to glutamic acid to form ornithine. This pathway, first discovered in *C. glutamicum* (Udaka and Kinoshita, 1958), is thought to be more economical for arginine formation than other arginine biosynthetic pathways (Umbarger, 1975). The second enzyme of arginine biosynthesis, N-acetyl-γ-glutamate kinase (4 in Fig. 5.5), is susceptible to strong feedback inhibition by arginine. Since the accumulation of N-acetylglutamate by an arginine auxotroph that was devoid of N-acetyl-γ-glutamate kinase was almost completely depressed by excessive arginine, it is thought that the former enzyme, N-acetylglutamate synthetase (3 in Fig. 5.5), is affected by feedback inhibition by arginine (Yoshida et al., 1980). In addition, all the enzymes except for the first (3 in Fig. 5.5) and the last (10 in Fig. 5.5) are repressed by arginine.

As for the tyrosine biosynthetic route, two pathways have been detected in microorganisms (Fig. 5.7). In the studies using *C. glutamicum* ATCC 13032 or *B. flavum* 2247, a transaminase (4) that has remarkable affinity to prephenate

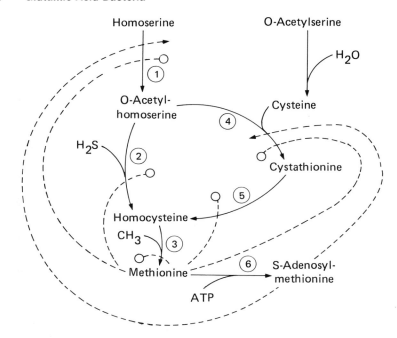

FIGURE 5.6   Biosynthetic pathways of L-methionine and its regulation in glutamic acid bacteria. 1, homoserine-o-acetyltransferase; 2, acetylhomoserine sulfhydrase; 3, homocysteine methylase complex; 4, cystathionine-γ-synthetase; 5, β-cystathionase; 6, S-adenosylmethionine synthetase.

and pretyrosine dehydrogenase (5) was found, but prephenate dehydrogenase (6) was not detected. In addition, a mutant requiring tyrosine owing to a defect of pretyrosine dehydrogenase accumulated pretyrosine. From these results it appears that tyrosine is formed via pretyrosine in glutamic acid bacteria (Fazel and Jensen, 1979). In relation to the biosynthesis of other aromatic amino acids, the following enzymes are present in the glutamic acid bacteria: DAHP synthetase (1), shikimate dehydrogenase (2), prephenate dehydratase (7), chorismate mutase (3), anthranilate synthetase (8), and tryptophan synthetase (9). Chorismate mutase and DAHP synthetase form a complex. The proposed regulatory mechanisms working on the biosynthesis of aromatic amino acids are also indicated in Fig. 5.7 (Hagino and Nakayama, 1975).

The biosynthetic pathways of the branched amino acids have been studied in *B. lactofermentum* 2256 and *B. flavum* 2247 (Fig. 5.8). Threonine dehydratase (1), acetohydroxy acid synthetase (2), isopropylmalate synthetase, and valine transaminase are present, and the regulatory mechanisms are summarized in Fig. 5.8 (Tsuchida and Momose, 1975).

Concerning histidine biosynthesis, phosphoribosyl-ATP pyrophosphorylase (the first enzyme) is subject to both feedback inhibition and repression by histidine in the glutamic acid bacteria. Histidinol phosphate-α-ketoglutarate

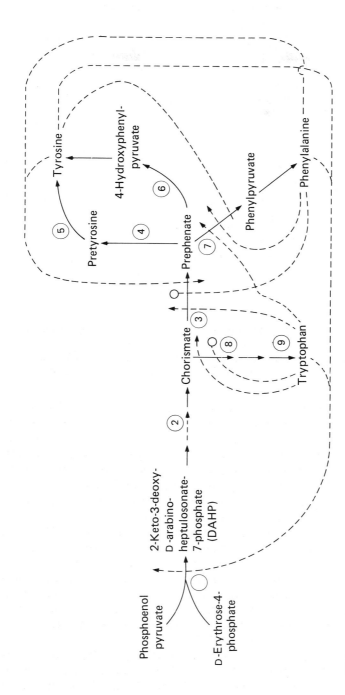

FIGURE 5.7 Regulation of aromatic amino acid biosynthesis in *Corynebacterium glutamicum*. 1, DAHP synthetase; 2, shikimate dehydrogenase; 3, chorismate mutase; 4, transaminase; 5, pretyrosine dehydrogenase; 6, prephenate dehydrogenase; 7, prephenate dehydratase; 8, anthranilate synthetase; 9, tryptophan synthetase. ——→, inhibition; ----○, repression; ----▲, activation.

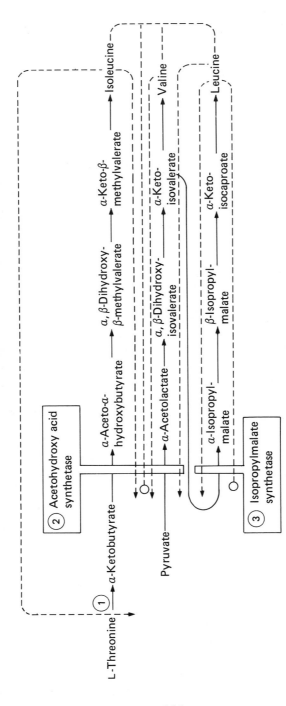

FIGURE 5.8 Regulation patterns of branched-chain amino acid biosynthesis in *Brevibacterium lactofermentum* 2256. 1, threonine dehydratase. ——▷, inhibition, ——○, repression.

transaminase and histidinol dehydrogenase are present in these organisms. Even though histidinol phosphatase has not been detected, it could be responsible for histidine biosynthesis in the glutamic acid bacteria as in *Salmonella typhimurium*, since the enzymes catalyzing the previous step and the next step do exist (Araki and Nakayama, 1974).

B. Decomposition of Amino Acids

Amino acid–degrading activities in glutamic acid bacteria are generally much lower than in other microorganisms. This is one of the main reasons for efficient accumulation of various amino acids by these microorganisms. Among amino acids, serine and threonine tend to be degraded, but histidine is only weakly degraded to imidazole and lactic acid. Glutamic acid is degraded to some extent, depending on culture conditions; it is transformed to α-ketoglutaric acid by transaminase, followed by further degradation through the TCA cycle (Shiio et al., 1982). Lysine is metabolized to only a small extent in *C. hydrocarboclastus* KY 8837 or in *C. glutamicum* ATCC 13032; the pathway seems to be via N-acetyl-lysine (Tomita et al., 1975).

C. Formation of γ-Glutamylpeptides

In glutamic acid fermentation broths, small amounts of γ-glutamylpeptides (γ-glu-val, γ-glu-leu, γ-glu-glu, etc.) are coproduced with glutamic acid, often disturbing the crystalization of glutamic acid in the purification process. The formation of these peptides is catalyzed by a single enzyme, γ-glutamyltranspeptidase in *C. glutamicum* KY 9909 (Hasegawa and Matsubara, 1978). The enzyme has hydrolytic activity toward the peptides but, in the presence of large amounts of glutamic acid, also catalyzes the reverse reaction to form γ-glu-glu, where γ-linked derivatives of glutamate or ATP are not required. In peptide formation by this enzyme, either L- or D-glutamate is essential as the N-terminal substrate, but other amino acids and dipeptides can act as a C-terminal substrate. Since the concentrations of amino acids other than glutamic acid in the fermentation broth are too low to act as substrate for the enzyme, it is likely that peptides such as γ-glu-val, γ-glu-leu, and γ-glu-glu are formed from γ-glu-glu by transpeptidation (Fig. 5.9).

VIII. CELL MEMBRANES

A. Excretion of Glutamic Acid

A considerable accumulation of glutamic acid occurs only when the microorganism is cultivated with a limiting supply of biotin (Tanaka et al., 1960). The excretion mechanism of glutamic acid has been studied (Kinoshita and Tanaka, 1972) and is closely related to the decrease in phospholipid content of cell membranes (Kikuchi and Nakao, 1973; Takinami et al., 1968). The

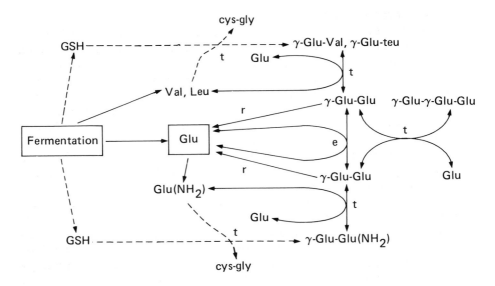

FIGURE 5.9 Proposed mechanism of the formation of γ-L-glutamyl-peptides in L-glutamic acid fermentation. ----, functional pathway at the initial stage of fermentation. e, exchange of C-terminal glutamic acid; r, reversal of hydrolysis; t, transpeptidation. All of these reactions are catalyzed by the γ-glutamyltranspeptidase.

decrease in phospholipid contents probably causes changes in membrane structure to increase excretion of glutamate from cells. With acceleration of glutamate excretion, intracellular concentration of the amino acid is lowered enough to relax feedback regulation by glutamate (Nunheimer et al., 1970).

From studies on *B. lactofermentum* 2256, *B. flavum* 2247, *B. divaricatum* NRRL-2312, *C. glutamicum* ATCC 13032, and *M. ammoniaphilum* ATCC 15354 it appears that phospholipid content represents more than 90% of the total lipids in the cells (Takinami et al., 1968, 1969) and varies from 2% to 4.5% of dry cell weight, depending on the biotin concentration of the culture medium. Favorable content of phospholipid for glutamic acid production is 2.4–3.4%. The main component of the phospholipid is cardiolipin (60%). Phosphatidylinositol, phosphatidylglycerol, and an unidentified sugar phospholipid are also present, while phosphatidylethanolamine, which was found in nocardiaform bacteria (Komura et al., 1975) and *) B. alkanolyticum* (Kikuchi and Nakao, 1973), is not. With respect to the fatty acid component in the phospholipids, palmitic and oleic acid are present in *C. glutamicum* ATCC 13032 and *Microbacterium ammoniaphilum* ATCC 15354, but their contents depend on culture conditions.

B. Transport Systems

In contrast to excretion, uptake of glutamate is energy-dependent in *B. lacto-fermentum* 2256. In the cells of glutamic acid bacteria cultivated under biotin limitation, permeability barriers for excretion of amino acids are fairly relaxed, and glutamate is easily excreted (Demain and Birnbaum, 1968). The uptake of glutamate into the cells of *B. ammoniagenes* ATCC 31169 requires Na^+, K^+, or Mg^{2+} and is inhibited by aspartate. The uptake activity is decreased when cells are cultured under biotin limitation (Oishi et al., 1970). In *B. ammoniagenes* ATCC 31169, *C. glutamicum* ATCC 13032, and *B. flavum* 2247, at least two sodium-dependent systems were observed for amino acid uptake; one related to threonine and serine, and the other to isoleucine, valine, and leucine. The requirement of sodium ion can be partially replaced by lithium ion in these systems (Araki et al., 1973). The uptake of tyrosine and phenylalanine compete with each other but compete only partially with tryptophan uptake (Sugimoto and Shiio, 1981). The above findings are also supported by studies on growth of auxotrophic mutants of *C. glutamicum* (Nakayama et al., 1960). Arginine and lysine appear to be taken up by a single system, but histidine uptake is independent of the uptake system for aromatic amino acids. Uptake of biotin is energy-dependent, the maximum accumulation being 130 mμg per g of dry cell weight, corresponding to 40 times the amount required for maximum growth (Oishi et al., 1970).

C. Respiratory Chain

Glutamic acid bacteria such as *C. glutamicum* No. 541 and *B. ammoniagenes* possess considerable NADH-oxidizing activity. In cells cultivated under biotin limitation, oxidizing activities toward organic acids such as succinate and pyruvate are low but are reactivated by the addition of NAD (Kinoshita and Tanaka, 1972; Oishi, 1967).

The following components of the respiratory chain are known in *B. flavum* 22 LD: three kinds of cytochrome (c_{555}, b_{567}, a_{604}), two kinds of cytochrome-oxidizing enzyme [Cyt O (420, 569 nm) and Cyt a_3 (435, 596 nm)], and menaquinone. A NADH oxidase that has been obtained from membrane fractions is resistant to cyanide or ultraviolet light irradiation and does not couple with the oxidative phosphorylation system. The respiratory chain of the microorganism has been proposed as indicated in Fig. 5.10 (Shvinka et al., 1979).

In advance of the studies described above, the respiratory chain in *Brevibacterium thiogenitalis* D248 has been investigated (Sugiyama et al., 1973). The work was initiated by the finding that the yield of glutamate from acetate was considerably increased by the addition of copper ion, but similar effects were not obtained when glucose was the substrate. Copper ion was found to markedly stimulate succinate-oxidizing activity. The cells grown with excess copper ion contained cytochrome *a*, *b*, and *c*, the contents of *a* and *c*

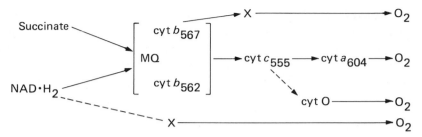

FIGURE 5.10 Proposed respiratory chain in *Brevibacterium flavum.* MQ, menaquinone; X, unidentified transmitters. (From Shvinka et al., 1979.)

being increased in parallel with copper concentration. In the absence of copper ion, cytochrome *d* was formed in place of *a* and *c*; the formation of *b* was not affected by the metal ion. The major component of menaquinone was found to be MQ-9 (II H). In oxidative phosphorylation a higher P/O ratio was observed with succinate than with NADH, but the ratio increased to the same level when an NADH-regenerating system was used in place of NADH. This suggested the existence of an NADH-oxidizing route that did not couple with the oxidative phosphorylation system.

IX. SUBSTRATES OTHER THAN CARBOHYDRATES

A. Taxonomy

It is known that acetate or ethanol serves as a good source of carbon for glutamic acid production by the microorganisms described above (Tsunoda et al., 1960; Oki et al., 1969). However, more potent glutamate-producing bacteria from acetate, *Corynebacterium acetoacidophilum*, *Corynebacterium acetoglutamicum*, and *Corynebacterium acetophilum*, have been reported; details of their taxonomic characters have not been described. *Brevibacterium* sp. B 136 was reported to produce glutamate efficiently from ethanol. More detailed information on these strains can be found in the review by Tsunoda and Okumura (1970).

   Corynebacterium hydrocarboclastus ATCC 15961, *Arthrobacter paraffineus* KY4303, *Corynebacterium alkanolyticum*, and *Corynebacterium petrophilum* have been studied as hydrocarbon-assimilating glutamic acid bacteria; all require thiamine for growth (Abe and Takayama, 1972).

   Extensive taxonomic studies have been conducted with *C. hydrocarboclastus* ATCC 15961. Having 61% G + C content in its DNA, containing *meso*-diaminopimelic acid in its cell wall, and possessing other properties, the microorganism was classified in Group I coryneform bacteria by K. Yamada and Komagata (1972). However, it was found later that the microorganism shows cell division of the bending type and complex morphological changes such as multibranching forms and short rods by fragmentation. From these results the microorganism has been reclassified as *Nocardia erythropolis* (Komura et al., 1973). The cell wall fraction of this microorganism contains arabinose, galactose, and mannose as main sugar components. Mycolic acid is also present in its cell walls and MK-8(H_2)/MK-7(H_2) are the menaquinone components.

B. Metabolism of Hydrocarbons

When hydrocarbon-assimilating bacteria are cultured, the medium becomes emulsified, indicating that surface-active agents are being produced. *A. paraffineus* KY4303 and *Corynebacterium* sp. KY4336 accumulate a trehalose ester of an α-branched-β-hydroxy fatty acid in the emulsion layer when cultured in hydrocarbon media. The sugar ester shows high surfactant activity in a mixture of water and *n*-paraffin. It appears that the sugar ester, located on the surface of the cells, participates in distribution of aqueous nutrients and *n*-paraffin and plays an important role in assimilation of hydrocarbon. Similar phenomena have been observed with other *n*-paraffin-assimilating microorganisms. Instead of trehalose, other sugars are present in the sugar ester components of *Nocardia* sp. KY4339 and *Corynebacterium fasciens* KY3541 (T. Suzuki et al., 1969).

Glutamic acid produced by *Corynebacterium* S10B1 grown with air containing $^{18}O_2$ yielded ^{18}O atoms in an α- and γ-carbonyl residue at a 45:55 ratio. This indicates that an oxygenase is involved in oxidation of *n*-dodecane to form dodecanoic acid via the alcohol and aldehyde forms. The acid would be further metabolized by β-oxidation to form acetyl-CoA, which is finally introduced into glutamic acid via the glyoxylate pathway (Imada et al., 1966).

X. CONCLUSION

During the past 25 years a vast volume of information on the biochemistry, physiology, and genetics of the glutamic acid bacteria has been accumulated. However, this information is still insufficient from the standpoint of biology because the studies were aimed mainly at establishing and improving industrial production of amino acids. The usefulness of the natural L-amino acids is recognized in a variety of fields, especially those concerned with food, pharmaceuticals, and animal feed. More fundamental and comprehensive studies on these microorganisms would bring further advancement to science and technology. In addition to conventional mutation methods, new genetic methods such as cell fusion and recombinant DNA are now beginning to be applied to these bacteria. The metabolic flow of substrates can now be quickly analyzed in situ by using NMR spectrometry. These new techniques will be of great help in accumulating and expanding new knowledge on glutamic acid bacteria and will accelerate the industrial development of this field.

REFERENCES

Abe, S., and Takayama, K. (1972) in *The Microbial Production of Amino Acids* (Yamada, K., et al., eds.), pp. 3–38, Kodansha, Tokyo; John Wiley and Sons, New York.

Abe, S., Takayama, K., and Kinoshita, S. (1967) *J. Gen. Appl. Microbiol. 13*, 279–301.

Aiba, S., Tsunekawa, H., and Imanaka, T. (1982) *Appl. Environ. Microbiol. 43*, 289–297.

Araki, K., and Nakayama, K. (1974) *Agr. Biol. Chem. 38*, 2219–2225.

Araki, K., Oka, T., and Nakayama, K. (1973) *Agr. Biol. Chem. 37*, 1357–1366.

Debabov, V. G., Zhdanova, N. I., Sokolov, A. K., Livshits, V. A., Kozlov, J. I., Khurges, E. M., Yankovsky, N. K., Gusyatiner, M. M., Sholin, A. F., Antipov, V. P., and Pozdnyakova, T. M. (1982) U.S. patent 4,321,325.

Demain, A. L., and Birnbaum, J. (1968) *Curr. Top. Microbiol. Immunol. 46*, 1–25.

Fazel, A. M., and Jensen, R. A. (1979) *J. Bacteriol. 138*, 805–815.

Hagino, H., and Nakayama, K. (1975) *Agr. Biol. Chem. 39*, 351–361.

Hasegawa, M., and Matsubara, K. (1978) *Agr. Biol. Chem. 42*, 383–391.

Imada, Y., Takahashi, T., Yamada, K., Uchida, K., and Aida, K. (1966) *Amino Acid Nucl. Acid 13*, 95–103.

Izumi, Y., Kono, Y., Inagaki, K., Kawase, N., Tani, Y., and Yamada, H. (1981) *Agr. Biol. Chem. 45*, 1983–1989.

Kaneko, H., and Sakaguchi, K. (1979) *Agr. Biol. Chem. 43*, 1007–1013.

Kase, H., and Nakayama, K. (1974) *Agr. Biol. Chem. 39*, 153–160.

Katsumata, R., and Furuya, A. (1982) *Jpn. Kokai Tokkyo Koho* JP 81 134,500.

Katsumata, R., and Furuya, A. (1983) Paper presented at the Annual Meeting of the Agricultural Chemical Society of Japan.

Katsumata, R., Takayama, K., and Furuya, A. (1980) *Jpn. Kokai Tokkyo Koho* JP 80 109,587.

Keddie, R. M., and Bousfield, I. J. (1980) in *Microbiological Classification and Identification* (Goodfellow, M., and Boad, R. G., eds.), pp. 167–188, Academic Press, New York.

Keddie, R. M., and Cure, G. M. (1978) in *Coryneform Bacteria* (Bousfield, I., and Calley, A. G., eds.), pp. 1–84, Academic Press, New York.

Kikuchi, M., and Nakao, Y. (1973) *Agr. Biol. Chem. 37*, 515–519.

Kinoshita, S. (1959) *Adv. Appl. Microbiol. 1*, 201–214.

Kinoshita, S., and Nakayama, K. (1978) *Economic Microbiology 2*, 209–261.

Kinoshita, S., and Tanaka, K. (1972) in *The Microbial Production of Amino Acids* (Yamada, K., et al., eds.), pp. 263–324, Kodansha, Tokyo; John Wiley and Sons, New York.

Kinoshita, S., Udaka, S., and Shimono, M. (1957a) *J. Gen. Appl. Microbiol. 3*, 193–205.

Kinoshita, S., Nakayama, K., and Udaka, S. (1975b) *J. Gen. App. Microbiol. 3*, 276–282.

Kinoshita, S., Nakayama, K., and Kitada, S. (1958) *J. Gen. Appl. Microbiol. 4*, 128–135.

Kinoshita, S., Itagaki, S., and Nakayama, K. (1960a) *Amino Acid 2*, 42–57.

Kinoshita, S., Samejima, H., Nara, T., and Fujita, C. (1960b) *Amino Acid 2*, 125–132.

Komatsu, Y. (1979) *J. Gen. Microbiol. 113*, 407–408.

Komatsu, Y., and Kaneko, T. (1980) *Rep. Ferment. Res. Inst. 55*, 1–5.

Komatsubara, S., Kisumi, M., and Chibata, I. (1980) *J. Gen. Microbiol. 119*, 51–61.

Komura, I., Komagata, K., and Mitsugi, K. (1973) *J. Gen. Appl. Microbiol. 19*, 161–170.

Komura, I., Yamada, K., Otsuka, S., and Komagata, K. (1975) *J. Gen. Appl. Microbiol. 21*, 251–261.

Minnikin, D. E., Goodfellow, M., and Collins, M. D. (1978) in *Coryneform Bacteria* (Bousfield, I., and Callely, A. G., eds.), pp. 85–160, Academic Press, New York.

Miwa, R., et al. (1981) Paper presented at the Annual Meeting of the Agricultural Chemical Society of Japan.

Momose, H., Miyashiro, S., and Oba, M. (1976) *J. Gen. Appl. Microbiol. 22*, 119–129.

Mori, M., Ozaki, H., and Shiio, I. (1981) *Seikagaku 53*, 984.

Nakayama, K. (1982) in *Prescott & Dunn's Industrial Microbiology* (Reed, G., ed.), pp. 748–801, Avi Publishing, Westport, Conn.

Nakayama, K., Sato, Z., and Kinoshita, S. (1960) *Nippon Nogei Kagaku Kaishi 34*, 934–942.

Nakayama, K., Kitada, S., and Kinoshita, S. (1961) *J. Gen. Appl. Microbiol. 7*, 145–154.

Nakayama, K., Tanaka, H., Hagino, H., and Kinoshita, S. (1966) *Agr. Biol. Chem. 30*, 611–616.

Nunheimer, T. D., Birnbaum, J., Ihnen, E. D., and Demain, A. (1970) *Appl. Microbiol. 20*, 215–217.

Ochiai, K., and Takayama, K. (1982) Paper presented at the Annual Meeting of the Agricultural Chemical Society of Japan.

Oishi, K. (1967) *Nippon Nogei Kagaku Kaishi 41*, R35–R44.

Oishi, K., and Aida, K. (1965) *Agr. Biol. Chem. 29*, 83–89.

Oishi, K., Otawa, M., Aida, K., and Uemura, T. (1970) *J. Gen. Appl. Microbiol. 16*, 259–277.

Oki, S., Nishimura, Y., Sayama, Y., Kitai, A., and Ozaki, A. (1969) *Amino Acid Nucl. Acid 19*, 73–87.

Okumura, S., Tokawa, R., Tsunoda, T., and Motozaki, S. (1962) *Nippon Nogei Kagaku Kaishi 36*, 204–211.

Oshima, K., Tanaka, K., and Kinoshita, S. (1965) *Amino Acid Nucl. Acid 11*, 112–118.

Ozaki, H., and Shiio, I. (1982) *J. Biochem. 91*, 1163–1171.

Sakaguchi, K., and Tanaka, A. (1979) *Jpn. Kokai Tokkyo Koho* JP 78 41,392.

Shapiro, J. A. (1976) *Appl. Environ. Microbiol. 32*, 179–182.

Shiio, I., and Ozaki, H. (1968) *J. Biochem. 64*, 45–53.

Shiio, I., and Sugimoto, S. (1981) *Agr. Biol. Chem. 45*, 2197–2207.

Shiio, I., and Ujigawa, K. (1978) *J. Biochem. 84*, 647–657.

Shiio, I., and Ujigawa-Takeda, K. (1980) *Agr. Biol. Chem. 44*, 1897–1904.

Shiio, I., Otsuka, S., and Takahashi, M. (1961) *J. Biochem. 49*, 398–403.

Shiio, I., Momose, H., and Oyama, A. (1969) *J. Gen. Appl. Microbiol. 15*, 27–40.

Shiio, I., Kawaoka, A., and Tsuchiya, T. (1970) *Nippon Nogei Kagaku Kaishi 44*, 245–251.

Shiio, I., Ozaki, H., and Mori, M. (1982) *Agr. Biol. Chem. 46*, 493–500.

Shvinka, Yu. É., Viestur, U. É., and Toma, M. K. (1979) *Microbiologiya 48*, 10–16.

Somerson, N. L., and Phillips, T. (1962) U.S. patent 3,080,297.

Sugimoto, S., and Shiio, I. (1981) *Seikagaku 53*, 984.

Sugiyama, Y., Kitano, K., and Kanzaki, T. (1973) *Agr. Biol. Chem. 37*, 1837–1847.

Sung, K., et al. (1982) Paper presented at the Annual Meeting of the Agricultural Chemical Society of Japan.

Suzuki, K., Kaneko, T., and Komagata, K. (1981) *Int. J. Syst. Bacteriol. 31*, 131–138.

Suzuki, T., Tanaka, K., Matsubara, I., and Kinoshita, S. (1969) *Agr. Biol. Chem. 33*, 1619–1627.

Takinami, K., Yoshii, H., Yamada, Y., Okada, H., and Kinoshita, K. (1968) *Amino Acid Nucl. Acid 18*, 120–160.

Takinami, K., Yamada, Y., and Shiro, T. (1969) *Proc. of Jap. Conf. on the Biochem. Lipids 11*, 49–53.

Tanaka, K., Iwasaki, A., and Kinoshita, S. (1960) *Nippon Nogei Kagaku Kaishi 34*, 593–600.

Tomita, F., Suzuki, T., and Furuya, A. (1975) *Amino Acid Nucl. Acid 32*, 1–6.

Tosaka, O., Takinami, K., and Hirose, Y. (1978a) *Agr. Biol. Chem. 42*, 745–752.

Tosaka, O., Hirakawa, H., Takinami, K., and Hirose, Y. (1978b) *Agr. Biol. Chem. 42*, 1501–1506.

Tosaka, O., Morioka, H., and Takinami, K. (1979) *Agr. Biol. Chem. 43*, 1513–1519.

Tsuchida, T., and Momose, H. (1975) *Agr. Biol. Chem. 39*, 2193–2198.

Tsunoda, T., and Okumura, S. (1970) in *Sekiyu Hakko* (Yamada, K., ed.), pp. 123–191, Saiwai Shobo, Tokyo.

Tsunoda, T., Shiio, I., and Mitsugi, K. (1960) *J. Gen. Appl. Microbiol. 7*, 18–40.

Uchida, K., and Aida, K. (1979) *J. Gen. Appl. Microbiol. 25*, 169–182.

Udaka, S., and Kinoshita, S. (1958) *J. Gen. Appl. Microbiol. 4*, 272–282.

Umbarger, H. E. (1975) in *Biochemistry Series One*, Vol. VII (Arnstein, H. R.V., ed.), pp. 1–56, Butterworths, London.

Vandecasteele, J. P., Lemal, J., and Coudert, M. (1975) *J. Gen. Microbiol. 90*, 178–180.

Walker, T. E., Han, C. H., Kollman, V. H., London, R. E., and Matwiyoff, N. A. (1982) *J. Biol. Chem. 257*, 1189–1195.

Yamada, H., and Kumagai, H. (1978) *Pure Appl. Chem. 50*, 1117–1127.

Yamada, K., and Komagata, K. (1972) *J. Gen. Appl. Microbiol. 18*, 417–431.

Yoshida, H., Araki, K., and Nakayama, K. (1980) *Agr. Biol. Chem. 44*, 361–365.

Biology of the Lactic, Acetic, and Propionic Acid Bacteria

Albert G. Moat

I. INTRODUCTION

The acid-forming bacteria, that is, those bacteria that produce lactic, acetic, or propionic acids as major end products of metabolism, constitute a wide assortment of gram-positive cocci (Streptococcaceae), gram-positive bacilli (Lactobacillaceae), gram-negative aerobic rods (*Acetobacter* and *Gluconobacter*), and one genus (*Propionibacterium*) from the coryneform group. They are considered together for reasons of their metabolic similarities or their industrial potential rather than any phylogenetic relationship. The importance of some of these organisms in other regards, for example, the role of many streptococci in disease states, is generally considered in the entirely different context of medical microbiology despite the obvious industrial applications in the development of chemotherapeutic agents, vaccines, and diagnostic aids. To keep the ensuing discussion within bounds, an arbitrary limit has been established to include primarily the genera *Streptococcus*, *Leuconostoc*, *Pediococcus*, *Lactobacillus*, *Acetobacter*, and *Propionibacterium*.

II. CLASSIFICATION AND ECOLOGY

The genera *Streptococcus*, *Leuconostoc*, *Pediococcus*, *Aerococcus*, and *Gemella* are gram-positive cocci that belong to the family Streptococcaceae. The genus *Streptococcus* includes 22 named species and six Lancefield groups that have not been given species names. The streptococci are found in a wide variety of habitats. *Streptococcus pyogenes* and *Streptococcus pneumoniae* are found in or on the human body and are the cause of serious infectious diseases of man. *Streptococcus faecalis* and *Streptococcus faecium* occur in the feces of humans and animals but may be found in many food products unrelated to direct fecal contamination. A number of species, for example, *Streptococcus equi*, *Streptococcus dysgalactiae*, *Streptococcus agalactiae*, *Streptococcus bovis*, *Streptococcus equinus*, and *Streptococcus uberis*, are found in various animals. Some of these play a major role in rumen fermentation. *Streptococcus thermophilus*, *Streptococcus lactis* and its subspecies *diactylactis*, and *Streptococcus cremoris* are found in milk and milk products (Deibel and Seeley, 1974). Lancefield's development of serological grouping brought order out of chaos in the identification of the streptococci. The increased frequency of isolation of group B and serological groups other than group A from human infections makes Lancefield grouping of streptococci more important than ever. However, the original precipitation technique is time-consuming and difficult to perform satisfactorily. Several serogrouping kits (Phadebact, SeroSTAT, and Streptex) have become available and are reported to be simple to perform and quite reliable (Burdash et al., 1981). An extensive numerical taxonomic study of strains of *Streptococcus* by Bridge and Sneath (1983) defined major areas within the genus that largely correspond to the traditional sections described by Deibel and Seeley (1974). The distinctness of the enterococcal species revealed in their study provides support for the concept of a separate genus for the enterococci. Their results also suggest that it may be useful to distinguish between a paraviridans species group that includes *S. bovis*, *S. equinus*, *Streptococcus salivarius*, and *Streptococcus mutans* and a viridans species group that includes *Streptococcus mitis*, *Streptococcus sanguis*, and a proposed new species, *Streptococcus oralis*. Bridge and Sneath (1983) also propose that a parapyogenic species group (*Streptococcus uberis*, Lancefield groups R, S, and T) be distinguished from the classical pyogenic streptococci (*S. agalactiae*, *S. pyogenes*, *S. equi*, and Group B strains of human origin). The extremely close similarity of *S. cremoris*, *S. lactis*, and its subspecies indicates that these probably all belong in one species.

Leuconostoc contains six recognized species: *Leuconostoc mesenteroides*, *Leuconostoc dextranicum*, *Leuconostoc paramesenteroides*, *Leuconostoc lactis*, *Leuconostoc cremoris*, and *Leuconostoc oenos* (Garvie, 1974). Garvie (1981) has shown by RNA/DNA hybridization studies that all but *L. oenos* form a natural group. This separation corresponds with their habitat, *L. oenos* being found in wine and having the ability to grow at a pH of 4.8–4.2 or lower.

The other genera are less tolerant of a low pH and are commonly present in fermenting vegetables and in milk and dairy products.

   Pediococcus contains five species: *Pediococcus cerevisiae, Pediococcus acidilactici, Pediococcus pentosaceus, Pediococcus halophilus,* and *Pediococcus urinae-equi. P. acidilactici* and *P. pentosaceus* are widely distributed in sauerkraut and fermenting materials such as pickles, silage, and cereal mashes. *P. cerevisiae* has been isolated from spoiled beer and brewer's yeast. *P. urinae-equi,* originally isolated from horse urine, has also been isolated from brewer's yeast (Kitahara, 1974). Back (1978) suggests the elevation of *P. cerevisiae* subsp. *dextranicus* to species status, *P. dextranicus.* DNA–DNA homology studies indicate that *Pediococcus* is a well-circumscribed genus (Dellaglio et al., 1981). *Aerococcus* contains a single species, *Aerococcus viridans.* However, *Gaffkya homari* apparently belongs in this species or in *Pediococcus. A. viridans* has been isolated from human infections and may be pathogenic for lobsters, since it is indistinguishable from *G. homari,* a known pathogen for these arthropods (Evans, 1974). *Gemella haemolysans,* a single genus that does not fit into any of the other prescribed genera of the Streptococcaceae, occurs in the respiratory tract of man (Reyn, 1974).

   The Micrococcaceae, which include *Micrococcus, Staphylococcus,* and *Planococcus,* are a rather diverse group of bacteria (Baird-Parker, 1974). As will be discussed below, there is reason to suspect that the Micrococcaceae do not represent a cohesive family group, and their classification status should be reexamined. The Peptococcaceae (which include *Peptococcus, Peptostreptococcus, Ruminococcus,* and *Sarcina*) are all anaerobic and represent a reasonably valid classification group. E. M. Barnes et al. (1977) suggest that some lactic acid–producing Peptococcaceae should be included in the genus *Streptococcus.*

   The family Lactobacillaceae contains gram-positive, asporogenic, rod-shaped bacteria in a single genus, *Lactobacillus.* Twenty-seven species are recognized (Rogosa, 1974). Although classified separately, primarily on the basis of their rod shape, the lactobacilli are more closely related to the streptococci than are some of the Micrococcaceae. Lactobacilli are found in fermenting animal and plant products wherever carbohydrates are available. They may also be present in the mouth, vagina, and intestinal tracts of various warm-blooded animals, including humans. *Lactobacillus lactis, Lactobacillus bulgaricus, Lactobacillus helveticus,* and possibly others are commonly present in mixed starter cultures used in the manufacture of cheese. Serological grouping has been useful in the classification of the lactobacilli. Seven serological groups, A to G, have been identified. The group G antigen is a negatively charged wall polysaccharide containing glucose, galactose, and lesser amounts of rhamnose, N-acetylglucosamine, and phosphate. Rhamnose is the immunodominant component. The phosphorylated polysaccharide nature of the group G antigen distinguishes it from the polysaccharides of groups B and C. Group A and C antigens are glycerol teichoic acids, that of group D is a ribitol teichoic acid, and that of group F a lipoteichoic acid (Knox et al., 1980).

Acetobacter is included as "Incertae sedis" under part 7 of the eighth edition of *Bergey's Manual of Derminative Bacteriology*, gram-negative, aerobic rods and cocci. The genus has been divided into three major groups: *Acetobacter aceti* (subspecies *aceti*, *orleanensis*, *xylinum*, and *liquefaciens*); *Acetobacter pasteurianus* (subspecies *pasteurianus*, *lovanensis*, *estunensis*, *ascendens*, and *parodoxus*); and *Acetobacter peroxidans* with no subspecies designations. *Gluconobacter oxydans* contains several subspecies: *oxydans*, *industrius*, *suboxydans*, and *melanogenes*. *Gluconobacter*, if motile, have polar flagella, whereas *Acetobacter* have peritrichous flagella. *Gluconobacter* do not oxidize acetic or lactic acid to CO_2, whereas *Acetobacter* are "overoxidizers" that oxidize these acids to CO_2 and H_2O via the Krebs (TCA) cycle. *Acetobacter* are often cited as producing cellulose, but *A. aceti* subsp. *xylinum* and *A. pasteurianus* subsp. *estunensis* are the only cellulose producers. *Gluconobacter oxydans* subsp. *suboxydans* is used in the manufacture of vinegar. *Acetobacter* are frequently found as contaminants in wine, beer, and other alcoholic beverages and on fruits and vegetables (DeLey and Frateur, 1974).

The family Propionibacteriaceae, genera *Propionibacterium* and *Eubacterium*, are included in part 17 of the eighth edition of *Bergey's Manual of Determinative Bacteriology* under Actinomycetes and related organisms, along with *Corynebacterium* and other corynebacteria (Moore and Holdeman, 1974). They are not, however, classified as *Actinomycetales*. Eight species of *Propionibacterium* are currently recognized: *Propionibacterium freudenreichii* (subspecies: *freudenreichii*, *globosum*, and *shermanii*), *Propionibacterium theonii*, *Propionibacterium acidi-propionici*, *Propionibacterium jensenii*, *Propionibacterium avidum*, *Propionibacterium acnes*, *Propionibacterium lymphophilum*, and *Propionibacterium granulosum*. *P. freudenreichii* subsp. *shermanii* is used in the manufacture of Swiss cheese and is responsible for the characteristic flavor and the prominent holes. Other species may be isolated from dairy products. *P. acnes* has been isolated from the normal skin as well as from soft tissue abscesses and other infections. It is found frequently as a contaminant in clinical specimens. The genus *Eubacterium* contains 28 recognized species (Holdeman and Moore, 1974). The majority of strains originally designated as *Corynebacterium parvum* are indistinguishable biochemically, serologically, and by DNA homology from *P. acnes* (Cummins and Johnson, 1974), and their transfer to the genus *Propionibacterium* seems well justified. The nutritional requirements of *P. acnes*, *P. avidum*, and *P. granulosum* (Ferguson and Cummins, 1978) are similar to those of other propionibacteria (Delwiche, 1949), a finding that provides support for their classification in this genus. A numerical taxonomic study of members of the Actinomycetaceae and related genera showed that reference cultures of *Propionibacterium* clustered together according to present species designations and were unrelated to any of the actinomycetes tested (Schofield and Schaal, 1981).

Although it would appear that most of the well-recognized groups of the acid-forming bacteria have been thoroughly categorized, many changes in their classification may be forthcoming. Recent developments in molecular sequencing techniques have provided a means of direct measurement of genealogical relationships. Comparative analysis of the nucleotide sequences of ribosomal RNA (rRNA) has established that rRNA's, particularly 16S rRNA, serve as phylogenetic markers for prokaryotic organisms. Since a typical bacterial cell contains from 10,000 to 20,000 ribosomes, sufficient rRNA is available to make direct nucleotide sequence analysis feasible. Comparison of the nucleotide sequences of the 16S rRNA of a variety of bacteria has shown that certain criteria currently used in bacterial classification distinguish valid phylogenetic relationships (e.g., the gram reaction), whereas others (e.g., morphology and mode of cell division) do not. Such studies delineated the *Bacillus–Lactobacillus–Streptococcus* (B–L–S) cluster containing gram-positive bacteria with a low DNA G + C content. Some of the evidence suggests that the family Micrococcaceae should be dissected (Fox et al., 1980). The anaerobic species *Peptococcus saccharolyticus* belongs to the B–L–S cluster, whereas some species of *Micrococcus* probably do not. The micrococci appear to be more closely related to the coryneform bacteria, whereas *Planococcus* and *Staphylococcus* share some relationship with the B–L–S cluster. Some members of *Peptostreptococcus* probably belong to the genus *Streptococcus* (Love et al., 1979). The lactobacilli are much more closely related phylogenetically to the streptococci and other gram-positive cocci than the current classification schemes indicate. *Eubacterium*, currently placed in the Propionibacteriaceae, are more closely related to the clostridia and hence to the B–L–S cluster. Transferability of antibiotic resistance traits between group A, B, and D streptococci, *S. pneumoniae*, and *Staphylococcus aureus* suggests a close relationship between these organisms (Clewell, 1981). Cataloging of the nucleotide sequences of the 16S rRNA of a larger number of strains of bacteria is likely to result in sweeping changes in our current classification system (Woese, 1981).

Nucleotide sequence data relate the mycoplasmas to the gram-positive bacteria. Lack of production of a cell wall apparently does not hold phylogenetic validity (Fox et al., 1980). Acholeplasmas possess FDP-activated lactate dehydrogenases that are strikingly similar to the unique lactate dehydrogenases found in streptococci. Comparative enzyme protein studies using specific antisera against purified *S. faecalis* enzymes related seven species of *Acholeplasma* phylogenetically to streptococci (Neimark and London, 1982). Solution DNA hybridization between 18 species of *Mycoplasma* and three species of *Acholeplasma* revealed that the mycoplasmas are quite heterogeneous. Several species within each genus showed 13–15% homology, but there was no detectable homology between species from the two genera. *Myco-*

plasma pneumoniae and *Mycoplasma neurolyticum* appeared to be unrelated to any of the other mycoplasma species or to each other. These findings suggest that the evolution of each wall-less species from gram-positive bacteria may have been an independent event (Sugino et al., 1980).

III. CELL STRUCTURE

A. Cell Envelope

Gram-positive and gram-negative bacteria have markedly different cell envelopes (Fig. 6.1). Nevertheless, both types perform the same essential function of maintaining the vitality of the cell. The cytoplasmic membrane is responsible for retention of the intracellular contents and controlling entrance and exit of metabolites. The cytoplasmic membranes of gram-positive and gram-negative cells contain phospholipids, glycolipids, and phosphatidyl glycolipids and appear quite similar under the electron microscope. Embedded in the lipid bilayer are various proteins and lipoproteins that function in metabolite transport (permeases) and the synthesis of macromolecular wall components. Certain proteins serve to maintain the integrity of the membrane–wall interface, while others provide protection against phagocytosis or act as attachment sites for bacteriophage or chemical agents such as antibiotics. In gram-positive cells, lipoteichoic acids are anchored in the cytoplasmic membrane.

The peptidoglycan of gram-positive cells is a thick, multilayered structure capable of maintaining the shape of the cell and providing physical protection for the cytoplasmic membrane. Teichoic acids (TA's) are present within the peptidoglycan matrix of the gram-positive wall and extend to the cell surface. Both wall and membrane TA's are important antigenic determinants (Ward, 1981). Peptidoglycan and the TA's and the acidic polysaccharides covalently attached to it should probably be considered as components of the same macromolecule and not as separate polymers.

In gram-negative bacteria the peptidoglycan is a monolayer. Between the cytoplasmic membrane and the peptidoglycan monolayer is a periplasmic space that extends beyond the monolayer to the outer membrane. It seems unlikely that the peptidoglycan monolayer, by itself, can provide shape and protective functions. The outer membrane, covalently attached to the underlying peptidoglycan layer, apparently reinforces the cell envelope and provides an additional permeability barrier. Some components of the outer membrane are qualitatively similar to those of the cytoplasmic membrane, but the ratio of phosphatidylglycerol to phosphatidylethanolamine and of cardiolipin to phosphatidylethanolamine is much lower in the outer membrane. The outer membrane also has a lower ratio of total phospholipid to protein. In addition, the outer membrane contains lipopolysaccharides (LPS) with the general structure:

(oligosaccharide)–(core polysaccharide)–(Lipid A)
Region I   Region II   Region III

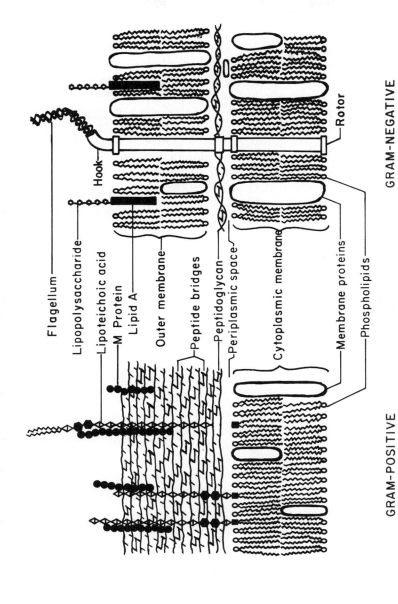

Flagellum

Lipopolysaccharide

Lipoteichoic acid

M Protein

Lipid A

Outer membrane

Peptide bridges

Peptidoglycan

Periplasmic space

Cytoplasmic membrane

Membrane proteins

Phospholipids

Hook

Rotor

GRAM-POSITIVE

GRAM-NEGATIVE

FIGURE 6.1 Comparison of the cell surfaces of gram-positive and gram-negative bacteria.

LPS is anchored in the outer membrane by lipid A, the endotoxic component. The terminal polysaccharide components of Region I are important antigenic markers, the O-antigens. They occur in a wide variety of combinations accounting for the antigenic diversity of gram-negative organisms. Region II is a "core" polysaccharide containing the group antigens of closely related organisms such as the salmonellae. Also present in the outer membrane are proteins called porins, which serve as a molecular sieve to regulate, to some extent, the access of certain metabolites to the cytoplasmic membrane. Although the outer and inner membranes appear to be similar, the fluid mosaic model of a true membrane does not seem to hold for the outer membrane. The term membrane is used primarily because of its lipid content and its appearance under the electron microscope.

Peptidoglycans are composed of linear strands of alternating β-1,4-linked N-acetylglucosamine (NaG) and N-acetylmuramic acid (NaM). The glycan linkages are considered to be uniform in all bacteria. Every D-lactyl group of the NaM is peptide-substituted. All glycans have short tetrapeptide units terminating with D-alanine or occasionally tripeptide units lacking the terminal D-alanine. The L-alanine at the N-terminus can be replaced by L-serine or glycine.

Ghuysen (1977) has classified peptidoglycans into four main types on the basis of their interpeptide bridges (Fig. 6.2).

Type I. Direct D-alanyl-R_3 peptide bond. This type of bridge is found in *Escherichia coli*, in most other gram-negative bacteria, and in many bacilli.

Type II. A single additional amino acid residue (glycine or another L- or D-amino acid). This type of interpeptide bridge, commonly found in the organisms under discussion here, varies from organism to organism as shown in the following examples:

Amino acid(s)	Organism	
-[Gly]$_5$-	*Staphylococcus aureus*	(1)
-[L-Ala]$_3$-L-Thr-	*Micrococcus roseus*	(2)
-[Gly]$_3$-[L-Ser]$_2$-	*Staphylococcus epidermidis*	(3)
-L-Ser-L-Ala-	*Lactobacillus viridescens*	(4)
-L-Ala-L-Ala-	*Streptococcus pyogenes*	(5)
-L-Ala-	*Arthrobacter crystallopoietes*	(6)
-D-Asp-NH$_2$	*Streptococcus faecalis*	(7)
	Lactobacillus casei	

Type III. A bridge composed of one to several peptides, each having the same amino acid sequence as the peptide unit attached to muramic acid. This occurs in *Micrococcus luteus (Micrococcus lysodeikticus).*

Type IV. A bridge extending between two carboxyl groups belonging to D-alanine and D-glutamic acid and a diamino acid residue or a diamino acid–containing short peptide. This is found in *Butyribacterium rettgeri.*

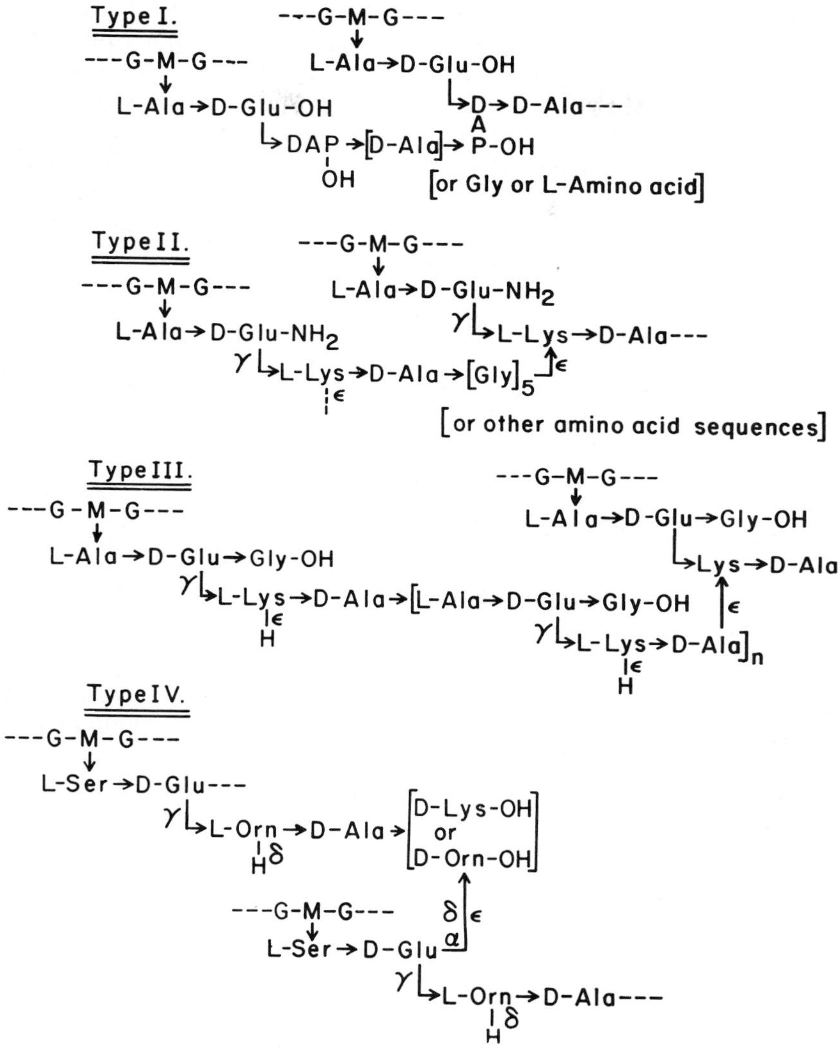

<u>Type I.</u>   ---G-M-G---
                                    ↓
---G-M-G---   L-Ala→D-Glu-OH
         ↓   ↳D→D-Ala---
    L-Ala→D-Glu-OH   A
         ↳DAP→[D-Ala]→ P-OH
              |
             OH   [or Gly or L-Amino acid]

<u>Type II.</u>   ---G-M-G---
                                    ↓
---G-M-G---   L-Ala→D-Glu-NH₂
         ↓   γ↳L-Lys→D-Ala---
    L-Ala→D-Glu-NH₂
         γ↳L-Lys→D-Ala→[Gly]₅ ↑ε
              |ε
              |
              [or other amino acid sequences]

<u>Type III.</u>   ---G-M-G---
                                            ↓
---G-M-G---   L-Ala→D-Glu→Gly-OH
       ↓   ↳Lys→D-Ala---
   L-Ala→D-Glu→Gly-OH
       γ↳L-Lys→D-Ala→[L-Ala→D-Glu→Gly-OH ↑ε
            |ε   γ↳L-Lys→D-Ala]ₙ
            H |ε
                                        H

<u>Type IV.</u>
---G-M-G---
       ↓
   L-Ser→D-Glu---
       γ↳L-Orn→D-Ala→[D-Lys-OH
            |δ   or
            H   D-Orn-OH]
                              ↑
       ---G-M-G---   δ|ε
           ↓   α
       L-Ser→D-Glu
                 γ↳L-Orn→D-Ala---
                      |δ
                      H

FIGURE 6.2 Major types of interpeptide bridges in bacterial peptidoglycans. G, N-acetylglucosamine; M, N-acetylmuramic acid; DAP, diaminopimelic acid. Other amino acid sequences replacing [Gly]₅ (pentaglycine) in Type II bridges are shown in tabular form in the text.

The proportion of peptide cross-linking varies from species to species but is usually rather low. The most highly cross-linked peptides are found in *S. aureus* (> 90%), whereas those of *E. coli* and most other gram-negative bacteria show approximately 50% cross-linking. In either case a continuous baglike sacculus completely surrounds the cell. Ghuysen (1977) states that all of the peptide chains are located on the same side of the glycan, whereas the 0–6 positions of the NaM residues are exposed on the other side of the structure and are readily available for substitution with acetyl or phosphodiester groups. However, Mendelson (1982) cites evidence suggesting that the glycans are helically twisted chains from which the peptide bridges radiate in all directions from the axis of the backbone. This latter model would permit interpeptide bridging between many different neighboring chains, forming a supramolecular network (mosaic) of peptidoglycan such as that observed in gram-positive organisms.

All bacteria produce autolysins, enzymes that hydrolyze bonds in the peptidoglycan structure. These show three types of activity:

1. Glycan strand hydrolysis
   a. Endo-N-acetylmuramidases
   b. Endo-N-acetylglucosaminidases
2. Endopeptidases hydrolyzing
   a. Peptide bonds in the interior of the peptide bridges
   b. Bonds involving the C-terminal D-alanine residue
3. N-Acetylmuramyl-L-alanine amidase acting at the junction between the glycan strands and the peptide units

Cell lysis may occur from within, by autolysins, or from without, by addition of lytic enzymes to cell suspensions. Autolysins may be found in the cytoplasm, associated with the membrane, in the periplasmic region, fixed on the wall, or in the culture medium as excretion products. Certain autolysins are localized in the region of the growing septum. The number of detectable autolysins varies from species to species. In *S. faecalis* and in *Lactobacillus acidophilus* the only autolytic activity detectable is an endo-N-acetylmuramidase, but the localization of the enzyme differs in the two organisms. In *S. pneumoniae* and other gram-positive organisms the major, if not the only, autolysin appears to be an N-acetylmuramyl:L-alanine amidase. By comparison, *E. coli* has at least six different hydrolases for peptidoglycan linkages. Although most gram-negative organisms are considered to contain autolysins, relatively few specific studies have been conducted with organisms other than the Enterobacteriaceae. Autolysins play an important role in septum and wall extension during cell growth as well as in cell separation, wall turnover, sporulation, competency for transformation, and excretion of toxins and exoenzymes.

Cell membranes of gram-positive bacteria can be isolated by using any of the enzymes that selectively degrade the peptidoglycan and digest away the wall structure. If cells are treated with lysins while suspended in a solute to which the cell is impermeable (e.g., sucrose) at a concentration that approximately balances the high osmotic pressure of the cell, then an osmotically fragile body (protoplast) can be formed. Osmotic shock will then cause lysis, and the membrane can be isolated. Mesosomes, membrane invaginations into the cytoplasm of the cell, are more difficult to isolate, but they have been isolated from a number of bacteria. These internal structures are often observed in physical association with developing cross-walls and with the bacterial nucleoid. In *S. aureus* the general composition of mesosomes is essentially the same as that of the plasma membrane, but the fatty acid content of mesosomes is 48% greater than that of the plasma membrane (Theodore and Panos, 1973). The cytoplasmic membrane contains numerous enzymes that perform various functions. Very few functions involving enzyme activities are localized in the outer membrane of gram-negative bacteria. Terminal electron transport enzymes and metabolite transport functions are found in the cytoplasmic membrane but are entirely lacking in the outer membrane. However, porins, proteins that facilitate transfer of metabolites from the exterior to the surface of the inner membrane, are prominent features of the outer membrane structure. A number of proteins associated with the cell surface provide a protective advantage to the cell. The M protein of *S. pyogenes* enables the organism to resist ingestion and killing by phagocytic cells. Type-specific antibodies against the M protein neutralize the antiphagocytic effect and enhance elimination of the invading organisms. M proteins contain repeating covalent structures with antigenic determinants that can be used in synthetic form to invoke protective immunity in experimental animals (Beachey et al., 1981).

Teichoic acids (TA's) are polymers of either ribitol phosphate or glycerol phosphate in which the repeating units are joined together through phospho-diester linkages (Fig. 6.3). This term has been extended to include all polymers containing glycerol phosphate or ribitol phosphate associated with the membrane, cell wall or capsule. TA's are apparently present in all gram-positive bacteria but are absent from gram-negative bacteria. They are effective antigens and serve as the group- or type-specific substances of many organisms.

Wall TA's are covalently linked to the peptidoglycan. Membrane-associated TA's are always glycerol phosphate polymers and are covalently linked to a glycolipid that is part of the cytoplasmic membrane. For this reason they are also termed lipoteichoic acids (LTA's). Although both wall and membrane TA's often occur in the same organism, they are apparently formed by independent mechanisms.

In *S. pneumoniae* the wall ribitol TA's that serve as the C-antigen are more complex. Choline is present along with glucose, ribitol, phosphorus, galactosamine, and 2,3,6-trideoxy-2,4-diaminohexose. The walls of *Lactobacillus plantarum* contain a mixture of one polymer of glucosylglycerol phosphate and two polymers of isomeric diglucosylglycerol phosphates (Fig. 6.3). In *L. acidophilus* the TA is a mixture of α- or β-1,6-linked polyglucose polymers with monomeric α-glycerol phosphate side chains attached on the C_2 or C_4 position. Other anionic polymers that lack polyol phosphate and may be present in cell walls are not, by definition, teichoic acids. Nevertheless, such components are important. For example, the group-specific polysaccharide of *S. pyogenes* group A contains phosphate, glycerol, rhamnose, and NaG in a molar ratio of 1:1:2:1 (Munoz et al., 1967). Rhamnose and NaG are incor-

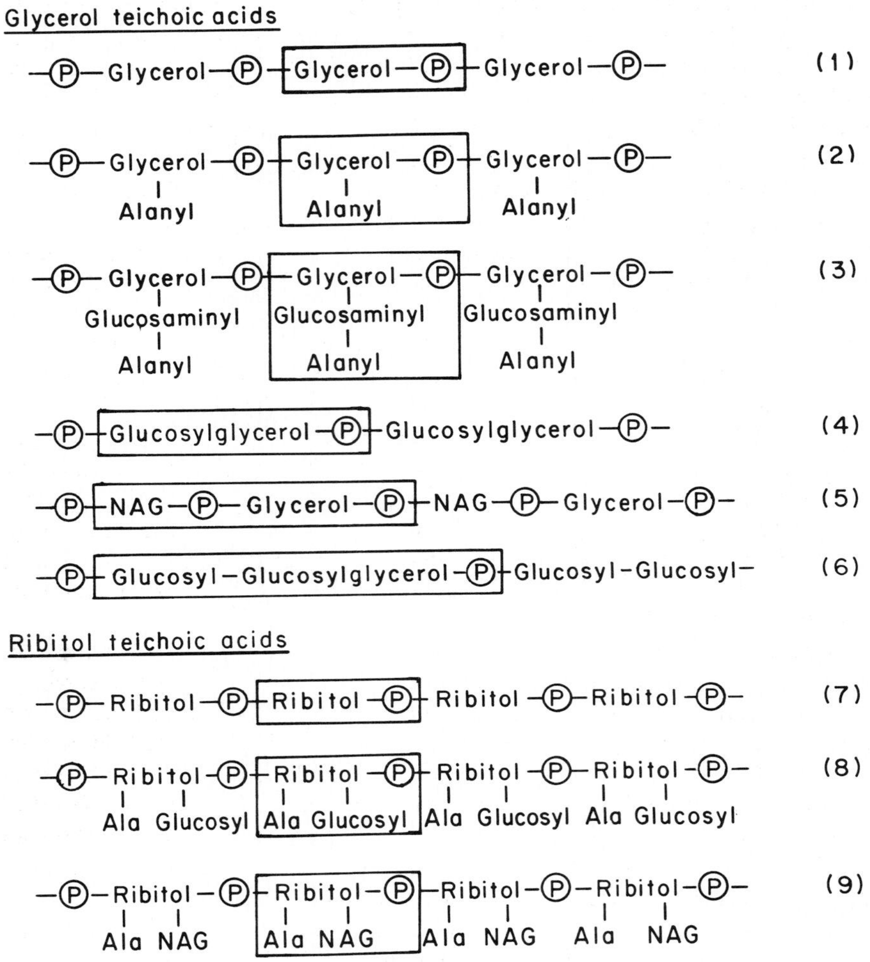

Glycerol teichoic acids

$$-\text{\textcircled{P}}-\text{Glycerol}-\text{\textcircled{P}}-\boxed{\text{Glycerol}-\text{\textcircled{P}}}-\text{Glycerol}-\text{\textcircled{P}}- \qquad (1)$$

$$-\text{\textcircled{P}}-\underset{\underset{\text{Alanyl}}{|}}{\text{Glycerol}}-\text{\textcircled{P}}-\boxed{\underset{\underset{\text{Alanyl}}{|}}{\text{Glycerol}}-\text{\textcircled{P}}}-\underset{\underset{\text{Alanyl}}{|}}{\text{Glycerol}}-\text{\textcircled{P}}- \qquad (2)$$

$$-\text{\textcircled{P}}-\underset{\underset{\underset{\text{Alanyl}}{|}}{\underset{\text{Glucosaminyl}}{|}}}{\text{Glycerol}}-\text{\textcircled{P}}-\boxed{\underset{\underset{\underset{\text{Alanyl}}{|}}{\underset{\text{Glucosaminyl}}{|}}}{\text{Glycerol}}-\text{\textcircled{P}}}-\underset{\underset{\underset{\text{Alanyl}}{|}}{\underset{\text{Glucosaminyl}}{|}}}{\text{Glycerol}}-\text{\textcircled{P}}- \qquad (3)$$

$$-\text{\textcircled{P}}-\boxed{\text{Glucosylglycerol}-\text{\textcircled{P}}}-\text{Glucosylglycerol}-\text{\textcircled{P}}- \qquad (4)$$

$$-\text{\textcircled{P}}-\boxed{\text{NAG}-\text{\textcircled{P}}-\text{Glycerol}-\text{\textcircled{P}}}-\text{NAG}-\text{\textcircled{P}}-\text{Glycerol}-\text{\textcircled{P}}- \qquad (5)$$

$$-\text{\textcircled{P}}-\boxed{\text{Glucosyl}-\text{Glucosylglycerol}-\text{\textcircled{P}}}-\text{Glucosyl}-\text{Glucosyl}- \qquad (6)$$

Ribitol teichoic acids

$$-\text{\textcircled{P}}-\text{Ribitol}-\text{\textcircled{P}}-\boxed{\text{Ribitol}-\text{\textcircled{P}}}-\text{Ribitol}-\text{\textcircled{P}}-\text{Ribitol}-\text{\textcircled{P}}- \qquad (7)$$

$$-\text{\textcircled{P}}-\underset{\underset{\text{Ala Glucosyl}}{|\quad|}}{\text{Ribitol}}-\text{\textcircled{P}}-\boxed{\underset{\underset{\text{Ala Glucosyl}}{|\quad|}}{\text{Ribitol}}-\text{\textcircled{P}}}-\underset{\underset{\text{Ala Glucosyl}}{|\quad|}}{\text{Ribitol}}-\text{\textcircled{P}}-\underset{\underset{\text{Ala Glucosyl}}{|\quad|}}{\text{Ribitol}}-\text{\textcircled{P}}- \qquad (8)$$

$$-\text{\textcircled{P}}-\underset{\underset{\text{Ala NAG}}{|\quad|}}{\text{Ribitol}}-\text{\textcircled{P}}-\boxed{\underset{\underset{\text{Ala NAG}}{|\quad|}}{\text{Ribitol}}-\text{\textcircled{P}}}-\underset{\underset{\text{Ala NAG}}{|\quad|}}{\text{Ribitol}}-\text{\textcircled{P}}-\underset{\underset{\text{Ala NAG}}{|\quad|}}{\text{Ribitol}}-\text{\textcircled{P}}- \qquad (9)$$

FIGURE 6.3 Basic structures of teichoic acids. In each structure the repeating unit is enclosed in a rectangle. Encircled P, interconnecting phosphate group; NAG, N-acetylglucosamine.

porated from thymidine 5′-diphosphorhamnose and UDP-NaG into the group A polysaccharide of *S. pyogenes*. Assembly of the group A polysaccharide occurs at the cell membrane with participation of a lipoid anchor or acceptor molecule (Reusch and Panos, 1977).

Membrane LTA's of the glycerolphosphate polymer type occur in many gram-positive bacteria. Their presence is not dependent upon growth conditions as is the case with wall-associated teichoic acids. The LTA's have a long polar glycerolphosphate chain linked to a small hydrophobic glycolipid. The LTA of *S. pyogenes* is an amphipathic molecule composed of a polymer of glycerophosphate linked to a glycerophosphoryldiglucosyl diglyceride (Slabj and Panos, 1976). In *S. faecalis* the glycolipid has been shown to be a phosphatidylkojibiosyl diglyceride. The structure of the LTA's of other streptococci and lactobacilli has been partially characterized. In *S. pneumoniae* the Forssman antigen is a LTA-like component. It is found in the cytoplasmic membrane and contains lipids and choline (Briles and Tomasz, 1975). LTA's are exposed at the cell surface in many organisms. In *L. plantarum*, specific antiserum to the glycerolphosphate sequence shows the label extending from the outer surface through the wall and even outside the boundary of the cell. Spontaneous release of TA and LTA has been described in streptococci (Joseph and Shockman, 1975) and lactobacilli (Markham et al., 1975). Release of LTA from *S. sanguis* and other streptococci is greatly stimulated during penicillin treatment. In contrast to spontaneous release, the LTA's are not replenished by synthesis during antibiotic treatment (Horne and Tomasz, 1979). In *S. pyogenes*, LTA binds via its polyanionic backbone to positively charged residues of surface M proteins (see Fig. 6.1). This orientation would leave the lipid moiety free to interact with fatty acid binding sites on host cell membranes (Ofek et al., 1982). *Lactobacillus fermenti*, but not *L. casei*, can be agglutinated by antisera to LTA. The long polar glycerolphosphate chains of LTA's probably extend through the network of the wall to evoke an immune response (Ward, 1981).

Several physiological roles for LTA have been proposed. These include: regulation of autolysin activity, scavenging of divalent cations such as Mg^{2+}, and interaction of bacteria with cells of involved hosts. When cells are grown under the conditions of phosphate limitation, no wall TA is formed. However, membrane LTA is still produced. TA's bind divalent cations. Hence they may serve in concentrating Mg^{2+} or other ions at the cell surface. In *Lactobacillus buchneri*, one Mg^{2+} ion is bound for every two phosphate groups in the wall TA. Release of TA and LTA may influence the interaction of bacterial pathogens with cells of the invaded host. In addition to their serological reactivity, TA's may activate the alternative pathway of complement and may play a role in the specific adhesion of bacteria to host epithelial surfaces (Winkelstein and Tomasz, 1978).

B. Nucleoid

With the aid of the electron microscope and the development of improved fixative techniques it has been possible to demonstrate a discrete nuclear region in bacterial cells. The absence of a nuclear membrane sets the DNA-containing region of bacterial cells apart from higher organisms and has led to a variety of terms to describe this area of the cell. In current usage the term "nucleoid," "bacterial nucleus," or "prokaryotic nucleus" is often applied to distinguish the nucleoid of bacteria from the "true" nucleus of eukaryotic cells. The bacterial nucleoid appears to be discrete even in the absence of a nuclear membrane. The DNA must be maintained in a more or less fixed spatial relationship with respect to the rest of the cytoplasm (Woldringh and Nanninga, 1976; Kleppe et al., 1979). Fixatives used in preparing cells for visualization under the electron microscope can alter the appearance of the nucleoid. When cells of *S. faecalis* are initially fixed in osmium tetroxide (OS), the nucleoid of exponential-phase cells is usually much more centralized than when they are fixed in glutaraldehyde (GA). In stationary-phase cells the nucleoid appears to be centralized regardless of whether OS or GA is used (Higgins and Daneo-Moore, 1974). The more dispersed configuration of the exponential-phase nucleoid can be preserved by fixation in GA but not in formalin or OS. Apparently, GA can rapidly cross-link the amino groups of macromolecules in cells (Daneo-Moore et al., 1980). The nucleoid of exponential-phase cells is concentrated in the center of cells but is much more dispersed by membrane attachment points or by coupled transcription–translation elements (e.g., ribonucleosomes) or both than is the nucleoid in stationary-phase cells (Fig. 6.4). When exponential-phase cultures enter the stationary phase of growth, a redistribution of cytoplasmic macromolecules takes place wherein nucleoid fibers become enriched in the center of the cytoplasm and other cytoplasmic components, such as ribosomes, accumulate at the periphery. Many of the procedures routinely used to study the morphology–physiology of cells (chilling, filtration, and fixation) result in reorganization of the cytoplasm, leading to an increase in the centralization of nuclear material (Edelstein et al., 1981).

C. Other Structures

Cores.   Cores are large, hollow, cylindrical structures that have been found almost exclusively in group D streptococci. These structures extend the width or length of a dividing cell and measure 0.1–0.16 μm in diameter, with a tube wall that is between 0.01 and 0.03 μm thick. These core structures appear in early stationary phase after a period of extensive mesosomal formation. Autolysis also results in core formation as does treatment with penicillin. Chemically, cores are composed of protein with little or no carbohydrate present. Cores are formed only after the pH of the culture medium drops below 6.5 and do not form in the presence of chloramphenicol. It has been postulated that cores represent a labile repository for cellular protein (Coleman and Bleiweis, 1977).

Pili.   Gram-negative bacteria have been shown to possess pili or fimbriae which are of two types: (1) common pili that serve a variety of functions such as bacteriophage absorption and (2) sex or F pili, which are produced by F^+ strains and are involved in the conjugation process. By comparison, very few gram-positive bacteria have been shown to have pili. Thin polar pili have been observed on *S. sanguis* (Henriksen and Henrichsen, 1975) and on some strains of *Streptococcus mitior* (Handley and Carter, 1979). Handley and Jacob

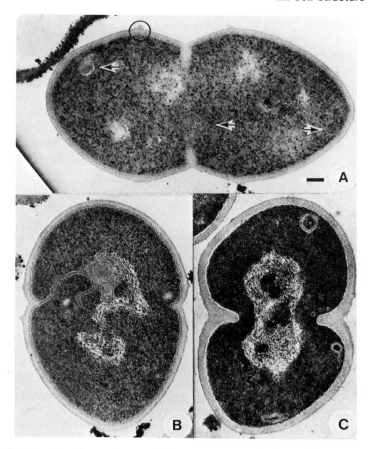

FIGURE 6.4 Comparison of the appearance of the stationary- and exponential-phase nucleoids of *Streptococcus faecalis*. (a) Longitudinal section of an exponential phase cell, showing external wall bands (circle), polar and septal mesosomes (arrows), and a dispersed nucleoid. Bar equals 100 nm. (b) Section of a cell after 60 min of threonine starvation (stationary-phase), showing thickened wall and nuclear and mesosomal pooling. Bar equals 100 nm. (c) Section of a cell starved of threonine for 20 h. The wall is greatly thickened and the nuclear pool greatly expanded; however, there is an apparent decreased area of internal membrane. Bar equals 100 nm. Reprinted with permission from Higgins and Shockman (1970).

(1981) compared the pili observed on a plasmid-free strain of *S. faecalis* and a strain carrying a self-transmissible plasmid. The number of piliated cells in each strain was dependent upon growth conditions. Plasmid transfer occurred between the two strains used in this study, but there was no direct association between the presence of pili and plasmid transfer.

D. Capsules

Extracellular polysaccharides or polypeptides are produced by a variety of bacteria. If these extracellular substances are of sufficiently high molecular weight and viscosity and are poorly soluble in water, they adhere to the cell and appear as a capsule or slime layer surrounding the cell. Capsules afford protection from the external environment and have been shown to aid bacteria in adhering to surfaces.

S. pneumoniae produces polysaccharide capsules. Type III pneumococci synthesize a polysaccharide capsule composed of glucopyranose (GP) and glucuronic acid (GUA) in alternating β-1,3- and β-1,4-linkages:

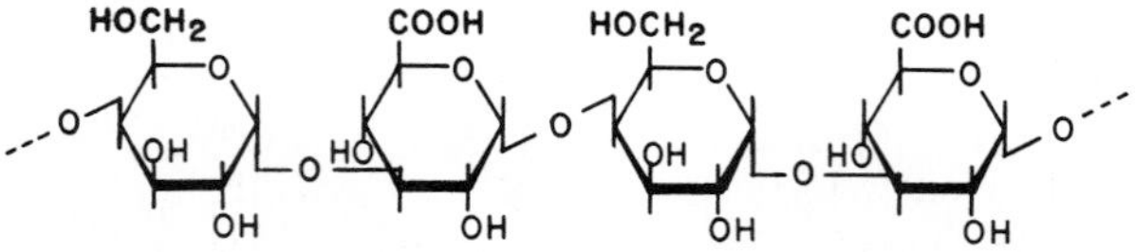

$$(1)$$

Pneumococcal polysaccharide confers antigenic specificity to each strain due to the occurrence of other sugars or substituted sugars such as galacturonic acid, rhamnose, uronic acid, or N-acetyl-amino sugars in varying proportions. Capsules contribute to the virulence of the pneumococcus by interfering with phagocytosis; unencapsulated strains lack virulence. Specific immunity to the capsular substance enhances phagocytosis and confers protective immunity. Vaccination against the more prevalent strains is now recommended as a preventive measure against pneumococcal pneumonia.

Group B streptococci also produce capsular polysaccharides, which can be used to identify four antigenic types (Mackie et al., 1979). The soluble group B streptococcal type III polysaccharide can be prepared directly from supernatant culture medium by using column chromatography. This polysaccharide is immunogenic in humans and gives rise to IgG antibody. Thus it may be possible to evoke an immune response in the mother against group B polysaccharide, which can cross the placental barrier to provide protection of the fetus from infection by group B streptococci (Eisenstein et al., 1982, 1983).

The hyaluronic acid capsule of group A streptococci has been shown to provide a protective shield from the destructive effects of atmospheric oxygen by virtue of its ability to aid in the formation of cell aggregates. Disruption of the aggregates with hyaluronidase results in an increased oxygen uptake and the production of toxic levels of hydrogen peroxide. Unencapsulated variants are sensitive to oxygen under similar conditions (Cleary and Larkin, 1979).

Many streptococci, lactobacilli, and *Leuconostoc* species produce dextrans or levans in response to growth on sucrose. Dextrans are α-1,6-linked glucose polymers with the following structure:

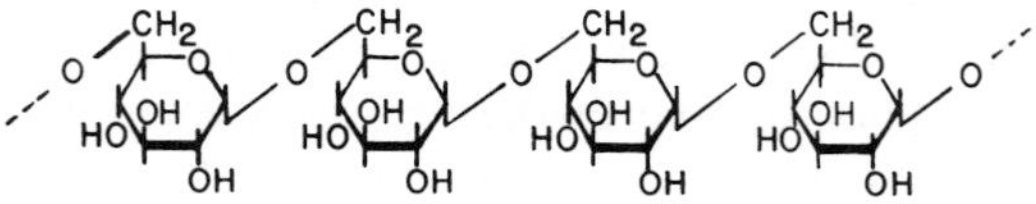

$$(2)$$

Mutans are branched glucose polymers with linear α-1,6-linked glucose units and branched α-1,3-linked glucose units:

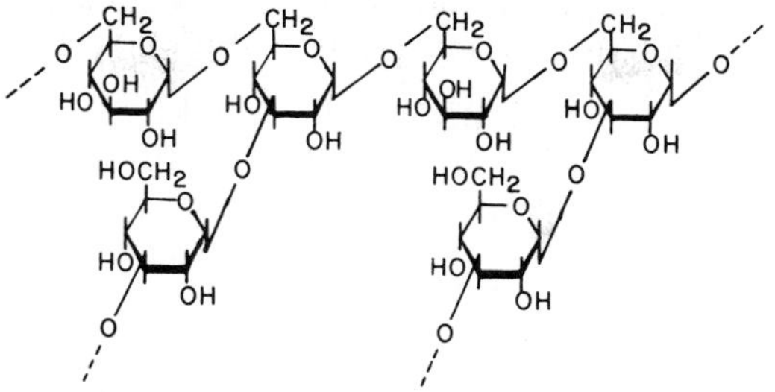

(3)

Fructans or levans are polymers of fructose with 2,6-linked fructose units:

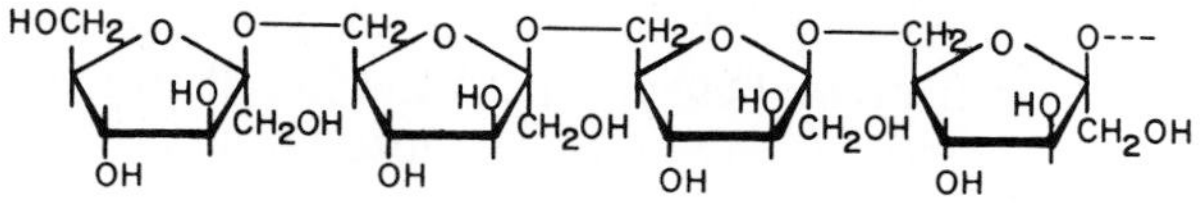

(4)

Depending on the enzymes present in a given strain, sucrose is cleaved by specific glucosyltransferases to yield free fructose and a polymer of glucose (dextransucrase) or free glucose and a polymer of fructose (levansucrase):

Dextran sucrase:

$$\begin{array}{c} \text{Glucose-fructose} \longrightarrow \text{Fructose} \; + \; \text{Glucose polymer} \\ \text{(Sucrose)} \qquad\qquad\qquad\qquad \text{(Dextran)} \\ \Big\downarrow \text{Fermentation} \\ \text{Lactic acid} \end{array}$$

(5)

Levan sucrase:

$$\begin{array}{c} \text{Glucose-fructose} \longrightarrow \text{Glucose} \; + \; \text{Fructose polymer} \\ \text{(Sucrose)} \qquad\qquad\qquad\qquad \text{(Levan)} \\ \Big\downarrow \text{Fermentation} \\ \text{Lactic acid} \end{array}$$

(6)

Certain members of the streptococci, notably *S. mutans*, produce both soluble and insoluble forms of dextrans. The insoluble form, called a mutan, is considered to be responsible for adherence of this organism to the enamel surfaces of teeth and the formation of bacterial aggregates (plaque). The genetics of adherence, aggregation, and glucosyltransferase production has been investigated in *S. mutans* (Murchison et al., 1981). Glucan-binding proteins are produced by this organism. Two of these proteins exhibit glucosyltransferase activity (Russell, 1979).

Cellulose is a polymer of β-1,4-linked glucose units:

(7)

Acetobacter xylinum synthesizes cellulose, which accumulates as an extracellular aggregate of crystalline microfibrils. Intertwined ribbons from large numbers of cells form a tough pellicle on the surface of the culture medium (R. M. Brown et al., 1976). Although cellulose synthesis has been studied extensively in higher plants as well as in bacteria, many questions remain regarding the intermediary steps involved in the bacterial pathway. Swissa et al. (1980) found that resting cells, and particulate membrane-bound preparations of *A. xylinum*, incorporated [1-^{14}C]-glucose into glucose 6-phosphate, glucose 1-phosphate, uridine glucose 5′-phosphate (UDPG), and cellulose. Labeling studies and the demonstration of enzyme activities in cell-free extracts indicate that the sequence of reactions leading to cellulose synthesis is:

$$\text{Glucose} \rightarrow \text{G-6-P} \rightarrow \text{G-1-P} \rightarrow \text{UDPG} \rightarrow \text{Cellulose} \qquad (8)$$

The evidence suggests that lipid- and protein-linked cellodextrins may function as intermediates between UDPG and cellulose. Elucidation of the details of cellulose biosynthesis has been hampered by the low rates of synthesis obtained using cell-free preparations. Aloni et al. (1983) developed a procedure for solubilization of cellulose synthetase (UDP-glucose:1,4-β-D-glucan 4-β-D-glucosyltransferase) by treatment of membranes from *A. xylinum* with digitonin (1–10%). The digitonin-solubilized enzyme is specifically activated by GTP in the presence of a protein factor, which can be removed from the enzyme by washing the membranes prior to solubilization. Association of the protein factor with the membrane-bound enzyme is promoted by polyethylene glycol or by Ca^{2+}; however, these compounds are ineffective in enhancing association of the protein factor with the solubilized enzyme. Calcofluor white ST (4,4′-bis[4-anilino-6-bis(2-hydroxyethyl)amino-s-triazin-2-ylamino]-2,2′-stilbendisulfonic acid), a fluorescent brightener used to whiten textiles and paper, is also used to identify cellulose in biological materials. Calcofluor white ST alters cellulose synthesis in *A. xylinum* by separating the terminal stages of polymerization and crystallization. Cellulose is synthesized intracellularly by a multienzyme complex. Extrusion pores in the LPS layer of the outer membrane facilitate aggregation of bundles of cellulose, which undergo crystallization by self-assembly at the cell surface (Haigler et al., 1980).

IV. CARBOHYDRATE METABOLISM

A. Lactic Acid Bacteria

Lactic acid fermentations.   In 1931, Kluyver proposed that bacteria fermenting glucose primarily to lactic acid be referred to as *homofermentative* and those that produced a mixture of products as *heterofermentative*. Gibbs et al. (1950) showed by means of ^{14}C-labeling studies that homofermentative species utilized the Embden-Meyerhof-Parnas (EMP) pathway for the production of lactic acid. Heterofermentative species utilize a pentose phosphate (phosphoketolase) pathway to yield approximately equimolar amounts of lactate, ethanol, and CO_2 (Fig. 6.5). Within the genus *Lactobacillus*, *L. casei*, and *Lactobacillus pentosus* are homofermentative, whereas *Lactobacillus lycopersici*, *Lactobacillus pentoaceticus*, and *Lactobacillus brevis* are heterofermentative. The group N lactic streptococci (*S. cremoris*, *S. lactis*, and *S. diacetylactis*) are regarded as homofermentative. For more extensive discussions of fermentative pathways, see Wood (1961), Doelle (1975), and Moat (1979).

DeMoss et al. (1951) found that heterofermentative species of lactic-producing bacteria lacked aldolase, the enzyme responsible for splitting fructose 1,6-diphosphate (FDP), a result suggesting that heterolactic species could not utilize the classic EMP pathway. Later studies showed that some species contained both the EMP and pentose pathways but that the pentose phosphate pathway was under close regulation by intermediates in the EMP pathway. Wolin (1964) found that the lactic dehydrogenase (LDH) of *S. bovis* is specifically activated by FDP. It has since been demonstrated that FDP-activated LDH's are characteristic of many of the lactic acid–producing bacteria. In apparently homolactic fermenters of glucose and lactose it has been found that they become heterolactic when grown with limiting carbohydrate concentrations in a chemostat. At high dilution rates with an excess of glucose present, about 95% of the fermented sugar is converted to L-lactate. At lower dilution rates, glucose becomes limiting, and some strains of lactic streptococci change to a heterolactic fermentation pattern. The products formed after this phenotypic change are formate, acetate, and ethanol. The level of LDH, which is dependent upon FDP for activity, decreases as fermentation becomes heterolactic. If cells are transferred from the chemostat to buffer containing glucose, the nongrowing cells revert to converting nearly 80% of the glucose to lactate. These findings suggest a close regulation of the enzymes of the two pathways by the intracellular concentration of FDP and the level of LDH (Thomas et al., 1979, 1980). A. T. Brown and Wittenberger (1972) found that *S. faecalis* produced a single NADP-linked 6-PGDH when grown with glucose as the primary energy source. Gluconate-adapted cells contained two separate 6-PGDHs, one specific for NADP and the other specific for NAD. The NADP-linked enzyme is specifically inhibited by FDP but is insensitive to ATP or other nucleotides. The NAD-linked enzyme is insensitive to FDP inhibition but is inhibited by ATP and other nucleotides. These findings suggest that the NAD enzyme is involved primarily in the catabolism of gluconate, whereas the NADP enzyme appears to function in the production of reducing equivalents (as NADPH) for use in various reductive biosynthetic reactions. The regulatory activities suggested by these observations are shown in Fig. 6.6.

In earlier studies on homolactic organisms it was noted that at alkaline pH the production of formate, acetate, and ethanol increased at the expense of

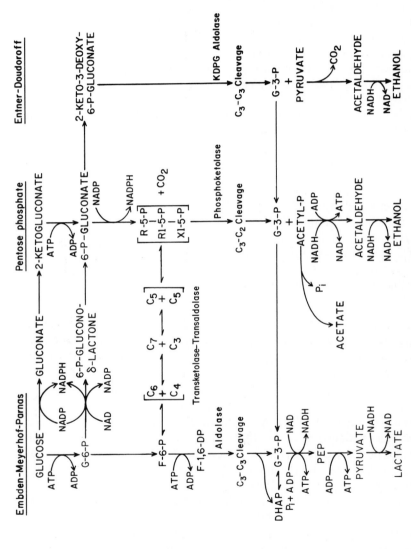

FIGURE 6.5 Composite diagram of the Embden-Meyerhof-Parnas (EMP), pentose phosphate, Entner-Doudoroff, and transketolase-transaldolase pathways. DHAP, dihydroxyacetonephosphate; G-3-P, glyceraldehyde 3-phosphate; PEP, phosphoenolpyruvate; KDPG, 2-keto-3-deoxy-6-phosphogluconate; R-5-P, ribose 5-phosphate; R1-5-P, ribulose 5-phosphate; X1-5-P, xylulose 5-phosphate.

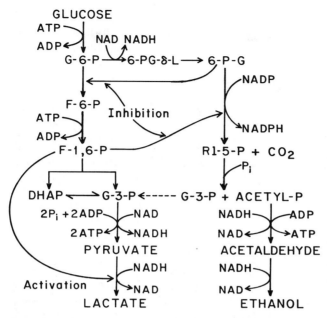

FIGURE 6.6   Regulatory activities in the fermentation pathways of lactic acid bacteria. 6-PG-δ-L, 6-phosphoglucono-δ-lactone; DHAP, dihydroxyacetonephosphate; 6-P-G, 6-phosphogluconate; G-3-P, glyceraldehyde-3-phosphate; R1-5-P, ribulose 5-phosphate.

lactate (W. A. Wood, 1961). Since labeling studies showed that the lactate arises via the EMP pathway, the diversion of fermentation must occur at a step after pyruvate formation. Under alkaline conditions a dismutation of pyruvate may give rise to lactate, acetate, and formate through the following sequence:

$$CH_3COCOOH + CoASH \rightarrow CH_3CO\text{-}SCoA + HCOOH \qquad (8)$$
$$CH_3CO\text{-}SCoA + P_i \rightarrow CH_3COPO_3H_2 + CoASH \qquad (9)$$
$$CH_3COCOOH + 2H \rightarrow CH_3CHOHCOOH \qquad (10)$$
$$CH_3COPO_3H_2 + ADP \rightarrow CH_3COOH + ATP \qquad (11)$$

Net: ───

2 Pyruvate + 2H + ADP + P_i → Lactate + Acetate
                                    + Formate + ATP   (12)

In *S. mutans*, triose phosphates (glyceraldehyde 3-phosphate or dihydroxyacetone phosphate) strongly inhibit pyruvate formate-lyase, the first step in the dismutation sequence. Inhibition by triose phosphates in cooperation with a reactivating effect of ferridoxin may regulate pyruvate formate-lyase activity in vivo (Takahashi et al., 1982). Homolactic and heterolactic strains appear to form ethanol not via pyruvate decarboxylase but via reduction of acetyl phosphate. When appreciable quantities of CO_2 appear in the reaction products rather than formate, it is apparent that the organism is utilizing the pentose phosphate pathway as a result of the decreased levels of FDP (A. T. Brown and Wittenberger, 1971). If formate is the major C_1 product, then a dismutation of pyruvate arising via the EMP pathway is more likely.

The transport of glucose, lactose, and hexitols by lactic streptococci and other lactic acid producers generally occurs via the phosphoenolypyruvate (PEP)-dependent phosphotransferase system (PTS), in which the sugar becomes phosphorylated simultaneously with translocation:

$$\text{PEP + enzyme I} \xrightarrow{\text{Mg}^{2+}} \text{pyruvate + enzyme I}{\sim}\text{P} \qquad (13)$$

$$\text{enzyme I}{\sim}\text{P + HPr} \longleftrightarrow \text{enzyme I + HPr}{\sim}\text{P} \qquad (14)$$

$$\text{HPr}{\sim}\text{P + enzyme III} \longleftrightarrow \text{HPr + enzyme III}{\sim}\text{P} \qquad (15)$$

$$\text{enzyme III}{\sim}\text{P sugar}_{\text{outside}} \xleftrightarrow{\text{enzyme II}} \text{enzyme III + sugar-P}_{\text{inside}} \qquad (16)$$

Enzyme I and HPr (heat-stable protein) are sugar-nonspecific and are synthesized constitutively. Enzymes II and III are inducible and are specific for the sugar added to the growth medium and are usually designated as enzyme II^{lac}, enzyme III^{gal}, and so on in referring to a specific PTS (see T. D. Thompson (1980); Park and McKay (1982); Dills and Seno (1983) for earlier references). Galactose may be transported via the PTS and utilized in homolactic fashion. Alternatively, galactose may be transported via an ATP-energized permease system (galP). The free sugar is then metabolized via the Leloir pathway (T. D. Thompson, 1980; Park and McKay, 1982). Methyl-β-D-thiogalactopyranoside (TMG) transport is mediated by the glucose:PTS and lactose:PTS. Accumulation of TMG-6P in *S. pyogenes* and *S. lactis* is regulated by exclusion and expulsion mechanisms (Reizer and Panos, 1980; J. Thompson and Saier, 1981). PTS sugars (glucose, mannose, 2-deoxyglucose, or lactose), when added simultaneously with TMG, exclude TMG from the cells, whereas galactose enhances TMG-6P accumulation. These same sugars (glucose, mannose, lactose) cause rapid expulsion of TMG from cells preloaded with TMG-6P. A variety of evidence has been accumulated to show that exclusion and expulsion are mediated by two independent mechanisms. In a survey of fermentative bacteria, Romano et al. (1979) observed that the PEP-dependent glucose:PTS is present in all homofermentative lactic acid bacteria that ferment glucose via the Embden-Meyerhof-Parnas pathway. The glucose:PTS is absent in heterofermentative species of *Lactobacillus* and *Leuconostoc*, which ferment glucose via the phosphoketolase pathway.

S. lactis, *S. cremoris*, *S. diacetylactis*, and various *Leuconostoc* and *Lactobacillus* species commonly produce acetoin and diacetyl. It has been considered that there were two main routes for the production of acetoin, one proceeding via condensation of pyruvate and hydroxyethylthiamine pyrophosphate (C_2-TPP) to form the intermediate α-acetolactate (AAL):

$$\text{Pyruvate + TPP} \rightarrow C_2\text{-TPP} + CO_2 \qquad (17)$$

$$\text{Pyruvate} + C_2\text{-TPP} \rightarrow \text{AAL + TPP} \qquad (18)$$

$$\text{AAL} \rightarrow \text{Acetoin} + CO_2 \qquad (19)$$

and the other directly via C_2-TPP condensation:

$$\text{Pyruvate} \rightarrow 2\ C_2\text{-TPP} \qquad (20)$$
$$2\ C_2\text{-TPP} \rightarrow \text{acetoin} \qquad (21)$$

Acetoin could then be oxidized to diacetyl or reduced to 2,3-butanediol:

$$\text{Diacetyl} \xleftarrow{-2H} \text{Acetoin} \xrightarrow{+2H} \text{2,3-butanediol} \qquad (22)$$

Condensation of acetyl-CoA with C_2-TPP can yield diacetyl directly:

$$C_2\text{-TPP} + \text{Acetyl-CoA} \rightarrow \text{Diacetyl} \qquad (23)$$

The diacetyl can then be reduced to acetoin by diacetyl reductase:

$$\text{Diacetyl} + 2H \rightarrow 2\ \text{Acetoin} \qquad (24)$$

In a medium containing acetate, acetyl-CoA is formed directly from acetate without the intermediary formation of pyruvate. *S. diacetylactis*, *L. casei*, and other lactic organisms that require lipoic acid for growth activate acetate via acetate kinase and phosphotransacetylase. The acetyl-CoA incorporated into diacetyl by organisms requiring lipoic acid should not be derived from pyruvate or pyruvate precursors in media devoid of lipoic acid. In a lipoic acid–free medium containing glucose and acetate, acetoin is formed via both the acetate + C_2-TPP and AAL routes (Speckman and Collins, 1973).

Oxygen relationships.   Organisms that metabolize glucose or other sugars via fermentative pathways do not normally utilize molecular oxygen as a final hydrogen acceptor and do not generate ATP via oxidative phosphorylation. Such organisms are, metabolically, anaerobes. However, in practical usage, if an organism grows in the presence of oxygen, it is referred to as being aerobic or aerotolerant.

Superoxide and other toxic products arise through reaction of oxidative enzymes with molecular oxygen, most commonly through autooxidation of flavins, quinones, etc.:

$$O_2 + e^- + \text{oxidative enzymes} \rightarrow O_2^{-}\cdot \qquad (25)$$
$$O_2^{-} + H_2O_2 \xrightarrow{\text{nonenzymatic}} OH + OH^- + O_2 \qquad (26)$$

Many aerobic and facultative organisms are protected from the toxic action of superoxide by the production of the enzyme superoxide dismutase (SOD):

$$O_2^{-}\cdot + O_2^{-}\cdot + 2H^+ \xrightarrow{\text{SOD}} O_2 + H_2O_2 \qquad (27)$$

Hydrogen peroxide is dissipated by the enzyme catalase:

$$2\ H_2O_2 \xrightarrow{\text{catalase}} 2\ H_2O + O_2 \qquad (28)$$

It has been observed that some fermentative organisms such as *L. plantarum* are aerotolerant but produce neither SOD nor catalase. These organisms do not reduce oxygen and therefore do not produce superoxide or hydrogen peroxide. Some organisms that produce soluble FAD-linked oxidases consume oxygen nonenzymatically under certain conditions to yield hydrogen peroxide:

$$\text{FAD-2H} + O_2 \rightarrow H_2O_2 + \text{FAD} \tag{29}$$

Archibald and Fridovitch (1981) reported that *L. plantarum* required a high concentration of Mn^{2+} and rapidly accumulated Mn^{2+} intracellularly. Extracts of these organisms contained 9 μg of Mn per milligram of protein and an ability to scavenge superoxide anion. This superoxide-scavenging activity was due to the dialyzable Mn^{2+} and not to other metals such as iron.

Peptococcus anaerobius is an anaerobic lactate-producing organism that some authors feel should be classified with the streptococci (E. M. Barnes et al., 1977). This organism has considerable tolerance to oxygen as a result of its ability to produce high levels of NADH oxidase, which reduces oxygen to water (Hoshino et al., 1978):

$$\text{NADH} + H^+ + \tfrac{1}{2} O_2 \rightarrow \text{NAD}^+ + H_2O \tag{30}$$

An NADPH oxidase that interacts with molecular oxygen to produce superoxide is also produced:

$$\text{NADPH} + H^+ + 2O_2 \rightarrow \text{NADP}^+ + O_2^- + H_2O_2 \tag{31}$$

However, the activity of this latter enzyme is much lower than the NADH oxidase, and it does not appear to produce toxic levels of superoxide.

Some streptococci produce functional cytochromes when grown in an aerobic environment under the appropriate nutritional conditions. In the presence of hematin, aerobically grown *S. faecalis* produces functional cytochromes that yield additional ATP. In the absence of heme, only a flavin system of electron transport is formed (Ritchey and Seeley, 1974). Distribution of cytochromelike respiration among representatives of the genus *Streptococcus* seems to be limited to enterococci and a few strains of *S. lactis* and its subspecies (Ritchey and Seeley, 1976). Collins and Jones (1981) surveyed a number of species of streptococci of serological groups D and N as well as some representative genera of *Aerococcus*, *Gemella*, *Leuconostoc*, and *Pediococcus* for the presence of isoprenoid quinones (menaquinones and ubiquinones), which play an important role in electron transport and possibly oxidative phosphorylation in other organisms. A large number of the organisms surveyed were found to lack both menaquinones and ubiquinones. *S. faecalis* and its subspecies contained demethylmenaquinones. Menaquinones with eight isoprene units were found in *S. faecium* and its subspecies, and menaquinones with nine isoprene units were present in *S. cremoris* and *S. lactis* and some of their subspecies.

B. Acetic Acid Bacteria

Gluconobacter oxydans (*Acetobacter suboxydans*) lacks a functional TCA cycle. However, all enzymes of this cycle except succinate dehydrogenase are

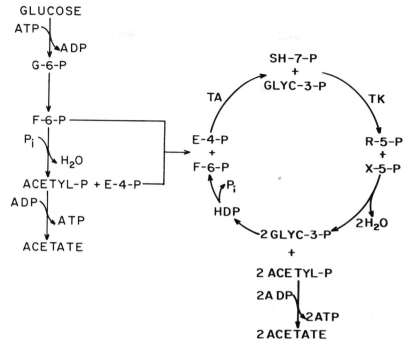

NET: Glucose + 2P$_i$ + 2 ADP $\longrightarrow$ 3 Acetate + 2 ATP + 3H$_2$O

FIGURE 6.7   Modified pentose cycle for *Acetobacter* and *Gluconobacter*. TA, transaldolase; TK, transketolase; E-4-P, erythrose 4-phosphate; SH-7-P, sedoheptulose 7-phosphate; R-5-P, ribose 5-phosphate; X-5-P, xylulose 5-phosphate; HDP hexodiphosphate; glyc-3-P, glyceraldehyde 3-phosphate.

present. Glutamate is synthesized from acetate via citrate, isocitrate, and α-ketoglutarate. *G. oxydans* cannot ferment glucose or other carbohydrates, since neither the EMP nor the Entner-Doudoroff pathway is present. It is dependent upon a modified pentose cycle for the metabolism of glucose under aerobic conditions (Greenfield and Claus, 1972). Acetate is a major end product of glucose metabolism (Fig. 6.7). Ethanol can be oxidized directly via acetaldehyde and acetate:

$$CH_3CH_2OH \rightarrow CH_3CHO + 2\,H \rightarrow CH_3COOH + 2\,H \qquad (32)$$

Other primary alcohols such as ethylene glycol and 1,3-butanediol are also oxidized to their corresponding carboxylic acids:

$$CH_2OHCH_2OH \rightarrow CH_3CHO + 2\,H$$
$$CH_3CHO + 2\,H \rightarrow CH_2OHCOOH + 2\,H \qquad (33)$$

$$CH_3CHOHCH_2CH_2OH \rightarrow CH_3CHOHCH_2CHO + 2\,H$$
$$CH_3CHOHCH_2CHO + 2\,H \rightarrow CH_3CHOHCH_2COOH + 2\,H \qquad (34)$$

Secondary alcohols (polyols) such as glycerol, 2,3-butanediol and 1,2-propanediol are oxidized to ketones:

$$CH_2OHCHOHCH_2OH \rightarrow CH_2OHCOCH_2OH + 2\ H \qquad (35)$$

$$CH_3CHOHCH_2OH \rightarrow CH_3COCH_2OH + 2\ H \qquad (36)$$

Sorbitol and other sugar alcohols (e.g., mannitol) are oxidized to their corresponding sugars:

$$
\begin{array}{ccc}
CH_2OH & & CH_2OH \\
| & & | \\
HOCH & & O{=}C \\
| & & | \\
HOCH & {\scriptstyle -2\,H} & HOCH \\
| & \xrightarrow{\hspace{2cm}} & | \qquad (37) \\
HCOH & & HCOH \\
| & & | \\
HOCH & & HOCH \\
| & & | \\
CH_2OH & & CH_2OH \\
\text{D-Sorbitol} & & \text{L-Sorbose}
\end{array}
$$

The products of these limited oxidations accumulate in the growth medium. *G. oxydans* increases polyol oxidation rates when cultures enter the maximum stationary phase (Claus et al., 1975). These single-step oxidations are catalyzed by membrane-bound dehydrogenases that increase in activity in parallel with the formation of intracytoplasmic membranes near the end of the growth cycle. Intracytoplasmic membrane formation in sorbitol-grown cells results in greater rates of oxidation of other polyols such as mannitol, *meso*-erythritol, and *meso*-inositol. The increased synthesis of intracytoplasmic membranes and the associated increase in respiratory capacity is independent of the polyol used to grow the cells or to measure respiration (White and Claus, 1982).

Acetobacter, by comparison, have a functional TCA cycle. As a result, the acetyl phosphate formed from the C_2–C_3 cleavage of pentose is oxidized to CO_2, and energy is produced via oxidative phosphorylation. Apparently, under anaerobic or partially anaerobic conditions, acetyl phosphate is not utilized through the TCA cycle, and acetate accumulates as an end product (Fig. 6.7). *Gluconobacter* lack pyridine nucleotide transhydrogenase and NADPH oxidase and suffer from a deficit of $NADP^+$, especially when utilizing a substrate such as fructose. When grown on fructose, *Gluconobacter cerinus* accumulates 5-ketofructose (5-KF) in significant amounts through the action of a non-NAD-dependent fructose 5-dehydrogenase (FDH). The resultant 5-KF is then reduced to fructose by an NADPH-linked to 5-KF reductase. This oxidation–reduction cycle apparently serves as a mechanism for the regeneration of $NADP^+$ during the growth cycle:

$$Fructose \rightarrow 5\text{-}KF + 2\,H \qquad (38)$$
$$5\text{-}KF + NADPH + H^+ \rightarrow Fructose + NADP^+ \qquad (39)$$
$$2\,H + \tfrac{1}{2}\,O_2 \rightarrow H_2O \qquad (40)$$

Net: _____

$$NADPH + H^+ + \tfrac{1}{2}\,O_2 \rightarrow NADP^+ + H_2O \qquad (41)$$

Mutants lacking FDH are dependent on 5-KF for growth (Mowshowitz et al., 1974). The FDH has been shown to exist as a membrane-bound cytochrome complex in *Gluconobacter industrius* (Ameyama et al., 1981).

C. Propionic Acid Bacteria

The major products of fermentation of glucose, glycerol, and lactate by propionibacteria are propionate, acetate, and carbon dioxide (H. G. Wood, 1982). The general reaction sequence for the formation of these compounds from glucose is usually given as

$$1.5\,Glucose + 3\,ADP + 3\,P_i \rightarrow 3\,Pyruvate + 3\,ATP$$
$$+ 3(2\,H) \qquad (42)$$
$$Pyruvate + ADP + P_i \rightarrow Acetate + ATP + 2\,H \qquad (43)$$
$$2\,Pyruvate + 2\,ADP + 2\,P_i + 4(2\,H) \rightarrow 2\,Propionate + 2\,ATP$$
$$+ 2\,H_2O \qquad (44)$$

Net: _____

$$1.5\,Glucose + 6\,ADP + 6\,P_i + \rightarrow Acetate + 2\,Propionate$$
$$+ 6\,ATP + 2\,H_2O \quad (45)$$

The details of the pathways by which propionate and acetate are formed by the propionibacteria are shown in Fig. 6.8. A major factor in the elucidation of the route of propionate formation was the discovery of the mechanism of transcarboxylation by methylmalonyl-oxaloacetate transcarboxylase. In this reaction, biotin plays an important catalytic role. Cobalamin (vitamin B_{12}) also serves as a cofactor in the formation of methylmalonyl-CoA from succinyl-CoA as shown in the following series of reactions (H. G. Wood, 1982):

$$Enz\text{-}Biotin\text{-}CO_2 + Pyruvate \rightarrow Oxaloacetate + Enz\text{-}Biotin \qquad (46)$$
$$Oxaloacetate + 4\,H \rightarrow Succinate + H_2O \qquad (47)$$
$$Succinyl\text{-}CoA\text{-}B_{12}\text{-}enz \rightarrow Methylmalonyl\text{-}CoA \qquad (48)$$
$$Methylmalonyl\text{-}CoA + Enz\text{-}Biotin \rightarrow Propionyl\text{-}CoA$$
$$+ Enz\text{-}Biotin\text{-}CO_2 \qquad (49)$$
$$Propionyl\text{-}CoA + Succinate \rightarrow Succinyl\text{-}CoA + Propionate \qquad (50)$$

Net: _____

$$Pyruvate + 4\,H \rightarrow Propionate + H_2O \qquad (51)$$

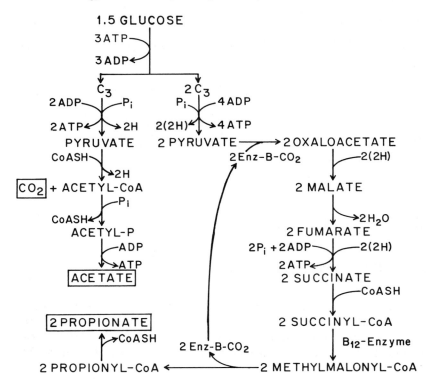

NET:
1.5 GLUCOSE + 6 ADP + 6 P_i $\longrightarrow$ 2 PROPIONATE + ACETATE + CO_2 + 6 ATP + 2 H_2O

FIGURE 6.8 Pathways to propionate, acetate, and CO_2 in *Propionibacterium*. Enz-B-CO_2, enzyme-biotin-CO_2 complex.

As shown in Fig. 6.8, the hydrogens required for the reduction of oxaloacetate to succinate are obtained from reactions in the conversion of glucose to acetate and CO_2 via pyruvate.

In practice, ideal carbon and C_1 balances were not always obtained for propionic acid fermentations. This turned out to be due in part to the production of appreciable amounts of succinate as a final product. Also, in many instances it was observed that the ratio of propionate:acetate was often higher than the expected 2:1 and that the ATP yields were higher than that shown in Eq. (45). H. G. Wood (1977) has made an interesting discovery with regard to the replacement of ATP by inorganic pyrophosphate (PP_i) as an energy source in fermentations by *P. shermanii* and other organisms that would explain the higher energy yields observed. The enzymes pyruvate, phosphate dikinase; carboxytransphosphorylase; and PP_i-phosphofructokinase catalyze reactions in which PP_i replaces ATP:

Pyruvate, phospate dikinase:

$$\text{Pyruvate} + \text{ATP} + \text{P}_i \leftrightarrow \text{Phosphoenolpyruvate} + \text{AMP} + \text{PP}_i \quad (52)$$

Carboxytransphosphorylase:

$$\text{Phosphoenolpyruvate} + \text{CO}_2 + \text{P}_i \leftrightarrow \text{Oxaloacetate} + \text{PP}_i \quad (53)$$

Net: ───

$$\text{Pyruvate} + \text{ATP} + \text{CO}_2 + 2\,\text{P}_i \leftrightarrow \text{Oxaloacetate} + \text{AMP} + 2\,\text{PP}_i \quad (54)$$

PP_i–*Phosphofructokinase*:

$$\text{PP}_i + \text{Fructose-6-P} \leftrightarrow \text{Fructose 1,6-diP} + \text{P}_i \quad (55)$$

Under appropriate conditions and in the presence of the requisite enzymes, yields of 11 mols of ATP per 3 mols of glucose could be achieved as shown by the following equation:

$$3\,\text{Glucose} + 3\,\text{PP}_i \rightarrow 4\,\text{Propionate} + 2\,\text{Acetate} + 2\,\text{CO}_2$$
$$+ 11\,\text{ATP} + 4\sim\text{P} \quad (56)$$

The participation of PP_i in the propionate fermentation serves to explain the high cell yield observed. Discovery of these reactions in which the energy inherent in PP_i may be utilized and not wasted through hydrolysis may have far-reaching implications, since many reactions in biological systems result in the formation of PP_i.

V. FATTY ACIDS AND LIPIDS

As was discussed in previous sections, many different lipid-containing components are present in membranes, mesosomes, and surface structures of bacterial cells. Lipid-linked sugars play a major role in the biosynthesis of glycoproteins. Isoprenoid alcohols function as carriers for polysaccharide molecules in the assembly of the complex glycans of the cell envelope. C_{55}-isoprenols are involved in the synthesis of cell surface structures. These compounds, referred to as bactoprenols, are present in mesosomes and have the following general structure:

$$\underset{|}{\overset{\text{CH}_3}{}} \qquad \underset{|}{\overset{\text{CH}_3}{}} \qquad \underset{|}{\overset{\text{CH}_3}{}}$$
$$\text{CH}_3\text{C}=\text{CHCH}_2(\text{CH}_2\text{C}=\text{CHCH}_2)_9\text{CH}_2\text{C}=\text{CHCH}_2\text{OPO}_3\text{H}_2$$

Fatty acid biosynthesis has been studied extensively in bacteria. However, the enzymes of the biosynthetic pathway may not have been directly demonstrated in many organisms. Nutritional studies have provided evidence that many lactobacilli and streptococci synthesize fatty acids as long as biotin is present in the culture medium. In the absence of biotin, unsaturated, and often saturated, fatty acids are required for growth (Broquist and Snell, 1951). Acetyl-CoA carboxylase, a biotin-containing enzyme that catalyzes the ATP-dependent fixation of CO_2 into acetyl-, propionyl-, and butyryl-CoA, catalyzes the first committed reaction in the de novo synthesis of fatty acids. Its presence in propionibacteria has been demonstrated (Stirling et al., 1981). The reaction occurs in two stages:

$$\text{Enz-B} + \text{ATP} + \text{HCO}_3^- \leftrightarrow \text{Enz-B-CO}_2 + \text{ADP} + \text{P}_i \qquad (57)$$

$$\text{Enz-B-CO}_2 + \text{Acetyl-CoA} \leftrightarrow \text{Enz-B} + \text{Malonyl-CoA} \qquad (58)$$

Net:

$$\text{Acetyl-CoA} + \text{ATP} + \text{HCO}_3^- \leftrightarrow \text{Malonyl-CoA} + \text{ADP} + \text{P}_i \qquad (59)$$

Biosynthesis of long-chain fatty acids from acetyl-CoA, malonyl-CoA, and reduced NADP is catalyzed by fatty acid synthetases that contain 4'-phosphopantetheine as the prosthetic group. A readily dissociable type of fatty acid synthetase complex has been isolated from *P. shermanii* and partially characterized (Ahmad et al., 1981). The partially purified synthetase requires acetyl-CoA, malonyl-CoA, NADH, NADPH, and a small acyl carrier protein (ACP) similar to that found in *E. coli*. The reactions following acetyl-CoA carboxylase in the biosynthetic sequence leading to fatty acid synthesis involve the additional reactions shown below:

Acetyl transferase:

$$\text{ACP}-\text{SH} + \text{CH}_3\text{CO}-\text{S}-\text{CoA} \leftrightarrow \text{CH}_3\text{CO}-\text{S}-\text{ACP} + \text{CoASH} \qquad (60)$$

Malonyl transferase:

$$\text{HOOCCH}_2\text{CO}-\text{S}-\text{CoA} + \text{ACP}-\text{SH} \leftrightarrow$$
$$\text{HOOC}-\text{CH}_2\text{CO}-\text{S}-\text{ACP} + \text{CoASH} \qquad (61)$$

Condensing enzyme:

$$\text{CH}_3-\text{CO}-\text{S}-\text{ACP} + \text{HOOCCH}_2\text{CO}-\text{S}-\text{ACP} \leftrightarrow$$
$$\text{CH}_3\text{COCH}_2\text{CO}-\text{S}-\text{ACP} + \text{ACP}-\text{SH} + \text{CO}_2 \qquad (62)$$

Ketoacyl-ACP reductase:

$$\text{CH}_3\text{COCH}_2\text{CO}-\text{S}-\text{ACP} + \text{NADPH} + \text{H}^+ \leftrightarrow$$
$$\text{CH}_3-\text{CHOH}-\text{CH}_2\text{CO}-\text{S}-\text{ACP} + \text{NADP} \qquad (63)$$

Enoyl-ACP hydrase:

$$\text{CH}_3\text{CHOH}-\text{CH}_2-\text{CO}-\text{S}-\text{ACP} \leftrightarrow$$
$$\text{CH}_2-\text{CH}=\text{CH}-\text{CO}-\text{S}-\text{ACP} + \text{H}_2\text{O} \qquad (64)$$

Continued chain elongation:

$$CH_3CH=CH-CO-S-ACP + NADPH\ H^+ \rightleftharpoons$$
$$CH_3CH_2CH_2CO-S-ACP + NADP^+ \quad (65)$$

$$CH_3CH_2CH_2CO-S-ACP + CoASH \rightleftharpoons$$
$$CH_3CH_2CH_2CO-SCoA + ACP-SH \quad (66)$$

Long-chain fatty acids are synthesized by continuation of this process until the desired chain length is achieved. In most bacteria there is a predominance of C_{16} (palmitic) and C_{18} (stearic) acids.

The biosynthesis of long-chain unsaturated fatty acids follows one of two distinct pathways. In eukaryotes (mammals, yeast, algae, protozoa) and some highly aerobic bacteria (*Alkaligenes*, *Mycobacterium*, *Corynebacterium*, *Bacillus*) the synthesis of long-chain unsaturated fatty acids involves an oxidative desaturation that introduces double bonds after the fatty acid has been synthesized:

$$
\underset{\substack{|\quad| \\ H\quad H}}{\overset{\substack{H\quad H \\ |\quad|}}{CH_3(CH_2)_7-C-C}}-(CH_2)_7CO-CoA \xrightarrow{1/2\ O_2}
$$

Stearoyl$-$CoA (18:0) $\hspace{6cm}$ (67)

$$
\underset{}{\overset{\substack{H\quad H \\ |\quad|}}{CH_3(CH_2)_7-C=C}}-(CH_2)_7CO-CoA
$$

Oleoyl$-$CoA $(18:1^{\Delta 11})$

The anaerobic pathway of fatty acid biosynthesis follows the same series of reactions to the C_{10} stage with the formation of β-hydroxydecanoyl-ACP. At this stage there is a branch point. Action of β-hydroxydecanoyl thioester dehydrase leads to the formation of a double bond between the β and γ carbons. Continued addition of malonyl-ACP leads to the formation of *cis*-9-hexadecenoic acid $(16:1^{\Delta 9}$ or palmitoleic acid) and *cis*-11-octadecenoic acid $(18:1^{\Delta 11}$ or *cis*-vaccenic acid):

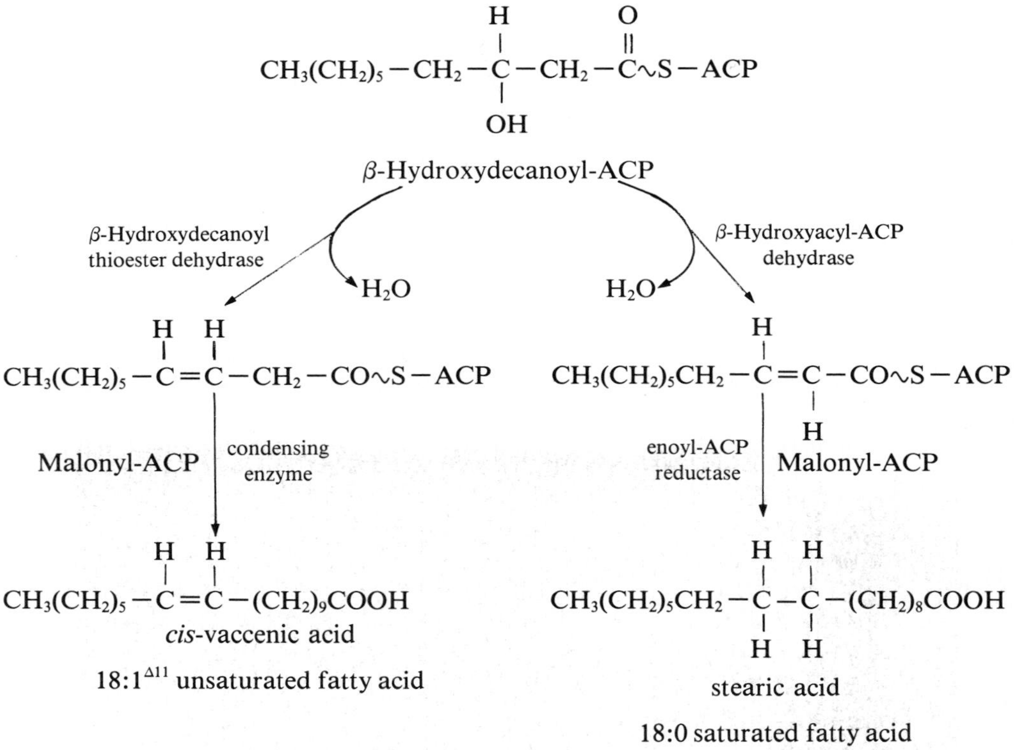

$$CH_3(CH_2)_5 - CH_2 - \overset{\overset{\displaystyle H}{|}}{\underset{\underset{\displaystyle OH}{|}}{C}} - CH_2 - \overset{\overset{\displaystyle O}{\|}}{C} {\sim} S - ACP$$

β-Hydroxydecanoyl-ACP

β-Hydroxydecanoyl thioester dehydrase

β-Hydroxyacyl-ACP dehydrase

H_2O H_2O

$$CH_3(CH_2)_5 - \overset{\overset{\displaystyle H}{|}}{C} = \overset{\overset{\displaystyle H}{|}}{C} - CH_2 - CO{\sim}S - ACP$$

$$CH_3(CH_2)_5CH_2 - \overset{\overset{\displaystyle H}{|}}{C} = C - CO{\sim}S - ACP$$
$$\overset{\overset{\displaystyle H}{|}}{}$$

Malonyl-ACP   condensing enzyme

enoyl-ACP reductase   Malonyl-ACP

$$CH_3(CH_2)_5 - \overset{\overset{\displaystyle H}{|}}{C} = \overset{\overset{\displaystyle H}{|}}{C} - (CH_2)_9COOH$$
cis-vaccenic acid

$18:1^{\Delta 11}$ unsaturated fatty acid

$$CH_3(CH_2)_5CH_2 - \overset{\overset{\displaystyle H}{|}}{\underset{\underset{\displaystyle H}{|}}{C}} - \overset{\overset{\displaystyle H}{|}}{\underset{\underset{\displaystyle H}{|}}{C}} - (CH_2)_8COOH$$

stearic acid

$18:0$ saturated fatty acid

The action of β-hydroxyacyl-ACP dehydrase leads to the formation of an α,β-*trans* double bond. Continued elongation then leads to the formation of saturated fatty acids as described above. As a result of this mode of unsaturated fatty acid synthesis, only monounsaturated fatty acids are formed, and *cis*-vaccenic acid ($18:1^{\Delta 11}$) is the major fatty acid formed by microorganisms that utilize the anaerobic pathway (Finnerty and Makula, 1975; Moat, 1979).

VI. NUTRITION AND NITROGEN ASSIMILATION

Historically, several B vitamins and some of the amino acids were discovered through studies designed to develop completely chemically defined culture media for the lactic acid bacteria as listed below (Lichstein, 1982, 1983):

Growth factor	*First identified as growth factor for*
Cobalamin (Vitamin B$_{12}$)	*Lactobacillus lactis*
p-Aminobenzoic acid	*Streptococcus haemolyticus*
	Lactobacillus plantarum
	Acetobacter suboxydans

Growth factor	*First identified as growth factor for*
Lipoic acid	*Lactobacillus casei*
	Streptococcus faecalis
Spermidine	*Lactobacillus casei*
Methionine	Streptococci
Pimelic acid	*Corynebacterium diphtheriae*

Many of the lactic and propionic acid bacteria are nutritionally fastidious and require virtually all of the B vitamins, purines, pyrimidines, and amino acids for growth under defined chemical conditions (Koser, 1968). However, a sampling of the nutritional requirements of these organisms reveals considerable variation in the minimal medium required for support growth (Table 6.1). *Staphylococcus aureus*, which is genetically related to the streptococci, is capable of growing on a defined medium containing glucose, ammonium and phosphate salts, other inorganic salts, nicotinic acid, thiamine, and ten amino acids (Bondi et al., 1954). The minimal nutritional requirements of *Streptococcus bovis* include biotin, pantothenate, thiamine, adenine, xanthine, arginine, glutamate, inorganic salts, and glycerol as a carbon source if CO_2 (as $NaHCO_3$) is supplied (I. J. Barnes et al., 1961). For *S. faecium*, CO_2 stimulates growth but is not an essential growth factor (Cook, 1976). This organism requires 12 amino acids, since it is unable to assimilate ammonia. *S. faecium* is responsible for a major portion of the ruminal ureolytic activity in sheep.

S. bovis can utilize ammonia as a sole source of nitrogen. It produces an NADP-linked glutamate dehydrogenase that appears to be the only mechanism of ammonia assimilation (Burchall et al., 1964):

$$\alpha\text{-Ketoglutarate} + NH_4^+ + NADPH + H^+ \leftrightharpoons \text{L-Glutamate} + NADP^+ \quad (68)$$

Although glutamate dehydrogenase is a major route for ammonia assimilation in a variety of organisms, the tandem operation of glutamine synthetase and glutamate synthase may also be an important mechanism of ammonia assimilation in some organisms. These reactions provide an effective scavenging system under the conditions of low ammonia concentration:

Glutamine synthetase:

$$\text{L-Glutamate} + NH_4^+ + ATP \rightarrow \text{L-Glutamine} + ADP + P_i \quad (69)$$

Glutamate synthase:

$$\alpha\text{-Ketoglutarate} + \text{L-glutamine} + NADPH + H^+ \rightarrow 2 \text{ L-glutamate} + NADP^+ \quad (70)$$

Net:

$$\alpha\text{-Ketoglutarate} + NH_4^+ + ATP + NADPH + H^+ \rightarrow \text{L-Glutamate} + ADP + P_i + NADP^+ \quad (71)$$

Griffith and Carlsson (1974) measured the activity of ammonia-assimilating enzymes in *S. sanguis*, *S. bovis*, *S. mutans*, and *S. salivarius*. Under the condi-

TABLE 6.1 Growth Factor Requirements of Acid-forming Bacteria

Nutrient	Streptococcus pyogenes[a]	Streptococcus agalactiae[b]	Streptococcus faecium[c]	Streptococcus salivarius[d]	Streptococcus bovis[e]	Lactobacillus plantarum[f]	Leuconostoc paramesenteroides[g]	Staphylococcus aureus[h]	Propionibacterium avidum[i]	Gluconibacter oxidans[j]
Glucose[k]	+	+	+	+	+	+	+	+	+	+
Na acetate	−	−	+	−	−	−	+	−	+	−
Phosphate	+	+	+	−	+	+	+	+	+	+
Inorganic salts[l]	+	+	+	+	+	+	+	+	+	+
NaHCO₃	+	+	+	−	+	−	−	−	−	−
Ammonium sulfate	−	−	−	+	+	+	+	+	+	+
Urea	−	−	+	−	−	−	−	−	−	−
Biotin	+	+	+	+	+	+	+	−	+	−
Cobalamin	−	−	−	−	−	−	+	−	−	−
Folate	−	+	+	−	−	+	+	−	−	−
PAB	+	−	+	+	−	+	+	−	−	+
Nicotinate	+	+	+	+	−	+	+	+	+	+
Pantothenate	+	+	+	+	+	+	+	−	+	+
Pyridoxine	+	+	+	+	−	+	+	−	−	−
Riboflavin	+	+	+	+	−	+	+	−	−	−
Thiamine	+	+	+	+	+	+	+	+	S	−
Adenine	+	+	+	−	+	+	+	−	+	−
Guanine	+	+	+	−	−	+	+	−	+	−
Uracil	+	+	+	−	−	+	+	−	+	−
Xanthine	−	+	−	−	+	−	+	−	+	−
L-Alanine	+	+	+	−	−	+	−	−	+	−
L-Arginine	+	+	+	−	+	+	+	+	±	−
L-Aspartate	+	+	+	−	−	+	+	+	−	−
L-Asparagine	−	−	+	−	−	−	−	−	−	−
L-Cysteine	+	+	−	+	−	+	+	+	+	−
L-Glutamate	+	+	+	−	+	+	+	−	+	−
L-Glutamine	−	+	−	−	−	−	−	−	+	−
Glycine	+	+	+	−	−	+	+	+	+	−
L-Histidine	+	+	+	−	−	+	+	+	+	−
L-Isoleucine	+	+	+	−	−	+	+	−	−	−
L-Leucine	+	+	+	−	−	+	+	+	−	−
L-Lysine	+	−	+	−	−	+	−	−	−	−
L-Methionine	+	+	+	−	−	+	+	−	+	−
L-Phenylalanine	+	+	+	−	−	+	+	+	+	−
L-Proline	+	+	+	−	−	−	+	+	+	−
L-Serine	+	+	+	−	−	+	+	−	+	−
L-Threonine	+	+	+	−	−	+	+	−	−	−

TABLE 6.1   (Continued)

Nutrient	Streptococcus pyogenes[a]	Streptococcus agalactiae[b]	Streptococcus faecium[c]	Streptococcus salivarius[d]	Streptococcus bovis[e]	Lactobacillus plantarum[f]	Leuconostoc paramesenteroides[g]	Staphylococcus aureus[h]	Propionibacterium avidum[i]	Gluconobacter oxydans[j]
L-Tryptophan	+	+	+	−	−	+	+	−	+	−
L-Tyrosine	+	+	+	−	−	+	+	−	+	−
L-Valine	+	+	+	−	−	+	+	+	−	−
Tween 80	−	−	−	−	−	S	+	−	S	−

S, stimulatory but not required by most species.
[a] Dassy and Alouf (1983). These authors cite previous studies on the development of chemically defined medium for *Streptococcus pyogenes*.
[b] Willett and Morse (1966).
[c] Cook (1976).
[d] Carlsson (1971).
[e] I. J. Barnes et al. (1961).
[f] Broquist and Snell (1951). The nutritional requirements shown for *Lactobacillus plantarum* also support optimal growth of *Lactobacillus casei* and *Streptococcus faecalis*. Tween 80 or oleic acid partially replaced the biotin requirement for *L. plantarum*.
[g] Garvie (1967).
[h] Bondi et al. (1954).
[i] Delwiche (1949), Ferguson and Cummins (1978). The nutritional requirements shown are for *Propionibacterium avidum*. Other species of propionibacteria often require additional amino acids for optimal growth.
[j] Greenfield and Claus (1972). The nutritional requirements shown for *Gluconobacter oxydans* also support optimal growth of *Acetobacter aceti*.
[k] Glucose will usually serve as the sole source of carbohydrate for the organisms shown here. In some cases, other carbohydrate sources such as glycerol, sucrose, or mannitol can be substituted or may be preferred.
[l] The inorganic salts most commonly required are $MgSO_4$, $MnCl_2$, and $FeSO_4$. Other inorganic salts may be required by individual organisms.

tions of ammonia limitation in a chemostat they found glutamate dehydrogenase to be the only route of ammonia assimilation in these organisms. Although glutamine synthetase was detected in *S. bovis*, it was absent in the other species, and glutamate synthase activity could not be detected in any.

A number of lactic acid–producing organisms are able to generate ATP from the degradation of arginine via the arginine deiminase (dihydrolase) pathway. This system involves the activity of three enzymes:

Arginine deiminase:

$$\text{L-Arginine} + H_2O \rightarrow \text{L-citrulline} + NH_3 \tag{72}$$

Ornithine transcarbamylase:

$$\text{L-citrulline} + P_i \leftrightarrow \text{L-Ornithine} + \text{Carbamyl phosphate} \tag{73}$$

Carbamate kinase:

$$\text{Carbamyl phosphate} + ADP \leftrightarrow ATP + CO_2 + NH_3 \qquad (74)$$

Production of the enzymes of this pathway in *S. lactis* is regulated by growth conditions. The specific activities of arginine deiminase and ornithine transcarbamylase are higher in galactose-grown cells than in glucose- or lactose-grown cells. Arginine added during growth increases the activity of these two enzymes regardless of the sugar added as carbon source. Carbamate kinase–specific activity is not altered by changing the composition of the growth medium. Arginine deiminase is induced by arginine, but not until glucose limitation has occurred. Crow and Thomas (1982) were unable to detect arginine deiminase activity in any of nine strains of *S. cremoris*. Some strains of *S. cremoris* displayed ornithine transcarbamylase activity, but all showed carbamate kinase activities comparable to those observed in *S. lactis*.

Pyridoxal 5′-phosphate (PLP) is the coenzyme of most of the general reactions of decarboxylation, transamination, deamination, and racemization of amino acids. Many PLP-dependent reactions involving amino acids have been demonstrated in lactic acid bacteria. However, in a strain of *Lactobacillus* an unusual histidine decarboxylase contains bound pyruvate at the active catalytic site. The pyruvate apparently arises from deamination of a serine residue originally incorporated at this site in the enzyme structure. The pyruvoyl residue apparently functions by a Schiff's base mechanism in a manner analogous to that proposed for PLP-dependent reactions (Vaaler et al., 1982).

L. plantarum, *L. casei*, and *S. faecalis* require all B vitamins except cobalamin and lipoate, purines, pyrimidines, and 18 amino acids for good growth in defined medium. In the absence of biotin, *L. plantarum* requires aspartate and an unsaturated fatty acid supplied as either oleic acid or Tween 80. CO_2 is stimulatory for growth (Broquist and Snell, 1951). The biotin:oleate relationship reported by these and other workers provided the background that led to the demonstration of the requirement for biotin in CO_2-fixing enzymes involved in fatty acid synthesis (H. G. Wood, 1982). It seems likely that CO_2 may be required by many of the more fastidious organisms in the lactic group, but the presence of CO_2 in the environment could preclude proper control of the CO_2 concentration for accurate assessment of such a requirement. In a fermentor in which the gaseous atmosphere could be carefully regulated, Repaske et al. (1974) found that the presence of CO_2 completely eliminated the lag period for *S. sanguis*.

Members of the genus *Leuconostoc* display considerable variation in their amino acid requirements (Garvie, 1967). Valine and glutamic acid are required by all strains. Methionine is stimulatory for most strains, but alanine is not required by any. Each of 14 other amino acids are required by, or stimulatory to, some strains. Most of the dextran-forming strains require fewer than eight amino acids, whereas strains that do not produce dextran require most of the 14 additional amino acids. *L. paramesenteroides* appears to have the most extensive requirements for B vitamins, purines, pyrimidines, and amino acids (Table 6.1). This species also exhibits a requirement for fatty acids (as supplied by Tween 80).

S. agalactiae (group B) displays extensive nutritional requirements. Nineteen amino acids, purines, pyrimidines, and all B vitamins except cobalamin, PABA, and lipoate are required for growth (Willett and Morse, 1966).

The propionibacteria (Table 6.1) display nutritional requirements similar to those of many of the lactic acid bacteria (Delwiche, 1949; Ferguson and Cummins, 1978).

The acetic acid bacteria have, on occasion, been cited as having complex nutritional requirements. However, Greenfield and Claus (1972) were able to grow *G. oxydans* and *A. aceti* on a simple medium containing inorganic salts, PABA, nicotinate, pantothenate, ammonium sulfate as the nitrogen source, and glycerol as the carbon source (Table 6.1). These simple growth requirements compare favorably with the less exacting nutritional requirements of many other gram-negative organisms.

Bacteria conserve and transduce metabolic energy by means of an electrochemical gradient of hydrogen ions across the cytoplasmic membrane. This electrochemical gradient gives rise to a proton motive force (PMF) and consists of a transmembrane potential and a transmembrane pH gradient. The PMF available is the net result of extrusion of hydrogen ions and their reentry into the cell. In an aerotolerant anaerobe such as *S. lactis*, H^+ efflux is catalyzed by the membrane ATPase complex, and H^+ influx is mediated by a number of proton-linked transport systems (Kashet et al., 1980).

The streptococci have been useful for the study of energy coupling to the transport of amino acids and B vitamins. Metabolite transport is more amenable to study if its subsequent intracellular utilization can be restricted. The nonmetabolizable analog of alanine, α-aminoisobutyric acid (AIB), has been useful as a model amino acid for the study of amino acid transport (J. Thompson, 1976; Reizer and Panos, 1982). As was mentioned earlier, *S. lactis* can generate energy via glycolysis or the arginine deiminase system. Galactose-grown cells of *S. lactis* accumulate AIB by using energy from glycolysis and arginine catabolism. Proton-conducting ionophores abolished uptake and induced AIB efflux under conditions in which glycolysis and arginine catabolism continued at high levels. These findings suggest that a PMF is most likely involved in the active transport of AIB. Acceleration of the rate of efflux of AIB from preloaded cells in the presence of metabolizable energy sources and the induction of rapid efflux by K^+ and Rb^+ suggest that energy is coupled to AIB exit (J. Thompson, 1976). In *S. pyogenes* the rate and extent of AIB accumulation increased significantly in the presence of glucose. The effect of various inhibitors on AIB uptake provides evidence that a proton motive force is involved in AIB accumulation. AIB was not accumulated by the physiologically isotonic L-form. A deficiency in the coupling of energy to AIB transport is responsible for the lack of active accumulation by the L-form (Reizer and Panos, 1982).

The active transport of folate and thiamine into *Lactobacillus casei* is accomplished by two independent systems. In conjunction with the transport process, appreciable amounts of folate and thiamine are bound to cellular components expressed only in cells propagated under conditions of vitamin limitation. Two separate proteins, one specific for folate and the other for thiamine, have been solubilized from membrane preparations of *L. casei*, purified to homogeneity, and shown to be extremely hydrophobic, water-insoluble proteins (Henderson et al., 1977). Although the folate-binding protein has no measurable affinity for thiamine, folate transport is strongly inhibited by thiamine (Henderson et al., 1979). Cells preloaded with thiamine can transport folate at a normal rate, a fact indicating that inhibition is the result of thiamine entry rather than its presence in the cell. Folate does not interfere with the binding of thiamine to its transport protein but does interfere with thiamine transport. Thiamine or folate uptake also inhibits biotin uptake, but biotin has only a minimal effect on the transport of either of these vitamins. The competition of these vitamins for transport via their specific binding proteins suggests that an additional component, present in limiting quantities, is a common requirement for all three transport activities. The ability of the specific transport protein to bind folate is dependent upon the presence of cations in the external medium, a fact suggesting that a cation-folate symport mechanism is involved (Henderson and Potuznik, 1982).

VII. GENETICS

Discovered as a "phenomenon" in *S. pneumoniae* by Griffith in 1928, the transformation of pneumococcal types was ultimately shown to be mediated by the transfer of DNA isolated from encapsulated strains to unencapsulated strains. (For an historical account of the early work of Griffith and others, see Downie (1972).) The elucidation of the mechanism of transformation is now recognized as the turning point in the establishment of DNA, rather than protein, as the bearer of hereditary information. This discovery also served as the starting point for the field of bacterial genetics. Since that time, transformation has been demonstrated in many other species of streptococci. In addition to *S. pneumoniae*, *S. sanguis* (group H, strains Challis and Wicky), *S. pyogenes*, and members of serological groups B and D have been studied in some detail. Intergeneric transformation by DNA from donor staphylococci to recipient *S. sanguis*, *S. pneumoniae*, and streptococci of groups A, B, and D has been achieved and provides an important basis for a close genetic relationship between staphylococci and streptococci (Engel et al., 1980). Detailed studies have established that competence for tranformation requires precise physiological conditions as well as the presence of specific competence factor(s) as described by Lacks (1977).

For many years it appeared that the conjugative transmission of genetic material was restricted to gram-negative bacteria. The first suggestion that conjugal transfer of genetic material might occur in gram-positive bacteria was made by Raycroft and Zimmerman (1964) to explain the transfer of chloramphenicol resistance in *S. faecalis*. Jacob and Hobbs (1974) provided specific proof of conjugal transfer of plasmid-borne multiple antibiotic resistance traits in *S. faecalis*. Other investigators have since demonstrated that transfer of plasmid-borne traits by a conjugative process occurs in *S. faecalis*, group D streptococci, and oral streptococci (Clewell, 1981). Intergeneric conjugal tranfer of plasmid-mediated antibiotic resistance from two species of *Streptococcus* to *Lactobacillus* at similar frequencies to interspecific streptococcal matings provides a basis for the genetic relatedness of lactobacilli and streptococci. Donor cell aggregation and transfer functions of conjugative plasmids of *S. faecalis* are induced in donor strains by sex pheromones (clumping-inducing agents) produced by recipient strains. Only strains preexposed to the sex pheromone behave as donors. A mutant plasmid that exhibits high donor potential in the absence of sex pheromone has been isolated. Optimal mating conditions independent of the sex pheromone have been defined (Dunny et al., 1982). Use of early exponential-phase donors at cell concentrations of 10^7 donor cells per milliliter and 10^8 recipient cells per milliliter and a short mating period of 10 min gave a high frequency of conjugal transfer. Krogstad et al. (1980) have provided direct visualization of cell–cell contact during conjugation in *S. faecalis*. High frequency variants of a lactose plasmid have been selected from *S. lactis*. These variants exhibit an unusual cell aggregation phenotype (Gasson and Davies, 1980). Walsh and McKay (1981) found that donor cell aggregation and high-frequency conjugation of *S. lactis* are associated with a 60-megadalton plasmid. Plasmid and "nonplasmid" tetracycline resistance determinants of *S. faecalis*, *S. agalactiae*, and *S. pneumoniae* have been cloned into *E. coli* (Burdett et al., 1982). Both PEP phosphotransferase and P-β-galactosidase activity are associated with a 32-megadalton plasmid in *S. lactis*. A chromosome-borne resistance transposon (Tn*916*) has been found in *S. faecalis*. The tetracycline resistance determinant located on this transposon is capable of transposing to several conjugative plasmids, often affecting the expression of a plasmid-encoded hemolysin determinant. A resistance plasmid can be transferred between streptococci of groups A, B, D, *S. pneumoniae*, and *Staphylococcus aureus*. Another resistance plasmid (pAMB) originating from *S. faecalis* is transferable by mating to *S. mutans*, *S. sanguis*, and *S. salivarius*. It can also be transferred to members of the Lactobacillaceae (see Clewell (1981) for earlier references).

Most strains of *S. cremoris*, *S. lactis*, and its subspecies *diacetylactis* carry a large complement of plasmids. The presence of multiple plasmids complicates the analysis of plasmid transfer experiments and the molecular study of individual plasmids. Development of a procedure for protoplast production and regeneration in lactic streptococci permits protoplast-promoted plasmid curing for a variety of plasmid phenotypes. This technique has been used to produce a plasmid-free strain of *S. lactis* subsp. *diacetylactis* and plasmid-free derivatives. A 33-megadalton plasmid (712) encoding genes for lactose utilization and protease production has been identified. This plasmid displays a high degree of molecular instability, a characteristic that accounts for the previously observed complexity of the *S. lactis* 712 plasmid. The ability to cure cryptic molecules from multiple plasmid complements by protoplast regeneration should prove useful for genetic studies in lactic streptococci and other gram-positive organisms (Gasson, 1983).

Transduction by temperate bacteriophage of a portion of the host chromosome to a recipient bacterium and its incorporation into the chromosome of the new host have been described in a number of streptococci. A temperate phage from lactose-positive (lac$^+$) *S. lactis* has been shown to transduce lactose-fermenting ability to lac$^-$ recipient strains. Treatment of the lysate with DNAse has no effect on this conversion, a finding indicating that the transfer was not mediated by free DNA. Lysates with 100-fold higher transducing ability were obtained from lac$^+$ transductants. These findings demonstrate the transduction of carbohydrate markers by a streptococcal phage and establish a temperate phage-mediated genetic transfer system in group N streptococci. Genetic analysis of antibiotic resistance markers of chromosomal origin has revealed the presence of at least three linkage groups in *S. pyogenes* (see Clewell, 1981).

VIII. GROWTH

Growth of a bacterial cell is usually perceived as an orderly but complex series of events involving: (1) replication of the chromosome, (2) increase in size and mass of the cell, and (3) extension of the cell surface and septation to form two daughter cells, each containing a copy of the chromosome and other vital components. Considerable evidence has been accumulated to support the view that the circular chromosome is attached to replication sites in the cell membrane or an extension of it (mesosome). Replication is initiated at a specific locus on the chromosome (origin) and DNA replicating forks proceed bidirectionally at a constant rate toward the terminus, a point equidistant from the origin. The terminus of the chromosome is also attached to the membrane. As chromosome replication proceeds, the attachment sites for the two chromosomes are separated by extension of the cell surface (membrane and wall) as shown in Fig. 6.9. The complex of membrane attachment sites and associated replication enzymes may be considered as a primitive mitotic apparatus that aids in the distribution of copies of the chromosome into the developing daughter cells. (For references supporting this view, see Moat (1979), Mendelson (1982), Parks et al. (1982), Schlaeppi and Karamata (1982).)

Growth of the cell envelope in *S. faecalis* and other gram-positive bacteria occurs primarily at the septal region. At this growing point there is an increase in area accomplished by progressive autolytic cleavage of the developing crosswall into two layers of peripheral wall. The two peripheral wall segments extend in both directions. Autolytic enzymes regulate both the rate at which the

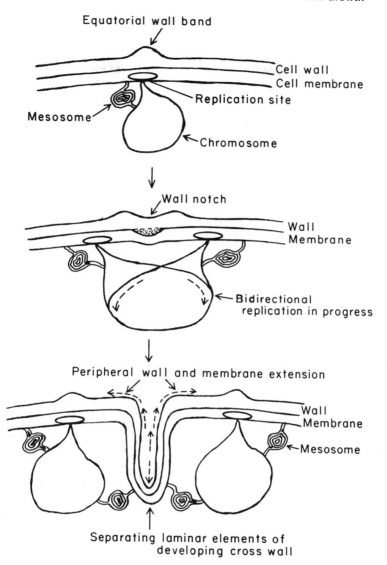

Equatorial wall band

Cell wall
Cell membrane
Replication site
Mesosome
Chromosome

Wall notch

Wall
Membrane

Bidirectional
replication in progress

Peripheral wall and membrane extension

Wall
Membrane
Mesosome

Separating laminar elements of
developing cross wall

FIGURE 6.9 Schematic diagram of the growth stages of a gram-positive bacterial cell. The bacterial chromosome is attached to a replication site embedded in the cell membrane. A mesosome is also attached to the chromosome. As the cell wall and cell membrane are extended, a notch develops in the wall. Extension of the surface layers of the cell results in separation of the replication sites as bidirectional replication of the chromosome continues. Mesosomes apparently aid in guiding the replicating chromosomes in the direction of cell extension. When chromosomal replication is complete, peripheral extension of the wall and membrane layers continues, and the laminar elements of the developing cross-wall are separated.

cross-wall layers are separated and the amount of surface remodeling occurring in the expanding surface layers. At early stages of growth, when autolytic capacity is maximal, incorporation of wall precursors into growing cross-walls and intercalation of precursors into separating peripheral wall surfaces are also maximal. As the cell cycle progresses and autolytic capacity decreases, the rate of cross-wall separation and expansion declines and precursors are directed into closing uncompleted septa. Reduction of autolytic capacity by antibiotic inhibition of protein synthesis results in preferential direction of cell wall synthesis toward closing of open cross-walls. Inhibition of DNA synthesis inhibits cell septation, a property suggesting that a signal indicating completion of a round of DNA replication is required to permit closing of the cross-walls (Higgins and Daneo-Moore, 1980).

Autolysins play an important role in the lysis accompanying inhibition by penicillin and other β-lactam antibiotics. In autolytic-defective pneumococci the dose response to penicillin is identical to that of wild-type cells, but lysis is replaced by inhibition of growth. Mutants lacking autolytic activity simultaneously develop resistance to all other cell wall inhibitors as well as to detergents and bacteriophage infection. These findings reflect the presence of a single autolysin in *S. pneumoniae* that is active in all lytic phenomena. These and other observations regarding the action of penicillin and other β-lactam agents have given rise to the development of new concepts regarding the mode of action of these agents on bacterial cells (Tomasz, 1979).

IX. SUMMARY

From the details presented here, it is obvious that considerable progress has been made in elucidating many facets of the physiology of the lactic, acetic, and propionic acid bacteria. In fact, apologies are hereby made to the many authors whose work has not been quoted owing to limitations of space. On the other hand, there are still many gaps in our knowledge of the specific metabolic and genetic capabilities of specific groups of organisms. I hope that this resume of what has transpired within the past few years will provide an indication as to areas of research that still remain to be undertaken.

ACKNOWLEDGMENTS

I would like to thank Drs. F. Ahmad, A. J. L. Cooper, G. W. Claus, J. W. Foster, L. Daneo-Moore, and C. Panos for their helpful comments on preliminary drafts of this manuscript and Drs. M. L. Higgins and D. Dicker for supplying the original photographs for Fig. 6.4.

REFERENCES

Ahmad, P. M., Stirling, L. A., and Ahmad, F. (1981) *J. Gen. Microbiol. 127*, 121–129.
Aloni, Y., Cohen, R., Benziman, M., and Delmer, D. (1983) *J. Biol. Chem. 258*, 4419–4423.

Ameyama, M., Shinagawa, E., Matsushita, K., and Adachi, O. (1981) *J. Bacteriol.* *145*, 814–823.

Archibald, F. S., and Fridovich, I. (1981) *J. Bacteriol.* *146*, 928–936.

Back, W. (1978) *Int. J. System. Bact.* *28*, 523–527.

Baird-Parker, A. C. (1974) in *Bergey's Manual of Determinative Bacteriology*, 8th ed. (Buchanan, R. E., and Gibbons, N. E., eds.), pp. 478–490, Williams and Wilkins, Baltimore.

Barnes, E. M., Impey, C. S., Stevens, B. J. H., and Peel, J. L. (1977) *J. Gen. Microbiol.* *102*, 45–53.

Barnes, I. J., Seeley, H. W., and VanDemark, P. J. (1961) *J. Bacteriol.* *82*, 85–93.

Beachey, E. H., Seyer, J. M., Dale, J. B., Simpson, W. A., and Kang, A. H. (1981) *Nature 292*, 457–459.

Bondi, A., Kornblum, J., and de St. Phalle, M. (1954) *J. Bacteriol.* *68*, 617–621.

Bridge, P. D., and Sneath, P. H. A. (1983) *J. Gen. Microbiol.* *129*, 565–597.

Briles, E. B., and Tomasz, A. (1975) *J. Gen. Microbiol.* *86*, 267–274.

Broquist, H. P., and Snell, E. E. (1951) *J. Biol. Chem.* *188*, 431–444.

Brown, A. T., and Wittenberger, C. L. (1971) *J. Bacteriol.* *106*, 456–467.

Brown, A. T., and Wittenberger, C. L. (1972) *J. Bacteriol.* *109*, 106–115.

Brown, R. M., Jr., Willison, J. H. M., and Richardson, C. L. (1976) *Proc. Nat. Acad. Sci. U.S.A. 73*, 4565–4569.

Burchall, J. J., Niederman, R. A., and Wolin, M. J. (1964) *J. Bacteriol.* *88*, 1038–1044.

Burdash, N. M., West, M. E., Newell, R. T., and Teti, G. (1981) *Amer. J. Clin. Pathol.* *76*, 819–822.

Burdett, V., Inamine, J., and Rajagopalan, S. (1982) *J. Bacteriol.* *149*, 995–1004.

Carlsson, J. (1971) *J. Gen. Microbiol.* *67*, 69–76.

Claus, G. W., Batzing, B. L., Baker, C. F., and Goebel, E. M. (1975) *J. Bacteriol.* *123*, 1169–1183.

Cleary, P. P., and Larkin, A. (1979) *J. Bacteriol.* *140*, 1090–1097.

Clewell, D. G. (1981) *Microbiol. Rev. 45*, 409–436.

Coleman, S. E., and Bleiweis, A. S. (1977) *J. Gen. Microbiol.* *129*, 445–456.

Collins, M. D., and Jones, D. (1981) *Microbiol. Rev. 45*, 316–354.

Cook, A. R. (1976) *J. Gen. Microbiol.* *97*, 235–240.

Crow, V. L., and Thomas, T. D. (1982) *J. Bacteriol.* *150*, 1024–1032.

Cummins, C. S., and Johnson, J. C. (1974) *J. Gen. Microbiol.* *80*, 435–442.

Daneo-Moore, L., Dicker, D., and Higgins, M. L. (1980) *J. Bacteriol.* *141*, 928–937.

Dassy, B., and Alouf, J. E. (1983) *J. Gen. Microbiol.* *129*, 643–651.

Deibel, R. H., and Seeley, H. W., Jr. (1974) in *Bergey's Manual of Determinative Bacteriology*, 8th ed. (Buchanan, R. E., and Gibbons, N. E., eds.), pp. 490–509, Williams and Wilkins, Baltimore.

DeLey, J., and Frateur, J. (1974) in *Bergey's Manual of Determinative Bacteriology*, 8th ed. (Buchanan, R. E., and Gibbons, N. E., eds.), pp. 251–253; 276–278, Williams and Wilkins, Baltimore.

Dellaglio, F., Trovatelli, L. D., and Sarra, A. G. (1981) *Zbl. Bakt. Hyg., 1. Abt. Orig. C2*, 140–150.

Delwiche, E. A. (1949) *J. Bacteriol. 58*, 395–398.

DeMoss, R. D., Bard, R. C., and Gunsalus, I. C. (1951) *J. Bacteriol. 62*, 499–511.

Dills, S. S., and Seno, S. (1983) *J. Bacteriol. 153*, 861–866.

Doelle, H. W. (1975) *Bacterial Metabolism*, 2nd ed., Academic Press, New York.

Downie, A. W. (1972) *J. Gen. Microbiol. 73*, 1–11.

Dunny, G., Yuhasz, M., and Ehrenfeld, E. (1982) *J. Bacteriol. 151*, 855–859.

Edelstein, E., Parks, L., Tsein, H.-C., Daneo-Moore, L., and Higgins, M. L. (1981) *J. Bacteriol. 146*, 798–803.

Eisenstein, T. K., Carey, R. B., Shockman, G. D., Smith, S. M., and Swenson, R. M. (1982) in *Seminars in Infectious Disease*, Vol. IV (Weinstein, L., and Fields, B. L., eds.), pp. 279–284, Georg Thieme, New York.

Eisenstein, T. K., DeCueninck, B. J., Resavy, D., Shockman, G. D., Carey, R. B., and Swenson, R. M. (1983) *J. Infect. Dis. 147*, 847–856.

Engel, H. W. B., Soedirman, N., Rost, J. A., Van Leeuwen, W. J., and Van Embden, J. D. A. (1980) *J. Bacteriol. 142*, 407–413.

Evans, J. B. (1974) in *Bergey's Manual of Determinative Bacteriology*, 8th ed. (Buchanan, R. E., and Gibbons, N. E., eds.), pp. 515–516, Williams and Wilkins, Baltimore.

Ferguson, D. A., Jr., and Cummins, C. S. (1978) *J. Bacteriol. 135*, 858–867.

Finnerty, W. R., and Makula, R. A. (1975) *CRC Crit. Rev. Microbiol. 4*, 1–40.

Fox, G. E., Stackbrandt, E., Hespell, R. B., Gibson, J., Maniloff, J., Dyer, T. A., Wolfe, R. S., Balch, W. E., Tanner, R. S., Magrum, L. J., Zablen, L. B., Blakemore, R., Gupta, R., Bonen, L., Lewis, B. J., Stahl, D. A., Luehrsen, K. R., Chen, K. N., and Woese, C. R. (1980) *Science 209*, 457–463.

Garvie, E. I. (1967) *J. Gen. Microbiol. 48*, 439–447.

Garvie, E. I. (1981) in *Bergey's Manual of Determinative Bacteriology*, 8th ed. (Buchanan, R. E., and Gibbons, N. E., eds.), pp. 510–513, Williams and Wilkins, Baltimore.

Garvie, E. I. (1981) *J. Gen. Microbiol. 127*, 209–212.

Gasson, M. J. (1983) *J. Bacteriol. 154*, 1–9.

Gasson, M. J., and Davies, F. L. (1980) *J. Bacteriol. 143*, 1260–1264.

Ghuysen, J.-M. (1977) in *Cell Surface Reviews*, Vol. IV, *Membrane Assembly and Turnover* (Poste, G., and Nicholson, G. L., eds.), ASP Biological and Medical Press, Amsterdam.

Gibbs, M., Dumrose, R., Bennett, F. A., and Bubeck, M. R. (1950) *J. Biol. Chem. 184*, 545–549.

Greenfield, S., and Claus, G. W. (1972) *J. Bacteriol. 112*, 1295–1301.

Griffith, C. J., and Carlsson, J. (1974) *J. Gen. Microbiol. 82*, 253–260.

Haigler, C. H., Brown, R. M., Jr., and Benziman, M. (1980) *Science 210*, 903–906.

Handley, P. S., and Carter, P. (1979) in *Pathogenic Streptococci* (Parker, M. T., ed.), pp. 241–242, Reedbooks, Chertsey, England.

Handley, P. S., and Jacob, A. E. (1981) *J. Gen. Microbiol. 127*, 289–293.

Henderson, G. B., and Potuznik, S. (1982) *J. Bacteriol. 150*, 1098–1102.

Henderson, G. B., Zevely, E. M., Kadner, R. J., and Huennekens, F. M. (1977) *J. Supramol. Struc. 6*, 239–247.

Henderson, G. B., Zevely, E. M., and Huennekens, F. M. (1979) *J. Bacteriol. 137*, 1308–1314.

Henriksen, D., and Henrichsen, J. (1975) *Acta Pathol. Microbiol. Scandinavica B83*, 133–140.

Higgins, M. L., and Daneo-Moore, L. (1974) *J. Cell Biol. 61*, 288–300.

Higgins, M. L., and Daneo-Moore, L. (1980) *J. Bacteriol. 141*, 938–945.

Higgins, M. L., and Shockman, G. D. (1970) *J. Bacteriol. 103*, 244–254.

Holdeman, L. V., and Moore, W. E. C. (1974) in *Bergey's Manual of Determinative Bacteriology*, 8th ed. (Buchanan, R. E., and Gibbons, N. E., eds.), pp. 641–657, Williams and Wilkins, Baltimore.

Horne, D., and Tomasz, A. (1979) *J. Bacteriol. 137*, 1180–1184.

Hoshino, E., Frolander, F., and Carlsson, J. (1978) *J. Gen. Microbiol. 107*, 235–248.

Jacob, A. E., and Hobbs, S. J. (1974) *J. Bacteriol. 117*, 360–372.

Joseph, R., and Shockman, G. D. (1975) *Infect. Immun. 12*, 333–338.

Kashet, E. R., Blanchard, A. G., and Metzger, W. C. (1980) *J. Bacteriol. 143*, 128–134.

Kitahara, K. (1974) in *Bergey's Manual of Determinative Bacteriology*, 8th ed. (Buchanan, R. E., and Gibbons, N. E., eds.), pp. 513–515, Williams and Wilkins, Baltimore.

Kleppe, K., Ovrebo, S., and Lossius, J. (1979) *J. Gen. Microbiol. 112*, 1–13.

Knox, K. W., Campbell, L. K., Evans, J. D., and Wicken, A. J. (1980) *J. Gen. Microbiol. 119*, 1203–1209.

Koser, S. A. (1968) *Vitamin Requirements of Bacteria and Yeasts*. Charles C. Thomas, Springfield, Ill.

Krogstad, D. J., Smith, R. M., Moellering, R. D., Jr., and Parquette, A. R. (1980) *J. Bacteriol. 141*, 963–967.

Lacks, S. (1977) in *Microbial Interactions*, Series B, *Receptors and Recognition*, Vol. II (Ressig, J. L., ed.), pp. 177–232, Chapman and Hall, Ltd., London.

Lichstein, H. C. (1982) in *Experiences in Biochemical Perception*, pp. 349–360, Academic Press, New York.

Lichstein, H. C. (1983) *Bacterial Nutrition*, Benchmark Papers in Microbiology, Vol. XIX (Umbreit, W. W., series ed.), Hutchison Ross, Stroudsburg, Penn.

Love, D. N., Jones, R. F., and Bailey, M. (1979) *J. Gen. Microbiol. 112*, 401–403.

Mackie, E. B., Brown, K. N., Lam, J., and Costerton, J. W. (1979) *J. Bacteriol. 138*, 609–617.

Markham, J. L., Knox, K. W., Wicken, A. J., and Hewett, J. J. (1975) *Infect. Immun. 12*, 378–386.

Mendelson, N. H. (1982) *Microbiol. Rev. 46*, 341–375.

Moat, A. G. (1979) *Microbiol Physiology*, John Wiley and Sons, New York.

Moore, E. E. C., and Holdeman, L. V. (1974) in *Bergey's Manual of Determinative Bacteriology*, 8th ed. (Buchanan, R. E., and Gibbons, N. E., eds.), pp. 633–641, Williams and Wilkins, Baltimore.

Mowshowitz, S., Englard, S., and Avigad, G. (1974) *J. Bacteriol. 119*, 363–370.

Munoz, E., Ghuysen, J.-M., and Heyman, H. (1967) *Biochemistry 6*, 3659–3670.

Murchison, H., Larrimore, S., and Curtiss, R., III (1981) *J. Bacteriol. 34*, 1044–1055.

Neimark, H., and London, J. (1982) *J. Bacteriol. 150*, 1259–1265.

Ofek, I., Simpson, W. A., and Beachey, E. H. (1982) *J. Bacteriol. 149*, 426–433.

Park, Y. H., and McKay, L. L. (1982) *J. Bacteriol. 149*, 420–425.

Parks, L. C., Rigney, D., Daneo-Moore, L., and Higgins, M. L. (1982) *J. Bacteriol. 152*, 191–200.

Raycroft, R. E., and Zimmerman, L. N. (1964) *J. Bacteriol. 87*, 799–801.

Reizer, J., and Panos, C. (1980) *Proc. Nat. Acad. Sci. U.S.A. 77*, 5497–5501.

Reizer, J., and Panos, C. (1982) *J. Bacteriol. 149*, 211–220.

Repaske, R., Repaske, A. C., and Mayer, R. D. (1974) *J. Bacteriol. 117*, 652–657.

Reusch, V. M., Jr., and Panos, C. (1977) *J. Bacteriol. 129*, 1407–1414.

Reyn, A. (1974) in *Bergey's Manual of Determinative Bacteriology*, 8th ed. (Buchanan, R. E., and Gibbons, N. E., eds.), pp 516–517, Williams and Wilkins, Baltimore.

Ritchey, T. W., and Seeley, H. W., Jr. (1974) *J. Gen. Microbiol. 85*, 220–228.

Ritchey, T. W., and Seeley, H. W., Jr. (1976) *J. Gen. Microbiol. 93*, 195–203.

Rogosa, M. (1974) in *Bergey's Manual of Determinative Bacteriology*, 8th ed.

(Buchanan, R. E., and Gibbons, N. E., eds.), pp. 576–593, Williams and Wilkins, Baltimore.

Romano, A. H., Trifone, J. D., and Brustolon, M. (1979) *J. Bacteriol. 139*, 93–97.

Russell, R. (1979) *J. Gen. Microbiol. 112*, 197–201.

Schlaeppi, J.-M., and Karamata, D. (1982) *J. Bacteriol. 152*, 1231–1240.

Schofield, G. M., and Schaal, K. P. (1981) *J. Gen. Microbiol. 127*, 237–259.

Slabj, B. M., and Panos, C. (1976) *J. Bacteriol. 127*, 855–862.

Speckman, R. A., and Collins, E. B. (1973) *Appl. Microbiol. 26*, 744–746.

Stirling, L. A., Ahmad, P. M., and Ahmad, F. (1981) *J. Bacteriol. 148*, 933–940.

Sugino, W. M., Wek, R. C., and Kingsbury, D. T. (1980) *J. Gen Microbiol. 121*, 333–338.

Swissa, M., Aloni, T., Weinhouse, H., and Benziman, M. (1980) *J. Bacteriol. 143*, 1141–1150.

Takahashi, S., Abbe, K., and Yamada, T. (1982) *J. Bacteriol. 149*, 1034–1040.

Theodore, T. S., and Panos, C. (1973) *J. Bacteriol. 116*, 571–576.

Thomas, T. D., Ellwood, D. C., and Longyear, V. M. C. (1979) *J. Bacteriol. 138*, 109–117.

Thomas, T. D., Turner, K. W., and Crow, V. L. (1980) *J. Bacteriol. 144*, 672–682.

Thompson, J. (1976) *J. Bacteriol. 127*, 719–730.

Thompson, J., and Saier, M. H., Jr. (1981) *J. Bacteriol. 146*, 885–894.

Thompson, T. D. (1980) *J. Bacteriol. 144*, 683–691.

Tomasz, A. (1979) *Ann. Rev. Microbiol. 33*, 113–137.

Vaaler, G. L., Recsei, D. A., Fox, J. L., and Snell, E. E. (1982) *J. Biol. Chem. 257*, 12770–12774.

Walsh, P. M., and McKay, L. L. (1981) *J. Bacteriol. 146*, 937–944.

Ward, J. B. (1981) *Microbiol. Rev. 45*, 211–243.

White, S. A., and Claus, G. W. (1982) *J. Bacteriol. 150*, 934–943.

Willett, N. P., and Morse, G. E. (1966) *J. Bacteriol. 91*, 2245–2250.

Winkelstein, J. A., and Tomasz, A. (1978) *J. Immunol. 120*, 174–178.

Woese, C. R. (1981) *Scientific American 244*, 98–122.

Woldringh, C. L., and Nanninga, N. (1976) *J. Bacteriol. 127*, 1455–1464.

Wolin, M. J. (1964) *Science 146*, 775–777.

Wood, H. G. (1977) *Fed. Proc. 36*, 2197–2205.

Wood, H. G. (1982) in *Of Oxygen, Fuels, and Living Matter*, Part 2 (Semenza, G., ed.), pp. 173–250, John Wiley and Sons, New York.

Wood, W. A. (1961) in *The Bacteria*, Vol. XI, *Metabolism* (Gunsalus, I. C., and Stanier, R. Y., eds.), pp. 59–149, Academic Press, New York.

Methanogens

Meyer J. Wolin
Terry L. Miller

I. INTRODUCTION

Bioconversion of organic substrates to CH_4 and CO_2 occurs in anaerobic environments and requires a complex microbial community. Biopolymers, for example, cellulose, hemicellulose, starch, and proteins, are common substrates for bioconversion. In complete bioconversion systems, nonmethanogenic populations of the microbial community ferment organic substrates to acetate, H_2, and CO_2, and methanogenic populations convert these products to CH_4. H_2 is used to reduce CO_2 to CH_4, and acetate is decarboxylated to CH_4 and CO_2. Common environments for complete bioconversion include swamps, muds, and anaerobic waste decomposition systems. Partial bioconversion of biopolymers to short-chain volatile fatty acids, CH_4 and CO_2 is common in intestinal tract environments, for example, the rumen of ruminating herbivores, the large intestine of monogastric animals, and the intestine of termites.

The methanogens of these important anaerobic systems constitute a phylogenetically diverse group of bacteria despite their significant physiological and biochemical similarities. This article describes the similarities and differences between methanogens. References are mainly to recent articles. The review by Balch et al. (1979) contains an extensive list of references to earlier publications.

TABLE 7.1   Key to Species of Methanogenic Bacteria Based on Simple Phenotypic Characters[a]

I. Gram-positive to gram-variable rods or lancet-shaped cocci often forming chains and filaments

                  Order I.   *Methanobacteriales*
                  Family I.   *Methanobacteriaceae*

  A. Slender, straight to irregularly crooked long rods often occurring in filaments

                  Genus I.   *Methanobacterium*

    1. Mesophilic
      a. Methane produced from formate
          Methanobacterium formicicum
      b. Methane not produced from formate
          Methanobacterium bryantii
    2. Thermophilic
          Methanobacterium thermoautotrophicum

  B. Short rods or lancet-shaped cocci that often occur in pairs or chains

                  Genus II.   *Methanobrevibacter*

    1. Cells form short, nonmotile rods that do not utilize formate
          Methanobrevibacter arboriphilus
    2. Chain-forming, lancet-shaped cocci that produce methane from formate and require acetate as a carbon source
      a. Growth requirement for 2-mercaptoethanesulfonic acid and D-α-methyl butyrate
          Methanobrevibacter ruminantium
      b. Do not have an obligate growth requirement for 2-mercaptoethanesulfonic acid or D-α-methyl butyrate
          Methanobrevibacter smithii

II. Gram-negative cells or gram-positive cocci occurring in packets

  A. Gram-negative, regular to slightly irregular cocci often forming pairs

                  Order II.   *Methanococcales*
                  Family I.   *Methanococcaceae*
                  Genus I.   *Methanococcus*

    1. Cells inhibited by addition of 5% NaCl to medium
          Methanococcus vannielii
    2. Cells not inhibited by addition of 5% NaCl to medium
          Methanococcus voltae

  B. Gram-negative rods or highly irregular cocci occurring singly

                  Order III.   *Methanomicrobiales*
                  Family I.   *Methanomicrobiaceae*

    1. Straight to slightly curved, motile, short rods
                Genus I.   *Methanomicrobium*
           Methanomicrobium mobile
    2. Irregular coccoid cells
                Genus II.   *Methanogenium*
      a. Cells require acetate
          Methanogenium cariaci

TABLE 7.1 (Continued)

b. Cells do not require acetate
Methanogenium marisnigri
3. Regularly curved, slender, motile rods, often forming continuous spiral filaments
Genus III.   *Methanospirillum*
Methanospirillum hungatei

C. Gram-positive coccoid cells that usually occur in packets and ferment methanol, methylamine, and acetate
Family II.   *Methanosarcinaceae*
Genus I.   *Methanosarcina*
Methanosarcina barkeri

[a] Reproduced with permission from Balch et al. (1979).

Nomenclature

The names of methanogens used in this article are based on the classification by Balch et al. (1979). The determinative key from that review is reproduced here as Table 7.1. The reader is referred to the review for relationships between the names of organisms listed in Table 7.1 and previously assigned names. Publications about new species that have appeared since the review are cited at appropriate points in the chapter.

Archaebacteria

Methanogens are members of a newly defined kingdom of prokaryotes, the Archaebacteria (Woese, 1982). Most of the more common bacteria, for example, *Streptococcus*, *Escherichia*, *Clostridium*, and *Pseudomonas*, are in the kingdom Eubacteria. Archaebacteria do not contain the peptidoglycan polymer that is common to the cell walls of Eubacteria, and the lipids of Archaebacter are distinctly different from the lipids of Eubacteria (Langworthy et al., 1982). The nonmethanogenic members of the Archaebacteria are either thermophilic or halophilic and are found in hot or high-salt environments.

II. NUCLEOTIDE SEQUENCE

An important method for determining whether particular species of bacteria are Archaebacteria or Eubacteria depends on comparisons of the degree of nucleotide sequence homology between the 16S ribosomal ribonucleic acids (16S rRNA's) of species of these kingdoms (Woese, 1982). The method is also extremely useful for showing relationships between species within kingdoms. Woese and his colleagues have demonstrated that comparison of nucleotide sequence homologies is a powerful method for establishing phylogenetic relationships. The method involves isolation of ^{32}P-labeled 16S rRNA, digestion of the RNA with T-1 ribonuclease and separation of the resulting oligonucleotides by two-dimensional electrophoresis. A fingerprint of oligonucleotide spots is obtained. The spots of the electrophoretic fingerprint are separated according to (a) the number of uridine residues and (b) the number and sequence of nucleotides in each oligonucleotide.

If two organisms are closely related, for example, two species of *Streptococcus*, their fingerprints will be very similar, whereas those of a *Streptococcus* and a *Pseudomonas* or *Salmonella* will be very different. The relationships can be expressed quantitatively by a similarity index (S_{AB}). The S_{AB} of organisms A and B is equal to 2 times the number of nucleotides in all identical oligonucleotides of hexamer and larger size, divided by the sum of all nucleotides found in hexamer and larger oligonucleotides in each organism. A S_{AB} of 1.0 indicates complete sequence homology between the two organisms, while a S_{AB} of 0.02 represents random coincidence between the sequences. More detailed descriptions of the sequencing procedures and their use in estimating phylogenetic relatedness are found in references cited by Balch et al. (1979) and Woese (1982).

S_{AB} comparisons show very little similarity between all Archaebacteria and all Eubacteria. This lack of homology suggests early divergence of the two kingdoms during evolution (Woese, 1982). Similar analyses of methanogens show considerable differences between some species, indicating significant phylogenetic diversity despite the common characteristic of CH_4 production (Fig. 7.1). The degree of similarity and differences between strains and species of methanogens, as measured by S_{AB} values, was a major tool in establishing the taxonomic scheme shown in Table 7.1.

III. CELL SHAPE AND PHYSIOLOGY

Morphology

Phylogenetic diversity among species of methanogens is also revealed by the variety of their cell shapes (Fig. 7.2). The family Methanobacteriaceae contains rod-shaped bacteria that range from small coccobacillary forms to long filaments, depending on the species. The members of the family Methanococcaceae are cocci, as are species of the genus *Methanogenium* of the family Methanomicrobiaceae. The latter family also includes two other genera: *Methanospirillum*, characterized by curved rods that form long, spiral filaments, and *Methanomicrobium*, whose members form short rods. The members of the family Methanosarcinaceae are large, spherical cocci, which often form packets of varying size. More recently described methanogens include a species that forms plates of cells with sharp edges, *Methanoplanus limicola* (Wildgruber et al., 1982) and a species of irregular cocci, *Methanolobus tindarius* (König and Stetter, 1982).

Motility

Methanogens may be motile or nonmotile. Motile genera include *Methanospirillum*, *Methanococcus*, *Methanomicrobium*, and *Methanogenium*.

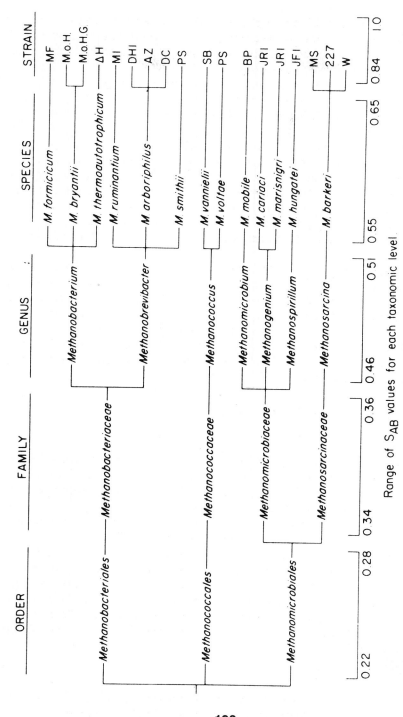

FIGURE 7.1 Taxonomic treatment of methanogenic bacteria based on 16-S rRNA cataloging. (Reproduced with permission from Balch et al., 1979.)

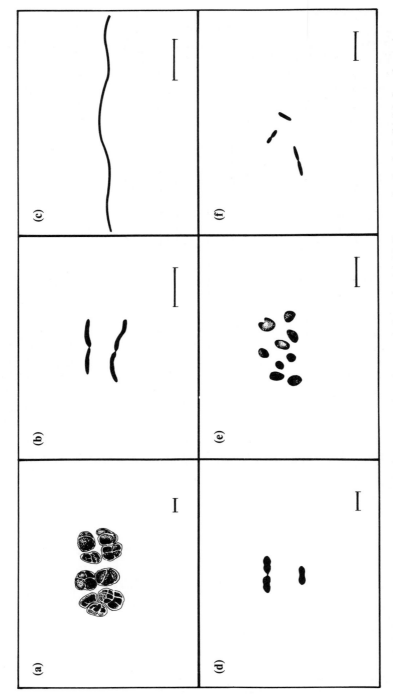

FIGURE 7.2 Some (not all) morphological forms of methanogenic bacteria. (a) *Methanosarcina barkeri*, (b) *Methanobacterium thermoautotrophicum*, (c) *Methanospirillum hungatei*, (d) *Methanobrevibacter ruminantium*, (e) *Methanogenium cariaci*, and (f) *Methanobrevibacter arboriphilus*. Bars in (a)–(e) = 5 μm. Bar in (f) = 8 μm.

Growth Temperatures

All methanogens isolated thus far are either mesophiles or thermophiles. *Methanothermus fervidus* is an extreme thermophile that grows optimally at 83°C (Stetter et al., 1981). *Methanobacterium thermoautotrophicum* (Zeikus and Wolfe, 1972), *Methanococcus thermolithotrophicus* (Huber et al., 1982), *Methanogenium thermophilicum* (Rivard and Smith, 1982), and a thermophilic *Methanosarcina* (Zinder and Mah, 1979) are moderate thermophiles that have optima at 65–70°C, 65°C, 55°C, and 50°C, respectively. No psychrophilic methanogens have been described.

pH

All methanogens grow in the range of about pH 6–8. No methanogen has yet been isolated that is capable of growth at extremely high or low pH values, even though CH_4 is produced in the low-pH environments of peat bogs.

Spores

Thus far no methanogen has been reported to produce spores. There are no reports of attempts to isolate heat-resistant forms of mesophilic species.

Anaerobiosis

All methanogens are obligate anaerobes. Growth is not initiated unless the oxidation–reduction potential of the medium is poised below about −300 mV, and several species are killed by exposure to small amounts (about 5 ppm) of O_2.

Isolation and cultivation procedures almost always make use of the techniques developed by Hungate (1969). Modifications have been developed (Miller and Wolin, 1974), but all of these methods use the same principles. Media are prepared under O_2-free gas prior to sterilization, and reducing agents and inocula are added to sterile media in the absence of O_2. The introduction of anaerobic glove boxes, based on the design described by Aranki and Freter (1972), has helped to facilitate the handling of methanogens. The Hungate technique and the glove boxes are also useful for studies of O_2-sensitive subcellular material, for example, for enzyme studies.

Substrates for Methanogenesis

The only known substrates for the production of CH_4 are H_2 and CO_2, formate, methanol, methylamine, dimethylamine, trimethylamine, and acetate. The equations for the various methanogenic reactions are:

$$4\ H_2 + CO_2 \longrightarrow CH_4 + 2\ H_2O$$
$$4\ HCO_2H \longrightarrow CH_4 + 3\ CO_2 + 2\ H_2O$$
$$4\ CH_3OH \longrightarrow 3\ CH_4 + CO_2 + 2\ H_2O$$
$$4\ CH_3NH_2Cl + 2\ H_2O \longrightarrow 3\ CH_4 + CO_2 + 4\ NH_4Cl$$
$$2\ (CH_3)_2NHCl + 2\ H_2O \longrightarrow 3\ CH_4 + CO_2 + 2\ NH_4Cl$$
$$4\ (CH_3)_3NCl + 6\ H_2O \longrightarrow 9\ CH_4 + 3\ CO_2 + 4\ NH_4Cl$$
$$CH_3CO_2H \longrightarrow CH_4 + CO_2$$

Production of CH_4 from any of these substrates provides the energy required for the growth of methanogens. No methanogen is known to use all of these substrates, but almost all methanogens can produce CH_4 from H_2 and CO_2. *Methanothrix soehngenii*, which produces CH_4 from acetate, was reported to be unable to use H_2 and CO_2 for growth (Huser et al., 1982), as was the thermophilic methanosarcina strain TM1 (Zinder and Mah, 1979), which produces CH_4 from acetate and methanol. However, incubation of TM1 for some months with H_2 and CO_2 yielded growth, and subsequent subcultures with H_2 and CO_2 produced good growth in a few days (unpublished results from this laboratory; confirmed by S. Zinder, Cornell University, personal communication). These results suggest that a mutant of TM1 was selected that could use H_2 and CO_2 in addition to acetate and methanol.

The only methanogens that produce CH_4 from acetate are *Methanosarcina* (Balch et al., 1979) and *Methanothrix soehngenii* (Huser et al., 1982). *Methanosarcina* also produces CH_4 from methanol and methylamines (Balch et al., 1979; Hippe et al., 1979). *Methanothrix soehngenii* does not use methanol or methylamines. None of the acetate-using methanogens appears to be able to use formate (but see the qualification below in connection with *Methanogenium*).

Methanolobus tindarius produces CH_4 from methanol and methylamines but not from acetate (König and Stetter, 1982). All other methanogens except one use only H_2 and CO_2 or formate in addition to H_2 and CO_2. The exception is a coccus isolated from human feces that produces CH_4 only from H_2 and methanol: $H_2 + CH_3OH \longrightarrow CH_4 + H_2O$ (Miller and Wolin, 1983). It does not use H_2 and CO_2, formate, methanol alone, or acetate and methylamines without or with H_2. The organism obtains energy for growth solely by using H_2 to reduce methanol to CH_4.

Formate utilization is variable among methanogens that are otherwise restricted to the use of H_2 and CO_2. Thus *Methanobacterium formicicum* uses formate, whereas *Methanobacterium bryantii* does not (Balch et al., 1979). *Methanogenium cariaci* and *Methanogenium marisnigri* did not use formate when grown in a medium buffered with bicarbonate but did use formate when piperazine-N,N'-bis-(2-ethanesulfonic acid) (PIPES) was substituted for bicarbonate (Romesser et al., 1979). Other methanogens have been tested for formate utilization only in media buffered with bicarbonate. *Methanothrix soehngenii* converts formate to H_2 and CO_2 but cannot use these substrates for production of CH_4 (Huser et al., 1982).

IV. CELL WALLS AND LIPIDS

Pseudomurein

None of the methanogens contain peptidoglycan (murein), but members of the family Methanobacteriaceae contain a cell wall polymer called pseudomurein, which resembles peptidoglycan (Kandler, 1982; König et al., 1982). Like murein, pseudomurein is the polymer responsible for cell shape. The adjacent glycan strands of pseudomurein consist of repeating units of D-N-acetylglucosamine and/or D-N-acetylgalactosamine and L-N-acetyltalosaminuronic acid, whereas murein contains repeating units of D-N-acetylglucosamine and D-N-acetylmuramic acid. Short peptide chains of amino acids that cross-link the glycan strands have the L-configuration in pseudomurein and the D-configuration in murein. The primary structures of the pseudomurein of *Methanobacterium thermoautotrophicum* and the murein of a typical Eubacterium are compared in Figure 7.3. Pseudomureins from various Methanobacteriaceae differ with respect to (a) whether the glycan contains N-acetylglucosamine or N-acetylgalactosamine or both and (b) the kinds of amino acids in the peptide chain (König et al., 1982).

Nonpseudomurein Walls

Methanobacteriaceae and *Methanosarcina* are the only methanogens that yield rigid sacculi when subjected to the usual techniques for preparation of bacterial cell walls (Kandler, 1982). They also are the only gram-positive methanogens. However, the thick cell walls of species of *Methanosarcina* do not contain pseudomurein. Their sacculi contain a heteropolysaccharide consisting of galactosamine, glucose, mannose, and glucuronic acid or galacturonic acid (Kandler, 1982). The envelopes of all other methanogens are composed of protein-containing subunits (Kandler, 1982). Other constituents, which have not been fully characterized, may be covalently or noncovalently linked to sacculi or envelopes. For example, the sacculi of Methanobacteriaceae contain neutral polysaccharides, which in some cases may be attached to the glycan of pseudomurein by phosphodiester linkages (König et al., 1982). Teichoic and teichuronic acids are not present in the sacculi of Methanobacteriaceae (König et al., 1982).

Antibiotic Sensitivity

Because cell walls of methanogens do not contain peptidoglycan, they are insensitive to some of the antibiotics that inhibit peptidoglycan synthesis in Eubacteria, for example, penicillins and cephalosporins (Hilpert et al., 1981). The methanogens are also insensitive to several antibiotics that inhibit other eubacterial biosynthetic processes, for example, rifampin, which inhibits RNA synthesis, and tetracyline, erythromycin, and kanamycin, which inhibit protein synthesis. Antibiotics have therefore been used to selectively isolate and enumerate methanogens from microbial ecosystems (Godsey, 1980; Miller and Wolin, 1982).

Immunology

Rabbit antisera, prepared against formalinized cell suspensions of almost all known species of methanogens, have been used to determine immunological

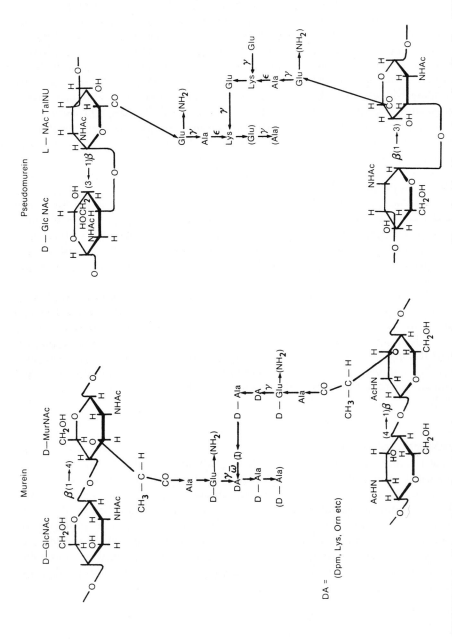

FIGURE 7.3 Comparison of murein and pseudomurein primary structures. (Reproduced with permission from Kandler, 1982.)

relatedness between species (Conway de Macario et al., 1982a). Immunological cross-reactions have been determined mainly by indirect immunofluorescence techniques. In brief, the highest dilution of an antiserum that produces maximal fluorescence with the homologous antigen is mixed with heterologous strains, and the fluorescence that results is scored. Fluorescence is actually measured after the antigen–antiserum complex is treated with goat anti-rabbit gamma globulin labeled with fluorescein isothiocyanate. There are no cross-reactions between members of different families of methanogens (Conway de Macario et al., 1981, 1982b). Cross-reactions of varying degrees occur between many antigen–antiserum combinations within families. Indirect enzyme immunoassay techniques have also been used to measure reactions between homologous and heterologous antigens and antibodies.

Immunotyping results can be used to reveal the relationships of strains within families, as well as to relate new isolates taxonomically to known species and strains (Miller et al., 1982). Estimates of relatedness made by immunofluorescence and immunoenzymatic techniques are essentially the same. Immunotyping is also potentially useful for direct identification of methanogens in microbial communities that contain populations of nonmethanogenic organisms and populations of different types of methanogens (Strayer and Tiedje, 1978).

Monoclonal antibodies prepared against methanogens have been used to examine the immunochemical specificity of reactions with pseudomurein (Conway de Macario et al., 1982c). Individual monoclonal antibody preparations have been obtained which specifically recognize the D-N-acetylglucosamine, D-N-acetylgalactosamine or gamma-glutamyl alanine residues of pseudomurein (E. Conway de Macario, New York State Health Department, personal communication).

Lipids

The lipids of Archaebacteria contain polyisoprenoid chains that are ether-linked to glycerol (Langworthy et al., 1982). In contrast, the lipids of Eubacteria are composed of fatty acid esters of glycerol. About 20–30% of the lipids of methanogens are neutral, and 70–80% are polar (Tornabene and Langworthy, 1979). The hydrophobic portions of the polar lipids are mainly C_{20} phytanyl and C_{40} biphytanyl diethers of glycerol. Both types of lipids are found in *Methanobacterium* and *Methanospirillum*, whereas *Methanosarcina* and *Methanococcus* contain none or only a trace of the biphytanyl lipids. Two polar glycophospholipids account for 64% of the total cellular lipids of *Methanospirillum hungatei* (Kushwaha et al., 1981). They are derivatives of dibiphytanyl diglycerol tetraether. One of the free hydroxyls of glycerol is esterified with glycerophosphoric acid; the other is glycosidically linked to a disaccharide. The disaccharide of one of the glycophospholipids contains glucose and galactose; the other contains a disaccharide of galactose.

The major neutral lipids of methanogens are C_{30} (squalene), C_{25}, and C_{20} hydrocarbons and their corresponding hydroisoprenoid derivatives (Tornabene et al., 1979). The distribution of chain lengths differs, depending on the genus examined. The C_{30} chain length predominates in *Methanobacterium*, *Methanobrevibacter*, *Methanococcus*, and *Methanospirillum*, whereas *Methanosarcina* species contain only C_{25} compounds. The smaller amounts of C_{20} isoprenoids and lack of C_{40} compounds suggest that they may be used for biosynthesis of the polar lipids.

Although isoprenoid compounds are found in higher organisms, they are not significant components of the lipids of Eubacteria. Squalene is found in animal cells and is a precursor of cholesterol. The phytane structure is found in the phytol group of chlorophyll. Polar lipids that are typical of Archaebacteria are found in sediments and petroleum, which indicates that a significant portion of the isoprenoids were synthesized directly by Archaebacteria (Chappe et al., 1982). No studies have been reported on the role of lipids in the structure and functions of membranes of methanogens.

V. NUTRITION

Cell Carbon

The substrates used for growth and methanogenesis have been discussed above. No other compounds can be used to supply energy for growth.

Other requirements for growth are not complex. Some methanogens are autotrophic or essentially autotrophic, i.e., they can be grown with $H_2 + CO_2$ and CO_2 is not only the precursor of CH_4, but the source of almost all cell carbon, except for requirements for trace amounts of organic nutrients. *Methanobacterium bryantii* (Bryant et al., 1971), *Methanobacterium thermoautotrophicum* (Zeikus and Wolfe, 1972), *Methanobrevibacter arboriphilus* (Zeikus and Henning, 1975), *Methanococcus vannielii* (J. B. Jones and Stadtman, 1977), *Methanococcus thermolithotrophicus* (Huber et al., 1982), *Methanococcus deltae* (Corder et al., 1983), and *Methanococcus maripaludis* (J. W. Jones et al., 1983) grow autotrophically.

Although some methanogens require organic compounds for growth, their requirements are relatively simple. Acetate is required by members of the Methanobacteriaceae, Methanococcaceae, and Methanomicrobiaceae (Whitman et al., 1982). These organisms do not produce CH_4 from acetate. Acetate is a major source of cell carbon for *Methanobrevibacter ruminantium* and *Methanococcus voltae* (Bryant et al., 1971; Whitman et al., 1982). Acetate is incorporated into about 2% of the cell carbon of *Methanobacterium thermoautotrophicum* grown with CO_2 as the only other source of carbon, although acetate is not required for growth (Fuchs and Stupperich, 1980). *Methanosarcina barkeri*, which produces CH_4 from acetate, incorporates acetate into significant amounts (60%) of cell carbon (Weimer and Zeikus, 1978a, b).

The use of acetate as a major carbon source and requirements for acetate have not been investigated for all methanogens, but the ability to use acetate as a carbon source may be a characteristic of most methanogens. Acetate and/or CO_2 probably are the sources of most of the cell carbon of most methanogens.

A few methanogens require short, branched-chain fatty acids or branched-chain amino acids for growth (Bryant et al., 1971; Whitman et al., 1982). *Methanococcus voltae* requires leucine and isoleucine, although these can be replaced by isovalerate and 2-methylbutyrate respectively. The amino acids are not used as sources of nitrogen. Most of the carbon of the amino acids or the branched-chain fatty acids is incorporated into cell protein. The amino acids contribute 55% of the protein carbon of *Methanococcus voltae*, but the pathway of incorporation is unknown. This organism derives the rest of its carbon from acetate and CO_2. *Methanobrevibacter ruminantium* requires short, branched-chain fatty acids and appears to assimilate these acids and acetate in a manner similar to that of *Methanococcus voltae*.

Nitrogen and Sulfur

All methanogens appear to be able to use NH_4^+ and S^{2-} as sources of nitrogen and sulfur. None use more oxidized forms of nitrogen, and members of the *Methanobacteriaceae* and *Methanococcus voltae* are incapable of using organic sources of nitrogen (Bryant et al., 1971; Whitman et al., 1982). More oxidized forms or organic sources of sulfur do not substitute for inorganic sulfide.

Some methanogens can reduce elemental sulfur to sulfide (Stetter and Gaag, 1983). The physiological significance of sulfur reduction is unclear.

Ions

Ion requirements have been demonstrated with several methanogens. *Methanococcus voltae* requires magnesium, calcium, iron, and nickel; its growth is stimulated by selenium (Whitman et al., 1982). *Methanococcus maripaludis* requires magnesium, and growth is stimulated by sodium and seleniun (J. W. Jones et al., 1983). Selenium and tungsten stimulate the growth of *Methanococcus vannielii* (J. B. Jones and Stadtman, 1977). Nickel, cobalt, molybdate (Schönheit et al., 1979), and sodium (Perski et al., 1981) are required by *Methanobacterium thermoautotrophicum*, and cobalt stimulates the growth of *Methanococcus voltae* (Whitman et al., 1982). Nickel requirements have also been shown for *Methanobacterium bryantii* (Jarrell et al., 1982), *Methanobrevibacter smithii*, and *Methanosarcina barkeri* (Diekert et al., 1981). Requirements for NaCl have been demonstrated for marine forms, including species of *Methanococcus* and *Methanogenium* (Balch et al., 1979). Some halophilic methanogens require very high concentrations of NaCl for growth (personal communications from R. Mah, University of California–Los Angeles and P. Smith, University of Florida).

Vitamins

No requirements for vitamin K or heme have been reported for methanogens. However, there are reports of requirements for and growth stimulation by B vitamins; for example, *Methanobrevibacter smithii* requires B vitamins (Bryant et al., 1971), and *Methanococcus voltae* is stimulated by pantothenic acid (Whitman et al., 1982). *Methanobrevibacter ruminantium* requires vitamin-like concentrations of coenzyme M (CoM), i.e., 2-mercaptoethanesulfonic acid ($HSCH_2CH_2SO_3H_2$), for growth (Taylor et al., 1974).

VI. BIOCHEMISTRY OF CH₄ FORMATION

Reduction of CO₂

Reduction of the most oxidized 1-carbon compound, CO_2, to the most reduced 1-carbon compound, CH_4, requires eight electrons. Early studies indicated that free one-carbon compounds at intermediate stages of reduction, i.e., formate, formaldehyde and methanol, were not intermediates. Barker (1956) postulated that the pathway of CH_4 formation was initiated by attachment of CO_2 to a carrier molecule and that all subsequent reduction steps involved one-carbon intermediates attached to a carrier molecule (Scheme 7.1). Contemporary studies of the biochemistry of methanogenesis support Barker's scheme, and the postulated carriers are coenzymes found uniquely in methanogens.

The last step of methanogenesis involves the reduction of methyl-coenzyme M (methyl-CoM) to CH_4 and CoM. The methyl group of the methylated coenzyme is attached to the sulfur of the sulfhydryl group (Taylor and Wolfe, 1974). Coenzyme M has been found only in methanogens (Balch and Wolfe, 1979). The reduction step requires H_2 and a catalytic amount of ATP (Gunsalus and Wolfe, 1978). Another coenzyme that was first discovered in methanogens, factor₄₂₀ (Eirich et al., 1978), participates in the enzymatic transfer of electrons from H_2, which are used to reduce the methyl group to CH_4. The coenzyme is a deazaflavin (Formula I), and its oxidized form emits a green fluorescence when illuminated at 420 nm. Epifluorescence microscopy of intact cells can be used to examine the factor₄₂₀ fluorescence of methanogens in pure cultures and in mixed microbial communities (Miller and Wolin, 1982). The purified methyl reductase protein of *Methanobacterium thermoautotrophicum* contains a chromophore that is a unique nickel-containing compound (Ellefson et al., 1982), whose organic component has a structure (Pfaltz et al., 1982) similar to that of a porphyrin (Formula II). The nickel compound is called factor₄₃₀ and is thought to play a role in the electron transfer reactions involved in the reduction of the methyl group to CH_4. Reduction of the methyl group is obviously a complex reaction involving several coenzymes, proteins, and ATP. The details of the mechanism of reduction have not yet been elucidated.

The immediate precursor of methyl-CoM is not completely clear. A methyl transferase in *Methanosarcina barkeri* transfers the methyl group of methanol to the sulfhydryl group of CoM to produce methyl-CoM, which is reduced to

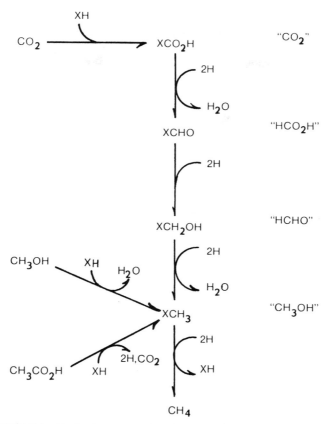

SCHEME 7.1 Barker's proposed pathway of CH_4 formation. The free one-carbon compounds whose oxidation state is equivalent to the C_1-intermediate attached to one or another coenzyme (x) are in quotes. (Adapted from Barker, 1956.)

CH_4 (Shapiro and Wolfe, 1980). However, *Methanosarcina barkeri* produces CH_4 from methanol, whereas *Methanobacterium thermoautotrophicum* does not. Transferases have been demonstrated in methanogens that form methyl-CoM from methylcobalamin, but this is believed to be nonphysiological (Shapiro, 1982). Chemically prepared hydroxymethyl-CoM is converted to CH_4 by extracts of *Methanobacterium thermoautotrophicum*, a result suggesting that the hydroxymethyl compound may be an intermediate that is reduced to methyl-CoM (Romesser and Wolfe, 1981). However, chemically prepared formyl-CoM is not converted to CH_4 (Romesser, 1978; Hutten et al., 1981). If hydroxymethyl-CoM is an intermediate, there is still a major question of how it might be formed from a C-1 carrier with the C-1 at the oxidation level of formate. A unique pteridine compound found in methanogens, called methano-

FORMULA I Structure of factor$_{420}$. (Reproduced with permission from Balch et al., 1979.)

FORMULA II Structure of the product of methanolysis of factor$_{430}$. (Reproduced with permission from Pfaltz et al., 1982.)

pterin (Formula III), is involved in the formation of the formyl carrier that is eventually reduced to methyl-CoM (Vogels et al., 1982). CO_2 is joined to the methanopterin (Keltjens et al., 1983), and the CO_2-methanopterin is reduced intramolecularly to formyl-methanopterin.

FORMULA III Structure of methanopterin. (Reproduced with permission from Vogels et al., 1982.)

Electron Transfer

Enzymes involved in the methanogens' use of electron donors have been studied. The hydrogenase of *Methanobacterium thermoautotrophicum* contains nickel (Graf and Thauer, 1981). The presence of nickel in factor$_{430}$, hydrogenase, and possibly other proteins of methanogens explains the requirement for nickel for growth. Selenium is present as selenocysteine in a hydrogenase purified from *Methanococcus vannielii* and also in the organism's formic dehydrogenase (Yamazaki, 1982; J. B. Jones et al., 1979). The latter enzyme requires zinc for activity.

Methanobrevibacter smithii contains enzymes that catalyze the reversible reduction of factor$_{420}$ by H_2 and by NADPH (Tzeng et al., 1975a, b). A formic dehydrogenase:factor$_{420}$ oxidoreductase was also demonstrated. These experiments provided the first evidence for a role for factor$_{420}$ in electron transport. A formic dehydrogenase:factor$_{420}$ oxidoreductase, an iron-sulfur molybdoenzyme, and a reduced factor$_{420}$:NADP oxidoreductase were found in *Methanococcus vannielii* (J. B. Jones and Stadtman, 1980). *Methanobacterium formicicum* has a formic dehydrogenase:factor$_{420}$ oxidoreductase that requires FAD for activity (Schauer and Ferry, 1983). The reduction of $NADP^+$ may be important for generation of NADPH, which is used in biosynthetic reactions and possibly in steps of CO_2 reduction. NADPH plus factor$_{420}$ or a high-molecular-weight, hydrogenase-containing protein plus H_2 can be used as electron donors for reduction of methyl-CoM by the methyl reductase of *Methanobacterium thermoautotrophicum* (Ellefson and Wolfe, 1980).

Methanosarcina barkeri is the only methanogen reported to contain the acidic iron-sulfur protein ferredoxin (Blaylock, 1968). *Methanosarcina barkeri* ferredoxin has been isolated and characterized (Moura et al., 1982; Hatchikian et al., 1982), but its physiological function is not yet known.

Sodium Requirement

Complete reduction of CO_2 to CH_4 with H_2 by cell suspensions of several methanogens requires Na^+ (Perski et al., 1982). Methanogens isolated from marine and nonmarine environments require Na^+ for production of CH_4. The specific site or sites where Na^+ functions in methanogenesis are not known.

Methyl Reduction

Although methyl transfer from methylamine, as well as from methanol, to CoM has been shown with extracts of *Methanosarcina barkeri*, no transfer was obtained from the methyl group of acetate. The transferase activity provides a mechanism for the entry of methanol and methylamine methyl groups into a terminal stage of methyl reduction that is common to the CO_2 reduction pathway. Several studies implicate the participation of a cobalamin coenzyme in the methylation of CoM by methanol. The transmethylation was shown to consist of two reactions (Van der Meijden et al., 1983). The evidence suggested that the two steps were a transmethylation from methanol to 5-hydroxyben-zimidazolylcobamide and transmethylation of CoM by the methylated cobal-amin. The paper reviews earlier evidence for a cobalamin requirement for methane formation. The corrinoids of *Methanosarcina barkeri* contain 5-hy-droxybenzimidazole as the alpha ligand (Scherer and Sahm, 1981; Pol et al., 1982). Corrinoids were found in species of Methanobacteriaceae as well as in *Methanosarcina* (Krzycki and Zeikus, 1980). The enzymology of conversion of the important substrate acetate is unknown.

Earlier studies with deuterated compounds showed that during the conversion of methyl groups of acetate and methanol to CH_4, none of the attached protons are exchanged with protons of H_2O (Barker, 1956). Whereas production of CH_4 from acetate is formally a decarboxylation, production from methanol or methylamine proceeds via oxidation of the methyl group to CO_2. The electrons removed during the oxidation are used to reduce the methyl groups to CH_4. *Methanosarcina* strains, which oxidize these methyl groups, contain cytochromes that probably participate in the oxidation reactions (Kühn et al., 1979; Kühn and Gottschalk, 1983). Cytochromes have not been reported in other methanogens. The coccus that has recently been isolated from human feces appears to be incapable of oxidizing the methyl group of methanol to CH_4 and requires exogenous H_2 for the reduction of methanol to CH_4 (Miller and Wolin, 1983).

Energy

Precise information about how CH_4 production is coupled to the generation of ATP is not yet available. There is some evidence that protonmotive mechanisms and membrane-bound ATPase are involved in ATP synthesis accompanying the reduction of CO_2 by H_2. There is essentially no information about how ATP synthesis is coupled to methanogenesis from acetate. References to studies of energy production mechanisms can be found in the review by Kell et al. (1981).

Nuclear magnetic resonance studies showed the presence of a unique phosphate storage compound in *Methanobacterium thermoautotrophicum* (Kanodia and Roberts, 1983; Seely and Fahrney, 1983). The compound cyclic-2,3-diphosphoglycerate is the intramolecular pyrophosphate derivative of glycerate 2,3-bisphosphate. It is present at an intracellular concentration of 10–12 mM.

VII. DNA, RNA, AND PROTEIN SYNTHESIS

DNA

There is no information about the subcellular organization of DNA of methanogens and little information about the physical structure of their DNA. The DNA of *Methanobacterium thermoautotrophicum* strain ΔH was calculated to be $\sim 10^9$ daltons, or 2.5 times smaller than the *Escherichia coli* chromosome (Mitchell et al., 1979). The DNA's of methanogens show a broad range of mole % guanine + cytosine (G + C), ranging from 27% to 61% (Table 7.2).

RNA

A DNA-dependent RNA polymerase has been purified from *Methanobacterium thermoautotrophicum* (Stetter et al., 1980). The polymerase is O_2-sensitive and composed of eight polypeptides (designated O, A through G) with a molar ratio of $O:A:B:C:D:E:F:G_6$. The polymerase was more active with a synthetic template, but native DNA from eubacterial phages or a *Halobacterium* phage served as a template for RNA synthesis, which required ATP, GTP, CTP, and UTP. Activity was inhibited by actinomycin D but resistant to rifampin and streptolydigin.

The tRNA and rRNA of methanogens contain posttranscriptional modifications of nucleotides that are characteristic of Archaebacteria and are distinctly different from Eubacteria or eukaryotes. The modified regions of nucleotides in the tRNA's from *Methanobacterium bryantii*, *Methanobrevibacter smithii*, *Methanococcus vannielii*, *Methanococcus voltae*, *Methanomicrobium mobile*, and *Methanosarcina barkeri* contain pseudouridine but lack ribothymidine and 7-methylguanosine (Gupta and Woese, 1980). The latter two nucleotides are usually present in eubacterial tRNA's. Dihydrouridine, which is also commonly found in Eubacteria, was present only in the modified region of *Methanosarcina barkeri*. The methylated nucleotides 2-o-methylcytidine, 1-methylguanosine, N^2,N^2-dimethylguanosine, and 1-methyl-adenosine were found in the modified region of all methanogens. All except *Methanosarcina barkeri* contain N^2-methylguanosine. The common arm sequence GTΨCG found in most eubacterial and eukaryotic tRNA's is not present in methanogen tRNA's.

Ribosomes and Protein Synthesis

The methanogens examined have eubacteria-like 70S ribosomes that dissociate in low $MgCl_2$ to 30 and 50S subunits (Schmid and Böch, 1982). The ribosomal proteins of methanogens are more acidic than those of Eubacteria, and the total numbers of proteins range from 54 to 60. Ribosomes of methanogens are insensitive to many antibiotics that specifically inhibit protein synthesis by eubacterial 70S ribosomes (Schmid et al., 1982). The ribosomal A protein from *Methanobacterium thermoautotrophicum* has a different amino acid composition than that of eubacterial ribosomal A proteins (Matheson et al., 1980). The N-terminal sequence of the methanogen's ribosomal A protein has 20–39% amino acid sequence homology with eukaryotic A protein and 6–16% homology with eubacterial A protein. The 30S and 50S ribosomal subunits have rRNA which sediments at 5, 16, and 23S. The primary sequence and secondary structure of 5S rRNA from some methanogens have recently been

TABLE 7.2 G + C (%) in DNA of methanogens

Methanogen	G + C (%)	References
Methanobacterium		
M. bryantii	33	a
M. formicicum	41	a
M. thermoautotrophicum	50	a
Methanobrevibacter		
M. arboriphilus	28–32	a
M. smithii	31	a
M. ruminantium	31	a
Methanococcus		
M. vannielii	31	a
M. voltae	31	a
M. thermolithotrophicus	31	Huber et al. (1982)
M. maripaludis	33	J. W. Jones et al. (1983)
M. deltae	41	Corder et al. (1983)
Methanococcoides		
M. methylutens	42	Sowers and Ferry (1983)
Methanothermus		
M. fervidus	33	Stetter et al. (1981)
Methanosarcina		
M. barkeri	40–42	b
Methanospirillum		
M. hungatei	45–47	a
Methanolobus		
M. tindarius	46	König and Stetter (1982)
Methanoplanus		
M. limicola	48	Wildgruber et al. (1982)
Methanomicrobium		
M. mobile	49	a
M. paynteri	45	Rivard et al. (1983)
Methanothrix		
M. soehngenii	52	Huser et al. (1982)
Methanogenium		
M. cariaci	52	a
M. thermophilicum	59	Rivard and Smith (1982)
M. marisnigri	61	a
M. olentangyi	54	Corder et al. (1983)

[a] Original references are given in the work of Balch et al. (1979).
[b] I. Yu and R. Mah, personal communication.

determined and used in taxonomic comparisons among methanogens and other Archaebacteria (Fox et al., 1982).

The present classification of the methanogens is heavily based on comparisons of the sequences of 16S rRNA oligonucleotides. The 16S rRNA molecule functions in binding mRNA to ribosomes. Although nucleotide sequence changes in parts of the 16S rRNA molecule have occurred during evolution, the molecule has retained the same function. The size of the molecule (~ 1600 nucleotides) makes it amenable to cataloging of oligonucleotide sequences but not to primary sequencing. The 16S rRNA's of methanogens have a number of unique posttranscriptionally modified sequences (Fox et al., 1977; Balch et al., 1979).

Methanogens have all of the machinery to synthesize proteins, but the details of transcription and translation may be somewhat different in methanogens, in comparison to either Eubacteria or eukaryotes. A protein in *Methanobacterium thermoautotrophicum* reacts in the eukaryotic-specific diptheria toxin reaction, which catalyzes ADP-ribosylation of a modified histidine in the eukaryotic protein synthesis elongation factor, EF2 (Kessel and Klink, 1980). Similarly reacting proteins have been found in *Methanosarcina barkeri*, *Methanobrevibacter arboriphilus*, *Methanococcus vannielii*, *Methanospirillum hungatei*, and *Methanothermus fervidus* (Kessel and Klink, 1982). However, the proteins of methanogens have not yet been shown to participate in in vitro protein synthesis. A similar protein from the archaebacterium *Halobacterium cutirubrum* was active in an in vitro protein assay (Kessel and Klink, 1980).

Genetic Analysis

The development of genetic systems for transfer of DNA from one methanogen to another has been hampered by the lack of an appropriate vector system and the difficulty in obtaining appropriate mutants of methanogens. A covalently closed circular DNA plasmid was isolated from *Methanobacterium thermoautotrophicum* strain Marburg (Meile et al., 1983). It is composed of ~ 4500 base pairs. A plasmid has also been isolated from an unidentified methanogen (Thomm et al., 1983). The functions of the plasmids are not known.

DNA's from methanogens serve as templates for mRNA synthesis when cloned into *E. coli*. DNA from *Methanobrevibacter arboriphilus* has been cloned into *E. coli* and shown to direct synthesis of a methanogen-specific polypeptide of unknown function (Bollschweiler and Klein, 1982). Recombinant plasmids containing DNA from *Methanosarcina barkeri* complemented mutations in biosynthetic genes of *E. coli* and directed polypeptide synthesis in minicells (Reeve et al., 1982). Recombinant plasmids containing *Methanococcus voltae* DNA complemented arginine or histidine gene mutations in *E. coli* and directed polypeptide synthesis in maxicells (Wood et al., 1983).

VIII. BIOSYNTHETIC PATHWAYS

Most of the information about pathways of incorporation of CO_2, acetate, or methanol into cell carbon has been obtained from enzymatic and pulse radioactive labeling studies with *Methanobacterium thermoautotrophicum* or *Methanosarcina barkeri*.

CO$_2$ Incorporation

Methanogens do not use a Calvin cycle (reductive pentose phosphate cycle) or a reductive tricarboxylic acid cycle for CO_2 assimilation. Fuchs and Stupperich (1982) recently summarized evidence from studies with *Methanobacterium thermoautotrophicum* that support the pathway of CO_2 assimilation shown in Scheme 7.2. Acetyl-CoA is synthesized by the condensation of two one-carbon units derived from two molecules of CO_2 (Stupperich and Fuchs, 1983). The one-carbon carrier molecules are not yet known, but evidence suggests the carboxyl group of acetyl-CoA is derived by the reduction of a bound carbonyl that exchanges with CO (Stupperich et al., 1983). Acetyl-CoA is reductively carboxylated with CO_2 and reduced factor$_{420}$ to form pyruvate. Pyruvate is the direct precursor of the alanine family of amino acids and phosphoenolpyruvate (PEP). Hexose and pentose phosphates and hexosamines are synthesized from PEP via triosephosphates and gluconeogenic enzymes (Jansen et al., 1982). Phosphoenolpyruvate is carboxylated to oxaloacetate, which is the direct precursor of the aspartate family of amino acids. Oxaloacetate is also reduced with NADH to form malate, which is then dehydrated to fumarate. Succinate is formed by reduction of fumarate, but the physiological electron donor is not known. Succinate, δ-aminolevulinic acid and the methyl group of methionine are incorporated into the tetrapyrrole structure of factor$_{430}$ (Thauer, 1982). The CoA thioester of succinate is formed by succinyl-CoA synthetase. Succinyl-CoA is reductively carboxylated with CO_2 and reduced factor$_{420}$ to form α-ketoglutarate, the direct precursor of the glutamate family of amino acids.

Methanosarcina barkeri uses a similar pathway for CO_2 incorporation into pyruvate, PEP and oxaloacetate. However, the organism synthesizes α-ketoglutarate by a pathway that involves condensation of acetyl-CoA and oxaloacetate to form citrate and oxidative decarboxylation of isocitrate with NADP to form α-ketoglutarate (Weimer and Zeikus, 1979).

Acetate Incorporation

Acetate is incorporated into cell carbon by *Methanobacterium thermoautotrophicum* and *Methanosarcina barkeri* and activation to acetyl-CoA, which then is metabolized to pyruvate and other key intermediates of cell carbon by the reactions demonstrated for CO_2 assimilation (Fuchs and Stupperich, 1982; Weimer and Zeikus, 1979). In *Methanobacterium thermoautotrophicum*, acetyl-CoA is synthesized from acetate, ATP, and CoA by acetyl-CoA synthetase (Oberlies et al., 1980):

$$\text{Acetate} + \text{ATP} + \text{CoA} \longrightarrow \text{Acetyl-CoA} + \text{AMP} + \text{PP}_i$$

In *Methanosarcina barkeri*, acetyl-CoA is formed by reactions involving acetate kinase and phosphotransacetylase (Kenealy and Zeikus, 1982):

$$\text{Acetate} + \text{ATP} \longrightarrow \text{Acetyl-P} + \text{ADP}$$

$$\text{Acetyl P} + \text{CoA} \longrightarrow \text{Acetyl-CoA} + \text{P}_i$$

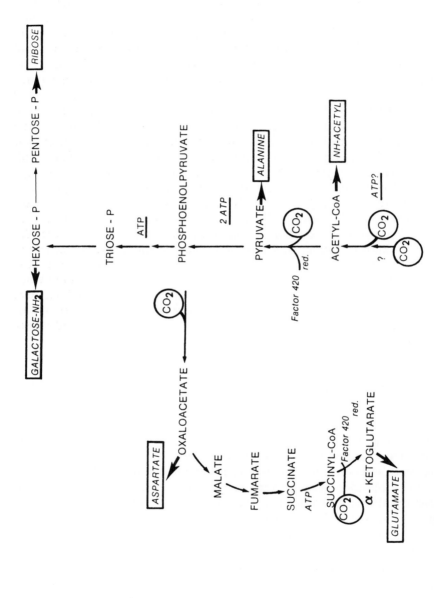

SCHEME 7.2 Autotrophic CO_2 assimilation pathway in *Methanobacterium thermoautotrophicum*. (Reproduced with permission from Fuchs and Stupperich, 1982.)

^{13}C Studies

Nuclear magnetic resonance analysis of cell constituents of a strain of *Methanospirillum hungatei* that requires acetate and CO_2 showed labeling patterns that are consistent with pathways deduced from the results of enzymatic and radioisotope studies (Ekiel et al., 1983). In addition, nucleoside labeling is consistent with conventional pathways for purine and pyrimidine synthesis and analysis of ribose indicated it is formed by decarboxylation of a hexose. Carbons 2 and 8 of purines are formed from the methyl group of acetate which suggests that the 1 carbon precursor of these purine carbon atoms is derived from the beta carbon of serine and not from glycine or formate obtained from the reduction of CO_2. Serine appears to be formed from 3-phosphoglycerate and glycine from serine. The labeling of threonine and methionine is consistent with formation from aspartate. Lysine appears to be formed by the diaminopimelic acid pathway from pyruvate and aspartate, and arginine and proline appear to be formed from glutamate. Labeling of phenylalanine and tyrosine is consistent with formation by the shikimic and chorismic acid pathway. Leucine and valine appear to be synthesized by the conventional acetolactate pathways, but isoleucine is formed from the condensation of 3 acetate molecules and a pathway involving the formation of precursor α-ketobutyrate from citramalate was suggested. Labeling studies with ^{14}C precursors also indicate that this pathway of isoleucine synthesis is operative in *Methanobacterium thermoautotrophicum* (Eikmanns et al., 1983). The methyl group of methionine is formed from CO_2. Labeling of the phytanyl chains of lipids is consistent with a condensation of three molecules of acetate to form mevalonate and subsequent formation of the phytanyl chains.

Methanol Incorporation

Studies of the incorporation of $[^{14}C]$-methanol into the carbons of alanine by *Methanosarcina barkeri* indicate that the methyl group can be a direct precursor of the methyl group of acetyl-CoA (Kenealy and Zeikus, 1982). Either exogenous CO_2 or CO_2 produced by the oxidation of methanol can be the precursor of the carboxyl group of acetyl-CoA. The enzymatic steps and C_1 carriers have not been elucidated. Methanol carbon is incorporated into acetyl-CoA, which is reductively carboxylated to pyruvate.

Reductions

NADH and reduced factor$_{420}$ participate as sources of reducing power for carbon assimilation in *Methanobacterium thermoautotrophicum* and *Methanosarcina barkeri*. Both organisms contain H_2:factor$_{420}$ oxidoreductase (hydrogenase), and both have reduced factor$_{420}$:NADP$^+$ oxidoreductase (Fuchs and Stupperich, 1982; Weimer and Zeikus, 1978a).

NH₃ Incorporation

Methanosarcina barkeri and *Methanobacterium thermoautotrophicum* have glutamine synthetase, glutamate synthase, glutamate:oxaloacetate transaminase, and glutamate:pyruvate transaminase (Kenealy et al., 1982). Neither organism has glutamate dehydrogenase. Synthesis of glutamate from α-ketoglutarate and glutamine by the glutamate synthase of *Methanosarcina barkeri* requires either reduced factor$_{420}$ or reduced flavin mononucleotide as electron donor. Alanine dehydrogenase activity is present in *Methanobacterium thermoautotrophicum* but was not detected in *Methanosarcina barkeri*.

IX. ECOSYSTEMS

Global Production

About 530–810 teragrams are biologically produced each year (Ehhalt, 1976). About 75% of the total is from methanogenesis in swamps, muds, and anaerobic sediments. Of these ecosystems, rice paddies are major contributors to CH_4 formation. About 25% of global CH_4 is produced in intestinal tract systems, especially the rumen of ruminating herbivores. Recent studies suggest that large amounts of CH_4 are also produced by the earth's population of termites (Zimmerman et al., 1982). Acetate decarboxylation contributes approximately 67% of the CH_4 of swamps, muds, and sediments; the rest is mainly produced by reduction of CO_2 by H_2. Therefore half of the annual production of CH_4 is from acetate decarboxylation, and half is from CO_2 reduction.

Influence of H₂

Production of CH_4 from H_2 and CO_2 influences the fermentations of the non-methanogenic bacteria that produce H_2, CO_2, and acetate, the major substrates used by methanogens (Wolin, 1982). Use of H_2 to reduce CO_2 to CH_4 keeps the partial pressure of H_2 of methanogenic ecosystems very low, despite the fact that large amounts of H_2 are produced. Since half of the annual production of CH_4 comes from reduction of CO_2, about 500–800 Tg of H_2 must be produced and used for methanogenesis each year.

Some of the H_2-evolving biochemical reactions are not inhibited by H_2, but others are. Removal of H_2 permits the latter reactions to proceed. Inhibition is due to reversal of H_2-producing reactions whose equilibrium constants favor the use rather than the production of H_2. The most important of these H_2-inhibited reactions is formation of H_2 from reduced pyridine nucleotides. Continuous reductions and oxidations of pyridine nucleotides are key parts of any fermentation. Reoxidation is normally accomplished by reduction of a fermentation intermediate to a fermentation product. Formation of products such as lactate, ethanol, propionate, succinate, and butyrate requires reoxidation of NAD(P)H. The use of NAD(P)H to reduce protons to H_2 can prevent the formation of these products, and their precursors can then be converted to acetate and CO_2. Shifts of electrons to H_2-dependent methanogenesis and the accompanying increase in acetate formation occur when H_2-producing, fermentative organisms are co-cultured with methanogens (see references in Wolin, 1982). Formation of the reduced products of monocultures of the fermentative species — for example, lactate, ethanol and succinate — decreases in these co-cultures.

Intestinal Tract

The increase in acetate formation caused by the use of H_2 for methanogenesis is reflected in the high concentrations of acetate that accumulate in methanogenic intestinal tract ecosystems, such as the rumen (Wolin, 1981). The other major products of intestinal tract fermentation are propionate and butyrate. Some fermentative organisms in these ecosystems produce the latter two products but do not have enzymes that reduce protons with reduced pyridine nucleotides. These fermentations are not altered by the use of H_2 by methanogens, and propionate and butyrate can accumulate.

Complete Bioconversion

Intestinal tract ecosystems turn over more rapidly than systems in which organic substrates are completely converted to CH_4 and CO_2 (McInerny and Bryant, 1981). Acetate is produced in the complete bioconversion systems in the same general way as in the intestinal tract systems. However, the turnover time of complete bioconversion systems is long enough to permit the establishment of high concentrations of the slow-growing methanogens that decarboxylate acetate to CH_4 and CO_2. In addition, very slow-growing bacteria that oxidize propionate (Boone and Bryant, 1980) and butyrate (McInerny et al., 1979) to acetate and H_2 can become established. Their growth is absolutely dependent on methanogens that use H_2 to reduce CO_2 to CH_4 because any accumulation of H_2 completely prevents the production of H_2 from propionate and butyrate.

The essential association of a H_2 producer with a H_2 user is an example of a syntrophic association, that is, one organism's growth depends upon the use of its products by a second organism. The shifting of all carbon flow to acetate and CO_2 through fermentation shifts and syntrophic associations and the decarboxylation of acetate to CH_4 and CO_2 are key features of complete bioconversion systems.

Methanogen Types in Ecosystems

Methanogens have been isolated from a variety of terrestrial and marine anaerobic ecosystems (Table 7.3). However, there is little information on the predominant types in most ecosystems or on variations in methanogen populations within and between similar ecosystems. *Methanosarcina* species have been

TABLE 7.3 Sources of Methanogens

Methanogen	Sources
Methanobacterium	
M. bryantii	Marine mud
M. formicicum	Sewage sludge
M. thermoautotrophicum	Sewage sludge, thermal springs, compost
Methanobrevibacter	
M. arboriphilus	Sewage sludge, decaying trees
M. smithii	Sewage sludge, human feces
M. ruminantium	Rumen
Methanococcus	
M. vannielii	Marine mud
M. voltae	Marine mud
M. thermolithotrophicus	Heated sea sediment
M. maripaludis	Marine mud
M. deltae	Marine mud
Methanococcoides	
M. methylutens	Marine mud
Methanothermus	
M. fervidus	Thermal spring
Methanosarcina	
M. barkeri	Sewage sludge, cow feces, rumen
Methanospirillum	
M. hungatei	Sewage sludge, pear waste
Methanolobus	
M. tindarius	Lake sediment
Methanoplanus	
M. limicola	Swamp mud
Methanomicrobium	
M. mobile	Rumen
M. paynteri	Marine mud
Methanothrix	
M. soehngenii	Sewage sludge
Methanogenium	
M. cariaci	Marine mud
M. thermophilicum	Heated sea sediment
M. marisnigri	Marine mud
M. olentangyi	River sediment

isolated from several terrestrial complete bioconversion systems, where they almost certainly have a significant role in production of CH_4 from acetate. They also can be sustained in the rumen, where they probably grow on methanol produced by hydrolysis of methylated polysaccharides by nonmethanogenic bacteria or on methylamines produced by the nonmethanogenic fermentation of choline (Patterson and Hespell, 1979). *Methanosarcina* grow more rapidly with CH_3OH and methylamines than with acetate. *Methanobrevibacter smithii* appears to be the predominant methanogen in the human large intestine

(Miller and Wolin, 1982). Various species of methanogens are found in high concentrations in the rumen.

Practical Aspects

Practical interest in methanogenesis has focused on both the intestinal tract and complete bioconversion systems. There has been considerable interest in altering the rumen fermentation to prevent the loss of the energy in dietary substrates that occurs when CH_4 is produced and eructated (Wolin, 1981). Retention of substrate electrons in the reduced fermentation products propionate and butyrate, rather than CH_4, is a benefit to the ruminant because the acids are used as major sources of the animal's carbon and energy. Various compounds have been examined for their ability to cause the desired alterations in the fermentation. Monensin, which is used extensively as a feed additive, decreases the production of CH_4, increases the production of propionate, and increases the efficiency of feed utilization of beef cattle.

Interest in complete bioconversion stems from the relatively long-standing practice of fermenting sewage solids to CH_4 and CO_2 in order to reduce their volume and to gain some energy benefit from the combustible gas produced. Investigations of strategies for improving the treatment of domestic sewage have expanded to include other wastes, such as animal manure, solid wastes (garbage), and wastes from the manufacture of chemicals. The goals are the same as for sewage solids, that is, reduction or elimination of a pollution problem and generation of a useful form of energy. Contemporary concerns about environmental pollution and the energy crisis are obviously important stimuli for investigations of complete bioconversion systems. The energy crisis has also stimulated investigations of possible cheap sources of biomass other than wastes as substrates for methanogenesis.

Reviews can be consulted for details about these practical fermentations (see references in McInerny and Bryant (1981), Mosey (1982), and Wolin, (1981)) to supplement the discussion in this article of the bacteria responsible for the terminal steps, the methanogens.

REFERENCES

Aranki, A., and Freter, R. (1972) *Amer. J. Clin. Nutr. 25,* 1329–1334.

Balch, W. E., and Wolfe, R. S. (1979) *J. Bacteriol. 137,* 256–263.

Balch, W. E., Fox, G. E., Magrum, L. J., Woese, C. R., and Wolfe, R. S. (1979) *Microbiol. Rev. 43,* 260–296.

Barker, H. A. (1956) *Bacterial Fermentations*, pp. 1–95, John Wiley and Sons, New York.

Blaylock, B. A. (1968) *Arch. Biochem. Biophys. 124,* 314–324.

Bollschweilier, C., and Klein, A. (1982) *Z. Bakt. Hyg., I. Abt. Orig. C3,* 101–109.

Boone, D. R., and Bryant, M. P. (1980) *Appl. Environ. Microbiol. 40,* 626–632.

218   Methanogens

Bryant, M. P., Tzeng, S. F., Robinson, I. M., and Joyner, A. E. (1971) *Adv. Chem. Ser. 105*, 23–40.

Chappe, B., Albrecht, P., and Michaelis, W. (1982) *Science 217*, 65–66.

Conway de Macario, E., Macario, A. J. L., and Wolin, M. J. (1981) *Science 214*, 74–75.

Conway de Macario, E., Macario, A. J. L., and Wolin, M. J. (1982a) *J. Bacteriol. 149*, 320–328.

Conway de Macario, E., Wolin, M. J., and Macario, A. J. L. (1982b) *J. Bacteriol. 149*, 316–319.

Conway de Macario, E., Macario, A. J. L., and Kandler, O. (1982c) *J. Immunol. 129*, 1670–1674.

Corder, R. E., Hook, L. A., Larkin, J. M., and Frea, J. L. (1983) *Arch. Microbiol. 134*, 28–32.

Diekert, G., Konheiser, U., Piechulla, K., and Thauer, R. K. (1981) *J. Bacteriol. 148*, 459–464.

Ehhalt, D. H. (1976) in *Symposium on Microbial Production and Utilization of Gases* (Schlegel, H. G., Gottschalk, G., and Pfennig, N., eds.), pp. 13–22, E. Golzer, Göttingen, West Germany.

Eikmanns, B., Linder, D., and Thauer, R. K. (1983) *Arch. Microbiol. 136*, 111–113.

Eirich, L. D., Vogels, G. D., and Wolfe, R. S. (1978) *Biochemistry 17*, 4583–4593.

Ekiel, I., Smith, I. C. P., and Sprott, G. D. (1983) *J. Bacteriol. 156*, 316–326.

Ellefson, W. L., and Wolfe, R. S. (1980) *J. Biol. Chem. 255*, 8388–8389.

Ellefson, W. L., Whitman, W. B., and Wolfe, R. S. (1982) *Proc. Nat. Acad. Sci. U.S.A. 79*, 3703–3710.

Fox, G. E., Magrum, L. J., Balch, W. E., Wolfe, R. S., and Woese, C. R. (1977) *Proc. Nat. Acad. Sci. U.S.A. 74*, 4537–4541.

Fox, G. E., Luehrsen, K. R., and Woese, C. R. (1982) *Z. Bakt. Hyg., I. Abt. Orig. C3*, 330–345.

Fuchs, G., and Stupperich, E. (1980) *Arch. Microbiol. 127*, 267–272.

Fuchs, G., and Stupperich, E. (1982) *Z. Bakt. Hyg., I. Abt. Orig. C3*, 277–288.

Godsey, E. M. (1980) *Appl. Environ. Microbiol. 39*, 1074–1075.

Graf, E., and Thauer, R. K. (1981) *FEBS Lett. 136*, 165–169.

Gunsalus, R. P., and Wolfe, R. S. (1978) *J. Bacteriol. 135*, 851–857.

Gupta, R., and Woese, C. R. (1980) *Curr. Microbiol. 4*, 245–249.

Hatchikian, E. C., Brauschi, M., Forget, N., and Scandellari, M. (1982) *Biochem. Biophys. Res. Comm. 109*, 1316–1323.

Hilpert, R., Winter, J., Hammes, W., and Kandler, O. (1981) *Z. Bakt. Hyg., I. Abt. Orig. C2*, 11–20.

Hippe, H., Caspari, D., Fiebig, K., and Gottschalk, G. (1979) *Proc. Nat. Acad. Sci. U.S.A. 76*, 494–498.

Huber, H., Thomm, M., König, H., Thies, G., and Stetter, K. O. (1982) *Arch. Microbiol. 132*, 47–50.

Hungate, R. E. (1969) in *Methods in Microbiology*, Vol IIIB (Norris, J. R., and Ribbons, D. W., eds.), pp. 117–132, Academic Press, New York.

Huser, B. E., Wuhrmann, K., and Zehnder, A. J. B. (1982) *Arch. Microbiol. 132*, 1–9.

Hutten, T. J., de Jong, M. H., Peeters, B. P. H., van der Drift, C., and Vogels, G. D. (1981) *J. Bacteriol. 145*, 27–34.

Jansen, K., Stupperich, E., and Fuchs, G. (1982) *Arch. Microbiol. 132*, 355–364.

Jarrell, K. F., Colvin, J. R., and Sprott, G. D. (1982) *J. Bacteriol. 149*, 346–353.

Jones, J. B., and Stadtman, T. C. (1977) *J. Bacteriol. 130*, 1404–1406.

Jones, J. B., and Stadtman, T. C. (1980) *J. Biol. Chem. 255*, 1049–1053.

Jones, J. B., Dilworth, G. L., and Stadtman, T. C. (1979) *Arch. Biochem. Biophys. 195*, 255–260.

Jones, J. W., Paynter, M. J. B., and Gupta, R. (1983) *Arch. Microbiol. 135*, 91–97.

Kandler, O. (1982) *Z. Bakt. Hyg., I. Abt. Orig. C3*, 149–160.

Kanodia, S., and Roberts, M. F. (1983) *Proc. Nat. Acad. Sci. U.S.A. 80*, 5217–5221.

Kell, D. B., Doddema, H. J., Morris, J. G., and Vogels, G. D. (1981) in *Microbial Growth on C1 Compounds* (Dalton, H., ed.), pp. 159–170, Heyden & Son Ltd., London.

Keltjens, J. T., Daniels, L., Jannsen, H. G., Borm, P. J., and Vogels, G. D. (1983) *Eur. J. Biochem. 130*, 545–552.

Kenealy, W. R., and Zeikus, J. G. (1982) *J. Bacteriol. 151*, 932–941.

Kenealy, W. R., Thompson, T. E., Schubert, K. R., and Zeikus, J. G. (1982) *J. Bacteriol. 150*, 1357–1365.

Kessel, M., and Klink, F. (1980) *Nature 287*, 250–251.

Kessel, M., and Klink, F. (1982) *Z. Bakt. Hyg., I. Abt. Orig. C3*, 140–148.

König, H., and Stetter, K. O. (1982) *Z. Bakt. Hyg., I. Abt. Orig. C3*, 478–490.

König, H., Kralik, R., and Kandler, O. (1982) *Z. Bakt. Hyg., I. Abt. Orig. C3*, 228–244.

Krzycki, J., and Zeikus, J. G. (1980) *Curr. Microbiol. 3*, 243–245.

Kühn, W., and Gottschalk, G. (1983) *Eur. J. Biochem. 135*, 89–94.

Kühn, W., Fiebig, K., Walther, R., and Gottschalk, G. (1979) *FEBS Lett. 105*, 271–274.

Kushwaha, S., Kates, M., Sprott, G. D., and Smith, I. C. P. (1981) *Science 211*, 1163–1164.

Langworthy, T. A., Tornabene, T. G., and Holzer, G. (1982) *Z. Bakt. Hyg., I. Abt. Orig. C3*, 228–244.

Matheson, A. T., Yaguchi, M., Balch, W. E., and Wolfe, R. S. (1980) *Biochim. Biophys. Acta 626*, 162–169.

McInerny, J. I., and Bryant, M. P. (1981) in *Biomass Conversion Processes for Energy and Fuels* (Sofar, S. S., and O. R. Zaborsky, eds.), pp. 277–296, Plenum, New York.

McInerny, J. I., Bryant, M. P., and Pfennig, N. (1979) *Arch. Microbiol. 122*, 129–135.

Meile, L., Kiener, A., and Leisinger, T. (1983) *Mol. Gen. Genet. 191*, 480–484.

Miller, T. L., and Wolin, M. J. (1974) *Appl. Microbiol. 27*, 985–987.

Miller, T. L., and Wolin, M. J. (1982) *Arch. Microbiol. 131*, 14–18.

Miller, T. L., and Wolin, M. J. (1983) *J. Bacteriol. 153*, 1051–1055.

Miller, T. L., Wolin, M. J., Conway de Macario, E., and Macario, A. J. L. (1982) *Appl. Environ. Microbiol. 43*, 227–232.

Mitchell, R. M., Loeblich, L. A., Klotz, L. C., and Loeblich, A. R. (1979) *Science 204*, 1082–1084.

Mosey, F. E. (1982) *Wat. Poll. Cont. 81*, 540–552.

Moura, I., Moura, J. J. G., Huynh, B.-H., Santes, H., LeGall, J., and Xavier, A. V. C. (1982) *Eur. J. Biochem. 126*, 95–98.

Oberlies, G., Fuchs, G., and Stupperich, E. (1980) *Arch. Microbiol. 128*, 248–252.

Patterson, J. A., and Hespell, R. B. (1979) *Curr. Microbiol. 3*, 79–83.

Perski, H. J., Moll, J., and Thauer, R. K. (1981) *Arch. Microbiol. 130*, 319–321.

Perski, H. J., Schönheit, P., and Thauer, R. K. (1982) *FEBS Lett. 143*, 323–326.

Pfaltz, A., Bernhard, J., Fassler, A., Eschenmosser, A., Jaenchen, R., Gilles, H. H., Diekert, G., and Thauer, R. K. (1982) *Helv. Chim. Acta. 65*, 828–865.

Pol, A., Van der Drift, C., and Vogels, G. D. (1982) *Biochem. Biophys. Res. Comm.* *108*, 731–737.

Reeve, J. N., Trun, N. J., and Hamilton, P. T. (1982) in *Genetic Engineering of Microorganisms for Chemicals*, (Hollaender, A., DeMoss, R. D., Kaplan, S., Konisky, J., Savage, D., and Wolfe, R. S., eds.), Vol. XIX, *Basic Life Sciences*, pp. 233–244, Plenum Press, New York.

Rivard, C. J., and Smith, P. L. (1982) *Int. J. Syst. Bacteriol. 32*, 430–436.

Rivard, C. J., Hensen, J. M., Thomas, M. V., and Smith, P. H. (1983) *Appl. Environ. Microbiol. 46*, 484–490.

Romesser, J. A. (1978) "The Activation and Reduction of Carbon Dioxide to Methane in *Methanobacterium thermoautotrophicum*." Ph.D. thesis, University of Illinois, Urbana.

Romesser, J. A., and Wolfe, R. S. (1981) *Biochem. J. 197*, 565–571.

Romesser, J. A., Wolfe, R. S., Mayer, F., Spiess, E., and Walther-Mauruschat, A. (1979) *Arch. Microbiol. 121*, 147–153.

Schauer, N., and Ferry, J. G. (1983) *J. Bacteriol. 155*, 467–472.

Scherer, P., and Sahm, H. (1981) *Acta Biotechnol. 1*, 57–65.

Schmid, G., and Böch, A. (1982) *Z. Bakt. Hyg., I. Abt. Orig. C3*, 347–353.

Schmid, G., Pecher, T., and Böch, A. (1982) *Z. Bakt. Hyg., I. Abt. Orig. C3*, 209–217.

Schönheit, P., Moll, J., and Thauer, R. K. (1979) *Arch. Microbiol. 123*, 105–107.

Seely, R. J., and Fahrney, D. E. (1983) *J. Biol. Chem. 258*, 10835–10838.

Shapiro, S. (1982) *Can. J. Microbiol. 28*, 629–635.

Shapiro, S., and Wolfe, R. S. (1980) *J. Bacteriol. 141*, 728–734.

Sowers, K. R., and Ferry, J. G. (1983) *Appl. Environ. Microbiol. 45*, 684–690.

Stetter, K. O., and Gaag, G. (1983) *Nature 305*, 309–311.

Stetter, K. O., Winter, J., and Hartlieb, R. (1980) *Z. Bakt. Hyg., I. Abt. Orig. C1*, 210–214.

Stetter, K. O., Thomm, M., Winter, J., Wildgruber, G., Huber, H., Zillig, W., Jané-Covic, D., König, H., Palm, P., and Wunderl, S. (1981) *Z. Bakt. Hyg., I. Abt. Orig. C2*, 166–178.

Strayer, R. F., and Tiedje, J. M. (1978) *Appl. Environ. Microbiol. 35*, 192–198.

Stupperich, E., and Fuchs, G. (1983) *FEBS Lett. 156*, 345–348.

Stupperich, E., Hammel, K. E., Fuchs, G., and Thauer, R. K. (1983) *FEBS Lett. 152*, 21–23.

Taylor, C. D., and Wolfe, R. S. (1974) *J. Biol. Chem. 249*, 4879–4885.

Taylor, C. D., McBride, B. C., Wolfe, R. S., and Bryant, M. P. (1974) *J. Bacteriol. 120*, 974–975.

Thauer, R. K. (1982) *Z. Bakt. Hyg., I. Abt. Orig. C3*, 265–270.

Thomm, M., Altenbuchner, J., and Stetter, K. O. (1983) *J. Bacteriol. 153*, 1060–1062.

Tornabene, T. G., and Langworthy, T. A. (1979) *Science 203*, 51–53.

Tornabene, T. G., Langworthy, T. A., Holzer, G., and Oro, J. (1979) *J. Mol. Evol. 13*, 73–78.

Tzeng, S. F., Bryant, M. P., and Wolfe, R. S. (1975a) *J. Bacteriol. 121*, 192–196.

Tzeng, S. F., Wolfe, R. S., and Bryant, M. P. (1975b) *J. Bacteriol. 121*, 184–191.

Van der Meijden, P., Hythuysen, H. J., Pouwels, A., Houwen, F., Van der Drift, C., and Vogels, G. D. (1983) *Arch. Microbiol. 134*, 238–242.

Vogels, G. D., Keltjens, J. T., Hutten, T. J., and van der Drift, C. (1982) *Z. Bakt. Hyg., I. Abt. Orig. C3*, 258–264.

Weimer, P. J., and Zeikus, J. G. (1978a) *Arch. Microbiol. 119*, 49–57.

Weimer, P. J., and Zeikus, J. G. (1978b) *Arch. Microbiol. 119*, 175–182.

Weimer, P. J., and Zeikus, J. G. (1979) *J. Bacteriol. 137*, 332–339.

Whitman, W. B., Ankwanda, E., and Wolfe, R. S. (1982) *J. Bacteriol. 149*, 852–863.

Wildgruber, G., Thomm, M., König, H., Ober, K., Ricchiuto, T., and Stetter, K. O. (1982) *Arch. Microbiol. 132*, 31–36.

Woese, C. R. (1982) *Z. Bakt. Hyg., I. Abt. Orig. C3*, 1–17.

Wolin, M. J. (1981) *Science 213*, 1463–1468.

Wolin, M. J. (1982) in *Microbial Interactions and Communities*, Vol. I (Bull, A. T., and Slater, J. H., eds.), pp. 323–356, Academic Press, London.

Wood, A. G., Redborg, A. H., Cue, D. R., Whitman, W. B., and Konisky, J. (1983) *J. Bacteriol. 156*, 19–29.

Yamazaki, S. (1982) *J. Biol. Chem. 257*, 7926–7929.

Zeikus, J. G., and Henning, D. L. (1975) *Anton van Leeuwenhoek J. Microbiol. and Serol. 41*, 543–552.

Zeikus, J. G., and Wolfe, R. S. (1972) *J. Bacteriol. 109*, 707–713.

Zimmerman, P. R., Greenberg, J. P., Wandiga, S. O., and Crutzen, P. J. (1982) *Science 218*, 563–565.

Zinder, S. H., and Mah, R. A. (1979) *Appl. Environ. Microbiol. 38*, 996–1008.

Biology of the Methylotrophs

Israel Goldberg

I. INTRODUCTION

Aerobic growth of nonautotrophic microorganisms on reduced carbon compounds lacking carbon–carbon bonds is referred to as methylotrophic growth (Colby et al., 1979). With obligate methylotrophs, one-carbon (C_1) compounds are their sole carbon-energy substrates, while facultative methylotrophs are able to grow on a variety of other organic compounds. Because one-carbon compounds such as methane, methanol, methylamine(s), formaldehyde, and formate occur in abundance throughout nature (Hanson, 1980), it is not surprising that a wide variety of microorganisms that utilize such compounds have been isolated (Anthony, 1982; Colby et al., 1979; Hanson, 1980; Higgins et al., 1981; Wolfe and Higgins, 1979). Methylotrophic microorganisms, especially the methane oxidizers, are primarily responsible for recycling of methane back to carbon dioxide, either directly by respiration or indirectly as a result of subsequent degradation of their biomass by aerobic and anaerobic heterotrophs (Hanson, 1980; Wolfe and Higgins, 1979).

Methylotrophs also play a significant role in the nitrogen cycle. Many of them oxidize ammonia to nitrite or nitrate (Hutton and Zobell, 1953; Wilkinson, 1971) and may contribute to the oxidative segment of the nitrogen cycle (Whittenbury et al., 1976). There have been a number of reports demonstrating that methane utilizers fix atmospheric nitrogen (deBont and Mulder, 1974; Whittenbury et al., 1970, 1976) and that methanol utilizers can be used efficiently in biological denitrification processes (Tani et al., 1978).

In recent years, C_1 compounds, especially methane and methanol, have attracted much attention as convenient raw materials for industrial fermentations (Goldberg, 1977; Senior and Windass, 1980). Interest in the microbial utilization of C_1 compounds initially focused on commercial exploitation of these microorganisms for single-cell protein (SCP) production and only more recently as sources of useful metabolites, for example, amino acids (L-glutamate, L-valine, L-alanine, L-serine, L-lysine, L-methionine, L-tryptophan, L-phenylalanine, and L-tryosine), organic acids (citric acid, fumaric acid, and α-ketogluconic acid), polysaccharides, β-hydroxybutyric acid, coenzyme Q_{10}, vitamin B_{12}, and FAD (Anthony, 1982; Higgins et al., 1981; Keune et al., 1976; Oki et al., 1973; Senior et al., 1982; Tanaka et al., 1980; Tani et al., 1978; Toraya et al., 1975; Yamada et al., 1982). Although the amounts of metabolites produced by methylotrophs are still low, there is the hope that increased knowledge on both the metabolism of C_1 compounds and genetics of methylotrophs will enhance development in this field. Important new possibilities for commercial exploitation of methylotrophs include (1) cooxidation of nongrowth substrates, particularly multicarbon compounds, to useful products (Anthony, 1982; Higgins et al., 1980; Stirling and Dalton, 1979b), (2) the use of immobilized organisms or enzymes in low-temperature processes for conversion of methane to methanol and methanol to formaldehyde or formate, and (3) detoxification of carbon monoxide and a variety of other oxidative processes (Anthony, 1982; Baratti et al., 1978; Couderc and Baratti, 1980; Higgins et al., 1980; Wolfe and Higgins, 1979).

The purpose of this chapter is to highlight some biological features of methylotrophic microorganisms that are able to produce energy and synthesize cell material from reduced C_1 compounds. Because most of the recent reviews on methylotrophs have been concerned with methane-utilizing bacteria (Chandra and Shethna, 1977; Colby et al., 1979; Higgins et al., 1981; Quayle, 1980a, b; Quayle and Ferenci, 1978; Wolfe and Higgins, 1979) and since in the author's view methanol-utilizing microorganisms have more potential application than other methylotrophs, emphasis will be directed to methanol-ultilizing bacteria and yeasts. For more details see the recent book *The Biochemistry of Methylotrophs* by Anthony (1982).

II. PHYSIOLOGICAL ASPECTS OF THE METHYLOTROPHS

Taxonomy

The present position on nomenclature and taxonomy of methylotrophs is confusing and there exists a need for a more detailed taxonomic investigation of these microorganisms (Anthony, 1982). For convenience the following discussion is divided into methane-utilizing bacteria, obligate methanol-utilizing bacteria, facultative methanol-utilizing bacteria, and methanol-oxidizing yeasts.

Methane-Utilizing Bacteria.   The most extensive study on the isolation of these bacteria was made by Whittenbury et al. (1970), who classified the isolates into two groups, Type I and Type II, according to several characteristics. Type I bacteria have the following properties: formation of intracytoplasmic membranes arranged as bundles of vesicular discs, formation of cysts as resting stages, assimilation of methane and methanol via the ribulose monophosphate (RMP) pathway, the presence of an incomplete tricarboxylic acid (TCA) cycle, and the predominant biosynthesis of C_{16} fatty acids in the membranes.

By comparison, Type II bacteria have the following characteristics: formation of intracytoplasmic paired membranes around cell periphery, formation of exospores or lipid cysts as resting stages, assimilation of C_1 compounds via the serine pathway, the presence of a complete TCA cycle, and the predominant biosynthesis of C_{18} fatty acids in the membranes (Colby et al., 1979; Higgins et al., 1981).

On the basis of a detailed survey of obligate methane utilizers by Romanovskaya et al. (1978) and the subsequent isolation of facultative methane-utilizing bacteria (Patel et al., 1978; Patt et al., 1974), Colby et al. (1979) suggested a tentative classification scheme that includes both obligate and facultative methane-utilizing bacteria. The scheme divides the bacteria into two major groups (Type I and Type II), but it adds subgroup A (*Methylomonas methanica* and *Methylomonas albus*) and subgroup B (*Methylococcus capsulatus*) to Type I and subgroup obligate (*Methanomonas methanooxidans* and *Methylosinus trichosporium*) and subgroup facultative (*Methylobacterium organophilum*) to Type II (Anthony, 1982).

The validity of this classification scheme is questionable, owing to recent findings showing little homology between the DNAs of *M. organophilum* XX and *M. trichosporium* OB3b (Hanson, 1980) and the finding that methane-utilizing bacteria (i.e., *M. capsulatus*) contain more than one pathway for the assimilation of C_1 units (Hanson, 1980; Higgins et al., 1981).

Obligate Methanol-Utilizing Bacteria.   The classification of obligate methanol- and methylamine-utilizing bacteria is incomplete (Loginova and Trotsenko, 1981). These bacteria were originally classified as belonging to the genus *Pseudomonas* or *Methylomonas*, and only recently were *Methylobacterium* (facultatives), *Methylobacillus*, and *methylophilus*—new generic terms for methylotrophs—introduced (Loginova and Trotsenko, 1981; Yordy and Weaver, 1979). The genus *Methylobacillus* includes obligate methylotrophic nonmotile rods that are able to grow on C_1 compounds but not on methane, while the genus *Methylophilus* was proposed to differentiate between motile-obligate bacteria utilizing methanol and bacteria able to grow on methane. Obligate methanol and methylamine utilizers assimilate C_1 compounds via the RMP pathway.

Facultative Methanol-Utilizing Bacteria. The majority of facultative methylotrophs isolated were pink-pigmented gram-negative bacteria belonging to *Pseudomonas*, *Vibrio*, *Protaminobacter*, *Flavobacterium*, *Achromobacter*, and *Hyphomicrobium* genera (Green and Bousfield, 1982). A few Gram-positive methanol-assimilating bacteria belonging to *Arthrobacter*, *Bacillus*, and *Corynebacterium* genera have been isolated (Green and Bousfield, 1982). One hundred and fifty pink-pigmented and 28 other facultative methylotrophs were compared in a numerical phenetic study consisting of 140 unit characters (Green and Bousfield, 1982). All the pink-pigmented bacteria were grouped in two closely related major clusters, and they were tentatively included in the genus *Methylobacterium* (see above) (Green and Bousfield, 1982). All these bacteria assimilate C_1 compounds via the serine pathway.

Methano-Assimilating Yeasts. A comprehensive taxonomical study on methanol-assimilating yeasts was prepared by Lee and Komagata (1980). They divided the methanol-utilizing yeasts into four groups: *Candida boidinii* group, methanol-assimilating *Hansenula* group (including closely related yeasts), *Pichia cellobiosa* group, and the *Hansenula capsulata* group.

Growth on C_1 Compounds

Methanol, formaldehyde, and formic acid, intermediary metabolites in the oxidation pathway of methane to CO_2, are very toxic to microorganisms. In order to obtain a large amount of biomass per liter of medium, the culture broth concentration of the toxic C_1 compounds should be minimal. This can be achieved by two methods (Goldberg, 1981): (1) with a continuous culture under conditions in which the carbon source is the only limiting nutrient (chemostat) and (2) with a mixed population of microorganisms, to be discussed later.

The advantages (Goldberg and Er-el, 1981) of continuous culture compared to batch culture facilitate the propagation of microorganisms on toxic C_1 compounds (Table 8.1), and these advantages have been applied by ICI in an industrial SCP process using a methylotrophic bacterium in continuous culture (Senior and Windass, 1980). The continuous culture (chemostat) technique was also used: to optimize growth medium for different methylotrophs (Goldberg and Er-el, 1981), to select for fast growing methylotrophs (Minami et al., 1978a, Smirnova et al., 1977), to adapt methylotrophs to grow in an acidic medium (Smirnova et al., 1977) or at relatively high temperatures

TABLE 8.1   Growth Parameters of Bacteria Grown on C_1 Compounds in a Continuous Culture

	C_1 Growth Substrate	Assimilation Pathway	μ_{max}[a] (h^{-1})	Y^b_M	M^c_s	$Y^{max d}_M$
Methylococcus NC1B 11083[e]	CH_4	RMP[f]	—	—	—	—
Mixed culture[g]	CH_4	—	0.143	10.8	0.48	16.0
Methylococcus NC1B 11083	CH_3OH	RMP	—	12.5	—	—
Pseudomonas C	CH_3OH	RMP	0.49	17.3	2.5	19.5
Pseudomonas 135	CH_3OH	Serine	0.14	12.1	2.6	12.5
Pseudomonas 135	CH_3NH_2	Serine	—	10.8	—	—
Pseudomonas 135	HCHO	Serine	—	8.1	2.7	11.2
Pseudomonas 135	HCOOH	Serine	—	6.9	4.4	7.9
Paracoccus denitrificans[h]	HCOOH	Calvin	0.04	—	0.7	2.9

[a] Maximum specific growth rate.
[b] Molar yield coefficient (g cell dry wt/mol substrate utilized).
[c] Maintenance coefficient (mmol substrate/g cell dry wt·h).
[d] Maximum molar yield coefficient corrected for maintenance energy ($m_s = 0$).
[e] Results are from Linton and Vokes (1978).
[f] Ribulose monophosphate pathway.
[g] Results are from Nagai et al. (1973).
[h] Results are from van Verseveld and Stouthamer (1978); other results are from Goldberg et al. (1976) and Rokem et al. (1978) (see Goldberg (1981) for more details).

(Minami et al., 1978a, b; Snedecor and Cooney, 1974), and to isolate both single-strain and mixed populations of methylotrophic microorganisms from different sources (Hazeu et al., 1975; Rock et al., 1976).

Microorganisms: Methane Utilizers.   Whittenbury et al. (1970) have isolated over 100 strains of methane-utilizing bacteria from various sources. The bacteria found were obligate methylotrophs capable of growth only on methane and methanol (Whittenbury et al., 1970) and not on dimethyl ether (Meyers, 1982). *Methylobacterium organophilum* (Patt et al., 1974, 1976) is the only facultative methane utilizer. The majority of methane utilizers are mesophiles (Whittenbury et al., 1970), some being able to grow at 50°C (thermotolerant) but with an optimum growth temperature near 37°C (Foster and Davis, 1966; Malashenko, 1976). Two thermophilic methane-utilizing bacteria, *Methylococcus thermophilus* and strain H2 (Malashenko, 1976; Shen et al., 1982), have an optimum temperature for growth of 50–55°C. A unique morphological feature of methanotrophs is their ability to develop extensive intracytoplasmic structures, which has been described in detail in the review by Higgins et al. (1981).

Five facultative yeast strains capable of oxidizing methane (Wolf and Hanson, 1978, Wolf et al., 1980) contain microbodies during growth on methane (O'Keeffe and Anthony, 1978). Several algae and fungi may have a limited capacity to utilize methane, but the evidence is not conclusive (Wolfe and Higgins, 1979).

Methanol and Methylamine-Utilizers. Since the isolation of Sohngen's strain (Sohngen, 1906), the first described methanol-utilizing bacterium, in 1906 and its reisolation by Dworkin and Foster in 1956, a large number of methylotrophic bacteria that utilize reduced C_1 compounds have been isolated and characterized (Anthony, 1975; Colby et al., 1979; Hanson, 1980; Quayle, 1972). The majority of the bacteria isolated from enrichments using methanol and methylamine as substrates also grow on other carbon substrates (facultative methylotrophs), and the rest are obligate methylotrophs. In addition, thermophilic methanol utilizers (Snedecor and Cooney, 1974) as well as bacteria capable of growth in methanol plus seawater have been isolated (Yamamoto et al., 1980). Ogata et al. (1969) first described the isolation of numerous methanol-utilizing yeasts from natural sources, and many more strains were subsequently isolated (Tani and Yamada, 1980). One mycelial fungus, *Trichoderma lignorium* (Tye and Willets, 1973), and one actinomycete, *Streptomyces* sp. 239 (Hanson, 1980), are also capable of utilizing methanol as a sole carbon source.

Formaldehyde- and Formate-Utilizing Methylotrophs. Facultative methanol-utilizing bacteria can grow on formaldehyde and formic acid as sole carbon sources in chemostat cultures (Goldberg et al., 1976; Rock et al., 1976). Two soil fungi, *Gliocladium deliquescens* and *Paecilomyces varioti* (Hanson, 1980), were able to grow on 0.1–0.2% formaldehyde or 0.5% sodium formate as the sole carbon sources.

Mixed Cultures. During batch or continuous cultivation of methane- or methanol-utilizers, formaldehyde, formate (Papoutsakis et al., 1978), and other organic compounds (Harrison, 1976) are excreted into the growth medium. The excretion of organic matter lowers the cellular yield (which is calculated on the basis of the total C_1 substrate utilized), results in the accumulation of inhibitory substances (such as methanol, formaldehyde, and formate), and increases the possibility of contamination by other bacteria able to utilize the excreted organic compounds. These problems do not occur in mixed cultures where soluble organic products are scavenged by associated bacteria (Goldberg, 1981).

Three classes of mixed cultures have been described for methylotrophic bacteria propagated on C_1 compounds. In the first class, component microorganisms compete for the primary carbon substrate. Under certain conditions these mixed cultures are unstable because one or the other of the component bacteria predominates, leading to the disappearance of the other component. An example of this class was described by Rokem et al. (1980) with methylotrophic bacteria in which either *Pseudomonas* C, *Pseudomonas* 1 or *Pseudomonas* 135 dominated in the continuous culture, depending on the dilution rate chosen. The yields of these mixed cultures were never higher than that obtained by growing single-strain cultures of the RMP pathway bacterium at optimal conditions (Papoutsakis et al., 1978; Rokem et al., 1980). In the second class of mixed cultures, one bacterium that grows on methanol produces organic by-products that serve as substrates for growth of other bacteria (Ballerini et al., 1977; Cremieux et al., 1977; Haggstrom, 1969; Harrison, 1976, 1978; Lamb and Garver, 1980; Rokem et al., 1980). A mixed culture composed of one methanol-utilizing bacterium and four non-methanol-utilizing bacteria (two *Pseudomonas* sp., a *Curtobacterium* sp., and an *Acinetobacter* sp.) and propagated on methanol gave higher growth yields and rates than those obtained with the methanol-utilizer alone (Harrison, 1976, 1978). In contrast, when different heterotrophs were added to a culture of *Pseudomonas* C grown in a chemostat (Rokem et al., 1980), there was no increase in the yield of biomass, calculated on the basis of the methanol utilized. Although changes in the proportion of the heterotrophic bacteria were noted in the culture, their total number remained low and fairly constant throughout the fermentation (less than 1% of the total population), resulting in no increase in total biomass.

The third class of mixed cultures is exemplified by a methane-utilizer that showed little (Malashenko et al., 1973; Shen et al., 1982) or no growth (Lamb and Garver, 1980; Linton and Buckee, 1977; Sheehan and Johnson, 1971; Wilkinson and Harrison, 1973; Wilkinson et al., 1974) on methane because of the toxicity of methanol derived from methane oxidation, and only the addition of one or more methanol utilizers supported growth. Examples include growth of a methane-utilizing *Pseudomonas* sp. on methane only in the presence of a *Hyphomicrobium* sp., which utilized the methanol produced, and an *Acinetobacter* sp. and a *Flavobacterium* sp. that utilized the organic intermediates excreted into the medium (Wilkinson and Harrison, 1973; Wilkinson et al., 1974). The addition of methanol did not affect the stability of this mixed culture (Wilkinson et al., 1974). When methane oxidation by the primary *Pseudomonas* sp. ceased, there was an increase (from 4% to 25%) in the proportion of *Hyphomicrobium* sp. This resulted in a rapid reduction of the methanol concentration, a resumption of methane oxidation by *Pseudomonas* sp., and a gradual restoration of the original community composition. In another mixed culture (Shen et al., 1982), the interaction of a methane-utilizer (strain H2) and a heterotroph (strain S) resulted from the production of a growth stimulant by strain S for strain H2. This latter system is similar to that reported by Lamb and Garver (1980), in which a heterotrophic bacterium produced an essential growth factor (not a stimulant) for a methane-utilizing bacterium.

III. CARBON METABOLISM

All C_1-utilizing methylotrophs have a need to biosynthesize cell material entirely from one-carbon compounds. As with CO_2-fixing autotrophic bacteria, a three- or four-carbon skeleton, such as pyruvate or succinate is synthesized from C_1 units. When such an intermediate is formed, the biochemical pathways by which other cell constituents synthesized are similar to those found in heterotrophic microorganisms (Quayle, 1980a; Wolfe and Higgins, 1979).

In autotrophic bacteria, synthesis of anabolic intermediates from CO_2 is affected by the ribulose diphosphate cycle of carbon dioxide fixation initially described by Calvin and his group (Bassham and Calvin, 1957). In methylotrophs, C_1 anabolism has been elucidated primarily by Quayle and his colleagues and consists of two new assimilation pathways for reduced C_1 compounds, the serine pathway (Fig. 8.1) (Quayle, 1972) and the ribulose monophosphate (RMP) pathway (Fig. 8.2) (Strom et al., 1974).

The Calvin, serine, and RMP pathways differ in two aspects, namely, the nature of the acceptor molecule and the oxidation level of the entering C_1 units. The acceptor molecule in the Calvin cycle is ribulose diphosphate, in the serine pathway glycine and phosphoenolpyruvate (PEP), and in the RMP pathway ribulose monophosphate.

The second difference among the three assimilation pathways is the oxidation level of the entering C_1 unit and therefore the ATP requirements for C_1

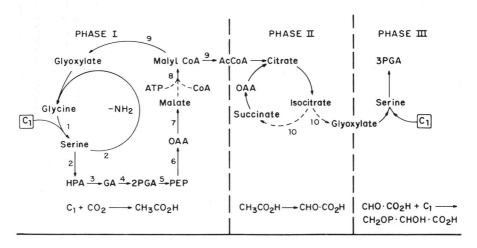

Sum: $2C_1 + CO_2 \longrightarrow CH_2OP \cdot CHOH \cdot CO_2H$

FIGURE 8.1 The isocitrate lyase$^+$-serine pathway. The abbreviations used are: PEP, phosphoenolpyruvic acid; PGA, 3-phosphoglyceric acid; HPA, hydroxypyruvic acid; OAA, oxaloacetic acid; GA, glyceric acid; 2PGA, 2-phosphoglyceric acid. 1, serine transhydroxymethylase; 2, serine-glyoxylate aminotransferase; 3, hydroxypyruvate reductase; 4, glycerate kinase; 5, enolase; 6, PEP-carboxylase; 7, malate dehydrogenase; 8, malate thiokinase; 9, malyl-CoA-lyase; 10, isocitrate lyase. (Reproduced with permission from *Biochemical Society Transactions.*)

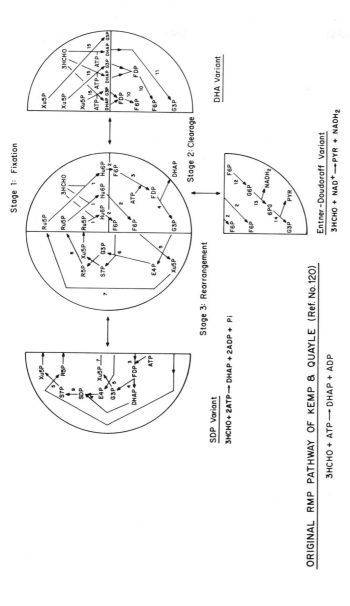

Stage 1: Fixation

DHA Variant

Stage 2: Clearage

Stage 3: Rearrangement

SDP Variant

3HCHO + 2ATP ⟶ DHAP + 2ADP + Pi

3HCHO + ATP ⟶ DHAP + ADP

Entner–Doudoroff Variant

3HCHO + NAD⁺ ⟶ PYR + NADH₂

ORIGINAL RMP PATHWAY OF KEMP & QUAYLE (Ref. No.120)

FIGURE 8.2 The ribulose monophosphate pathway. The abbreviations used are: Ru5P, Ribulose-5-P; Hu6P, D-erythro-L-glyc-ero-3-hexulose-6-P; F6P, fructose-6-P; FDP, fructose-1,6-diphosphate; G3P, glyceraldehyde-3-P; DHA, dihydroxyacetone; DHAP, DHA-P; E4P, erythrose-4-P; Xu5P, xylulose-5-P; S7P, sedoheptulose-7-P; SDP, sedoheptulose-1,7-diphosphate; R5P, ribose-5-P; G6P, glucose-6-P; 6PG, 6-phosphogluconate; PYR, pyruvic acid. 1, hexulose phosphate synthase; 2, phospho-3-hexu-loisomerase; 3, 6-phosphofructokinase; 4, fructose diphosphate aldolase; 5, transketolase; 6, transaldolase; 7, ribulose phosphate epimerase; 8, phosphoriboisiomerase; 9, sedoheptulose diphosphatase; 10, fructose diphosphatase; 11, triosephosphate isomerase; 12, phosphoglucoisomerase; 13, glucose-6-phosphate dehydrogenase; 14, 6-phosphogluconate dehydrogenase; and phospho-2-keto-3-deoxygluconate aldolase; 15, transketolase and triokinase. (Reproduced with permission from *Microbiological Reviews*, Vol. 45, 1981, by Annual Reviews, Inc.)

231

assimilation. Carbon dioxide alone is incorporated to form a C_3 compound by the Calvin cycle, formaldehyde and CO_2 are incorporated by the serine pathway, whereas only formaldehyde is fixed via the RMP pathway. Because the ATP requirements vary for these pathways, significant differences are observed in cellular yields (Table 8.1) (Anthony, 1975; Fujii and Tonomura, 1975; Harder and van Dijken, 1976).

A few microorganisms can use the Calvin cycle during growth on reduced C_1 compounds (Colby et al., 1979; Hanson, 1980; Quayle, 1980a). Examples include the aerobic-autotrophic metabolism of formate by *Pseudomonas oxalaticus* (Quayle and Keech, 1959a, b) and *Thiobacillus novellus* (Chandra and Shethna, 1977); aerobic growth on methanol by *Paracoccus denitrificans* (Cox and Quayle, 1975; Shively et al., 1978; van Verseveld, and Stouthamer, 1978), *Achromobacter, Pseudomonas,* and *Mycobacterium* (Loginova et al., 1979b); and photosynthetic bacteria (Douthit and Pfennig, 1976; Quayle and Pfennig, 1975). Because the definition of methylotrophs used in the introduction excludes bacteria that grow autotrophically on reduced C_1 compounds, the above bacteria utilizing the Calvin pathway will not be discussed further.

Obligate methane-utilizers (Higgins et al., 1981) that use the serine pathway have been isolated less frequently than those employing the RMP pathway. This may reflect the greater efficiency of the latter cycle with respect to growth rates and yields on methane, although it does not necessarily reflect the abundance of either type of organism in the environment (Goldberg, 1981; Harder and van Dijken, 1976; Higgins et al., 1981).

The Serine Pathway

The serine pathway (Fig. 8.1) was first described in 1961 by Large, Peel, and Quayle from short-term isotope incubation studies using the methanol-grown facultative methylotrophs *Pseudomonas* AM1 and *Hyphomicrobium vulgare* (see also Kaneda and Roxburgh, 1959). It was subsequently observed that all facultative methylotrophs possess high levels of the serine pathway enzymes. The list of methylotrophic species utilizing the serine pathway has been reviewed (Anthony, 1982; Colby et al., 1979; Higgins et al., 1981).

Short-term incubations of *Pseudomonas* AM1 cells with [^{14}C] methanol resulted in early labeling of serine followed by glycine, malate, and aspartate (Large et al., 1961). This study together with results (Lawrence et al., 1970) obtained on the early labeling of C_4-carboxylic acids, serine, glycine, and "unknown compounds" in methane-grown *M. methanooxidans* led to the description of the C_1 fixation reactions as: (1) the fixation of formaldehyde by serine transhydroxymethylase (serine hydroxymethyltransferase) (Harder and Quayle, 1971) and (2) the formation of malate by carboxylation of a C_3 compound derived from serine (Large et al., 1962a). Subsequently, the latter reaction was found to be catalyzed by PEP carboxylase (Large et al., 1962b), and many of the intervening reactions between serine and oxaloacetate were identified by enzymological studies with cell-free extracts of *Pseudomonas* AM1

(Blackmore and Quayle, 1970; Large and Quayle, 1963). Further labeling experiments using $[2,3\text{-}^{14}C]$ succinate (Salem et al., 1972) confirmed that the C_2 acceptor molecule (glycine) is regenerated by cleavage of a C_4 compound (presumably derived from oxaloacetate), thus completing a cyclic serine pathway for the formation of a C_3 compound from two molecules of formaldehyde and one molecule of CO_2 (Quayle, 1980a).

Formaldehyde enters the serine pathway by reacting chemically with tetrahydrofolic acid (THF) to produce N^5,N^{10}-methylene-THF. The subsequent enzymatic step, catalyzed by serine transhydroxymethylase, involves the condensation of N^5,N^{10}-methylene-THF with glycine to form serine. In most bacteria, serine transhydroxymethylase is required for serine and purine biosynthesis (Huennekens, 1968); but in serine pathway methylotrophs the enzyme assumes a central role in carbon assimilation. O'Connor and Hanson (1975) purified two serine transhydroxymethylase activities from a facultative methane-utilizer, *M. organophilum* XX. One activity predominated during growth on methane or methanol, and the other predominated during growth on succinate. The methanol-induced enzyme had twice the molecular weight of the succinate-induced enzyme, possessed different electrophoretic and sedimentation properties, and, unlike the other isoenzyme, was not stimulated by Mg^{2+} and Zn^{2+}. Both activities were competitively inhibited by glycine, but only the methanol-induced enzyme was stimulated by glyoxyalate. This result is not surprising in view of the fact that glyoxyalate is the precursor of glycine via the serine-glyoxyalate-aminotransferase reaction (Fig. 8.1).

Previously, it was thought that the presence of high activities of hydroxypyruvate reductase, a key enzyme in the serine pathway, indicated the pathway's operation in methylotrophs. However, it has been shown (Bamforth and Quayle, 1977) that RMP bacteria contain low activity of this enzyme, and therefore such evidence alone is inadequate. Evidence for the operation of the serine pathway in facultative methylotrophs was shown by the induction of the following key serine pathway enzymes during growth on methanol: glyoxyalate-stimulated serine transhydroxymethylase, serine-glyoxyalate aminotransferase, and glycerate kinase (Boulton and Large, 1977; McNerney and O'Connor, 1980; O'Connor and Hanson, 1975). Mutants of *Pseudomonas* AM1 lacking hydroxypyruvate reductase (Quayle, 1972), or serine-glyoxyalate aminotransferase (Quayle, 1972), or glycerate kinase (Dunstan et al., 1972), or malyl-CoA lyase (Anthony, 1975) do not grow on C_1 compounds. In this bacterium the four enzymes are thought to be regulated coordinately, and their synthesis repressed by succinate or one of its metabolites (Anthony, 1975; Dunstan et al., 1972).

In the serine pathway, half the 2-phosphoglycerate generated by formaldehyde fixation is converted to 3-phosphoglycerate, the net product of the pathway. Phosphoglycerate mutase, the enzyme that catalyzes this reaction, has been purified from *Hyphomicrobium* X and *Pseudomonas* AM1 (Hill and Attwood, 1976). The other half is converted to PEP, the acceptor molecule for the second carbon fixation step in this pathway. Phosphoenolpyruvate is carboxylated by the irreversible reaction of PEP carboxylase (Large et al., 1962b; Large and Quayle, 1963) to form oxaloacetate, which is then reduced to malate by malate dehydrogenase (Large et al., 1962a). Carbon dioxide is derived mainly from the oxidation of reduced C_1 compounds, and it has been calculated (Large et al., 1961) that approximately 50% of the carbon derived from methanol is assimilated as CO_2 by different serine pathway facultative methylotrophs. Strom et al. (1974) predicted higher CO_2 requirements for serine pathway organisms than for RMP pathway organisms, and their prediction was confirmed by the observation (Loginova and Trotsenko, 1979a; Trotsenko, 1976) that carboxylase activities (PEP carboxylase; see Loginova and Trotsenko (1979a)) in the former are an order of magnitude higher than those in the latter (PEP carboxylase and pyruvate carboxylase; see Babel and Loffhagen (1977) and Loginova and Trotsenko (1979a)). In addition, it was shown (Ben-Bassat et al., 1980; Goldberg, 1981; Samuelov and Goldberg, 1982a) that in the RMP pathway–obligate methylotroph *Pseudomonas* C (Samuelov and Goldberg, 1982b), only 7% of the carbon derived from methanol is assimilated as CO_2, and that the cellular yield of *Pseudomonas* C is higher than that of *Pseudomonas* AM1.

Phosphoenolpyruvate carboxylase plays a key role in methylotrophic metabolism. The studies of Newaz and Hersh (1975) suggested that the enzyme is found at the branch point between carbon assimilation and energy generation. The PEP carboxylase of methylamine-grown *Pseudomonas* MA (Newaz and Hersh, 1975; Stirling and Dalton, 1979b) is activated by NADH and inhibited by ADP, effects indicating that in this bacterium PEP carboxylase is under metabolic regulation.

In contrast, O'Connor (1981) described two isoenzymes of PEP carboxylase in *M. organophilum*, which fractionated at different ammonium sulfate concentrations. The isoenzyme of methanol-grown cells was insensitive to both acetyl-CoA and NADH, while succinate-grown cells contained acetyl-CoA-stimulated activity.

Malate dehydrogenase is ubiquitous in both obligate and facultative methylotrophs (Patel et al., 1978; Patt et al., 1974).

Until 1972 a basic unsolved problem was how glycine, necessary for each passage through the serine pathway, is regenerated from C_4 compounds. The mechanism for this regeneration was first elucidated in two methylamine-utilizing methylotrophs, *Pseudomonas* MA (Hersh and Bellion, 1972) and bacterium 5H2 (Cox and Zatman, 1973, 1976). These bacteria synthesized high levels of an ATP- and CoA-dependent malate-cleavage enzyme system and isocitrate lyase (Fig. 8.1). Malate cleavage is catalyzed by malate thiokinase: Malate + ATP + CoA → Malyl-CoA + ADP + P_i and malyl-CoA lyase: Malyl-CoA → Glyoxylate + Acetyl-CoA (Hersh, 1973; Tuboi and Kikuchi, 1963). These reactions together with isocitrate lyase and the enzymes involved in the serine pathway constitute a biochemical pathway, which has been named the isocitrate lyase (icl$^+$)-serine pathway (Fig. 8.1) (Hersh and Bellion, 1972). This pathway was found in *Pseudomonas* MA (Hersh and Bellion, 1972),

Pseudomonas aminovorans (Bamforth and O'Connor, 1979; Large and Carter, 1973), bacterium $5H_2$ (Cox and Zatman, 1973), and *Pseudomonas* MS (Wagner and Levitch, 1975) and is reviewed elsewhere (Anthony, 1982; Colby et al., 1979; Higgins et al., 1981; Quayle, 1972, 1980a; Wolfe and Higgins, 1979). Because there are many methane-utilizers (Colby et al., 1979; Patel et al., 1978; Trotsenko, 1976; Trotsenko and Shishkina, 1980) and methanol-utilizers (Anthony, 1975; Bellion and Kim, 1979; Colby et al., 1979; Cox and Zatman, 1973; Leadbetter and Gottlieb, 1967; Peel and Quayle, 1961; Quayle, 1975) that lack the activity of isocitrate lyase, it was clear that there are two variants of the serine pathway. Only malyl-CoA-lyase activity, but not malate thiokinase and isocitrate lyase, was detected in *Pseudomonas* AM1 (Salem et al., 1973). Biochemical studies with wild-type and mutant methylotrophs (Anthony, 1975; Colby et al., 1979; Higgins et al., 1981; Quayle, 1972, 1980a; Ribbons et al., 1970) suggested that the carbon assimilation pathway in isocitrate lyase–less organisms is similar to the pathway shown in Fig. 8.1. However, despite the efforts of several groups (Bellion et al., 1981; Kortstee, 1980), they have been unable to determine how the icl⁻-serine pathway bacteria regenerate glycine.

The RMP Pathway (Quayle Cycle)

The ribulose monophosphate (RMP) pathway of formaldehyde fixation (Fig. 8.2) was proposed by Quayle and co-workers (Johnson and Quayle, 1965; Kemp and Quayle, 1966, 1967) from studies with *Pseudomonas methanica (Methylomonas methanica)* and *Methylococcus capsulatus*. Short-term incubations of the methane-utilizing bacteria with [^{14}C]-labeled methane, methanol, or formaldehyde resulted in early labeling of sugar phosphate, particularly hexose phosphates. Subsequent work (Kemp, 1974; Lawrence et al., 1970) showed two key reactions catalyzed by 3-hexulose phosphate synthase and phospho-3-hexuloisomerase. These two enzymes, together with appropriate labeling patterns, have been found in many other methylotrophs (Colby et al., 1979). The net result of the RMP pathway is the condensation of three molecules of formaldehyde to one molecule of triose phosphate, an intermediate required for synthesis of cell constituents.

The RMP pathway is conveniently divided into three stages (Fig. 8.2) (Anthony, 1982; Quayle, 1980a). Stage 1, the aldol condensation of three molecules of formaldehyde with three molecules of ribulose-5-phosphate to yield three molecules of fructose-6-phosphate is common to all methylotrophic bacteria using this pathway (Colby et al., 1979). Attempts to purify 3-hexulose phosphate synthease, the enzyme responsible for the condensation reaction, from *M. capsulatus* yielded two protein fractions (Kemp, 1972), a soluble fraction that catalyzed the isomerization of ribose-5-phosphate into ribulose-5-phosphate, and a particulate fraction that catalyzed the condensation of ribulose-5-phosphate with formaldehyde. Kemp (1974) reported that the condensation product is not allulose-6-phosphate but rather D-arabino-3-hexulose-6-phosphate (D-erythro-L-glycero-3-hexulose-6-phosphate). Both the synthase and the isomerase have been purified from *M. capsulatus* (Ferenci et al., 1974); they have molecular weights of 310,000 and 67,000, respectively. The synthase dissociates into subunits of molecular weight 49,000 under conditions of low pH and low ionic strength. The purified synthase from methanol-grown *Methylomonas* M15 (Sahm et al., 1976), *Pseudomonas oleovorans* (Multer and Sokolov, 1979; Sokolov and Trotsenko, 1978b), and *Methylomonas aminofaciens* 77a (Kato et al., 1977, 1978) differs from that of *M. capsulatus* by the molecular weight and intracellular distribution.

Stage 2 of the RMP pathway involves the splitting of one molecule of fructose-6-phosphate to produce two triose phosphates. The cleavage of hexose phosphate into two such C_3 compounds is accomplished in the following ways: (1) by fructose diphosphate cleavage as in Embden-Meyerhof pathway, for example, in *Bacillus* sp. PM6 and S2A1 (Colby and Zatman, 1975b) and *Arthrobacter globiformis* B175 (Loginova and Trotsenko, 1977), and (2) by the Entner-Doudoroff pathway, for example, in bacterial strains 4B6, C2A1, and W3A1 (Colby and Zatman, 1975b), *Pseudomonas* W6 (Babel and Hofman, 1975; Babel and Miethe, 1974, Hofman et al., 1975), *Pseudomonas oleovorans* (Loginova and Trotsenko, 1977), and *Pseudomonas* C (Samuelov and Goldberg, 1982a). The methane utilizers *P. methanica* and *M. capsulatus* (Strom et al., 1974) contain enzymes of both pathways, although their relative importance is not known. The two routes of cleavage can also be coupled to two methods of rearrangement (see below).

Stage 3 of the RMP pathway involves the regeneration of three molecules of ribulose-5-phosphate from two molecules of fructose-6-phosphate and one molecule of glyceraldehyde-3-phosphate produced in stages 1 and 2 (Fig. 8.2). These sugar phosphate interconversions are catalyzed by transaldolase and transketolase enzymes in bacterial strains 4B6, C2A1, and W3A1 (Colby and Zatman, 1975b), *P. methanica* and *M. capsulatus* (Strom et al., 1974). However, *Bacillus* sp. PM6 and S2A1 lack transaldolase activity, and their sugar phosphate interconversions involve sedoheptulose-1,7-diphosphate and fructose diphosphate aldolase (Colby and Zatman, 1975b).

There are thus four variants (permutations) of the cycle based on two modes of cleavage with two modes of rearrangement. Further studies are needed to establish correlations among individual organisms, their growth physiology, and the cycle variants employed (Anthony, 1982; Quayle, 1980a).

Originally, it was thought that a particular C_1-utilizing microorganism possesses either the Calvin cycle, the serine pathway, or the RMP pathway to the exclusion of the other two (Anthony, 1975; Quayle, 1972). However, evidence has accumulated that several microorganisms may contain more than one mechanism for C_1 assimilation either constitutively or induced under different growth conditions (Colby et al., 1979; Higgins et al., 1981; Quayle, 1980a). Taylor (1977) has demonstrated the presence of ribulose-1,5-diphosphate carboxylase and phosphoribulokinase, key enzymes of the Calvin pathway, in extracts of the RMP pathway bacterium *M. capsulatus* (Bath). Radioactive studies demonstrated early incorporation of label from the C_1 unit into 3-phosphoglycerate, aspartate, citrate, and malate, the latter three compounds arising from carboxylation of C_3 metabolites. Ribulose diphosphate carboxylase thus functions in vivo in this methane utilizer, generating labeling patterns similar to those observed for autotrophic bacteria grown heterotrophically. Despite the presence of a complete Calvin cycle, the rate of CO_2 fixation by intact cells of *M. capsulatus* was only 25% of the total cell carbon (Taylor et al., 1980), and this explains *M. capsulatus*'s (Bath) inability to grow autotrophically on CO_2 (Higgins et al., 1981; Quayle, 1980a).

Methanol-grown *Streptomyces* sp. or methane-grown *M. capsulatus* (Texas) and *P. methanica* (Salem et al., 1973) that use the RMP pathway as their major mechanism for C_1 assimilation contain low to moderate specific activities of hydroxypyruvate reductase and other enzymes of the serine pathway. In *M. capsulatus* (Bath) and in other methane-utilizing bacteria (Malashenko, 1976), hydroxypyruvate reductase activity was detected at 45°C, but not at 30°C (Whittenbury et al., 1975). Pulse labeling studies in continuous cultures of *M. capsulatus* (Bath) using [^{14}C] methanol demonstrated the incorporation of label into sugar phosphates at both 30°C and 45°C, but label was incorporated into serine and glycine only at 45°C (Whittenbury et al., 1976).

A Pentose Phosphate Cycle of Formaldehyde Fixation in Yeasts

Sugar phosphates were postulated as early intermediates in the methanol assimilation pathway in yeasts. The incorporation of ^{14}C-labeled C_1 compounds to phosphate esters of fructose and glucose was first reported by Fujii and his colleagues in methanol-grown yeasts (Fujii et al., 1974; Fujii and Tonomura, 1975). The absence of hydroxypyruvate reductase (serine pathway) and ribulose diphosphate carboxylase (Calvin cycle) and the labeling pattern was evidence for the operation of a RMP pathway in yeasts. The finding from several laboratories (Diel et al., 1974; Fujii et al., 1974; Sahm, 1977) that cell-free extracts of methanol-grown yeasts catalyzed a pentose phosphate–dependent fixation of [^{14}C] formaldehyde into hexose phosphates supported this view. However, the specific activity of hexulose phosphate synthase was too low to account for growth of yeast on methanol; and unlike the bacterial enzyme, the yeast enzyme was stimulated by ATP (Fujii et al., 1974; Fujii and Tonomura, 1975). Furthermore, Kato et al. (1977) could not detect the RMP pathway enzyme phosphohexulose isomerase in cell-free extracts of *Kloeckera* sp. 2201. This dilemma was solved by Quayle's group when they demonstrated that the condensation of formaldehyde and xylulose monophosphate catalyzed by dihydroxyacetone (DHA) synthase yielded DHA and glyceraldehyde-3-phosphate.

Mutants of *Hansenula polymorpha* and *Candida boidinii* lacking triokinase, and a mutant of *H. polymorpha* with low levels of fructose diphosphatatase (O'Connor and Quayle, 1979) provided further evidence for the operation of the DHA pathway. These mutants could not grow on methanol, while revertants to wild-type growth on methanol had retained the ability to synthesize wild-type levels of these enzymes. Recently, DHA synthase was purified from *C. boidinii* 2201 and found to be a new type of transketolase with respect to its specificity for ketol acceptors (Kato et al., 1982).

IV. ENERGY METABOLISM

Methylotrophs oxidize C_1 growth substrates to CO_2 via a series of special oxidation pathways shown in Fig. 8.3. From these oxidative reactions, ATP for biosynthesis of cell constituents is derived. Formaldehyde occupies a central position in the metabolism of methane and methanol because at the formaldehyde oxidation level carbon is both assimilated into biomass and oxidized to CO_2 to provide energy for growth.

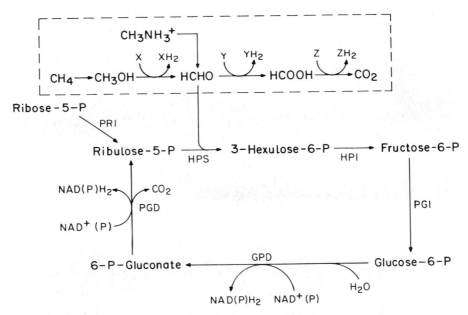

FIGURE 8.3 Pathways for direct and cyclic oxidation of reduced C_1 compounds to CO_2. The abbreviations used are: GPD, glucose-6-phosphate dehydrogenase; HPS, hexulose phosphate synthase; PGD, 6-phosphogluconate dehydrogenase; PGI, phosphoglucoisomerase; HPI, hexulose isomerase; PRI, phosphoriboisomerase; X, Y, Z represent electron acceptors. The broken line is the direct oxidation pathway.

The Direct Oxidation Pathway

The oxidation of C_1 compounds in most of methylotrophs proceeds via a series of two-electron oxidation steps: methane to methanol to formaldehyde to formate and then to CO_2 (the direct oxidation pathway in Fig. 8.3). Each oxidation step is sufficiently exergonic to permit, in theory, the generation of one or more ATP molecules (Wolfe and Higgins, 1979).

Evidence for operation of the direct oxidation pathway in methylotrophs is supported by the following data (Wolfe and Higgins, 1979):

1. cell suspensions of methylotrophs oxidized methane, methanol, formaldehyde, and formate to CO_2;
2. high specific activities of enzymes catalyzing each step were detected in cell-free extracts and extensively purified;
3. during oxidation of methane or methanol to CO_2 by cell suspensions, oxidative intermediates were shown to accumulate in the presence or absence of trapping agents or inhibitors; and
4. methanol oxidation by particulate fractions derived from the intracytoplasmic membranes of *M. capsulatus* yielded formate as the major product.

Oxidation of Methane to Methanol. In methylotrophs the enzyme system responsible for the oxidation of methane to methanol is a methane monooxygenase (MMO), which catalyzes the following reaction:

$$CH_4 + O_2 + NAD(P)H_2 \rightarrow CH_3OH + NAD(P)^+ + H_2O$$

The properties of purified enzymes from *Methylosinus trichosporium* OB3b (a facultative methane-utilizer) (Tonge et al., 1975, 1977b) and *M. capsulatus* (Bath) (an obligate methane-utilizer) (Colby and Dalton, 1978, 1979) have been reviewed previously (Colby et al., 1979; Higgins et al., 1981; Quayle, 1980a; Wolfe and Higgins, 1979). The enzyme from *M. capsulatus* consists of three component proteins of molecular weights 220,000 (the hydroxylase component), 15,000, and 44,000 (the reductase component). This MMO is similar to that found in *M. trichosporium* (Scott et al., 1981; Stirling et al., 1979), and close functional similarity was demonstrated (Stirling and Dalton, 1979a) by cross-reactivity between two of the components of the enzyme complex from *M. capsulatus* (A and C fractions) and fraction 1 of the MMO complex of *M. trichosporium*.

Although several methane-utilizing yeasts were isolated, the enzyme responsible for methane oxidation to methanol has not been detected in vitro.

Oxidation of Methanol to Formaldehyde. All methylotrophic bacteria oxidize methanol to formaldehyde by a soluble methanol dehydrogenase, first studied in the facultative methylotroph *Pseudomonas* M27 (Anthony, 1982;

Anthony and Zatman, 1964, 1965, 1967a, b). The enzyme has been purified from *Methylomonas* J (Tobari et al., 1981), *Pseudomonas* M27 (Anthony and Zatman, 1964; Patel et al., 1972), *Hyphomicrobium* X (Duine et al., 1978), *Pseudomonas* C (Goldberg, 1976), *M. organophilum* (Wolf and Hanson, 1978), *M. capsulatus* (Texas) (Patel et al., 1972), *Methylomonas methylovora* (Patel et al., 1979), *Methylosinus sporium* (Patel and Felix, 1976), *Pseudomonas* strains TP1 and W1 (Sperl et al., 1974), *Pseudomonas* RJ1 (Mehta, 1973), and a facultative methylotroph PAR (Bellion and Wu, 1978). The enzyme from different bacteria is characterized by low substrate specificity for alcohols, a requirement for high pH, and the ability, in vitro, to couple methanol oxidation to phenazine methosulfate in the presence of ammonia or a primary amine activator. By the use of X-ray crystallography a novel methanol dehydrogenase prosthetic group was found in numerous methylotrophs and named "methoxatin" (Salisbury et al., 1979) or "pyrrolo-quinoline quinone" (Duine and Frank, 1980). The coenzyme in *Hyphomicrobium* methanol dehydrogenase was shown to act as an electron carrier and a functional coupler to cytochrome C (Duine et al., 1979). All methanol dehydrogenases examined also oxidize formaldehyde to formate, presumably because hydrated formaldehyde, which predominates in aqueous solutions ($> 99\%$), is similar in structure to methanol (Quayle, 1980b; Sperl et al., 1974). The physiological significance of this dual substrate specificity is not known.

Several groups (Patel and Felix, 1976; Tonge et al., 1975, 1977b; Wadzinski and Ribbons, 1975a) purified methanol dehydrogenase from methane-grown bacteria and have shown that it is a particulate enzyme located on the intracytoplasmic membranes. However, the membrane-bound enzyme proved to be indistinguishable from the soluble methanol dehydrogenase of other methylotrophs.

In yeasts, growth on methanol is accompanied by a pronounced increase in the number of microbodies known as peroxisomes (van Dijken et al., 1975b, 1978; Fukui et al., 1975a, b; Hazeu et al., 1975; Sahm et al., 1975; Tanaka et al., 1976). These organelles contain high levels of flavin-dependent alcohol oxidase (Tani et al., 1972, 1978; Tani and Yamada, 1980; Veenhuis et al., 1976) that catalyzes the reaction:

$$CH_3OH + O_2 \rightarrow HCHO + H_2O_2$$

The bulk of formaldehyde formed in the peroxisomes is exported to the cytoplasm, where it is oxidized to CO_2 by NAD^+-dependent dehydrogenase. Under certain conditions a portion of the formaldehyde is oxidized to formate in the peroxisomes either by methanol oxidase itself (Sahm, 1975) or by the reaction of H_2O_2 and calalase, an enzyme also present in methanol induced peroxisomes (van Dijken et al., 1975a; Fujii and Tonomura, 1975; Roggenkamp et al., 1974; Roggenkamp et al., 1975).

Oxidation of Methylamine(s) to Formaldehyde. Numerous enzyme systems responsible for the oxidation of methylamine(s) to formaldehyde have been found in different methylotrophs propagated on methylamine(s) (Colby et al., 1979; Eady and Large, 1968; Hersh et al., 1971; Matsumoto (1978); Matsumoto et al., 1978; Matsumoto and Tobari, 1978a, b; McFadden, 1973; Mehta, 1976, 1977; Meiberg and Harder, 1979; Quayle, 1961, 1980b; Shirai et al., 1978; Steenkamp and Mallison, 1976) (Fig. 8.3). In bacteria, several dehydrogenases, oxygenases, and oxidases that catalyze methylamine oxidation to formaldehyde were isolated. Yeasts are able to utilize methylamine(s) only as a nitrogen source (but not as a carbon source) by oxidizing it to formaldehyde and ammonia with at least two methylamine oxidases present in peroxisomes (Colby et al., 1979; van Dijken et al., 1979; Green et al., 1982).

Oxidation of Formaldehyde to Formate.   In a previous review, Quayle (1980b) enumerated the following enzyme systems for the oxidation of formaldehyde to formate:

1. A $NAD^+(P)$-independent phenazine methosulphate-linked formaldehyde dehydrogenase of *Pseudomonas* AM1 (Johnson and Quayle, 1964), *Pseudomonas* C (Ben-Bassat and Goldberg, 1980), *Hyphomicrobium* X (Marison and Attwood, 1980), and *Pseudomonas* RJ1 (Mehta, 1975).
2. Bacterial methane dehydrogenase or yeast methanol oxidase possessing dual substrate specificity (Quayle, 1980b; Sahm, 1975; Sperl et al., 1974).
3. A NAD^+-dependent dehydrogenase from *Bacillus* 4B6 (Colby and Zatman, 1973), *Pseudomonas* MS (Kung and Wagner, 1970), *M. capsulatus* (Bath) (Stirling and Dalton, 1978), and *M. trichosporium* OB3b (Stirling and Dalton, 1979a).
4. A NAD^+-linked and reduced glutathione-dependent formaldehyde dehydrogenase in bacteria (Johnson and Quayle, 1964) and yeasts (Kato et al., 1974; Tani et al., 1972).
5. A THF-mediated enzyme cycle to be discussed later (see Section V).

Recently, it was suggested (Marison and Attwood, 1980) that the dye-linked formaldehyde dehydrogenase is probably a general aldehyde dehydrogenase not directly involved in the dissimilation of C_1 compounds.

Oxidation of Formate to CO_2.   Formate is oxidized to CO_2 by NAD^+-dependent formate dehydrogenase. The enzyme was purified from methylotrophic bacteria (Egorov et al., 1979) and yeasts (van Dijken et al., 1976a; Kato et al., 1980).

The Dissimilatory Ribulose Monophosphate Pathway (The Cyclic Pathway)

Colby and Zatman (1975a, b) found low or negligible activities of formaldehyde dehydrogenase and NAD^+-dependent formate dehydrogenase in extracts of several methylamine-grown bacteria. However, these extracts contained high specific activities of phosphoglucoisomerase, glucose-6-phosphate dehydrogenase, and 6-phosphogluconate dehydrogenase, and Colby and Zatman proposed that these enzymes together with hexulose phosphate synthase and phosphohexuloisomerase participate in a cyclic sequence for complete oxidation of formaldehyde to CO_2 (Fig. 8.3). Since the first two enzymes of the oxidation cycle (i.e., hexulose phosphate synthase and phosphohexuloisomerase) are common to the assimilatory RMP pathway, such a formaldehyde dissimilatory pathway is limited to microorganisms that possess the RMP assimilatory cycle (Anthony, 1982; Babel and Mothes, 1978; Ben-Basset and Goldberg, 1977, 1980; Quayle, 1980a, b; Sokolov and Trotsenko, 1977; Strom et al., 1974).

Both in the direct oxidation pathway (via formic acid) and the cyclic oxidation pathway, two mols of reduced pyridine nucleotides and one mol of CO_2 are formed for each mol of formaldehyde oxidized (Fig. 8.3). However, the cyclic oxidation pathway yields both NADPH and NADH, whereas the direct oxidation pathway generates only NADH (Ben-Bassat and Goldberg, 1977) with the exception of a $NADP^+$-linked formaldehyde dehydrogenase described in *M. capsulatus* (Bath) (Stirling and Dalton, 1978). Evidence that both the cyclic oxidation and the direct oxidation pathways operate in methanol-grown *Pseudomonas* C was shown by the following (Ben-Bassat and Goldberg, 1977):

1. Detection in cell extracts of high specific activities of enzymes of the cyclic oxidation pathway and low activities of formaldehyde and formate dehydrogenases.

2. Stimulation of NAD^+ or $NADP^+$ formaldehyde–dependent reduction in cell extracts by D-ribulose-5-phosphate.

3. Oxidation of [^{14}C] formaldehyde to CO_2 by cell extracts required the presence of NAD^+ (or $NADP^+$) and was greatly increased by the addition of D-ribulose-5-phosphate.

4. Demonstration of the cyclic oxidation pathway by incubating cell extracts, labeled glucose-6-phosphate (1-^{14}C and U-^{14}C), and formaldehyde resulting in an initial ratio of $^{14}CO_2$ derived from [1-^{14}C]glucose-6-phosphate/[U-^{14}C]glucose-6-phosphate of 6.8, which subsequently decreased upon a prolonged incubation owing to interconversion of ribulose-5-phosphate and unlabeled formaldehyde to glucose-6-phosphate.

From enzymological studies it was found that *P. methanica* (Strom et al., 1974) can oxidize C_1 compounds via both the cyclic and the direct oxidation pathways. However, in *Pseudomonas* C, about half of the methanol carbon that is converted to CO_2 is oxidized via the cyclic pathway during growth on methanol (Samuelov and Goldberg, 1982a), but in *P. methanica*, most of the methanol carbon is oxidized to CO_2 via formate (Johnson and Quayle, 1964; Strom et al., 1974). In bacterium W6A (Colby and Zatman, 1975a, b) and *Arthrobacter* P1 (Levering et al., 1981) C_1 compounds are oxidized only via the cyclic oxidation pathway (see Section V).

Glucose-6-phosphate dehydrogenase (G6PDH) and 6-phosphogluconate dehydrogenase (6PGDH) have been purified from methanol-grown *Pseudomonas* C (Ben-Basset and Goldberg, 1980), *Pseudomonas oleovorans* (Sokolov et al., 1980), and *Candida boidinii* (Kato et al., 1979). The above enzymes from *Pseudomonas* C and *C. boidinii* differ in their specificity for pyridine nucleotides (Ben-Bassat and Goldberg, 1980). In *Pseudomonas* C, G6PDH and 6PGDH react at comparable rates with both $NADP^+$ and NAD^+, but the *C. boidinii* enzymes utilize only $NADP^+$. Similar NAD^+ and $NADP^+$ utilization was reported with purified and partially purified G6PDH of *Methylomonas* M15 (Sahm and Steinbach, 1977) and *Pseudomonas* W6 (Miethe and Babel, 1976), respectively; and in *Pseudomonas* W6 (Miethe and Babel, 1976), 6PGDH also utilized both coenzymes. In these bacteria, both G6PDH and 6PGDH have higher affinities for $NADP^+$ than for NAD^+, but the intracellular NAD^+ concentration (2–6 mM; see Andersen and Meyenburg (1977), Bowien et al. (1974), Opheim and Bernlohr (1973), Setlow and Setlow (1977), Wimpenny and Firth (1972)) is higher than that of $NADP^+$ (0.3–1.8 mM; see Andersen and Meyenburg (1977), Opheim and Bernlohr (1973), Setlow and Setlow (1977)). It appears that the lower affinities of the enzymes for NAD^+ are compensated by higher intracellular NAD^+ concentrations.

Reduced pyridine nucleotides and ATP inhibit the activity of $NAD^+/NADP^+$-dependent G6PDH of *Pseudomonas* C (Ben-Bassat et al., 1980), *Pseudomonas* W6 (Miethe and Babel, 1976), *Methylomonas* M15 (Sahm and Steinbach, 1977), and *Methylophilus methylotrophus* (Quayle, 1980b). Reduced NAD^+ and $NADPH^+$ and ATP also inhibit the NAD^+- and $NADP^+$-linked 6-phosphogluconate dehydrogenases of *M. methylotrophus* (Quayle, 1980a) and the single $NAD^+/NADP^+$-dependent 6PGDH of *Pseudomonas* C (Ben-Bassat and Goldberg, 1980), while NADH had no effect on the enzyme in *C. boidinii* (Kato et al., 1979). The physiological importance of these findings are discussed later in this chapter (see Section V).

The Tricarboxylic Acid (TCA) Cycle

Glucose-6-phosphate is an important intermediate in methylotrophic bacteria that contain both the assimilatory and dissimilatory RMP pathway (for example *Pseudomonas* C; see Ben-Bassat and Goldberg (1977)). Glucose-6-phosphate may be oxidized to CO_2 via the Embden-Meyerhof pathway and the tricarboxylic acid (TCA) cycle. However, it was found that methylotrophic bacteria that operate the RMP pathway possess an incomplete TCA cycle (Anthony, 1982; Colby et al., 1979; Higgins et al., 1981). A key enzyme indicating the existence of the TCA cycle is α-ketoglutarate dehydrogenase. Many obligate RMP pathway bacteria (Davey et al., 1972) (for example, *Pseudomonas* C (Ben-Bassat and Goldberg, 1977), *Bacillus* spp. W3A1, W6A, S2A1, PM6 and 5B1 (Colby and Zatman, 1975a), and *M. capsulatus* (Patel et al., 1975)) have no detectable activity of α-ketoglutarate dehydrogenase, although all other TCA enzymes were detected at low activities. In methylotrophs the incorporation of radioactive acetate into cell constituents was used to study the significance of the TCA cycle. With *M. capsulatus* (Texas), the endogenous rate of acetate uptake was stimulated by methane, methanol, and formate (Patel et al., 1969, 1975). Label from $[1\text{-}^{14}C]$ acetate was incorporated only into proline, arginine, and leucine and the C5, but not C1, of glutamate, indicating the absence of α-ketoglutarate dehydrogenase. Unlike RMP pathway bacteria, methylotrophic bacteria using the serine pathway possess a complete TCA cycle (Colby et al., 1979; Higgins et al., 1981). In the serine pathway bacterium *M. methano-oxidans*, the addition of acetate enhanced growth, and label from $[1\text{-}^{14}C]$ acetate was incorporated into amino acids of the glutamate, aspartate, and pyruvate families, indicating a complete and functional TCA cycle (Wadzinski and Ribbons, 1975b). For some time the obligatory growth of methylotrophs on C_1 compounds was explained by the absence of α-ketoglutarate dehydrogenase (Smith and Hoare, 1977; Taylor and Anthony, 1975, 1976). However, because "restricted methylotrophs," which utilize glucose and C_1 compounds, lack α-ketoglutarate dehydrogenase activity (Colby and Zatman, 1975a), the absence of this enzyme alone cannot explain the inability of obligate methylotrophs to grow on non-C_1 compounds (Colby and Zatman, 1975a; Smith and Hoare, 1977). The role of the TCA cycle in methylotrophs is not fully elucidated, and more detailed knowledge in this area is required (Anthony, 1982).

Electron Transport Systems and Energy Derived from C_1 Compounds Oxidation

A scheme for electron transport in methylotrophs including cytochromes a/a_3, c/c_0, and b, ubiquinone-10, and flavoprotein is discussed elsewhere (Anthony, 1982; O'Keeffe and Anthony, 1978; Wolfe and Higgins, 1979). The presence of different electron transport system components has been shown in *M. capsulatus* (Ribbons et al., 1970), *M. albus* and *M. trichosporium* (Davey and Mitton, 1973; Tonge et al., 1974, 1977a, b; Weaver and Dugan, 1975), *P. methanica* (Ferenci, 1974, 1976; Tonge et al., 1974, 1977b), *Pseudomonas* P11 (Brabikowska, 1977), *P. extorquens* (Higgins et al., 1976), *M. methylotrophus* (Dawson and Jones, 1981), and *Pseudomonas* AM1 (O'Keeffe and Anthony, 1978; Widdowson and Anthony, 1975) from different spectra with or without carbon monoxide as well as electron transport inhibition studies using whole cells and cell-free extracts.

Despite the presence of electron transport systems, the exact sites of ATP synthesis in methylotrophs are not known. With *Methylococcus* NC1B 11083, methane to methanol oxidation does not produce energy that can be coupled to growth (Linton and Vokes, 1978). This is in agreement with the observation (Linton and Vokes, 1978) that the same molar yield is obtained with growth on methane and methanol (Table 8.1). Furthermore, it indicates not only that there is an increased oxygen demand for growth on methane as compared to methanol (Drozd and Wren, 1980; Linton and Vokes, 1978), but that the potential energy available in methane to methanol oxidation (39.9 kcal/mol) is lost as heat.

Measurements of respiration-driven proton translocation in different methylotrophs (Dawson and Jones, 1981; Drozd and Wren, 1980; Netrusov et al., 1977; Tonge et al., 1977a; van Verseveld and Stouthamer, 1978) indicate a P/O ratio of 1 at the oxidation step of methanol to formaldehyde. From molar growth yields on methanol or formaldehyde a P/O ratio of 1 was calculated for *Pseudomonas* strains 1, 135, AM1, M27, and *rosea* (Goldberg et al., 1976; Rokem et al., 1978), all serine pathway methanol-utilizing bacteria. Methylotrophic oxidation of methanol to formaldehyde has a total potential energy equivalent to at least 3 ATP (36.8 kcal/mol) (Drozd and Wren, 1980), and the above finding that only 1 mol of ATP is produced indicates a very low coupling of electron transfer to ATP production. The NAD-dependent oxidation steps of formaldehyde and formic acid in an obligate methane utilizer *M. trichosporium* OB3b yielded a P/O ratio of 1 for each of the two steps (Tonge et al., 1977a). This contrasts with results obtained with the facultative methanol-grown *Pseudomonas* AM1, in which a maximum P/O ratio of 2 for NADH oxidation was noted (O'Keefe and Anthony, 1978; see also Kell et al., 1978). Recently, in *Candida boidinii* 2201 a novel method for demonstrating ATP formation from AMP during oxidation of methanol and formate was described (Tani et al., 1982). This technique may be applicable to study the efficiency of ATP formation in C_1-utilizing bacteria.

In facultative methylotrophs the relationship between ATP derived from oxidation of C_1 compounds and the actual molar yield values will be discussed in Section VI.

V. REGULATION OF CARBON METABOLISM

Regulation of Initial Pathway Enzymes

Changes in the activity of methane monooxygenase (MMO) activity were noted in whole cells of *M. trichosporium* OB3b grown under different conditions (Scott et al., 1981). Bacteria propagated in a batch culture had a particulate MMO activity (associated with intracytoplasmic membranes) sensitive to different inhibitors, while bacteria grown in continuous culture contained a soluble MMO activity insensitive to chelating and thiol agents. The synthesis of MMO was repressed by substrates other than methane and was high during growth under O_2-limiting conditions but decreased at higher O_2 tensions (Patt and Hanson, 1978).

Comprehensive studies were reported on the regulation of the dissimilatory enzymes in yeasts grown on methanol. Van Dijken et al. (1976b) found in methanol-limited chemostat cultures of *H. polymorpha* operated at dilution rates of 0.16 and 0.03 h^{-1} that the level of methanol oxidase increased from 7% to 20% of the total soluble protein, respectively. Induction by methanol and repression by glucose of methanol oxidase, catalase, formaldehyde, and formate dehydrogenases, the four methanol dissimilatory enzymes, have been reported (Egli et al., 1982a, b, c). During growth under a nitrogen-limited environment, peroxisomal enzymes, methanol oxidase, and catalase were degraded (catabolite inactivation) (Egli, 1982).

Much remains to be learned about the control of the flux of carbon from methane and methanol to formaldehyde in methylotrophs (Anthony, 1982). As formaldehyde is found at the branch point between dissimilatory and assimilatory pathways in methylotrophs, control mechanism(s) are required to regulate the flow of carbon between these pathways.

Serine Pathway Microorganisms.   As was discussed earlier, formaldehyde is oxidized to CO_2 in serine pathway methylotrophs by the direct oxidation pathway. Formaldehyde is incorporated into cell constituents via the serine pathway by the following reactions:

$$\text{Formaldehyde} + \text{THF} \rightarrow N^5,N^{10}\text{-Methylene-THF}$$

$$N^5,N^{10}\text{-Methylene-THF} + \text{Glycine} \rightarrow \text{Serine} + \text{THF}$$

In serine pathway methylotrophs there is no evidence that the first reaction is catalyzed by an enzyme, and it is not known how the flux of formaldehyde is diverted between dissimilation and assimilation pathways (Quayle, 1980b). Harder and Attwood (1978) suggested that only when all the THF in the cell is in the form of C_1 derivatives would formaldehyde be available for oxidative energy generation via the formaldehyde dehydrogenase reaction. Thus formaldehyde flux is controlled by the relative levels of reduced pyridine nucleotide (NADH), ATP, and THF. However, Quayle (1980b) suggested that in addition to the direct oxidation pathway, two alternative oxidation pathways might operate in serine pathway bacteria. The first one involves THF-mediated reactions in which formaldehyde is oxidized to formic acid and bypass the possible assimilative control mechanism. The second alternative of formaldehyde oxidative pathway, described by Newaz and Hersh (1975) in *Pseudomonas* MA, involves the oxidation of formaldehyde to CO_2 by the serine pathway and the TCA cycle enzymes. In this pathway, carboxylation of PEP (the branch point metabolite) to oxaloacetate by PEP carboxylase diverts the flow of carbon into assimilation, whereas dephosphorylation of PEP to pyruvate by pyruvate kinase diverts the carbon flux into dissimilation via glyoxyalate and the TCA cycle. It was found (Millay and Hersh, 1976; Millay et al., 1978; Newaz and Hersh, 1975) that PEP carboxylase activity is stimulated by NADH; therefore at high NADH concentrations, oxaloacetate production is favored. The activation by NADH occurs by transformation of the enzyme from a less active high-molecular-weight form (320,000) to a more active low-molecular-weight form (180,000). However, at low levels of NADH, PEP carboxylase is inactive, and this permits the bulk of carbon to be oxidized for energy.

A different control mechanism in facultative methylotrophs is demonstrated by the presence of isoenzymes of serine transhydroxymethylase (O'Connor and Hanson, 1975), isocitrate lyase (Bellion and Kim, 1978; Bellion and Woodson, 1975), and PEP carboxylase (Newaz and Hersh, 1975). One isoenzyme of the above three enzymes is formed during growth on C_1 compounds, while the other isoenzyme is synthesized only during growth on non-C_1 compounds (Quayle, 1980b).

RMP Pathway Bacteria.   The oxidation of methanol to CO_2 in the obligate methanol-grown *Pseudomonas* C (Samuelov and Goldberg, 1982b) is carried out via two oxidation pathways (the direct and cyclic pathways) in which formaldehyde is situated at the branch point (Ben-Bassat et al., 1980). Formaldehyde (or methanol) is assimilated into cell material via the KDPG-aldolase (phospho-2-keto-3 deoxygluconate aldolase) reaction of the RMP pathway (Ben-Bassat et al., 1980). In *Pseudomonas* C the flux of carbon from methanol via the different biochemical pathways leading to the production of either cell biomass or CO_2 was studied (Ben-Bassat et al., 1980; Samuelov and Goldberg, 1982a) by measuring the distribution between CO_2 and cell constituents of different ^{14}C-labeled intermediates after their injection into a chemostat culture. Two conclusions derived from the results were the following:

1. Under different growth conditions, distribution changes of carbon from methanol between various biochemical pathways were noted. The bacterium controlled the flux of methanol carbon via formaldehyde dehydrogenase, hexulose phosphate synthase, 6-phosphogluconate dehydrogenase, and KDPG-aldolase according to its cellular requirements and to prevent accumulation of toxic intermediates (Ben-Bassat et al., 1980; Samuelov and Goldberg, 1982a).
2. In cells grown with excess amounts of carbon and energy the importance of the direct oxidation pathway increased, while in cells grown with limited supply of either carbon or energy the importance of the cyclic oxidation pathway increased together with an increase in the biomass yield (Samuelov and Goldberg, 1982a).

From measurements of enzyme activities (Ben-Bassat et al., 1980; Samuelov and Goldberg, 1982a) it appears that control of carbon flow is regulated by in vivo activities of these enzymes and not by varying the rates of synthesis of the various enzymes (which is reflected in their steady state concentrations). When methanol growth of *Pseudomonas* C was carried out in an excess of carbon and energy (limitation by iron or nitrogen), the molar yield based on the methanol utilized decreased by about 60% and only 44–48% of the carbon from methanol was diverted via the hexulose phosphate synthase reaction (Samuelov and Goldberg, 1982a). Also, 54% of the methanol carbon was diverted via the formaldehyde dehydrogenase reaction (the direct oxidation pathway) as compared to 25% during growth on methanol alone. In cells grown with excess amounts of carbon and energy, the calculated in vivo activity of formaldehyde dehydrogenase was 2.9–3.8 times higher than methanol-limited-grown cells. The absence of changes in hexulose phosphate synthase activity during growth of *Pseudomonas* C implies that changes in formaldehyde dehydrogenase activity affects the flow of carbon into the direct oxidation pathway or the RMP assimilatory and dissimilatory (cyclic) pathways. Small molecule modulation of the activity of hexulose phosphate synthase, a branch point enzyme in obligate methylotrophs, has not yet been encountered (Quayle, 1980b), while in the facultative *Pseudomonas oleovorans* the activity of the enzyme is inhibited by NADH and NADPH (Sokolov and Trotsenko, 1978a).

Anthony (1982) questioned the ^{14}C-distribution patterns in *Pseudomonas* C described above on the basis that the use of uniformly labeled [^{14}C]-glucose cannot be taken as a tracer of the 6-phosphogluconate metabolism occurring during carbon-limited growth in continuous culture on methanol containing a low concentration of glucose. This is because the 6-phosphogluconate will have relatively less label at the C-1 position owing to the condensation of uniformly labeled ribulose-5-phosphate with nonradioactive formaldehyde, and hence measurement of ^{14}CO$_2$ will give a low value for estimating the proportion of formaldehyde being oxidized by the cyclic pathway. Anthony (1982) estimated that more than 90% of the oxidation of methanol is by way of the cyclic pathway in *Pseudomonas* C supporting his generalization that "all obligate methylotrophs (except methanotrophs) having the RMP pathway oxidize formaldehyde by the cyclic route." The situation is more complicated by the presence of rearrangement reactions leading to regeneration of ribulose-5-phosphate from glyceraldehyde phosphate. These reactions, which were not considered by Anthony, will give, on the other hand, a high value for estimating the proportion of formaldehyde being oxidized by the cyclic pathway. Clearly, more experimental work is needed in order to understand how obligate methylotrophs control the flux of methanol (or formaldehyde) carbon via the RMP assimilatory pathway and the cyclic and direct dissimilatory pathways.

The enzyme 6-phosphogluconate dehydrogenase is another branch point of metabolism in obligate methylotrophs; decarboxylation of 6-phosphogluconate to ribulose-5-phosphate directs the flow of carbon into dissimilation via the cyclic oxidation pathway, whereas cleavage to pyruvate and glyceraldehyde-3-phosphate commits the flow into assimilation. Changes in the distribution of methanol carbon between different pathways of C$_1$-metabolism were explained (Ben-Bassat et al., 1980) in part by in vitro studies on both the inhibitory effects of ATP and reduced coenzymes (NADH and NADPH) on the activities of glucose-6-phosphate dehydrogenase and 6-phosphogluconate dehydrogenase. Inhibition by ATP, NADH, or NADPH of the dissimilatory cyclic pathway of formaldehyde oxidation suggests a control mechanism for carbon flux (Ben-Bassat and Goldberg, 1980; Newaz and Hersh, 1975; Quayle, 1980a). To confirm the in vivo regulation of enzyme activities involved in the C$_1$ carbon flux, both the intracellular levels of the metabolites during growth and the effect of small molecules on 6-phosphogluconate dehydratase and phospho-2-keto-3-deoxygluconate aldolase activities must be determined. A study on carbon distribution between the different C$_1$ metabolic pathways would be facilitated by the isolation of mutants impaired in the direct oxidation pathway or the cyclic dissimilatory pathway.

VI. GROWTH YIELDS OF METHYLOTROPHIC MICROORGANISMS

Because of commercial interest in using methanol- or methane-grown bacteria for single-cell protein (SCP) production and the influence of yield on the economics of SCP production (Goldberg, 1977, 1981; Moo-Young, 1977), it is important to know the theoretical maximum cellular yield coefficient for growth on C_1 compounds for comparison with experimental molar growth yields. The latter must be determined under optimal conditions, which can be defined and determined only in a continuous chemostat culture. For the calculation of C_1 theoretical yields, both the ATP yield from and the ATP requirement for dissimilation and assimilation pathways, respectively, as well as the composition of the bacterial cell and its maintenance energy requirement have to be known.

Theoretical yield values of methylotrophs grown on C_1 compounds were first calculated by van Dijken and Harder (van Dijken and Harder, 1975; Harder and van Dijken, 1976) and later elaborated by Anthony (1978, 1982). However, these calculations did not consider the following (Mateles, 1979): (a) the excretion of partly oxidized intermediates into the culture broth by bacteria grown in a chemostat on limiting C_1 substrate (Goldberg and Mateles, 1975; Brabikowska, 1977; Papoutsakis et al., 1978), (b) the existence of two oxidation pathways in several methylotrophs as discussed above (see the subsection entitled "The Dissimilatory Ribulose Monophosphate Pathway"), (c) the changes in cellular composition under different growth conditions, and (d) the maintenance energy requirements (Rokem et al., 1978). Nevertheless, three generalizations are evident from theoretical data:

1. Bacteria growing on C_1 compounds via the RMP pathway exhibit a higher theoretical biomass yield than serine pathway bacteria. This was confirmed by experimental molar yield values (Goldberg, 1977, 1981; Goldberg et al., 1976) and true molar yield (maintenance coefficient = 0) values (Rokem et al., 1978) determined for different methylotrophs (Table 8.1). In general, the RMP pathway bacteria also exhibit higher maximum specific growth rates (Goldberg, 1977, 1981; Goldberg et al., 1976; Rokem et al., 1978), which would result in higher process productivity.

2. A general assumption in the calculation of theoretical C_1 growth yields was that 1 mol of NADH is produced per mol of formaldehyde oxidized to CO_2. It follows that bacteria that produce higher amounts of NADH during the oxidation of formaldehyde to CO_2 would result in higher biomass yields. This conclusion was supported by results showing that with *Pseudomonas* C (Samuelov and Goldberg, 1982a) grown with limited carbon but not energy, the relative activity of the cyclic oxidation pathway in-

creased, resulting in an increase in cellular yield obtained. However, with cells grown in excess carbon and energy the activity of the direct oxidation pathway increased, and the cellular yield decreased.

3. Linton and Stephenson (1977) showed that experimental cellular yields from C_1 substrates were lower than yields calculated on the basis of chemical potential energy available from C_1 substrate oxidation. This suggested (Drozd and Wren, 1980) that growth was reductant or carbon-limited and there might be poor coupling of ATP production to potential energy obtained from oxidation of C_1 compounds. Anthony (1978, 1982) concluded that methanol-grown RMP pathway bacteria become internally energy limited, but serine pathway methylotrophs become internally reductant limited. If RMP pathway bacteria are internally energy limited during growth on methanol, an increase in either ATP yield from methanol or the addition of an exogenous energy source would result in an increase in cellular yield. This was elegantly demonstrated with the RMP pathway bacterium *M. methylophilus* (Senior and Windass, 1980). The parent strain was genetically modified to assimilate nitrogen by an enzyme (glutamate dehydrogenase) that required less ATP than the parent nitrogen assimilation pathway (the GOGAT pathway). This resulted in a growth yield increase of 5%. In addition, formic acid (an energy source but not a carbon source) added to methanol-grown *Pseudomonas* C, resulted in a molar growth yield increase of 12% (Ben-Bassat et al., 1980).

With *Pseudomonas* 135 (Table 8.1) and other serine pathway bacteria (Goldberg et al., 1976; Rokem et al., 1978), growth yield values were determined for steady state growth on methanol, methylamine, formaldehyde, and formate; the molar yield values decreased as the oxidation level of the C_1-growth substrate increased. The finding that the molar yield values of bacteria grown on methylamine were higher than the values obtained for formaldehyde-grown bacteria suggests that the oxidation step from methylamine to formaldehyde might be energy yielding (Table 8.1) (Goldberg et al., 1976).

VII. GENETICS OF METHYLOTROPHS

Advances in genetics of methylotrophs have lagged behind our knowledge of the biochemistry and physiology of these microorganisms. This can be attributed in part to the difficulty of obtaining mutants from obligate methylotrophs. Spontaneous antibiotic resistant *M. capsulatus* mutants were obtained at frequencies of one per 10^7 to 4×10^8 cells (Harwood et al., 1972), but higher frequencies were obtained with facultative methylotrophs (one per 10^4 cells for *Pseudomonas* AM1 and *Pseudomonas extorquens* and one per 10^6 cells for *Pseudomonas* 3A2 and *M. organophilum* (Newaz and Hersh, 1975)). Induction of mutation in *M. capsulatus* with a variety of mutagens yielded few mutants that reverted at high frequency (Harwood et al., 1972; Williams et al., 1977). Similar attempts with several other obligate methane-utilizing bacteria were unsuccessful. Treatment of *M. organophilum*, a facultative methylotroph with various mutagenic agents, yielded several mutants unable to grow on C_1 compounds at a frequency of one per 5×10^5 cells. Also, drug-resistant mutants were obtained at a frequency of one per 10^7 cells (O'Connor et al., 1977).

The following mutant types were isolated from methylotrophs: drug-resistant mutants (Harwood et al., 1972; Jeyaseelan and Guest, 1979; Warner et al., 1980), auxotrophs (Jeyaseelan and Guest, 1979), mutants which accumulate O-methyl-L-homoserine (Tanaka et al., 1980) or L-methionine (Yamada et al., 1982), amino acid analog–resistant mutants (Yamada, 1982), a mutant with an increased yield on methanol (Senior and Windass, 1980), cytochrome-deficient mutants (O'Keefe and Anthony, 1978; Widdowson and Anthony, 1975), and mutants unable to grow on C_1 compounds (O'Connor and Hanson, 1977; O'Connor and Quayle, 1979; O'Connor et al., 1977; Taylor and Anthony, 1975). Of particular importance are the last type of mutants, since they enable us to understand the organization and regulation of C_1 metabolism (Colby et al., 1979; Higgins et al., 1981; Wolfe and Higgins, 1979). These mutants have been discussed in several sections of this chapter.

Transformation occurs in *M. capsulatus* as a *p*-aminobenzoic acid–requiring auxotroph was transformed at a frequency of 0.17% using DNA extracted from wild-type strain (Williams et al., 1971). Similar results were obtained (O'Connor et al., 1977) with a glutamate auxotroph of *M. organophilum* (isocitrate dehydrogenase-mutant), which was transformed at a frequency of 0.5% using wild-type DNA.

Plasmids isolated from facultative methane utilizers (Hanson, 1980) may code for methane monooxygenase, the initial enzyme in methane oxidation (Wopat and O'Connor, 1980). The facultative methanol utilizer *Pseudomonas* AM1 contains three plasmids (Warner et al., 1980), but their function is not clear. Plasmids containing antibiotic-resistance markers were successfully transformed into the following facultative methylotrophs: *Pseudomonas extorquens* (Jeyaseelan and Guest, 1979; Warner et al., 1980), *Pseudomonas* 3A2 (Jeyaseelan and Guest, 1979), *Pseudomonas* AM1 (Jeyaseelan and Guest, 1979), and *M. organophilum* (Jeyaseelan and Guest, 1979), and into the obligate methylotroph *M. trichosporium* OB3b (Jeyaseelan and Guest, 1979). A hybrid plasmid consisting of R1162 and *Pseudomonas* AM1 chromosomal DNA was constructed and mobilized into *Pseudomonas* AM1/M15a, a methanol dehydrogenase minus mutant (methanol minus phenotype), resulting in colonies that were able to grow on methanol. This suggested that the gene for methanol dehydrogenase was present on the vector and cloned in *Pseudomonas* AM1/M15a (Gautier and Bonewald, 1980).

Reliable recombinant DNA techniques are necessary for genetic manipulation of methylotrophs to enhance production of industrial and medical biochemicals, biotransformation reactions, and overproduction of different enzymes. Recently (Hennan et al., 1982), two eukaryotic cDNA's encoding chicken ovalbumin and mouse dihydrofolate reductase were expressed in *Methylophilus methylotrophus*, demonstrating the potential of this organism for production of mammalian peptides.

VIII. CONCLUSIONS

Interest in methylotrophs began in 1956 when Dworkin and Foster reisolated Sohngen's methanol-utilizing strain and noted some physiological and biochemical peculiarities of these microorganisms. Since Dworkin's and Foster's publication (1956) the large research described above has unraveled many of the basic biochemical problems of C_1 compounds metabolism. Much of this research was undoubtedly stimulated by interest in SCP derived from methanol and methane as a nonagricultural means of producing foods or feeds. Biological and biochemical studies on C_1 metabolism greatly benefited the development of the ICI's methanol-SCP process, the first industrial application utilizing methylotrophs. The ICI's process has opened the way for further exploitation of other methylotrophs for industrial processes from this fascinating group of microorganisms.

REFERENCES

Andersen, K. B., and Meyenburg, K. J. (1977) *Biol. Chem. 252*, 4151–4156.
Anthony, C. (1975) *Sci. Prog. (Oxford) 62*, 167–206.
Anthony, C. (1978) *J. Gen. Microbiol. 104*, 91–104.
Anthony, C. (1982) *The Biochemistry of Methylotrophs*. Academic Press, London.
Anthony, C., and Zatman, L. J. (1964) *Biochem. J. 92*, 614–621.
Anthony, C., and Zatman, L. J. (1965) *Biochem. J. 96*, 808–812.
Anthony, C., and Zatman, L. J. (1967a) *Biochem. J. 104*, 953–959.
Anthony, C., and Zatman, L. J. (1967b) *Biochem. J. 104*, 960–969.
Babel, W., and Hofmann, K. (1975) *Z. Allg. Mikrobiol. 15*, 53–57.
Babel, W., and Loffhagen, N. (1977) *Z. Allg. Mikrobiol. 17*, 75–79.
Babel, W., and Miethe, D. (1974) *Z. Allg. Mikrobiol. 15*, 153–156.
Babel, W., and Mothes, G. (1978) *Z. Allg. Mikrobiol. 18*, 17–26.
Ballerini, D., Parlouar, D., Lapeyronnie, M., and Sri, K. (1977) *Eur. J. Appl. Microbiol. 4*, 11–19.
Bamforth, C. W., and O'Connor, M. L. (1979) *J. Gen. Microbiol. 110*, 143–149.
Bamforth, C. W., and Quayle, J. R. (1977) *J. Gen. Microbiol. 101*, 259–267.
Baratti, J., Couderc, R., Cooney, C. L., and Wang, D. I. C. (1978) *Biotech. Bioeng. 20*, 333–348.
Bassham, J. A., and Calvin, M. (1957) *The Path of Carbon in Photosynthesis*, Prentice-Hall, Englewood Cliffs, N.J.
Bellion, E., and Kim, Y. S. (1978) *Biochim. Biophys. Acta 541*, 425–434.
Bellion, E., and Kim, Y. S. (1979) *Curr. Microbiol. 2*, 31–34.
Bellion, E., and Woodson, J. (1975) *J. Bacteriol. 122*, 557–564.
Bellion, E., and Wu, C. T. S. (1978) *J. Bacteriol. 135*, 251–258.
Bellion, E., Bolbot, J. A., and Lash, T. D. (1981) *Curr. Microbiol. 6*, 367–372.

Ben-Bassat, A., and Goldberg, I. (1977) *Biochim. Biophys. Acta 497*, 586–597.

Ben-Bassat, A., and Goldberg, I. (1980) *Biochim. Biophys. Acta 611*, 1–10.

Ben-Bassat, A., Goldberg, I., and Mateles, R. I. (1980) *J. Gen. Microbiol. 116*, 213–223.

Blackmore, M. A., and Quayle, J. R. (1970) *Biochem. J. 118*, 53–59.

Boulton, C. A., and Large, P. J. (1977) *J. Gen. Microbiol. 101*, 151–156.

Bowien, B., Cook, A. M., and Schlegel, H. G. (1974) *Arch. Microbiol. 97*, 273–281.

Brabikowska, A. K. (1977) *Biochem. J. 168*, 171–178.

Chandra, T. S., and Shethna, Y. I. (1977) *J. Bacteriol. 13*, 383–398.

Colby, J., and Dalton, H. (1978) *Biochem. J. 171*, 461–468.

Colby, J., and Dalton, H. (1979) *Biochem. J. 177*, 903–908.

Colby, J., and Zatman, L. J. (1973) *Biochem. J. 132*, 101–112.

Colby, J., and Zatman, L. J. (1975a) *Biochem. J. 148*, 505–511.

Colby, J., and Zatman, L. J. (1975b) *Biochem. J. 148*, 513–520.

Colby, J., Dalton, H., and Whittenbury, R. (1979) *Ann. Rev. Microbiol. 33*, 481–517.

Couderc, R., and Baratti, J. (1980) *Biotech. Bioeng. 22*, 1155–1173.

Cox, R. B., and Quayle, J. R. (1975) *Biochem. J. 150*, 569–571.

Cox, R. B., and Zatman, L. J. (1973) *Biochem. Soc. Trans. 1*, 669–671.

Cox, R. B., and Zatman, L. J. (1976) *J. Gen. Microbiol. 93*, 397–400.

Cremieux, A., Chevalier, J., Combet, M., Dumenil, G., Parlouar, D., and Ballerini, D. (1977) *Eur. J. Appl. Microbiol. 4*, 1–9.

Davey, J. F., and Mitton, J. R. (1973) *FEBS Lett. 37*, 335–337.

Davey, J. F., Whittenbury, R., and Wilkinson, J. F. (1972) *Arch. Microbiol. 87*, 357–366.

Dawson, M. J., and Jones, C. W. (1981) *Eur. J. Biochem. 118*, 113–118.

deBont, J. A. M., and Mulder, E. G. (1974) *J. Gen. Microbiol. 83*, 113–121.

Diel, F., Held, W., Schlanderer, G., and Dellweg, H. (1974) *FEBS Lett. 38*, 274–276.

van Dijken, J. P., and Harder, W. (1975) *Biotech. Bioeng. 17*, 15–30.

van Dijken, J. P., Otto, R., and Harder, W. (1975a) *Arch. Microbiol. 106*, 221–226.

van Dijken, J. P., Veenhuis, M., Vermeulen, C. A., and Harder, W. (1975b) *Arch. Microbiol. 105*, 261–267.

van Dijken, J. P., Oostra-Demkes, G. J., Otto, R., and Harder, W. (1976a) *Arch. Microbiol. 111*, 77–83.

van Dijken, J. P., Otto, R., and Harder, W. (1976b) *Arch. Microbiol. 111*, 137–144.

van Dijken, J. P., Harder, W., Beardsmore, A. J., and Quayle, J. R. (1978) *FEMS Microbiol. Lett. 4*, 97–102.

van Dijken, J. P., Veenhuis, M., Zwart, K., Large, P. J., and Harder, W. (1979) *Soc. Gen. Microbiol. Quart. 6*, 70.

Douthit, H. A., and Pfennig, N. (1976) *Arch. Microbiol. 107*, 233–234.

Drozd, J. W., and Wren, S. J. (1980) *Biotech. Bioeng. 22*, 353–362.

Duine, J. A., and Frank, J. (1980) *Biochem. J. 187*, 221–226.

Duine, J. A., Frank, J., and Westerling, J. (1978) *Biochim. Biophys. Acta 524*, 277–287.

Duine, J. A., Frank, J., and deRuiter, L. G. (1979) *J. Gen. Microbiol. 115*, 523–526.

Dunstan, P. M., Anthony, C., and Drabble, W. T. (1972) *Biochem. J. 128*, 107–115.

Dworkin, M., and Foster, J. W. (1956) *J. Bacteriol. 75*, 646–659.

Eady, R. R., and Large, P. J. (1968) *Biochem. J. 106*, 245–255.

Egli, T. (1982) *Arch. Microbiol. 131*, 95–101.

Egli, T., Haltmeier, T., and Fiechter, A. (1982a) *Arch. Microbiol. 131*, 174–175.

Egli, T., Kappeli, O., and Fiechter, A. (1982b) *Arch. Microbiol. 131*, 1–7.

Egli, T., Kappeli, O., and Fiechter, A. (1982c) *Arch. Microbiol. 131*, 8–13.

Egorov, A. M., Avilova, T. V., Dikov, M. M., Popov, V. O., Rodionov, Y. V., and Berezin, I. V. (1979) *Eur. J. Biochem. 99*, 569–576.

Ferenci, T. (1974) *FEBS Lett. 41*, 94–98.

Ferenci, T. (1976) *Arch. Microbiol. 108*, 217–219.

Ferenci, T., Strom, T., and Quayle, J. R. (1974) *Biochem. J. 144*, 477–486.

Foster, J. W., and Davis, R. H. (1966) *J. Bacteriol. 91*, 1924–1931.

Fujii, T., and Tonomura, K. (1973) *Agr. Biol. Chem. 37*, 447–449.

Fujii, T., and Tonomura, K. (1975) *Agr. Biol. Chem. 39*, 2325–2330.

Fujii, T., Asada, Y., and Tonomura, K. (1974) *Agr. Biol. Chem. 38*, 1121–1127.

Fukui, S., Tanaka, A., Kawamoto, S., Yusahara, S., Tanaka, A., Osumi, M., and Imaizumi, F. (1975a) *Eur. J. Biochem. 59*, 561–566.

Fukui, S., Tanaka, A., Kawamoto, S., Yusahara, S., Terashimi, Y., and Osumi, M. (1975b) *J. Bacteriol. 123*, 317–328.

Gautier, F., and Bonewald, R. (1980) *Mol. Gen. Genet. 178*, 375–380.

Goldberg, I. (1976) *Eur. J. Biochem. 63*, 233–240.

Goldberg, I. (1977) *Proc. Biochem. 12*, 12–18.

Goldberg, I. (1981) in *Advances in Biotechnology*, Vol. II (Moo-Young, M., and Robinson, C. W., eds.), pp. 419–424, Pergamon Press, New York.

Goldberg, I., and Er-el, Z. (1981) *Process Biochem. 10*, 1–8.

Goldberg, I., and Mateles, R. I. (1975) *Appl. Microbiol. 122*, 47–53.

Goldberg, I., Rock, J. S., Ben-Bassat, A., and Mateles, R. I. (1976) *Biotech. Bioeng. 18*, 1657–1668.

Green, J., Haywood, G. W., and Large, P. J. (1982) *J. Gen. Microbiol. 128*, 991–996.

Green, P. N., and Bousfield, I. J. (1982) *J. Gen. Microbiol. 128*, 623–638.

Haggstrom, L. (1969) *Biotech. Bioeng. 11*, 1043–1054.

Hanson, R. S. (1980) *Adv. Appl. Microbiol. 26*, 3–39.

Harder, W., and Attwood, M. M. (1978) *Adv. Microb. Physiol. 17*, 303–359.

Harder, W., and van Dijken, J. P. (1976) in *Microbial Production and Utilization of Gases (H₂, CH₄, CO)* (Schlegel, H. G., Gottschalk, G., and Pfennig, N., eds.), pp. 403–418, Erich Goltze, Göttingen, West Germany.

Harder, W., and Quayle, J. R. (1971) *Biochem. J. 121*, 753–762.

Harrison, D. E. F. (1976) *Chemtech. 6*, 570–574.

Harrison, D. E. F. (1978) *Adv. Appl. Microbiol. 24*, 129–164.

Harwood, J. H., Williams, E., and Bainbridge, B. W. (1972) *J. Appl. Bacteriol. 35*, 99–108.

Hazeu, W., Batenburg-van den Vegte, W. H., and Nieuwdorp, P. J. (1975) *Experientia 31*, 926–927.

Hazeu, W., Batenburg-van den Vegte, W. H., and de Bruyn, J. C. (1980) *Arch. Microbiol. 124*, 211–220.

Hennan, J. F., Cunningham, A. E., Sharpe, G. S., and Atherton, K. T. (1982) *Nature 297*, 80–82.

Hersh, L. B. (1973) *J. Biol. Chem. 248*, 7295–7303.

Hersh, L. B., and Bellion, E. (1972) *Biochem. Biophys. Res. Comm. 48*, 712–719.

Hersh, L. B., Peterson, J. A., and Thompson, A. A. (1971) *Arch. Biochem. Biophys. 145*, 115–120.

Higgins, I. J., Taylor, S. C., and Tonge, G. M. (1976) *Proc. Soc. Gen. Microbiol. 3*, 179.

Higgins, I. J., Best, D. J., and Hammond, R. C. (1980) *Nature 280*, 561–564.

Higgins I. J., Best, D. J., Hammond, R. C., and Scott, D. (1981) *Microbiol. Rev. 45*, 556–590.

Hill, B., and Attwood, M. M. (1976) *J. Gen. Microbiol. 96*, 185–193.

Hofmann, K., Sawistowsky, J., and Babel, W. (1975) *Z. Allg. Mikrobiol. 15*, 599–604.

Huennekens, F. M. (1968) in *Biological Oxidation*, (Singer, T. P., ed.), Interscience, New York.

Hutton, W. E., and Zobell, C. E. (1953) *J. Bacteriol. 65*, 216–219.

Jeyaseelan, K., and Guest, J. R. (1979) *FEMS Microbiol. Lett. 6*, 87–89.

Johnson, P. A., and Quayle, J. R. (1964) *Biochem. J. 93*, 281–290.

Johnson, P. A., and Quayle, J. R. (1965) *Biochem. J. 95*, 859–867.

Kaneda, T., and Roxburgh, J. M. (1959) *Biochim. Biophys. Acta 33*, 106–110.

Kato, N., Tani, Y., and Ogata, K. (1974) *Agr. Biol. Chem. 38*, 675–677.

Kato, N., Ohashi, H., Hori, T., Tani, Y., and Ogata, K. (1977) *Agr. Biol. Chem. 41*, 1133–1140.

Kato, N., Ohashi, H., Tani, Y., and Ogata, K. (1978) *Biochim. Biophys. Acta 523*, 236–244.

Kato, N., Sahm, H., Schutte, H., and Wagner, F. (1979) *Biochim. Biophys. Acta 566*, 1–11.

Kato, N., Sakazawa, C., Nishizawa, T., Tani, Y., and Yamada, H. (1980) *Biochim. Biophys. Acta 611*, 323–332.

Kato, N., Higuchi, T., Sakazawa, C., Nishizawa, T., Tani, Y., and Yamada, H. (1982) *Biochim. Biophys. Acta 715*, 143–150.

Kell, D. B., John, P., and Ferguson, S. J. (1978) *Biochem. J. 174*, 257.

Kemp, M. B. (1972) *Biochem. J. 127*, 64P.

Kemp, M. B. (1974) *Biochem. J. 139*, 123–134.

Kemp, M. B., and Quayle, J. R. (1966) *Biochem. J. 99*, 41–48.

Kemp, M. B., and Quayle, J. R. (1967) *Biochem. J. 102*, 94–102.

Keune, H., Sahm, H., and Wagner, F. (1976) *Eur. J. Appl. Microbiol. 2*, 175–184.

Kortstee, G. J. J. (1980) *FEMS Microbiol. Lett. 8*, 59–65.

Kung, H. F., and Wagner, C. (1970) *Biochem. J. 116*, 357–365.

Lamb, S. C., and Garver, J. C. (1980) *Biotech. Bioeng. 22*, 2119–2135.

Large, P. J., and Carter, R. H. (1973) *Biochem. Soc. Trans. 1*, 1291–1293.

Large, P. J., and Quayle, J. R. (1963) *Biochem. J. 87*, 386–396.

Large, P. J., Peel, D., and Quayle, J. R. (1961) *Biochem. J. 81*, 470–480.

Large, P. J., Peel, D., and Quayle, J. R. (1962a) *Biochem. J. 82*, 483–488.

Large, P. J., Peel, D., and Quayle, J. R. (1962b) *Biochem. J. 85*, 243–250.

Lawrence, A. J., Kemp, M. B., and Quayle, J. R. (1970) *Biochem. J. 116*, 631–639.

Leadbetter, E., and Gottleib, T. A. (1967) *Arch. Microbiol. 59*, 211–217.

Lee, J. D., and Komagata, K. (1980) *J. Gen. Appl. Microbiol. 26*, 133–158.

Levering, P. R., van Dijken, J. P., Veenhuis, M., and Harder, W. (1981) *Arch. Microbiol. 129*, 72–80.

Linton, J. D., and Buckee, J. C. (1977) *J. Gen. Microbiol. 101*, 219–225.

Linton, J. D., and Stephenson, R. J. (1977) *FEMS Microbiol. Lett. 2*, 95–98.

Linton, J. D., and Vokes, J. (1978) *FEMS Microbiol. Lett. 4*, 125–128.

Loginova, N. V., and Trotsenko, Yu. A. (1977) in *Microbial Growth on C₁ Compounds* (Skryabin, G. K., Ivanov, M. V., Kandratieva, E. N., Zavarzin, G. A., Trotsenko, Yu. A., and Nesterov, A. I., eds.), pp. 37–39, USSR Academy of Sciences, Puschino.

256   **Biology of the Methylotrophs**

Loginova, N. V., and Trotsenko, Yu. A. (1979a) *Microbiology (USSR) 48*, 202–207.
Loginova, N. V., and Trotsenko, Yu. A. (1979b) *FEMS Microbiol. Lett. 5*, 239–243.
Loginova, N. V., and Trotsenko, Yu. A. (1981) *Microbiology (USSR) 50*, 13–18.
Malashenko, Yu. B. (1976) in *Microbial Production and Utilization of Gases (H₂, CH₄, CO)* (Schlegel, H. G., Gottschalk, G., and Pfennig, N., eds.), pp. 293–300, Erich Goltze, Göttingen, West Germany.
Malashenko, Yu. B., Romanovskaya, V. A., Bogachenko, V. N., Khotian, L. V., and Voloshin, N.V. (1973) *Microbiology (USSR) 42*, 403.
Marison, I. W., and Attwood, M. M. (1980) *J. Gen. Microbiol. 117*, 235–313.
Mateles, R. I. (1979) in *Microbial Technology: Current State, Future Prospects* (Bull, A. T., Ellwood, D. C. E., and Ratledge, C., eds.), pp. 29–52, Cambridge University Press, Cambridge, England.
Matsumoto, T. (1978) *Biochim. Biophys. Acta 522*, 291–302.
Matsumoto, T., Hiraoka, B. Y., and Tobari, J. (1978) *Biochim. Biophys. Acta 522*, 303–310.
Matsumoto, T., and Tobari, J. (1978a) *J. Biochem. 83*, 1591–1597.
Matsumoto, T., and Tobari, J. (1978b) *J. Biochem. 84*, 461–465.
McFadden, B. A. (1973) *Bacteriol. Rev. 37*, 289–319.
McNerney, T., and O'Connor, M. L. (1980) *Appl. Environ. Microbiol. 40*, 370–375.
Mehta, R. J. (1973) *Antonie van Leeuwenhoek 39*, 303–312.
Mehta, R. J. (1975) *Antonie van Leeuwenhoek 41*, 89–95.
Mehta, R. J. (1976) *J. Ferm. Technol. 54*, 596–602.
Mehta, R. J. (1977) *Can. J. Microbiol. 23*, 402–406.
Meiberg, J. B. M., and Harder, W. (1979) *J. Gen. Microbiol. 15*, 49–58.
Meyers, A. J. (1982) *J. Bacteriol. 150*, 966–968.
Miethe, D., and Babel, N. (1976) *Z. Allg. Microbiol. 16*, 289–299.
Millay, R. H., and Hersh, L. B. (1976) *J. Biol. Chem. 251*, 2754–2760.
Millay, R. H., Schilling, H., and Hersh, L. B. (1978) *J. Biol. Chem. 253*, 1371–1377.
Minami, K., Yamamura, M., Shimizu, S., Ogawa, K., and Sekine, N. (1978a) *J. Gen. Appl. Microbiol. 24*, 155–164.
Minami, K., Yamamura, M., Shimizu, S., Ogawa, K., and Sekine, N. (1978b) *J. Ferm. Technol. 56*, 1–7.
Moo-Young, M. (1977) *Process Biochem. 5*, 6–10.
Multer, R., and Sokolov, A. P. (1979) *Z. Allg. Microbiol. 19*, 261–267.
Nagai, S., Mori, T., and Aiba, S. (1973) *J. Appl. Chem. Biotechnol. 23*, 549–562.
Netrusov, A. I., Rodionov, Y. V., and Kondratieva, E. N. (1977) *FEBS Lett. 76*, 56–58.
Newaz, S. S., and Hersh, L. B. (1975) *J. Bacteriol. 124*, 825–833.
O'Connor, M. L. (1981) *Appl. Environ. Microbiol. 41*, 437–441.
O'Connor, M. L., and Hanson, R. S. (1975) *J. Bacteriol. 124*, 985–996.
O'Connor, M. L., and Hanson, R. S. (1977) *J. Gen. Microbiol. 101*, 327–332.
O'Connor, M. L., and Quayle, J. R. (1979) *J. Gen. Microbiol. 113*, 203–208.
O'Connor, M. L., Wopat, A., and Hanson, R. S. (1977) *J. Gen. Microbiol. 98*, 265–272.
Ogata, K., Nishikawa, H., and Ohsugi, M. (1969) *Agr. Biol. Chem. 33*, 1519–1520.
O'Keeffe, D. T., and Anthony, C. (1978) *Biochem. J. 170*, 561–567.
Oki, T., Kitai, A., Kouno, K., and Ozaki, A. (1973) *J. Gen. Appl. Microbiol. 19*, 79–83.
Opheim, D., and Bernlohr, R. W. (1973) *J. Bacteriol. 116*, 1150–1159.
Papoutsakis, E., Lin, H. C., and Tsao, G. T. (1978) *AIChE J. 24*, 406–417.
Patel, R. N., and Felix, A. (1976) *J. Bacteriol. 128*, 413–424.

Patel, R. N., Hoare, D. S., and Taylor, B. F. (1969) *Bacteriol. Proc.* p. 128.

Patel, R. N., Bose, H. R., Mandy, W. J., and Hoare, D. S. (1972) *J. Bacteriol. 110,* 570–577.

Patel, R., Hoare, S. L., Hoare, D. S., and Taylor, B. F. (1975) *J. Bacteriol. 123,* 382–384.

Patel, R. N., Hou, C. T., and Felix, A. (1978) *J. Bacteriol. 136,* 352–358.

Patel, R. N., Hou, C. T., and Felix, A. (1979) *Arch. Microbiol. 122,* 241–247.

Patt, T. E., Cole, G. C., Bland, J., and Hanson, R. S. (1974) *J. Bacteriol. 120,* 955–964.

Patt, T. E., Cole, G. C., and Hanson, R. S. (1976) *Int. J. System. Bacteriol. 26,* 226–229.

Patt, T. E., and Hanson, R. S. (1978) *J. Bacteriol. 134,* 636–644.

Peel, D., and Quayle, J. R. (1961) *Biochem. J. 81,* 465–469.

Quayle, J. R. (1961) *Ann. Rev. Microbiol. 15,* 120–152.

Quayle, J. R. (1972) *Adv. Microbiol. Physiol. 7,* 119–203.

Quayle, J. R. (1975) in *Proceedings of the International Symposium on Microbial Growth on C₁-Compounds* (Terui, G., ed.), pp. 59–65, Tokyo Society of Fermentation Technology, Osaka, Japan.

Quayle, J. R. (1980a) *Biochem. Soc. Trans. 8,* 1–10.

Quayle, J. R. (1980b) *FEBS Lett. 117,* K16–27.

Quayle, J. R., and Ferenci, T. (1978) *Microbiol. Rev. 42,* 251–273.

Quayle, J. R., and Keech, D. B. (1959a) *Biochem. J. 72,* 623–630.

Quayle, J. R., and Keech, D. B. (1959b) *Biochem. J. 72,* 631–637.

Quayle, J. R., and Pfennig, N. (1975) *Arch. Microbiol. 102,* 193–198.

Ribbons, D. W., Harrison, J. E., and Wadzinski, A. M. (1970) *Ann. Rev. Microbiol. 24,* 135–158.

Rock, J. S., Goldberg, I., Ben-Bassat, A., and Mateles, R. I. (1976) *Agr. Biol. Chem. 40,* 2129–2135.

Roggenkamp, R., Sahm, H., and Wagner, F. (1974) *FEBS Lett. 41,* 283–286.

Roggenkamp, R., Sahm, H., Hinkelmann, W., and Wagner, F. (1975) *Eur. J. Biochem. 59,* 231–236.

Rokem (Rock), J. S., Goldberg, I., and Mateles, R. I. (1978) *Biotech. Bioeng. 20,* 1557–1564.

Rokem, J. S., Goldberg, I., and Mateles, R. I. (1980) *J. Gen. Microbiol. 116,* 225–232.

Romanovskaya, V. A., Malashenko, Yu. B., and Bogachenko, V. M. (1978) *Microbiology (USSR) 47,* 120–130.

Sahm, H. (1975) *Arch. Microbiol. 105,* 179–181.

Sahm, H. (1977) *Adv. Biochem. Eng. 6,* 77–103.

Sahm, H., and Steinbach, R. (1977) in *Microbial Growth on C₁-Compounds* (Skryabin, G. K., Ivanov, M. V., Kandratieva, E. N., Zavarzin, G. A., Trotsenko, Yu. A., and Nesterov, A. I., eds.), pp. 50–51, USSR Academy of Sciences, Puschino.

Sahm, H., Roggenkamp, R., and Wagner, F. (1975) *J. Gen. Microbiol. 88,* 218–222.

Sahm, H., Schutte, H., and Kula, M. R. (1976) *Eur. J. Biochem. 66,* 591–596.

Salem, A. R., Hacking, A. J., and Quayle, J. R. (1973) *Biochem. J. 136,* 83–96.

Salem, A. R., Large, P. J., and Quayle, J. R. (1972) *Biochem. J. 128,* 1203–1211.

Salisbury, S. A., Forrest, H. S., Cruse, W. B. T., and Kennard, O. (1979) *Nature 280,* 843–844.

Samuelov, N., and Goldberg, I. (1982a) *Biotech. Bioeng. 24,* 731–736.

Samuelov, N., and Goldberg, I. (1982b) *Biotech. Bioeng. 24,* 731–736.

Scott, D., Brannan, J., and Higgins, I. J. (1981) *J. Gen. Microbiol. 125,* 63–72.

Senior, P. J., and Windass, J. (1980) *Biotechnol. Lett.* 2, 205–210.

Senior, P. J., Wright, L. F., and Alderson, B. (1982) U.S. Patent No. 4,324,907.

Setlow, B., and Setlow, P. (1977) *J. Bacteriol.* 129, 857–865.

Sheehan, B. T., and Johnson, M. (1971) *J. Appl. Microbiol.* 21, 511–515.

Shen, G., Kodama, T., and Minoda, Y. (1982) *Agr. Biol. Chem.* 46, 191–197.

Shirai, S., Matsumoto, T., and Tobari, J. (1978) *J. Biochem.* 83, 1599–1607.

Shively, J. M., Saluja, A., and McFadden, B. A. (1978) *J. Bacteriol.* 134, 1123–1132.

Smirnova, Z. S., Gorskaya, L. A., Dibstov, V. P., and Osokima, N. V. (1977) in *Microbial Growth on C1-Compounds* (Skryabin, G. K., Ivanov, M. V., Kandratieva, E. N., Zavarzin, G. A., Trotsenko, Yu. A., and Nesterov, A. I., eds.), p. 179, USSR Academy of Sciences, Puschino.

Smith, A. J., and Hoare, D. S. (1977) *Bacteriol. Rev.* 41, 419–448.

Snedecor, B., and Cooney, C. L. (1974) *Appl. Microbiol.* 27, 1112–1117.

Sohngen, N. L. (1906) *Centr. Bakt. Parasitenk 15*, 513–517.

Sokolov, A. P., and Trotsenko, Yu. A. (1977) *Microbiology (USSR) 46*, 1119–1121.

Sokolov, A. P., and Trotsenko, Yu. A. (1978a) *Biochem.* 43, 673–680.

Sokolov, A. P., and Trotsenko, Yu. A. (1978b) *Biochem.* 43, 782–787.

Sokolov, A. P., Luchin, S. V., and Trotsenko, Yu. A. (1980) *Biochem.* 45, 1371–1378.

Sperl, G. T., Forrest, H. S., and Gibson, D. T. (1974) *J. Bacteriol.* 118, 541–550.

Steenkamp, D. J., and Mallison, J. (1976) *Biochim. Biophys. Acta 429*, 705–719.

Stirling, D. I., and Dalton, H. (1978) *J. Gen. Microbiol.* 107, 19–29.

Stirling, D. I., and Dalton, H. (1979a) *Eur. J. Biochem.* 96, 205–212.

Stirling, D. I., and Dalton, H. (1979b) *FEMS Microbiol. Lett.* 5, 315–318.

Stirling, D. I., Colby, J., and Dalton, H. (1979) *Biochem. J.* 177, 361–364.

Strom, T., Ferenci, T., and Quayle, J. R. (1974) *Biochem. J.* 144, 465–476.

Tanaka, A., Yasuhura, S., Kawamoto, S., Fukui, S., and Osumi, M. (1976) *J. Bacteriol.* 126, 919–927.

Tanaka, A., Araki, K., and Nakayama, K. (1980) *Biotechnol. Lett.* 2, 67–74.

Tani, Y., Miya, T., Nishikawa, H., and Ogata, K. (1972) *Agr. Biol. Chem.* 36, 68–75.

Tani, Y., Kato, N., and Yamada, H. (1978) *Adv. Appl. Microbiol.* 24, 165–186.

Tani, Y., and Yamada, H. (1980) *Biotech. Bioeng.* 22, 163–175.

Tani, Y., Mitani, Y., and Yamada, H. (1982) *Agr. Biol. Chem.* 46, 1097–1099.

Taylor, I. J., and Anthony, C. (1975) *Proc. Soc. Gen. Microbiol.*, vol. 2, p. 46.

Taylor, I. J., and Anthony, C. (1976) *J. Gen. Microbiol.* 93, 259–265.

Taylor, S. (1977) *FEMS Microbiol. Lett.* 2, 305–307.

Taylor, S., Dalton, H., and Dow, C. (1980) *FEMS Microbiol. Lett.* 8, 157–160.

Tobari, J., Ohat, S., and Fujita, T. (1981) *J. Biochem.* 90, 205–213.

Tonge, G. M., Knowles, C. J., Harrison, D. E. F., and Higgins, I. J. (1974) *FEBS Lett.* 44, 106–110.

Tonge, G. M., Harrison, D. E. F., Knowles, C. J., and Higgins, I. J. (1975) *FEBS Lett.* 58, 293–299.

Tonge, G. M., Drozd, J. W., and Higgins, I. J. (1977a) *J. Gen. Microbiol.* 99, 229–232.

Tonge, G. M., Harrison, D. E. F., and Higgins, I. J. (1977b) *Biochem. J.* 161, 333–344.

Toraya, T., Yongsmith, B., Tanaka, A., and Fukui, S. (1975) *Appl. Microbiol.* 30, 477–479.

Trotsenko, Yu. A. (1976) in *Microbial Production and Utilization of Gases (H2, CH4, CO)* (Schlegel, H. G., Gottschalk, G., and Pfennig, N., eds.), pp. 329–336, Erich Goltze Verlag, Göttingen, West Germany.

Trotsenko, Yu. A., and Shishkina, Y. A. (1980) in *Abstracts of Posters, Third International Symposium on Microbial Growth on C₁-Compounds* (Tonge, G., ed.), pp. 64–65, University of Sheffield Press, Sheffield, England.

Tuboi, S., and Kikuchi, G. (1963) *J. Biochem. 53*, 364–373.

Tye, R., and Willets, A. (1973) *J. Gen. Microbiol. 77*, 1P.

Veenhuis, M., van Dijken, J. P., and Harder, W. (1976) *Arch. Microbiol. 111*, 123–135.

van Verseveld, H. W., and Stouthamer, A. H. (1978) *Arch. Microbiol. 118*, 21–26.

Wadzinski, A. M., and Ribbons, D. W. (1975a) *J. Bacteriol. 122*, 1364–1374.

Wadzinski, A. M., and Ribbons, D. W. (1975b) *J. Bacteriol. 123*, 380–381.

Wagner, C., and Levitch, M. E. (1975) *J. Bacteriol. 122*, 905–910.

Warner, P. J., Higgins, I. J., and Drozd, J. W. (1980) *FEMS Microbiol. Lett. 7*, 181–185.

Weaver, T. L., and Dugan, P. R. (1975) *J. Bacteriol. 122*, 433–436.

Whittenbury, R., Phillips, K. C., and Wilkinson, J. F. (1970) *J. Gen. Microbiol. 61*, 205–218.

Whittenbury, R., Dalton, H., Eccleston, M., and Reed, H. L. (1975) in *Proceedings of the International Symposium on Microbial Growth on C₁-Compounds* (Terui, G., ed.), pp. 1–9, Tokyo Society of Fermentation Technology, Osaka, Japan.

Whittenbury, R., Colby, J., Dalton, H., and Reed, H. L. (1976) in *Microbial Production and Utilization of Gases (H₂, CH₄, CO)* (Schlegel, H. G., Gottschalk, G., and Pfennig, N., eds.), pp. 281–292, Erich Goltze, Göttingen, West Germany.

Widdowson, D., and Anthony, C. (1975) *Biochem. J. 152*, 349–356.

Wilkinson, J. F. (1971) in *Microbes and Biological Productivity, 21st Symposium*, (Hughes, D. E., and Rose, A. H. eds.), p. 15, Society of General Microbiology, Cambridge University Press, Cambridge, England.

Wilkinson, T. F., and Harrison, D. E. F. (1973) *J. Appl. Bacteriol. 36*, 309–313.

Wilkinson, T. G., Topiwala, H. H., and Hamer, G. (1974) *Biotech. Bioeng. 16*, 41–59.

Williams, E., and Bainbridge, B. W. (1971) *J. Appl. Bacteriol. 34*, 683–687.

Williams, E., Shimin, M. A., and Bainbridge, B. W. (1977) *FEMS Microbiol. Lett. 2*, 293–296.

Wimpenny, I. W. T., and Firth, A. (1972) *J. Bacteriol. 111*, 24–34.

Wolf, H. J., and Hanson, R. S. (1978) *Appl. Environ. Microbiol. 36*, 105–114.

Wolf, H. J., Christiansen, M., and Hanson, R. S. (1980) *J. Bacteriol. 141*, 1340–1349.

Wolfe, R. S., and Higgins, I. J. (1979) *Int. Rev. Biochem. 21*, 267–353.

Wopat, A. E., and O'Connor, M. L. (1980) in *Abstracts of Posters, Third International Symposium on Microbial Growth on C₁-Compounds*, (Tonge, G. M., ed.), p. 16, University of Sheffield Press, Sheffield, United Kingdom.

Yamada, H., Morimaga, Y., and Tani, Y. (1982) *Agr. Biol. Chem. 46*, 47–55.

Yamamoto, M., Iwaki, H., Kouno, K., and Inui, T. (1980) *J. Ferm. Technol. 58*, 99–106.

Yordy, J. R., and Weaver, T. L. (1979) *Int. J. Syst. Bacteriol. 27*, 247–255.

Zymomonas, a Unique Genus of Bacteria

Bland S. Montenecourt

I. INTRODUCTION

Zymomonas is a unique genus of bacteria in many respects, not the least of which is its rapid rise to prominence as a microorganism of potential industrial application. For centuries, *Zymomonas* has been utilized in the tropical areas of the world as a natural fermentative agent for home-brewed alcoholic beverages produced from plant saps. A well-known example of a national alcoholic beverage derived in part through *Zymomonas* fermentation is pulque. Pulque is a milky viscous beverage containing approximately 3–5% alcohol. It is classically produced from the sap of the *Agave* plant through successive fermentation by *Zymomonas*, yeast, and *Leuconostoc*. The application of *Zymomonas* in the fermentation of a variety of local beverages has been extensively reviewed (Dadds and Martin, 1973; Swings and De Ley, 1977; Okafor, 1978; Stokes et al., 1981).

As was pointed out by Swings and De Ley (1977) in their extensive monograph on the biology of *Zymomonas*, this organism had to date been overlooked as an agent for the industrial-scale production of ethanol. As a result of their review and also partially owing to the energy crisis of the late 1970's, *Zymomonas* has become a popular microorganism for study, especially in the fields of engineering, biochemistry, and genetics. Its superior ethanol productivity in comparison to yeast has been confirmed (Rogers et al., 1979).

Several recent reviews have dealt primarily with ethanol production by *Zymomonas* (Lawford et al., 1983; Rogers et al., 1982b). In addition, a comprehensive treatise on recombinant DNA approaches for enhancing the ethanol productivity of *Zymomonas* has been included in this series (Dally et al., 1983). It is not the intent of the present overview to reiterate these excellent reviews directed at ethanol production but rather to cover in detail the biological description of the genus *Zymomonas* with special emphasis on recent contributions in this area that have occurred since the monograph of Swings and De Ley (1977).

II. DESCRIPTIVE BIOLOGY AND TAXONOMY

Zymomonas was first described by Barker and Hillier (1912) as the causative agent in "cinder sickness", which they designated Strain A. Barker and Hillier did not affix a scientific name to this isolate, and according to Swings and De Ley (1977), this strain was not maintained. Although it cannot be proven, the taxonomic description of this microorganism resembled that of the subspecies *pomaceae*, and thus it is presumed to be of this subspecies. A number of other isolates of the cider sickness agent have been studied (Barker, 1948; Millis, 1956). All strains produced the typical aroma and flavor characteristic of spoiled cider. In general, the cider sickness organisms are pleomorphic gram-negative motile rods that form neither spores nor capsules. They produce 1.5–1.9 moles of ethanol per mole of glucose. In addition to ethanol, CO_2, H_2S, acetaldehyde, and lactic acid are also formed (Millis, 1951, 1956). It is H_2S, acetaldehyde, and lactic acid that are responsible for the offensive aroma and taste of the infected cider. The strains studied by Millis fermented only glucose, fructose, and sucrose. Other sugars such as lactose, raffinose, maltose, sucrose, galactose, rhamnose, arabinose, xylose, dulcitol, mannitol, and sorbitol were not utilized.

The discovery of the nominate variety of *Zymomonas*, subspecies *mobilis*, is generally attributed to Lindner (1928), who isolated a number of microorganisms from Mexican pulque. Lindner named the isolate *Thermobacterium mobile* (Lindner, 1931). This isolate has survived and is now known as *Z. mobilis* subspecies *mobilis* (ATCC 10988). This was the beginning of a long list of taxonomic identifications that this organism was to have between 1928 and 1976. In the intervening years, *Zymomonas* was to be reclassified into five different genera and to have on the order of 20 different species designations (Table 9.1).

A listing of the phenotypic characteristics of *Zymomonas* is given (Table 9.2). *Bergey's Manual* (8th edition) places *Zymomonas* in part 8 in the genera of uncertain affiliation. This classification is principally due to the fact that *Zymomonas* is the only genus identified to date that exclusively utilizes the

TABLE 9.1   Nomenclatures Assigned to *Zymomonas* **(1928–1971)**

Thermobacterium mobile	Lindner (1928)
Pseudomonas lindneri	Kluyver and Hoppenbrouwers (1931)
Zymomonas mobile	Kluyver and van Niel (1936)
Zymomonas mobilis	Kluyver and van Niel (1936)
Achromobacter anaerobium	Shimwell (1937)
Saccharobacter	Shimwell (1937)
Saccharomonas anaerobia	Shimwell (1950)
Saccharomonas lindneri	Shimwell (1950)
Zymomonas anaerobia	Carr (1974)
Zymomonas mobilis var. *recifensis*	Gonçalves de Lima et al. (1970)
Zymomonas congolensis	Van Pee and Swings (1971)

Entner-Doudoroff pathway anaerobically. *Zymomonas* also possesses a number of additional unique features, which will be discussed below.

In the 1970's, Swings and De Ley applied modern molecular biological techniques to compare approximately 40 different strains of *Zymomonas* for their biochemical relatedness. The parameters studied were deoxyribonucleic acid (DNA) base composition, genome size and similarity (Swings and De Ley, 1975; De Ley et al., 1981), relatedness of electrophoretic protein patterns (Swings et al., 1976), numerical analysis of phenotypic characteristics (De Ley and Swings, 1976), and infrared spectroscopy (Swings and Van Pee, 1977). On the basis of the data obtained, Swings and De Ley (Swings and De Ley, 1977; Swings et al., 1977) concluded that all of the strains of *Zymomonas* described to date belong to a single species, *Z. mobilis*, with two subspecies — *Z. mobilis* subsps. *mobilis*, the organism currently envisioned for industrial ethanol production, and *Z. mobilis* subsps. *pomaceae*, the agent responsible for beer and cider spoilage. Recent data (to be discussed in detail below) describing the intersubspecies relatedness of the various *Zymomonas* plasmids (Dally et al., 1982; Stokes et al., 1983) and the similarity of phospholipid content (Tornabene

TABLE 9.2   Phenotypic Description of the Genus *Zymomonas*

1. gram-negative rods, 2–6 μm length 1–1.5 μm width
2. either motile or nonmotile; motility can be easily lost; 1–4 lophotrichous flagella
3. pleomorphic cell arrangement, rosettes, chains, filaments
4. spores, capsules, intracellular storage compounds (lipids, glycogen and poly-β-hydroxybutyrate absent)
5. catalase positive, oxidase negative
6. anaerobic and microaeroduric
7. ferment glucose and fructose $\geq$ 1.5 mole ethanol and CO_2
8. sucrose utilization inducible, may be accompanied by levan production
9. no other monosaccharides, disaccharides, polysaccharides, or fatty acids metabolized
10. G + C content 47.5–49.5%
11. genome size 1.5 $\times$ 10^9, approximately 1500 cistrons

From Carr (1974) and Swings and De Ley (1977).

et al., 1982; Carey and Ingram, 1983) support this conclusion. Swings and De Ley have proposed that since all *Zymomonas* strains examined, with the exception of those in the subspecies *pomaceae*, are phenotypically and genetically identical, the genus has evolved recently and has not as yet had time to diversify. Alternatively, there may be something in the genetic makeup of the genus which confers unusual stability, in spite of its worldwide distribution. A survey of the antibiotic sensitivity of *Zymomonas* (Swings and De Ley, 1977; Dally, 1982; Stokes et al., 1982, 1983) indicates that the genus is resistant to many of the more commonly used antibiotics and antimicrobial agents (tetracycline, novobiocin, and chloramphenicol are the exceptions). Gonçalves de Lima (1972) examined his strains (Table 9.1) for antibiotic production in order to explain the antagonistic effect of *Zymomonas* against other bacteria. However, no antibiotics were detected. In addition, broad-host-range plasmids are difficult to maintain stably in *Zymomonas* (Skotnicki et al., 1980, 1982a, b; Dally et al., 1983; Stokes et al., 1982, 1983). These two characteristics taken together could form the basis for the genetic stability of the genus.

III. INDUSTRIALLY USEFUL STRAINS

Even after the sound rational and supportive data compiled by Swings and De Ley (Swings and De Ley, 1977; Swings et al., 1977) for classification of *Zymomonas* into a single species, the burst of enthusiasm over *Zymomonas* in the late 1970's and 1980's has resulted in a myriad of new nomenclatures for the various laboratory strains. A comparative listing of the new strain designations as well as the old (Table 9.3) is given in order for the scientific community to be able to recognize and evaluate the industrially useful variants. An example of the confusion abounding in the literature is the exact strain designation of Z3 and Z4. The Z designation was originally assigned to isolates from *Elaeis* sap, Kinshasa, Zaire (1967). However, Saddler et al. (1981, 1982) have renamed two culture collection strains (ATCC 29192 and ATCC 29191) Z3 and Z4, respectively. Although Z4 may be related to ATCC 29191, since both were isolated in Zaire, it is unlikely that Z3 and ATCC 29192 are the same strain, since the latter (ATCC 29192) was isolated in the United Kingdom in the early 1950's and the former (Z3) in Zaire in the late 1960's. Although secret code numbers for production strains are ubiquitous in industry, they serve only to obscure relationships and identities in academic research. It would be useful if strain designations could be limited to culture collection acquisition numbers or to the original strain designations with appropriate literature citations. As was pointed out recently, strains CP3 and CP4 appear to be identical on the basis of their natural plasmids (Dally, 1982; Dally et al., 1983).

As interest in *Zymomonas* increases, new isolates will most certainly be obtained. Several have already been reported and appear to have superior characteristics when compared to at least one culture collection strain (Viikari et al., 1981, 1982), for example, greater ethanol production and temperature

TABLE 9.3 Nomenclature and Origin of Industrially Useful Strains of *Z. mobilis*

Original Isolation/yr. Subspecies mobilis	ATCC	NRRL	NCIB	UQM	Rogers et al. (1982b)	Lyness and Doelle (1980)	Saddler et al. (1981)	Gonçalves de Lima et al. (1972)	Swings and De Ley (1977)
Lindner, Mexico (1924)	10988	B-806	8938	410	ZM1	Z10	Z1	—	—
Gonçalves de Lima, Mexico (1950)	—	B-4576	—	—	Ag11	—	—	Ag11	
Lovanium University, Zaire (1967)	29191	B-4490	11199	—	ZM6	Z7	Z4	—	Z6
Gonçalves de Lima, Brazil (1970)	—	B-14022	—	—	—	—	—	CP3	
Gonçalves de Lima, Brazil (1970)	—	B-14023	—	—	ZM4	—	—	CP4	
Lovanium University, Zaire (1967)	—	—	—	—	—	—	—	—	Z1 Z5 Z2 Z7 Z3 Z8
subspecies *pomaceae*									
Barker, United Kingdom (1951)	29192	B-4491	11200	—	—	—	Z3	—	—

ATCC: American Type Culture Collection, Washington, D.C., U.S.A.
NRRL: National Regional Research Laboratory, U.S.D.A., Peoria, Illinois, U.S.A.
NCIB: National Collection of Industrial Bacteria, Aberdeen, Scotland.
UQM: University of Queensland, Microbiology, Queensland, Australia.

265

tolerance. As new isolates emerge, it will be of interest to note whether the close genetic and phenotypic similarities among the members of the genus are maintained.

IV. SPECIFIC GROWTH CONDITIONS

A. Growth Medium

In their extensive review, Swings and De Ley (1977) outline some of the more commonly used culture media. Recently, several semidefined media have been developed for continuous culture studies (Cromie and Doelle, 1982; Rogers et al., 1982a; Lawford et al., 1983). Finally, two minimal media have been described for use in the selection of auxotrophic strains of *Zymomonas* (Goodman et al., 1982; Lawford et al., 1983). The various complex, semi-defined and defined media are summarized in Table 9.4. It should be noted that in the semidefined formulations of Lawford and co-workers (1983), technical-grade yeast extract may be substituted for the more expensive laboratory grade (Oxoid or Difco) without a significant effect on specific sugar uptake rate (q_s)[1] or ethanol yield (Y_p/s),[2] although growth yield is reduced by one half. Similarly, in a minimal medium, growth yield (as measured by OD660) is halved (Belaïch and Senez, 1965), but the final ethanol concentration is unchanged.

B. Growth Requirements

As was reported by Swings and De Ley (1977), no strains of *Zymomonas* were able to grow on the defined medium of Kluyver and Hoppenbrouwers (1931), a result indicating a requirement for organic nitrogen or some growth factors or possibly both. Evidence also exists that magnesium ions are required to stabilize RNA and prolong cell viability (Dawes and Large, 1970). There has been one report that molybdenum stimulated growth of certain strains (Dadds and Martin, 1973).

Vitamin requirements of the two subspecies appear to be an anomaly. As was first pointed out by Swings and De Ley (1977), results with the same strains varied among the different laboratories and appeared to fluctuate seasonally. Belaïch et al. (1969) identified pantothenate as a vitamin require-ment. Goodman et al. (1982) have defined the following vitamin requirements (mg/l) for strains CP4, AgII, and ATCC 10988: calcium pantothenate, 5; thia-mine hydrochloride, 1; nicotinic acid, 1; pyridoxine hydrochloride, 1; and biotin, 1. In addition, they have defined a trace metal solution containing magnesium, calcium, molybdenum and iron (Table 9.4). As their defined minimal medium contains only inorganic nitrogen $(NH_4)_2SO_4$, it would appear

---

[1] q_s specific sugar uptake rate gg^{-1}h^{-1}

[2] Y_p/s ethanol yield gg^{-1}

TABLE 9.4 Complex, Semisynthetic and Minimal Media Formulations for *Z. mobilis*

Component (g/l)	Complex[1,2]	Semidefined[3]	Semidefined[4]	Semidefined[5]	Defined Mineral Salts[6]	Minimal[7]
glucose	20	10	100	100	100	20
tryptone		10				
yeast extract	5	10	2.5	5.0		
KH_2PO_4		5		2.0	3.48	1.0
K_2HPO_4					1.0	1.0
$(NH_4)_2SO_4$			2.0	1.0	1.98	1.0
NH_4Cl						
$MgSO_4 \cdot 7H_2O$		0.16	6.0	1.0	1.0	0.2*
$MnSO_4 \cdot 4H_2O$		0.032				
NaCl		0.08				0.5
$FeSO_4 \cdot 7H_2O$		0.08				0.025*
$CaCl_2 \cdot 7H_2O$						0.2*
$Na_2MoO_4 \cdot 2H_2O$						0.025*
Ca-pantothenate mg/l					1.5	
biotin mg/l					1.0	5.0*
trace minerals ml/l					1.0[a]	1.0*

[1] Swings and De Ley (1977)
[2] ATCC catalogue (1982)
[3] Gibbs and DeMoss (1951)
[4] Cromie and Doelle (1982)
[5] Rogers et al. (1982b)
[6] Lawford et al. (1983)
[7] Goodman et al. (1982)

[a] composition of trace minerals stock solution

$CaCl_2 \cdot 2H_2O$	100μM
$FeCl_3 \cdot 6H_2O$	90μM
$MnCl_2 \cdot 4H_2O$	50μM
$ZnSO_4 \cdot 7H_2O$	25μM
$CoCl_2 \cdot 6H_2O$	10μM
$CuCl_2 \cdot 2H_2O$	5μM
H_3BO_3	5μM
H_2MoO_3	10μM

dissolved in approx. 3M HCl

* Filter sterilized.

267

that *Zymomonas* contains all of the metabolic pathways necessary to synthesize all the amino acids and does not require preformed organic nitrogen.

C. Semidefined Media for Specific Product Formation

Three semidefined media have been described for continuous culture of *Zymomonas* for ethanol production (Table 9.4). Cromie and Doelle (1981, 1982) have indicated that high magnesium concentration resulted in a threefold increase in the specific ethanol production rate (q_p).[3] This observation has not been confirmed by others (Rogers et al., 1982b; Lawford et al., 1983), who maintain that elevated magnesium actually depresses alcohol production. It should also be noted that in the semidefined medium of Cromie and Doelle (1982) there is an extremely high sulfate-to-phosphate ratio. It may be that under these conditions, phosphate is limiting, an effect that results in an uncoupling of growth and product formation (see below for discussion). As was pointed out by Cromie and Doelle, these conditions reduce the maintenance energy coefficient to zero, which implies a nongrowing condition.

D. Effects of Temperature on Growth

In the survey of Swings and De Ley (1977), 100% of the 43 strains tested grew at 30°C, 75% at 38°C, and 5% at 40°C. Members of the subspecies *pomaceae* do not appear to grow above 34°C. Therefore a growth temperature test is suggested as a good method of distinguishing between the two subspecies. The new strains recently isolated by Viikari et al. (1981) are reported to be more temperature-tolerant than the culture collection strains. Two culture collection strains CP3 and CP4 originally isolated by Gonçalves de Lima et al. (1970) were reported to grow at 42°C (Forrest, 1967). This observation has recently been substantiated by Lee et al., (1981b) with strain CP4 and Chase and Pincus (personal communication) for strain CP3. Elevated temperature (34.5°C) has also been reported to increase the intracellular ethanol concentration and the maintenance energy coefficient (Laudrin and Goma, 1982).

E. Tolerances

Sugar Tolerance.   *Zymomonas* is a highly osmotolerant microorganism. In their survey of the various subspecies, Swings and De Ley (1977) reported that 88% of those tested could grow in 30% glucose and 54% in 40% glucose with a variable lag time. Certain subspecies appear to be more osmotolerant than others. In general, the subspecies *mobilis* will grow at higher sugar concentrations than members of the subspecies *pomaceae*. Although *Zymomonas* grows in these elevated sugar solutions, the rate of growth is generally slower. CP4 is

---

[3] q_p specific ethanol production rate $gg^{-1}h^{-1}$

reported to be the best strain for alcohol production at 200 gl^{-1} glucose or sucrose (Skotnicki et al., 1981). This strain also appears to be the most temperature tolerant (Lee et al., 1981b).

Alcohol Tolerance.   Many strains of *Zymomonas* are reported to grow in 55 gl^{-1} ethanol (Swings and De Ley, 1977). Since *Zymomonas* is traditionally isolated from fermenting saps, it would seem reasonable that it should tolerate elevated ethanol levels. Recently, *Zymomonas* alcohol tolerance has been compared with that of yeast (Ohta et al., 1981). *Z. mobilis* subsp. *mobilis* was found to be considerably more alcohol-tolerant than subsp. *pomaceae*. A titer of 14% (v/v) ethanol was achieved at 30°C with stepwise addition of glucose. Saké yeasts can tolerate higher amounts of ethanol at low temperatures (20°C) (Hayashida and Ohta, 1981). On the other hand, *Z. mobilis* showed a sensitive response to temperature with the optimum at 30°C. Thus, lower temperature did not afford protection to *Zymomonas*. Additionally, Ohta et al. (1981) demonstrated a titer of 16.3% (v/v) of ethanol in a liver supplement medium, while no increased tolerance was observed with the addition of lipoprotein (albumin). The same workers also analyzed the fatty acid composition of *Zymomonas* and found high levels of C18:1 fatty acids (63.6%). Ingram and coworkers have extensively studied the effect of growth in ethanol on *Escherichia coli* cell membranes. Ethanol causes alteration in the composition of the plasma membrane by increasing the proportion of C18:1 fatty acids and decreasing C16:0 content (Ingram, 1976, 1977, 1981; Ingram et al., 1980a, b, 1982). A similar response has been found with yeast (Rose and Beavan, 1981; Beavan et al., 1982). More recently, Carey and Ingram (1983) have turned their attention to *Zymomonas*. They corroborated the results of Ohta et al. (1981) and Tornabene et al. (1982) with respect to the high concentration of C18:1 and identified this component as vaccenic acid (C18:1Δ11) in the *Zymomonas* plasma membrane. Moreover, they found no major changes in the fatty acid composition at high glucose or ethanol concentrations. They concluded that the increased amounts of vaccenic acid in the membrane of *Zymomonas*, in comparison to other gram-negative bacteria, may be an evolutionary adaptation for survival in the presence of ethanol. In contrast to the stability of the membrane fatty acid composition, both high glucose and high ethanol caused a major shift in the phospholipid composition as well as the lipid-to-protein ratios found in crude membranes. Specifically, a decrease in phosphatidylethanolamine and phosphatidylglycerol and an increase in cardiolipin and phosphatidylcholine were reported (Carey and Ingram, 1983). Intracellular ethanol concentration has been measured in ATCC 10988 during the time course of a fermentation (Laudrin and Goma, 1982). The peak intracellular ethanol concentration (118.3 gl^{-1}) occurred about 5 h into the fermentation, when the external ethanol concentration was low (less than 5 gl^{-1}). The cell growth rate was not affected by the peak in intracellular ethanol, but the specific rate of ethanol production was greatly repressed at the maximum intracellular ethanol concentration. Millar et al. (1982) have examined the effect of ethanol on the glycolytic enzymes (Fig. 9.1) in *Zymomonas*. They found that these enzymes were not denatured by ethanol but were inhibited noncompetitively. At ethanol concentrations roughly equivalent to the peak intracellular ethanol levels the activity of the metabolic enzymes is inhibited, but this inhibition is reversible.

Salt and pH Tolerance. Most of the *Zymomonas* strains tested by Swings and De Ley (1977) were able to tolerate 1% NaCl. However, none could grow in 2% NaCl. Members of the subsp. *pomaceae* appear to be less halotolerant and will not grow in standard medium containing 0.5%–0.7% NaCl. One commercial process envisioned for *Zymomonas* is ethanol production from molasses syrups. However, inhibitory substances are present in molasses that result in a dramatic reduction in ethanol productivity. It has been suggested (Skotnicki et al., 1982a) that inhibition may be due to high concentrations of Mg^{2+} and K^+. This group reports that the difficulty may be overcome by membrane desalting of the molasses and the selection of salt tolerant mutants (Rogers et al., 1982b). *Zymomonas* typically tolerates a wide range of pH (3.5–7.5) (Swings and De Ley, 1977). Most standard growth media start with a neutral pH, which drops during the course of the fermentation due to the evolution of CO_2. King and Hossain (1982) have recently shown that the final yield of the product, ethanol, is insensitive to pH in the range between 5.0 and 7.5.

F. Enrichment Cultures for *Zymomonas*

As was mentioned in the introduction, *Zymomonas* may be isolated from a wide variety of tropical locations, as well as breweries. Since it is likely that the search for new strains that exhibit increased ethanol tolerance, temperature tolerance, and product formation will continue, a section on isolation and detection of *Zymomonas* is included. Swings and De Ley (1977) outlined a selection medium containing malt extract, 0.3%; yeast extract, 0.3%; glucose, 2%; peptone, 0.5% and cycloheximide, 0.002%, with 3% ethanol added after autoclaving. The presence of *Zymomonas* is indicated by abundant gas production. Recently, two rapid methods for the detection of *Zymomonas* have been described (Dennis and Young, 1982; Woodward, 1982). The test is based on detection of acetaldehyde with Schiff's reagent after growth on a selective medium containing cycloheximide. Yeasts were generally unable to grow in the selective medium owing to the antibiotic. The low pH of the selective medium suppressed the growth of most bacteria. Those species that could grow did not produce acetaldehyde and did not react with the Schiff's reagent. *Zymomonas* strains, on the other hand, grew well and produced gas, and the medium developed a deep purple color when reacted with the Schiff's reagent. As was pointed out by Sanchez-Marroquin et al. (1967) and later by Swings and De

Ley (1977), only young fermenting saps (<24 h) should be used as a source of inocula, since the viability of *Zymomonas* rapidly decreases in aged fermentation broths.

V. BIOCHEMISTRY OF GLUCOSE, FRUCTOSE, AND SUCROSE UTILIZATION

A. Substrate Utilization

The biochemistry of *Zymomonas* is unique in many respects. First, *Zymomonas* can utilize only three carbohydrates: glucose, fructose, and sucrose. Growth on sucrose appears to be an inducible phenomenon in most strains (Dadds et al., 1973; Richards and Corbey, 1974; Swings and De Ley, 1977; Lavers et al., 1981; Lyness and Doelle, 1981a; Skotnicki et al., 1981), and certain strains are reported not to utilize sucrose at all (Swings and De Ley, 1977). The ability to utilize sucrose is presumably due to the presence of invertase (3.2.1.26) or levan sucrase (2.4.1.10) (Ribbons et al., 1962; Dawes et al., 1966). One recent report (Lyness and Doelle, 1981b) suggests that strain NCIB 11199 possesses increased amounts of invertase activity. However, the actual activity of the enzyme was not measured, but its presence was implied by the amounts of free glucose and fructose in the medium. Many of the strains that can grow on sucrose produce an extracellular fructose polymer, levan (Dawes et al., 1966). This polymer is synthesized only during growth on sucrose and not on fructose (Ribbons et al., 1962), and its production is reported to be reduced at higher temperatures (Skotnicki et al., 1981).

The mechanism of sucrose utilization and levan formation remains elusive. Several important questions remain to be resolved. Which of the two enzymes (invertase and levan sucrase) is involved in sucrose breakdown? Perhaps there is a third possibility, a sucrose glucosyltransferase (2.4.1.4). This activity was analyzed in one strain by Dawes et al. (1966) and was found to be lacking. What controls the synthesis of these enzymes at the gene level (constitutive or inducible) and what regulates their activity? Several strains of *Zymomonas* are reported to utilize raffinose (Millis, 1956; Dadds et al., 1973) and sorbitol (Millis, 1956) without the formation of gas. If the sucrose-utilizing strains of *Zymomonas* possess an invertase similar to yeast, it is surprising that they are unable to utilize raffinose or other disaccharides or trisaccharides that contain the β-fructofuranoside linkage. No other carbohydrates tested will support growth of *Zymomonas*. In addition, amino acids as a sole carbon source will not support growth without glucose, fructose, or sucrose and are not fermented (Belaïch, 1963). Thus *Zymomonas* is unique in that it is extraordinarily limited in its metabolic capabilities.

B. Metabolic Pathway

Initially, *Zymomonas* was thought to closely resemble yeast in its fermentation of sugars. Since it fermented glucose, fructose, and sucrose to ethanol and

CO_2, it was assumed that *Zymomonas* utilized the normal glycolytic pathway and had a yeast-type pyruvate decarboxylase. In the early 1950's, Gibbs and De Moss discovered that *Zymomonas* utilized a modified Entner-Doudoroff pathway for glucose and fructose metabolism (Gibbs and De Moss, 1951, 1954). This was later confirmed radiorespirometrically (Stern et al., 1960).

A detailed scheme of the pathway including the presumed enzymes is given (Fig. 9.1). Few of the enzymes in the pathway have been studied in detail. Recently, Doelle (1982a, b) has isolated and purified a constitutive glucokinase and a fructokinase from *Zymomonas*. The glucokinase showed two pH optima, pH 7.0 and pH 8.2, gave hyperbolic saturation kinetics and reacted equally well with ATP, UTP, ITP, and CTP. Inhibition of activity occurred at high levels of the nucleoside triphosphates and with ADP, AMP, and the end product glucose-6-phosphate. No inhibition was evident with sucrose, glucose, fructose, ethanol, mannose, ribose, galactose, deoxyglucose, lactose, or gluconate. The enzyme was absolutely specific for glucose. The fructokinase has also been purified by Doelle and co-workers. This enzyme exhibits a pH optimum at pH 7.4 and gives hyperbolic saturation kinetics. However, unlike the glucokinase, the fructokinase is absolutely specific for ATP and fructose. Mannose, glucose, GTP, CTP, UTP, and ITP do not react with the enzyme. Inhibition of fructokinase activity occurs with glucose, glucose-6-phosphate, and slightly with the end product fructose-6-phosphate. In addition, slight inhibition occurs with ATP and high concentrations of ADP and AMP. Doelle (1982a, b) proposes that these two enzymes are exclusively responsible for sugar uptake and phosphorylation and that no broad specificity hexokinase is present in *Zymomonas*. The regulatory properties of these enzymes will be discussed below.

Glucose-6-phosphate dehydrogenase has been studied by Sly and Doelle (1968). They found the optimum pH to be 8.7 and the optimal $MgCl_2$ concentration to be $10^{-2} M$. The K_m for glucose-6-phosphate was $5 \times 10^{-4} M$, and that for $NADP^+$ was $3.6 \times 10^{-5} M$. However, Swings and De Ley (1977) suggest that the oxidation–reduction balance during fermentation is probably maintained through NAD^+ rather than $NADP^+$, since no NADP:NAD oxidoreductase (E.C. 1.6.1.1) has yet been detected in *Zymomonas*. More recently, the glucose-6-phosphate dehydrogenase from *Zymomonas* has been purified and studied with respect to its allosteric regulation (Opitz and Schlegel, 1978). These workers reported a molecular weight of 260,000 daltons and confirmed a pH optimum between pH 8.2 and 8.8. The enzyme showed negative allosteric regulation by phosphoenolypyruvate (for discussion, see below). They also reported that the glucose-6-phosphate dehydrogenase from *Zymomonas* showed dual specificity for NAD and NADP. However, the NAD dependent reaction was more strongly inhibited by phosphoenolpyruvate.

Thus there is conflicting opinion in the literature as to the preferred co-substrate, NAD or NADP. Since the maximum Hill coefficient was the same with both co-substrates (Opitz and Schlegel, 1978), it may turn out that the *Zymomonas* glucose-6-phosphate dehydrogenase may use both co-substrates. In addition, inhibition of the enzyme by nucleotides (ATP, ADP, and AMP) was

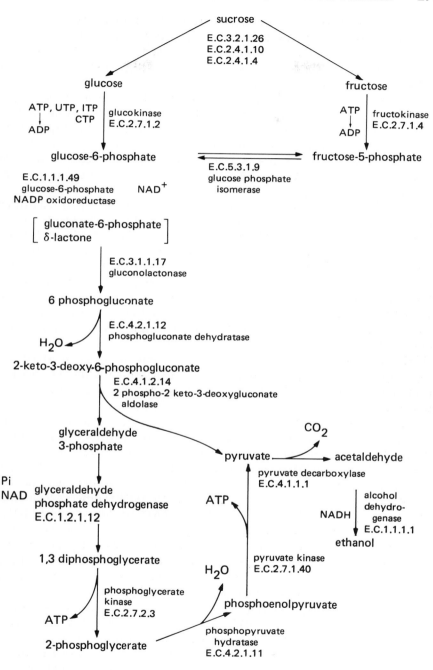

FIGURE 9.1 Schematic representation of the Modified Entner-Doudoroff pathway utilized by *Zymomonas*.

found not to be allosteric regulation but rather competition by ATP for the co-substrate binding site. This inhibition could not be relieved by addition of magnesium. Surprisingly, they found that 10 mM by $MgCl_2$ inhibited the enzyme, a finding that conflicts with the earlier optimal Mg^{2+} concentration reported by Sly and Doelle (1968).

If magnesium does inhibit glucose-6-phosphate dehydrogenase, it is a possible explanation for the poor growth and product formation on molasses media (Skotnicki et al., 1982a) and the zero order maintenance energy coefficient for *Zymomonas* at high magnesium concentration (Cromie and Doelle, 1981, 1982). With respect to end product inhibition of glucose-6-phosphate dehydrogenase, Opitz and Schlegel (1978) reported that both subspecies of *Zymomonas* exhibited noncompetitive inhibition by NADH.

Initially, phosphohexoisomerase was reported to be absent in cell extracts of *Zymomonas* (Raps and De Moss, 1962), but later an enzyme with very low specific activity was detected by McGill and co-workers (McGill et al., 1965; McGill and Dawes, 1971). An isomerase of this type must be present during fermentation of fructose, since neither phosphofructokinase nor fructose diphosphate aldolase has been detected in *Zymomonas*. Phosphogluconate dehydratase and 2-phospho-2-keto-3-deoxygluconate aldolase have been detected in crude extracts of *Zymomonas* by the same workers, but neither enzyme has been purified. Gluconolactonase does not appear to have been studied in *Zymomonas*.

Two enzymes are involved in the conversion of pyruvate to ethanol, pyruvate decarboxylase, and alcohol dehydrogenase. Although pyruvate decarboxylase has been found in crude extracts (McGill et al., 1965; McGill and Dawes, 1971), no detailed biochemistry of this enzyme has been reported. Recently, Wills et al. (1981) have characterized two alcohol dehydrogenases from *Zymomonas* (ZADH-I and ZADH-II). One of the isozymes, ZADH-I, is a tetramer with a subunit molecular weight of 34,700 daltons. It closely resembles the yeast alcohol dehydrogenase (ADH). The second, ZADH-II, is a dimer with a subunit molecular weight of 31,100 daltons. Both enzymes contain one atom of zinc per subunit. ZADH-I and ZADH-II differ in their amino acid composition, apparent K_m, and kinetics. ZADH-I shows kinetics that are very similar to the yeast ADH-II and can oxidize a broad range of alcohols. The ZADH-II isozyme appears to be irreversibly lost during growth on a medium containing low sucrose concentration. The authors (Wills et al., 1981) suggest that the genes for ZADH-II may be plasmid encoded. The absolute function of the two ZADH's in *Zymomonas* is not known. Since ZADH-II can be irreversibly lost, its function does not appear to be critical. In yeast, expression of the two ADH's is regulated by the growth conditions. One isozyme is responsible for ethanol production and the other for ethanol oxidation (Lutstorf and Megnet, 1968). Since ZADH-II has a very high K_m for ethanol, it is unlikely that its key activity is the oxidation of ethanol. Wills and co-workers grew their *Zymomonas* cultures aerobically with shaking; thus it is possible that they induced the genes for a vestigial enzyme that is normally not expressed under anaerobic conditions. Only a few *Zymomonas* strains are able to oxidize ethanol further (Swings and De Ley, 1977).

The fate of glyceraldehyde-3-phosphate depends on the demands of the culture for four- and five-carbon precursors of RNA and DNA biosynthesis. The enzymes utilized in the conversion of glyceraldehyde 3-phosphate to pyruvate appear to be similar to those of glycolysis, although they have not been purified and studied. Clearly, a large portion of the glyceraldehyde-3-phosphate must be converted to pyruvate considering the near theoretical ethanol yield. Additionally, these are the only energy-generating steps in the pathway, and their regulation must play an important role in cell energetics.

In summary, two unique features arise from a study of the metabolism of *Zymomonas*. First, the metabolic pathway itself is unique, since *Zymomonas* is the only genus known to utilize the Entner-Doudoroff pathway anaerobically. This is not a very efficient pathway for the microorganism, since only one mole of ATP is synthesized per mole of hexose fermented. However, it is of special importance to the potential industrial application of *Zymomonas*, since as a result the yield of the desired product, ethanol, approaches the theoretical 2 moles/mole. As the catabolic activity has been implicated as the rate-limiting step in *Zymomonas* fermentation (Cromie and Doelle, 1982), it would seem that a detailed study of all the enzymes involved in the conversion of hexose to ethanol is of utmost importance. The pyruvate decarboxylase and the glycolytic enzymes involved in the conversion of glyceraldehyde-3-phosphate appear to have been overlooked. The latter may turn out to be critical in regulation, since the energy level of the cell is reported to determine the rate of sugar uptake. (Belaïch et al., 1968; Cromie and Doelle, 1980). Second, two unique highly specific constitutive enzymes, glucokinase and fructokinase, have been shown to be the sole mode of sugar transport, a finding that explains the strikingly limited substrate utilization pattern characteristic of *Zymomonas*.

C. Product Formation

The main products formed by *Zymomonas* are ethanol and CO_2. A number of extensive studies on the kinetics of ethanol production and optimization have been carried out. These have been summarized in recent comprehensive reviews (Rogers et al., 1982a, b; Lawford et al., 1983). *Zymomonas* is very inefficient in converting glucose, fructose and sucrose to biomass. It has been found that 2–2.6% of the sugar fermented is used for cell growth (McGill et al., 1965; McGill and Dawes, 1971), the remainder is quantitatively converted to ethanol and CO_2. The molar conversion efficiency appears to be strain-specific and varies between 1.5 and 1.9 mole ethanol/mole glucose. Acetaldehyde, acetate, lactate, acetylmethylcarbinol, and glycerol have been reported to form in trace quantities. These end products are principally formed under aerobic conditions (McGill et al., 1965; McGill and Dawes, 1971). Higher alcohols, principally *n*-propanol and iso-amyl alcohol, have been detected with some strains of *Zymomonas*, but they appear to be present in trace amounts (Bevers and Verachtert, 1976).

One important product formed by *Zymomonas* during growth on sucrose is levan, a polymer of fructose. Levan production by *Zymomonas* was first studied by Dawes et al. (1966). They found that roughly 10% of the sucrose in the medium was converted into levan; the remainder was converted into fructose and glucose. They postulated that an acceptor was necessary in the hydrolysis of sucrose. If the acceptor was water, then glucose and fructose were formed. If the acceptor was fructose, then levan was formed. Levan sucrase was implicated as the enzyme responsible for both sucrose hydrolysis and levan production. However, as was mentioned earlier, the true identity of the enzymes involved in sucrose utilization remains to be elucidated. Levan production has been reported to show a temperature response. Certain strains do not produce levan at elevated temperatures (Skotnicki et al., 1981), and one strain, Ag11, is reported not to produce any levan during growth on sucrose. Lavers et al. (1981) have suggested that sucrose utilization and levan production might be plasmid encoded in *Zymomonas*. However, this has been refuted by Skotnicki et al. (1983a, b). Their laboratory strain of ATCC 10988 (ZM1) is reported to be devoid of plasmids yet still synthesizes levan.

D. Glycolysis and Tricarboxylic Acid Cycle Enzymes

As was mentioned previously, *Zymomonas* appears to lack several key enzymes in glycolysis, in particular, a broad specificity hexokinase (Doelle, 1982b) as well as phosphoenolpyruvate phosphotransferase (Romano et al., 1979) and fructose-1-6 diphosphatealdolase (Dawes et al., 1966). In addition, *Zymomonas* is reported to contain detectable levels of only certain tricarboxylic acid cycle enzymes: citrate synthase, aconitase, isocitrate dehydrogenase, and malate dehydrogenase. The remainder of the enzymes in the TCA cycle were not detected. Studies of the cytochromes of several strains of *Zymomonas* indicate that this bacterium contains both a *c*-type cytochrome and a *b*-type cytochrome (Belaïch and Senez, 1965) In addition, ATCC 10988 is reported to contain pigment P-503 (Kropinski et al., 1973). Since *Zymomonas* is not a true facultative anaerobe but microaeroduric, it is likely that some key component of the electron transport chain is missing.

E. Energy Balances and Uncoupled
Growth and Metabolism

Since only a single net ATP is generated by fermentation employing the Entner-Doudoroff pathway, the molar growth yield coefficient (Y_s),[4] which is equivalent to grams dry weight of cells per mole of substrate, is always equal to the ATP yield coefficient (Y_{ATP})[5] expressed as grams dry weight mole^{-1} ATP.

---

[4] Y_s growth yield coefficient g(dry weight) mole^{-1} substrate
[5] Y_{ATP} energy yield coefficient g(dry weight) mole^{-1} ATP

The growth yield coefficient appears to vary with the culture conditions and the strain and ranges from $Y_s = 3.5$ to 9.3 (Swings and De Ley, 1977). These values reflect the low efficiency of *Zymomonas* in converting energy sources into cell material. As was pointed out earlier, 98% of the glucose consumed is converted into products, ethanol, and CO_2, and only 2% is conserved for cell maintenance and growth.

Further uncoupling of growth and glucose catabolism may be achieved by vitamin starvation (Belaïch et al., 1969, 1972; Lazdunski and Belaïch, 1972) and elevated temperature (Forrest, 1967). In particular, the vitamin pantothenate has been implicated in the uncoupling process. Pantothenate is required for coenzyme A formation, which is, in turn, involved in the activity of the various transferase reactions. Belaïch et al. (1972) reported that pantothenate starvation resulted in a simultaneous decrease in both the molar growth yield and the specific growth rate. However, the rate of glucose fermentation per unit of dry weight remained constant. These authors suggested that the excess ATP was dissipated by direct or indirect ATPase activity.

VI. BIOCHEMICAL COMPOSITION OF THE CELL

A. Cell Envelope

It is assumed that *Zymomonas* possess a normal gram-negative cell wall consisting of a peptidoglycan monolayer with lipoprotein covalently linked to the peptide units of the peptidoglycan. The fact that spheroplasts may be formed by treatment with lysozyme and protease (Dally et al., 1982) supports this conclusion. However, no detailed biochemical analysis of the *Zymomonas* cell wall was been reported.

The lipopolysaccharide outer membrane of *Z. mobilis* has recently been examined in considerable detail (Tornabene et al., 1982). These authors report that the outer membrane lipopolysaccharide fraction of *Zymomonas* is distinctly different from that of other gram-negative bacteria. The principal components were found to be deoxyhexose, pentose, hexose, aminopentose, uronic acid, phosphate, and myristic acid. Ketodeoxyoctulonic acid, heptoses, and hydroxy fatty acids, which are reported to be constituents of the lipopolysaccharide of most gram-negative bacteria, were not detected in *Zymomonas*.

B. Phospholipids and Fatty Acids

The extractable lipids of *Zymomonas* are reported to be 6.3% of the cell on a dry weight basis. Three recent reports on the fatty acid composition of *Zymomonas* concur that this bacterium contains uniquely high levels of C18:1 fatty acids in its membrane (Ohta et al., 1981; Tornabene et al., 1982; Carey and Ingram, 1983). The fatty acid composition is summarized (Table 9.5). There does not appear to be a shift in the fatty acid composition during adaptation to growth in high concentrations of ethanol as has been reported for other

TABLE 9.5   Fatty Acid Composition of *Zymomonas mobilis* subsp. *mobilis*

Fatty Acid	Ohta et al. (1981)	Percentage (%) Total Fatty Acids Tornabene et al. (1982)[a]	Carey and Ingram (1983)
14:0	25.0	17.4	11.0
14:1	0.5	0.1	trace
16:0	8.2	9.9	12
16:1	2.7	1.8	7
18:1	63.6	66.8	69

[a] Data calculated from the reported fatty acid composition of neutral, methanol-insoluble polar, and methanol-soluble polar lipid fractions (Tornabene et al., 1982).

bacteria and yeast (Ingram, 1976; Rigomier et al., 1980; Beavan et al., 1982), since the fatty acid composition of log phase cells closely resembles that of stationary phase cells (Carey and Ingram, 1983). In addition, no branched-chain or cyclopropane fatty acids were detected.

The major phospholipids were found to be cardiolipin, phosphatidylethanolamine, phosphatidylglycerol, phosphatidylcholine, and lysophosphatidylethanolamine. Phosphatidylethanolamine is reported to make up the bulk of the phospholipid, 55–60% (Tornabene et al., 1982; Carey and Ingram, 1983). Thus the phospholipid composition of *Zymomonas* is not uniquely different from that of other bacteria.

Tornabene et al. (1982) have examined the glycolipoprotein fraction of the *Zymomonas* outer membrane and found it to be compositionally similar to lipid A, which is commonly found in gram negative bacteria. The material consisted of 11 different amino acids, two hexosamines, and a range of fatty acids. The major amino acids were glycine and alanine; glucosamine was the predominant hexosamine; and myristic acid was the major fatty acid. This glycolipoprotein fraction differed from the lipopolysaccharide fraction.

In their detailed chemical analysis of the neutral lipid fraction of *Zymomonas*, Tornabene et al. (1982) have detected squalene and 21 pentacyclic triterpene hydrocarbons. The pentacyclic triterpenes fall into two classes, five six-carbon ringed components and four six-carbon rings plus one five-carbon ring. The authors claim that these cyclic triterpenes are unique to *Zymomonas*, while the polyol form is commonly found in other bacteria. In addition, they imply that the presence of these cyclic triterpenes may have a bearing on the ethanol tolerance of *Zymomonas*, since they resemble sterols, and the amount of sterols with unsaturated side chains (ergosterol and stigmasterol) has been directly implicated in yeast alcohol tolerance (Mayashida and Ohta, 1980; Rose and Beavan, 1981). However, these speculations must be viewed with caution. The neutral lipid fractions of many bacteria have not been extensively studied, and thus it is a bit premature to deem *Zymomonas* unique. Second, the cyclic triterpenes are present in such minute quantities that it seems unlikely that they would have a major effect on membrane integrity.

TABLE 9.6   Macromolecular Composition of *Zymomonas* and *E. coli*

Macromolecule	Weight % (g/100 g dry weight)	
	E. coli[a]	*Zymomonas*
Protein	52.4	54[b]
Polysaccharide	16.6	4–5[b]
Lipid	9.4	6.3[c]
RNA	15.7	17–22[b]
DNA	3.2	2.7[b]

[a] Gottschalk (1979).
[b] Dawes and Large (1970).
[c] Tornabene et al. (1982).

C. Other Cellular Constituents

The cellular components of *Zymomonas* appear reasonably typical in comparison to *E. coli* (Table 9.6). The only obvious difference is in carbohydrate content. It is clear that there is very little reserve material in *Zymomonas*, consistent with the lack of glycogen, poly-β hydroxybutyrate, or starch.

VII. FLOCCULATION

Flocculation is a useful phenotype for industrial microorganisms, since it allows facile cell concentration and cell recycle. Flocculant strains have been described (Swings and De Ley, 1977; deBoks and van Eybergen, 1981), and spontaneous and induced flocculant mutants have been isolated (Skotnicki et al., 1981; Lee et al., 1982; Fein et al., 1983a, b). There is one report that flocculation can be induced in ATCC 10988 (Prince and Barford, 1982). At the recent International Symposium on the Genetics of Industrial Microorganisms, Skotnicki et al. (1982b) reported that flocculation in one of their mutant strains was the result of production of sheets of cellulose, which could be visualized by electron and fluorescent light microscopy. These flocs could be dispersed by treatment with crude *Trichoderma* cellulase. However, these results should be viewed with caution, since the crude extracellular protein fraction of *T. reesei* contains at least 60 proteins in addition to cellulase (Sheir-Neiss and Montenecourt, unpublished observation) and any of those activities could be dissolving the flocs.

Cellulose biosynthesis has been elegantly elucidated in *Acetobacter xylinum* by Brown and coworkers (Benziman et al., 1980; Haigler et al., 1980). Since *Zymomonas* exhibits close ribosomal ribonucleic acid cistron homology with *Acetobacter* species, it is possible that they share also the ability to synthesize cellulose (Gillis and De Ley, 1980). The two genera also share the Entner-Doudoroff pathway and many phenotypic characteristics such as flag-gelation, tolerance to acid pH, tolerance toward high glucose and ethanol concentrations, and limited carbon assimilation.

VIII. ENZYME REGULATION

A. Transport System

It has been demonstrated that *Zymomonas* lacks a PEP glucose phosphotransferase system (Romano et al., 1979) and a permease system. Earlier evidence of Belaïch et al. (1968) indicated that glucose transport was twofold greater than glucose utilized. Therefore the rate of glucose catabolism was not limited by glucose permeation, and they suggested that a mediated or facilitated diffusion transport system was the only method of sugar uptake. Lyness and Doelle (1980, 1981a) have recently observed that during growth on sucrose, fructose is always taken up by the cell more slowly than glucose. This observation led them to a more intensive study of the glucokinase and fructokinase enzymes of *Zymomonas*. The glucokinase is inhibited by ADP, AMP, glucose-6-phosphate, and high concentrations of nucleoside triphosphates (ATP, UTP, ITP, and CTP). Thus the enzyme is regulated by the energy status of the cell. Fructokinase reacts only with fructose and ATP and is inhibited by glucose and glucose-6-phosphate. Thus the presence of glucose regulates the uptake of fructose (Doelle, 1982a, b). From this information, Doelle (1982b) proposed that these two enzymes for sugar transport were independently controlled. Glucokinase is controlled by ATP and ADP (e.g., cellular energetics), and fructokinase, which is unaffected by the nucleotides, is severely repressed by glucose. They suggested that this differential control system could explain the existence of two independent constitutive enzymes. In most bacteria the group translocation and transport system for glucose is constitutive, and the fructose transport system is inducible (Dills et al., 1980). These data support the earlier conclusions of Belaïch (1968) that the rate of catabolic activity and the energy balance of the cell control glucose uptake. Fructose uptake is less dependent on the internal cell energetics. One very surprising observation (Doelle, 1982b) is the lack of end product inhibition of the transport enzymes by ethanol. The inhibitory effects of ethanol on growth (Lee et al., 1979, 1981a) must be due to perturbation of the cell membrane or inhibition of one of the catabolic enzymes. In *E. coli* a delicate balance between ADP/AMP and the end products of glycolysis regulate the system. Phosphofructokinase is activated by ADP/AMP and inhibited by phosphoenolpyruvate. Pyruvate kinase, on the other hand, is activated by AMP and fructose-1-6 diphosphate. *Zymomonas* clearly lacks these regulatory mechanisms. It has been suggested that when growth and metabolism are uncoupled, the excess ATP that is not channeled into biosynthetic pathways is removed by an ATPase (Lazdunski and Belaïch, 1972). It would seem that removal of the high-energy intermediates is a prerequisite for continued glucose transport.

B. Catabolic Enzymes

The regulatory properties of phosphoenolpyruvate with respect to glucose-6-phosphate dehydrogenase have recently been elucidated (Opitz and Schlegel, 1978). Phosphoenolpyruvate has been shown to be a strong allosteric inhibitor of this enzyme in *Zymomonas*. In contrast, the enzyme from *E. coli* was unaffected, and that of *Leuconostoc* and *Pseudomonas* only weakly inhibited. Species of *Azotobacter*, *Caulobacter*, and *Corynebacterium* showed negative allosteric regulation similar to *Zymomonas*. The authors suggest that glucose-6-phosphate dehydrogenase may be a key regulating enzyme in mediating between catabolism and anabolism.

Regulation of the other catabolic enzymes in the Entner-Doudoroff Pathway has not been studied. Although the kinetics of the forward and backward reactions of the two alcohol dehydrogenases have been elucidated (Wills et al., 1981), inhibition kinetics have not been reported. More important, few studies have been carried out on the pyruvate decarboxylase or the regulatory function of phosphoenylpyruvate other than its effect on glucose-6-phosphate dehydrogenase. Delineation of the regulatory mechanisms of the two ATP-generating enzymes, phosphoglycerate kinase and pyruvate kinase, should help to determine the relationship between catabolism and the energy balance of the cell.

IX. GENETICS OF *ZYMOMONAS*

A. Classical Mutation

Until recently, few mutants of *Zymomonas* have been described. Optimal conditions for mutagenesis and selection of auxotrophs have only recently been reported (Goodman et al., 1982). These authors have described a minimal medium for auxotrophic screening (Table 9.4). Ultraviolet light was not found to be a suitable mutagen for *Zymomonas*, at least in the selection of rifampicin-resistant mutants (Skotnicki et al., 1982a). Nitrosoguanidine (NTG) has been suggested as a better mutagen for *Zymomonas*. The susceptibility of the various strains of *Z. mobilis* to NTG mutagenesis varies and the range of NTG concentration suggested was 25–100 μg/ml (Skotnicki et al., 1982a). It should also be noted that the best conditions for mutagenesis involve treatment of lag phase cells with NTG in a rich medium under active growth conditions rather than in buffer. Subsequent incubation for expression is for only a few hours rather than the usual overnight incubation.

B. Transposon Mutagenesis

Skotnicki et al. (1980) have apparently developed a system for transposon mutagenesis of *Z. mobilis* (Rogers et al., 1982a, b). The Tn-5 containing plasmid pJ4JI, which has been utilized for transposon mutagenesis in

Rhizobium (Beringer et al., 1978), can be stably maintained in *Z. mobilis*. Rogers et al. (1982a) reported that subsequent superinfection with plasmid pJB3JI, a kanamycin-sensitive derivative of the broad-host-range R plasmid R68.45, results in the transposition of the Tn-5 transposon into the chromosome and loss of the pJ4JI plasmid. Although they speculate that this method of mutagenesis will help locate the genes for levan production as well as other critical chromosomal genes, mutants developed employing this technique have not been described to date.

C. Mutants of *Z. mobilis*

With the recent surge of interest in *Zymomonas*, a range of mutants have been isolated that show superior characteristics of industrial importance in comparison to the wild-type. These mutants, their phenotype, and the parent strains are listed (Table 9.7). The special feature of the ZM481 ethanol-tolerant variant appears to be a higher cell viability at the elevated ethanol concentration. Mutants with greater flocculation characteristics and tolerance to molasses have also been described, but the exact nature of the genetic lesions has not been defined. Auxotrophic mutants of *Z. mobilis* have been described (Goodman et al., 1982) that will greatly facilitate future genetic studies.

Isolating mutants exhibiting increased temperature tolerance and decreased levan production are goals for the future. Although temperature-tolerant mutants have been described (Skotnicki et al., 1982a, b), the increased tolerance is only a few degrees (40°C) above the temperature for optimal growth. Levan production during industrial ethanol production is inefficient use of the feedstock. Mutants unable to synthesize levan and yet maintain high ethanol productivity will enhance the industrial potential of *Zymomonas*.

D. Natural Antibiotic Resistance of *Z. mobilis*

Zymomonads exhibit an unusual array of antibiotic resistances. Swings and De Ley (1977) tested 38 strains for susceptibility to 20 antibiotics; many of the

TABLE 9.7   Mutant Strains of *Z. mobilis*

Parent	Isolate	Phenotype	Reference
CP4	ZM 481	Ethanol tolerant	Skotnicki et al. (1982a)
CP4	ZM 401	Flocculent	Lee et al. (1983)
ATCC 10988	Undesignated	Flocculent	Prince and Barford (1982)
ATCC 29191	Undesignated	Flocculent	Fein et al. (1983)
CP4	ZM 482	Molasses tolerant	Skotnicki et al. (1982a)
CP4	Undesignated	Temperature tolerant	Skotnicki et al. (1982a)
Ag11			
CP4	Undesignated	Auxotrophic	Goodman et al. (1982)
ATCC 10988			
ATCC 10988	Undesignated	ZADH negative	Wills et al. (1981)

strains were resistant to at least 15 of these antibiotics. All strains tested were resistant to the following antibiotics: bacitracin, gentamicin, kanamycin, lincomycin, nalidixic acid, neomycin, penicillin, polymyxin, and streptomycin. In addition, Stokes et al. (1983) have shown strains ATCC 10988, CP3, CP4, and Ag11 to also be resistant to trimethoprim and mercuric chloride. A list of antimicrobial agents tested against *Zymomonas* is given (Table 9.8). All strains screened were susceptible to tetracycline at a concentration of 10 μg/ml. In addition, all strains thus far tested are also sensitive to rifampicin (Skotnicki et al., 1982a, b; Stokes et al., 1983). Mutation to rifampicin resistance was utilized as a test genetic marker in the development of a mutagenesis procedure for *Zymomonas* (Skotnicki et al., 1982a, b). Resistance to ampicillin is variable according to the strain. Swings and De Ley (1977) reported that 50% of the 38 strains that they tested were resistant to 10 μg/ml of this antibiotic. Stokes et al. (1983) recently reported a variation in the minimal inhibitory concentration for the four purported industrially useful strains CP3, CP4, Ag11, and ATCC 10988. They found that there was a tenfold difference in the concentration of ampicillin needed to kill 50% of the cells (range 20–300 μg/ml), Ag11 being the most sensitive and CP4 the most resistant. It is likely that the vast array of antibiotics to which *Zymomonas* is resistant may be a result of a permeability barrier. Anaerobes are reported to be resistant to aminoglycoside antibiotics owing to inadequate electrical potential across the membrane, which is required for efficient antibiotic uptake (Bryan and Kwan, 1981). This idea is consistent with the recent observations (Romano et al., 1979; Doelle, 1982b) that indicate that neither the proton motive force nor a phosphoenolpyruvate glucose transferase is involved in group translocation in *Zymomonas*. It would appear that only glucose and fructose are transported into this microorganism and that they are specifically transported by independent enzymes.

TABLE 9.8   Effect of Antimicrobial Agents on *Zymomonas*

Resistant	Sensitive
Ampicillin	Chloramphenicol
Bacitracin	Fusidic acid
Erythromicin	Novobiocin
Gentamycin	Rifampicin
Kanamycin	Sulfafurazole
Lincomycin	Tetracycline
Methicillin	
Nalidixic acid	
Neomycin	
Penicillin	
Polymyxin	
Streptomycin	
Trimethoprim	

E. Plasmids in *Zymomonas*

Native plasmids in *Zymomonas* were recently reported by a number of laboratories (Dally et al., 1983; Skotnicki et al., 1982a; Rogers et al., 1982b; Tonomura et al., 1982). As with many of the characteristics described for *Zymomonas*, the number and size of the plasmid varies with the strains. There has been considerable confusion in interpreting some of the data (Skotnicki et al., 1982a; Rogers et al., 1982b). The authors claim that their variant of ATCC 10988 (ZM1) does not contain plasmids. Part of the confusion stems from the fact that the supporting gel in one publication (Rogers et al., 1982b) was printed in the improper orientation so that the lanes should have been identified from right to left instead of from left to right in the legend (P. Rogers, personal communication). In general, the plasmids range in molecular weight from 1 to 40 megadaltons, and both ends of the molecular weight spectrum may be represented in the same strain (Dally et al., 1982; Skotnicki et al., 1982a). Tonumura et al. (1982) have found that the largest plasmids are on the order of 25 megadaltons.

The molecular weights of the plasmids from four potential industrial strains have been more definitively analyzed both by agarose gel electrophoresis with appropriate molecular weight standards and by measurement of the contour lengths via electron microscopy (Stokes et al., 1983); they are summarized (Table 9.9). All strains tested, including ATCC 10988, seem to have in common a large 46-megadalton plasmid. Restriction endonuclease digestion of the plasmids of CP3 and CP4 yielded identical maps (Dally et al., 1982), suggesting that they are the same strain. More recently, the smaller plasmids have been partially mapped by restriction endonuclease analysis, and the 1.1-megadalton plasmid has been shown to contain a single Sau31 recognition site (Dally et al., 1983) and may be useful as a cloning vector.

To date, no phenotypic characteristics have been attributed to the plasmids from these zymomonads. It has been suggested that sucrose utilization and levan production may be plasmid encoded. This hypothesis has been refuted by Skotnicki and co-workers, who claim that their variant of ATCC 10988 contains no plasmids and yet utilizes sucrose and produces levan. On the other hand, strain Ag11, which is reported not to produce any detectable levan (Skotnicki et al., 1981), contains at least three plasmids and appears to be missing only the small (1.1–1.65 megadalton) plasmids.

TABLE 9.9 Plasmids of *Z. mobilis*

| | *Molecular Weight (daltons)* | | |
ATCC 10988	*AG 11*	*CP3*	*CP4*
46×10^6	46×10^6	46×10^6	46×10^6
16×10^6	19×10^6	21×10^6	21×10^6
1.65×10^6	6.6×10^6		
1.1×10^6			

Data from Stokes et al. (1983).

A second speculation (Willis et al., 1981) for a plasmid-encoded phenotype is the alcohol dehydrogenase (ZADHII) gene. The activity of this enzyme may be irreversibly lost during continued cultivation on sucrose. If the variant of ATCC 10988 (ZM1) is truly free of native plasmids, it should be a useful organism with which to test this hypothesis.

F. Genetic Recombination

Conjugal transfer of four plasmids, three from the IncP1 incompatibility group and one from the IncFII incompatibility group, into *Zymomonas* have been reported (Skotnicki et al., 1980). It was shown that *Zymomonas* can participate in intrageneric transfer as the plasmid recipient with *Pseudomonas aeruginosa* and *E. coli* and in intraspecies transfer with other *Z. mobilis* strains. These results were confirmed by Stokes et al. (1983) employing the IncP1 broad host range plasmids R1 and R68.45. The efficiency of plasmid transfer, as with many other characteristics, appears to show considerable variation. One problem with the broad-host-range plasmids is that they are frequently not stably maintained in *Zymomonas*. To date, no transformation or transduction systems have been described for *Zymomonas*.

X. SUMMARY AND CONCLUSIONS

In summary, *Zymomonas* appears to be unique in many respects. Taxonomically, it is a genus with a single representative species and only two identifiable subspecies. Its pattern of carbon assimilation is extraordinarily limited, glucose and fructose being the only identifiable carbon sources transported into the cell. These sugars are uniquely metabolized via the Entner-Doudoroff pathway under anaerobic conditions. Physiologically, the cell membrane contains abnormally high levels of C18:1 fatty acids and small amounts of cyclic triterpenes which distinguish it from other prokaryotes. In addition, it shows unusually high tolerances to acid pH, sugar concentrations, and ethanol.

As was pointed out by Swings and De Ley (1977) and Gillis and De Ley (1980), *Zymomonas* is genetically about equidistant from *Gluconobacter*, *Acetobacter*, *Rhodopseudomonas*, *Rhizobium*, and *Agrobacterium*. Phenotypically, it more closely resembles *Gluconobacter* than *Acetobacter* (polar flagellation, incomplete Krebs Cycle, occurrence of the Entner-Doudoroff pathway, and ravenous consumption of glucose). It would be of interest to know whether the phospholipid and fatty acid contents of the cell membrane of *Zymomonas* and *Gluconobacter* are also similar.

Swings and De Ley (1977) have proposed that *Zymomonas* is a newly evolved genus that has not yet had time to develop major species variation or, alternatively, is inherently a very stable genus in spite of its worldwide distribution. Certainly, *Zymomonas* has evolved rapidly as a microorganism of potential industrial application. Increasing interest in *Zymomonas* will undoubtedly uncover new unique features of this genus and better define its relationship to other genera. The notion that the high levels of C18:1 in the fatty acids of *Zymomonas* membranes are an evolutionary adaptation to allow growth and survival in high concentrations of ethanol suggests that at least some evolutionary change has occurred. The development of a genetic transfer system for *Zymomonas* (Dalley et al., 1983) is exciting and will, it is hoped, result in strains capable of using a much broader range of substrates (e.g., cellulose and starch) increase with industrial potential.

ACKNOWLEDGMENTS

The author wishes to thank T. Chase, Jr., D. E. Eveleigh, and J. A. Sands for their helpful critiques of this manuscript and E. Dally, D. E. Eveleigh, L. O. Ingram, and H. Stokes for allowing the inclusion of their unpublished results.

REFERENCES

Barker, B. T. P. (1948) *Ann. Rep. Agr. Hort. Res. Stn.* Long Ashton Bristol, pp. 174–181.

Barker, B. T. P., and Hillier, V. F. (1912) *J. Agr. Sci. 5*, 67–85.

Beavan, M. C., Carpentier, C., and Rose, A. H. (1982) *J. Gen. Microbiol. 128*, 1447–1455.

Belaïch, J. P. (1963) *C. R. Soc. Biol. 157*, 316–322.

Belaïch, J. P., and Senez, J. C. (1965) *J. Bacteriol. 89*, 1195–1200.

Belaïch, J. P., Senez, J. C., and Murgier, M. (1968) *J. Bacteriol. 95*, 1750–1757.

Belaïch, J. P., Simonpietri, P., and Belaïch, A. (1969) *J. Gen. Microbiol. 58*, vii.

Belaïch, J. P., Belaïch, A., and Simonpietri, P. (1972) *J. Gen. Microbiol. 70*, 179–185.

Benziman, M., Haigler, C. H., Brown, R. M., White, A. R., and Cooper, K. M. (1980) *Proc. Nat. Acad. Sci. U.S.A. 77*, 6678–6682.

Beringer, J. E., Beynon, J. L., Buchanan-Wollaston, A. V. and Johnston, A. W. B. (1978) *Nature* (London) *276*, 633–634.

Bevers, J., and Verachtert, H. (1976) *J. Inst. Brew. 82*, 35–40.

Bryan, L. E., and Kwan, S. (1981) *J. Antimicrobial Chemotherapy 8*, Suppl. D, 1–8.

Carey, V. C., and Ingram, L. O. (1983) *J. Bacteriol. 154*, 1291–1300.

Carr, J. G. (1974) in *Bergey's Manual of Determinative Bacteriology*, 8th ed. (Buchanan, R. E., and Gibbons, N. E., eds.), Williams and Wilkins, Baltimore.

Cromie, S., and Doelle, H. W. (1980) *Biotechnol. Lett. 2*, 357–362.

Cromie, S., and Doelle, H. W. (1981) *Eur. J. Appl. Microbiol. Biotechnol. 11*, 116–119.

Cromie, S., and Doelle, H. W. (1982) *Eur. J. Appl. Microbiol. Biotechnol. 14*, 69–73.

Dadds, M. J. S., and Martin, P. A. (1973) *J. Inst. Brew.* (London) *79*, 386–391.

Dadds, M. J. S., Martin, P. A., and Carr, J. G. (1973) *J. Appl. Bacteriol. 36*, 531–539.

Dally, E. L., (1982) "Plasmid studies in Zymomonas mobilis," M.S. thesis, Rutgers University; New Brunswick, N. J.

Dally, E. L., Stokes, H. W., and Eveleigh, D. E. (1982) *Biotechnol. Lett. 4*, 91–96.

Dally, E. L., Stokes, H. W., and Eveleigh, D. E. (1983) in *Organic Chemicals from Biomass* (Wise, D. L., ed.), Addison-Wesley, Reading, Mass.

Dawes E. A., and Large, P. J. (1970) *J. Gen. Microbiol. 60*, 31–42.

Dawes, E. A., Robbins, D. W., and Rees, D. A. (1966) *Biochem. J. 98*, 804–812.

deBoks, P. A., and van Eybergen, G. C. (1981) *Biotechnol. Lett. 3*, 577–582.

De Ley, J., and Swings, J. (1976) *Int. J. Syst. Bacteriol. 26*, 146–157.

De Ley, J., Gillis, M., and De Vos, P. (1981) *Z. Bakt. Hyg., I Abt. Orig. C2*, 263–268.

Dennis, R. T., and Young, T. W. (1982) *J. Inst. Brew. 88*, 25–29.

Dills, S. S., Apperson, A., Schmidt, M. R., and Daier, M. H., Jr. (1980) *Microbiol. Rev. 44*, 385–418.

Doelle, H. W. (1982a) *Eur. J. Appl. Microbiol. Biotechnol. 15*, 20–24.

Doelle, H. W. (1982b) *Eur. J. Appl. Microbiol. Biotechnol. 14*, 241–246.

Fein, J., Zawadzki, B., Lawford, G. R., and Lawford, H. G. (1983a) *Appl. Environ. Microbiol. 45*, 1899–1904.

Fein, J., Lawford, H. G., Lawford, G. R., Zawadzki, B. and Charley, R. C. (1983b) *Biotechnol. Lett. 5*, 19–24.

Forrest, W. W. (1967) *J. Bacteriol. 94*, 1459–1463.

Gibbs, M., and De Moss, R. D. (1951) *Arch. Biochem. Biophys. 34*, 478–479.

Gibbs, M., and De Moss, R. D. (1954) *J. Biol. Chem. 207*, 689–694.

Gillis, M., and De Ley, J. (1980) *Int. J. Syst. Bacteriol. 30*, 7–27.

Gonçalves de Lima, O., De Aránjo, J. M., Schumacher, I. E., and Cavalcanti Da Silva, E. (1970) *Rev. Inst. Antibiot. Univ. Recife 10*, 3–15.

Gonçalves de Lima, O., Schumacher, I. E., and De Aránjo, J. M. (1972) *Rev. Inst. Antibiot. Univ. Recife 12*, 57–69.

Goodman A. E., Rogers, P. L., and Skotnicki, M. L. (1982) *Appl. Environ. Microbiol. 44*, 496–498.

Gottschalk, G. (1979) *Microbial Metabolism*, Springer, New York.

Haigler, C. H., Brown, R. M., and Benziman, M. (1980) *Science 210*, 903–906.

Hayashida, S., and Ohta, K. (1981) *J. Inst. Brew. 87*, 42.

Ingram, L. O. (1976) *J. Bacteriol. 125*, 670–678.

Ingram, L. O. (1977) *Can. J. Microbiol. 23*, 779–789.

Ingram, L. O. (1981) *J. Bacteriol. 146*, 331–336.

Ingram, L. O., Dickens, B. F., and Buttke, T. M. (1980a) *Adv. Exp. Med. Biol. 126*, 299–337.

Ingram, L. O., Vreeland, N. S., and Eaton, L. C. (1980b) *Pharmacol. Biochem. Behavior 13*, 191–195.

Ingram, L. O., Eaton, L. C., Erdos, G. W., Tedder, T. F., and Vreeland, N. L. (1982) *J. Memb. Biol. 65*, 31–40.

King, F. G., and Hossain, M. A. (1982) *Biotechnol. Lett. 4*, 531–536.

Kluyver, A. J., and Hoppenbrouwers, W. J. (1931) *Arch. Mikrobiol. 2*, 245–260.

Kluyver, A. J., and van Niel, K. (1936) *Hyg. Abt. 94*, 369–403.

Kropinski, A. M. B., Boon, J., Poulson, R., and Polglase, W. J. (1973) *Can. J. Microbiol. 19*, 1235–1238.

Laudrin, I., and Goma, G. (1982) *Biotechnol. Lett. 4*, 537–542.

Lavers, B. H., Pang, P., MacKenzie, G. R., Lawford, G. R., Pik, J., and Lawford, H. G. (1981) in *Advances in Biotechnology* Vol. II, (Moo-Young, M., and Robinson, C. W., eds.), pp. 195–200, Pergamon Press, Canada, Ltd.

Lawford, G. R., Lavers, B. H., Good, D., Charley, R., Fein, J., and Lawford, H. G. (1983) in *Proceedings of the International Symposium on Ethanol from Biomass, Winnipeg, Canada* (Duckworth, E. H., ed.), pp. 482–507, Royal Society of Canada, Ottawa.

Lazdunski, A., and Belaïch, J. P. (1972) *J. Gen. Microbiol. 70*, 187–197.

Lee, K. J., Tribe, D. E., and Rogers, P. L. (1979) *Biotechnol. Lett. 1*, 421–426.

Lee, K. J., Skotnicki, M., Tribe, D. E., and Rogers, P. L. (1981a) *Biotechnol. Lett. 3*, 207–212.

Lee, K. J., Skotnicki, M., Tribe, D. E., and Rogers, P. L. (1981b) *Biotechnol. Lett. 3*, 291–296.

Lee, K. J., Skotnicki, M. L., and Rogers, P. L. (1982) *Biotechnol. Lett. 4*, 615–620.

Lindner, P. (1928) *Bot. Zool. Ver. 50*, 253–255.

Lindner, P. (1931) *Z. Ver. Dsch. Zuckerind. 81*, 25–36.

Lutstorf, U., and Megnet, R. (1968) *Arch. Biochem. Biophys. 126*, 933–944.

Lyness, E., and Doelle, H. W. (1980) *Biotechnol. Lett. 2*, 549–554.

Lyness, E., and Doelle, H. W. (1981a) *Biotechnol. Bioeng. 23*, 1449–1460.

Lyness, E., and Doelle, H. W. (1981b) *Biotechnol. Lett. 3*, 257–260.

Mayashida, S., and Ohta, K. (1980) *Agr. Biol. Chem. 44*, 2561–2567.

McGill, D. J., and Dawes, E. A. (1971) *Biochem. J. 125*, 1059–1068.

McGill, D. J., Dawes, E. A., and Ribbons, D. W. (1965) *Biochem. J. 97*, 44P–45P.

Millar, D. G., Griffiths-Smith, K., Cugar, E., and Scopes, R. K. (1982) *Biotechnol. Lett. 4*, 601–606.

Millis, N. F. (1951) "Some Bacterial Fermentations of Cider," Ph.D. thesis, University of Bristol, Bristol, England.

Millis, N. F. (1956) *J. Gen. Microbiol. 15*, 521–528.

Ohta, K. Supanwong, K., and Hayashida, S. (1981) *J. Ferment. Technol. 59*, 435–439.

Okafor, N. (1978) *Adv. Appl. Microbiol. 24*, 237–256.

Opitz, R., and Schlegel, H. G. (1978) *Biochem. Syst. Ecol. 6*, 149–155.

Prince, I. G., and Barford, J. P. (1982) *Biotechnol. Lett. 4*, 525–530.

Raps, S., and De Moss, R. D. (1962) *J. Bacteriol. 84*, 115–118.

Ribbons, D. W., Dawes, E. A., and Rees, D. A. (1962) *Biochem. J. 82*, 45P.

Richards, M., and Corbey, D. A. (1974) *J. Inst. Brew. 80*, 241–244.

Rigomier, D., Bohin, J., and Lubochinsky, B. (1980) *J. Gen. Microbiol. 121*, 139–149.

Rogers, P. L., Lee, K. J., and Tribe, D. E. (1979) *Biotechnol. Lett. 1*, 165–170.

Rogers, P. L., Lee, K. J., Skotnicki, M. L., and Tribe, D. E. (1982a) *Adv. Biotechnol. 2*, 189–194.

Rogers, P. L., Lee, K. J., Skotnicki, M. L., and Tribe, D. E. (1982b) in *Advances in Biochemical Engineering*, Vol. XXIII (Fiechter, A., ed.), pp. 37–84, Springer, New York.

Romano, A. H., Trifone, J. D., and Brustolon, M. (1979) *J. Bacteriol. 139*, 93–97.

Rose, A. H., and Beavan, M. J. (1981) in *Trends in the Biology of Fermentations for Fuels and Chemicals*, Vol. XVIII, Basic Life Sciences, (Hollaender, A., Rabson, R., Rogers, P., San Pietro, A., Valentine, R., and Wolfe, R., eds.), pp. 513–531, Plenum, New York.

Saddler, J. N., Chan, M. K. H., and Louis-Seize, G. (1981) *Biotechnol. Lett. 3*, 321–326.

Saddler, J. N., Hogan, C., Chan, M. K. H., and Louis-Seize, G. (1982) *Can. J. Microbiol. 28*, 1311–1319.

Sanchez-Marroquin, A., Larios, C., and Vierna, L. (1967) *Rev. Latinoamer. Microbiol. Parasitol. 9*, 83–85.

Shimwell, J. L. (1937) *J. Inst. Brew. London 43*, 507–509.

Shimwell, J. L. (1950) *J. Inst. Brew. London 56*, 179–182.

Skotnicki, M. L., Tribe, D. E., and Rogers, P. L. (1980) *Appl. Environ. Microbiol. 40*, 7–12.

Skotnicki, M. L., Lee, K. J., Tribe, D. E., and Rogers, P. L. (1981) *Appl. Environ. Microbiol. 41*, 889–893.

Skotnicki, M. L., Lee, K. J., Tribe, D. E., and Rogers, P. L. (1982a) in *Genetic Engineering of Microorganisms for Chemicals, Basic Life Sciences* pp. 271–290 (Hollaender, A., De Moss, R. D., Kaplan, S., Konisky, J., Savage, D., and Wolfe, R. S., eds.), Plenum, New York.

Skotnicki, M. L., Warr, R. G., Goodman, A. E., and Rogers, P. L. (1982b) *Proc. 4th Int. Congress Genetics Ind. Microbiol.* pp. 361–365 (Y. Ikeda and T. Beppu, eds.), Kodansha Ltd., Tokyo, Japan, 1982.

Sly, L. I., and Doelle, H. W. (1968) *Arch. Mikrobiol. 63*, 197–213.

Stern, I. J., Wang, C. H., and Gilmour, C. M. (1960) *J. Bacteriol. 79*, 601–611.

Stokes, H. W., Dally, E., Spaniel, D. T., Williams, R. L., Montenecourt, B. S., and Eveleigh, D. E. (1981) *Biomass Digest 3*, 124–132.

Stokes, H. W., Dally, E., Williams, R. L., Montenecourt, B. S., and Eveleigh, D. E. (1982) in *Chemistry in Energy Production* (Keller, O. L., Wymer, R. G., eds.) (American Chemical Society Southern-Southwest Regional Meeting), pp. 115–121, Oak Ridge National Laboratory, Oak Ridge, Tenn.

Stokes, H. W., Dally, E. L., Yablonski, M. D., and Eveleigh, D. E. (1983) *Plasmid 9*, 138–146.

Swings, J., and De Ley, J. (1975) *Int. J. Syst. Bacteriol. 25*, 324–328.

Swings, J., and De Ley, J. (1977) *Bacteriol. Rev. 41*, 1–46.

Swings, J., and Van Pee, W. (1977) *J. Gen. Appl. Microbiol. 23*, 297–301.

Swings, J., Kersters, K., and De Ley, J. (1976) *J. Gen. Microbiol. 93*, 266–271.

Swings, J., Kersters, K., and De Ley, J. (1977) *Int. J. Syst. Bacteriol. 27*, 271–273.

Tonomura, K., Naotaka, K., Konishi, S., Kawaski, H. (1982) *Agr. Biol. Chem. 46*, 2851–2853.

Tornabene, T. G., Holzer, G., Bittner, A. S., and Grohmann, K. (1982) *Can. J. Microbiol. 28*, 1107–1118.

Van Pee, W., and Swings J., (1971) *E. Afr. Agr. For. J. 36*, 311–314.

Viikari, L., Nybergh, P., and Linko, M. (1981) in *Advances in Biotechnology* Vol. II, *Fuels, Chemicals, Foods and Waste Treatment* (Moo-Young, M., and Robinson, C. W., eds.), pp. 137–142, Pergamon Press, New York.

Viikari, L., Lindberg, K., Linko, M., and Parkinen, E. (1982) in *Abstracts of the XIIIth International Congress of Microbiology*, p. 72.7, American Society for Microbiology, Washington, D.C.

Wills, C., Kratofil, P., Londo, D., and Martin, T. (1981) *Arch. Biochem. Biophys. 210*, 775–785.

Woodward, J. D. (1982) *J. Inst. Brew. 88*, 84–85.

Biology of *Streptomyces*

C. F. Hirsch
P. A. McCann-McCormick

I. INTRODUCTION

Streptomyces clearly ranks near the top among microorganisms of industrial importance. They are outstanding as producers of antibiotics and other useful and interesting compounds, a feature that no doubt has done much to attract researchers. Despite the considerable body of knowledge that has accumulated over the past 30 or so years on the secondary metabolism of *Streptomyces*, comparatively little is known about other aspects of their biology. This is unfortunate, since the streptomycetes are among some of the most exquisite and fascinating bacteria to be found in the microbial world. Over the past few decades a handful of investigators have provided what insight we presently have concerning the genetics, developmental biology, structure, physiology and metabolism, and ecology of *Streptomyces*. However, much remains to be learned before we can begin to fully comprehend these bacteria. Such a dearth of information may cause some to despair over the state of our understanding of *Streptomyces* biology, but we suggest that it be viewed rather as an exciting and challenging area for study, or, as put by Ensign (1978), "virgin territory awaiting the curious investigator."

In our approach to this chapter we have tried to sketch an overall picture of *Streptomyces* for those investigators now working with them and especially for those workers entering the realm of *Streptomyces* for the first time. We have purposely avoided lengthy discussions of products, industrial processes, and secondary metabolism of these bacteria and refer the interested reader to other volumes in this series for detailed treatment of these topics.

II. ECOLOGY

Streptomyces strains are widely distributed in nature (Lechevalier, 1981; Kutzner, 1981). Their primary niche is the soil ecosystem (Williams, 1978), but they also occur in freshwater and marine environments (Cross, 1981; Weyland, 1981; Okami and Okazaki, 1978), in salt marsh areas (Hunter et al., 1981), in fodders and related materials (Lacey, 1978), and in the air (Lloyd, 1969b). The populations of *Streptomyces* present in soils usually range from 10^4 to 10^7 colony forming units per gram of dry soil. The physical state of streptomycetes in soil generally is spores rather than vegetative hyphae (Mayfield et al., 1972; Lloyd, 1969a). A variety of soil types are inhabited by streptomycetes, and their presence is positively correlated with the organic matter and water content of the soil (Williams, 1978; Kutzner, 1981; Lacey, 1973).

Streptomycetes appear late in the natural microbial successions that colonize and degrade organic matter. This results in a fairly restricted spectrum of nutrients available to support their growth. These nutrients are mostly high-molecular-weight resistant polymers such as chitin (Williams and Robinson, 1981). Streptomycetes appear to occupy an important position in the turnover of chitin in the soil (Mitchell, 1963; Okafor, 1966). Other polymers degraded by streptomycetes include starch (Williams, 1978; Lacey, 1973), pectin (Kaiser, 1971), and certain hemicelluloses (Waksman and Diehm, 1931). Cellulose does not play a major role in streptomycete nutrition, nor do humic acids (Williams, 1978). Streptomycetes apparently are involved in the biodegradation of lignin in nature (Crawford et al., 1982).

In addition to their degradative role in the soil, streptomycetes interact with other soil microorganisms through competition for nutrients and the production of lytic and inhibitory factors (Lloyd and Lockwood, 1966; Lockwood and Lingappa, 1963). Antibiotic production by streptomycetes actively growing in normal soil is believed to occur but has yet to be clearly demonstrated (Williams and Khan, 1974).

Streptomycetes are obligate aerobes and prefer soils of moderate moisture content over those which are waterlogged (Williams et al., 1972b). Very dry soils are inhibitory to streptomycete growth. In such soils, however, the portion of the total measurable microbial population identified as streptomycetes may increase from the 1–20% usually found (Kutzner, 1981) to 70% (Williams et al., 1972b) and up to 98% in soils under drought conditions (Meiklejohn, 1957). The enrichment of streptomycetes in dry soils is believed to be due to the desiccation resistance of their spores.

Streptomycetes can be grouped into two general categories based on pH requirements for growth (Williams, 1978; Kutzner, 1981). Acidophilic streptomycetes grow in the pH range of 3.5–6.5, with optima near pH 5.0 (Khan and Williams, 1975; Williams et al., 1971), while neutrophilic streptomycetes grow from 5.0 to 9.0, with optima around pH 7.0. Some streptomycetes have been isolated from soils of pH greater than 9.0 (Johnstone, 1947; Taber, 1960); however, they seem to occur only rarely in such alkaline environments.

Most streptomycetes grow optimally in a mesophilic temperature range. Some species are known, however, that grow at temperatures well above the mesophilic range, that is, 50–60°C. Whether these species represent true thermophiles or thermotolerant mesophiles is not clear (Kutzner, 1981).

In the soil, streptomycetes generally are found in greatest numbers near the surface and in decreasing number with increasing depth down to about 10–15 cm (Davies and Williams, 1970; Hagedorn, 1976). They may occur in greater numbers in the rhizosphere of plants, depending on the type of plant and its age, older plants having much higher rhizosphere numbers than young plants (Williams, 1978). There do not appear to be any significant qualitative differences between rhizosphere and nonrhizosphere streptomycetes. Indirect evidence suggests that some rhizosphere streptomycetes may interact in a beneficial way with plants (Katznelson and Cole, 1965; Walker et al., 1966; Mitchell, 1963; Whaley and Boyle, 1967). Conversely, a few streptomycetes are known to be pathogenic to certain plants, being the causative agents of scab disease of potato and sugar beet (Lapwood, 1973) and soil rot of sweet potato (Person and Martin, 1940).

The classic procedures used for enrichment and isolation of *Streptomyces* from nature have been comprehensively reviewed by Kutzner (1981). A useful new procedure has been described that selectively isolates streptomycetes on the basis of their ability to grow through membrane filters (Hirsch and Christensen, 1983).

III. ULTRASTRUCTURE AND COMPOSITION

Streptomycetes possess a prokaryotic cellular ultrastructure typical of other gram-positive bacteria (Williams et al., 1973). The streptomycete cell wall consists of a peptidoglycan network cross-linked by peptide subunits of L-alanine, L-glutamic acid, and LL-diaminopimelic acid joined by glycine bridges (Leyh-Bouille et al., 1970; Arima et al., 1968; Nakamura et al., 1967, 1977). In contrast to the vegetative hyphal wall, the spore walls of streptomycetes are resistant to lysozyme or 15% KOH (Kats, 1963; Sohler et al., 1958; DeJong and McCoy, 1966). Qualitative analysis of the spore walls of different streptomycetes show them to contain essentially the same kinds of amino acids, amino sugars, and sugars as vegetative walls (DeJong and McCoy, 1966; Barabas and Szabo, 1965; Cummins and Harris, 1958). Dipicolinic acid has not been detected in spores of *Streptomyces* (Kalakoutskii and Agre, 1976; Kalakoutskii et al., 1969). Teichoic acids are associated with the cell wall (Naumova et al., 1980; Bekesy et al., 1966). Studies of mutants defective in teichoic acid biosynthesis suggest that these wall components play an important part in the normal development and differentiation of aerial hyphae (Chater and Merrick, 1979). A polysaccharide composed primarily of glucose and galactose is a part of the cell wall of *Streptomyces chrysomallus* (Streshinskaya et al., 1979), and a glycocalyx surrounds the hyphal wall of *Streptomyces aureofaciens* (Hostalek et al., 1981). Certain antibiotics have been claimed to be structural components in the cell walls of some streptomycetes (Barabas et al., 1980; Kalakoutskii and Agre, 1976), but the evidence for this is not strong.

The cytoplasmic membrane has a unit membrane structure. Mesosomes are routinely associated with the membrane in thin sections (Williams et al., 1973), but it is not clear what function mesosomes might have in streptomycetes. Recent findings suggest, however, that these structures may be artifacts of the fixation methods used for electron microscopy (Ebersold et al., 1981).

The nuclear material of streptomycetes is situated centrally in the cells and is distributed in discrete packets along the entire length of the hyphae (Chen, 1966; Hopwood and Glauert, 1960; Shamina, 1964a). Ribosomal particles are numerous and distributed uniformly through the cytoplasm. The cytoplasm contains areas that appear characteristically transparent in electron micrographs of thin sections. These areas apparently represent inclusions of carbon reserve materials, including lipids, polysaccharide, and poly-β-hydroxybutyrate (Williams et al., 1972a; Manzanal et al., 1981; Kannan and Rehacek, 1970).

Three distinct types of cross-wall formation occur in *Streptomyces* (Hardisson and Manzanal, 1976). One kind seems to be restricted to septum formation in vegetative hyphae (Williams et al., 1973), while all three types are involved in spore formation.

Spore formation is accompanied by a series of morphological events (Williams et al., 1973; Wildermuth and Hopwood, 1970; Hardisson and Manzanal, 1976). Sporogenesis begins with the simultaneous septation of aerial mycelium at regular intervals of 1–2 μm. The nuclear material becomes constricted at the septum sites, divides, and assumes a compact structure as septum formation proceeds. Polysaccharide granules similar in structure to glycogen (Brana et al., 1982) appear and then disappear as new wall material is added to the septa and inner spore wall, resulting in an overall thickened spore wall (Manzanal et al., 1981). The delineated spores assume a more ellipsoidal shape, and the outer hyphal wall disintegrates. The spores are maintained in chains by a fibrous outer sheath, which is synthesized by aerial mycelia prior to sporulation (Williams et al., 1972a; Rancourt and Lechevalier, 1964). The sheath confers on the spores a distinctive surface ornamentation, that is, spines, hairs, warts, or smoothness (Wildermuth, 1970b; Williams et al., 1972a; Dietz and Mathews, 1971; Hopwood and Glauert, 1961). Tenuously associated with the outer surface of the sheath is an unidentified lipid component, which appears to be responsible for the hydrophobicity exhibited by spores (R. A. Smucker, unpublished results).

The sheath is a fibrillar material (Wildermuth, 1970b), arranged in at least two distinct patterns (Matselyukh, 1978). It is composed of rodlike or spherical subunits that may or may not be soluble in different organic solvents, depending on the source of the sheath material (Kalakoutskii and Agre, 1976; Smucker and Pfister, 1978). The sheath composition also varies. In general, it contains only small amounts of lipid or protein (Cherny et al., 1974b; Bradshaw and Williams, 1976). Amino sugars, carbohydrates, fatty acids, and various inorganic elements have been reported (Bradshaw and Williams, 1976; Pozharitskaja et al., 1976), as well as a polyene antibiotic (Pozharitskaja et al., 1974) and chitin (Smucker and Pfister, 1978, and unpublished results). A self-assembly process may be involved in sheath formation (Cherny et al., 1974a).

There does not appear to be any remarkable difference in the fine structure of streptomycete spores and their parent hyphae (Chater and Hopwood, 1973; Kalakoutskii and Agre, 1976). The spore walls usually are 1.5–2.0 times thicker than the vegetative wall and may have ornamentation as noted earlier. The amount of water found associated with spores varies with the particular environmental conditions of the spores. Desiccated spores are reported to have 2–4% of their dry weight as firmly bound water (Lapteva et al., 1972). *Streptomyces streptomycini* spores contain more calcium, manganese, and potassium than vegetative mycelia (Pozharitskaja et al., 1973). Spores of *Streptomyces viridochromogenes* also have higher calcium levels than are normally found in vegetative bacterial cells (Eaton and Ensign, 1980). The amino acid composition of total protein hydrolysates of spores and vegetative mycelia of *Streptomyces venezuelae* are similar except for lower amounts of arginine and leucine detected in spores (Bradley and Ritzi, 1968). Trehalose and glycogen occur in *Streptomyces hygroscopicus* spores (Hey-Ferguson et al., 1973), and trehalose has been detected in spores of *S. viridochromogenes* (Ensign, 1978).

The DNA content of *Streptomyces* spores is somewhat higher than that of vegetative hyphae (Kalakoutskii and Agre, 1976), and each spore appears to contain 1–2 copies of the genome (Shamina, 1964b). The bouyant density of spore DNA is reported to be less than that of the DNA present in vegetative hyphae of *S. venezuelae* (Tewfik and Bradley, 1967). Spore DNA in this streptomycete is complexed with a yellow pigment and contains 10–15% protein.

The RNA content of spores may be lower or higher than that of vegetative mycelia (Kalakoutskii and Agre, 1976; Kalakoutskii and Pozharitskaja, 1973). The RNA of newly formed *S. venezuelae* spores hybridizes 100% with the RNA of vegetative mycelia; however, this decreases to only 30–60% with spores that are 8–10 days old (Enquist and Bradley, 1971). *Streptomyces griseus* and *Streptomyces granaticolor* spore ribosomes are more stable in salt solutions than those of the vegetative mycelia (Valu and Szabo, 1973; Mikulik et al., 1975). In both cases the spore ribosomes are found to be associated with tightly bound pigments. The pigment complexes with spore ribosomes and DNA may occur in vivo or may be artifacts of the extraction procedures used. Pigments do appear to be localized in the spore walls or outer sheath of most *Streptomyces* species. These pigments belong to a variety of chemical classes (Kutzner, 1981), and little is understood about their function in the spores.

During germination the spores swell; their nuclear material, plasma membrane, and ribosomes become more well defined; and the spore walls differentiate into two distinct layers (Bradley and Ritzi, 1968; Sharples and Williams, 1976; Hardisson et al., 1978). Germ tube emergence occurs by rupture and penetration of the outer wall layer. Following germ tube emergence, two nuclear regions are observed, one in the mother spore and the other in the germ tube (Hardisson et al., 1978).

IV. PHYSIOLOGY AND METABOLISM

Nutritionally, streptomycetes are classified as chemoheterotrophs. In general, they do not exhibit any special nutritional requirements, most being able to grow on media consisting of inorganic salts and glucose.

Nitrate and ammonia are utilized by many streptomycetes. Only sparse information exists concerning inorganic nitrogen assimilation and the regulatory mechanisms involved. Both glutamine synthetase and glutamate synthase have been reported in *Streptomyces clavuligerus* (Aharonowitz and Friedrich, 1980), and glutamine synthetase has been demonstrated in *Streptomyces cattleya* (Streicher and Tyler, 1981). The activity of the *S. cattleya* enzyme was observed to rapidly decline following exposure of cells to high exogenous ammonia levels (Wax et al., 1982). Further study showed that glutamine synthetase activity in *S. cattleya* is regulated via adenylylation of the enzyme in a manner similar to that found in other bacteria (Streicher and Tyler, 1981).

Glutamate dehydrogenase was not detected in extracts of *S. clavuligerus* cells grown at high ammonia levels; however, alanine dehydrogenase, which catalyzes the reductive amination of pyruvate to L-alanine, was found (Aharonowitz and Friedrich, 1980). Synthesis of this enzyme is induced by ammonia, and its specific activity is highest under growth conditions with excess ammonia. The alanine dehydrogenase of *S. clavuligerus* differs in its kinetic properties and molecular weight from the alanine dehydrogenases of other microorganisms.

Streptomycetes produce proteolytic exoenzymes (Nomoto et al., 1960) and readily grow on complex proteins. Amino acid transport in *Streptomyces hydrogenans* was found to be under both positive and negative feedback control (Gross et al., 1970; Ring et al., 1970). The modulation of amino acid transport activity in this streptomycete was shown to be related to the physiological age of the cells and the intracellular level of the amino acid pool (Langheinrich and Ring, 1976). Continuous culture experiments revealed that amino acid transport activity is closely correlated to growth rate and is regulated, at least partially, by changes in the concentrations of the components of the transport systems (Alim and Ring, 1976). Exogenous cyclic AMP was found to stimulate active amino acid uptake in resting cells of *S. hydrogenans*, apparently through an enhancement of de novo synthesis of certain transport enzymes (Ring et al., 1977a, b).

It is important to note, with regard to transport in streptomycetes, the existence of a "cold shock" phenomenon. Ring (1981) showed that mycelia subjected to a rapid shift to low temperature, for example, from 30°C down to -2°C within a few seconds, experience a reversible loss of control of membrane permeability as evidenced by massive leakage or uptake of small molecules. Analysis of the membrane lipid and phospholipid composition suggested that the loss of permeability control is the result of a lipid phase transition in the membrane to the solid gel state at low temperatures.

Little is known about nitrogen metabolism in *Streptomyces*. Kendrick and Wheelis (1982) showed that histidine dissimilation in *Streptomyces coelicolor* is inducible and occurs via N-formyl-L-glutamic acid using the same pathway as is used by some *Pseudomonas* species.

Sabater and Asensio (1973) were unable to detect the phosphoenolpyruvate-phosphotransferase system for the active transport of sugars in *Streptomyces violaceoruber*. Studies showed that glucose, mannose, and fructose are each transported by a specific, constitutive transport system and that the transport of mannose and fructose is of the passive type. Phosphorylation of these sugars occurs via specific kinases (Sabater et al., 1972a). The glucokinase of *S. violaceoruber* is a constitutive enzyme and similar in properties to the glucokinases known in other bacteria. The kinases for fructose and mannose are inducible and were characterized as unique enzymes (Sabater et al., 1972a, b; Sabater and Delafuente, 1975). *Streptomyces coelicolor* A3(2) possesses inducible active transport systems for arabinose and glycerol and constitutive active transport systems for glucose, galactose, and fructose. Glucose represses synthesis of the arabinose and glycerol active transport systems and inhibits the transport of galactose and fructose, as well as the passive diffusion of glycerol in this streptomycete (Hodgson, 1982).

Carbon-energy reserve materials are known to occur in a variety of streptomycetes (Elbein, 1967a; Kannan and Rehacek, 1970). Trehalose synthesis in nine different *Streptomyces* species was found to occur via a unique enzyme (Elbein, 1967b, 1968). The mobilization of trehalose occurs during spore germination of *S. hygroscopicus* (Hey-Ferguson et al., 1973). Trehalase has been characterized in *S. hygroscopicus* and shown to be present in seven other *Streptomyces* species (Hey and Elbein, 1968).

As in other bacteria, the transport of ions in streptomycetes may involve the activity of ATPase. Bansal et al. (1979) demonstrated a calcium activated ATPase in *S. griseus*. This enzyme exhibits a pH optimum of 8.5 at 37°C and its Ca^{2+} requirement can be met by Cd^{2+}, Zn^{2+}, and Mn^{2+}. Mg^{2+} inhibits the enzyme noncompetitively.

Early work using manometric and radiotracer techniques showed the presence of an operative tricarboxylic acid cycle in cells of *S. griseus* and demonstrated its central role in biosynthesis and respiration of this streptomycete (Gilmour et al., 1955). Anaplerotic fixation of CO_2 also occurs in *S. griseus* (Butterworth et al., 1955). Cytochromes *c*, *b*, and *a* were detected in mycelia of *Streptomyces fluorescens* and *Streptomyces albus* (Taptykova et al., 1969); however, little else is known about the composition or operation of the terminal respiratory pathways in streptomycetes.

Phosphate metabolism is intimately involved in the regulation of secondary metabolism (Martin et al., 1978; Ragan and Vining, 1978; Gersch et al., 1979b; Terry and Springham, 1981; Vu-Trong et al., 1980). How this regulation is mediated is still not clear. Even less is understood about phosphate metabolism in growing mycelia not producing secondary metabolites. Highly phosphorylated guanine and adenine nucleotides may have a role in the development of some streptomycetes (Stastna and Mikulik, 1981; Simuth et al., 1979). Phospholipid metabolism has been studied in *S. griseus* (Verma et al., 1980) and shown to involve an intracellular phospholipase C (Verma and Khuller, 1983).

Mizukami and Bradley (1966) described the first cell-free protein synthesis system in *Streptomyces*; since then, other in vitro systems have been developed (Jones, 1975; Valu and Szabo, 1978). Jones (1975) compared the efficiency of polyphenylalanine synthesis by preparations from young (12 h) and old (48 h) populations of *Streptomyces antibioticus*. He found that preparations from older cultures were much less efficient owing to decreased activity of the ribosomes and lower amino acid acceptor activity of the t-RNA. Valu and Szabo (1981), working with an in vitro system from *S. griseus*, also found that ribosomes isolated from older cultures are less efficient at polypeptide synthesis than those from young mycelia. Elongation factor EFTu of *Streptomyces collinus* has a molecular weight of 52,500 and shares some common immunochemical determinants with EFTu of *S. aureofaciens* and *Escherichia coli* (Mikulik et al., 1982). RNA polymerase of *S. antibioticus* is composed of subunits corresponding to the β, β', and α subunits of *E. coli* RNA polymerase; however, no band or activity corresponding to the σ subunit is detected in the streptomycete polymerase (Jones, 1979).

V. DEVELOPMENTAL BIOLOGY

A. Formation and Differentiation of Aerial Hyphae

The *Streptomyces* life cycle is shown schematically in Fig. 10.1. Upon encountering a suitable environment a spore germinates to produce a germ tube. The germ tube then develops into a branched mycelial network, which may penetrate and become firmly attached to a solid supporting matrix, for example, soil or agar. From this mycelial network arise hyphae, which grow up into the air and differentiate to form chains of spores. Wildermuth (1970a) has observed that the production of aerial hyphae and sporulation continue in waves until the mature colony consists of sporulated and unsporulated aerial hyphae, the remaining mycelia having lysed.

The distinguishing developmental features of streptomycetes are their growth as mycelial networks and reproduction via the formation of chains of aerial spores. The branched mycelia that comprise a young streptomycete colony generally are called vegetative, substrate, or primary mycelia. They give the colony a compact appearance and a tough, leathery consistency. Extension

Streptomyces Life Cycle

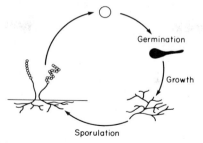

FIGURE 10.1   Schematic representation of the *Streptomyces* life cycle.

of the vegetative mycelia occurs by growth of the apical regions (Locci, 1981). The factors that govern branch formation by vegetative hyphae are not understood.

The aerial hyphae, also known as secondary or reproductive hyphae, arise from those vegetative mycelia nearest the surface of the colony (Wildermuth, 1970a). Aerial hyphae generally are thicker and exhibit much less branching than vegetative mycelia. They usually are gray, pigmented, and enclosed in a hydrophobic outer sheath (Higgins and Silvey, 1966). The latter feature is thought to protect aerial hyphae from desiccation (Kalakoutskii and Agre, 1976).

There is evidence that aerial hyphae are physiologically different from vegetative hyphae (Francisco and Silvey, 1971; Hopwood and Glauert, 1960; Giolitti, 1960). Functionally, aerial mycelia are very different from vegetative mycelia because they differentiate to form chains of spores. The formation of aerial hyphae and their differentiation to spores is greatly influenced by the nutritional conditions used to grow streptomycetes (Aharonowitz and Demain, 1979; Redshaw et al., 1976; Foor et al., 1982). Coleman and Ensign (1982) studied the nutritional control of aerial mycelia formation and sporulation in *S. viridochromogenes*. They found that 0.5% or higher concentrations of casein hydrolysate or various individual amino acids added to a glycerol-based chemically defined medium repressed aerial mycelium formation. The repression by casein hydrolysate was reversed by adding adenine, dimethyladenine, or 8-azaguanine and by the use of low levels (≤ 3 mM) of phosphate in the media.

The physiological basis for the nutritional regulation of aerial mycelia and spore formation in *Streptomyces* is not understood. It is clear that other factors besides nutritional ones are involved in regulating these developmental processes. Khokhlov et al. (1973) described a compound, A-factor (autoregulatory factor), produced by *S. griseus* that is required for the expression of the normal life cycle of this streptomycete (Kornitskaya and Tovarova, 1976). Low concentrations of A-factor, for example, 0.1 μg/ml, restore sporulation and antibiotic production to *S. griseus* mutants blocked in these processes

(Hara and Beppu, 1982a) and induce the synthesis of streptomycin-6-phosphotransferase (Hara and Beppu, 1982b). A-factor or A-factor-like substances are produced by a number of different species of *Streptomyces, Actinomyces,* and *Nocardia* (Hara and Beppu, 1982a; Efremenkova et al., 1979). A-factor is a low-molecular-weight (mw 242), heat-stable compound identified as 2S-isocaprylolyl-3R-oxymethyl-8-butyrolactone (Pliner et al., 1975; Kleiner et al., 1976) and has been synthesized chemically (Mori, 1981). Its mechanism of action is not known, although it has been shown to inhibit the activity of glucose-6-phosphate dehydrogenase (Voronia et al., 1975). Evidence suggests that the genes responsible for A-factor formation may be plasmid borne (Hara and Beppu, 1982a) in some species and chromosomal in others.

Factor C (cyto-differentiation factor) is another substance involved in the life cycle of *S. griseus.* Factor C induces sporulation of a nonsporulating mutant (Szabo et al., 1967a), binds to DNA, and stimulates (Szabo et al., 1967b) or inhibits (Szeszak and Szabo, 1973) RNA transcription by inhibiting repressor production or the binding of RNA polymerase. Factor C is a polypeptide of molecular weight 34,500 (Biro et al., 1980) and is calculated to be active at a level of about 10 molecules per genome.

Streptomyces alboniger produces a number of compounds that stimulate or inhibit aerial mycelium formation (Pogell et al., 1976; McCann et al., 1977). The stimulatory compound is an antibiotic called pamamycin (McCann and Pogell, 1979). It is a water-insoluble, heat-labile molecule of molecular weight 621 with the empirical formula $C_{36}H_{63}NO_7$. The antibiotic activity of this compound against *Staphylococcus aureus* is due to specific inhibition of the active transport of nucleosides, inorganic phosphate, and purine and pyrimidine bases (Chou and Pogell, 1981). The mechanism by which pamamycin stimulates aerial mycelia formation in *S. alboniger* is not known. Some results suggest the involvement of extrachromosomal DNA in the production of pamamycin and its regulation of aerial mycelia formation (Redshaw et al., 1979).

The occurrence of a high percentage of bald (no aerial mycelia) mutants following ethidium bromide treatment of *S. alboniger* suggested that plasmid DNA may carry the genes for aerial mycelia and spore production in this streptomycete (Redshaw et al., 1979). In contrast, bald mutants of *S. coelicolor* were all found to map on the chromosome (Merrick, 1976). Stable mutants of *S. coelicolor* that produce aerial mycelia but fail to sporulate also have been described. These mutants fall into six phenotypic categories, and at least eight genetic loci are identified, most being clustered in the same region of the chromosome (Chater, 1972, 1975; McVittie, 1974).

B. Sporulation

An interesting and perhaps useful system for studying developmental biology in *Streptomyces* has been described by Koepsel and Ensign (1980). They found conditions that allow *S. viridochromogenes* to complete its entire developmental cycle in liquid medium within 24 hours of spore germination and without extensive mycelial growth. In general, sporulation of streptomycetes in liquid submerged culture is not observed. It is clear, however, that some streptomycetes do sporulate under such conditions. The spores formed in submerged cultures are analogous to spores produced on aerial hyphae. Submerged spores are reported to be less resistant to high temperature (Kuimova, 1980) and to have different levels of some enzymes than aerial spores (Kalakoutskii and Douzha, 1967). Kendrick and Ensign (personal communication) made a systematic study of the formation and properties of spores produced in submerged cultures of S. griseus. They found that aerial and submerged spores were similar in trehalose content, endogenous respiration, spore proteins, germination, and resistance to sonication. Submerged spores were more sensitive to lysozyme than were aerial spores. The onset of sporulation of S. griseus in submerged culture correlated with either phosphate or ammonia depletion in the presence of exogenous glucose. Antibiotic production and extracellular protease activity were not required for sporulation, and glutamine synthetase activity was proposed as a sporulation marker under conditions of phosphate limitation.

C. Spore properties

Streptomycete spores serve the dual functions of dispersal and survival during times unfavorable for vegetative growth. Their resistant properties have long been indicated by the observation that well-sporulated cultures survive long periods of storage (Kalakoutskii and Pozharitskaja, 1973; Pridham et al., 1965, 1973). Not surprisingly, the ability of spores to survive is affected by the environmental conditions to which they are subjected. Moisture seems to be a critical factor; desiccated spores or spores stored at low relative humidities survive for years, whereas spores stored at high relative humidities suffer a rapid loss of viability (Lapteva et al., 1972). The effect of moisture on the survival of streptomycete spores may be related to its effect on their endogenous metabolism. Dry spores of S. streptomycini and S. viridochromogenes exhibit a very low level of respiration (Kalakoutskii et al., 1970; Hirsch and Ensign, 1978). Wetting the spores with buffer results in an increased respiration rate and a 4.5-fold increase in ATP content in the case of S. viridochromogenes. Thus the deleterious effect of moisture on spore survival may be due to the increased metabolism of endogenous spore components.

Spores of Streptomyces are only slightly more resistant to heat than vegetative mycelia (Ebner and Frea, 1970). It is not clear that water enhances the lethal effects of heat on spores. Aqueous suspensions of spores of four different streptomycetes were killed by heating to 60–70°C for 10 min (Dorokhova et al., 1970). However, spores of S. streptomycini survived 30 min of exposure to dry heat or Vaseline oil at 120°C and undecane at 110°C (Aslanyan et al., 1971). Most experiments designed to examine the effect of heat on spores do so by measuring the difference in colony-forming units between unheated and heated spores. Lapteva et al. (1976) found that *S. strep-tomycini* spores heated to 100°C for 10 min in water do not produce colonies when plated immediately after heating. After incubation of the heated spore suspension at 28°C for several days, however, almost all of the original colony-forming units were recovered. Aerial spores of *S. chrysomallus* behaved similarly following heating and storage at 28°C (Kuimova, 1980). Recovery of heated *S. streptomycini* spores occurred if they were stored at 28° or 37°C but not at 9°, 10°, or 45°C. The basis for the recovery from heat exposure is not understood.

The metabolic activity of *Streptomyces* spores is much lower than that of vegetative mycelia (Ensign, 1978). Spores contain certain enzymes involved in the catabolism of glucose and various phosphorus- and nitrogen-containing compounds (Kalakoutskii and Douzha, 1967; Kalakoutskii and Agre, 1976) as well as cytochromes *a*, *b*, and *c* (Taptykova et al., 1969). *S. viridochromogenes* spores appear to possess a complete and functional respiratory system, since wetting the spores results in increased respiration and the generation of ATP (Hirsch and Ensign, 1978). Moreover, dormant spores can oxidize certain sugars, fatty acids, and amino acids (J. C. Ensign, personal communication) and incorporate exogenous precursors into protein and RNA (Kalakoutskii et al., 1966; Hardisson et al., 1980). Dormant streptomycete spores thus appear to be freely permeable to water and various small molecules, to have the capacity to oxidize endogenous and exogenous substrates with the formation of ATP, and to possess a functional system for RNA and protein synthesis.

D. Germination

Studies of the transition from the dormant spore to the vegetative state have been complicated by the use of different definitions and terms to describe the stages in the transition process (Ensign, 1978). For purposes of this discussion, we have chosen to separate the transition into the distinct stages of activation, initiation, germination, and outgrowth.

Activation of streptomycete spores poises them for initiation. Activation is manifested as the loss of a lag period preceding the onset of germination events and as increased synchrony of initiation in the spore population (Ensign, 1978). Activation of *S. viridochromogenes* spores occurs by heating (Hirsch and Ensign, 1976b) or by treating with certain detergents (Grund and Ensign, 1982). Both heat- and detergent-activated spores show increased rates of endogenous respiration and are not affected by a germination inhibitor (see below). Activated spores are deactivated, that is, exhibit a lag period and asynchronous initiation, by incubation in buffer for 8–12 hours at 25°C. The mechanism of activation of *S. viridochromogenes* spores is not understood. Germination of *S. granaticolor* spores was inhibited by the heat activation conditions described for *S. viridochromogenes* (Mikulik et al., 1977), while the germination of *S. antibioticus* spores was not affected by heat treatment (Hardisson et al., 1978).

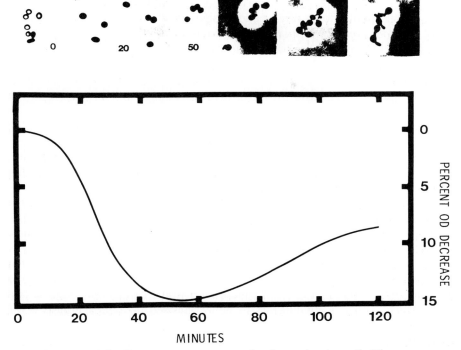

FIGURE 10.2   *Streptomyces* spore germination and outgrowth. The curve shows the percent decrease in absorbance measured at 600 nm of a *S. viridochromogenes* spore suspension incubated at 30 C in 0.1% yeast extract. The series of photographs shows the sequential morphological changes undergone by the spores during germination and outgrowth. The numbers on the photographs correspond to the minutes of incubation in 0.1% yeast extract at 30 C.

Dormant streptomycete spores appear refractile when examined with phase contrast optics. As germination proceeds, the appearance of the spores changes as they lose refractility and swell slightly (Fig. 10.2) (Atwell and Cross, 1973; Hirsch and Ensign, 1976a; Hardisson et al., 1978). These morphological changes are accompanied by a 15–30% decrease in the absorbance of a spore suspension (Fig. 10.2). Outgrowth of spores is signaled by the appearance of germ tubes and is accompanied by a slight increase in the absorbance of a germinated spore suspension (Fig. 10.2).

Media that support spore germination can be simple, consisting of a few amino acids, a carbon energy source, and some inorganic salts. Little is known about the events involved in triggering germination in streptomycete spores. Initiation of *S. viridochromogenes* spores is absolutely dependent on calcium ions and carbon dioxide (Hirsch and Ensign, 1976a). Spores of *S. antibioticus* initiate germination in the presence of only calcium, magnesium, or iron ions

(Hardisson et al., 1978). The role of metal ions in triggering germination in these spores is not known. Eaton and Ensign (1980) studied calcium initiation of *S. viridochromogenes* spores. They found that nearly all the calcium is bound external to the spore membrane and that it must be continuously present following initiation for germination to proceed normally. The carbon dioxide requirement for initiation of *S. viridochromogenes* spores was traced to the need for anaplerotic reactions to maintain operation of the TCA cycle during germination (Grund and Ensign, 1978).

Germination is an energy-requiring process (Hirsch and Ensign, 1978; Hardisson et al., 1978). Endogenous reserves of trehalose as well as exogenous substrates are oxidized during germination (Ensign, 1978; Hey-Ferguson et al., 1973). The ATP content of germinating spores can be as great as 20 times that found in dormant spores (Hirsch and Ensign, 1978).

During germination of *S. viridochromogenes* spores, about 25% of the total spore carbon is released into the medium (Hirsch and Ensign, 1978). Nothing is known about the nature of this material except that it contains a compound(s) which inhibits germination.

It has been shown with at least four different *Streptomyces* species that germination is a time of rapid biosynthesis of RNA and protein (Hirsch and Ensign, 1978; Mikulik et al., 1977; Hardisson et al., 1978, 1980; Nagatsu and Matsuyama, 1970). RNA appears to be synthesized first, its synthesis being detected at the onset or within a few minutes of germination. Guijarro et al. (1982) found that as much as 80% of the RNA synthesized by spores of *S. antibioticus* during the first 15 min of germination is stable. This percentage decreased to 60% after one hour. The half-life of the unstable RNA (m-RNA) also was found to be greatest early in germination and then to decline as germination progressed.

Depending on the germination system studied, protein synthesis begins at the same time as RNA synthesis or anywhere from a few minutes to as late as five hours after the onset of RNA synthesis. The variation in the timing of protein synthesis most likely reflects poor permeability of some of the radioactive precursors used as well as failure to account for metabolism of the precursors (Ensign, 1978). The composition of the germination medium can have profound effects on the kinetics of precursor incorporation (Hardisson et al., 1980).

Net DNA synthesis starts at the beginning of the outgrowth (germ tube emergence) stage. Spores incubated with inhibitors of DNA synthesis undergo normal germination but cease development when they have short germ tubes.

Dormant streptomycete spores demonstrate a capacity for rather active metabolism (see above). Germination of the spores does not begin, however, until they are provided with the proper initiation conditions. It would seem then that some type of endogenous mechanism is in place to maintain the spores in a dormant condition. The existence, let alone nature, of such a mechanism has not been shown. There is evidence, however, that certain small molecules may be involved in maintaining dormancy or regulating germination. Spores of *S. hygroscopicus* are inhibited from germinating by adding 5–10 mM cyclic AMP (Gersch and Strunk, 1980). Electron micrographs of these spores showed that the surface sheath and outer layers of the spore wall had fractured but that no further transition to the vegetative form had occurred. Cyclic AMP is well known as a biological regulator and has been implicated in the regulation of metabolism and differentiation in various streptomycetes (Gersch, 1980; Gersch et al., 1979a; Ragan and Vining, 1978). Perhaps it also is involved in regulating streptomycete spore germination.

The germination inhibitor associated with *S. viridochromogenes* spores is another example of a possible regulator of dormancy in spores (Ensign, 1978). This compound is excreted from germinating spores and inhibits germination when added to spores. The structure of the inhibitor is not known, but it is a low-molecular-weight molecule that is unaffected by proteases or nucleases and cannot be extracted from dormant spores. It is calcium activated and inhibits membrane-bound ATPase activity and respiration. Activated spores or spores that have initiated germination are not affected by the inhibitor, a result suggesting that it acts on a target that is involved in initiation or at a very early time in germination. The germination inhibitor has antibiotic activity against some gram-positive bacteria.

VI. GENETICS

The genetic material of *Streptomyces* is arranged on a single circular chromosome (Hopwood and Merrick, 1977). In *S. coelicolor* A3(2), the only species with well-established genetics, approximately 100 genes have been recognized, representing auxotrophic, drug-resistant, temperature-sensitive, and morphological mutations. The linkage map of *S. coelicolor* A3(2) is shown in Fig. 10.3. Many streptomycetes also contain one or more pieces of extrachromosomal DNA. These plasmids exist in a wide range of size and copy number and have been implicated in the control of many diverse functions, including fertility, antibiotic biosynthesis and resistance, pigment and exoenzyme production, and differentiation (Chater, 1979; Baumann and Kocher, 1976).

Genetic exchange in *Streptomyces* is thought to be accomplished primarily through conjugation. In *S. coelicolor* A3(2), conjugation is facilitated by sex factor plasmids SCP1 (Hopwood et al., 1973), and SCP2 (Bibb et al., 1977). Other known fertility plasmids include the SLP1 family of *S. lividans* (Bibb et al., 1980) and SRP1 of *S. rimosus* (Friend et al., 1978). These fertility plasmids, and many other streptomycete plasmids, exhibit a phenomenon called lethal zygosis (Bibb et al., 1977). The transfer of plasmid into a recipient strain results in a clear zone (pock) of growth inhibition in the recipient lawn. This property has proven very useful in genetic studies of streptomycetes because it allows detection of plasmid transformation events.

Because of the mycelial growth pattern of streptomycetes, most systems for interrupting mating or isolating zygotes are not very useful. Chromosomal

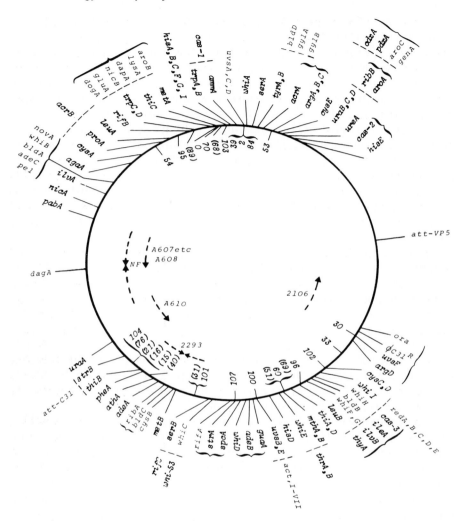

FIGURE 10.3   Linkage map of *Streptomyces coelicolor* A3(2). Arrows indicate donation of markers by bidirectional or unidirectional donors. (Used by permission of Prof. D. A. Hopwood.)

linkage mapping is accomplished by analyzing the progeny of crosses carried out on solid substrates. The parental strains are mixed and allowed to grow and sporulate, usually on a nonselective medium. The haploid recombinant spores are then analyzed to determine genotypes. The original linkage maps of *S. coelicolor* (Hopwood, 1959), *S. rimosus* (Friend and Hopwood, 1971), *S. glaucescens* (Baumann et al., 1974), and *S. acrimycini* (Wright and Hopwood, 1977) were constructed by this technique. The progeny of four-factor crosses were analyzed by selecting colonies on four different media, each selecting for a different pair of markers, one from each parent. This procedure leaves two markers (or four genotypes) unselected and allows the recovery of nine out of 14 genotypes of recombinant progeny, including two complementary pairs. Comparison of the frequencies of recombinant genotypes provides enough information for the estimation of relative linkage distances between all pairs of markers on a circular linkage map (Hopwood, 1959).

Genetic analysis of streptomycetes can be complicated by the frequent occurrence of heterokaryons and heteroclones. The heterokaryons are not very useful, since they constantly segregate into the parental genotypes during sporulation. Heteroclones are essentially partial diploids, containing mixtures of recombinant and parental genotypes. Subsequent growth allows crossing over at various points within the duplicated segment, generating a family of recombinant spores that can be useful in calculating map intervals (Hopwood and Sermonti, 1962).

In other species, especially those exhibiting an apparent lack of natural fertility, recently developed techniques of protoplast fusion (Hopwood et al., 1977; Baltz, 1978) should aid in the development of genetic systems. The cell walls of most streptomycetes can be weakened by growth in a medium supplemented with glycine and then readily digested with lysozyme. The resulting protoplasts can be mixed, treated briefly with polyethylene glycol, and allowed to regenerate on a hypertonic medium. Studies by Hopwood and Wright (1978) concluded that the primary fusion products contain essentially complete genomes of both parents. During the growth on regeneration plates, several rounds of recombination can occur; but by the time sporulation occurs, the process is complete, and all the spores behave as pure haploids.

The availability of protoplasts has also made systems for transformation and transfection available, thus opening the way for gene amplification and cloning systems and more sophisticated fine structure mapping. Vector DNA can be introduced into streptomycete protoplasts by the use of polyethylene glycol 1000 (Bibb et al., 1978) resulting in high-frequency transformation. Since the development of this technique, a number of potential plasmid vectors have been developed. To be useful, the plasmids must have a broad host range or be readily modified to broad host range. There is now available in streptomycete plasmids the potential to construct vectors with selectable markers, high or low copy number, narrow or broad host range, and conjugal or nonconjugal transfer. Vectors currently in use include SCP2* (Bibb et al., 1977), SLP1.2 (Bibb et al., 1980), and pIJ101 (Kieser et al., 1982). Chromosomal DNA can also be introduced into streptomycete protoplasts by polyethylene glycol–mediated fusion with lipid vesicles (Makins and Holt, 1981).

Streptomycete phage occur widely in nature and can be readily isolated from soil (Lanning and Williams, 1982). Transfection and bacteriophage-mediated transduction of protoplasts promoted by polyethylene glycol is emerging as an important genetic tool. Temperate phage with broad host range and a limited number of restriction enzyme target sites in their DNA are the best potential cloning vectors. Several described phage, including ϕC31, R4, and SH 10 meet these criteria (Lomovskaya et al., 1980).

Protoplast-facilitated recombination, gene cloning, and conjugation are complementary tools in the study of streptomycete genetics and will be useful in unraveling the mysteries of differentiation and antibiotic production in these remarkable bacteria.

ACKNOWLEDGMENTS

We wish to thank Dr. Kathy Kendrick, Prof. Jerry Ensign, and Prof. R. A. Smucker for kindly providing us with some of their unpublished results.

REFERENCES

Aharonowitz, Y., and Demain, A. L. (1979) *Can. J. Microbiol. 25*, 61–67.

Aharonowitz, Y., and Friedrich, C. G. (1980) *Arch. Microbiol. 125*, 137–142.

Alim, S., and Ring, K. (1976) *Arch. Microbiol. 111*, 105–110.

Arima, K., Nakamura, T., and Tamura, G. (1968) *Agr. Biol. Chem. 32*, 530–531.

Aslanyan, R. R., Agre, N. S., Kalakoutskii, L. V., and Kirillova, I. P. (1971) *Mikrobiologiya 40*, 293–296.

Atwell, R. W., and Cross, T. (1973) in *Actinomycetales: Characteristics and Practical Importance* (Sykes, G., and Skinner, F. A., eds.), pp. 197–206, Academic Press, New York.

Baltz, R. H. (1978) *J. Gen Microbiol. 107*, 93–102.

Bansal, V. S., Verma, J. N., Mahmood, A., and Khuller, G. K. (1979) *J. Gen. Microbiol. 112*, 393–395.

Barabas, G., and Szabo, G. (1965) *Arch. Mikrobiol. 50*, 156–163.

Barabas, G., Szabo, I., Ottenberger, A., Zs-Nagy, V., and Szabo, G. (1980) *Can. J. Microbiol. 26*, 141–145.

Baumann, R., and Kocher, H. P. (1976) in *Second International Symposium on the Genetics of Industrial Microorganisms* (MacDonald, K. D., ed.), pp. 535–551, Academic Press, London.

Baumann, R., Hutter, R., and Hopwood, D. A. (1974) *J. Gen. Microbiol. 81*, 463–474.

Bekesy, J., Szabo, G., and Valu, G. (1966) *Arch. Mikrobiol. 54*, 125–128.

Bibb, M. J., Freeman, R. F., and Hopwood, D. A. (1977) *Mol. Gen. Genet. 154*, 155–166.

Bibb, M. J., Ward, J. M., and Hopwood, D. A. (1978) *Nature* (London) *274*, 398–400.

Bibb, M. J., Ward, J. M., and Hopwood, D. A. (1980) *Dev. Ind. Microbiol. 21*, 55–64.

Biro, S., Bekesi, I., Vitalis, S., and Szabo, G. (1980) *Eur. J. Biochem. 103*, 359–363.

Bradley, S. G., and Ritzi, D. (1968) *J. Bacteriol. 95*, 2358–2364.

Bradshaw, R. M., and Williams, S. T. (1976) *Microbios 15*, 57–65.

Brana, A. F., Manzanal, M. B., and Hardisson, C. (1982) *Can. J. Microbiol 28*, 1320–1323.

Butterworth, E. M., Gilmour, C. M., and Wang, C. H. (1955) *J. Bacteriol. 69*, 725–727.

Chater, K. F. (1972) *J. Gen. Microbiol. 72*, 9–28.

Chater, K. F. (1975) *J. Gen. Microbiol. 87*, 312–325.

Chater, K. F. (1979) in *Genetics of Industrial Microorganisms* (Sebek, O. K., and Laskin, A. I., eds.), pp. 123–133, American Society for Microbiology, Washington, D.C.

Chater, K. F., and Hopwood, D. A. (1973) *Symp. Soc. Gen. Microbiol. 23*, 143–160.

Chater, K. F., and Merrick, M. J. (1979) in *Developmental Biology of Prokaryotes* (Parish, J. H., ed.), pp. 93–114, University of California Press, Berkeley.

Chen, P. L. (1966) *Am. J. Bot. 53*, 291–298.

Cherny, N. E., Tikhonenko, A. S., and Kalakoutskii, L. V. (1974a) *Arch. Microbiol. 101*, 71–82.

Cherny, N. E., Tikhonenko, A. S., Kuroda, S., and Kalakoutskii, L. V. (1974b) *Microbios 10*, 7–14.

Chou, W.-G., and Pogell, B. M. (1981) *Biochem. Biophys. Res. Comm. 100*, 344–350.

Coleman, R. H., and Ensign, J. C. (1982) *J. Bacteriol. 149*, 1102–1111.

Crawford, D. L., Barder, M. J., Pometto, A. L., and Crawford, R. L. (1982) *Arch. Microbiol. 131*, 140–145.

Cross, T. (1981) *J. Appl. Bacteriol. 50*, 397–423.

Cummins, C. S., and Harris, H. (1958) *J. Gen. Microbiol. 18*, 173–189.

Davies, F. L., and Williams, S. T. (1970) *Soil Biol. Biochem. 2*, 227–238.

DeJong, P. J., and McCoy, E. (1966) *Can. J. Microbiol. 12*, 985–994.

Dietz, A., and Mathews, J. (1971) *Appl. Microbiol. 21*, 527–533.

Dorokhova, L. A., Agre, N. S., Kalakoutskii, L. V., and Krassilnikov, N. A. (1970) in *The Actinomycetales* (Prauser, H., ed.), pp. 227–232, Gustav Fischer, New York.

Eaton, D., and Ensign, J. C. (1980) *J. Bacteriol. 143*, 377–382.

Ebersold, H. R., Cordier, J. L., and Luthy, P. (1981) *Arch. Microbiol. 130*, 19–22.

Ebner, J. B., and Frea, J. I. (1970) *Microbios 5*, 43–48.

Efremenkova, O. V., Anisova, L. N., and Khokhlov, A. S. (1979) *Mikrobiologiya 48*, 999–1003.

Elbein, A. D. (1967a) *J. Bacteriol. 94*, 1520–1524.

Elbein, A. D. (1967b) *J. Biol. Chem. 242*, 403–406.

Elbein, A. D. (1968) *J. Bacteriol. 96*, 1623–1631.

Enquist, L. W., and Bradley, S. G. (1971) *Dev. Ind. Microbiol. 12*, 225–234.

Ensign, J. C. (1978) *Ann. Rev. Microbiol. 32*, 185–215.

Foor, F., Tyler, B., and Morin, N. (1982) *Dev. Ind. Microbiol. 23*, 305–313.

Francisco, D. E., and Silvey, J. K. G. (1971) *Can. J. Microbiol. 17*, 347–351.

Friend, E. J., and Hopwood, D. A. (1971) *J. Gen. Microbiol. 68*, 187–197.

Friend, E. J., Warren, M., and Hopwood, D. A. (1978) *J. Gen. Microbiol. 106*, 201–206.

Gersch, D. (1980) *Proc. Biochem. 15*, 21–25.

Gersch, D., and Strunk, C. (1980) *Curr. Microbiol. 4*, 271–275.

Gersch, D., Romer, W., and Krugel, H. (1979a) *Experientia 35*, 749–751.

Gersch, D., Skurk, A., and Romer, W. (1979b) *Arch. Microbiol. 121*, 91–96.

Gilmour, C. M., Butterworth, E. M., Noble, E., and Wang, C. H. (1955) *J. Bacteriol. 69*, 719–724.

Giolitti, G. (1960) *J. Gen. Microbiol. 23*, 83–86.

Gross, W., Ring, K., and Heinz, E. (1970) *Arch. Biochem. Biophys. 137*, 253–261.

Grund, A. D., and Ensign, J. C. (1978) *Arch. Microbiol. 118*, 279–288.

Grund, A. D., and Ensign, J. C. (1982) *Curr. Microbiol. 7*, 223–228.

Guijarro, J. A., Suarez, J. E., Salas, J. A., and Hardisson, C. (1982) *FEMS Microbiol. Lett. 14*, 205–209.

Hagedorn, C. (1976) *Appl. Environ. Microbiol. 32*, 368–375.

Hara, O., and Beppu, T. (1982a) *J. Antibiot. 35*, 349–356.

Hara, O., and Beppu, T. (1982b) *J. Antibiot. 35*, 1208–1214.

Hardisson, C., and Manzanal, M. B. (1976) *J. Bacteriol. 127*, 1443–1454.

Hardisson, C., Manzanal, M. B., Salas, J. A., and Suarez, J. E. (1978) *J. Gen. Microbiol. 105*, 203–214.

Hardisson, C., Salas, J. A., Guijarro, J. A., and Suarez, J. E. (1980) *FEMS Microbiol. Lett. 7*, 233–235.

Hey, A. E., and Elbein, A. D. (1968) *J. Bacteriol. 96*, 105–110.

Hey-Ferguson, A., Mitchell, M., and Elbein, A. D. (1973) *J. Bacteriol. 116*, 1084–1085.

Higgins, M. L., and Silvey, J. K. G. (1966) *Trans. Amer. Microsc. Soc. 85*, 390–395.

Hirsch, C. F., and Christensen, D. L. (1983) *Appl. Environ. Microbiol. 46*, 925–929.

Hirsch, C. F., and Ensign, J. C. (1976a) *J. Bacteriol. 126*, 13–23.

Hirsch, C. F., and Ensign, J. C. (1976b) *J. Bacteriol. 126*, 24–30.

Hirsch, C. F., and Ensign, J. C. (1978) *J. Bacteriol. 134*, 1056–1063.

Hodgson, D. A. (1982) *J. Gen. Microbiol. 128*, 2417–2430.

Hopwood, D. A. (1959) *Ann. N.Y. Acad. Sci. 81*, 887–898.

Hopwood, D. A., and Glauert, A. M. (1960) *J. Biophys. Biochem. Cytol. 8*, 267–278.

Hopwood, D. A., and Glauert, A. M. (1961) *J. Gen. Microbiol. 26*, 325–330.

Hopwood, D. A., and Merrick, M. J. (1977) *Bacteriol. Rev. 41*, 595–635.

Hopwood, D. A., and Sermonti, G. (1962) *Adv. Genet. 11*, 273–342.

Hopwood, D. A., and Wright, H. M. (1978) *Mol. Gen. Genet. 162*, 307–317.

Hopwood, D. A., Chater, K. F., Dowding, J. E., and Vivian, A. (1973) *Bacteriol. Rev. 37*, 371–405.

Hopwood, D. A., Wright, H. M., Bibb, M. J., and Cohen, S. N. (1977) *Nature* (London) *268*, 171–174.

Hostalek, Z., Jechova, V., Curdova, E., and Vorisek, J. (1981) in *Actinomycetes* (Schaal, K. P., and Pulverer, G., eds.), 281–285, Gustav Fischer, New York.

Hunter, J. C., Eveleigh, D. E., and Casella, G. (1981) in *Acetinomycetes* (Schaal, K. P., and Pulverer, G., eds.), pp. 195–200, Gustav Fischer, New York.

Johnstone, D. B. (1947) *Soil Sci. 64*, 453–458.

Jones, G. H. (1975) *J. Bacteriol. 124*, 364–372.

Jones, G. H. (1979) *Arch. Biochem. Biophys. 198*, 195–204.

Kaiser, P. (1971) *Ann. Inst. Pasteur, Paris 121*, 389–404.

Kalakoutskii, L. V., and Agre, N. S. (1976) *Bacteriol. Rev. 40*, 469–524.

Kalakoutskii, L. V., and Douzha, M. I. (1967) *Mikrobiologiya 36*, 279–283.

Kalakoutskii, L. V., and Pozharitskaja, L. M. (1973) in *Actinomycetales: Characteristics and Practical Importance* (Sykes, G., and Skinner, F. A., eds.), pp. 155–171, Academic Press, New York.

Kalakoutskii, L. V., Bobkova, E. A., and Krassilnikov, N. A. (1966) *Dokl. Akad. Nauk SSSR 170*, 705–707.

Kalakoutskii, L. V., Agre, N. S., and Aslanjan, R. R. (1969) *Dokl. Akad. Nauk SSSR 184*, 1214–1216

Kalakoutskii, L. V., Mukhin, L. M., Lapteva, E. A., Taptykova, S. D., Skovortsova, I. N., and Douzha, M. V. (1970) *Izv. Akad. Nauk SSSR Ser. Biol. Nauk. 4*, 593–599.

Kannan, L. V., and Rehacek, Z. (1970) *Ind. J. Biochem. 7*, 126–129.

Kats, L. N. (1963) *Mikrobiologiya 32*, 459–464.

Katznelson, H., and Cole, S. E. (1965) *Can. J. Microbiol. 11*, 733–742.

Kendrick, K. E., and Wheelis, M. L. (1982) *J. Gen. Microbiol. 128*, 2029–2040.

Khan, M. R., and Williams, S. T. (1975) *Soil Biol. Biochem. 7*, 345–348.

Khokhlov, A. S., Anisova, L. N., Tovarova, I. I., Kleiner, E. M., Kovalenko, I. V., Krasil'nikova, O. I., Kornitskaya, E. Ya., and Pliner, S. A. (1973) *Z. Allg. Mikrobiol. 13*, 647–655.

Kieser, T., Hopwood, D. A., Wright, H. M., and Thompson, C. J. (1982) *Mol. Gen. Genet. 185*, 223–228.

Kleiner, E. M., Pliner, S. A., Soifer, V. S., Onoprienko, V. V., Balashova, T. A., Rozynov, B. V., and Khokhlov, A. S. (1976) *Bioorg. Chem. 2*, 1142–1147.

Koepsel, R., and Ensign, J. C. (1980) in *Abstracts of the Annual Meeting of the American Society for Microbiology*, I 130, p. 106, American Society for Microbiology, Washington, D.C.

Kornitskaya, E. Ya., and Tovarova, I. I. (1976) *Mikrobiologiya 45*, 302–305.

Kuimova, T. F. (1980) *Mikrobiologiya 49*, 74–77.

Kutzner, H. G. (1981) in *The Prokaryotes* (Starr, M. P., Stolp, H., Truper, H. G., Balows, A., and Schlegel, H. G., eds.), pp. 2028–2082, Springer, New York.

Lacey, J. (1973) in *Actinomycetales: Characteristics and Practical Importance* (Sykes, G., and Skinner, F. A., eds.), pp. 231–246, Academic Press, New York.

Lacey, J. (1978) in *Nocardia and Streptomyces* (Mordarski, M., Kurylowicz, W., and Jeljaszewicz, J., eds.), pp. 161–168, Gustav Fischer, New York.

Langheinrich, W., and Ring, K. (1976) *Arch. Microbiol. 109*, 227–235.

Lanning, S., and Williams, S. T. (1982) *J. Gen. Microbiol. 128*, 2063–2071.

Lapteva, E. A., Kusnetsov, V. D., and Kalakoutskii, L. V. (1972) *Mikrobiologiya 41*, 845–848.

Lapteva, E. A., Agre, N. S., and Kalakoutskii, L. V. (1976) *Mikrobiologiya 45*, 559–561.

Lapwood, D. H. (1973) in *Actinomycetales: Characteristics and Practical Importance* (Sykes, G., and Skinner, F. A., eds.), pp. 253–259, Academic Press, New York.

Lechevalier, M. P. (1981) in *Actinomycetes* (Schaal, K. P., and Pulverer, G., eds.), pp. 159–164, Gustav Fischer, New York.

Leyh-Bouille, M., Bonaly, R., Ghuysen, J. M., Tinelli, R., and Tipper, D. (1970) *Biochemistry* (New York) *9*, 2944–2952.

Lloyd, A. B. (1969a) *J. Gen. Microbiol. 56*, 165–170.

Lloyd, A. B. (1969b) *J. Gen. Microbiol. 57*, 35–40.

Lloyd, A. B., and Lockwood, J. L. (1966) *Phytopathol. 53*, 917–920.

Locci, R. (1981) in *Actinomycetes* (Schaal, K. P., and Pulverer, G., eds.), pp. 119–128, Gustav Fischer, New York.

Lockwood, J. L., and Lingappa, B. T. (1963) *Phytopathology 53*, 917–920.

Lomovskaya, N. D., Chater, K. F., and Mkrtumian, N. M. (1980) *Microbiol. Rev. 44*, 206–229.

Makins, J. F., and Holt, G. (1981) *Nature 293*, 671–673.

Manzanal, M. B., Brana, A. F., and Hardisson, C. (1981) in *Actinomycetes* (Schaal, K. P., and Pulverer, G., eds.), pp. 147–151, Gustav Fischer, New York.

Martin, J. F., Liras, P., and Demain, A. L. (1978) *Biochem. Biophys. Res. Comm. 83*, 822–828.

Matselyukh, B. P. (1978) in *Nocardia and Streptomyces* (Mordarski, M., Kurylowicz, W., and Jeljaszewicz, J., eds.), pp. 235–239, Gustav Fischer, New York.

Mayfield, C. I., Williams, S. T., Ruddick, S. M., and Hatfield, H. L. (1972) *Soil Biol. Biochem. 4*, 79–89.

McCann, P. A., and Pogell, B. M. (1979) *J. Antibiot. 32*, 673–678.

McCann, P., Redshaw, P. A., and Pogell, B. M. (1977) in *Abstracts of the Annual Meeting of the American Society of Microbiology*, I 129, p. 176, American Society of Microbiology, Washington, D.C.

McVittie, A. (1974) *J. Gen. Microbiol. 81*, 291–302.

Meiklejohn, J. (1957) *J. Soil Sci. 8*, 240–247.

Merrick, M. J. (1976) *J. Gen. Microbiol. 96*, 299–315.

Mikulik, K., Janda, I., Ricicova, A., and Vinter, V. (1975) in *Spores VI* (Gerhardt, P., Costilow, R. N., and Sadoff, H. L., eds.), pp. 15–27, American Society of Microbiology, Washington, D.C.

Mikulik, K., Janda, I., Maskova, H., Stastna, J., and Jiranova, A. (1977) *Folia Microbiol. 22*, 252–261.

Mikulik, K., Weiser, J., and Haskova, D. (1982) *Biochem. Biophys. Res. Comm. 108*, 861–867.

Mitchell, R. (1963) *Phytopathol. 53*, 1068–1071.

Mizukami, I., and Bradley, S. G. (1966) *Dev. Ind. Microbiol. 7*, 326–334.

Mori, K. (1981) *Tetrahedron Lett. 22*, 3431–3432.

Nagatsu, C., and Matsuyama, A. (1970) *Agr. Biol. Chem. 34*, 860–869.

Nakamura, T., Tamura, G., and Arima, K. (1967) *J. Ferment. Technol. 45*, 869–878.

Nakamura, T., Tamura, G., and Arima, K. (1977) *Agr. Biol. Chem. 41*, 763–768.

Naumova, I. B., Kuznetsov, V. D., Kudrina, K. S., and Bezzubenkova, A. P. (1980) *Arch. Microbiol. 126*, 71–75.

Nomoto, M., Narahashi, Y., and Murakami, M. (1960) *J. Biochem. 48*, 593–601.

Okafor, N. (1966) *Plant and Soil 43*, 211–235.

Okami, Y., and Okazaki, T. (1978) in *Nocardia and Streptomyces* (Mordarski, M., Kurylowicz, W., and Jeljaszewicz, J., eds.), pp. 145–151, Gustav Fischer, New York.

Person, L. H., and Martin, W. J. (1940) *Phytopathol. 30*, 913–926.

Pliner, S. A., Kleiner, E. M., Kornitskaya, E. Ya., Tovarova, I. I., Rozynov, B. V., Smirnova, G. M., and Khokhlov, A. S. (1975) *Bioorg. Chem. 1*, 70–76.

Pogell, B. M., Sankaran, L., Redshaw, P. A., and McCann, P. A. (1976) in *Microbiology 1976* (Schlessinger, D., ed.), pp. 543–547, American Society of Microbiology, Washington, D.C.

Pozharitskaja, L. M., Aslanjan, R. R., Lebedev, V. I., and Kalakoutskii, L. V. (1973) *Mikrobiologiya 42*, 354–355.

Pozharitskaja, L. M., Taptykova, S. D., Sarkisijan, S. T., Tulskii, S. V., Cherny, N. E., and Kalakoutskii, L. V. (1974) *Antiobiotiki 19*, 963–966.

Pozharitskaja, L. M., Taptykova, S. D., Cherny, N. E., Tulskii, S. V., Lebedev, V. I., and Kalakoutskii, L. V. (1976) *Antibiotiki 21*, 787–791.

Pridham, T. G., Lyons, A. G., and Seckinger, H. L. (1965) *Int. Bull. Bacteriol. Nomencl. Taxon. 15*, 191–237.

Pridham, T. G., Lyons, A. J., and Phrompatima, B. (1973) *Appl. Microbiol. 26*, 441–442.

Ragan, C. M., and Vining, L. C. (1978) *Can. J. Microbiol. 24*, 1012–1015.

Rancourt, M. W., and Lechevalier, H. A. (1964) *Can. J. Microbiol. 10*, 311–316.

Redshaw, P. A., McCann, P. A., Sankaran, L., and Pogell, B. M. (1976) *J. Bacteriol. 125*, 698–705.

Redshaw, P. A., McCann, P. A., Pentella, M. A., and Pogell, B. M. (1979) *J. Bacteriol. 137*, 891–899.

Ring, K. (1981) in *Actinomycetes* (Schaal, K. P., and Pulverer, G., eds.), pp. 265–278, Gustav Fischer, New York.

Ring, K., Gross, W., and Heinz, E. (1970) *Arch. Biochem. Biophys. 137*, 243–252.

Ring, K., Foit, B., and Ehle, H. (1977a) *FEMS Microbiol. Lett. 2*, 27–30.

Ring, K., Langheinrich, W., Ehle, H., and Foit, B. (1977b) *Arch. Microbiol. 115*, 199–205.

Sabater, B., and Asensio, C. (1973) *Eur. J. Biochem. 39*, 201–205.

Sabater, B., and Delafuente, G. (1975) *Biochim. Biophys. Acta 377*, 258–270.

Sabater, B., Sebastian, J., and Asensio, C. (1972a) *Biochim. Biophys. Acta 284*, 406–413.

Sabater, B., Sebastian, J., and Asensio, C. (1972b) *Biochim. Biophys. Acta 284*, 414–420.

Shamina, Z. B. (1964a) *Izv. USSR Acad. Sci., Biol. Ser. N 3*, 460–463.

Shamina, Z. B. (1964b) *Mikrobiologiya 33*, 831–835.

Sharples, G. P., and Williams, S. T. (1976) *J. Gen. Microbiol. 96*, 323–332.

Simuth, J., Hudec, J., Hoang, T. C., Danyi, O., and Zelinka, J. (1979) *J. Antibiot. 32*, 53–57.

Smucker, R. A., and Pfister, R. M. (1978) *Can. J. Microbiol. 24*, 397–408.

Sohler, A., Romano, A. H., and Nickerson, W. J. (1958) *J. Bacteriol. 75*, 283–290.

Stastna, J., and Mikulik, K. (1981) in *Actinomycetes* (Schaal, K. P., and Pulverer, G., eds.), pp. 481–485, Gustav Fischer, New York.

Streicher, S. L., and Tyler, B. (1981) *Proc. Nat. Acad. Sci. U.S.A. 78*, 229–233.

Streshinskaya, G. M., Naumova, I. B., and Panina, L. I. (1979) *Mikrobiologiya 48*, 814–819.

Szabo, G., Valyi-Nagy, T., and Vitalis, S. (1967a) *Acta. Biol. Acad. Sci. Hung. 18*, 237–243.

Szabo, G., Bekesi, I., and Vitalis, S. (1967b) *Biochim. Biophys. Acta. 145*, 159–165.

Szeszak, F., and Szabo, G. (1973) *Acta. Biol. Acad. Sci. Hung. 24*, 11–17.

Taber, W. A. (1960) *Can. J. Microbiol. 6*, 503–514.

Taptykova, S. D., Kalakoutskii, L. V., and Agre, N. S. (1969) *J. Gen. Appl. Microbiol. 15*, 383–386.

Terry, J., and Springham, D. G. (1981) *Can. J. Microbiol. 27*, 1044–1047.

Tewfik, E. M., and Bradley, S. G. (1967) *J. Bacteriol. 94*, 1994–2000.

Valu, G., and Szabo, G. (1973) *Acta. Biol. Acad. Sci. Hung. 24*, 171–174.

Valu, G., and Szabo, G. (1978) in *Nocardia and Streptomyces* (Mordarski, W., Kurylowicz, W., and Jeljaszewicz, J., eds.), pp. 409–412, Gustav Fischer, New York.

Valu, G., and Szabo, G. (1981) in *Actinomycetes* (Schaal, K. P., and Pulverer, G., eds.), pp. 487–493, Gustav Fischer, New York.

Verma, J. N., and Khuller, G. K. (1983) *IRCS Med. Sci. Biochem. 11*, 18–19.

Verma, J. N., Khera, A., Khuller, G. K., and Subrahmanyam, D. (1980) *Curr. Microbiol. 4*, 13–15.

Voronia, O. I., Tovarova, I. I., and Khokhlov, A. S. (1975) *Bioorg. Chem. 1*, 985–990.

Vu-Trong, K., Bhuwapthanapun, S., and Gray, P. P. (1980) *Antimicrob. Agents Chemother. 17*, 519–525.

Waksman, S. A., and Diehm, R. A. (1931) *Soil Sci. 32*, 97–117.

Walker, J. T., Specht, C. H., and Bekker, J. F. (1966) *Can. J. Microbiol. 12*, 347–353.

Wax, R., Snyder, L., and Kaplan, L. (1982) *Appl. Environ. Microbiol. 44*, 1004–1006.

Weyland, H. (1981) in *Actinomycetes* (Schaal, K. P., and Pulverer, G., eds.), pp. 185–192, Gustav Fischer, New York.

Whaley, J. W., and Boyle, A. M. (1967) *Phytopathol 57*, 347–351.

Wildermuth, H. (1970a) *J. Gen. Microbiol. 60*, 43–50.

Wildermuth, H. (1970b) *J. Bacteriol. 101*, 318–322.

Wildermuth, H., and Hopwood, D. A. (1970) *J. Gen. Microbiol. 60*, 51–59.

Williams, S. T. (1978) in *Nocardia and Streptomyces* (Mordarski, M., Kurylowicz, W., and Jeljaszewicz, J., eds.), pp. 137–142, Gustav Fischer, New York.

Williams, S. T., Sharples, G. P., and Bradshaw, R. M. (1973) in *Actinomycetales: Characteristics and Practical Importance* (Sykes, G., and Skinner, F. A., eds.), pp. 113–126, Academic Press, New York.

Williams, S. T., Davies, F. L., Mayfield, C. I., and Kahn, M. R. (1971) *Soil Biol. Biochem. 3*, 187–195.

Williams, S. T., Bradshaw, R. M., Costerton, J. W., and Forge, A. (1972a) *J. Gen. Microbiol. 13*, 349–358.

Williams, S. T., Shameemullah, M., Watson, E. T., and Mayfield, C. I. (1972b) *Soil Biol. Biochem. 4*, 215–225

Williams, S. T., and Khan, M. R. (1974) *Post. Hig. Med. Dosw. 28*, 395–408.

Williams, S. T., and Robinson, C. S. (1981) *J. Gen. Microbiol. 127*, 55–63.

Wright, H. M., and Hopwood, D. A. (1977) *J. Gen. Microbiol. 102*, 417–421.

Biology of Actinomycetes Not Belonging to the Genus *Streptomyces*

Mary P. Lechevalier
Hubert Lechevalier

I. INTRODUCTION

Actinomycetes are filamentous, branching bacteria that are widely distributed in nature (M. P. Lechevalier, 1981). They can be separated into two broad groups of unequal size: the mainly fermentative forms that are most often associated with the natural cavities of man and animals and the much larger group of primarily oxidative organisms for which soil is the basic reservoir and from which they are disseminated to most other environments. The first group is composed of organisms that do not produce special morphological features other than a branching mass of filaments (mycelium) and that are chemically characterized by cell walls that do not contain diaminopimelic acid (DAP). The second group harbors organisms that may be more intricate morphologically than the first and that contain DAP in their peptidoglycan. We will be mainly concerned here with the second group, since it contains the actinomycetes important in applied microbiology.

The knowledge of the physiology and biochemistry of actinomycetes is almost exclusively the fruit of studies in the fields of microbial taxonomy and industrial microbiology. In the first case, our knowledge is often limited to "plus" and "minus" answers: Does the organism degrade, transform, utilize, or produce acid from a given substrate? These data are suitable for tabular presentation and for phenetic analysis; however, rarely is detailed attention given to intermediates. Industrial research that is oriented toward products such as antibiotics and enzymes or toward processes that might degrade a waste product, transform a compound, or accumulate a metal has given us far more information on intermediate metabolism, albeit of a rather esoteric nature. Thus in this account of the biology of the nonstreptomycetes (rare actinomycetes), we have relied heavily on industrially oriented studies basically because they are the only ones available.

A. Morphological and Chemical Properties of Soil Actinomycetes

On the basis of morphological and chemical characteristics, actinomycetes can be separated into comparatively large groups that can be called genera. Rational separation into families must await the development of our knowledge of these organisms at the molecular biology level (Stackebrandt and Woese, 1981). In any case the separation into families is not essential to the development of a satisfactory system for the identification of these organisms. Useful morphological characters for this purpose include the presence or absence of aerial hyphae, the mode of division of hyphae, the presence and arrangement of various types of spores, the motility of various elements, and the production of specialized structures such as sclerotia or synnemata. Chemical properties that have been found to be useful in the taxonomy of these organisms include cell wall, whole cell, and lipid composition (M. P. Lechevalier and Lechevalier, 1980).

The G + C content of the DNA of actinomycetes is high. The mycobacteria and the nocardiae are on the low side of this spectrum (60–70 mol %), and the streptomycetes are on the high side with 70–75 mol %. Some thermophilic actinomycetes that produce true bacterial endospores and that might be considered filamentous, branching bacilli, have DNA's with low G + C percentages (44–54 mol %) (genus *Thermoactinomyces*). Members of a second genus, *Dermatophilus*, which cause dermatoses in animals and man, have G + C's of about 55 mol %.

Some of the actinomycetes lacking DAP are soil inhabitants. These include *Actinomyces humiferus*, *Agromyces ramosus*, *Promicromonospora* spp., both species of *Oerskovia*, and a group of related bacteria that are called NMO's (nonmotile oerskoviae) (Table 11-1). All these organisms are aerobic to microaerophilic with rather undistinguished bacteroid morphology, forming only a substrate (primary) mycelium, which may break up into rod-shaped elements. In the case of the oerskoviae these elements are motile, having flagella. The agromycetes are catalase-negative, and the oerskoviae are catalase-positive when grown aerobically. The cell walls of agromycetes contain diaminobutyric acid and glycine, while those of oerskoviae and promicro-

TABLE 11-1   Some Diagnostic Properties of Soil Actinomycetes That Do Not Contain Diaminopimelic Acid

	Diagnostic Cell Wall Constituents							Other Characteristics					
	Ornithine	Lysine	Aspartic Acid	Glycine	DAB[1]	Galactose	Rhamnose	Fermentation of Sugars	Oxidation of Sugars	Catalase	Motility	Color	Reference
Actinomyces humiferus	+[2]	+	+	−	−	−	+	+	−	−	−	W[5]	Gledhill and Casida (1969a)
Agromyces ramosus	−[3]	−	−	+	+	−	+	−	+	−	−	W	Gledhill and Casida (1969b)
Oerskovia turbata	−	+	±[6]	−	−	+	−	+	+	V[4]	+	Y[7]	H. A. Lechevalier and Lechevalier (1981b)
Oerskovia xanthineolytica	−	+	±	−	−	+	−	+	+	V	+	Y	H. A. Lechevalier and Lechevalier (1981b)
N M O	−	+	+	−	−	−	+	+	+	V	−	Y/W	H. A. Lechevalier and Lechevalier (1981b)
Promicromonospora spp.	−	+	+	−	−	−	−	−	+	+	−	Y/W	H. A. Lechevalier and Lechevalier (1981b)

1. DAB = diaminobutyric acid.
2. + = positive.
3. − = negative.
4. V = variable.
5. W = White.
6. ± = Minor amounts.
7. Y = Yellow.

monosporae contain lysine and aspartic acid. Cell walls of *Actinomyces humiferus* strains contain ornithine, lysine, and aspartic acid. The NMO's and promicromonosporae lack the galactose found in the walls of the true oerskoviae.

The remainder of the soil actinomycetes, which are oxidative, have cell walls with mucopeptides containing DAP. These can be separated into two large groups: those containing mainly the L form of DAP and those with the *meso* form of this diamino acid. The streptomycetes, which are the subject of Chapter 10, form the largest group of the L-DAP-containing actinomycetes. The morphological properties of the actinomycetic genera with this composition are found in Table 11.2. Unfortunately, the various strains of a given genus may not always exhibit all the typical morphological features of the

TABLE 11.2 Morphological Properties of Soil Actinomycetes Containing L-DAP (Type I Cell Wall)

Genus	Morphology	Reference
Streptomyces	Production of both substrate and aerial hyphae. Long chains of arthrospores are formed on the aerial hyphae and, more rarely, on both mycelia.	Williams et al. (1981)
Streptoverticillium	Similar to *Streptomyces*, but the aerial hyphae bear sporophores in verticils and umbels of spore chains.	Locci et al. (1982)
Elytrosporangium	Similar to *Streptomyces* but with single or short chains of large spores on the substrate mycelium.	Cross and Al-Diwany (1981)
Microellobosporia	Production of both substrate and aerial hyphae bearing merosporangia (sporangia containing a single row of spores).	Cross et al. (1963)
Actinosporangium	Similar to *Streptomyces* but with formation of pseudosporangia (masses of spores enclosed within a mucilaginous mass).	Hussein and Krassilnikov, (1969)
Actinopycnidium	Similar to *Streptomyces* but with the formation of pseudopycnidial accumulation of spores.	Williams (1970)
Chainia	Primary mycelium giving rise to sclerotial bodies that are filled with lipids at maturity. Aerial mycelium may be formed that bears chains of arthrospores	M. P. Lechevalier et al. (1973)
Kitasatoa	Similar to *Streptomyces* but with formation of merosporangia with motile sporangiospores on both mycelia.	Matsumae et al. (1968)
Kitasatosporia	Similar to *Streptomyces*, but the cell wall contains both L- and *meso*-DAP in about equal amounts.	Omura et al. (1982)
Nocardioides	Both substrate and aerial hyphae fragment into rod-shaped or coccoid elements. Aerial mycelium scanty and formed of hyphae that either do not branch or branch sparsely.	Prauser (1976a)

genus. A streptomycete may become bald, and a chainia may lose the property to form sclerotia. Variation in biological material must always be kept in mind.

Actinomycetes containing major amounts of *meso*-DAP in their cell walls can be separated into four main groups on the basis of their cell wall composition and the sugar composition of the hydrolysates of their whole cells (H. A. Lechevalier and Lechevalier, 1981a). The morphological characteristics of the genera falling into these four groups are presented in Tables 11.3, 11.4, 11.5, and 11.6.

In Table 11.3 the morphological properties of actinomycetes with a cell wall (CW) of the so-called Type II are given. Most of the organisms of this group are sporangia-forming actinomycetes with flagellated sporangiospores. The exception is the micromonosporae, which form single conidia on their substrate mycelium. Interestingly, all the organisms with a cell wall of Type II either lack an aerial mycelium or have a very rudimentary one. One should also note that among nonstreptomycetes the members of the genera *Micromonospora* and *Actinoplanes* have been the most important producers of secondary metabolites and thus are among the most important actinomycetes from an industrial point of view.

TABLE 11.3 Morphological Properties of Soil Actinomycetes Containing Major Amounts of *meso*-DAP and Glycine (Type II Cell Wall) and Whole-cell Hydrolysates with Xylose and Arabinose (Type D Whole-cell Sugar Pattern)

Genus	Morphology	Reference
Micromonospora	Aerial mycelium absent or scanty. Single conidia formed on the substrate mycelium.	Luedemann (1970)
Actinoplanes	Aerial mycelium absent or scanty. Formation of globose to lageniform sporangia containing globose spores with one polar tuft of flagella.	Bland and Couch (1981)
Amorphosporangium	Same as *Actinoplanes*, but sporangia are highly irregular and larger than those of *Actinoplanes*.	Bland and Couch (1981)
Ampullariella	Same as *Actinoplanes*, but sporangiospores rod-shaped with one polar tuft of flagella.	Bland and Couch (1981)
Dactylosporangium	Same as *Actinoplanes*, but fingerlike merosporangia are formed, each with one chain of spores each bearing one polar tuft of flagella. Single vesicles are also formed on the vegetative mycelium.	Sharples and Williams (1974)

TABLE 11.4   Morphological Properties of Soil Actinomycetes Containing Major Amounts of *meso*-DAP (Type III Cell Wall) and Having No Diagnostic Sugar in Their Whole-cell Hydrolysates

Genus	Morphology	Reference
Thermoactinomyces	Single spores formed on both the aerial and the substrate mycelia. The spores are true heat-resistant bacterial endospores.	Cross (1981)
Thermomonospora	Single spores formed on the aerial mycelium or on both the aerial and the substrate hyphae. Spores are not heat-resistant bacterial endospores.	Cross (1981)
Nocardiopsis	Substrate mycelium fragmenting into coccoid elements. Aerial hyphae mature into long chains on arthrospores.	J. Meyer (1976)
Actinosynnema	Substrate mycelium forms synnemata or domelike bodies on which aerial hyphae are formed. The hyphae break up into motile rods that are peritrichously flagellated.	Hasegawa et al. (1978)
Geodermatophilus	Aerial mycelium not formed. Substrate mycelium very limited and dividing in all planes to form a mass of cuboid to coccoid nonmotile cells and elliptical to lanceolate flagellated zoospores.	Luedemann (1968)
Streptoalloteichus	Both substrate and aerial mycelia formed. Small spherical to elongated sporangia formed on the substrate mycelium bearing a few spores, each with a single flagellum. Aerial hyphae bearing chains of conidia.	Tomita et al. (1978)

The morphological properties of the genera of actinomycetes with a Type III cell wall are summarized in Tables 11.4, and 11.5. Table 11.4 gives the properties of the genera whose members do not contain major amounts of madurose in their whole-cell hydrolysates. Of these, the thermoactinomycetes differ from the other actinomycetes not only in their DNA G + C %, as previously discussed, but also because they form true bacterial endospores complete with a high content of calcium and dipicolinic acid. These thermophilic organisms are important medically as one of the causes of an allergic pneumonitis called "farmers lung" and industrially as potential sources of heat-stable enzymes.

The status of the genus *Streptoalloteichus* is still uncertain, since in our laboratory a strain of the type species was not found to form sporangia and no motile elements have been observed.

TABLE 11.5 Morphological Properties of Soil Actinomycetes Containing Major Amounts of *meso*-DAP (Type III cell wall) and with madurose (3-O-methyl-D-galactose) in Their Whole-cell Hydrolysates (Type B Whole-cell Sugar Pattern)

Genus	Morphology	Reference
Actinomadura	Both substrate and aerial mycelia formed. Short chains of conidia on aerial hyphae.	Goodfellow et al. (1979)
Microtetraspora	Similar to *Actinomadura* but with 1–6 (average 4) spores per conidial chain.	Thiemann et al. (1968)
Microbispora	Similar to *Microtetraspora* but with only 2 spores per conidial chain.	Nonomura and Ohara (1971)
Streptosporangium	Both substrate and aerial mycelia formed. Globose sporangia containing round nonmotile spores formed on the aerial mycelium.	Bland and Couch (1981)
Spirillospora	Similar to *Streptosporangium*, but the sporangia contain rod-shaped spores each with a subpolar tuft of flagella.	Bland and Couch (1981)
Planomonospora	Both substrate and aerial mycelia formed. The aerial mycelium bears rows of small sporangia each containing one flagellated spore.	Thiemann et al. (1967)
Planobispora	Both substrate and aerial mycelia formed. The aerial hyphae bear two-chambered merosporangia, each chamber of which contains a flagellated zoospore.	Thiemann and Beretta (1968)

In Table 11.5 are grouped the genera of actinomycetes with a Type III cell wall, the cells of which contain the diagnostic sugar, madurose (3-O-methyl-D-galactose) (CW III B type). Within this chemotype are found the actinomadurae, organisms long confused with the streptomycetes and the nocardiae. Some actinomadurae can cause human mycetomas, but it seems that their natural habitat is the soil, and many are antibiotic producers. The genera *Microtetraspora* and *Microbispora* might be considered as a crystallization of some of the morphological stages found in the genus *Actinomadura*. The sporangiate organisms with this cell wall composition are of two types: (1) the streptosporangia and spirillosporae with multispored sporangia containing nonmotile and motile spores, respectively, and (2) the planomonosporae and planobisporae, both forming sporangia containing only one motile spore per chamber. The mode of sporulation of sporangiate actinomycetes has been the subject of several studies. Sharples et al. (1974) differentiated between the formation of sporangiospores by actinomycetes that, like actinoplanetes, form their spores by the ingrowth of hyphae within the outer sheath of the parent hypha (H. A. Lechevalier and Holbert, 1965), and those that, like dactylosporangia and the planomonosporae, form their sporangiospores "en-

TABLE 11.6   Morphological and Chemical Properties of Soil Actinomycetes Containing Major Amounts of *meso*-DAP, Arabinose, and Galactose (Type IV cell wall, Type A Whole-cell Sugar Pattern)

Genus	Morphology	Reference
Mycobacterium	Filamentation usually very limited to nil; aerial mycelium usually not formed; hyphae when formed easily fall apart into rods and cocci, which may be pleomorphic. Mycolic acids with about 80 atoms of carbon present. Rarely pink to red.	Barksdale and Kim (1977)
Rhodococcus	Morphology as for *Mycobacterium* but mycelium more developed and more persistent. Mycolic acids with about 35–65 atoms of carbon present. Often pink to red. Colonies often entire.	Goodfellow and Alderson (1977)
Nocardia	Morphology very variable. Similar to *Rhodococcus* at one end of the spectrum or *Streptomyces* at the other. Mycolic acids with about 50 atoms of carbon.	M. P. Lechevalier (1976)
Micropolyspora	Formation of substrate and aerial mycelia. Short chains of spherical conidia formed on both mycelia. Mycolic acids as for nocardiae.	Goodfellow and Pirouz (1982)
Faenia	Morphology as in *Micropolyspora*. No mycolic acids.	Kurup and Agre (1983)
Pseudonocardia	Substrate and aerial mycelia produced. Long cylindrical blastospores formed in chains on the aerial hyphae. No mycolates.	Henssen and Schafer (1971)
Saccharomonospora	Both substrate and aerial mycelia formed. Single conidia densely produced on aerial mycelium only. No mycolates.	Cross (1981)
Saccharopolyspora	Morphology basically as for *Nocardiopsis*: fragmenting substrate mycelium, aerial mycelium segmenting into spores contained within a sheath. No mycolates.	Lacey and Goodfellow (1975)
Actinopolyspora	Both substrate and aerial mycelia formed. Substrate hyphae may fragment. Aerial hyphae with long chains of spores produced basipetally. No mycolates. Extreme halophiles.	Gochnauer et al (1975)

dogenously" by the development of wall material within the parent hypha. In our estimation, both types of spores are formed endogenously, since they are both formed within a preexisting envelope. Although it is valuable to note these cytological differences, one is tempted to believe that they may rather be linked to the number of spores formed within the sporangium in question than to basic differences within the organisms themselves. No doubt, further investigation of the fundamental properties of these organisms will throw light on this subject.

The genera listed in Table 11.6 consist of organisms with a Type IV cell wall. These contain *meso*-DAP, arabinose, and galactose as diagnostic constituents. These sugars can also be found in whole-cell hydrolysates, a characteristic that makes the identification of this cell wall type especially easy. Although this cell wall type is shared by the mycobacteria and nocardiae, they can be differentiated from each other and from their relatives, the corynebacteria, by the size of the mycolates that they contain. The methods for carrying out these lipid analyses and for cell wall and whole-cell analyses have been detailed by M. P. Lechevalier and Lechevalier (1980). A group of bacteria long referred to as the "rhodochrous" group that is intermediate between the corynebacteria and the nocardiae is now located in the genus *Rhodococcus*. The separation of the rhodococci from the nocardiae is not easy, but sets of physiological characteristics to help in this endeavor have been proposed by Tsukamura (1982). Some mycobacteria and nocardiae are serious pathogens. However, the nocardiae that are important antibiotic producers have never been shown to be pathogenic.

The genus *Micropolyspora* was proposed before chemotaxonomy was introduced into the classification of actinomycetes. The type species of this form-genus, *Micropolyspora brevicatena*, has been found to contain nocardomycolates and thus may be considered a species of *Nocardia* by present day criteria (Goodfellow and Pirouz, 1982). The most important species of this form-genus is *M. faeni*, one of the causes of farmer's lung. Kurup and Agre (1983) have proposed to remove this ecologically and medically important species from the form-genus *Micropolyspora* and to create the genus *Faenia* to accommodate it.

A more elaborate discussion of the actinomycete genera known to date, along with details of their morphology, chemistry, physiology, and ecology, will be found in the work of Starr et al. (1981).

II. HABITATS AND ISOLATION OF NONSTREPTOMYCETES

A. Introduction

The microflora of a soil sample contains mainly bacteria, including actinomycetes, and fungi. Various methods have been devised for the selective isolation of a soil population. As a matter of fact, any procedure is selective for the isolation of part of the microflora, since no single method will permit one to isolate representatives of all the microorganisms of a soil sample.

Basically, there are three methods for the selective isolation of specific groups of organisms and often all three methods are used together.

1. The substrate from which isolation is made can be selected on the basis of its potential richness in the group of interest.
2. It can be treated to increase the relative abundance of the desired organisms (pretreatment).
3. The media used can be selective for the coveted microflora.

The first method requires a knowledge of the ecology and habitats of the various groups of actinomycetes; the second and the last depend on our knowledge of the nutrition and physiology of these organisms and on those of the other organisms that might be present in the substrate. Unfortunately, we know precious little about all these things, and the selective isolation of actinomycetes is at present probably more an art than a science. The methods proposed for the preferential isolation of specific actinomycete genera and species have been reviewed by Cross (1982).

In general, it is considered that drying a sample and bringing its pH mildly onto the alkaline side will favor the isolation of most actinomycetes. The soil sample can be mixed with calcium carbonate and dried in this form with good results (El-Nakeeb and Lechevalier, 1963).

Unfortunately actinomycetologists do not have at their disposal a set of procedures that will permit them to isolate selectively members of all the groups of actinomycetes that have been recognized so far. The streptoverticillia have been particularly elusive, but the following is a summary of proposed methods for the selective isolation of the groups of actinomycetes that have been the most cooperative, together with a short discussion of the ecology of various actinomycetes. In the case of the rarer genera, so few strains have been isolated that generalizations on ecological niches cannot be made.

B. DAP-lacking Soil Actinomycetes

Actinomyces humiferus seems to favor organically rich soils, whereas *Agromyces ramosus*, although inhabiting the same substrates, can also accommodate itself to desert soils. Oerskoviae and NMO's are found in soils rich in decaying plant materials but also in aluminum hydroxide gels and blood samples. *A. ramosus* grows slowly on ordinary media, but its growth is stimulated by the presence of peroxide-removing substances such as bovine catalase or manganese dioxide. Under these conditions the cells are pigmented yellow because of carotenoid production (Jones et al., 1970).

C. L-DAP-containing Soil Actinomycetes

It is assumed that organisms representing variations on the *Streptomyces* theme, such as the actinopycnidia or the microellobsporiae, are widely distributed in soils but are rarer than the streptomycetes. Chainiae and streptoverticillia are suspected to be more abundant in arid exotic soils than in those from the backyards of New Jersey or England; however, quantitative data are lacking.

Strains of *Nocardioides* are widely distributed in soil and other sources such as kaolin. Successful isolation by plating out on an oatmeal-salt medium has been reported (Prauser, 1976a). On such a medium the surface of the colonies is pasty and whitish to faint yellow-brown; if aerial mycelium is formed, it is thin and chalky.

D. Actinomycetes with Cell Wall Type II

Strains of *Micromonospora* dominate the actinomycetic flora of lacustrine environments. This is especially marked in mud deposits (Johnston and Cross, 1976). They have been selectively isolated by incorporating gentamicin into the isolation media; however, Wakisaka et al. (1982) obtained better results by using a selective medium containing tunicamycin. When large amounts of gram-negative bacteria were present in the sample, their number could be reduced by treating the sample for a few minutes with 0.01 N NaOH. With the tunicamycin medium an average of four strains of *Micromonospora* per soil sample were obtained.

Strains of *Actinoplanes* are found in soil, including some rather arid sandy ones, and on plant materials located at the edge of freshwater bodies. These organisms were first isolated by Couch through the use of a baiting procedure that he had perfected for the isolation of chytrids; this method is still used. It consists of placing soil in a petri dish and adding water passed through charcoal to cover the sample completely. Various sterile materials are then floated on the water as baits. These may be grains of pollen or various other pieces of vegetable matter. Hair and cellulose fibers are also effective baits. After incubation the pieces of bait are examined with a microscope for the detection of sporangia of actinoplanetes. These can be picked up and streaked on a nutritive medium (Couch, 1949).

In the case of actinoplanetes and relatives, plant material can be collected from the shores of lakes and streams and incubated in humid chambers to help the formation of sporangia. Dehydration of the samples, before they are suspended in water for dilution and plating out, helps to reduce the number of contaminating gram-negative bacteria. Once the sample has been suspended in water, one should wait about an hour before picking up the supernate to make the serial dilutions, thus giving time for the sporangia to release their zoospores (Makkar and Cross, 1982).

Palleroni found that spores of actinoplanetes are attracted by chloride ions and described a method for their isolation that takes advantage of this property. The method consists of flooding the soil sample with water and waiting about one hour in order to give time for the release of the zoospores. The spores are attracted to a capillary filled with a phosphate buffer containing

KCl. The orifice of the capillary is placed about one mm under the surface of the water and is left there for one hour. The capillary is then taken out, washed externally with a jet of sterile water, and its contents blown out for plating (Palleroni, 1980).

E. Actinomycetes with Cell Wall Type III

Members of the genus *Thermoactinomyces* grow in nature in warm habitats such as piles of plant debris. Human-made piles of such materials, such as stored hay, are favorite locations for these organisms. They can be isolated by incubating at 50°C plates of a Czapek-yeast extract medium containing Vitamin-free Casamino acids, with novobiocin and cycloheximide or nystatin added to prevent overgrowth by fungi (Cross and Attwell, 1974).

Strains of *Actinosynnema* have been obtained by placing pieces of vegetable matter such as grass blades on diluted potato-carrot agar in a moist chamber (Hasegawa et al., 1978). The synnemata can be observed microscopically, and the organisms can be transferred to a variety of media suitable for the growth of actinomycetes.

Nonomura and Ohara (1969a, b; 1971) advocated a procedure for the selective isolation of strains of *Microbispora* and *Streptosporangium* that consisted of heating the sample for one hour at 120°C and then plating it out on a chemically defined medium containing the antifungal antibiotics nystatin and cycloheximide. Variations on this method have also been useful in the isolation of strains of *Actinomadura* from soil.

F. Actinomycetes with a Type IV Cell Wall

Nocardiae can be isolated from soil by baiting with paraffin. However, these are not the only microorganisms capable of growing at the expense of this hydrocarbon. Orchard and Goodfellow (1974) recommended Oxoid Diagnostic Sensitivity Test Agar supplemented with cycloheximide, nystatin, and various antibacterial antibiotics such as demethylchlortetracycline, methacycline, and chlortetracycline for the selective isolation of nocardiae.

Strains of *Rhodococcus* have been isolated from a variety of habitats such as soil, the gut of various animals, and clinical specimens. One species, *Rhodococcus coprophilus*, can be used as an indicator of environmental pollution caused by farm animal feces. Isolation of rhodococci can be carried out by heating the samples for a few minutes at 55°C before plating out on a salt-propionate medium (Mara and Oragui, 1981).

III. CELLULAR COMPOSITION

Something of the composition of the walls, polysaccharides, lipids, and nucleic acids of the cells of actinomycetes is discussed under the sections on chemotaxonomy and genetics. A review of the fatty acids found in various genera has been published by Kroppenstadt and Kutzner (1978). The dominant types of fatty acids (FA's) in most actinomycetes are the branched-chain types; the *Mycobacterium–Nocardia–Rhodococcus* group and the dermatophili are the exceptions. Members of these genera contain largely normal and mono-unsaturated FA's, often accompanied by a 10-methyl FA such as tuber-culostearic acid. The more complex lipids include not only the mycolic acids, discussed in a previous section, but also many other compounds as well, in-cluding peptidolipids, glycolipids, and menaquinones (M. P. Lechevalier, 1977; Asselineau, 1978; Collins et al., 1982). Phospholipid composition of ac-tinomycetes has been extensively studied and found to be of value in the tax-onomy of these organisms (M. P. Lechevalier et al., 1977, 1981). The report of a phenol-soluble lipopolysaccharide containing lipid A from the cell walls of a *Micropolyspora faeni* strain (Hollingdale, 1975) may have important implica-tions in the study of the pathogenesis of farmer's lung. Finally, in one of the few analyses of the elements present in actinomycete cells, Quaroni and Locci (1978) found X-ray evidence for Ca, Mg, and Si in microcrystalline structures within cells of *N. farcinica.*

IV. PRODUCTION OF ANTIBIOTICS BY NONSTREPTOMYCETES

Among the "rare actinomycetes," some are rarer than others, and in general, the soil population is richer in strains of *Micromonospora, Actinoplanes, Ac-tinomadura,* and *Nocardia* than in the other nonstreptomycetes. Thus it is not surprising that strains of these more common genera have contributed more antibiotics than those of most of the other genera of rare actinomycetes.

Table 11.7 presents a tabulation of an estimate of the number of an-tibiotics that have been described as products of rare actinomycetes between the years 1975 and 1982. Antibiotics are usually produced by microorganisms as families of related substances. To illustrate the point, one actinomycete strain rarely produces only one actinomycin but rather a series of ac-tinomycins, some present as major components and some as minor com-ponents. In Table 11.7 the number of antibiotics reported to be produced by the rare actinomycetes is on the low side because if a strain was reported to produce more than one related antibiotic, the family was counted as 1. For ex-ample, *Ampullariella regularis* strain A11079 was reported in 1979 and 1981 to produce the nucleosides neplanocins A, B, C, and F. These were scored in the table as one antibiotic complex.

Thus using this rather conservative method of counting, one will note that the rare actinomycetes have yielded about 200 antibiotics worthy of publica-tion during the years 1975–1982. During that period, about 600 antibiotics have been reported as products of streptomycetes.

TABLE 11.7 Estimate of the Numbers of Antibiotics or Antibiotic Families Reported from Actinomycetes between 1975–1982.

Genus	Number of Antibiotics	Main Types of Antibiotics Reported
Actinomadura	16	4 anthracyclines
		3 polyethers
Actinoplanes	43	5 vancomycins
		4 polyenes
		18 peptides
Actinosporangium	6	6 anthracyclines
Actinosynnema	7	2 ansamycins
Ampullariella	2	
Chainia	3	
Dactylosporangium	3	2 aminoglycosides
Kitasatosporia	1	
Kitasatoa	1	quinoline derivative
Microellobosporia	1	thiazolyl peptide
Micromonospora	65	36 aminoglycosides
		5 macrolides
		8 peptides
		4 nucleosides
Microtetraspora	1	acyltetramic acid derivative
Nocardia	15	5 ansamycins
Nocardiopsis	2	
Pseudonocardia	2	2 peptides
Rhodococcus	2	
Saccharomonospora	1	vancomycin-type
Saccharopolyspora	3	2 aminoglycosides
Streptoalloteichus	2	
Streptosporangium	9	5 peptides
Streptoverticillium	22	3 aminoglycosides
		5 polyenes
		4 peptides
Thermoactinomyces	1	1 alkaloid
Thermomonospora	1	
Unidentified	2	
Total	211	
Streptomyces	about 600	

This estimate was obtained by reviewing the pages of the *Journal of Antibiotics* for these years and by consulting the monumental compilation of Berdy et al. (1980–1981).

As was mentioned previously, the number of antibiotics reported to be produced by strains of a given genus roughly follows the abundance of, or the ease of, isolation of these organisms from natural substrates. One exception is the genus *Streptoverticillium*. Although rarely isolated from soil, these organisms rank third among the rare actinomycetes as producers of antibiotics, and the variety of antibiotic substances that they produce rivals that of the two major producers, the micromonosporae and the actinoplanetes. One possible reason for the discrepancy between the rareness of the streptoverticillia and their comparative fertility as antibiotic producers might be that they are not recognized as different from streptomycetes until they are subject to further study because of their antibiotic activity. An inactive *Streptoverticillium* probably does not attract attention.

V. METABOLISM AND PHYSIOLOGY

Basic studies of actinomycete physiology, metabolism, and biochemistry, unrelated to some process or product of industrial interest or to pathogenic processes, are few. A beginning has been made in understanding the basic physiology of the nitrogen-fixing actinomycetes of the genus *Frankia* (see the subsection entitled "Nitrogen Fixation"). As with the genetic studies on nonstreptomycetes (see Section VI), the actinomycetes that have been most studied from the point of view of physiology are the rhodococci, probably for much the same reasons, notably their rapidity of growth and ease of manipulation because of their bacteroid cell morphology.

Regulation of nutrient uptake in rhodococci and nocardiae is through (1) passive diffusion, (2) facilitated diffusion (for example, glucose in *Nocardia asteroides* and *Nocardia brasiliensis*), (3) constitutive, carrier-mediated transport (glucose uptake in *Rhodococcus erythropolis*), or (4) inducible transport (succinate, mannitol, and asparagine in *R. erythropolis*). Cellular enzymes in these two genera are both constitutive and inducible. Glucose is dissimilated through the Embden-Meyerhof-Parnas pathway in *R. erythropolis*, but the Entner-Doudoroff pathway is not operational or present as it is in *Rhodococcus opacus* (Bradley, 1978). Bond and Bradley (1978) isolated and examined five enzymes of a strain of *R. erythropolis*, including isocitric dehydrogenase, 6-phosphogluconate dehydrogenase, malate dehydrogenase, mannitol dehydrogenase and glucose-6-phosphate dehydrogenase. It was determined that the first two were constitutive in this strain and the last three inducible. The two constitutive enzymes were heat-stable whereas the inducible enzymes were affected by heat. Differences were also found in the relative rapidity of the degradation of isocitric dehydrogenase and malate dehydrogenase within the cells (Bond and Bradley, 1978).

A. Enzymes in Transformations and Degradations of Hydrophobic Compounds

Actinomycetes of all taxa contribute a large share of the degradative enzymatic activity that goes on in nature (M. P. Lechevalier, 1981). Among those that have been studied in vitro there have been many reports dealing with the metabolism and transformation of great variety of chemical compounds by what used to be called "nocardiae," "rhodochrous," "gordonae," "jenseniae," or "mycobacteria." Many of these species have now been assigned to the genus *Rhodococcus*. Several reviews of the metabolic activities of rhodococci, nocardiae, and actinomadurae have been published. That of Tárnok (1976) covered growth requirements, production of pigments or enzymes, capacities to utilize, transform and/or degrade, often stereospecifically, alkanes, cycloaryl, cycloalkyl, heterocyclic compounds, steroids, pesticides, and antibiotics. Golovliev et al. (1978) discussed the degradation of alicyclics, aromatics, and heterocycles, while Raymond and Jamison (1971) focused on the decomposition of aliphatic compounds and steroids. Despite the impressive battery of enzymes demonstrable from rhodococci, interestingly enough, there have been few reports of antibiotics from this group (see Table 11.7).

The capacity of the rhodococci to elaborate enzymes with unusual activities, to tolerate, utilize and/or degrade relatively high concentrations of hydrophobic chemicals, is a result of their relatively rapid growth rate, their rapid production of a variety of enzymes and the lipophilic nature of their cell envelope. This last characteristic, coming largely from the nocardomycolates covalently linked to the peptidoglycan, undoubtedly enhance these organisms' capacity to solubilize or take up many of the highly hydrophobic compounds they are credited with transforming. Also important is that micelle-entrapped enzymes (hydrated reverse micelles of surfactant in organic solvent) may not only show increased activity, but may even change in their specificity depending on the location of their hydrophobic regions as one or another portion of the enzyme is exposed to substrate contact (Martinek et al., 1982). Surfactants are produced by rhodococci and other nocardiaceae in the form of nocardomycolate trehalose esters (usually α-α-trehalose-6,6$'$-dimycolates). In these amphipathic compounds the hydrophobic moiety (nocardomycolate), a β-OH, α-branched fatty acid that contains between 30 and 50 carbons, is optimally produced when the strains are grown on *n*-alkanes (Rapp et al., 1977; Margaritis et al., 1979; Akit et al., 1981; Kretschmer et al., 1982; Cairns et al., 1982; Kosaric et al., 1983). Other surfactants produced by *R. erythropolis* grown on hexadecane were found to be the methyl ester of hexadecanoic acid and a monoglyceride, probably of the same fatty acid (MacDonald et al., 1981).

Despite their preeminence, the discussion below is by no means limited to rhodococci. Species of other genera are beginning to be investigated. Even though these are more slow-growing than the rhodococci, they produce different enzymes and, often, novel transformations. What purpose the enzymatic batteries involved in these reactions serve in the cellular processes of the producer is often unknown or only guessed at. Generally, when enzymes have been purified, no data are available on transformations they might carry out that would be involved in the primary metabolism of the producer.

Hydrocarbons and Related Compounds.   The isolation of transforming or degradative actinomycetes from the natural environment is accomplished by a variety of means, but largely by selection through their capacity for growth in media where the compounds to be degraded, such as alkanes, straight-chained, alicyclic, or *n*-paraffinic, are used as sole sources of carbon or are co-metabolized in the presence of other carbon sources. Nitrogen sources may be as simple as ammonia or complex such as yeast extract or peptones (Jamison et al., 1976; Austin et al., 1977; Raymond et al., 1967; Gordon and Hagan, 1937). In a case in which petroleum products had leaked from storage into a local limestone aquifer and was contaminating the local water supplies, a successful stimulation of bacterial activity in situ was obtained by addition of ammonium sulfate and monobasic sodium phosphate (Raymond et al., 1977). A "*Nocardia*" sp. isolated, by enrichment with Bunker C fuel oil, from a sample collected at a marine spill site was also capable of growing on Venezuelan crude oil, hexadecane, and a hydrocarbon mixture at 5–28°C. It degraded largely the *n*-alkane portion and another unidentified fraction of the oils. Since only 0.05 mg of nitrogen was required to bring about degradation of 1 mg of hexadecane, nitrogen was not considered to be a limiting factor in the natural environment, whereas phosphorus and temperature were (Mulkins-Phillips and Stewart, 1974).

Alkane degraders, such as *Rhodococcus salmonicolor*, oxidize the terminal methyl to carboxyl, then degrade the chain piecemeal by β-oxidation. This capability also extends to the alkyl side chain of alkyl benzenes (Abbott, 1979). *N. asteroides*, *Nocardia flavescens*, and rhodococci showed wide strain variability in utilizing C_{6-10}, C_{11-17}, and C_{18-24} *n*-alkanes. Most can grow on C_{13-19}, whereas the other compounds are variably utilized (Nesterenko et al., 1979). 1-Alkenes (C_3, C_4, and C_{13-18}) were degraded by *Rhodococcus corallina* to 1,2-epoxyalkanes (Furuhashi et al., 1981). An unidentified *Nocardia* degrades butadiene to acetate via a monoepoxide intermediate (Watkinson and Somerville, 1976), and a different strain of the same genus transforms 2,4-dimethyl-1,4-nonadiene to 4,6-dimethyl-3,6-heptadienoic acid (Nakajima et al., 1980). Pristane, an isoprenoid hydrocarbon, is converted to pristanol and pristanic acid by a nocardia that can also grow on *n*-paraffins and monomethyl paraffins but not aromatic hydrocarbons (Nakajima and Sato, 1981). With urea as a source of nitrogen, the strain oxidizes squalene to squalendioic acid (Nakajima et al., 1981).

Gaseous hydrocarbons are also utilized. For example, acetylene serves as a sole source of carbon and energy for a rhodococcus that degrades it first to acetaldehyde and then to acetate. The acetylene hydrase, acetaldehyde dehydrogenase, and acetothiokinase are apparently constitutive (Kanner and Bartha, 1979, 1982). *N. ucrainica* has been reported to use such compounds as ethane, propane, butane, propanol, ethanol, and butanol in the vapor phase (Kvasnikov et al., 1971; Shchurova and Kirnitskaya, 1977). And finally, carbon monoxide can be oxidized to carbon dioxide by *R. salmonicolor* and slight activity of this type has been detected in strains of *Actinoplanes*, *Agromyces*, *Microbispora*, and *Mycobacterium* (Bartholomew and Alexander, 1979).

Alicyclic hydrocarbons are also degraded by various nocardiae and rhodococci. A bright yellow slime-forming bacterium identified tentatively as a *Nocardia* sp. degrades cyclohexane via cyclohexanol, cyclohexanone, and ε-caprolactone to acetate and succinate (Stirling et al., 1977). The same final product is produced by a strain of *Rhodococcus globerulus* but via cyclohexanone and 1-oxa-2-oxocycloheptane (Donoghue et al., 1976). The cyclohexanone monoxygenase isolated from the latter organism has a molecular weight (M.W.) of 53,000 and a flavin adenine dinucleotide (FAD) prosthetic group and contains a single polypeptide chain. The pure enzyme has broad ketone specificity (Trudgill, 1982). The cyclohexanol dehydrogenase is inducible, has a M.W. of 145,000 and is nicotinamide adenine dinucleotide (NAD)-linked (Stirling and Perry, 1980). Cycloheptanone is inducibly dissimilated by a nocardia to pimelic acid; 1-oxa-2-oxo-cyclooctane and 7-hydroxyheptanoic acid are intermediates. The strain can degrade other alicycles as well (Hamano et al., 1982). The so-called "*N.*" *petreophila* reported to degrade methylcyclohexane (Tonge and Higgins, 1974) is not an actinomycete (M. P. Lechevalier, unpublished).

Oxygenated compounds related to hydrocarbons are also attacked. Propylene oxide is degraded via 1,2-propanediol by a *Nocardia* sp. (DeBont et al., 1982). *Rhodococcus rhodochrous* can dissimilate camphor by first hydroxylating it at C_6, followed by dehydrogenation to a β-diketone and cleaving. The second ring is opened by the insertion of oxygen adjacent to the keto group, and the resultant lactone is then susceptible to further degradation (Chapman et al., 1966; Trudgill, 1982). Carvone is transformed by three different actinomycetes in different ways: *Nocardia* sp. 1-3-11 yields (−)-cis-carveol as the major metabolite, whereas *N. lurida* A-:0141 transforms it to (+)-neoisodihydrocarveol and + (−)-neodihydrocarveol. *Streptosporangium roseum* IFO 3776 accomplishes the last reaction stereospecifically. In the last two organisms, (+)-dihydrocarvone and (+)-isodihydrocarvone are intermediates (Noma, 1979a, b).

Nitrogen-Containing Nonaromatic Compounds.   Since the report of Winter (1962) of the isolation of two strains of nocardiae capable of degrading up to 90% of a 100-ppm potassium cyanide solution as well as 68–90% of complexed metal cyanides, other studies have shown that cyanide-substituted substances are also degraded by rhodococci. Among these are hydroacrylonitrile, butenenitrile, succinonitrile, acetonitrile and propionitrile (DiGeronimo and Antoine, 1976), benzonitrile (Harper, 1977), acrylonitrile and *n*-butyronitrile (Miller and Gray, 1982), acrylamide monomer (Arai et al., 1981), and acrylonitrile-methyl acrylate butadiene terpolymer (Antoine et al., 1980).

The acetonitrile, hydroacrylonitrile, and propionitrile are used as sources of carbon and nitrogen; acrylonitrile and acrylamide only as a nitrogen source. The acetonitrile was shown to be degraded by an intracellular, inducible enzyme to acetamide and acetic acid with release of ammonia. Propionitrile was converted to propionic acid and ammonia (DiGeronimo and Antoine, 1976). The benzonitrilase is activated by the substrate to cause the association of two 45,000 M.W. subunits to give the active 12-unit enzyme, M.W. 560,000, which is specific for all benzene ring-borne nitrile groups except those in the ortho position (Harper, 1977). The acrylamide monomer is deaminated to acrylic acid, ammonia and water (Arai et al., 1981).

A number of amides are degraded and serve as support for growth by a *Rhodococcus* strain including formamide, acetamide, propionamide, and *n*-butyramide. Interestingly, the nonsubstrates malonamide, benzamide, α-phenylacetamide, and 3-aminopropionitrile also support growth, whereas benzonitrile, phenylacetonitrile, malononitrile, and aminoacetonitrile do not (Miller and Gray, 1982).

Pyridine and methyl pyridines up to a concentration of 0.1% are used as a sole source of carbon, nitrogen and energy by a variety of nocardiae. Pyridine is degraded to glutarate semialdehyde in one case (Watson and Cain, 1975), but monohydroxypyridines, tetrahydroxypyridines, and piperidine are not attacked. A second nocardial strain transforms pyridine to 3-hydroxypyridine and 3-methylpyridine to 3-hydroxymethylpyridine and nicotinic acid (Golovlev et al., 1976; Korosteleva et al., 1981). 2,6-Dimethylpyridine-N-oxide, α-picoline and α-picoline are the products of a *Nocardia* sp. grown on 2,6-dimethylpyridine (Kost et al., 1977). Another nitrogen-containing heterocycle, atrazine (I), a toxic herbicide, is degraded by a *Nocardia* sp. to a novel, nontoxic product, 4-amino-2-chloro-1,3,5-triazine. It was also used as a source of N and C by the same strain (Giardina et al., 1980).

Although in most actinomycetes adenine is dissimilated via the usual pathway of hypoxanthine→xanthine→uric acid (Vitols et al., 1974), *Oerskovia xanthineolytica* strains and the related NMO's transform adenine to 8-hydroxyadenine, a compound previously unknown from microorganisms. Xanthine is oxidized by the oerskoviae to uric acid, but no intermediates in the breakdown of hypoxanthine can be detected (M. P. Lechevalier et al., 1982a; M. P. Lechevalier, unpublished).

Aromatic Compounds. The regulation of aromatic and hydroaromatic catabolic pathways in nocardioform organisms has recently been reviewed (Cain, 1981). Both oxidation and oxidative cleavage can occur. A *Nocardia* sp. oxidizes *p*-xylene to *p*-toluic acid (Ooyama and Foster, 1965). Another strain of the same genus likewise transforms *o*- and *m*-xylenes to the equivalent toluates, whereas a third cleaves *o*-xylene to 3,4-dimethyl catechol (Gibson, 1978). When hexadecane is provided as a carbon and energy source, *N. tartaricans* ATCC 31190 transforms ethylbenzene to 1-phenethanol and acetophenone; when glucose is substituted, the transformation does not take place (Cox and Goldsmith, 1979). 1-Phenylalkanes can serve as a source of carbon and energy for a strain of *R. salmonicolor*. In this degradation the side chain of 1-phenyldodecane is thought to be catabolized by ω-oxidation of the terminal methyl group, β-oxidation to 4-phenylbutyrate, β- and α-oxidation to phenylacetic acid, hydroxylation to homogentisate via *o*-hydroxyphenyl acetate, and ring cleavage to maleylacetoacetate (Sariaslani et al., 1974). In the case of 2-phenylbutane and 3-phenylpentane the aromatic ring is cleaved by a *Nocardia* sp. to hydroxymuconic semialdehyde via a diol; but with 3-phenyl-dodecane the side chain is oxidized to give phenyl-substituted fatty acids (Treccani, 1971).

Benzoate is catabolized by rhodococci via the 3-oxoadipate pathway, as are *p*- and *m*-toluates, despite the methyl substitutions (Miller, 1981). Phthalate esters are dissimilated by a similar strain by the protocatechuate pathway (Kurane et al., 1979, 1980), as is 4-chlorobenzoic acid after preliminary dechlorination (Klages and Lingens, 1979). Ten other aromatic carboxylic acids (ACAs) are cleaved by *Nocardia* sp. DSM 43251 via the three most common bacterial routes: (1) catecholate muconate, (2) protocatechuate, or (3) gentisate and maleylpyruvate (Engelhardt et al., 1979b). When the capacity of a number of rhodococcal strains to degrade ACA's as a sole source of carbon and energy was investigated, it was concluded that so many different degradative patterns exist that they might find use in the taxonomy of this group (Rast et al., 1980a).

Nitroaromatics are readily attacked by a variety of rhodococci and nocardiae (Cain, 1958; Villanueva, 1961). Some of the substrates that are utilized include *p*-, *o*, and *m*-nitrobenzoic, 2,5-dinitro,3,5-dinitrobenzoic acids, 2,4,6-trinitrobenzoic acids, *o*-, *m*-, and *p*-dinitrobenzene, and 1,3,5-trinitrobenzene. *p*-Nitrobenzoate is utilized as a source of carbon and nitrogen with the production of an arylamine, and *p*-nitrobenzene is reduced to *p*-nitroaniline (Villanueva, 1964; Garcia-Acha and Villanueva, 1962). The latter enzyme is constitutive and requires a sulphydryl compound for activity and pyrimidine nucleotides as electron donors. Strains of *Rhodococcus calcareus* and *Nocardia* sp. DSM 43251 degrade not only aniline and its chlorinated derivatives (Russel, 1978) but also substituted phenols such as 3-methyl-4-(methylthio)-, 4-(methylmercapto)-, and 4-(methylsulfinyl)-phenol (Rast et al., 1979; Englehardt et al., 1977), all via catechol or a substituted catechol. Phenol itself may be dissimilated by both rhodococci and nocardiae (Rizzuti et al., 1979; Ono, 1973). Other phenol-related substances co-metabolized by strain DM 43251 (in the presence of various carbon sources) include various isomers of cresol, hydroxyanisol, and 3,4-dimethylphenol (Engelhardt et al., 1979a). Biphenyl is meta-hydroxylated to 3-hydroxybiphenyl by a rhodococcus, and others of this taxon transform biphenyl to 2-hydroxy- or 4- hydroxy-biphenyl or 2,2'-dihydroxybiphenyl (Schwartz, 1981). A napththalene-grown *Nocardia* sp. can oxidize *cis*-dihydrodiols of monocyclic and polycyclic aromatic compounds. The isolated enzyme (M.W. 92,000; pH optimum 8.4) has no cross-reaction with antibodies developed against a similar enzyme from *Pseudomonas putida* (Patel and Gibson, 1976).

Aromatic compounds as diverse as the substituted coumarin and vitamin K antagonist, warfarin (Davis and Rizzo, 1982), cannabinoids such as nabilone (Abbott et al., 1977; Archer et al., 1979), the substituted quinoline, papaverine (Haase-Aschoff and Lingens, 1978), and alkylphenol polyethoxylates such as hexaethyleneglycol monophenyl ether (Baggi et al., 1978) are also degraded or transformed by rhodococci and nocardiae. In the case of warfarin the reduction to the corresponding alcohol was stereoselective (Davis and Rizzo, 1982). Papaverine was used as a sole source of carbon and nitrogen (Haase-Aschoff and Lingens, 1978). It was found that the nocardia responsible had two inducible proteins that were involved in the degradation, both bound to the cell wall membrane fraction and having M.W.'s of 29,000–60,000 daltons (Hauer et al., 1982).

Sterol and Steroid Transformation.   Sterol and steroid transformation by actinomycetes has had a long and honorable history of accomplishing in the fermentor what the chemist could not do readily or at all in the reaction flask. Although the enzymes responsible for these reactions are undoubtedly part of the basic cell armamentarium, little is known of their basic functions, and the steroids are rarely broken down and utilized as carbon or energy sources. In some cases the isolation of pure enzymes has made it possible to bypass the microbial fermentation process with all its potential for contamination and production of unwanted by-products, but little attention has been paid to these enzymes' true cellular function. Other methods that avoid the time-consuming steps involved in enzyme purification have found favor, including the use of microbial cells within a variety of organic solvent–water mixtures or entrapment of cells within various solid or semisolid matrices. Again here the vast majority of reports of transformations of these hydrophobic compounds deals with rhodococci and nocardiae, with an occasional *Streptoverticillium* or *Actinomadura*. In fact, in a study comparing the efficiency of 87 different microorganisms in the transformation of 5-cholestane, rhodococci, nocardiae and the related corynebacteria and mycobacteria stood out above the others (Mulheirn and Van Eyk, 1981).

Among the enzymes studied in some detail, one elaborated by a strain of *R. corallinus* that mediates the stereospecific reduction of cholest-4-en-3-one, has a K_m of 70 M at pH 6.5 and requires NADPH as its coenzyme (Lefebvre et al., 1980). A cholesterol oxidase from *R. rhodochrous* can be extracted from the cells by non-ionic surfactants such as Triton X-100 or proteases such as trypsin. However only the detergent-extracted material can be readsorbed by the cells; the trypsin-extracted crude remains unadsorbed. It was felt that the trypsin destroyed the hydrophobic region of the protein responsible for its anchorage to the membrane (Cheetham et al., 1980).

B. Metabolism of Simple Water-soluble Molecules of Biological Origin

Actinomycetes, in general, have the capacity to take up and utilize a broad spectrum of simple compounds of biological origin. This is nowhere more obvious than in the taxonomic literature, where the utilization or transformation of dozens of compounds form the basis of the classification of species of each genus. It is perhaps of general biological significance that actinomycetes that are pathogens, parasites, or symbionts (e.g., *N. asteroides, Nocardia vaccinii, Nocardia rhodnii, Actinomadura madurae, Actinomadura pelletieri, Dermatophilus congolensis*, and certain frankiae) are often significantly less active in this respect than saprophytes such as streptomycetes.

Enzymes Active on Sugars and Related Compounds.   Isomerases for simple sugars certainly abound among actinomycetes. The glucose isomerase of *Actinoplanes missouriensis* NRRL B-3342 has been more widely investigated than any other among nonstreptomycetes. The strain is an overproducer of the enzyme and is insensitive to catabolite repression (Farrés et al., 1982). The isomerase has been purified on DEAE cellulose, is composed of two identical subunits of ~42,000 daltons, and has a pH optimum of 7.0 and a K_m for glucose of 1.33M (Gong et al., 1980). Mg^{2+} serves to activate it, and Ca^{2+} serves to stabilize it (Hupkes and Van Tilburg, 1976).

Two nicotinamide adenine dinucleotide (NAD)-linked polyol dehydrogenases have been reported in *R. corallina*. One, which is labile, converts sorbitol to D-fructose. A second, more stable, converts D-mannitol to D-fructose and D-arabitol to D-xylose (Maurer and Batt, 1962). Enzymes in the hydrolase category, such as glycosidases, are quite widespread among nonstreptomycetes. Using such substrates as esculin or arbutin, activity was found in 13 of 14 genera examined (Goodfellow and Pirouz, 1982). From an *Actinoplanes* sp., Michalski and Domnas (1974) isolated two β-D-glucoside glucohydrolases (E.C. 3.2.1.21), one inducible and one constitutive. These have been purified and found to be identical: pH optima of 5.8–6.0, stability to 30°C at pH 5.5–7.3 for 2 h, and rapidly inactivated above 50°C. They are also inactivated by 100 μM of Cu^{2+}, Hg^{2+}, Pb^{2+}, and Ag^{2+}. The inhibitory effect of *p*-chloromercuribenzoate can be counteracted by cysteine or mercaptoethanol; thus they are felt to be sulfhydryl enzymes. The inducible enzyme has a M.W. of 165,000, a K_m of 2.5×10^{-4}, and an Arrhenius activation energy of 8.5 Kcal $mole^{-1}$.

Phosphatases.   A single acid (I) and three alkaline (II) phosphatases, all constitutive, have been reported from a strain of *T. vulgaris* (Sinha and Singh, 1980). The strain can use organic and inorganic phosphates with equal effectiveness for growth, and the enzymes are elaborated throughout the growth cycle. I is more stable than II, but neither is stable at 50°C, the optimum growth temperature of the producer. Mg^{2+} acts to stabilize I, but Ca^{2+} is inhibitory. It was felt that the cations might be differentially bound to the enzyme (Sinha and Singh, 1981; Singh and Sinha, 1982).

C. Enzymes Attacking Biopolymers

Amylases.   As with the transformation of simple sugars, starch hydrolysis figures in most physiological taxonomic analyses of actinomycetes. Thermophilic actinomycetes have been especially well studied because of their ease of isolation, the high temperature optima of their enzymes, and their rapidity of growth. This is true despite the observation that certain *Thermoactinomyces* enzymes in pure form are not completely stable at the optimum growth temperature of the producer, even with the addition of stabilizing ions (Singh and Sinha, 1982). Furthermore, a potential health problem exists here, since unlike the types of largely nonpathogenic nocardiae and rhodococci that have been the object of study in the preceding sections, these thermophilic actinomycetes can create problems for the personnel engaged in their large-scale fermentation. *Thermoactinomyces vulgaris, Thermomonospora curvata, Micropolyspora faeni*, and *M. sacchari* are all implicated as the causes of "farmer's lung" (extrinsic allergic alveolitis), a serious allergic reaction to the inhalation of spores of these organisms (Blyth, 1973). *T. curvata* produces substances activating the alternative pathway of complement, a factor thought to be the possible cause of these hypersensitivity reactions (Stutzenberger and Bowden, 1980).

Many, if not most, *Thermoactinomyces vulgaris* strains produce α-amylases, either simultaneously or sequentially. In one study (Allan and Hartman, 1972), two electrophoretically differing isoenzymes were found in a single strain; nevertheless, the products of their activities were the same: maltose, maltotriose and some maltotetraose, maltopentaose, and glucose. The larger oligosaccharides are formed first and broken down later. In a different strain, three isoenzymes are found at different stages of growth (Taufel et al., 1981). An α-amylase produced by still another member of this taxon has been purified to homogeneity by ammonium sulfate precipitation and chromatography on DEAE cellulose and CM-cellulose. Its optimum temperature is 70°C, and the pH optimum and isoelectric point are around 5.0. Stabilized by Ca^{2+}, it gives a yield of 70% panose from pullulan (I). The enzyme's activity is proposed to be through attack on some of the $(1 \rightarrow 6)$-α-D-glucosidic linkages of the partially hydrolyzed I as well as the $(1 \rightarrow 4)$-α-D-glucosidic links in I and starch. Kinetic evidence led to the conclusion that a single catalytic site is responsible for both activities (Shimizu et al., 1978; Sakano et al., 1978, 1982; Fukushima et al., 1982).

An extracellular α-amylase from another thermophile, *Thermomonospora curvata*, is produced maximally at 53°C in a chemically defined medium. It is unstable to heating to 65°C at pH 4–6 but can be activated by heating of pH 8. Purified by discontinuous gel electrophoresis, it has a pH optimum at 5.5–6.0 and an optimum temperature of 65°C. It has a M.W. of 62,000 daltons and a K_m for starch of 0.39 mg/ml. The principal products of the digestive process are maltotetraose and maltopentose (Glymph and Stutzenberger, 1977). A strain of *Saccharomonospora* ("*Thermomonospora*") *viridis* also has been reported to produce a thermostable amylase (Upton and Fogarty, 1977).

Dextranases, Xylanases, Chitinases, Pectinases and Cellulases. Dextranase activity was reported in a number of isolates of *Oerskovia xanthineolytica*

(Hayward and Sly, 1976) and in *Actinomyces israelii* (Staat and Schachtele, 1975). An isomaltodextranase releasing isomaltose and the trisaccharide α-D-Glc*p*-(1→3)-α-D-Glc*p*(1→6)-D-Glc from dextran was found in the broths of an *Actinomadura* strain. Its pH optimum is 5.0 (Sawai et al., 1981). Xylanases are elaborated by strains of *Microbispora*, *Micromonospora*, and *Thermomonospora fusca* (Goodfellow and Pirouz, 1982) and chitinase by many different actinomycetes including *Nocardia mediterranei* (Tominaga and Tsujisaka, 1976), *Actinomadura pelletieri*, *Microbispora*, *Micromonospora*, *Nocardiopsis*, *Planobispora*, *Planomonospora*, *Saccharomonospora viridis*, *Thermoactinomyces*, and *Thermomonospora* (Goodfellow and Pirouz, 1982). Pectin is degraded by members of *Micromonospora*, *Microbispora*, *Actinoplanes*, and *Streptosporangium*; however, this activity was not found in nocardiae and thermophilic actinomycetes (Kaiser, 1971).

Most natural cellulose exists in a crystalline form buried in a matrix of lignin (lignocellulose). Lignin is not readily attacked by microorganisms (see discussion below); however, cellulases abound. These are actually multienzyme systems composed variously of endoglucanase (hydrolyzes the high-molecular-weight polymers into oligosaccharides), exoglucanase and cellobiohydrolase (which split glucose or cellobiose, respectively, from the nonreducing end of cellulose), and cellobiase (Bellamy, 1979).

Among actinomycete cellulase producers, strains of *Thermomonospora* spp. have been the principal objects of the research reported over the last decade. Stutzenberger and his colleagues have studied various *T. curvata* strains, originally isolated to find the source of the cellulases observed in composted municipal waste. Ground cotton was found to support the highest cellulase levels (Stutzenberger, 1972), although the strain could digest most cellulosic substrates with the exception of highly lignified sawdust and barley straw (Stutzenberger, 1979). Of the toxic metals commonly found in municipal waste, Al^{2+} and Ca^{2+} were found to inhibit production but not activity of the cellulase (Stutzenberger, 1978). Although glucose and maltose were found to repress cellulase production, the products liberated by the inducible amylase were not repressive when the strain was grown on substrates with starch/cellulose ratios similar to those of compost (Stutzenberger and Carnell, 1977).

The cellulase of a strain of *Thermomonospora fusca* can be induced by a variety of cellulosic substrates including carboxymethylcellulose (CMC) and lignocellulose pulps (Crawford and McCoy, 1972; Crawford et al., 1973a; Crawford, 1974, 1975). The principal products from CMC are cellobiose, glucose, and oligosaccharides of intermediate size. Cellobiase is apparently absent. A partially purified preparation showed that the enzyme is optimally active at 65°C and pH 6.5 (Crawford and McCoy, 1972).

A strain (YX) that has been referred to as *Thermoactinomyces* sp., *T. vulgaris*, *Thermomonospora* sp., and finally identified as a *Thermomonospora fusca* (J. Ferchak, personal communication) has also been the subject of a number of studies. Enzymes attacking CMC and microcrystalline cellulose (Avicel) are produced concomitantly with cellular growth. Fifty percent of this extracellular activity is adsorbed on the residual cellulosic substrate at any given time. The two principal enzymes are an endo-β-1→4 glucanase and a cellobiosylhydrolase that act synergistically. Two minor components are an exo-β-1→4 glucanase and an extracellular cellobiase or β-glucosidase. The optimum pH for CMCase activity is 5.9, with an optimum temperature of 70°C. The optima for the "avicelase" are 7.0 and 65°C, respectively, and for the β-glucosidase (which also has cellobiase activity) they are pH 6.5 and 55°C. A pH of 7.3 destroys all three enzymatic activities. A xylanase is also produced (Humphrey et al., 1977; Hagerdal et al., 1978, 1979, 1980; H. P. Meyer and Humphrey, 1982).

Among mesophilic actinomycetes a report of cellulase production by *Micromonospora* strains is worth noting (Sandrak, 1977), since it is not a common characteristic of this taxon. Also, the little-studied, rapidly growing facultative NMO's (nonmotile oerskoviae) produce cellulases digesting Whatman filter paper in two to three days in stationary culture. These strains are thought to be intermediates between oerskoviae and cellulomonads (M. P. Lechevalier, 1972).

Ligninases.   Lignin, a natural polymer that occurs with cellulose, accounts for 25–30% of all wood. Since it is therefore second only to cellulose in abundance and, in addition, one of the most resistant natural polymers to both chemical and microbial attack known, it represents a serious disposal problem for the pulp industry. Nevertheless, it does disappear slowly under the sequential and combined attack of a variety of organisms and unquestionably serves some of them as a natural nutrient. Chemically, it is a three-dimensional amorphous copolymer of phenylpropanoid monomers. The monomers are derived from n-, monomethoxylated and dimethoxylated coumaryl alcohols linked by one or more random ether and carbon-to-carbon links. Microbial lignin degradation has recently been reviewed (Crawford and Sutherland, 1980). Certain nocardiae, rhodococci, and micromonosporae have been shown to degrade various lignin monomers or other lignin-related compounds (Crawford et al., 1973b; Gradziel et al., 1978; Eggeling and Sahm, 1980, 1981; Rast et al., 1980b; Haraguchi et al., 1980). A strain of *N. autotrophica* can degrade lignin and assimilate the breakdown products as a carbon source. It was shown to release $^{14}CO_2$ from ^{14}C-labeled methoxy groups, side chains or ring C of coniferyl alcohol dehydropolymers (Trojanowski et al., 1977). Carbon 13-labeled lignin is attacked via aromatic ring openings, cleavage of aryl alkyl ether bonds, and degradation of free side chains (Ellwardt et al., 1981).

Proteases and Peptidases.   Serine proteases are produced by strains of *Thermoactinomyces vulgaris* (Ruttloff et al., 1978). Proteases are also produced by members of many other actinomycete taxa (Ovcharov and Konovalov, 1968; Mishra et al., 1980; Goodfellow and Pirouz, 1982; Bhumibhamon, 1982), but the latter have not been as fully investigated, probably because most are less thermostable than those from the thermophiles.

Protease production in *T. vulgaris* strain PA II 4a is stimulated by oils that are 50% consumed during the fermentation and thought to control cell autolysis, a troublesome phenomenon that is very common among thermophilic actinomycetes (Leuchtenberger and Ruttloff, 1979, 1981). An alkaline cationic serine protease ("Thermitase") from a strain of the same taxon, isolated by isoelectric focusing, has a M.W. of 37,400, an isoelectric point of 9.0, and a single N-terminal group (Froemmel et al., 1978). Its temperature optimum ranges from 60°C to 90°C depending on the substrate. In general, the larger the substrate, the higher the temperature optimum (e.g., 60°C for esterolysis versus 90°C for proteinolysis) (Behnke et al., 1978; Kleine, 1982; Rothe et al., 1982). An anionic protease is elaborated concomitantly with thermitase. It shares certain characters such as size, sensitivity to Hg^{2+} and other inhibitors, ultraviolet spectrum, as well as pH and immunological characteristics. However, it differs in its isoelectric point (6.5), greater thermal stability, and decreased enzymatic activity. At pH 8 or 6 and 25°C or 4°C it is hydrolyzed completely by Thermitase and, like other protein substrates, serves to stabilize this enzyme (Kleine and Kettman, 1982). Thermitase binds two molecules of Ca^{2+}, the most weakly bound of which causes a quenching of the protein fluorescence emission, acts to stabilize the molecule against thermal denaturation and autolysis, and increases the esterolytic activity of the enzyme. The strongly bound calcium has none of these effects (Froemmel and Hoehne, 1981). Amino acid sequencing shows the enzyme to be of the subtilisin type (Hausdorf et al., 1980), an interesting observation in view of the many lines of evidence (G + C mol %, 16S RNA nucleotide catalogs and typical endospore formation) that show that despite their hyphal nature, thermoactinomycetes may be more closely related to *Bacillus* spp. than to other actinomycetes (Stackebrandt and Woese, 1981). A malate dehydrogenase (MDH) (E.C. 1.1.1.37) from another species of this genus, *T. thalpophilus*, although similar in stability and regulatory properties to MDH from thermophilic bacilli, has a M.W. only about 1/2 its size (Miller, 1982).

A serine protease from a different strain of *T. vulgaris* has a M.W. of 28,000, an isoelectric point of 8–9, and a temperature optimum of 55°C. It was stable to heating 1 h at 37°C and has a pH range of 7–9 (Stepanov et al., 1980). A comparison with a protease from *Bacillus thuringiensis* showed that both enzymes contain activity-essential cysteine residues and 10–14 homologous N-terminal sequences (Stepanov et al., 1981) and are thus considered very similar.

A heat-stable alkaline protease not further characterized was reported from a thermophilic *Micropolyspora* sp. It is part of a crude enzyme mixture that degraded walls of the yeast *Candida utilis* (Okazaki, 1972). Other yeast-lytic enzyme complexes (see discussion below) also contain proteases necessary for their action. A protease isolated from an *Oerskovia* has a M.W. of 30,000 and hydrolyzes Azocoll and a number of denatured proteins. It is insensitive to most protease inhibitors but is inhibited by yeast mannan and other polysaccharides (Scott and Schekman, 1980). Other proteases, such as keratinases, are produced by actinomadurae, micromonosporae, nocardiopsis, planomonosporae, and planobisporae. Elastinases are elaborated by strains of 12 of 14 genera of nonstreptomycetes studied (Goodfellow and Pirouz, 1982).

DD-Carboxypeptidase (DCP) activity refers to the D-alanyl-D-alanine peptidases implicated in the final steps of peptidoglycan synthesis. These enzymes split the terminal D-Ala-D-Ala link of the peptide, releasing free D-Ala. DCP activity was found in broths of *Nocardiopsis dassonvillei*, *"Nocardia" mediterranei*, *"Nocardia" piracicabensis*, and eight of nine actinomadurae tested, but not in any of the true (mycolate-containing) nocardiae examined. It is considered a taxonomic character of potential usefulness (Pellon et al., 1978). Among the strains looked at in greater detail, *Actinomadura madurae* R39 produces a 53,000-dalton DCP, which sequencing studies demonstrate to have a very uneven distribution of lysine and arginine residues in the molecule (Duez et al., 1981).

Lipases.   The lipases that permit the consumption of oils during fermentative growth of *T. vulgaris* (see the discussion of proteases) can be optimally produced by growth on corn oil. When fermented at 55°C, the strain utilizes 18 of 28 natural oils studied (Elwan et al., 1978a, b). Certain strains of this group isolated from milk, have both exocellular and endocellular lipases, the latter having less activity (Falkowski, 1979). Phospholipases are reported from a high percentage of *Streptoverticillium* spp. tested (Okawa and Yamaguchi, 1978) and from strains of *Micromonospora chalcea* (Ovcharov and Konovalov, 1968; Ko and Hora, 1970).

Lysins for Eukaryotes.   In the 1960's, before protoplast fusion gained the importance for genetics it has today, several groups reported that micromonosporae produced enzymes capable of yielding protoplasts from a variety of yeasts, molds, and bacteria, principally of the gram-positive types (Kotani et al., 1962; Ochoa et al., 1963, Gascon and Villanueva, 1964; Gascon et al., 1965; Novaes et al., 1966; Monreal and Reese, 1968). The crude preparations from *Micromonospora* sp. AS contain protease, glucanase, and chitinase activities (Novaes et al., 1966). A partially purified preparation is thermolabile and has a pH optimum of 5.8–8.8. Many commonly used osmotic substances such as NaCl and sucrose inhibit the activity; however, mannitol, dulcitol, and rhamnose do not (Gascon et al., 1965). Crude preparations from culture filtrates of *Thermoactinomyces vulgaris* yield, under appropriate stabilizing conditions, protoplasts of *Aspergillus niger* and various algae (Ismailova et al., 1977).

From *Micropolyspora* sp. 434, a heat-stable enzyme preparation active in lysing intact *Candida utilis* cells was shown to contain protease and

β-1,3-glucanase activities (Okazaki, 1972). More recently, attention has been directed toward the yeast-lytic activities of *Oerskovia* spp. and NMO's (nonmotile oerskoviae) (Mann et al., 1972). These actinomycetes (also under the name of *Arthrobacter luteus*) produce a variety of enzymes, including laminaranases, cellulases, proteases, β-1,3 glucanases, β-1,6-glucanases, and α-mannanases (Obata et al., 1975, 1977). Oerskoviae demonstrate chemotactic, predatory activity toward various yeasts, both alive and killed, and glucanase activity is optimally stimulated by growth on yeast walls or glucan. In one strain, at least four different β-1,3-glucanases have been found by isoelectric focusing, and these are thought to be the enzymes principally responsible for the yeast cell-lytic activity (Mann et al., 1978). Other workers feel that the glucanases act synergistically with the alkaline proteases present in the crude extract and only after the proteases have acted on the walls first (Obata et al., 1977; Scott and Schekman, 1980). A purified preparation of one glucanase has a M.W. of 55,000 and a K_m for yeast glycan of 0.4 mg/ml and 5.9 mg/ml for laminarin. The action on β-1,3-glucans is endolytic; the size of the oligomers released vary with the pH, the smaller ones predominating at low pH (Scott and Schekman, 1980).

D. Nitrogen Fixation

Biological nitrogen fixation holds hope of giving both agriculturalists and foresters breathing space in the coming years as petroleum stocks dwindle and fertilizer prices rise. Although known to exist for more than 100 years, the nitrogen-fixing actinomycete endophytes (actinorhiza) forming root-borne nodules on a variety of angiospermous plants were not isolated and maintained in pure culture until 1978 (Callaham et al., 1978). To date, more than 300 strains have been isolated (Baker, 1982; M. Lalonde, personal communication). These *Frankia* spp. are characterized by rather slow growth (2–5 days doubling time) and a tendency to microaerophily. All can grow to some degree on simple, defined media, but complex media are often used for their maintenance (M. P. Lechevalier et al., 1982b; Benson, 1982; Shipton and Burggraaf, 1982). Cultivated in vitro, they show morphological traits similar to those seen in planta (Newcomb et al., 1979; Van Dijk and Merkus, 1976) and can fix nitrogen at low levels in the presence of oxygen (Tjepkema et al., 1980; Gauthier et al, 1981; Torrey et al., 1981). The site of nitrogen fixation is believed to be within a specialized structure formed by the actinomycete called a vesicle (Torrey et al., 1981).

    Recent studies have shown that the strains presently in pure culture may be divided into at least two groups. Members of both groups have the same type of cell wall chemotype (III) and phospholipid pattern (PI). One group (A) is made up of strains that take up and utilize carbohydrates at low concentrations (0.5%), produce proteases, and have pigmented cells. They tend to be more aerobic than the second group and have very diverse, and often novel, whole-cell sugar patterns. They are ineffective (do not fix nitrogen) in their host plant and are heterogeneous serologically. In contrast, the members of the second group (B) do not appear to take up carbohydrates at a 0.5% concentration and probably lack an active carbohydrate transport system. Passive diffusion occurs at higher concentrations, and facilitated transport may occur in the presence of the amphipath, Tween 80. All are strictly microaerophilic and form a tight group serologically. Their whole-cell sugar pattern in all cases is of Type D, and all strains isolated thus far are effective (fix nitrogen) in the host plant. Both types of strains may be isolated from the same species of host plant and probably represent degrees of commitment to symbiosis (M. P. Lechevalier et al., 1983). In a study of over 100 isolates from speckled alder, five distinct groups were distinguished by SDS–polyacrylamide gel electrophoresis; 80% of the isolates fall into one gel group (Benson and Hanna, 1983).

Effectivity of frankiae in plants may vary from strain to strain (Dillon and Baker, 1982). Nodulation occurs in axenic culture (Lalonde et al., 1981) but in some cases may be "helped" by bacteria in the rhizosphere (Knowlton et al., 1980). The amount of nitrogen fixed compares favorably with the rhizobium-legume system (Torrey, 1978), and the energy requirement for nitrogen fixation in planta is also comparable (Tjepkema and Winship, 1980).

In *Frankia* sp. AvcII a study of nitrogen metabolism (with Tween 80 as the carbon source) showed that uptake of both ammonia and nitrate occurs and that there is selective uptake of glutamate and aspartate from a mixture of amino acids. Glutamine synthetase (GS) and glutamate oxaloacetate transaminase (GOT) are found in cell-free extracts and are probably involved in ammonia assimilation, but no glutamate dehydrogenase is present. Besides Tween 80, fatty acids and propionate can serve as carbon sources for this strain. Since AvcII belongs to group B, it is not surprising that these authors found no uptake of various sugars, alcohols and dicarboxylic acids at concentrations of 0.2%. Grown on Tween, there is no evidence of hexokinase, pyruvate kinase, or pyruvate dehydrogenase. Rather, glyoxylate cycle enzymes such as isocitrate lyase and malate synthetase are found and five citric acid cycle enzymes, which is logical in a fatty acid oxidizer grown on Tween. Phosphoglycerate kinase and 3-phosphoglyceraldehyde kinase were also reported to be present. Propionate is thought to be metabolized via succinate, although the latter is not taken up by the cells (Blom and Harkink, 1981; Akkermans et al., 1982).

The analysis of the base compositions of twelve strains of frankiae showed that the G + C mol % ranged from 68.4% to 72.1% DNA homology indicates that at least two distinct groups exist (An et al., 1983). Plasmids have been isolated from endophyte tissue taken from nodules (Dobritsa, 1982) and from strains grown in vitro (Marvel et al., 1983).

VI. GENETICS AND MOLECULAR BIOLOGY

Most genetic research on nonstreptomycetes has been carried out on species of rhodococci or nocardiae. *Rhodococcus* spp. are particularly amenable to genetic manipulation because they are among the most rapidly growing actino-mycetes (doubling time of 1/2–3/4 h) and their hyphae break down on aging into small coccoid units of relatively uniform size.

Genetic linkage maps have been published for *R. erythropolis*, *R. rhodochrous* and *Rhodococcus canicruria* (Adams, 1974; Brownell et al., 1981) and *Nocardia mediterranei* (Schupp et al., 1975). The map of *R. erythropolis* is linear and contains about 65 genetic traits (Brownell, 1978a), of which three are mating genes. Little information on zygote formation or phenotypic expression of the mating genes is available, although this has been studied from the point of view of the effects of various chemical and physical agents on the mating process. It has been hypothesized that one gene may code for specific cell surface antigens and one for the autolytic digestion of cell walls. The mating event is slow and probably involves cell fusion. Intermediate forms containing genomic material from both mating cells must be held in suboptional nutritional conditions for mating to take place (Adams and Adams, 1974), although limited metabolic activity is also necessary. Stationary phase cells were found to be best, and a ratio of 1:1 for the mating types and a cell density of one cell of each mating type per 100 μm^2 were optimal for recovery of recombinants of *R. erythropolis* (Kasweck, 1978). Colonial variants were found that produce recombinants at about 30-fold higher frequencies than the parent strains, but further efforts to improve these frequencies beyond this were unsuccessful (Brownell and Walsh, 1972; Adams and Adams, 1974). Evidence for "genetic drift" in vitro in a strain of *R. erythropolis* repeatedly cloned by single colony subculture is based on the detectable heterogeneity of the nucleotide sequences in the various substrains thus derived. These substrains did not vary when assessed by normal physiological testing or phage sensitivity, and thus the mutations might be considered "neutral" (Bradley, 1978).

Stable haploid recombinants, although at first believed interspecific (*erythropolis* × *canicruria*), were later shown to be *erythropolis* × *erythropolis* because of misclassification. These recombinants showed partial diploidy, probably because of the structure of their linkage maps. Such organisms, which are merozygotes, behave as though stable recombination has taken place in one region whereas in another heterogeneity is demonstrable. This phenomenon aids in mapping (Hopwood, 1973; Brownell, 1978a; Adams and Mayberry, 1978). Interestingly, recombination of mutants of *R. erythropolis* of homologous origin does not take place, and the reasons for this are not clear (Hopwood, 1973).

In contrast, there are reports of recombination of mutants of *N. opaca* (Schupp et al., 1975; Krallman-Wenzel, 1978; Krallman-Wenzel and Tarnok, 1978), *N. mediterranei* (Schupp et al., 1975, 1981), *N. asteroides* (Kasweck and Little, 1982), and *Micromonospora* (Beretta et al., 1971). In *N. opaca* the gene specifying hydrogen autotrophy (*aut*) has been transferred to *aut⁻* mutants (Reh and Schlegel, 1981). The marker can be cured by treatment with ultraviolet light or mitomycin C or transferred to other *R. opaca* strains as well as *R. erythropolis* and *Rhodococcus* (*"Corynebacterium"*) *hydrocarboclastus*, especially in late log phase (Reh, 1981). The recombination of *N. mediterranei* mutants differs from the rhodococcal system (as exemplified by *R. erythropolis*) and is more like that in *Streptomyces* (Hopwood, 1973). Genes coding for rifamycin production in this strain are clustered but are probably not plasmid-borne (Hopwood, 1981). An intergeneric cross between auxotrophic mutants of *N. mediterranei* and *Streptomyces griseus* carried out by conventional "mix and mate" techniques gave a recombinant frequency of 10^{-7}–10^{-6}, with 70–100% of the prototrophs being stable (Wesseling and Lago, 1981).

The transformation systems of *Thermoactinomyces vulgaris*, as first delineated by Hopwood and Wright (1972), show features like those of eubacteria in that recombination can take place both by mixing strains to be crossed or by incubating "competent" strains (the most common type) with donor DNA. Incompetent strains (not capable of receiving donor DNA but capable of acting as donors) have been found but are uncommon. A third type, which is represented by a single isolate, produces large amounts of DNase and therefore can apparently neither undergo nor bring about transformation (Hopwood, 1973). The presence of 0.025 M of Ca^{2+} and Mg^{2+} increases the transformation of a similar strain of *T. vulgaris* (Prasad and Sinha, 1982). Recombination of mutants of "*Streptosporangium*" *viridogriseum* subsp. *kofuense*, a chloramphenicol producer, is markedly increased by use of protoplast fusion (1% versus 5×10^{-5} by conventional techniques). Early- to mid-log phase cells are more resistant to protoplasting than late-log phase. Oh and co-workers used 2 mg/ml lysozyme stabilized in Okanishi's P solution and regenerated the strain on 0.3 M of sucrose agar containing Mg^{2+}, Ca^{2+}, and oatmeal. Polyethylene glycol (PEG) 6000 (40%) was used for fusion (Oh et al., 1980).

New rifamycins, not previously obtainable through mutation, can be obtained through fusion of various nonproducing mutant substrains of *N. mediterranei* (Schupp et al., 1981; Traxler et al., 1981). Fusion of two high-producing cephamycin mutants of *N. lactamdurans* increases production of this antibiotic complex by 10–15% (Wesseling and Lago, 1981). Protoplasts of *Micromonospora* mutants have been obtained by growth in high glycine media, followed by treatment with 1 mg/ml of lysozyme for 90 min at 37°C. Protoplast fusion occurred in 40% PEG 6000 at room temperature, and regeneration on appropriate media was successful (Szvoboda et al., 1980). When selected indirectly (on nutrient-sufficient media), only 1% of the recombinants were stable, whereas direct selection yielded 90% stability, but only if the selecting antibiotic were added after the phenotypic lag period (Ferenczy, 1981).

A. Phage

Phage-mediated gene transfers in actinomycetes have not been successful until recently (Hopwood, 1973). Recent intense interest in genetic engineering of these organisms has resulted in reports that some gene transfer can be accomplished by this means (Brownell, 1978a; Stuttard, 1983). In particular, transfection by deletion mutants of phage φEC in *R. erythropolis* is accomplished by use of protoplasts of the host (Brownell et al., 1982), this despite the fact that φEC-bearing lysogens of *R. erythropolis* yield phage that are relatively lytic and relysogenize at low frequencies (Adams and Brownell, 1976). The prophage can exist as either an integrated element or a plasmid and can be transferred by the lysogens not only to other related strains of the *R. erythropolis* mating complex, but also to strains of *Rhodococcus globerula* and *Rhodococcus calcarea* as well at better than 10^4 higher frequencies than the transfer of nonphage chromosomal markers (Brownell, 1978b). The sites of integration of the prophage into the chromosome have been determined (Brownell, 1978a), and restriction maps of the phage genome have been published (Brownell et al., 1981, 1982). The phage has a linear, ds-DNA (45 kbp) molecule with cohesive ends. Although the DNA is not restricted by the enzymes Bgl, II, EcoRI, Hind III, Pst I, Sal I, or Sma I, potential cloning sites include Bam HI, XbaI, and Pvu II. It was thought that this unusual pattern might reflect the presence of modified bases. Deletion mutants show that the genome can be reduced from 47 to 37 kbp (Brownell et al., 1982). A phage virulent for the strain of *T. vulgaris* (1227) used by Hopwood in his transformation studies was isolated by Sarfert et al. (1979). Analysis of the pure phage DNA shows it to have a G + C mol % of 41–42 and a M.W. of 28.8 × 10^6 daltons and to be capable of infecting the actinomycete and giving rise to normal phage progeny.

Studies with actinophage from a variety of actinomycete genera have shown that phage sensitivity of a given strain is in good part determined by the composition of its peptidoglycan (Prauser and Falta, 1968; Williams et al., 1980; Prauser, 1981). Nevertheless, similar peptidoglycan composition does not guarantee sensitivity to a phage lytic on a host with the same cell wall type. For example, phage active on mycolate-positive *Nocardiaceae* such as *N. asteroides* or *R. rhodochrous* are not active on mycolateless types such as *N. mediterranei* or *N. orientalis*, although they all have a cell wall chemotype of IV. Also phages may be species- or even strain-specific (Treuhaft, 1977; Prauser, 1982).

To date, phages have been isolated that are active on members of the following genera: *Actinoplanes, Actinopycnidium, Actinosporangium, Amorphosporangium, Chainia, Dactylosporangium, Elytrosporangium, Intrasporangium, Kitasatoa, Microellobosporia, Micromonospora, Micropolyspora, Mycobacterium, Nocardia, Nocardiodes, Nocardioipsis, Oerskovia, Promicromonospora, Rhodococcus, Saccharomonospora, Streptoverticillium,* and *Thermoactinomyces* (Higgins and Lechevalier, 1969; Prauser and Momirova, 1970; Prauser, 1976b; Kurup and Heinzen, 1978; Willoughby et al., 1972). Phages active on *Actinomadura* sp. have not been found (Prauser, 1982).

B. Plasmids

Although a strain of the hydrogen-oxidizing *R. opaca* was found to contain three plasmids with M.W.'s of 11, 22, and 90 Md, none is associated with the "aut" marker (ability to grow on a mineral agar medium without organic carbon under a mixed atmosphere of hydrogen-oxygen and carbon dioxide) (Reh, 1981).

Super-coiled plasmid DNA was detected in *Streptoverticillium mashuensis* KCC S-0059. It has a M.W. of 16×10^6 daltons and Bam HI and Sal I but not EcoRI or Hind III restriction sites (Okanishi et al., 1980).

A plasmid (25 Md) controlling the production of cephamycin C is absent in 5% of the population of the producer *Nocardia* (*Streptomyces*) *lactamdurans* NRRL 3802. Loss of the plasmid is not only accompanied by the loss of the production of the antibiotic and the capacity to form aerial mycelium, but also the loss of the ability to grow on lysine as the sole source of N and C (Castro et al., 1982). An attempt to clone into *Escherichia coli* genes coding for the production of the glucose isomerase of *Actinoplanes missouriensis* was not successful (Farrés et al., 1982).

C. DNA

DNA homology studies in actinomycetes often show disappointingly low ranges of reassociation in presumably closely allied species. For example, in various rhodococci, DNA:DNA pairing under stringent conditions with reference DNA from the nine type strains of the genus, shows homology values in a wide range (38–100%) (Mordarski et al., 1981). In a similar study investigating the relatedness of 21 representatives of 18 actinomycete genera (Stackebrandt et al., 1981) it was stated that DNA:DNA homology values of greater than 20%, obtained under optimal conditions of hybridization, can be considered evidence that species are closely related at the generic level. The authors felt that even species with homology values below 15% could be considered members of the same genus.

Stackebrandt et al. (1981) found three DNA homology clusters with reassociation values in a range of 18–49%. The first cluster (all cell wall chemotype II) contains *Actinoplanes*, *Ampullariella*, *Amorphosporangium*, *Micromonospora*, and *Dactylosporangium*; the second (all cell wall chemotype III) contains *Streptosporangium*, *Planomonospora*, and *Planobispora*; and the third (all cell wall chemotype I) group *Streptomyces*, *Chainia*, *Kitasatoa*, *Streptoverticillium*, *Microellobiosporia*, *Elytrosporangium*, and *Actinosporangium*. The isolated DNA's from the above strains were hybridized with three 16S rRNA preparations (from *Actinoplanes*, *Streptosporangium*, and *Kitasatoa*), and the results were analyzed in terms of both the percent of RNA bound to the DNA and the thermal stabilities of the duplexes. On this basis, none of the clusters were thought to be "specifically related." Representative species of *Actinomadura*, *Spirillospora*, and *Microtetraspora* do not fall clearly into any one of the three clusters.

Studies on the DNA homologies of various nocardial species show that the taxon *N. asteroides* consists of at least four homology groups. In some cases, carbohydrates contaminating the purified DNA of nocardiae caused aberrations in reassociation studies; however, even in uncontaminated material, some DNA's do not reassociate according to second-order kinetics (Bradley et al., 1978). The creation of a new genus of the Actinoplanaceae containing actinoplanetes with a peptidoglycan containing lysine rather than diaminopimelic acid is supported by DNA homology studies (Tille et al., 1982), the first time such a criterion has been used for generic differentiation.

GLOSSARY

1. *Arthrospores*: spores formed by the subdivision of an hypha.
2. *Basipetally*: growing from the base, with the apical portion the oldest.
3. *Blastospores*: spores formed by budding.
4. *Conidium* (*ia*): general term used to refer to most asexual spores of fungi and actinomycetes which are formed terminally, singly or in chains.
5. *Hypha* (*ae*): branching filament forming the thallus of fungi and actinomycetes.
6. *Mycelium* (*ia*): masses of hyphae.
7. *Mycolate*: α-branched, β-hydroxylated fatty acids formed by some actinomycetales with a type IV cell wall.
8. *Pycnidium* (*ia*): bottle-shaped structure containing spores.
9. *Sclerotium* (*ia*): spherical, hard structure formed by some fungi and actinomycetes. Sclerotia are considered to be forms resistant to adverse conditions.
10. *Sporangium* (*ia*): baglike structure containing spores.
11. *Synnema* (*ata*): fructification consisted of appressed, parallel hyphae at the tip of which spores are produced.

REFERENCES

Abbott, B. J. (1979) *Dev. Ind. Microbiol. 20*, 345–365.

Abbott, B. J. , Fukuda, D. S., and Archer, R. A. (1977) *Experientia 33*, 718–720.

Adams, J. N. (1974) in *Handbook of Microbiology* (Laskin, A. I., and Lechevalier, H. A., eds.), p. 661, CRC Press, Cleveland, Ohio.

Adams, J. N., and Adams, M. M. (1974) *J. Bacteriol. 119*, 646–649.

Adams, J. N., and Brownell, G. H. (1976) in *The Biology of the Nocardia* (Goodfellow, M., Brownell, G. H., and Serrano, J. A., eds.) pp. 285–309, Academic Press, London.

Adams, J. N., and Mayberry, K. J. C. (1978) in *Genetics of the Actinomycetales* (Freerksen, E., Tarnok, I. and Thumin, J. H. eds.), pp. 149–161, Gustav Fischer, New York.

Akit, J., Cooper, D. G., Manninen, K. I., and Zajic, J. E. (1981) *Curr. Microbiol. 6*, 145–150.

Akkermans, A. D. L., Blom, J., Huss-Danell, K., and Roelofsen, W. (1982) in *Proceedings of the Second National Symposium on the Biology of Nitrogen Fixation*, pp. 169–179, Helsinki.

Allen, M. J., and Hartman, P. A. (1972) *J. Bacteriol. 109*, 452-454.

An, C. S., Willis, J. W., Riggsby, W. S., and Mullin, B. C. (1983) *Can. J. Bot. 61*, 2856-2862.

Antoine, A. D., Dean, A. V., and Gilbert, S. G. (1980) *Appl. Environ. Microbiol. 39*, 777-781.

Arai, T., Kuroda, S., and Watanabe, I. (1981) *Z. Bakt. Suppl. 11*, 297-307.

Archer, R. A., Fukuda, D. S., Kossoy, A. D., and Abbott, B. J. (1979) *Appl. Environ. Microbiol. 37*, 965-971.

Asselineau, T. (1978) in *Actinomycetes* (Schaal, K. P., and Pulverer, G., eds.) pp. 391-400, Gustav Fischer, New York.

Austin, B., Colwell, R. R., Walker, J. D., and Calomiris, J. (1977) *Dev. Indust. Microbiol. 18*, 685-698.

Baggi, G., Beretta, L., Galli, E., Scolastico, C., and Treccani, V. (1978) in *Oil Ind. Microb. Ecosyst. Proc. Meet., 1977* (Chater, K. W. A., and Somerville, H. J., eds.), pp. 129-136, Heyden, London.

Baker, D. (1982) *The Actinomycetes 17*(1), 35-42.

Barksdale, L., and Kim, K. S. (1977) *Bact. Rev. 41*, 217-372.

Bartholomew, G. W., and Alexander, M. (1979) *Appl. Env. Microbiol. 37*, 932-937.

Behnke, U., Kleine, R., Ludewig, M., and Ruttloff, H. (1978) *Acta. Biol. Md. Ger. 37*, 1205-1214.

Bellamy, W. D. (1979) *ASM News 45*, 326-331.

Benson, D. R. (1982) *Appl. Environ. Microbiol. 44*, 461-465.

Benson, D. R., and Hanna, D. (1983) *Can. J. Bot. 61*, 2919-2923.

Berdy, J., Aszalos, A., Bostian, M., and McNitt, K. L. (1980-1981) *CRC Handbook of Antibiotic Compounds*, Vols. I to VII, CRC Press, Boca Raton, Fla.

Beretta, M., Betti, M., and Polsinelli, M. (1971) *J. Bacteriol. 107*, 415-419.

Bhuminbhamon, O. (1982) *Thai J. Agric. Sci. 15*, 85-94.

Bland, C. E., and Couch, J. N. (1981) in *The Prokaryotes* (Starr, M. P., Stolp, H., Trüper, H. G., Balows, A., and Schlegel, H. G. eds.), pp. 2004-2010, Springer, New York.

Blom, J., and Harkink, R. (1981) *FEMS Microbiol. Lett. 11*, 221-224.

Blyth, W. (1973) in *Actinomycetales* (Sykes, G. S., and Skinner, F. A., eds.), pp. 261-275, Academic Press, New York.

Bond, J. S., and Bradley, S. G. (1978) in *Nocardia and Streptomyces*, pp. 389-397, Gustav Fischer, New York. (Mordarski, M., Kurylowicz, W., and Jeljaszewicz J., eds.)

Bradley, S. G. (1978) in *Nocardia and Streptomyces*, pp. 287-302, Gustav Fischer, New York. (Mordarski, M., Kurylowicz, W. and Jeljaszewicz, J., eds.)

Bradley, S. G., Enquist, L. W., and Scribner, H. E. (1978) in *Genetics of the Actinomycetales*, pp. 207-224, Gustav Fischer, New York. (Freerksen, E., Tarnds, I. and Thumim, J. H. eds.)

Brownell, G. H. (1978a) in *Genetics of the Actinomycetales* (Freerksen, E., Tarnok, I. and Thumin, J. H. eds.), pp. 137-148, Gustav Fischer, New York.

Brownell, G. H. (1978b) *Zbl. Bakt. Abt. I. Suppl. 6*, 313-317.

Brownell, G. H., and Walsh, R. S., (1972) *Genetics 70*, 341-351.

Brownell, G. H., Enquist, L. W., and Denniston-Thompson, K. (1981) in *Actinomycetes* (Schaal, K. P., and Pulverer, G., eds.), pp. 563-575, Gustav Fischer, New York.

Brownell, G. H., Saba, J. A., Enquisdt, L. W., and Denniston, K. (1982) in *Proceedings of the Fifth International Symposium on Actinomycetes Biology*, August 1982, pp. 159–160, Oaxtepec, Mexico.

Cain, R. B. (1958) *J. Gen. Microbiol. 19*, 1–14.

Cain, R. B. (1981) in *Actinomycetes* (Schaal, K. P., and Pulverer, G., eds.), pp. 335–354, Gustav Fischer, New York.

Cairns, W. L., Cooper, D. G., Zajic, J. E., Wood, J. M., and Kosaric, N. (1982) *Appl. Environ. Microbiol. 43*, 362–366.

Callaham, D., Del Tredici, P., and Torrey, J. G. (1978) *Science 199*, 899–902.

Castro, J. M., Liras, P., Aquilar, A. and Martin, J. F. (1982) in *Fifth International Symposium on Actinomycetes Biology Resumés*, August 1982, pp. 203–204, Oaxtepec, Mexico.

Chapman, P. J., Meerman, G., Gunsalus, I. C., Srinivasan, R., and Reinhart, K. L. (1966) *J.A.C.S. 88*, 618–619.

Cheetham, P. S. J., Dunnill, P., and Lilly, M. D. (1980) *Enzyme Microb. Technol. 2*, 201–205.

Collins, M. D., McCarthy, A. J., and Cross, T. (1982) *Z. Bakt. Abt. I Orig. C.3*, 358–363.

Couch, J. N. (1949) *J. Elisha Mitchell Sci. Soc. 65*, 315–318.

Cox, D. P., and Goldsmith, C. D. (1979) *Appl. Environ. Microbiol. 38*, 514–520.

Crawford, D. L. (1974) *Can. J. Microbiol. 20*, 1069–1072.

Crawford, D. L. (1975) *Can. J. Microbiol. 21*, 1842–1848.

Crawford, D. L., and McCoy, E. (1972) *Appl. Microbiol. 24*, 150–152.

Crawford, D. L., and Sutherland, J. B. (1980) in *Lignin Biodegradation: Microbiology, Chemistry and Potential Applications* (Kirk, T. K., Higuchi, T., and Chang, H.-M., eds.), pp. 95–101, CRC Press, Boca Raton, Fla.

Crawford, D. L., McCoy, E., Harkin, J. M., and Jones, P. (1973a) *Biotech. Bioeng. 15*, 833–843.

Crawford D. L., McCoy, E., Harkin, J. M., Kirk, T. K., and Obst, J. R. (1973b) *Appl. Microbiol. 26*, 176–184.

Cross, T. (1981) in *The Prokaryotes* (Starr, M. P., Stolp, H., Trüper, H. G., Balows, A., and Schlegel, H. G., eds.), pp. 2091–2102, Springer, New York.

Cross, T. (1982) *Devel. Indust. Microbiol. 23*, 1–18.

Cross, T., and Al-Diwany, L. J. (1981) in *Actinomycetes* (Schaal, K. P., and Pulverer, G., eds.), pp. 59–65, Gustav Fischer, New York.

Cross, T., and Attwell, R. W. (1974) in *Spore Research, 1977* (Barker, A. N., Gould, G. W., and Wolf, J., eds.), pp. 11–20, Academic Press, London.

Cross, T., Lechevalier, M. P., and Lechevalier, H. A. (1963) *J. Gen. Microbiol. 31*, 421–429.

Davis, P. J., and Rizzo, J. D. (1982) *Appl. Environ. Microbiol. 43*, 884–890.

DeBont, J. A. M., Van Dijken, J. P., and Van Ginkel, K. G. (1982) *Biochim. Biophys. Acta. 714*, 465–470.

DiGeronimo, M. J., and Antoine, A. D. (1976) *Appl. Environ. Microbiol. 31*, 900–906.

Dillon, J. T., and Baker, D. (1982) *New Phytol. 92*, 215–219.

Dobritsa, S. V. (1982) *FEMS Microbiol. Lett. 15*, 87–91.

Donoghue, N. A., Griffin, M., Norris, D. B., and Trudgill, P. W. (1976) in *Proceedings of the Third International Biodegradation Symposium* (Sharpley, J. M., and Kaplan, A. M., eds.), pp. 43–56, Applied Science Publishers, Ltd., London.

Duez, C., Frère, J. M., Ghuysen, J. M., Van Beeumen, J., and Vandekerckhove, J. (1981) *Biochim. Biophys. Acta. 671*, 109–116.

Eggeling, L., and Sahm, H. (1980) *Arch. Microbiol. 126*, 141–148.

Eggeling, L., and Sahm, H. (1981) *Z. Bakt. Suppl. 11*, 361–365.

Ellwardt, P. C., Haider, K., and Ernst, L. (1981) *Holzforschung 35*, 103–109.

El-Nakeeb, M., and Lechevalier, H. A. (1963) *Appl. Microbiol. 11*, 75–77.

Elwan, S. H., Mostafa, S. A., Khodair, A. A., and Ali, O. (1978a) *Z. Bakt. Abt. 2 133*, 706–712.

Elwan, S. H., Mostafa, S. A., Khodair, A. A., and Ali, O. (1978b) *Z. Bakt. Abt. 2 133*, 713–722.

Engelhardt, G., Rast, H. G., and Wallnoefer, P. R. (1977) *Arch. Microbiol 114*, 25–33.

Engelhardt, G., Rast, H. G., and Wallnoefer, P. R. (1979a) *FEMS Microbiol. Lett. 5*, 377–383.

Engelhardt, G., Rast, H. G., and Wallnoefer, P. R. (1979b) *FEMS Microbiol. Lett. 5*, 245–251.

Falkowski, J. (1979) *Z. Bakteriol. Abt. 1. Orig. Reiche B 168*, 361–366.

Farrés, A., Bolivar, F., Calva, E., and Sanchez, S. (1982) in *Fifth International Symposium on Actinomycetes Biology, Résumés*, August 1982, pp. 44–45, Oaxtepec, Mexico.

Ferenczy, L. (1981) in *Genetics as a Tool in Microbiology* (Glover, S. W., and Hopwood, D. A., eds.), pp. 1–34, Cambridge University Press, Cambridge, England.

Froemmel, C., and Hoehne, W. E. (1981) *Biochim. Biophys. Acta. 670*, 25–31.

Froemmel, C., Hausdorf, G., Hoehne, W. E., Behnke, O., and Ruttloff, H. (1978) *Acta Biol. Med. Ger. 37*, 1193–1204.

Fukushima, J., Sakano, Y., Iwai, H., Itoh, Y., Tamura, M., and Kobayashi, T. (1982) *Agr. Biol. Chem. 46*, 1423–1424.

Furuhashi, K., Taoka, A., Uchida, S., Karube, I., and Suzuki, S. (1981) *Eur. J. Appl. Microbiol. Biotechnol. 12*, 39–45.

Garcia-Acha, I., and Villanueva, J. R. (1962) *Microbiol. Espanol 15*, 165–169.

Gascon, S., and Villanueva, J. R. (1964) *Can. J. Microbiol. 10*, 301–303.

Gascon, S., Ochoa, A. G., Novaes, M., and Villanueva, J. R. (1965) *Arch. Mikrobiol. 51*, 156–167.

Gauthier, D., Diem, H. G., and Dommerques, Y. (1981) *Appl. Environ. Microbiol. 41*, 306–308.

Giardina, M. C., Giardi, M. T., and Filacchioni, G. (1980) *Agr. Biol. Chem 44*, 2067–2072.

Gibson, D. T. (1978) *U.S. NTIS, AD Rep. Announce. Index (U.S.) 64*, CA 89: 1174849.

Gledhill, W. E. and Casida, L. E. (1969a) *Appl. Microbiol. 18*, 114–121.

Gledhill, W. E. and Casida, L. E. (1969b) *Appl. Microbiol. 18*, 340–349.

Glymph, J. L., and Stutzenberger, F. J. (1977) *Appl. Environ. Microbiol. 34*, 391–397.

Gochnauer, M. B., Leppard, G. G., Komaratat, P., Kates, M., Novitsky, T., and Kushner, D. J. (1975) *Can. J. Microbiol. 21*, 1500–1511.

Golovlev, E. L., Golovleva, L. A., Ananin, V. M., and Skryabin, G. K. (1976) *Izv. Akad. Nauk SSSR, Ser. Biol. 1976*, 834–939 (CA 86: 68102r).

Golovlev, E. L., Golovleva, L. A., Eroshina, N. V., and Skryabin, G. K. (1978) in *Nocardia and Streptomyces*, pp. 269–283, Gustav Fischer, New York. (Mordarski, M., Kurylowicz, W. and Jeljaszewicz, J., eds.)

Gong, C.-S., Chen, L.-F., and Tsao, G. T. (1980) *Biotechnol. Bioeng. 22*, 833–845 (CA 92: 176339f).

Goodfellow, M., and Alderson, G. (1977) *J. Gen. Microbiol. 100*, 99–122.

Goodfellow, M., and Pirouz, T. (1982) *J. Gen. Microbiol. 128*, 503–527.

Goodfellow, M., Alderson, G., and Lacey, J. (1979) *J. Gen. Microbiol. 112*, 95–111.

Gordon, R. E., and Hagan, W. A. (1937) *Amer. Rev. Tuberc. Pulm. Dis. 36*, 549–554.

Gradziel, K., Haider, K., Kochmanska, J., Malarczyk, E., and Trojanowski, J. (1978) *Acta Microbiol. Pol. 27*, 103–109.

Haase-Aschoff, K., and Lingens, F. (1978) *Hoppe-Seyler's Z. Physiol. Chem. 360*, 621–632 (CA 91: 71492b).

Hagerdal, B. G. R., Ferchak, J. D., and Pye, E. K. (1978) *Appl. Environ. Microbiol. 36*, 606–612.

Hagerdal, B. G. R., Harris, H., and Pye, E. K. (1979) *Biotechnol. Bioeng. 21*, 345–355.

Hagerdal, B. G. R., Ferchak, J. D., and Pye, E. K. (1980) *Biotechnol. Bioeng. 22*, 1515–1526.

Hamano, K., Obata, H., and Tokuyama, T. (1982) *Agr. Biol. Chem. 46*, 1139–1143 (CA 97: 35825c).

Haraguchi, T., Hayashi, E., Takahashi, Y., Tijima, Y., and Hatakeyama, H. (1980) in *Biodeterioration. Proceedings of the Fourth International Biodeterion Symposium* (Oxley, T. A., Becker, G., and Allsopp, D. eds.), pp. 123–126, Pitman Publishing, London.

Harper, D. B. (1977) *Biochem. J. 165*, 304–319.

Hasegawa, T., Lechevalier, M. P., and Lechevalier, H. A. (1978) *Int. J. System. Bacteriol. 28*, 304–310.

Hauer, B., Haase-Ashoff, K., and Lingens, F. (1982) *Hoppe-Seyler's Z. Physiol. Chem. 363*, 499–506 (CA 97: 5226b).

Hausdorf, G., Krueger, K., and Hoehne, W. E. (1980) *Int. J. Pept. Protein Res. 15*, 420–429.

Hayward, A. C., and Sly, L. I. (1976) *J. Appl. Bacteriol. 40*, 355–364.

Henssen, A., and Schafer, D. (1971) *Int. J. System. Bacteriol. 21*, 29–34.

Higgins, M., and Lechevalier, M. P. (1969) *J. Virol. 3*, 210–216.

Hollingdale, M. R. (1975) *J. Gen. Microbiol. 86*, 250–258.

Hopwood, D. A. (1973) in *Actinomycetales: Characteristics and Practical Importance* (Sykes, G., and Skinner, F. A., eds.), pp. 131–153, Academic Press, London.

Hopwood, D. A. (1981) in *Genetics as a Tool in Microbiology* (Glover, S. W., and Hopwood, D. A., eds.), pp. 187–218, Cambridge University Press, Cambridge, England.

Hopwood, D. A., and Wright, H. M. (1972) *J. Gen. Microbiol. 71*, 383–398.

Humphrey, A. E., Moreira, A., Arminger, W., and Zabriskie, D. (1977) in *Single Cell Protein from Renewable and Nonrenewable Resources* (Humphrey, A. E., and Gaden, E. L., eds.), pp. 45–64, *Biotechnology and Bioengineering Supplement 7*, John Wiley & Sons, New York.

Hupkes, J. V., and Van Tilburg, R. (1976) *Staerke, 28*, 356–360 (CA 86: 1729g).

Hussein, A., and Krassilnikov, N. A. (1969) *Microbiology USSR 38*, 748–753.

Ismailova, D. Yu., Yakovleva, M. B. Golovina, I. G., and Loginova, L. G. (1977) *Prikl. Biokhim. Mikrobiol 13*, 219–224 (CA 86: 167634x).

Jamison, V. W., Raymond, R. L., and Hudson, J. D. (1976) in *Biodegradation of High-Octane Gasoline): (Sharpley, J. M., and Kaplan, A. M., eds.), Proceedings of the Third International Biodegradation Symposium*, pp. 187–196, Applied Science Publishers, London.

Johnston, D. W., and Cross, T. (1976) *Freshwater Biol. 6*, 457–463.

Jones, D., Watkins, J., and Meyer, D. J. (1970) *Nature 226*, 1249–1250.

Kaiser, P. (1971) *Ann. Inst. Past. 121*, 389–404.

Kanner, D., and Bartha, R. (1979) *J. Bacteriol. 139*, 225–230.

Kanner, D., and Bartha, R. (1982) *J. Bacteriol. 150*, 989–992.

Kasweck, K. L. (1978) in *Genetics of the Actinomycetales* (Freerksen, E., Tarnok, I., and Thumin, J. H., eds.), pp. 163–185, Gustav Fischer, New York.

Kasweck, K. L., and Little, M. L. (1982) *J. Bacteriol. 149*, 403–406.

Klages, V., and Lingens F. (1979) *FEMS Microbiol. Lett. 6*, 201–203.

Kleine, R. (1982) *Acta Biol. Med. Ger. 41*, 89–102 (CA 97: 105982c).

Kleine, R., and Kettmann, U. (1982) *Hoppe-Seyler's Z. Physiol. Chem. 363*, 843–853.

Knowlton, S., Berry, A., and Torrey, J. G. (1980) *Can. J. Microbiol. 26*, 971–977.

Ko, W.-H., and Hora, F. K. (1970) *Soil Science 110*, 355–358.

Korosteleva, L. A., Kost, A. N., Vorob'eva, L. I., Modyanova, L. V., Terent'ev, P. B., and Kulikov, N. S. (1981) *Prikl. Biokhim. Microbiol. 17*, 380–388 (CA 95: 57803e).

Kosaric, N., Gray, N. C. C., and Cairns, W. L. (1983) in *Biotechnology*, Vol. III (Dellweg, H., ed.), pp. 576–592, Verlag Chemie, Weinheim, West Germany.

Kost, A. N., Vorob'eva, L. I., Terent'ev, P. B., Modyanova, L. V., Shibilkina, O. K., and Korosteleva, L. A. (1977) *Prikl. Biokhim. Mikrobiol. 13*, 696–703 (CA 87: 197036s).

Kotani, S., Harada, K., Kitaura, T., Hashimoto, Y., Matsubara, T., and Chimori, M. (1962) *Biken. J. 5*, 117–119.

Krallmann-Wenzel, U. (1978) in *Nocardia and Streptomyces* (Mordarski, M., Kurylowicz, W., and Jeljaszewicz, J., eds.), pp. 335–341, Gustav Fischer, New York.

Krallmann-Wenzel, U., and Tarnok, I. (1978) in *Genetics of the Actinomycetales* (Freerksen, E., Tarnok, I., and Thumin, J. H., eds.), pp. 187–191, Gustav Fischer, New York.

Kretschmer, A., Bock, H., and Wagner, F. (1982) *Appl. Environ. Microbiol. 44*, 864–870.

Kroppenstadt, R. M., and Kutzner, H. J. (1978) in *Nocardia and Streptomyces* (Kurylowicz, W., and Jeljaszewicz, J., eds.), pp. 125–133, Gustav Fischer, New York.

Kurane, R., Suzuki, T., Takahara, Y., Kurita, N., and Miyaji, M. (1979) *Agr. Biol. Chem. 43*, 2093–2098.

Kurane, R., Suzuki, T., and Takahara, T. (1980) *Agr. Biol. Chem. 44*, 523–527.

Kurup, V. P., and Agre, N. S. (1983) *Int. J. Syst. Bacteriol. 33*, 663–665.

Kurup, V. P., and Heinzen, R. J. (1978) *Can. J. Microbiol. 24*, 794–797.

Kvasnikov, E. I., Nesterenko, O. A., Romanovskaya, V. A., and Kasumova, S. A. (1971) *Microbiology USSR* (Engl. Trans.) *40*, 240–246.

Lacey, J., and M. Goodfellow (1975) *J. Gen. Microbiol. 88*, 75–85.

Lalonde, M., Calvert, H. E., and Pine, S. (1981) in *Current Perspectives in Nitrogen Fixation* (Gibson, A., and Newton, W., eds.), pp. 296–299, Australian Academy of Sciences, Canberra.

Lechevalier, H. A., and Holbert, P. E. (1965) *J. Bacteriol. 89*, 217–222.

Lechevalier, H. A., and Lechevalier, M. P. (1981a) in *The Prokaryotes* (Starr, M. P., Stolp, H., Trüper, H. G., Balows, A., and Schlegel, H. G., eds.), pp. 1915–1922, Springer, New York.

Lechevalier, H. A., and Lechevalier, M. P. (1981b) in *The Prokaryotes* (Starr, M. P., Stolp, H., Trüper, H. G., Balows, A., and Schlegel, H. G., eds.), pp. 2118–2123, Springer, New York.

Lechevalier, M. P. (1972) *Int. J. Syst. Bacteriol. 22*, 260–264.

Lechevalier, M. P. (1976) in *The Biology of the Nocardiae* (Goodfellow, M., Brownell, G. H., and Serrano, J. A., eds.), pp. 1–38, Academic Press, London.

Lechevalier, M. P. (1977) *Crit. Rev. Microbiol. 5*, 109–210.

Lechevalier, M. P. (1981) in *Actinomycetes* (Schaal, K. P., and Pulverer, G., eds.), pp. 159–166, Gustav Fischer, New York.

Lechevalier, M. P., and Lechevalier, H. A. (1980) in *Actinomycete Taxonomy*, SIM Special Publication No. 6 (Dietz, A., and Thayer, D. W., eds.), pp. 227–291, Society for Industrial Microbiology, Arlington, Va.

Lechevalier, M. P., Lechevalier, H. A., and Heintz, C. E. (1973) *Int. J. Syst. Bacteriol. 23*, 157–170.

Lechevalier, M. P., de Bièvre, C., and Lechevalier, H. A. (1977) *Biochem. Syst. Ecol. 5*, 249–260.

Lechevalier, M. P., Stern, A. E., and Lechevalier, H. A. (1981) in *Actinomycetes* (Schaal, K. P., and Pulverer, G., eds.), pp. 111–116, Gustav Fischer, New York.

Lechevalier, M. P., Gerber, N. N., and Umbriet, T. A. (1982a) *Appl. Environ. Microbiol. 43*, 367–370.

Lechevalier, M. P., Horrière, F., and Lechevalier, H. A. (1982b) *Dev. Indust. Microbiol. 23*, 51–60.

Lechevalier, M. P., Baker, D., and Horrière, F. (1983) *Can. J. Bot. 61*, 2826–2833.

Lefebvre, G., Germain, P., and Schneider, F. (1980) *Bull. Soc. Chim. Fr.* (1–2, Pt. 2), 96–97 (CA 92: 196365g).

Leuchtenberger, A., and Ruttloff, H. (1979) *Z. Allg. Mikrobiol. 19*, 609–627.

Leuchtenberger, A., and Ruttloff, H. (1981) *Abh. Akad. Wiss. DDR Abt. Math. Naturwiss. Tech. 1981*, 347–356 (CA 96: 102428v).

Locci, R., Rogers, J., Sardi, P., and Schofield, G. M. (1982) *The Actinomycetes 16*, 140–142.

Luedemann, G. M. (1968) *J. Bacteriol. 96*, 1848–1858.

Luedemann, G. M. (1970) *Adv. Appl. Microbiol. 11*, 101–133.

MacDonald, C. R., Cooper, D. G., and Zajic, J. E. (1981) *Appl. Environ. Microbiol. 41*, 117–123.

Makkar, N. S. and Cross, T. (1982) *J. Appl. Bacteriol. 52*, 209–218.

Mann, J. W., Heintz, C. E., and MacMillan, J. D. (1972) *J. Bacteriol 111*, 821–824.

Mann, J. W., Jeffries, T. W., and MacMillan, J. D. (1978) *Appl. Environ. Microbiol. 36*, 594–605.

Mara, D. D., and Oragui, J. I. (1981) *Appl. Environ. Microbiol. 42*, 1037–1042.

Margaritis, A., Kennedy, K., Zajic, J. E., and Gerson, D. F. (1979) *Dev. Indust. Microbiol. 20*, 623–630.

Martinek, K., Levashov, A. V., Khmelnitsky, Yu. L., Klyachko, N. L., and Berezin, I. V. (1982) *Science 218*, 889–891.

Marvel, D. J., Buikema, W. S., and Torrey, J. G. (1983) Abstr. The Biology of *Frankia* and Its Associations with Higher Plants. Univ. of Wisc. 1982, p. 114.

Matsumae, A., Ohtani, M., Takeshima, H., and Hata, T. (1968) *J. Antibiot. 21*, 616–625.

Maurer, R. R., and Batt, R. D. (1962) *J. Bacteriol. 83*, 1131–1139.

Meyer, J. (1976) *Int. J. Syst. Bacteriol. 26*, 487–493.

Meyer, H. P., and Humphrey, A. E. (1982) *Biotechnol. Bioeng. 24*(8), 1901–1904.

Michalski, C. J., and Domnas, A. (1974) *Neth. 33*. 153–170.

Miller, D. J., (1981) in *Actinomycetes* (Schaal, K. P., and Pulverer, G., eds.), pp. 355–360, Gustav Fischer, New York.

Miller, D. J. (1982) in *Proceedings of the Fifth International Symposium on Actinomycetes Biology*, August 1982, pp. 110–111, Oaxtepec, Mexico.

Miller, J. M., and Gray, D. O. (1982) *J. Gen. Microbiol. 128*, 1803–1809.

Mishra, S. K., Gordon, R. E., and Barnett, D. A. (1980) *J. Clin. Microbiol. 11*, 628–736.

Monreal, J., and Reese, E. T. (1968) *Arch. Biochem. Biophys. 126*, 960–962.

Mordarski, M., Kaszen, I., Tkacz, A., Goodfellow, M., Alderson, G., Schaal, K. P., and Pulverer, G. (1981) in *Actinomycetes* (Schaal, K. P., and Pulverer, G., eds.), pp. 25–31, Gustav Fischer, New York.

Mulheirn, L. J., and Van Eyk, J. (1981) *J. Gen. Microbiol. 126*, 267–275.

Mulkins-Phillips, G. J., and Stewart, J. E. (1974) *Appl. Microbiol. 28*, 915–922.

Nakajima, K., and Sato, A. (1981) *Nippon Nogei, Kagaku, Kaishi 55*, 825–828.

Nakajima, K., Sato, A., Takahara, Y., Hosaka, T., and Tamiguchi, M. (1980) *Nippon Nogei Kagaku Kaishi 54*, 27–29, 1980.

Nakajima, K., Sato, A., Misono, T., and Takeo, I. (1981) *Nippon Nogei Kagaku Kaishi 55*, 1187–1195.

Nesterenko, O. A., Kasumova, S. A., and Kvasnikov, E. E. (1979) *Mikrobiol. Zh. (Kiev) 41*, 110–114 (CA 91: 2290).

Newcomb, W., Callaham, D., Torrey, J. G., and Peterson, R. L. (1979) *Bot. Gaz. 140*, (Suppl.) S22–S34.

Noma, Y. (1979a) *Koen Yoshishu-Koryo, Terupen Oyobi Seiyu Kagaku ni Kansuru Toronkai, Chem. Soc. Tokyo, 23.* Japan 42–44.

Noma, Y. (1979b) *Nippon Nogei Kagaku Kaishi 53*, 35–39.

Nonomura, H., and Ohara, Y. (1969a) *J. Ferment. Technol. 47*, 463–469.

Nonomura, H., and Ohara, Y. (1969b) *J. Ferment. Technol. 47*, 701–709.

Nonomura, H., and Ohara, Y. (1971) *J. Ferment. Technol. 49*, 887–894.

Novaes, M., Gascon, S., Ochoa, A. G., and Villanueva, J. R. (1966) *Biochim. Biophys. Acta 115*, 486–488.

Obata, T., Kamashita, K., and Nunokawa, Y. (1975) *J. Ferment. Technol. 53*, 256–263.

Obata, T., Fujioka, K., Hara, S., and Namba, Y. (1977) *Agr. Biol. Chem.* (1977) *41*, 671–678.

Ochoa, A. G., Acha, I. G., Gascon, S., and Villanueva, J. R. (1963) *Experientia 19*, 581–582.

Oh, Y. K., Speth, J. L., and Nash, C. H. (1980) *Dev. Indust. Microbiol. 21*, 219–226.

Okanishi, M., Manome, T., and Umezawa, H. (1980) *J. Antibiot. 33*, 88–91.

Okawa, Y., and Yamaguchi, T. (1978) *Microbiol. Immunol. 22*, 233–234.

Okazaki, H. (1972) *J. Ferm. Technol. 50*, 405–413.

Omura, S., Takahashi, Y., Iwai, Y., and Tanaka, H. (1982) *J. Antibiot. 35*, 1013–1019.

Ono, H. (1973) *Kagaku Kogaku 37*, 451–456 (CA 79: 45372v).

Ooyama, S., and Foster, J. W. (1965) Antonie van Leeuwenhoek. *J. Microbiol. Serol. 31*, 45–65.

Orchard, V. A., and Goodfellow, M. (1974) *J. Gen. Microbiol. 85*, 160–162.

Ovcharov, A. K., and Konovalov, S. A. (1968) *Prikl. Biokhim. Microbiol. SSSR 4*, 499–505.

Palleroni, N. J. (1980) *Arch. Microbiol. 123*, 53–55.

Patel, T. R., and Gibson, D. T. (1976) *J. Bacteriol. 128*, 842–850.

Pellon, G., Pommier, M. T., Voiland, A., and Michel, G. (1978) *Actinomycetes Relat. Org. 13*, 31–34.

Prasad, U., and Sinha, U. (1982) in *Fifth International Symposium on Actinomycetes Biology Resumés*, pp. 175–176, Oaxtapec, Mexico.

Prauser, H. (1976a) *Int. J. Syst. Bacteriol. 26*, 58–65.

Prauser, H. (1976b) in *The Biology of the Nocardiae* (Goodfellow, M., Brownell, G. H., and Serrano, J. A., eds.), pp. 266–284, Academic Press, London.

Prauser, H. (1981) in *Actinomycetes* (Schaal, K. P., and Pulverer, G., eds.), pp. 88–92, Gustav Fischer, New York.

Prauser, H. (1982) *Fifth International Symposium on Actinomycetes Biology, Resumés*, August 1982, Oaxtepec, Mexico.

Prauser, H., and Falta, R. (1968) *Z. Allg. Mikrobiol. 8*, 39–46.

Prauser, H., and Momirova, S. (1970) *Zeit. Allg. Mikrobiol. 10*, 219–222.

Quaroni, S., and Locci, R. (1978) *Ann. Microbiol. Enzymol. 28*, 79–84 (CA 93: 18225c).

Rapp, P., Bock, H., Urbani, E., Wagner, F., Gebetsberger, W., and Schulz, W. (1977) *DECHEMA Monogr. 81* (1670–1692), 177–186 (CA 88: 61112b).

Rast, H. G., Engelhardt, G., Wallnoefer, P. R., Oehlmann, L., and Wagner, K. (1979) *J. Agr. Food. Chem. 27*, 699–702.

Rast, H. G., Engelhardt, G., and Wallnoefer, P. R. (1980a) *FEMS Microbiol. Lett. 7*, 1–6.

Rast, H. G., Englehardt, G., Ziegler, W., and Wallnoefer, P. R. (1980b) *FEMS Microbiol. Lett. 8*, 259–263.

Raymond, R. L., and Jamison, V. W. (1971) *Adv. Appl. Microbiol. 14*, 93–122.

Raymond, R. L., Jamison, V. W., and Hudson, J. O. (1967) *Appl. Microbiol. 15*, 857–869.

Raymond, R. L., Jamison, V. W., and Hudson, J. O. (1977) *AIChE Symp. Ser. 73*, 390–404.

Reh, M. (1981) in *Actinomycetes* (Schaal, K. P., and Pulverer, G., eds.), pp. 577–583, Gustav Fischer, New York.

Reh, M., and Schlegel, H. G. (1981) *J. Gen. Microbiol. 126*, 327–336.

Rizzuti, L., Augugliaro, V., Torregrossa, V., and Savarino, A. (1979) *Eur. J. Appl. Microbiol. Biotechnol. 8*, 113–118 (CA 92: 152527g).

Rothe, U., Broemme, D., Koennecke, A., and Kleine, R. (1982) *Acta. Biol. Med. Ger. 41*, 477–450.

Russel, S. (1978) *Zesz. Nauk. Szk. Gl. Gospod. Wiejsk. Aklad. Roln. Warsaqwie, Rozpr. Nauk 101*, 33 (CA 93: 1524k).

Ruttloff, H., Klingenberg, P., Behnke, U., Leuchtenberger, A., and Taeufel, A. (1978) *Z. Allg. Microbiol 18*, 437–445.

Sakano, Y., Kogure, M., Kobayashi, T., Tamura, M., and Suekane, M. (1978) *Carbohydr. Res. 61*, 175–179.

Sakano, Y., Hiraiwa, S., Fukushima, J., and Kobayashi, T. (1982) *Agr. Biol. Chem. 46*, 1121–1129.

Sandrak, N. A. (1977) *Mikrobiologiya 46*, 478–481.

Sarfert, E., Kretschmer, S., Triebel, H., and Luck, G. (1979) *Z. Allg. Mikrobiol 19*, 203–210.

Sariaslani, F. S., Harper, D. B., and Higgins, I. J. (1974) *Biochem. J. 140*, 31–45.

Sawai, T., Ohara, S., Ichimi, Y., Kaji, S., Hisada, K., and Fukaya, N. (1981) *Carbohydr. Res. 89*, 289–299.

Schupp, T., Hütter, R., and Hopwood, D. A. (1975) *J. Bacteriol. 121*, 128–136.

Schupp, T., Traxler, P., and Auden, J. A. L. (1981) *J. Antibiot. 34*, 965–970.

Schwartz, R. D. (1981) *Enzyme Microb. Technol. 3*, 158–159.

Scott, J. H., and Schekman, R. (1980) *J. Bacteriol. 142*, 414–423.

Sharples, G. P., and Williams, S. T. (1974) *J. Gen. Microbiol. 84*, 219–222.

Sharples, G. P., Williams, S. T., and Bradshaw, R. M. (1974) *Arch. Microbiol. 101*, 9–20.

Shchurova, Z. P., and Kirnitskaya, A. V. (1977) *Mikrobiol. Zh. (Kiev) 39*, 499 (CA 87: 186822u).

Shimizu, M., Kanno, M., Tamura, M., and Suekane, M. (1978) *Agr. Biol. Chem. 42*, 1681–1688.

Shipton, W. A., and Burggraaf, A. J. P. (1982) *Plant and Soil 69*, 149–161.

Singh, V. P., and Sinha, U. (1982) *Ind. J. Exp. Biol. 20*, 26–30.

Sinha, U., and Singh, V. P. (1980) *Biochem. J. 190*, 457–460.

Sinha, U., and Singh, V. P. (1981) *Ind. J. Exp. Biol. 19*, 453–457.

Staat, R. H., and Schachtele, C. F. (1975) *Infect. Immun. 12*, 556–563.

Stackebrandt, E., and Woese, C. R. (1981) *Curr. Microbiol. 5*, 197–202.

Stackebrandt, E., Wunner-Fuessl, B., Fowler, V. J., and Schleifer, K.-H. (1981) *Int. J. Syst. Bacteriol. 31*, 420–431.

Starr, M. P., Stolp, H., Trüper, H. G., Balows, A., and Schlegel, H. G. (eds.) (1981) *The Prokaryotes*, 2 vols, Springer, New York.

Stepanov, V.M., Rudenskaya, G. N., Nesterova, N. G., Kupriyanova, T. I., Khoklova, Y. M., Usaite, I., Loginova, L. G., and Timokhina, E. A. (1980) *Biokhimiya (Moscow) 45*, 1871–1880.

Stepanov, V. M., Chestukina, G. G., Rudenskaya, G. N., Epremayan, A. S., Osterman, A. L., Khodova, O. M., and Belyanova, L. P. (1981) *Biochem. Biophys. Res. Comm. 100*, 1680–1687.

Stirling, L. A., and Perry, J. (1980) *Curr. Microbiol.* (1980) *4*, 37–40.

Stirling, L. A., Watkinson, R. J., and Higgins, I. J. (1977) *J. Gen. Microbiol. 99*, 119–125.

Stuttard, C. (1983) in *83rd Annual Meeting of the American Society of Microbiology, March 6–11, 1983*, H29, p. 26.

Stutzenberger, F. J. (1972) *Appl. Microbiol. 24*, 77–82.

Stutzenberger, F. J. (1978) *Appl. Environ. Microbiol. 36*, 201–204.

Stutzenberger, F. J. (1979) *Biotechnol. Bioeng. 21*, 909–913.

Stutzenberger, F. J., and Bowden, M. W. (1980) *Biotechnol. Bioeng. 22*, 2443–2447.

Stutzenberger, F. J., and Carnell, R. (1977) *Appl. Envir. Microbiol. 34*, 234–236.

Szvoboda, G., Lang, T., Gado, I., Ambrus, G., Kari, C., Fodor, K., and Alföldi, L. (1980) in *Advances in Protoplast Research*, (Ferency, L., and Farkas, G. L., eds.), pp. 235–240, Budapest: Akademiai Kiado; Oxford: Pergamon Press.

Tárnok, I. (1976) in *The Biology of the Nocardiae* (Goodfellow, M., Brownell, G. H., and Serrano, J. A., eds.), pp. 451–500, Academic Press, London.

Taufel, A., Behnke, U., and Ruttloff, H. (1981) *Naturwiss. Tech. 1981*, 427–433.

Thiemann, J. E., and Beretta, G. (1968) *Arch. Mikrobiol. 62*, 157–166.

Thiemann, J. E., Pagani, H., and Beretta, G. (1967) *Giorn. Microbiol. 15*, 27–38.

Thiemann, J. E., Pagani, H., and Beretta, G. (1968) *J. Gen. Microbiol. 50*, 295–303.

Tille, D., Vetterman, R., and Prauser, H. (1982) in *Fifth International Symposium on Actinomycetes Biology, Resumés*, August 1982, p. 182, Oaxtepec, Mexico.

Tjepkema, J. D., and Winship, L. J. (1980) *Science 209*, 279–281.

Tjepkema, J. D., Ormerod, W., and Torrey, J. G. (1980) *Nature 287*, 633–635.

Tominaga, Y., and Tsujisaka, Y. (1976) *Agr. Biol. Chem. Jap. 40*, 2325–2333.

Tomita, K., Uenoyama, Y., Numata, K.-I., Sasahira, T., Hoshino, Y., Fujisawa, K.-I., Tukiura, H., and Kawaguchi, H. (1978) *J. Antibiot. 31*, 497–510.

Tonge, G. M., and Higgins, I. J. (1974) *J. Gen. Microbiol. 81*, 521–524.

Torrey, J. G. (1978) *Bioscience 28*, 586–592.

Torrey, J. G., Tjepkema, J. D., Turner, G. L., Bergersen, F. J., and Gibson, A. H. (1981) *Plant Physiol. 68*, 983–984.

Traxler, P., Schupp, T., Fuhrer, H., and Richter, W. J. (1981) *J. Antibiot. 34*, 971–979.

Treccani, V. (1971) *Riv. Ital. Sostanze Grasse 48*, 233–235 (CA 75: 126897g).

Treuhaft, M. W. (1977) *J. Clin. Microbiol. 6*, 420–424.

Trojanowski, J., Haider, K., and Sundman, V. (1977) *Arch. Microbiol. 114*, 149–153.

Trudgill, P. W. (1982) in *Experience in Biochemical Perception* (Ornston, L. N., and Sligar, S. W., eds.), pp. 59–73, Academic Press, New York.

Tsukamura, M. (1982) *J. Gen. Microbiol. 128*, 2385–2388.

Upton, M. E., and Fogarty, W. M. (1977) *Appl. Environ. Microbiol. 33*, 59–64.

Van Dijk, C., and Merkus, E. (1976) *New Phytol. 77*, 73–91.

Villanueva, J. R. (1961) *Microbiol. Espan. 14*, 157–162.

Villanueva, J. R. (1964) Antonie van Leeuwenhoek. *J. Microbiol. Serol. 30*, 17–32.

Vitols, M., Volsky, V., and Maurina, H. (1974) in *Proceedings of the First International Conference on Biology of the Nocardia* (Brownell, G. H., ed.) pp. 64–65, McGowan, Augusta, Ga.

Wakisaka, Y., Kawamura, Y., Yasuda, Y., Koizumi, K., and Nishimoto, Y. (1982) *J. Antibiot. 35*, 822–836.

Watkinson, R. J., and Somerville, H. J. (1976) in *Proceedings of the Third International Biodegradation Symposium* (Sharpley, J. M., and Kaplan, A. M., eds.), pp 35–42, Applied Science Publishers, Ltd., London.

Watson, G. K., and Cain, R. B. (1975) *Biochem. J. 146*, 157–172.

Wesseling, A. C., and Lago, B. D. (1981) *Dev. Indust. Microbiol. 22*, 641–651.

Williams, S. T. (1970) *J. Gen. Microbiol. 62*, 67–73.

Williams, S. T., Wellington, E. M. H., Goodfellow, M., Alderson, G., Sackin, M., and Sneath, P. H. A. (1981) in *Actinomycetes* (Schaal, K. P., and Pulverer, G., eds.), pp. 47–57, Gustav Fischer, New York.

Williams, S. T., Wellington, E. M. H., and Tippler, L. S. (1980) *J. Gen. Microbiol. 119*, 173–178.

Willoughby, L. G., Smith, S. M., and Bradshaw, R. M. (1972) *Freshwater Biol. 2*, 19–26.

Winter, J. A. (1962) *Purdue Univ. Eng. Bull. Ext. Ser. 112*, 703–716.

Taxonomy of Fungi and Biology of the Aspergilli

J. W. Bennett

I. INTRODUCTION

In the enormous literature about the aspergilli, certain words are encountered over and over. These include a humble noun: "mold" (not "mildew"); some marveling adjectives: "cosmopolitan," "ubiquitous," "successful"; and a series of rather maleficent verbs: "rot," "spoil," "injure," "decompose," "putrefy," "decay."

What is a mold? In the vernacular, mold (Brit. = mould) is used to mean both "a wooly or furry growth (consisting of minute fungi) which forms on vegetable and animal substances that lie for some time in moist warm air" (*Oxford English Dictionary*, 1933) and "a fungus producing such growth" (*Webster's Third New International Dictionary of the English Language,* 1964).

Prodigious numbers of minute mold spores are present in air and elsewhere. These spores, invisible to the naked eye, readily contaminate food and other organic matter, making it easy for people in earlier ages to believe in spontaneous generation and that molds were the consequence, not the cause, of decay. Modern usage of "mold" and association of molds with inanimate putrefaction reflect some of these outdated views:

> A fungus is a sterile kind of plant, that is to say destitute of flower and seed, arising from putrefactive fermentation (wherefore they arise chiefly during a moist and

rainy period and consist for the most part of a soft and spongy substance), yet retaining its characteristic look which it owes to a definite and specific juice of decay from which it originated. (Dillenius, 1719, quoted by Ainsworth, 1976, p. 17)

Biologists agree that members of this particular group of molds—the genus *Aspergillus*—are among the most widely distributed and abundant of living things. The aspergilli exist in astonishingly large numbers and display great physiological versatility. They are easily isolated from soil, air, stored foods, organic debris, and many other habitats. Several species have had their metabolic powers harnessed for commercial exploitation, and members of this genus figure prominently in the history of both mycology and biotechnology (Table 12.1).

In writing a selective and abbreviated review about such a complex and well-studied group it is necessary to pick and choose. I am choosing to emphasize three areas: taxonomy, genetics, and selected economic aspects. Even within these topics, coverage is not comprehensive, but an attempt has been made to give well-balanced literature citations.

The single most useful reference for identification, "care and feeding," and general background information is still Raper and Fennell's 1965 manual, *The*

TABLE 12.1 Historical Landmarks in Mycology and Biotechnology Involving the Genus *Aspergillus*

Date	Discoverer	Landmark
1729	P. A. Micheli	Publication of *Nova plantarum genera*, name *Aspergillus* given to group of molds with characteristic spore head
1827	J. Schilling	First continuous observation of the growth of a fungus from spore to spore, *A. glaucus*
1854	A. deBary	First association of a perfect state (*Eurotium herbariorum*) and an imperfect state (*A. glaucus*)
1867	P. Van Tieghem	First to establish importance of aspergilli in biochemical field with identification of *A. niger* in gallic acid fermentation
1869	J. Raulin	First defined medium for a microorganism, *A. niger*
1894	J. Takamine	First patent for a commercial enzyme from fungus, "Takadiastase" from *A. flavus-oryzae*
1917	J. N. Currie	Commercial citric acid production from *A. niger* demonstrated as feasible
1952	J. A. Roper	Parasexual cycle discovered in *A. nidulans*
1961	Sargeant et al.	Identification of *A. flavus* as mold associated with Turkey-X disease; "mycotoxin revolution" begins (Sargeant et al., 1961) (name aflatoxin given in 1962)
1983	Ballance et al.	Transformation system described in *A. nidulans*, second transformation system elucidated for a mold

After Prescott and Dunn (1940), Ramsbottom (1941), Lechevalier and Solotorovsky (1965), and Ainsworth (1976).

Genus Aspergillus. This manual is a revision of Thom and Raper's 1945, *A Manual of the Aspergilli*, which in turn was a revision of Thom and Church's 1926 classic monograph, *The Aspergilli*. There is no single contemporary work that gives as broad coverage of research as the 1965 manual. However, the proceedings of a 1976 symposium of the British Mycological Society held in Birmingham, England, about the group has been published under the title *Genetics and Physiology of Aspergillus* (J. E. Smith and Pateman, 1977) and contains many useful references. For other areas of study in which the aspergilli have figured prominently, the following reviews are recommended: ecological studies (Griffin, 1972; Wicklow and Carroll, 1981); developmental model studies (J. E. Smith et al., 1977; J. E. Smith and Berry, 1974; Turian, 1974; Turian and Hohl, 1981; Zonneveld, 1977); and fungal physiology (Wolf and Wolf, 1947; Foster, 1949; Cochrane, 1958; Ainsworth and Sussman, 1966; J. E. Smith and Berry, 1975, 1976, 1978; J. E. Smith et al., 1983).

Some species of *Aspergillus* are involved in diseases of humans and animals, with effects ranging from allergy to systemic opportunistic infections. Texts on medical mycology (Ainsworth, 1952; Vanbreuseghem and Wilkinson, 1958; Wilson and Plunkett, 1965; Baker, 1971; Emmons et al., 1977) contain introductory coverage. Especially useful is the chapter on aspergillosis in the work of Rippon (1982).

The aspergilli are easy to handle and grow in the laboratory. Recipes for the common media used for culturing species, as well as basic descriptions for isolation, storage, transfer, and other routine microbiological maneuvers, are given in the work of Raper and Fennell (1965) as well as in the work of Onions et al. (1981).

II. SOME SYSTEMATICS FOR THE NONMYCOLOGIST

What is *Aspergillus*? *Aspergillus* is the name of a taxon, a genus name for a group of fungi. What is a taxon? What is a genus? What is a fungus?

To answer these questions, one must enter the arcane world of classical biology, the domain of taxonomy, a discipline of legal labyrinths and evolutionary theories. Many modern biologists have not been taught any taxonomy and dismiss this field as a necessary but hopelessly old-fashioned, barely scientific endeavor, to the point that they do not even have a rudimentary vocabulary. What follows is a brief introduction to systematics for industrial microbiologists, molecular biologists, and others in related fields.

A. What Is Systematics?

Taxonomists worry a lot about terminology, not only for formal nomenclature, but also for the words they use to label their discipline. Taxonomy is usually subdivided into two parts: classification (the orderly arrangement of units) and nomenclature (the naming or labeling of units). The construction of classification systems is different from the act of classifying (the process of placing an individual into a given group), which is more properly referred to as identification (Dunn and Everitt, 1982). Bacteriologists in particular view identification as a worthy activity, and some speak of "The Trinity that is Taxonomy: classification, nomenclature and identification" (Cowan, 1971).

Taxonomy and systematics may be defined equivalently as the science of classification, or separate meanings may be assigned (Talbot, 1971). In this latter usage, systematics becomes "the scientific study of the kinds of and diversity of organisms and of any and all relationships among them" (Simpson, 1961, p. 7); taxonomy becomes "the theoretical study of classification, including its bases, principles, procedures and rules" (Simpson, 1961, p. 11); and classification becomes "the ordering of organisms into groups or sets on the basis of their relationships" (Sneath and Sokal, 1973, p. 3). Here the principles underlying classification are "taxonomy" and are clearly distinguished from nomenclature and identification, while the whole endeavor is subsumed under the heading "systematics." These nuances are lost on many of us, and a truly shocking number of my undergraduates have taxonomy confused with taxidermy.

Classification.   Classification is the systematic arrangement of organisms into groups or categories according to a definite sequence using certain underlying principles. The relationships used to classify organisms may be genetic and evolutionary (phylogenetic) or may simply refer to similarities of phenotype (phenetic) (Dunn and Everitt, 1982).

According to the modern version of the Linnaean scheme for classification, organisms are placed into a series of more comprehensive groups by assessing similarities and relationships. First the most similar organisms are grouped. Then the most similar groups of groups are in turn arranged in groups. The systematic groups so produced are taxa (singular = taxon). The arrangement continues until a hierarchical grouping has been achieved. The different levels are known as taxonomic ranks.

Every accepted species is classified within higher taxa: genus, family, order, class, and division (phylum). These categories are standardized by various international codes, are arranged in a fixed order, and have standardized endings. The botanists, zoologists, and bacteriologists have somewhat different conventions. International Codes guide the different disciplines and serve as statute books for biologists, giving the rules for attaching a name to an organism and ranking it in the hierarchy of classification. Taxonomists are supposed to conform to this overall framework. The headings used in the three systems are summarized in Table 12.2. An excellent, terse introduction to biological nomenclature and classification is given by Jeffrey (1977).

When the taxonomic categories represent groups of organisms that are presumably related by descent (phylogenetically), the classification is said to be "natural." The fossil record, supported by morphological, genetic, and physiological data, gives clues to such an ordering. When the classification is devised merely for convenience and makes no assumptions about phylogeny it is said

TABLE 12.2 English Language Categories of the Taxonomic Hierarchy and Standardized Endings for Ranks of Taxa. (No Entry Is Given When There Is No Recommended Standardized Ending. The Most Important Taxa Are Given in Capitals, Those Seldom Used Are Enclosed in Parentheses

Botanical	Bacteriological	Zoological
KINGDOM	KINGDOM	KINGDOM
		Subkingdom
		(Superphylum)
DIVISION (-*phyta/* -*mycota*)[a,b]	Division	PHYLUM
Subdivision (-*phytina/* -*mycotina*)	(Subdivision)	Subphylum
		Superclass
CLASS (*phyceae/-mycetes/* -*opsida*)[a,b]	CLASS	CLASS
Subclass (-*phycidae,* -*mycetidae, -idae*)[a,b]	(Subclass)	Subclass
		Infraclass
Superorder		Superorder
ORDER (-*ales*)	ORDER (-*ales*)	ORDER
(Suborder) (-*inease*)	(Suborder) (-*ineae*)	Suborder
		Infraorder
		Superfamily (-*oidea*)[a]
FAMILY (-*aceae*)	FAMILY (-*aceae*)	FAMILY (-*idae*)
Subfamily (-*oideae*)	(Subfamily) (-*oideae*)	Subfamily (-*inae*)
		Supertribe
Tribe (-*eae*)	Tribe (-*eae*)	Tribe (-*ini*)[a]
Subtribe (-*inae*)	Subtribe	(Subtribe) (-*inae*)
GENUS	GENUS	GENUS
Subgenus	(Subgenus)	Subgenus
Section		
Subsection		
Series		
Subseries		
SPECIES	SPECIES	SPECIES
Subspecies	(Subspecies)	Subspecies
Variety		
(Subvariety)		
Form		
(Subform)		

[a] Recommended but not mandatory under the respective Code.
[b] The endings -*phyta* and -*phytina* are used for the names of green plants; -*phyceae* and -*phycideae* for algae; -*opsida* and -*idae* for higher green plants; -*mycota*, -*mycotina*, -*mycetes*, and -*mycetiae* for taxa of fungi.
After Jeffrey (1977).

to be "artificial." There are many philosophical objections that can be raised to the terms "natural" and "artificial," and systematists are quick to concede that *all* classifications are to some degree "artificial."

Numerical taxonomy is the grouping by numerical means of taxonomic units into taxa based on a matrix of resemblances (Sneath and Sokal, 1973). A basic attitude of numerical taxonomy is the strict separation of phylogenetic speculation from taxonomic procedures; this separation of "overall similarity (phenetics) from evolutionary branching sequences (cladistics) is an important advance in taxonomic thinking" (Sneath and Sokal, 1973, p. 10). Despite this empiric bias, the field has not escaped the taxonomists' penchant for semantic proliferation. Other names suggested for numerical taxonomy include: "numerical systematics," "mathematical taxonomy," "taximetrics," "taxometrics," and (good grief!) "multivariate morphometrics." (Literature citations for these terms are given by Sneath and Sokal, 1973.) The fundamental position used in numerical taxonomy is associated with the ideas of a pre-Darwinian French botanist, Michel Adanson (1727–1809), so the approach is sometimes also called Adansonian or Neo-Adansonian.

The reliance on numerical taxonomy varies tremendously from discipline to discipline and from taxon to taxon. Traditionally, bacteriologists have had the hardest time constructing phylogenetic classifications; therefore they have been quick to accept empiric schemes (Mandel, 1969; LaPage et al., 1976; Colwell and Austin, 1980). Current texts are full of expressions about this approach: "It is futile to construct arbitrary phyletic classifications based on comparisons of the extremely diversified present day bacterial life forms" (Joklik et al., 1980, p. 13), and the Eighth Edition of *Bergey's Manual of Determinative Bacteriology* claims that "no attempt has been made to provide a complete hierarchy, as in previous editions, because a complete and meaningful hierarchy is impossible" (Buchanan and Gibbons, 1974, p. 1).

Fungal classification is based on the life cycle of species and the morphology of the spores produced. This emphasis on classical morphology and the use of microscopic characters relegates biochemical and other attributes to a secondary status. Only in the yeasts are cultural and physiological characteristics widely used as definitive characters (Rij, 1973; Von Arx, 1979). In the virtual absence of a fossil record, and using criteria that reflect only a fraction of the organism's total genome, the phylogenetic classification of fungi is speculative. Nevertheless, with the important exception of the Fungi Imperfecti (the Deuteromycetes or "conidial fungi"), many mycologists attempt to find "natural" relationships. This philosophy is expressed by Ainsworth (1973, p. 1):

> By supplementing the morphological approach to taxonomy with the results of nutritional and biochemical studies, it is possible to arrive at arrangements which, in general, correspond with current beliefs on possible evolutionary sequences.

Classification schemes are modified as biologists change their conceptions of groupings on the basis of new experimental approaches that add new data and reveal new relationships. Consensus changes with time. For fungi this has resulted in a gradual elevation of station through the hierarchical classification scheme.

Linnaeus divided the plant kingdom into 25 classes and called one of these groups the Cryptogamia, for plants with "concealed" reproductive organs (G. M. Smith, 1938). After Darwin, botanists proposed various phylogenetic systems; one such system, originally proposed by Eichler (1886) was widely adopted:

> *Kingdom:* Plant
> > *Subkingdom:* Cryptogamae
> > *Division:* Thallophyta
> > *Class:* Fungi

As time passed, the notion of first cryptograms and then thallophytes gradually went out of favor, and a new division, the Eumycetae (true fungi) was accepted. G. M. Smith (1938, p. 38) was able to write with assurance: "Mycologists are in universal accord in dividing the true fungi into four classes. . . : Phycomycetae, Ascomycetae, Basidiomycetae and Fungi Imperfect." A popular botany text of the 1950's described "the most nearly natural system of classification of living plants" (Fuller and Tippo, 1954, p. 35) as one in which the fungi were given the status of a Phylum:

> *Kingdom:* Plant
> > *Subkingdom:* Thallophyta
> > *Phylum:* Eumycota

The traditional two-kingdom classification of plants and animals has always caused microbiologists difficulties, and since the nineteenth century, alternative schemes have been proposed. Four kingdom classifications were outlined and justified by Copeland (1956) and Barkley (1968), but neither of these treatments has achieved the acceptance of Whittaker's (1969) five-kingdom system. Mycologists have been quick to adopt Whittaker's convention and elevate fungi to the status of a kingdom (Ainsworth and Sussman, 1966; Ainsworth 1971, 1973; Sieburth, 1978). The second edition of a widely used introductory mycology text (Alexopoulos, 1962) displayed this scheme:

> *Kingdom:* Plantae
> > *Division:* Mycota
> > *Subdivision:* Eumycotina

The third edition (Alexopoulos and Mims, 1979) omitted plants entirely:

> *Superkingdom:* Eukaryota
> > *Kingdom:* Mycetae (Fungi)
> > *Division:* Amastigomycota

Novices confused by the apparent lability of classification rank should take note of the date of publication of their reference and also recognize the coexistence of different ways of organizing information by different authors.

Nomenclature.   Formal taxonomic nomenclature is the allocation of names to taxa. Classification always precedes the giving of names; without adequate classification it is impossible to give rational names. In turn, nomenclature precedes identification, for without a system of labeled units it is impossible to identify and communicate the results.

In biology, nomenclature is an international vocabulary of Latin or Latinized names of organisms and groups of organisms. This vocabulary is sanctioned, regulated, and standardized under rules set down by International Commissions. These legalistic documents are supposed to stabilize nomenclature; but as we shall see, the statutes embodied in the International Codes may have the opposite effect. Moreover, because the Codes are subject to periodic revision, the rules change over time.

Names should provide a single unambiguous way of referring to a given organism. The uniform system for naming living things was introduced by Linnaeus with the publication of *Species Plantarum* in 1753. For Linnaeus, each species was properly designated by a generic name plus a diagnostic phrase by which one "can distinguish this species from all others of the same genus speedily, safely and pleasantly" (Ainsworth, 1976); Linnaeus also epigrammatized, "Ignorant persons impose absurd names." Modern mycologists have requested "that each name proposed be short, significant, euphonious, and both properly formed and transliterated" (Clements and Shear, 1931, p. 16). The genus name *Aspergillus* fulfills these criteria. The genus was first described by Pier' Antonio Micheli, an Italian priest and botanist who published *Nova Plantarum Genera* in 1729 and who is viewed as the "father of mycology" (Ainsworth, 1976). The spore heads of *Aspergillus* reminded Micheli of a liturgical device used for sprinkling holy water, the aspergillum (from the Latin, aspergere, to sprinkle), consisting of a perforated globe holding a sponge affixed to a short handle. Raper (personal communication, 1983) reports that *Aspergillus* was erroneously translated as meaning "rough head" in the early publications by Thom and Church (1926) and Thom and Raper (1945).

In the Linnaean system now familiar to all biologists, names for organisms consist of a double term called a binomial. A capitalized, Latin genus name is followed by a lowercase specific epithet, also in Latin, and agreeing with the Latin genus noun. The name is usually printed in italics (underlined), for example, *Aspergillus terreus*. On subsequent use in a paper the generic epithet is usually abbreviated, for example, *A. terreus*.

Fungal species are named in accordance with the rules described in the International Codes of Botanical Nomenclature. These rules are discussed and revised at the time of various International Botanical Congresses. According to these Codes, each valid name is determined by a type species and priority of publication.

The type specimen is the specimen on which the author based the original description. The type specimen is usually deposited in an herbarium. In theory, identifications can be made by comparison with type material. In practice, comparison of a living variable organism with a dead, static "type" is often of little value, especially when the type is a 150-year-old mold. Herbarium types are not necessarily "typical," nor are they appropriate for standardizing nomenclature for microscopic fungi. For *Aspergillus* the most useful materials are lyophilized cultures deposited in culture collections.

When giving the name of a genus or species, it is customary to add the name of the author who published the first description and bestowed the name, for example, *A. terreus* Thom. Sometimes the name of the author is abbreviated: Mich. stands for Micheli. See *Ainsworth & Bisby's Dictionary of the Fungi* (Ainsworth, 1971) for a list of abbreviations of authors' names commonly associated with fungal genera.

Each taxonomic group can have only one name, and that name must be in Latin. However, a given organism is often described in more than one place by different workers, who may then publish new names in ignorance of each other's work. When it is subsequently realized that this has happened, the oldest validly published name has priority; other names are called synonyms. Generic synonyms for *Aspergillus* are *Alliospora* Pim, *Aspergillopsis* Spegazzini, *Cladaspergillus* Ritg., *Cladosarum* Yuill & Yuill, *Euaspergillus* Ludwig, *Gutturomyces* Rivolta, *Raperia* Subramanian & Rajendran, *Redaellia* Ciferri, *Rhopalocystis* Grove, *Sceptromyces* Corda, *Spermatoloncha* Spegazzini, *Sphaeromyces* Montagne, *Sterigmatocystis* Cramer, and *Stilbothamnium* Hennings (Carmichael et al., 1980).

Under the rules enacted at the XIIIth International Botanical Congress in 1981, valid publication of all "plant" names, including the fungi, begins with *Species Plantarum* by Linnaeus in 1753 (Korf, 1982a, b). Linnaeus was notoriously vague, if not downright wrong, in his interpretation of fungi (Ramsbottom, 1941; Ainsworth, 1976), so special treatment was given to fungal species described in 1801 by Persoon and between 1821 and 1832 by Fries. These names "take priority over homonymous and synonymous names published earlier" (Korf, 1982b).

Because Micheli published the name *Aspergillus* in 1729 (prior to Linnaeus), even though his description was accurate, according to the Code his description had to be validated. Therefore the genus name is frequently written: *Aspergillus* Mich. *ex* Fries, meaning "*Aspergillus* described and named by Micheli, validated by Fries." In the new systems enacted in 1981, author citations using "*ex*" in this fashion are replaced by a colon, and the notation now reads: *Aspergillus* Mich.: Fries.

As originally used by Micheli in 1729, the name *Aspergillus* was applied to the asexual stage (conidial anamorph) of the common molds observed. Later, Link in 1809 described a fungus found on herbarium specimens. He called the asexual spored structures *Aspergillus glaucus* and, not realizing a common origin, applied the term *Eurotium herbarium* to the cleistothecia (sexual fruiting bodies) formed by the same fungus. These names were accepted in 1821 by Fries in his *Systema Mycologicum*, the official starting date for the names of fungi within the group with which *Aspergillus* is classified. It was not until 1854 that deBary demonstrated that the conidiophore of "*Aspergillus glaucus*" and the cleistothecium of "*Eurotium herbarium*" arose from the same vegetative mycelium (Raper and Fennell, 1965). Ever since then, there have been nomenclatural problems with the genus.

Only a few species of *Aspergillus* exhibit an ascosporic (sexual) stage. However, the rules of botanical nomenclature state that names for sexual (perfect) forms take precedence over asexual forms: "Generic and specific names of imperfect states may not be used to refer to the perfect states, nor do they compete with names given to the latter for purposes of priority" (Jeffrey, 1977, p. 41). This predicament is discussed in the sections on *Aspergillus* taxonomy below.

The intracacies of Latin symbols, abbreviations, special notations, and nomenclatural jargon are difficult for the nonspecialist. Glossaries of terms used in fungal nomenclature are provided by Ainsworth (1971) and Hawksworth (1974). The continually revised rules of the system can generate surprising nomenclatural tangles, especially concerning priority of accepted names. When taxonomists gather to revise their codes and overcome some of the problems inherent in a given system, violent disagreements often break out. Cowan claims that "Nomenclature is the supreme generator of heat, bad temper and ill-will among taxonomists and every kind of microbiologist" (Cowan, 1971, p. 4). Some useful guides for interpreting what is legal within the botanical codes are Bisby (1953), Talbot (1971), Hawksworth (1974), and Jeffrey (1977). For the *really* strong at heart, see Weresub (1979a, b), Demoulin et al. (1981), and Korf (1982a, b) for detailed expositions of mycological nomenclature. Nomenclature of antibiotic-producing fungi has been reviewed by Hesseltine and Ellis (1975).

Identification.   Identification is the recognition of identity; identity is "sameness in all that constitutes the objective reality of a thing" (*Webster's Third New International Dictionary of the English Language*, 1964). Those who belittle the ability to identify biological unknowns are conveniently ignoring an enormous scientific superstructure based on centuries of accumulated data from observation and experimentation on the living world.

When an identification of an unknown organism is made, it is related to a previously described organism. To do this, a number of theoretical conditions, usually taken for granted, must be met. Identification presupposes that one already has taxa with which to identify. Also implied is the existence of an adequate scheme for classification and a standardized scheme for naming. Having classified and named our organism, we can ascertain further information about it and can talk about it. "We may even, naively, consider that we know what it is" (Johnson, 1970, p. 205).

For many of us the main purpose and function of taxonomy is identification. We have an "unknown" organism isolated from nature, or perhaps as a contaminant in our laboratory, and we want to know what to call it. So we bring our unknown to a taxonomist who will look at it, study it, and if all goes well, will give us the "correct" name. Depending on the skill and expertise of the taxonomist, the identification will be more or less precise: The unknown is a fungus. It is a Deuteromycete. It is a species of *Aspergillus*. It is *Aspergillus niger*.

Over and over, the writers of identification manuals have discovered that it is not necessary to use the theoretical framework of the taxonomic hierarchy in the production of a useful identification key. It is merely necessary to come up with a practical way of distinguishing organisms that share a number of features and differ in a number of other features. Keys are constructed according to any desired sequence of categories that divide the taxa into successively smaller subsets. The main keys used for the identification of members of the genus *Aspergillus* are given by Raper and Fennell (1965).

B. What Is a Fungus?

General Overview.   What are fungi? How do they differ from bacteria, plants, and animals?

Fungus is the Latin word for mushroom (related to the verb, *fungor*, to flourish). The word for the study of fungi, mycology, is also derived from a classical word for mushroom, in this case Greek (Gr. mykes = mushroom + logos = discourse) (Bessey, 1950; Alexopoulos and Mims, 1979). Mushrooms are among the largest of fungi and are easily seen with the unaided eye. With the discovery of lenses and microscopes, naturalists discovered many other organisms with mushroomlike attributes, and the concept of fungus was enlarged to include these microscopic forms. However, there is no complete agreement on the exact definition of fungus, and different experts include or exclude different groups. There are a number of excellent mycology texts to guide the novice in studying fungi: Alexopoulos (1962), Hawker (1966), Burnett (1968), Alexopoulos and Mims (1979), Ross (1979), Deacon (1980), and Moore-Landecker (1982).

Fungi are eukaryotic. They have membrane-bound nuclei, mitochrondria, 80S ribosomes, and the other paraphernalia of the eukaryotic cell. Because fungi were once classified as plants without roots, stems, or leaves (Thallophytes), the basic vegetative body of the fungus is often termed a thallus. With the exception of the yeasts, the thallus of the "true fungus" grows as a mass of branched filaments or threads, the mycelium. The mycelium constitutes the assimilative part of the fungus. Each individual filament of the mycelium is a hypha. In the so-called "lower fungi" the hyphae do not possess cross-walls; they are aseptate. (Some older mycologists prefer to call these aseptate filaments "siphons" and reserve "hyphae" for filaments with cross-walls). In the "higher fungi" (Asomycotina, Basidiomycotina, and Deuteromycotina) the hyphae do possess cross-walls; they are septate. However, the septations are perforated, creating cytoplasmic continuity and allowing nuclear migrations within the mycelium. Thus the hyphae and mycelia of fungi function essentially as a coenocyte.

Hyphal walls of the higher fungi are rigid and contain chitin. As filamentous fungi grow, they penetrate the substrate by elongating at the apex and branching, and then they secrete extracellular enzymes that degrade macromolecules in the substrate. Nutrients are absorbed across the cell wall. Nutritionally, the fungi are heterotrophic. Depending on their ecological relationships, the nutritional mode can be further described as saprophytic, parasitic, or symbiotic, and fungi may spend different parts of their life cycles in different modes.

When the time for reproduction occurs, a specific portion of the mycelium differentiates and forms sexual or asexual spores, often with associated complex structures. These spores and associated structures form the major criteria for taxonomic schemes in which fungi are classified and named and by which species are identified.

It is common for one species of fungus to have two or more different kinds of spore (Gregory, 1966). The condition of having more than one spore stage in a life cycle is sometimes referred to as pleomorphism. When only asexual spores are known, the fungus is classified among the Fungi Imperfecti, a group also known as the Deuteromycetes or Deuteromycotina (Gr. deutero = second + mycetes).

A commonly used classification for the major groups of fungi is that presented by Ainsworth (1971, 1973):

Division: Eumycota
    Subdivisions: Mastigomycotina
                  Zygomycotina
                  Ascomycotina
                  Basidiomycotina
                  Deuteromycotina

The Mastigomycotina and Zygomycotina comprise what used to be called the Phycomycetes, now "best restricted as a trivial term, phycomycetes, an appropriate synonym of lower fungi"—those with aseptate mycelia (Ainsworth, 1971).

Fungi Imperfecti. Fungi with a septate mycelium and with no sexual phase are commonly called imperfect fungi and, technically, Fungi Imperfecti. Depending on the author, the group may be ranked as a class (Deuteromycetes), a subdivision (Deuteromycotina), or not granted equivalent hierarchical status at all.

The genera and species within the Fungi Imperfecti are sorted and arranged in various ways by different workers. As we shall see, some mycologists even refuse to designate the units in this group as "true" genera and species. The early history of classification was dominated by Saccardo and has been reviewed by Hughes (1953), Barron (1968), and Chesters (1968), who uniformly point out that Saccardo's system was "artificial" but useful.

The sixth edition of *Ainsworth & Bisby's Dictionary of the Fungi* (Ainsworth, 1971) and Ainsworth (1973) rank the Fungi Imperfecti as a subdivision, the Deuteromycotina, containing three classes: Blastomycetes (yeasts); Hyphomycetes (molds), and Coelomycetes (pycnidials and acervular conidial fungi). This classification has been widely adopted (see, for example, Carmichael et al., 1980; Alexopoulos and Mims, 1979). However, you may encounter other ways of dividing the group as you read the literature. These are found in the work of Bessey (1950), Alexopoulos (1962), and Stevens (1974) and in the fifth edition of *Ainsworth & Bisby's Dictionary of the Fungi* (Ainsworth, 1961). Here the group is divided into four orders: Sphaeropsidales, Melanconiales, Moniliales, and Mycelia Sterilia. The first two groups produce their conidia inside some kind of cavity; the Moniliales produce spores in other ways (budding, fragmentation of hyphae into oidia, etc.); the Mycelia Sterilia have no known spores or specialized reproductive cells. Clements and Shear (1931) recognized three orders: Phomales, Melanconiales, and Moniliales. In the current fashionable three-class system (Ainsworth, 1971, 1973) the Moniliales have been split and elevated into the Blastomycetes and the Hyphomycetes, while the Melanconiales and Sphaeropsidales have been subsumed in the Coelomycetes. Depending on the author and date of your source, the genus *Aspergillus* will be found classified among the Hyphomycetes, the Moniliales, or both. Rank (i.e., Moniliaceae or Moniliales) may also vary from work to work.

Hughes (1953) is credited with beginning the modern reassessment of the Hyphomycetes. He examined more than a thousand types and authentic collections of classical material in European herbaria. He produced a new classification of Hyphomycetes based on conidiophore and conidium development, giving sectional numbers to groups of spore types, thus avoiding a classification into "natural" taxa. Later works have adopted and modified this "Hughesian System" and given names and ranks to most of the numbered sections. The group in which *Aspergillus* is found is called Section IV by Hughes (1953), Tuberculariaceae by Subramanian (1961), and Phialosporae by Tubaki (1963) and Barron (1968). The Hyphomycetes are enjoying a renaissance among taxonomists; and there are many competent, current, detailed works that review classification, mechanisms of spore development, spore ornamentation, and, by no means last, spore nomenclature (Kendrick, 1971, 1979; Cole and Samson, 1979; Carmichael et al., 1980; Cole and Kendrick, 1981; Subramanian, 1983).

The Fungi Imperfecti is a taxonomically difficult group for a number of reasons, not the least of which concerns the rules of botanical nomenclature and the philosophy of theoretical fungal systematists who persist in seeking some kind of "natural" classification system. Nowhere else in mycology does the power of words and the conservatism of historical precedent have as much influence as it does in this group. First, let us examine the imposition of a dichotomy between "natural" and "artificial" classifications.

According to the phylogenetic philosophy, a "true" taxon is an assemblage of organisms that are postulated, explicitly or implicitly, to be related by descent, i.e., are members of a "natural" group based on sexual criteria. When a systematist of this school brings together a group of organisms on the basis of similarities that have nothing to do with such "natural" relationships, the resulting unit may end up not labeled a taxon at all "but simply a 'group', or an informal 'category' or, at most, a 'parataxon' or a 'form-taxon'" (Pirozynski and Weresub, 1979, p. 667). Since the Fungi Imperfecti are, by definition, based on nonsexual attributes, these orthodox taxonomists may take enormous care to segregate classification ranks with the group from other classification ranks. Carmichael et al. (1980), for example, reject the use of the name Deuteromycotina "since it implies a division equivalent" to other groups such as the Ascomycota and Basidiomycota, and just use Fungi Imperfecti.

A common convention adopted by many mycologists for classifying fungi known only by their asexual states is to affix the prefix "form" to all taxonomic ranks used within the group. (This can be confusing because the International Codes of Botanical Nomenclature uses "form genus" with reference to fossils.) In any event, excruciating care may be taken to distinguish "true" taxonomic ranks in "natural" classifications from taxonomic ranks in which the group contains members with no known sexual phase and hence are divided in an "artificial" fashion. Such punctiliousness is exemplified by Alexopoulos and Mims (1979):

> *Superkingdom:* Eukaryota
> *Kingdom:* Myceteae (Fungi)
> *Division:* Amastigomycota
> *Subdivision:* Deuteromycotina
> *Form class:* Deuteromycetes
> *Form-subclass:* Hypomycetidae
> *Form-order:* Moniliales
> *Form-family:* Moniliaceae
> *Form-genus:* *Aspergillus*

Because this system is ponderous and because it can be confused with the Botanical Code usage for fossils, most mycologists dispense with the "form" prefix, apologize for giving Imperfect Fungi equivalent ranking to "natural" groups, and divide the Deuteromycotina with the usual taxonomic headings: class, order, family, genus, and species. Ainsworth (1971) writes that it is convenient if incorrect "to treat the Fungi Imperfecti as a Subdivision . . . because they are both taxonomically and nomenclatorially subservient to the perfect fungi which comprise the other subdivisions of the Eumycota" (p. 158) but then goes ahead and bows to convenience. Virtually all biologists, with the exception of some traditional mycological systematists, agree with Ainsworth and use standard ranks for designations of taxa. I follow this convention and accept *Aspergillus* as a valid generic taxon.

Perfect and Imperfect States. Having dispensed with the concept of form taxa, we must now grapple with another taxonomic problem, the nomenclatural and theoretical conundrum that results from the fact that many fungi produce several different spore types: one or more mitotic spores (asexual or "imperfect" types) and sometimes an additional meiotic spore (the sexual or "perfect" type). Nineteenth century mycologists had to overthrow the Linnaean paradigm of "one kind of fructification, one kind of plant" to accept this fungal pleomorphism. DeBary's recognition that *Aspergillus glaucus* and *Eurotium herbarium* were the same organism was a major breakthrough in his time. Subsequently, many other fungal species named on the basis of their asexual spores have been "connected" to a perfect state, usually in the Ascomycotina and more rarely in the Basidiomycotina. What do you name such organisms? Which name—an older imperfect name or a more recent perfect name—should take precedence? Are all aspergilli potentially members of the ascomycetes?

Within the genus *Aspergillus*, some species regularly produce both sexual and asexual spores; in other species the sexual form is rarely seen and then only under special circumstances; finally, there are many asexual species that have never been (and, it is safe to hazard, never will be) "connected" to a sexual stage. Several other economically important fungal genera such as *Penicillium* and *Fusarium* share this property of sporadic sexual pleomorphism and thus also share the resultant nomenclatural and philosophical complexities. My remarks about *Aspergillus* pertain to these taxa as well.

When the international rules were first enunciated for fungal nomenclature, the problem of perfect and imperfect states was dealt with by proclaiming certain special arrangements. Mycologists could name imperfect states separately from perfect states, legalizing a system of dual nomenclature. In this system the name of the perfect state takes precedence over the name for the imperfect state. In practice, this does not always happen. Many of the imperfect binomial epithets are deeply entrenched in the literature, and workers trained outside of mycology and taxonomy are blithely unaware of the rules of botanical nomenclature.

One of the premier model systems in eukaryotic genetics is the species almost universally known as *Aspergillus nidulans*. This organism regularly undergoes a sexual cycle and therefore, according to the strict rules of Botanical Nomenclature, should be given a different genus name based on the sexual (perfect) stage. Yet in the thousands of publications in major journals throughout the world it is still called *Aspergillus nidulans*.

Article 59 of the Botanical Code governs the relationship of perfect and imperfect states of fungi, legitimizing a dual nomenclature and giving priority to the names of perfect states. Over the years the terms "perfect," "imperfect," and "state" have been used in such a convoluted fashion that even the systematists have had trouble with definitions. Therefore the new terms "teleomorph," "anamorph," and "holomorph" were introduced by Hennebert and Weresub

(1979) and further defined by Weresub and Hennebert (1979). In this vocabulary the nonsexual phase is referred to as the anamorph, the sexual phase is termed the teleomorph, and the fungus in all its forms, both sexual and asexual, is called the holomorph. These words were almost immediately embraced by taxonomists and have been used widely; see, for example, the two-volume work entitled *The Whole Fungus, The Sexual-Asexual Synthesis* (Kendrick, 1979). At the most recent Botanical Congress in Sydney in 1981 the "Subcommittee on Article 59" recommended that "teleomorph" and "anamorph" replace the older terms "perfect state" and "imperfect state" (Korf, 1982b). Unfortunately for the many people who have trouble remembering that to mycologists "perfect" means sexual and "imperfect" means asexual, the new jargon has made the taxonomic literature more inaccessible, although more rational.

In this parlance, *Aspergillus* is an anamorph. Some members of the group have teleomorphic states in the Ascomycotina, which are given separate names. According to the Rules of Botanical Nomenclature, the names for teleomorphic states are the names given to the holomorph.

Translated, this means that *Aspergillus* is a genus defined by a characteristic mode of asexual reproduction. Certain species within the group also reproduce sexually. These sexual forms are given separate binomial epithets, and in the formal rules of Botanical Nomenclature these latter names are the ones that should be used when referring to a species in all facets of its life cycle, that is, the legal name is the name of the sexual state, if one is known.

Why belabor these points? I do so because I am concerned that the power of nomenclature and the philosophy of Hyphomycete taxonomists may be counter to the trends of modern biology. Increased commercial exploitation of imperfect species with concomitant patent applications for new processes requires a sound taxonomy that is understandable to nontaxonomists and consistent with advances in molecular biology.

C. A Plea to "Hyphomycetologists"

So much for an introduction to systematics for industrial biologists. I have already taken much longer than is expected in an essay of this sort; it certainly has taken me longer to read and review the taxonomic literature than I would have liked. I learned to appreciate one taxonomist's self-commentary: "Taxonomy is written by taxonomists for taxonomists; in this form the subject is so dull that few, if any, non-taxonomists are tempted to read it, and presumably even fewer try their hand at it" (Cowan, 1971, p. 1).

In my self-appointed role as interpreter to nontaxonomists, I read this literature. Only some of it was dull because, as a group, taxonomists have better vocabularies and more vituperative debates than most scientists. Now that I have done my reading, I am going to "try my hand at it" and attempt to highlight a semantic-philosophical tradition that I perceive as isolating mycologists.

First is the semantic burden conveyed by the word "imperfect." Imperfect means incomplete, undeveloped, defective. The implication is that an imperfect fungus like *Aspergillus* is an incomplete fungus. What is lacking? — the "more noble" sexual stage, which would allow a "natural" classification (Weresub and Pirozynski, 1979). When imperfects are "connected" to sexual stages, they become mere parts or stages of the sexual species. Imperfect fungi not so connected are put into form taxa with inferior status.

This linguistic connotation of imperfect may have contributed to the widely held presumption that "the imperfect fungi represent conidial stages of Ascomycetes whose ascigerous stages either are rarely formed in nature and have not been found, or have been dropped from the life cycle in the evolution of these organisms" (Alexopoulos, 1962, p. 387). The notion that molds in the Deuteromycotina have secondary status is pervasive — the "true" evolutionary place of such asexually reproducing organisms can only be found in the sexual phases they are presumed once to have possessed. In this Platonic search for an ideal taxonomy, little if any effort is spent examining the proposition that speciation might be possible in the absence of meiosis. Sex does not have a monopoly in evolution.

Major empirical and theoretical strides have been made by bacteriologists who must contend with an entire kingdom of organisms that lack meiosis. These organisms have many "alternatives to sex" (Haldane, 1955) for generating recombination and variation. Speciation in asexual fungi presents many of the same theoretical problems as speciation within the bacteria, yet "Hyphomycetologists" seem intent on erecting a nomenclatural smoke screen to avoid facing the hard fact that these organisms cannot be classified by using the same principles that work for maples, mammals, and mushrooms. What mycologists disparagingly label demes of imperfect fungi, bacteriologists comfortably accept as species.

Although the species is the basic unit for taxonomy, no one has been able to come up with a single, succinct, satisfactory definition. Within eukaryotes the species concept recognizes barriers to gene flow between members of different species and the possibility of the exchange of genes within species due to the limitations imposed by sexual reproduction. For plant and animal species in which reproduction is not obligately sexual, complex reticulate phylogenies may result and mathematical taxonomy is useful (Johnson, 1970). In bacteria the modern species concept is based on the recognition of partial discontinuities; "each species represents a cluster of biotypes, more or less resembling a strain that is maintained as the type culture of the species" (Davis, 1980).

Bacteriologists regularly utilize biochemical characteristics and DNA relatedness in the determination of species. Mycologists depend largely on microscopic anatomy, often using dead specimens. Priority for bacterial species names now starts as of January 1, 1980, with publication of the "Approved Lists of Bacterial Names" (Skerman et al., 1980). Mycologists are still arguing about moving their starting date for nomenclature for Hyphomycetes from Fries in 1821 back to Persoon in 1801. Bacteriologists broke away from the Botanical Code in 1947, primarily because living cultures were not accepted as types. At the International Botanical Congress in Sydney in 1981, living types were again disallowed for fungi, despite the recommendations of a subcommittee. I found my exasperation rising to the boiling point when I read the following explanations: "Since bacteriologists have 'type cultures' it is understandable that mycologists and phycologists may feel 'left out' by this prohibition. . . . Living 'types' would always be subject to mutation and to unpredictable change, and are thus a poor nomenclatural base" (Korf, 1982b, p. 482). Feeling "left out" has little to do with the issue. The capacity for mutation and change is inherent in living systems. A nomenclature that ignores the spectrum of diversity is absurd.

Hyphomycetologists have made laudable progress in standardizing and regularizing nomenclature for pleomorphic fungi. Although they are far from unanimous in their philosophies and perspectives (see Carmichael (1979) for an honest display of differences), many of them are still guided by the philosophy that "anamorphs" are only "pseudo-species" and that it takes a "teleomorph" to make a "whole fungus." For these workers the new jargon only masks the old Platonism: that the "true" status of these asexually reproducing organisms can only be known by association with a sexual stage. This quest for perfection certainly guides the membership of the commissions that draft the sections of the Botanical Code dealing with pleomorphic fungi.

Taxonomists are fond of describing their work as more an art than a science. The current version of the Code on pleomorphic fungi is not either. It is a legal system that has evolved out of debate and compromise. Like all legal systems, it is imperfect.

There is an old story about a microscopist who pointed out to Pasteur that an organism that he had taken for a coccus was in reality a small bacillus. Pasteur replied, "If only you knew how little difference that makes to me" (quoted by Davis, 1980). I suspect that many of my readers feel the same way.

III. THE GENUS *ASPERGILLUS*

A. Morphology

Now we can ask again, what is an *Aspergillus*? An *Aspergillus* is a kind of mold that produces a characteristic asexual spore-bearing structure, the conidial head or conidial apparatus. A schematic diagram of the *Aspergillus* conidial apparatus is given in Fig. 12.1. A glossary of specialized terms commonly encountered in descriptions of *Aspergillus* species is presented in Table 12.3. Definitions and usage vary somewhat in the literature; the problematic vocabulary is indicated within the glossary (Table 12.3).

The characteristic asexual spore-bearing structure, the conidiophore, arises from the vegetative mycelium as an erect, elongated modified hypha, which enlarges at the top into a globose, multinucleate vesicle. The lower portion, which is not a separate cell, is nonetheless called a "foot cell." The entire conidiophore (foot cell, stalk, vesicle) resembles certain modern water towers.

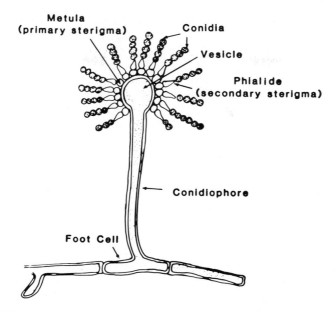

FIGURE 12.1 Schematic representation of an *Aspergillus* conidial apparatus.

The vesicle bears a single (uniseriate) or double (biseriate) row of fertile cells called phialides. The fertile area may cover the entire surface of the vesicle or may consist of a small area on the apex. Within the phialides, nuclei divide mitotically and migrate to the end. Then a wall is formed, the cytoplasm plus a nucleus or nuclei are cut off, and differentiation into mature conidiospores ensues. The process is repeated forming a long chain of conidiospores; the whole chain of conidia bears the genetic heritage of the phialide. In species with uninucleate spores the chain of conidia constitutes a clone.

How are species determined among these molds? Species are distinguished in the aspergilli on the basis of the microscopic character of the conidial apparatus and the culture characteristics of the colony. Four representative kinds of *Aspergillus* are shown in Fig. 12.2, illustrating the variation in microscopic appearance.

Modern classification requires observations made with pure cultures grown under standardized conditions of light, temperature, age, and substrate. The medium used by Raper and Fennell (1965) and most other laboratories for identification of *Aspergillus* species is Czapek-Dox Solution agar, which is formulated as follows (per liter of water): $NaNO_3$, 3.0 g; K_2HPO_4, 1.0 g; $MgSO_4 \cdot 7H_2O$, 0.5 g; KCl, 0.5 g; $FeSO_4 \cdot 7H_2O$, 0.01 g; sucrose, 30.0 g; and agar, 15.0 g. This medium is sometimes enriched or supplemented with various micronutrients. Addition of yeast extract, peptone, corn steep liquor, and

TABLE 12.3   Glossary of Morphological Terms Encountered in Descriptions of the Aspergilli

Term	Definition	Commentary
ascocarp	The ascospore-bearing structure of the ascomycetes; a sporocarp having asci, an ascoma	The ascocarp found in the genus *Aspergillus* is a cleistothecium.
ascogonium	The cell in ascomycetes fertilized by the sexual act	Recommended during nomenclature sessions of "Kananaskis II" (see Kendrick, 1979, p. 41).
ascoma	Synonym of ascocarp	
ascospore	The meiospore produced in an ascus	These are the haploid spores produced after karyogamy and meiosis.
ascus	The saclike reproductive cell of the sexual state of an ascomycete, generally containing eight ascospores; a meiosporangium	
cleistothecium	A fruit body having no special openings; a closed ascocarp; a cleistocarp	In Thom and Church (1926) and Thom and Raper (1945) this structure is called a perithecium.
conidium (pl. = conidia)	A conidiospore, a thin-walled asexual spore borne upon a specialized hyphae	A term first used by Link in 1807, now used loosely to mean almost any asexual spore among the higher fungi. The conidia of aspergilli are also called phialospores.
conidiophore	A specialized hypha bearing conidiogenous cells from which conidia are produced	Raper and Fennell (1965) state, "The erect perpendicular branch arising from the foot cell and ultimately producing the conidial head is known as the conidiophore or stalk."
foot cell	A basal cell supporting the conidiophore in *Aspergillus*	The presence of the foot cell is a diagnostic criterion in identifying members of the genus.
heterokaryon	A mycelium containing two or more genetically dissimilar nuclei	
Hülle cell	Terminal or intercalary thick-walled cells that are associated with and may surround cleistothecia in *Aspergillus*	The presence of Hülle cells is a useful diagnostic feature for identifying species in certain groups of the genus.
hypha	One of the filaments of the mycelium; the assimilative cell of the fungus	Vuillemin confined the term to septate filaments.

378

metula (pl. = metulae)	The short, sterile cells subtending phialides in some species of *Aspergillus* or *Penillium*; used for other kinds of conidiogenous cells in other genera of conidial fungi	Raper and Fennell (1965) use the term "primary sterigmata" for metulae of biseriate species.
mycelium (pl. = mycelia)	The collective term for a mass of hyphae; the thallus of a fungus	
phialide	A cell that develops one or more open ends from which develops a basipetal succession of conidia; in *Aspergillus* the phialide is an end cell of a conidiophore	See Kendrick (1971) for a "Report of the Committee on Phialides" in which it is recommended that "sterigma" should not be used for phialide.
sclerotium	A resting body of variable size, frequently rounded, composed of a hardened mass of thick walled hyphae, having no spores in or on it	The presence of sclerotia is a useful diagnostic feature for identifying species in certain groups of the genus.
sterigma (pl. = sterigmata)	A tiny spiculelike pedicel upon which a spore (esp. a basidiospore) is borne "inadvisedly used for somewhat analogous structures in groups other than basidiomycetes" (Snell and Dick, 1957)	Raper and Fennell (1965) use this term to refer to conidium-producing cells. When there is one layer of fertile cells, they call them primary sterigmata; when there are two layers of fertile cells, they call them primary sterigmata and secondary sterigmata. These are better called phialides (the actual conidiogenous cells) and metulae (the subtending cells, where present). See Fig. 12.1.
thallus	Body of a plant without roots, stems, or leaves; of fungi, the entire assimilative or vegetative phase	
vesicle	The swollen apex of the conidiophore	

After Snell and Dick (1957), Raper and Fennell (1965), and Ainsworth (1971).

FIGURE 12.2 Characteristic conidial heads from four species of *Aspergillus*. (a) *A. flavus*. (b) *A. fumigatus*. (c) *A. niger*. (d) *A. repens*.

other undefined supplements increases growth and sporulation. Raper and Fennell (1965) recommend the unsupplemented version because the "moderate growth of a majority of the species is commonly more useful than the great mass of mycelium and conidia which is readily obtained on enriched substrata" (p. 36).

Raper and Fennell (1965) recognized 132 species classified within 18 "groups." They provided two keys for identification to the group level. When an *Aspergillus* is assigned to a group, further keys are given to distinguish species within that group.

In the first key the primary characteristic is the "number of sterigmata — whether uniseriate, biseriate and uniseriate, or strictly biseriate." In the second key the primary characteristic is the color of the conidial heads. For example, members of the *Aspergillus flavus, A. nidulans,* and *Aspergillus fumigatus* groups are some shade of green, while the *A. niger* group is black or dark brown, and the *Aspergillus ochraceus* group is yellow. The ease with which colony color can be altered by both mutation and environment makes color an unreliable attribute unless it is rigorously associated with other characteristics.

Secondary diagnostic features for identifying species of *Aspergillus* include size, shape, and ornamentation of spores; rate of colony growth; texture of basal and aerial mycelium; character of colony margins; and production of specialized structures such as cleistothecia, sclerotia, and coremia. Trivial features such as zonation and odor may also be mentioned.

No one has attempted a thorough revision of the entire genus since Raper and Fennell recognized 132 species in 1965. However, Samson (1979) compiled and discussed the 90 additional species and varieties published after Raper and Fennell (1965). Subsequently, descriptions of new species have also been published in conjunction with synoptic keys for the *A. flavus* group (Christensen, 1981), the *A. ochraceus* group (Christensen, 1982a), and the *A. nidulans* group (Christensen and Raper, 1978; Christensen, 1982b).

B. Sexual Stages (Teleomorphs)

Sexual development in aspergilli results in the formation of cleistothecia. Asci within the cleistothecia contain eight unordered ascospores (Fig. 12.3). Thus the sexual stage is classified in the Ascomycotina.

A few aspergilli regularly produce cleistothecia and ascospores, while others do so only sporadically. Most have no known teleomorphs. With two

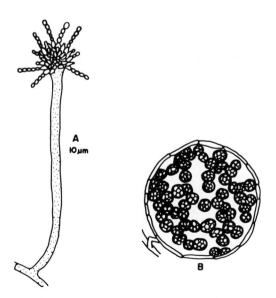

FIGURE 12.3 *Aspergillus repens.* (a) Conidial apparatus. (b) Cleistothecium containing asci with ascospores (after Webster, 1970).

exceptions, all known sexual aspergilli are homothallic. The discovery of true heterothallism for *Aspergillus heterothallicus* (Kwon and Raper, 1967) and for *Aspergillus fennelliae* (Kwon-Chung and Kim, 1974) raises the possibility that other aspergilli may possess incompatibility systems.

On a gross morphological level, sclerotia resemble cleistothecia, and some researchers consider sclerotia to be degenerate ascocarps (Rudolph, 1962). Raper et al. (1953) discovered ascospores within "aged sclerotia" of *Aspergillus citrisporus* leading Raper (1957) to predict that "in time and with patience" ascospores might be found in the sclerotia of the *Aspergillus candidus, A. flavus, A. niger* and *A. ochraceus* groups. Indeed, a teleomorphic stage for *A. ochraceus* has been observed (see Table 12.4), suggesting that the prediction should be reexamined for other groups as well. The widespread use of pure cultures derived from single-spore isolates may militate against the likelihood of finding sexual stages if heterothallism is at all common within the genus.

Formal botanical rules allowing dual nomenclature and giving precedence to sexual stages have been discussed already. Until 1973, each teleomorphic genus name within the aspergilli was associated with conidial species in only one of Raper and Fennell's (1965) "groups." Since then, the situation has become more complicated, and various mycologists have described new sexual forms and rearranged the classification of old ones. Fennell (1977) summarized this arcane literature and provided two keys for "orienting the reader in this monumental maze" (Fennell, 1977, p. 11). The generic names of the perfect states and the associated imperfect "groups" are listed in Table 12.4. For further clarification, see Benjamin (1955), Raper (1957), Malloch and Cain (1972a, b), Subramanian (1962, 1972, 1983), Wiley and Simmons (1973), Fennell (1977), and Samson (1979).

The original widely adopted classifications of Thom and Church (1926) and Thom and Raper (1945) were based on the asexual characteristics of the

TABLE 12.4 Names of Sexual Stages (Teleomorphs) within the Genus *Aspergillus*

Genus Name for Sexual Stage (Teleomorph)	Aspergillus *"Group"* (sensu *Raper and Fennell, 1965*)
Chaetosartorya	*A. cremeus*
Dichlaena	*A. ornatus*
Edyuilla	*A. glaucus*
Emericella	*A. nidulans* and *A. niveus*
Eurotium	*A. glaucus*
Fennellia	*A. flavipes*
Hemicarpenteles	*A. ornatus*
Neosartorya	*A. fumigatus*
Petromyces	*A. ochraceus*
Sclerocleista	*A. ornatus*
Warcupiella[a]	*A. ornatus*

[a] The conidial state of *Warcupiella* was transferred to the genus *Raperia* by Subramanian and Rajendran (1975).
After Fennell (1977) and Samson (1979).

genus. They felt strongly that nomenclature similarly based on the name given to an asexual stage was appropriate. Their devotion to this "illegal" nomenclature has provided decades of controversy. When Dodge (1945) reviewed the first *Manual of the Aspergilli* (Thom and Raper, 1945), he wrote, "It is gratifying to see that the authors have continued to maintain that the generic name *Aspergillus* should apply not only to the conidial stage but also to the ascosporic stage. The genus *Aspergillus* is here to stay, regardless of 'rules of nomenclature.'"

In a letter to *Science*, Martin (1946) remonstrated,

> This statement should not be allowed to pass unchallenged by those who believe that progress in any branch of science dealing with living organisms will be facilitated by precise designation of the organisms concerned, and that such precision can best be attained by conforming to an established procedure. . . . In any specific instance where a species of *Aspergillus* is known to be the imperfect stage of a species of *Eurotium*, then the species, as a unit, must be transferred to *Eurotium*. . . . Thom and Raper's volume constitutes one of the most outstanding mycological considerations of recent years . . . it is all the more to be regretted that the attitude toward nomenclature is so dogmatically asserted and so completely unsupported by examination of the considerations involved.

Thom and Raper (1946, p. 735) retorted that they had looked at more strains and reviewed the literature better than anyone else and were "convinced that more nomenclatural stability is to be attained by adding recognization of ascus formation in sections of Micheli's genus *Aspergillus* than could be reached by perpetuating the mistakes of Link and the negligence of Fries."

Why all this fuss? What's in a name? An *Aspergillus* by any other name would still form the same conidial head. This poses no problem in the laboratory. Here we can expect the taxonomic community to persist in giving ascosporous aspergilli "other names," while most scientists will recognize the members of the genus by the characteristic asexual spore structure. The problem arises in the literature. Shakespeare's rose will smell as sweet by any other name when we have the actual rose before us. When we have only the written word, then we hope that a "rose is a rose is a rose." For this reason I agree with Thom: ". . . one point is certain, a perfect strain in a well known imperfect series is entitled to appear only as completion of the species designation, not as a new name replacing the generic name already widely known (Thom, 1957, p. 26). Raper (1957) wrote in a similar vein on behalf of two great groups of molds:

> Conservation of *Aspergillus* and *Penicillium* for those species with a perfect state would stabilize the nomenclature now generally employed through the world; more importantly, it would leave together in the mycological literature the names of fungi that Nature never really separated (Raper, 1957, p. 660).

To the continued dismay of the nomenclatural priesthood, industrial microbiologists and most of the rest of us have adopted the Thom-Raper-Fennell convention. I concur; I have been calling, and will continue to call, both the perfect (teleomorphic) and imperfect (anamorphic) state of this fungus *Aspergillus*.

IV. GENETICS

A. Introduction to *A. nidulans*

Of the millions of cellular species in the world, very few are genetically well characterized. In this small group the prokaryote *Escherichia coli* is preeminent. Among eukaryotes we find a peculiar assortment of species: ourselves (*Homo sapiens*), corn (*Zea mays*), mouse (*Mus musculus*), the "fruit fly" (*Drosophila melanogaster*), and three species of fungi (*Neurospora crassa*, *Saccharomyces cerevisae*, and *A. nidulans*).

A. *nidulans* is one of the best model systems for examining gene expression and gene regulation in eukaryotes. The relative simplicity of the life cycle, the ease with which both mitotic and meiotic spores are obtained, and the presence of the parasexual cycle all contribute to the suitability of *A. nidulans* for genetical analysis. Overviews of formal genetics in this species are presented by Pontecorvo et al. (1953b), Clutterbuck (1974), and Cove (1977).

Asexually, *A. nidulans* reproduces by the formation of uninucleate conidia. Each conidiophore may bear thousands of identical spores which are readily mutagenized with standard chemical and physical mutagens. The mutant phenotypes available for study include auxotrophy, temperature sensitivity, drug resistance and sensitivity, spore color, mycelial color, variations in colony morphology, and alterations in catabolic pathways. Replica plating can be effected in *A. nidulans* by using a multineedle inoculating device for transfer of gridded colonies or damp velveteen to transfer colonies from a master plate on which growth is restricted by the presence of sodium desoxycholate (Mackintosh and Pritchard, 1963).

A. *nidulans* is homothallic, and cleistothecia develop from a single fertilization event. The absence of mating types means that any two strains can be crossed. If self-fertilization has occurred, all the ascospores will be of the same type as the single parental strain. If cross-fertilization has occurred, each ascus will segregate for the parental characteristics. Distinguishing hybrid from selfed cleistothecia is a bit of a nuisance. Usually, parental strains with contrasting spore color markers are used, and a small sample of ascospores from each putative hybrid cleistothecium is pretested for presence of both spore color markers.

Intact asci can be isolated from young cleistothecia and tetrad analysis on the unordered ascospores is possible. Usually, however, a whole hybrid cleistothecium is crushed, and dilute suspensions of ascospores are plated on unselective media. A single cleistothecium may contain up to half a million ascospores (Cove, 1977), allowing for fine structure analysis of meiotic events.

Most of the marked mutant strains used in *A. nidulans* genetics are derived from a single wild-type isolate, subcultures of which are variously designated A69 and NRRL 194 (Croft and Jinks, 1977). Pontecorvo's group in Glasgow did much of the original work isolating mutants from this strain, which were then distributed and redistributed to laboratories around the world where additional mutants were derived. These isogenic mutants are often called "the Glasgow strains." Generally, the Glasgow strains readily anastomose with one another to form heterokaryons. Heterokaryons are ephemeral in prototrophic strains but can be maintained between two strains with different auxotrophic requirements on a minimal medium. Complementation tests can be performed in heterokaryons.

Many wild-type isolates independent of the Glasgow strains have also been studied and mutagenized. These *A. nidulans* strains display considerable variation in growth rate, morphology, and biochemical properties such as penicillin production (Croft and Jinks, 1977). Heterokaryon incompatibility between prototrophic strains of these different wild-type isolates is common (Grindle, 1963). Despite the heterokaryon incompatibility, most strains of *A. nidulans* nevertheless will form cleistothecia when paired in the laboratory, allowing genetic analysis of natural polymorphisms. Population genetic studies for *A. nidulans* have been reviewed by Croft and Jinks (1977). They conclude that wild populations of *A. nidulans* consist of a number of clonally derived subpopulations with little or no gene exchange between them, with each incompatibility group possibly evolving into one of a group of sibling species. Interestingly, the Raper and Fennell (1965) classification scheme orders the various species into groups of sibling species.

B. The Parasexual Cycle

The parasexual cycle was discovered in *A. nidulans* (Roper, 1952; Pontecorvo et al., 1953a). This "alternative to sex" allows for somatic recombination between both linked and unlinked genes. Heterokaryosis is followed by the rare formation of a somatic diploid. Within the diploid, somatic crossing over may lead to formation of mitotic recombinants between linked genes or, alternatively, a series of nondisjunction events leads to the isolation of haploid segregants with new combinations of unlinked genes. The latter process is called haploidization and its frequency can be increased by the use of various haploidization agents such as benlate ("Benomyl") and para-fluorophenylalanine.

Haploidization is a powerful tool for mapping genes; it "is effectively like meiosis without crossing over" (Birkett and Roper, 1977). During haploidization, whole chromosomes segregate together. Appropriately constructed master strains will segregate randomly except the unmapped gene, which will be linked (in repulsion) to one of the eight markers (*A. nidulans* has eight chromosomes). Fungal parasexual analysis has been reviewed by Bradley (1962), Casselton (1965), and Roper (1966).

In mammalian systems a similar process, usually called somatic cell genetics, relies on interspecific protoplast fusion in vitro. Unstable rodent–human hybrids break down in a process analogous to fungal haploidization and allow mapping of human genes. In fact, this method derived from fungal parasexual genetics is the major way in which human genes are currently assigned to chromosomes (McKusick and Ruddle, 1977; Ruddle, 1981; Eustachio and Ruddle, 1983).

Once elucidated in *A. nidulans*, parasexual analysis was extended to other sexual species, as well as to asexual species within the genus (Table 12.5). Evidence for the parasexual cycle has also been found for members of genera of fungi other than *Aspergillus*, allowing recombinational analysis for several asexual fungal phytopathogens (Tinline and MacNeil, 1969; Webster, 1974) and antibiotic-producing industrial strains (Sermonti, 1959; Ball, 1973).

The use of the parasexual cycle for breeding industrial strains has had little impact in relation to its theoretical potential (Rowlands, 1983). Protoplast fusion is a relatively new technique that overcomes certain barriers between strains and species incompatible in traditional parasexual crosses and may extend the use of parasexual analysis. Interspecific protoplast fusion and the isolation of an unstable interspecific hybrid between *A. nidulans* and *Aspergillus rugulosus* has been achieved (Kevei and Peberdy, 1979), but an attempt to effect similar interspecific protoplast fusions between *A. flavus* and *Aspergillus parasiticus* has failed (Leong et al., 1981).

The role of the parasexual cycle in nature is not known, although it is generally thought to be minor. The discovery of naturally occurring diploid wild isolates of *A. niger* (Nga et al., 1975) and *A. nidulans* (Upshall, 1981) shows that somatic diploidy is not exclusively a laboratory phenomenon.

C. Molecular Genetics

Fungi are eukaryotes, but their small genome size and their ease of handling in the laboratory make them rivals to bacteria as convenient systems for detailed

TABLE 12.5   Parasexual Cycle Demonstrated for Species of *Aspergillus*

Species	Sexual Stage Present	Reference
A. nidulans	yes	Roper (1952); Pontecorvo et al. (1953b)
A. niger	no	Pontecorvo et al. (1953a), Lhoas (1967)
A. oryzae	no	Ishitani et al. (1956)
A. sojae	no	Ishitani et al. (1956)
A. fumigatus	no	Berg and Garber (1962); Stromnaes and Garber (1963)
A. amstelodami	yes	Lewis and Barron (1964)
A. rugulusis	yes	Coy and Tuveson (1964)
A. flavus	no	Papa (1973); Gussack et al. (1977)
A. parasiticus	no	Bennett (1973); Papa (1978)

genetic analysis. In fact, the birthday of molecular biology is sometimes equated with the publication of the "one gene–one enzyme" hypothesis by Beadle and Tatum (1941), which was based on research conducted on the filamentous ascomycete, *N. crassa*. For overviews on fungal genetics, see Esser and Kuenen (1967), Burnett (1975), and Fincham et al. (1979). The monograph on *Genetics and Physiology of Aspergillus* (J. E. Smith and Pateman, 1977) mentioned earlier gives background information on the many areas of genetic analysis where *A. nidulans* has served as an elegant model system. Space limitations here preclude anything but a cursory sampling from this enormous literature.

Three important generalizations can be made about gene action and gene organization in fungi:

1. Most fungal gene regulation occurs at the level of transcription (Cooper, 1980).

2. Genes coding for related or sequential functions are seldom clustered on chromosomes in fungi. In contrast, in *E. coli*, about 50% of the known structural genes are linked in groups related to their metabolic function (Pateman and Kinghorn, 1977; Arst, 1983).

3. In *A. nidulans* and *N. crassa*, but not in *S. cerevisiae*, there is a clear preponderance of positive control regulatory genes (Arst, 1983). The products of such regulatory genes in positive acting systems elicit rather than prevent the expression of structural genes.

With "positive control" the natural condition of structural genes is inactivity, and the positive action of a specific metabolite is needed to activate the system. Metzenberg (1979) has written about genetic control mechanisms in *N. crassa*, and much of what he says also pertains to *A. nidulans*. He points out that it is "intuitively easy to see" that it "demands less regulatory vigilance to have each gene inherently inactive, except when it is specifically called upon." The negative control systems well defined in *E. coli* require regulatory macromolecules to turn off the majority of genes most of the time. Metzenberg (1979) hypothesizes that there is probably some upper limit on genome size for which negative control is practical.

Two classes of positive-acting regulatory genes are characterized in *A. nidulans*; these may be dubbed "pathway-specific" and "wide-domain" regulatory genes (Arst, 1981). The best studied of the pathway-specific regulatory genes are *nirA* (involved in the induction of the synthesis of the enzymes of nitrate assimilation; see Cove (1979) and *uaY* (involved in the induction of the synthesis of enzymes of purine catabolism; see Scazzocchio and Gorton (1977) and Scazzocchio et al. (1982)). The *uaY* gene of *A. nidulans* affects the regulation of expression of at least eight unlinked genes specifying different steps in purine uptake and degradation. The probable product of the *uaY* gene is a protein isolated as a uric acid–binding fraction after DNA–cellulose column chromatography (Philippides and Scazzocchio, 1981). A fine structure map and complementation analyses of *uaY* are available (Scazzocchio et al., 1982).

Ammonium is involved in the regulation of many different enzyme and uptake systems in *A. nidulans*. In the presence of ammonium these various systems have little detectable enzyme activity, ensuring the preferential utilization of the favored nitrogen source (ammonium). This mechanism is called nitrogen metabolite repression, ammonium repression, or nitrogen catabolite repression and is analogous to catabolite repression found in the carbon metabolism of many bacteria. Systems repressed by ammonium include both nitrate reduction and purine degradation (Pateman and Kinghorn, 1977). All the enzymes and permeases whose synthesis is subject to nitrogen metabolite repression are under the control of a positive acting regulator gene designated *areA* (Arst and Cove, 1973; Arst and Scazzocchio, 1975; Arst and Bailey, 1977; Rand and Arst, 1977, 1978).

The best characterized wide-domain regulatory genes are *areA* of *A. nidulans*, and its probable equivalent in *N. crassa, nit-2*. Loss of function mutations designated *areA^r* result in inability to utilize nitrogen sources other than ammonium and low or undetectable levels of *areA*-controlled enzymes and permeases. Other, much rarer alleles, designated *areA^d*, lead to derepression of one or more ammonium-repressible activities (Shaffer et al., 1983). The probable *N. crassa nit-2* gene product has been isolated and characterized as a nonhistone nuclear protein (Grove and Marzluf, 1981). It is likely that a similar gene product of *areA* is necessary for the expression of structural genes whose products are subject to ammonium repression.

Gene cloning, restriction enzyme analysis, and the rest of the armamentary of "the new biology" are increasingly being brought to bear on aspects of *A. nidulans* biology. Mitochondrial DNA is particularly amenable to analysis. The coding section of the apocytochrome *b* gene (*cobA*) has been sequenced and found to contain one intron. The derived amino acid sequence shows 61% and 51% homology with the cognate genes in yeast and human mitochondria, respectively. Comparison of these sequences indicate that UGA, a nonsense codon in the "universal genetic code," codes for tryptophan in the *A. nidulans* mitochondrial system (Waring et al., 1981).

Restriction enzyme analysis of mitochondrial DNA has been used as a taxonomic aid for the genus *Aspergillus*. Restriction enzyme cleavage patterns of mitochondrial DNA from *A. nidulans, A. wentii, A. awamori, A. niger, A. oryzae, A. tamarii,* and *A. echinulatus* were compared. The two members of the Raper and Fennell (1965) "*A. nidulans* group" (*A. nidulans* and *A. echinulatus*) as well as the two members of the "*A. flavus* group" (*A. oryzae* and *A. tamarii*) showed close homology with each other. Unexpectedly, *A. wentii* showed an identical EcoRI and Hind III restriction pattern with *A. oryzae*, indicating a close phylogenetic relationship between these two species (Kozlowski and Stepien, 1982).

Studies of gene regulation are also benefiting from the new methods of gene manipulation. Approximately 1300 diverse polyA + RNA's that are not detectable in hyphae appear during conidiogenesis in *A. nidulans* (Timberlake, 1980). About 25% of these RNA's apparently accumulate selectively in conidia; chromosomal DNA's coding for RNA's abundant in conidia but absent in hyphae have been isolated (Zimmerman et al., 1980). One 13.3-kb region designated *SpoC* was selected from a recombinant phage library formed between *A. nidulans* DNA and λCharon4A. Restriction fragments were recloned into plasmids by ligation and transformation (Zimmerman et al., 1980). The *SpoC*$_1$ region encodes for one partial and five complete RNA coding sequences. The genes of the *SpoC*$_1$ cluster are expressed at the same time during conidiation, but the concentrations of the polyA + − RNA products vary considerably. The physiological function of the *SpoC*$_1$-encoded RNA's are not known (Timberlake and Barnard, 1981).

There is a recent report of the introduction and expression of a gene from *N. crassa* into *A. nidulans* protoplasts. Southern hybridization analysis indicated integration of the *N. crassa* DNA and some pBR322 sequences into the *A. nidulans* genome (Ballance et al., 1983). As transformation efficiencies are improved and transformation technologies are extended, gene manipulations may also become possible in other species of *Aspergillus* and may even permit strain improvement of industrial strains.

V. ECONOMIC ASPECTS

A. Secondary Metabolites

Secondary metabolites are an enormous group of chemical differentiation products. It is impossible to give a simple definition for secondary metabolites, so they are usually defined by a list of their attributes: low molecular weight, biosynthetic origin from a few simple precursors, sporadic taxonomic distribution, production in chemical families, and obscure function. Frequently, secondary metabolites are produced after active growth has ceased during a fermentation stage dubbed "idiophase" (Bu'Lock et al., 1965), which is roughly equivalent to stationary phase for bacterial cultures.

The dividing line between secondary metabolites and primary metabolites (compounds that are essential to growth) is not always clear. For example, citric acid (Fig. 12.4a) is sometimes viewed as a secondary metabolite when it is produced in unphysiological amounts as occurs during commercial citric acid fermentations (see below). Yet obviously the citric acid cycle is one of the foremost of the primary metabolic pathways of the cell.

The best known secondary metabolites are bioactive in one form or another. Antibiotics have received the most attention, and many microbial secondary metabolites have been discovered during large-scale screening for new antibiotics. Most of these compounds exhibit adverse side effects in mammalian systems, a result disqualifying them for drug use.

Turner (1971) divided fungal secondary metabolites into seven groups according to their biosynthetic origin: polyketides; terpenes and steroids; metabolites derived from fatty acids; metabolites derived from the tricarboxylic acid cycle; metabolites derived from amino acids; metabolites derived without acetate; and a final miscellaneous category.

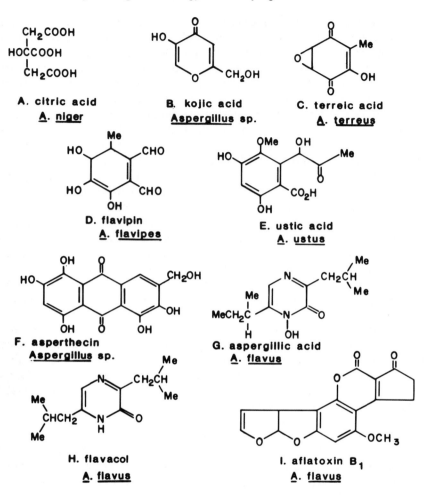

FIGURE 12.4   *Aspergillus* metabolites.

Polyketides are formed by the condensation of acetyl and malonyl units with concomitant decarboxylation. The polyketides may be classified according to the number of C_2 units contributing to the polyketide chain, for example, tetraketides and pentaketides. The polyketide pathway leads almost exclusively to secondary metabolites and is used mainly by fungi classified with the Ascomycotina and Deuteromycotina (Turner, 1971).

With the exception of citric acid (Fig. 12.4a), which is not a secondary metabolite, the compounds illustrated in Fig. 12.4 were chosen because they are all produced by members of the genus *Aspergillus* and because their trivial name was derived from the name of the original source of isolation. Kojic acid (Fig. 12.4b) is produced by many *Aspergillus* species and was first isolated from the koji molds of Oriental food fermentations. It arises by direct conversion of glucose (Arnstein and Bentley, 1956), a relatively rare pathway for secondary metabolism in fungi. Terreic acid (Fig. 12.4c) from *A. terreus* and flavipin (Fig. 12.4d) from *A. flavipes* are both tetraketides; ustic acid (Fig. 12.4e) from *A. ustus* is a pentaketide; asperthecin (Fig. 12.4f) from numerous *Aspergillus* species is an octaketide. Aflatoxin B_1 (Fig. 12.4i) from *A. flavus* and *A. parasiticus* is a C_{17} compound originally classified as a nonaketide; recent biosynthetic studies using blocked mutants and isotope labeling have now verified its biosynthetic origin from a decaketide anthraquinone precursor (Bennett and Christensen, 1983). The aflatoxins are toxigenic, carcinogenic, mutagenic, and teratogenic. Their discovery in the early 1960's heralded the beginning of the "mycotoxin revolution"—the realization that naturally occurring molds might not just be decomposing foodstuffs, but also contaminating those foods with highly potent toxins. Aflatoxins and other mycotoxins constitute a serious hazard to human and animal health. The mycotoxin problem has been reviewed many times (Goldblatt, 1969; Rodricks et al., 1977; Wyllie and Moorehouse, 1977, 1978; Heathcote and Hibbert, 1978; Ciegler and Bennett, 1980).

Two additional *A. flavus* metabolites are aspergillic acid (Fig. 12.4g) and flavacol (Fig. 12.4h). These are closely related diketoperazines, originally isolated because of their antibiotic activities (Turner, 1971).

In recent years there has been a growing awareness that secondary metabolites may exhibit pharmacological activity other than antibiosis and toxicosis. This topic has been reviewed by Demain (1981, 1983). Other recommended reviews on fungal secondary metabolism are those of Bentley and Campbell (1968), Bu'Lock (1961, 1965, 1967, 1975), Weinberg (1971), Demain (1974), Rose (1979), and Bennett and Ciegler (1983).

B. *Aspergillus* Metabolites Used in Food Processing

Organic acids are used in many aspects of modern food processing. They serve as flavoring agents, buffers, preservatives, synergists to antioxidants, and meat curing agents. By far, the acidulant with the widest commercial application is citric acid, and by far, the most common commercial fermentation for citric acid production employs selected strains of *A. niger*. In modern fermentations, 90% (w/w) of sucrose provided is recovered as citric acid (Kubicek and Röhr, 1982). Citric acid production may be considered an example of "overflow" metabolism (Ma et al., 1981).

Mold-produced citric acid was introduced by Currie (1917) and originally employed a surface culture process. Submerged fermentations are now used. Both processes rely on interruption of normal growth. Since commercial citric acid processes are usually kept secret by producing companies, the research on citric acid regulation in Vienna is welcome (Kubicek and Röhr, 1982). This group has also done work on a pilot scale to mimic industrial environments (Roehr et al., 1981).

Citric acid accumulation is triggered by trace metal deficiency, low pH, and phosphate exhaustion. Manganese deficiency is an essential condition for citric acid accumulation, resulting in a reduction of the activities of the enzymes of the hexose monophosphate pathway and the tricarboxylic acid cycle. The effect of manganese has been studied extensively (Kubicek and Röhr, 1977, 1978; Kubicek et al., 1979; Kisser et al., 1980). Citric acid secretion may be considered a way in which *A. niger* is able to handle overflow of glycolytic metabolites (Orthofer et al., 1979). Phosphofructokinase has been identified as an important regulatory enzyme in the citric acid process (Habison et al., 1979).

In addition to citric acid, other organic acids are used in many aspects of modern food and beverage manufacturing. Major ones produced by aspergilli are listed in Table 12.6. It should be noted, however, that many of these metabolites can also be produced by other genera of molds, as well as by yeasts and bacteria.

In addition to organic acids, numerous industrial enzymes are of fungal origin (Table 12.6). Given the widespread domestic use of *Aspergillus* fermen-

TABLE 12.6   *Aspergillus*-derived Metabolites Used in Food Processing

Compounds	Species	Use
Acids		
citric acid	*A. niger*	soft drinks, dairy products, jams, jellies, candies, frozen foods, fats, canned goods
gluconic acid	*A. niger*	baking powder, bread mixes, desserts, bottle-washing formulations
itaconic acid	*A. itaconicus,* *A. terreus*	shortenings, resin coatings in contact with food
malic acid	*Aspergillus* sp.	beverages, jams, jellies, syrups, candy, sour dough
oxalic acid	*A. niger*	hydrolysis of starch to glucose
tartaric acid	*A. niger,* *A. griseus*	carbonated beverages, desserts, jellies
Enzymes		
α-amylase	*A. oryzae,* *A. niger*	corn syrup, dextrose, baking, food dextrins, chocolate syrups
glucoamylase	*A. awamori,* *A. niger*	dextrose, dextrose syrup, baking
glucose oxidase, catalase	*A. niger*	powdered egg products, brewing, wines, mayonnaise
lactase	*A. niger*	dairy products
naringinase	*A. niger*	debittering grapefruit juice
pectinase	*A. niger*	clarifying fruit juice and wine
protease	*A. oryzae*	tenderizing meat, soy sauce, brewing, baking
Miscellaneous		
mannitol	*A. candidus*	bulking agent, humectant

After Beuchat (1978).

tations in Japan, it is not surprising that modern industrial enzyme technology began with a scientist of Japanese origin working with *A. oryzae* (Takamine, 1894).

C. Oriental Food Fermentations

Fermented foods have different qualities of flavor and improved digestibility. The most common Western food fermentations involve yeasts (bread, wine) and bacteria (yogurt, pickles, etc.). Molds are rarely encountered except in the manufacture of certain cheeses. In the Orient, on the other hand, filamentous fungi are employed in a number of complex fermentations involving soybeans, rice, wheat, fish, and other foodstuffs. Of these Oriental products, the best known in the West are soy sauce, saké (rice wine), and miso (fermented soybean paste). They are all made by a two-stage fermentation process in which the first step involves *Aspergillus* species.

The initial step in soy sauce, saké, and miso preparation is the production of a "koji." Koji comes from a Japanese word that roughly translates as "bloom of mold" (Yong and Wood, 1974). *Aspergillus sojae* or *A. oryzae* is inoculated on steamed rice, wheat, or other cereal under warm, humid conditions. The mold grows by infiltrating the substrate with a mass of hyphae while at the same time secreting hydrolytic enzymes that partially degrade this same substrate. The fungal proteases and amylases have been studied extensively (see Yong and Wood, 1974, for references). The resultant koji is a heterogeneous, fragrant, crumbly mass of mold and substrate. The koji serves both as a source of enzymes and as an inoculum. Sometimes the ash of woody plants is added to furnish salts that improve conidiation. Kojis can be classified into many types, distinguished by the particular features of mycelial development and sporulation, the kind of grain being used, and the purpose to which the koji will be put (Abiose et al., 1982). The term koji is used rather loosely to mean the mold, the mold substrate mixture used as an inoculum ("tane koji"), and the bulk component of the main fermentation mass. Tane koji can also be called mold seed or mold starter.

For soy sauce the tane koji is mixed with about equal proportions of soybeans and wheat. As the mold grows, it releases proteases and amylases that break down the beans and grain. The heterogenous hydrolyzed product is added to a salt solution (17–18%), and an anaerobic fermentation with lactobacilli and yeasts ensues. The final liquid is pressed and clarified as soy sauce (Yong and Wood, 1974; Pederson, 1979).

Miso has no counterpart in Western cuisine. Much of it is made at home in Japanese homes, much the way some Western households make their own yogurt, but on a much wider scale. Miso has the consistency of peanut butter and is used primarily as a flavoring agent. There are many kinds of miso; the different types are made by using different raw materials, mixed in different proportions, and varying the length of fermentation. For example, hatcho miso uses only soybeans, muji miso uses barley and soybeans, and kome miso uses rice and soybeans. Miso is further classified by color (red, light yellow, white), flavor (salty, mellow, sweet), texture (chunky, smooth), method of fermentation (traditional, mass produced) and region of origin. Certain flavors, colors, and types of fermentation are characteristic of different provinces, and the people in different geographic regions may have preferences molded by culinary tradition, social class, and urbanization. Although there is no hard and fast relationship, in general, saltier varieties tend to be darker in color, and depth of color is more or less related to the length of fermentation. Miso made with rice is the most common form (Shurtleff and Aoyagi, 1976). The chemical composition of several rice, barley, and soybean misos are given by Abiose et al. (1982).

Saké is an alcoholic drink made from rice. The history of saké brewing has been reviewed by Kodama (1970) and Kodama and Yoshizawa (1977). Rice koji is mixed with steamed rice and water and incubated with occasional stirring for 3–4 days. This is the moto stage, during which the koji enzymes partially degrade and saccharify the rice. A well-defined change in microorganisms occurs with fungi and bacteria being overgrown by the saké yeasts to yield the moroni stage. The moroni stage is the longest; during this time, yeast metabolism predominates to produce the alcoholic products which are filtered, bottled, and pasteurized (Abiose et al., 1982; Steinkraus, 1983).

Records of soy sauce, miso, and saké production date from Oriental classical antiquity. Most of this Chinese and Japanese literature is totally inaccessible to Westerners. An early monograph on soybeans described traditional soybean foods (Piper and Morse, 1923), and there are several reviews aimed at microbiologists (Yong and Wood, 1974; Kodama and Yoshizawa, 1977; Aidoo et al., 1982; Steinkraus, 1983). The popular work by Shurtleff and Aoyagi (1976) contains an excellent bibliography of books published in Japanese, a good history of the way in which Chinese fermented foods were introduced into Japan, and a wonderfully eclectic collection of miso recipes ranging from pedestrian-sounding dishes such as "miso with noodles" to particularly American hybrid concoctions such as "miso pizza."

The "health food" culture has embraced these Oriental food fermentations as wholesome and healthful, and Western culture in general has embraced Japanese products as having a particular chic. It is likely that molded foods will become more common in the West in the near future, and lyophilized molds will be available in trendy shops next to the brewer's yeasts and sourdough starters.

D. Taxonomy and the "Real World"

Only five microorganisms are "generally recognized as safe" (GRAS) by the U.S. Food and Drug Administration (FDA). These are one bacterium (*Bacillus subtilis*), two yeasts (*S. cerevisae, Kluyveromyces fragilis*), and two species of *Aspergillus* (*A. niger* and *A. oryzae*). Except for *A. niger*, *A. oryzae* is the most widely used mold in industry. When Oriental food fermentations are considered, *A. oryzae* may be the most important mold associated with human food use in the world. *A. oryzae* is classified in the "*A. flavus* group" (Raper and Fennell, 1965). The discovery of aflatoxins (the most potent natural carcinogens known) from *A. flavus* has made taxonomic distinctions within the *A. flavus* group far more than an academic matter.

Wicklow (1983) has commented that colony color is the taxonomic character of the longest standing stability in connection with this group and proclaimed that "the yellow-green aspergilli are spectacular molds" (Wicklow, 1983, p. 7). Spectacular or not, the yellow-green aspergilli are difficult for taxonomists. Variability, a common characteristic in the whole genus, seems to be particularly pronounced within this group. Thom and Raper (1945) considered the group as an aggregate of strains in which no sharp lines of demarcation between species existed. They separated the group into two major series centered upon *A. oryzae* and *A. flavus*. Most subsequent investigations have reconfirmed the difficulty of establishing sharp demarcations between species and the existence of subgroups within the "*A. flavus* group."

Japanese workers have been particularly concerned that aflatoxins might be present in their traditional foodstuffs. However, in surveys of hundreds of Japanese industrial and domestic strains, no evidence of aflatoxin production was found (Murakami et al., 1967, 1968a). In most cases there were clear mycological distinctions between koji molds and aflatoxigenic strains. Of 16 industrial strains that were similar to the *A. flavus* species concept, none showed exactly the same mycological characteristics, and more importantly, none produced aflatoxins (Murakami et al., 1968b). Murakami (1971) then studied 20 morphological, physiological, and cultural characteristics of over 400 strains within the *A. flavus* group. After computer analysis of the resultant data, all the strains were placed into one of two clusters, with all the koji molds in the *A. oryzae* cluster and all the other strains in the *A. flavus* cluster.

Despite the rigor of the studies from Murakami's laboratory, the taxonomic issue has not been laid to rest. El-Hag and Morse (1976) published a paper in *Science* claiming that a variant of *A. oryzae* produced aflatoxin on cowpeas but not on soybeans. In a later letter to *Science*, Fennell (1976) gave a well-supported diagnosis that the El-Hag and Morse strain had been contaminated through mite infestation and was not *A. oryzae* at all. Writing with barely veiled sarcasm, she said, "I am not surprised that the culture they returned to me produces aflatoxins B_1, B_2, G_1 and G_2. *Aspergillus parasiticus* is known for this ability."

Nomenclatural problems cannot all be settled by evidence of strain contamination. Some of them are intractable and have to do with the elusiveness of the species concept within the group itself.

The *A. flavus* group, *sensu* Raper and Fennell (1965) encompassed nine species (*A. flavus, A. parasiticus, A. oryzae, A. zonatus, A. tamarii, A. flavofurcatis, A. subolivaceus,* and *A. avenaceus*) and two varieties (*A. flavus* var. *columnaris* and *A. oryzae* var. *effusus*). Raper and Fennell (1965) viewed *A. sojae* as a probable synonym for a variety of *A. parasiticus*. Since *A. sojae* is a koji mold and *A. parasiticus* is usually a strong aflatoxin producer, the "what's in a name?" problem had immediate ramifications in the Japanese food industry. It is not surprising that Murakami (1971) elevated *A. sojae* back to species rank, where it has remained (Christensen, 1981; Wicklow, 1983).

In a reevaluation of the entire *A. flavus* group, Christensen (1981) examined new species and varieties described since the publication of Raper and Fennell (1965). She interpreted *Aspergillus leporis, A. avenaceus, A. clavatus-flavus, A. coremiformis, A. sojae, A. subolivaceus,* and *A. zonatus* as morphologically and ecologically distinct species. *A. flavus, A. oryzae,* and *A. parasiticus* were separable but demonstrated "broad and overlapping morphologies." Christensen (1981, pp. 1080–1081) interpreted this as being due to the long association of these molds with human activities:

> In *A. flavus, A. oryzae, A. parasiticus, A. sojae,* and *A. toxicarius* I suspect that we actually are dealing in part or wholly with members of the 'domesticated landscape.' In a domesticated mold, as in a domesticated plant or animal, it can be postulated that there exists a very large gene pool (expressed totally in haploid fungi) which has persisted because of man's behavior; we have spared those species the rigors of selection that exist in natural environments. Thus *A. leporis, A. zonatus,* and *A. clavos*-flavus (exclusively 'wild' species in the *A. flavus* group) like wild duck and wild species in *Canis*, are far more easily distinguished and characterized than are domesticated duck, dogs, or aspergilli.

Wicklow (1983) reviewed the entire literature on the taxonomic features of the group and pointed out that Murakami (1971) and Christensen (1981) differ sharply in their choice of stable and reliable taxonomic characters. Of Murakami's three most significant features—sclerotium production, seriation of sterigmata, and roughness of conidial walls, only sclerotia (under the heading sclerotium morphology)—is included in Christensen's (1981) calculated coefficients of similarity. According to Wicklow (1983), if Christensen had not included the koji molds in her sample, the cluster of "'partly or wholly' domesticated strains no longer is apparent." He concluded that "the clustering of domesticated koji molds and wild species (by Christensen, 1981) represents a clear case of guilt by previous association" (Wicklow, 1983, p. 12)—whatever that is. Wicklow (1983) presents the following classification:

"Wild" species	*Domesticated varieties*
A. parasiticus Speare	*A. parasiticus* var. *sojae*
Ancestral strain	
A. flavus Link ex Fr	*A. flavus* var. *oryzae* Saito

Needless to say this classification of a GRAS mold as a variety of an aflatoxigenic mold has the potential to cause problems for certain industries.

The definition of species remains problematic within the group. In one recent study, Kozakiewicz (1982) examined transfers of cultures on which some of the early descriptions of *A. flavus* and *A. parasiticus* were based and concluded that they might not be what they were purported to be. In the absence of "real" type material for *A. flavus*, Thom and Church (1921) used a culture designated Thom 108 (NRRL 482) as typical for that species. Raper and Fennell (1965) no longer considered this strain as typical but did not say why. Kozakiewicz (1982) reveals that the accession card for this isolate bears a handwritten note by Dorothy Fennell stating, "This appears to be *A. oryzae* in stocks now." Then, in the same paper, Kozakiewicz (1982) contructs a remarkable geneology of the "type" isolate of *A. parasiticus* Speare from 1912 through its distribution and transfer to culture collections throughout the world. She reaches the disturbing conclusion that Speare's original isolate was not a single entity, but rather a mixture of material representing two taxa, one like the modern concept of *A. flavus* and the other like *A. parasiticus.*

Secondary metabolites have not been considered as reliable taxonomic criteria by most systematists, yet in this example it is the production or nonproduction of a secondary metabolite (aflatoxin) that is the crux of the problem. It is to be hoped that future revisions of the genus will place greater emphasis on biochemical attributes and secondary metabolism. This has also been proposed for the genus *Penicillium* (Pitt, 1979; Ciegler et al., 1981; Frisvad, 1981). DNA annealing studies, restriction enzyme analysis, and eventually, base sequence data may also help resolve difficulties within the group.

VI. CONCLUSION

The aspergilli are at the interface of macroscopic and microscopic life. An entire colony is visible as a colored mold or a fuzzy discoloration on all sorts of organic matter. To see form, one must supplement the naked eye with a microscope. Thus, although "historically the Aspergilli, as a part of the moldiness of things have always been a factor in man's environment" (Thom and Raper, 1945, p. 3), it required a modern era to name and characterize these organisms in a scientific manner.

Taxonomic problems confronted within the genus are similar to those faced for bacterial taxonomy and have been compounded by forcing both nomenclature and classification into a scheme developed for vascular plants. The economic importance of the genus takes *Aspergillus* taxonomy out of the ivory tower and into the marketplace and courtroom. New biochemical diagnostic criteria to supplement traditional morphological attributes are needed.

With the penicillia, the aspergilli are the major molds of this world. They are typical filamentous fungi with the capacity to invade solid substrates, to secrete degradative enzymes, and thence to rot, to spoil, to putrefy, and to transform. They play a major role in nature, silently and relentlessly recycling the remains of dead organisms. In human commerce their degradative enzymes have been harnessed in food and industrial fermentations.

Unique anabolic pathways have evolved in many species within the genus. Polyketide-derived secondary metabolites are particularly abundant. Frequently, chemists endow these secondary metabolites with trivial names derived from the species of original isolation, so some secondary metabolites advertise the genus with names like aspergillic acid, aflatoxin, and asperthecin.

A. nidulans is one of the best models for eukaryotic genetics. Haploidization analysis via the parasexual cycle provided the theoretical framework for gene mapping in mammals by somatic cell hybridization. Genetic dissection of developmental and degradative pathways is progressing well. The availability of a transformation system in *A. nidulans* ensures a continued place for these versatile molds in contemporary biology.

ACKNOWLEDGMENTS

My thanks to the many taxonomists who helped guide me through their literature and reviewed this manuscript, especially Meredith Blackwell, Martha Christensen, Bryce Kendrick, Kenneth Raper, and Arthur Welden. Thanks also to Martha Christensen, Herb Arst, Max Röhr, Robert Samson, and Claudio Scazzocchio for reprints, Irving LaValle for manuscript preparation, and Nancy Meadow for illustrations. Ingrid Krohn, the editor who originally conceived this series, deserves special mention for her guidance and inspiration. Work in this laboratory has been supported by a Cooperative Agreement from the U.S. Department of Agriculture (58-7B30-3-556).

REFERENCES

Abiose, S. H., Allan, M. C., and Wood, B. J. B. (1982) *Adv. Appl. Microbiol. 28*, 239–265.

Aidoo, K. E., Hendry, R., and Wood, B. J. B. (1982) *Adv. Appl. Microbiol. 28*, 201–237.

Ainsworth, G. C. (1952) *Medical Mycology: An Introduction to Its Problems*, Pitman, New York.

Ainsworth, G. C. (1961) *Ainsworth & Bisby's Dictionary of Fungi*, 5th ed., Commonwealth Mycological Institute, Kew, Surrey, England.

Ainsworth, G. C. (1971) *Ainsworth & Bisby's Dictionary of the Fungi*, 6th ed., Commonwealth Mycological Institute, Kew, Surrey, England.

Ainsworth, G. C. (1973) in *The Fungi*, Vol. IVA, *Taxonomic Review with Keys, Ascomycetes and Fungi Imperfecti* (Ainsworth, G. C., Sparrow, F. K., and Sussman, A. S., eds.), pp. 1–7, Academic Press, New York.

Ainsworth, G. C. (1976) *Introduction to the History of Mycology*, Cambridge University Press, Cambridge, England.

Ainsworth, G. C., and Sussman, A. S. (eds.) (1966) *The Fungi: An Advanced Treatise*, Vol. II., *The Fungal Organism*, Academic Press, New York.

Alexopoulos, C. J. (1962) *Introductory Mycology*, 2nd ed., John Wiley, New York.

Alexopoulos, C. J., and Mims, C. W. (1979) *Introductory Mycology*, 3rd ed., John Wiley and Sons, New York.

Arnstein, H. R. V., and Bentley, R. (1956) *Biochem. J. 62*, 403–407.

Arst, H. (1981) in *Genetics as a Tool in Microbiology* (Glover, S. W., and Hopwood,

D. A., eds.), Society for General Microbiology Symposium 31, pp. 131–160, Cambridge University Press, Cambridge, England.

Arst, H. N., Jr. (1983) in *Eukaryotic Genes: Their Structure, Activity, and Regulation* (Maclean, N., Gregory, S. P., and Flavell, R. A., eds.), Butterworths, London.

Arst, H. N., Jr., and Bailey, C. R. (1977) in *Genetics and Physiology of Aspergillus* (Smith, J. E., and Pateman, J. A., eds.), pp. 131–146, Academic Press, New York.

Arst, H. N., Jr., and Cove, D. J. (1973) *Mol. Gen. Genet. 126*, 111–141.

Arst, H. N., Jr., and Scazzocchio, C. (1975) *Nature 254*, 31–34.

Baker, R. D. (ed.) (1971) *Human Infections with Fungi, Actinomycetes and Algae*, Springer, New York.

Ball, C. (1973) in *Genetics of Industrial Microorganisms* (Vanék, Z., Hostálek, Z., and Cudlin, J., eds.), pp. 227–237, Academia, Prague.

Ballance, D. J., Buxton, F. P., and Turner, G. (1983) *Biochem. Biophys. Res. Comm. 112*, 284–289.

Barkley, F. A. (1968) *Outline Classification of Organisms*, 2nd ed., J. Hopkins Press, Providence, Mass.

Barron, G. L. (1968) *The Genera of Hyphomycetes from Soil*, Williams and Wilkins, Baltimore.

Beadle, G. W., and Tatum, E. L. (1941) *Proc. Nat. Acad. Sci. U.S.A. 27*, 499–506.

Benjamin, C. R. (1955) *Mycologia 47*, 669–687.

Bennett, J. W. (1973) *Genetics 74*, 520.

Bennett, J. W., and Christensen, S. B. (1983) *Adv. Appl. Microbiol. 29*, 56–92.

Bennett, J. W., and Ciegler, A. (eds.) (1983) *Secondary Metabolism and Differentiation in Fungi*, Marcel Dekker, New York.

Bentley, R., and Campbell, I. M. (1968) in *Comprehensive Biochemistry, Metabolism of Cyclic Compounds* (Florkin, M., and Stotz, E. H., eds.), pp. 415–487. Elsevier, Amsterdam.

Berg, C. M., and Garber, E. D. (1962) *Genetics 47*, 1139–1146.

Bessey, E. A. (1950) *Morphology and Taxonomy of Fungi*, Blakiston Co., Philadelphia.

Beuchat, L. R. (1978) *Food and Beverage Mycology*. AVI, Westport, Conn.

Birkett, J. A., and Roper, J. A. (1977) in *Genetics and Physiology of Aspergillus* (Smith, J. E., and Pateman, J. A., eds.), pp. 293–303, Academic Press, New York.

Bisby, G. R. (1953) *An Introduction to the Taxonomy and Nomenclature of Fungi*, 2nd ed., Commonwealth Mycological Institute, Kew, Surrey, England.

Bradley, S. G. (1962) *Ann. Rev. Microbiol. 16*, 35–52.

Buchanan, R. E., and Gibbons, N. E. (eds.) (1974) *Bergey's Manual of Determinative Bacteriology*, 8th ed., Williams and Wilkins, Baltimore.

Bu'Lock, J. D. (1961) *Adv. Appl. Microbiol. 3*, 293–342.

Bu'Lock, J. D. (1965) *The Biosynthesis of Natural Products*, McGraw-Hill, London.

Bu'Lock, J. D. (1967) *Essays in Biosynthesis and Microbial Development*, Wiley, New York.

Bu'Lock, J. D. (1975) in *The Filamentous Fungi*, Vol. I (Smith, J. E., and Berry, D. R., eds.), pp. 33–58, Wiley, New York.

Bu'Lock, J. D., Hamilton, D., Hulme, M. A., Powell, A. J., Smalley, H. M., Shephard, D., and Smith, G. N. (1965) *Can. J. Microbiol. 11*, 765–778.

Burnett, J. H. (1968) *Fundamentals of Mycology*, Edward Arnold, Ltd., London.

Burnett, J. H. (1975) *Mycogenetics: An Introduction to the General Genetics of Fungi*, Wiley, New York.

Carmichael, J. W. (1979) in *The Whole Fungus: The Sexual-Asexual Synthesis*, Vol. I

(Kendrick, B., ed.), pp. 31–41, National Museum of Natural Sciences and Kananaskis Foundation, Ottawa, Canada.

Carmichael, J. W., Kendrick, W. B., Conners, I. L., and Sigler, L. (1980) *Genera of Hyphomycetes*, University of Alberta Press, Edmonton, Alberta, Canada.

Casselton, L. A. (1965) *Sci. Progr. 53*, 107–115.

Chesters, C. G. (1968) in *The Fungi*, Vol. III, *The Fungal Population* (Ainsworth, G. C., ed.), pp. 517–542, Academic Press, New York.

Christensen, M. (1981) *Mycologia 72*, 1056–1084.

Christensen, M. (1982a) *Mycologia 74*, 210–225.

Christensen, M. (1982b) *Mycologia 74*, 226–235.

Christensen, M., and Raper, K. B. (1978) *Trans. Br. Mycol. Soc. 71*, 177–191.

Ciegler, A., and Bennett, J. W. (1980) *Bioscience 30*, 512–515.

Ciegler, A., Lee, L. S., and Dunn, J. J. (1981) *Appl. Environ. Microbiol. 42*, 446–449.

Clements, F. C., and Shear, C. L. (1931) *The Genera of Fungi*, Hafner, New York.

Clutterbuck, A. J. (1974) in *Handbook of Genetics*, Vol. I (King, R. C., ed.), Plenum, New York.

Cochrane, V. W. (1958) *Physiology of Fungi*, John Wiley and Sons, New York.

Cole, G. T., and Kendrick, B. (1981) *Biology of Conidial Fungi*, Vols. I and II, Academic Press, New York.

Cole, G. T., and Samson, R. A. (1979) *Patterns of Development in Conidial Fungi*, Pitmans, London.

Colwell, R. R., and Austin, B. (1980) in *Manual of Methods for General Bacteriology* (Murray, R. G. E., Costilow, R. N., Nester, E. W., Wood, W. A., Krieg, N. R., and Phillips, G. B., eds.), pp. 444–449, American Society for Microbiology, Washington, D.C.

Cooper, T. G. (1980) *Trends in Biochem. Sci. 5*, 332–334.

Copeland, H. F. (1956) *The Classification of Lower Organisms*, Pacific Books, Palo Alto, Calif.

Cove, D. J. (1977) in *Genetics and Physiology of Aspergillus* (Smith, J. E., and Pateman, J. A., eds.), pp. 81–95, Academic Press, New York.

Cove, D. J. (1979) *Biol. Rev. 54*, 291–327.

Cowan, S. T. (1971) *J. Gen. Microbiol. 67*, 1–8.

Coy, D. O., and Tuveson, R. W. (1964) *Genetics 50*, 847–853.

Croft, J. H., and Jinks, J. L. (1977) in *Genetics and Physiology of Aspergillus* (Smith, J. E., and Pateman, J. A., eds.), pp. 339–360, Academic Press, New York.

Currie, J. N. (1917) *J. Biol. Chem. 31*, 15–37.

Davis, B. D. (1980) *Microbiology Including Immunology and Molecular Genetics*, 3rd ed. (Davis, B. D., Dulbecco R., Eisen, N. H., and Ginsberg, H. S.), pp. 1–14, Harper and Row, Hagerstown, Md.

Deacon, J. W. (1980) *Introduction to Modern Mycology*, Halsted Press, John Wiley and Sons, New York.

Demain, A. L. (1974) *Ann. N.Y. Acad. Sci. 235*, 601–612.

Demain, A. L. (1981) *Science 214*, 987–995.

Demain, A. L. (1983) *Science 219*, 709–714.

Demoulin, V., Hawksworth, D. L., Korf, P. L., and Pouzar, Z. (1981) *Taxon 30*, 52–63.

Dodge, B. O. (1945) *Science 102*, 460–461.

Dunn, G., and Everitt, B. S. (1982) *An Introduction to Mathematical Taxonomy*, Cambridge University Press, Cambridge, England.

Eichler, A. W. (1886) *Syllabus der Vorlesungen uber specielle und medicinisch-pharm-*

aceutische Botanik, 4th ed., Berlin, cited in G. M. Smith (1938) *Crytogamic Botany*, McGraw-Hill, New York.

El-Hag, N., and Morse, R. E. (1976) *Science 192*, 1345–1346.

Emmons, C. W., Binford, C. H., Utz, J. P., and Kwon-Chung, K. J. (1977) *Medical Mycology*, Lea and Febiger, Philadelphia.

Esser, K., and Kuenen, R. (1967) *Genetics of Fungi* (transl. E. Steiner), Springer, New York.

Eustachio, P. D., and Ruddle, F. H. (1983) *Science 220*, 919–924.

Fennell, D. I. (1976) *Science 194*, 1188.

Fennell, D. I. (1977) in *Genetics and Physiology of Aspergillus* (Smith, J. E., and Pateman, J. A., eds.), pp. 1–21, Academic Press, New York.

Fincham, J. R. S., Day, P. R., and Radford, A. (1979) *Fungal Genetics*, 4th ed., Blackwell, Oxford, England.

Foster, J. W. (1949) *Chemical Activities of Fungi*, Academic Press, New York.

Frisvad, J. C. (1981) *Appl. Environ. Microbiol. 41*, 568–579.

Fuller, H. J., and Tippo, O. (1954) *College Botany*, Holt, Rinehart and Winston, New York.

Goldblatt, L. A. (ed.) (1969) *Aflatoxin*, Academic Press, New York.

Gregory, P. H. (1966) in *The Fungus Spore* (Madelin, M. F., ed.), pp. 1–13, Butterworths, London.

Griffin, D. M. (1972) *Ecology of Soil Fungi*, Syracuse University Press, Syracuse, New York.

Grindle, M. (1963) *Heredity 18*, 397–405.

Grove, G., and Marzluf, G. A. (1981) *J. Biol. Chem. 256*, 463–470.

Gussack, G., Bennett, J. W., Cavalier, S., and Yatsu, L. (1977) *Mycopathologia 61*, 159–165.

Habison, A., Kubicek, C. P., and Röhr, M. (1979) *FEMS Microbiol. Lett. 5*, 39–42.

Haldane, J. B. S. (1955) *New Biol. 19*, 7–26.

Hawker, L. E. (1966) *Fungi: An Introduction*, Hutchinson University Library, London.

Hawksworth, D. L. (1974) *Mycologists Handbook*, Commonwealth Mycological Institute, Kew, Surrey, England.

Heathcote, J. G., and Hibbert, J. R. (1978) *Aflatoxins: Chemical and Biological Aspects*, Elsevier, Amsterdam.

Hennebert, G. L., and Weresub, L. K. (1979) in *The Whole Fungus: The Sexual-Asexual Synthesis*, Vol. I (Kendrick, B., ed.), pp. 27–30, National Museum of Natural Sciences and Kananaskis Foundation, Ottawa, Canada.

Hesseltine, C. W., and Ellis, J. J. (1975) *Adv. Appl. Microbiol. 19*, 47–57.

Hughes, S. J. (1953) *Can. J. Bot. 31*, 560–659.

Ishitani, C., Ikeda, Y., and Sakaguchi, I. I. (1956) *J. Gen. Appl. Microbiol. 2*, 401–430.

Jeffrey, C. (1977) *Biological Nomenclature*, 2nd ed., Edward Arnold, London.

Johnson, L. A. S. (1970) *System. Zool. 19*, 203–239.

Joklik, W. K., Willett, H. P., and Amos, D. B. (1980) *Zinsser Microbiology*, 17th ed., Appleton-Century-Crofts, New York.

Kendrick, B. (ed.) (1971) *Taxonomy of Fungi Imperfecti*, University of Toronto Press, Toronto.

Kendrick, B. (ed.) (1979) *The Whole Fungus: The Sexual-Asexual Synthesis*, Vols. I and II, National Museum of Natural Sciences and Kananaskis Foundation, Ottawa, Canada.

Kevei, F., and Peberdy, J. F. (1979) *Mol. Gen. Genet. 170*, 213–218.

Kisser, M., Kubicek, C. P., and Röhr, M. (1980) *Arch. Microbiol. 128*, 26–33.

Kodama, K. (1970) in *The Yeasts* (Rose, A. H., and Harrison, J. S., eds.), pp. 225–282, Academic Press, New York.

Kodama, K., and Yoshizawa, K. (1977) in *Economic Microbiology* (Rose, A. H., ed.), pp. 423–475, Academic Press, London.

Korf, R. P. (1982a) *Mycotaxon 14*, 476–490.

Korf, R. P. (1982b) *Mycologia 74*, 250–255.

Kozakiewicz, Z. (1982) *Mycotaxon 15*, 293–305.

Kozlowski, M., and Stepien, P. P. (1982) *J. Gen. Microbiol. 128*, 471–476.

Kubicek, C. P., Hampel, W., and Röhr, M. (1979) *Arch. Microbiol. 123*, 73–79.

Kubicek, C. P., and Röhr, M. (1977) *Eur. J. Appl. Microbiol. 4*, 167–175.

Kubicek, C. P., and Röhr, M. (1978) *Eur. J. Appl. Microbiol. Biotech. 5*, 263–271.

Kubicek, C. P., and Röhr, M. (1982) in *Overproduction of Microbial Products* (Krumphanzl, V., Sikyta, B., and Vanék, Z., eds.), pp. 253–262, Academic Press, London.

Kwon, K. J., and Raper, K. B. (1967) *Amer. J. Bot. 54*, 36–48.

Kwon-Chung, K. J., and Kim, S. J. (1974) *Mycologia 66*, 628–638.

LaPage, S. P., Sneath, P. H., Lessel, E. A., Skerman, V. B., Seeliger, H. P., and Clark, W. A. (1976) *International Code of Nomenclature of Bacteria*, 1975 rev., American Society of Microbiology, Washington, D.C.

Lechevalier, H. A., and Solotorovsky, M. (1965) *Three Centuries of Microbiology*, McGraw-Hill, New York.

Leong, P. M., Bennett, J. W., and Ciegler, A. (1981) *Dev. Ind. Microbiol. 22*, 661–668.

Lewis, L. A., and Barron, G. L. (1964) *Genet. Res. Camb. 5*, 162–163.

Lhoas, P. (1967) *Genet. Res. Camb. 10*, 45–61.

Ma, H., Kubicek, C. P., and Röhr, M. (1981) *FEMS Microbiol. Lett. 12*, 147–151.

Mackintosh, M. E., and Pritchard, R. H. (1963) *Genet. Res., Cambridge 4*, 320–322.

Malloch, D., and Cain, R. F. (1972a) *Can. J. Bot. 50*, 2613–2618.

Malloch, D., and Cain, R. F. (1972b) *Can. J. Bot. 51*, 1647–1648.

Mandel, M. (1969) *Ann. Rev. Microbiol. 23*, 239–274.

Martin, G. W. (1946) *Science 103*, 116–117.

McKusick, V. A., and Ruddle, F. H. (1977) *Science 196*, 390–405.

Metzenberg, R. L. (1979) *Microbiol. Rev. 43*, 361–383.

Moore-Landecker, E. (1982) *Fundamentals of the Fungi*, 2nd ed., Prentice-Hall, Englewood Cliffs, N.J.

Murakami, H. (1971) *J. Gen. Appl. Microbiol. 17*, 281–309.

Murakami, H., Sagawa, H., and Takase, S. (1968b) *J. Gen. Appl. Microbiol. 14*, 251–262.

Murakami, H., Takase, S., Ishii, T. (1967) *J. Gen. Appl. Microbiol. 13*, 323–334.

Murakami, H., Takase, S., and Kuwabara, K. (1968a) *J. Gen. Appl. Microbiol. 14*, 97–110.

Nga, B. H., Teo, S. P., and Lim, G. (1975) *J. Gen. Microbiol. 88*, 364–366.

Onions, A. H. S., Allsopp, D., and Eggins, H. O. W. (1981) *Smith's Introduction to Industrial Mycology*, Edward Arnold, London.

Orthofer, R., Kubicek, C. P., and Röhr, M. (1979) *FEMS Microbiol. Lett. 5*, 403–406.

Oxford English Dictionary (1933) Oxford University Press, London.

Papa, K. E. (1973) *Mycologia 65*, 1201–1205.

Papa, K. E. (1978) *Mycologia 70*, 766–773.

Pateman, J. A., and Kinghorn, J. R. (1977) in *Genetics and Physiology of Aspergillus* (Smith, J. E., and Pateman, J. A., eds.), pp. 203–241, Academic Press, New York.

Pederson, C. S. (1979) *Microbiology of Food Fermentations*, 2nd ed., AVI, Westport, Conn.

Philippides, D., and Scazzocchio, C. (1981) *Mol. Gen. Genet. 181*, 107–115.

Piper, C. V., and Morse, W. J. (1923) *The Soybean*, McGraw-Hill, New York.

Pirozynski, K. A., and Weresub, L. K. (1979) in *The Whole Fungus: The Sexual-Asexual Synthesis*, Vol. II (Kendrick, B., ed.), pp. 653–688, National Museum of Natural Sciences and Kananaskis Foundation, Ottawa, Canada.

Pitt, J. I. (1979) *The Genus Penicillium and Its Teleomorphic States Eupenicillium and Talaromyces*, Academic Press, London.

Pontecorvo, G., Roper, J. A., and Forbes, E. (1953a) *J. Gen. Microbiol. 8*, 198–210.

Pontecorvo, G., Roper, J. A., Hemmons, L. M., MacDonald, K. D., and Bufton, A. W. J. (1953b) *Adv. Genet. 5*, 141–238.

Prescott, S. C., and Dunn, C. G. (1940) *Industrial Microbiology*, McGraw-Hill, New York.

Ramsbottom, J. (1941) *Proc. Linnean Soc. London 151*, 280–367.

Rand, K. N., and Arst, H. N., Jr. (1977) *Mol. Gen. Genet. 155*, 67–75.

Rand, K. N., and Arst, H. N., Jr. (1978) *Nature 272*, 732–734.

Raper, K. B. (1957) *Mycologia 49*, 644–662.

Raper, K. B., and Fennell, D. I. (1965) *The Genus Aspergillus*, Williams and Wilkins, Baltimore.

Raper, K. B., Fennell, D. I., and Tresner, H. D. (1953) *Mycologia 45*, 671–692.

Rij, K. V. (1973) in *The Fungi*, Vol. IVA (Ainsworth, G. D., Sparrow, F. K., and Sussman, A. S., eds.), pp. 11–32, Academic Press, New York.

Rippon, J. W. (1982) *Medical Mycology*, 2nd ed., W. B. Saunders, Philadelphia.

Rodricks, J. V., Hesseltine, C., and Mehlman, M. A. (eds.) (1977) *Mycotoxins in Human and Animal Health*, Pathotox, Park Forest, Ill.

Roehr, M., Zehentgruber, O., and Kubicek, C. P. (1981) *Biotech. Bioeng. 23*, 2433–2455.

Roper, J. A. (1952) *Experientia 8*, 14–15.

Roper, J. A. (1966) in *The Fungi*, Vol II, *The Fungal Organism* (Ainsworth, G. C., and Sussman, A. S., eds.), pp. 589–617, Academic Press, New York.

Rose, A. H. (ed.) (1979) *Secondary Products of Metabolism*, Academic Press, London.

Ross, I. K. (1979) *Biology of the Fungi*, McGraw-Hill, New York.

Rowlands, R. T. (1983) in *The Filamentous Fungi*, Vol. IV, *Fungal Technology* (Smith, J. E., Berry, D. R., and Kristiansen, B., eds.), pp. 346–379, Edward Arnold, London.

Ruddle, F. H. (1981) *Nature 294*, 115–120.

Rudolph, E. O. (1962) *Amer. J. Bot. 49*, 71–78.

Samson, R. A. (1979) *A Compilation of the Aspergilli Described since 1965*, Studies in Mycology No. 18, Centraalbureau voor Schimmelcultures, Baarn, Netherlands.

Sargeant, K., Sheridan, A., O'Kelly, J., and Carnaghan, R. B. A. (1961) *Nature 192*, 1096–1097.

Scazzocchio, C., and Gorton, D. (1977) in *Genetics and Physiology of Aspergillus* (Smith, J. E., and Pateman, J. A., eds.), pp. 255–279, Academic Press, New York.

Scazzocchio, C., Sdrin, N., and Ong, G. (1982) *Genetics 100*, 185–208.

Sermonti, G. (1959) *Ann. N.Y. Acad. Sci. 81*, 950–966.

Shaffer, P. M., Arst, H. N., Jr., Cairns, J. R., Dorman, D. C., and Lott, A. M. (1984) *Mol. Gen. Genet. 198*, 139–145.

Shurtleff, W., and Aoyagi, A. (1976) *The Book of Miso*, Ballantine Books, New York.

Sieburth, J. (1978) in *CRC Handbook of Microbiology*, Vol. II, *Fungi, Algae, Protozoa, and Viruses* (Laskin, A. I., and Lechevalier, H. A., eds.), pp. 1-7, CRC Press, West Palm Beach, Fl.

Simpson, G. G. (1961) *Principles of Animal Taxonomy*, Columbia University Press, New York.

Skerman, V. D. B., McGowan, V., and Sneath, P. H. A. (eds.) (1980) *Int. J. Syst. Bacteriol. 30*, 225-420.

Smith, G. M. (1938) *Cryptogamic Botany*, Vol. I, *Algae and Fungi*, McGraw-Hill, New York.

Smith, J. E., and Berry, D. R. (1974) *An Introduction to Biochemistry of Fungal Development*, Academic Press, London.

Smith, J. E., and Berry, D. R. (eds.) (1975) *The Filamentous Fungi*, Vol. I, *Industrial Mycology*, John Wiley and Sons, New York.

Smith, J. E., and Berry, D. R. (eds.) (1976) *The Filamentous Fungi*, Vol. II, *Biosynthesis and Metabolism*, John Wiley and Sons, New York.

Smith, J. E., and Berry, D. R. (eds.) (1978) *The Filamentous Fungi*, Vol. III, *Developmental Mycology*, John Wiley and Sons, New York.

Smith, J. E., Anderson, J. G., Deans, S. G., and Davis, B. (1977) in *Genetics and Physiology of Aspergillus* (Smith, J. E., and Pateman, J. A. eds.), pp. 23-58, Academic Press, New York.

Smith, J. E., Berry, D. R., and Kristiansen, B. (eds.) (1983) *The Filamentous Fungi*, Vol. IV, *Fungal Technology*, John Wiley and Sons, New York.

Smith, J. E., and Pateman, J. A. (eds.) (1977) *Genetics and Physiology of Aspergillus*, Academic Press, New York.

Sneath, P. H. A., and Sokal, R. R. (1973) *Numerical Taxonomy: The Principles and Practice of Numerical Classification*, W. H. Freeman, San Francisco.

Snell, W. H., and Dick, E. A. (1957) *A Glossary of Mycology*, Harvard University Press, Cambridge, Mass.

Steinkraus, K. H. (1983) in *The Filamentous Fungi*, Vol. IV, *Fungal Technology* (Smith, J. E., Berry, D. R., and Kristiansen, B., eds.), pp. 171-189, John Wiley and Sons, New York.

Stevens, R. B. (ed.) (1974) *Mycology Guidebook*, University of Washington Press, Seattle.

Stromnaes, O., and Garber, E. D. (1963) *Genetics 48*, 653-662.

Subramanian, C. V. (1961) *Bull. Bot. Surv. Ind. 4*, 249-259 (cited in Barron, 1968).

Subramanian, C. V. (1962) *Curr. Sci. 31*, 409-411.

Subramanian, C. V. (1972) *Curr. Sci. 41*, 755-761.

Subramanian, C. V. (1983) *Hyphomycetes: Taxonomy and Biology*, Academic Press, New York.

Subramanian, C. V., and Rajendran, C. (1975) *Kavaka 3*, 129-133.

Takamine, J. (1894) U.S. Pat. 525,820 and 525,823.

Talbot, P. H. B. (1971) *Principles of Fungal Taxonomy*, Macmillan, London.

Thom, C. (1957) *Annals N.Y. Acad. Sci. 60*, 24-34.

Thom, C., and Church, M. B. (1921) *Amer. J. Bot. 8*, 103-125.

Thom, C., and Church, M. B. (1926) *The Aspergilli*, Williams and Wilkins, Baltimore.

Thom, C., and Raper, K. B. (1945) *A Manual of the Aspergilli*, Williams and Wilkins, Baltimore.

Thom, C., and Raper, K. B. (1946) *Science 103*, 735.

Timberlake, W. E. (1980) *Dev. Biol. 78*, 497–510.

Timberlake, W. E., and Barnard, E. C. (1981) *Cell 26*, 29–37.

Tinline, R. D., and MacNeil, B. H. (1969) *Ann. Rev. Phytopathol. 7*, 147–170.

Tubaki, K. (1963) *Ann. Rep. Inst. Fermentation* (Osaka) *1*, 25–54 (cited in Barron, 1968).

Turian, G. (1974) *Ann. Rev. Phytopathol. 12*, 129–137.

Turian, G., and Hohl, H. R. (1981) *The Fungal Spore: Morphogenetic Controls*, Academic Press, London.

Turner, W. B. (1971) *Fungal Metabolites*, Academic Press, London.

Upshall, A. (1981) *J. Gen. Microbiol. 122*, 7–10.

Vanbreuseghem, R., and Wilkinson, J. (1958) *Mycoses of Man and Animals*, Pitman and Sons, London.

Von Arx, J. A. (1979) in *The Whole Fungus: The Sexual-Asexual Synthesis,* Vol. II (Kendrick, B., ed.), pp. 555–571, National Museum of Sciences and Kananaskis Foundation, Ottawa, Canada.

Waring, R. B., Davies, R. W., Lee, S., Grisi, E., Berks, M. M., and Scazzocchio, C. (1981) *Cell 27*, 4–11.

Webster, J. (1970) *Introduction to Fungi*, Cambridge University Press, Cambridge, England.

Webster, R. K. (1974) *Ann. Rev. Phytopathol. 12*, 331–353.

Webster's Third New International Dictionary of the English Language, Unabridged (1964) G. & C. Merrian Co., Springfield, Mass.

Weinberg, E. D. (1971) *Perspect. Biol. Med. 14*, 565–577.

Weresub, L. K. (1979a) in *The Whole Fungus: The Sexual-Asexual Synthesis*, Vol. II, (Kendrick, B., ed.), pp. 689–709, National Museum of Natural Sciences and Kananaskis Foundation, Ottawa, Canada.

Weresub, L. K. (1979b) *Sydowia 8*, 416–431.

Weresub, L. K., and Hennebert, G. L. (1979) *Mycotaxon 8*, 181–186.

Weresub, L. K., and Pirozynski, K. A. (1979) in *The Whole Fungus: The Sexual-Asexual Synthesis*, Vol. II (Kendrick, B., ed.), pp. 17–25, National Museum of Natural Sciences and Kananaskis Foundation, Ottawa, Canada.

Whittaker, R. H. (1969) *Science 163*, 150–160.

Wicklow, D. T. (1983) in *Aflatoxin and Aspergillus flavus in Corn* (Davis, N. D., and Diener, U. L., eds.), pp. 6–12, Alabama Experimental Station, Opelika, Alabama.

Wicklow, D. T., and Carroll, G. C. (eds.) (1981) *The Fungal Community, Its Organization and Role in the Ecosystem*, Marcel Dekker, New York.

Wiley, B., and Simmons, E. G. (1973) *Mycologia 65*, 569–578.

Wilson, V. W., and Plunkett, O. A. (1965) *The Fungous Diseases of Man*, University of California Press, Berkeley and Los Angeles.

Wolf, F. A., and Wolf, F. T. (1947) *The Fungi*, Vols. I and II, John Wiley and Sons, New York.

Wyllie, T. D., and Morehouse, L. G. (eds.) (1977) *Mycotoxic Fungi, Mycotoxins, Mycotoxicoses*, Vol. I, Marcel Dekker, New York.

Wyllie, T. D., and Morehouse, L. G. (eds.) (1978) *Mycotoxic Fungi, Mycotoxins, Mycotoxicoses*, Vol. II, Marcel Dekker, New York.

Yong, F. M., and Wood, B. J. B. (1974) *Adv. Appl. Microbiol. 17*, 157–194.

Zimmerman, C. R., Orr, W. C., Leclerc, R. F., Bernard, E. C., and Timberlake, W. E. (1980) *Cell 21*, 709–715.

Zonneveld, B. J. M. (1977) in *Genetics and Physiology of Aspergillus* (Smith, J. E., and Pateman, J. A., eds.), pp. 59–80, Academic Press, New York.

The Biology of *Penicillium*

John F. Peberdy

I. INTRODUCTION

Species of *Penicillium* are employed in several industrial processes. The best known is the manufacture of the β-lactam antibiotic, penicillin, by *Penicillium chrysogenum*. Other important uses of these fungi are in enzyme production, for example, β-1,3-glucanase by *Penicillium emersonii* and in the production of cheeses, for example, *Penicillium roqueforti* and *Penicillium camembertii*.

Most species of this genus are saprophytic, occurring in soil and on decaying vegetation. They play an important role in the turnover of organic substrates. Some of them are also important as causative agents of biodeterioration, especially of cereal grains; and in these situations, mycotoxins may be produced. A few species are parasitic or commensal in their life style. The best known are those associated with plants and especially citrus and pome fruits, for example, *Penicillium digitatum*, *Penicillium expansum*, and *Penicillium italicum*. Economic losses arising from the activities of these fruit-infecting species can be very high. There is only one known report of a *Penicillium* species being an animal pathogen, namely, *Penicillium marneffei*, which caused the death of a Vietnamese bamboo rat (Segretain, 1959). Pathogenicity of this species to other animals was also described. It may not be surprising in future years to see penicillia being recorded as opportunistic pathogens in man, particularly as the use of immunosuppressant drugs increases. An indication of these fungi adopting this role is seen in the report of *P. chrysogenum* as the cause of endopthalmitis in man (Eschete et al., 1981).

The taxonomy of the genus has been the subject of debate and revision over the last few decades. The first major monograph (Raper and Thom, 1949) described 137 species. This treatment has been used almost exclusively by taxonomists over the past 30 years, although during this period, many "new" species were described (Kulik, 1968). Five years ago a totally revised classification of the genus was produced (J. I. Pitt, 1979). This recognized the new taxonomic conventions that had been adopted during the intervening period. More than 200 species were recognized, and physiological characteristics were used more extensively. Another feature is the separation of those species with both sexual (teleomorphic) and asexual (anamorphic) states (Hennebert and Weresub, 1977) into two genera: *Eupenicillium* and *Talaromyces*.

II. GENERAL CHARACTERISTICS

A. Vegetative Morphology

The penicillia are mycelial fungi comprised of septate hyphae, the septa having simple pores and the hyphal segments being multinucleate. In the apical compartment of *Penicillium cyclopium* these nuclei exhibit synchrony or near synchrony at mitosis (Rees and Jinks, 1952). As in all filamentous fungi, hyphal growth in *Penicillium* is a consequence of apical hyphal extension. A Spitzenkorper, described originally by Brunswick (1924) as an iron-haemotoxylin positive area in the apical tips of some fungal hyphae, has also been observed by phase contrast microscopy in *Penicillium* species (Girbardt, 1955, 1957).

The hyphal wall composition of several species has been analyzed, and these data are summarized in Table 13.1. Clear differences are seen in the galactose, glucosamine, protein, lipid, and ash contents of walls of the different species. The consequences of this are not yet clear. It can be concluded, however, that the hyphal walls have a high glucan content, most probably a β-1,3-glucan, with chitin as the skeletal component. More recently, Bobbit and Nordin (1978) have demonstrated the presence of nigeran, and α-D-glucan with alternating 1,3 and 1,4 linkages, in wall preparations from *P. chrysogenum*, *P. expansum* and *P. roqueforti*. However, the polymer is not universally present; several species belonging to the *Penicillium javanicum* series have been found to be lacking in this polymer.

B. Reproductive Morphology

The majority of the penicillia are anamorphic, that is, asexual, mitotic, imperfect species. Asexual reproduction involves the production of conidia (phialoco-

TABLE 13.1 Chemical Composition of Hyphal Walls of *Penicillium* Species

Component	P. chrysogenum[1]	P. notatum[2]	P. roqueforti[3] (component mg 100 mg^{-1} wall)	P. digitatum[4]	P. italicum[4]	P. rubrum[5]
Neutral sugars	55.2	51.0	56.0	49.2	55.4	52.0
Glucose	36.0	43.0	41.0	45.4	51.6	44.6
Galactose	11.0	7.0	14.0	3.8	3.8	5.4
Mannose	4.0	1.0	1.0	ND	ND	0.6
Hexosamine	20.1	18.5	13.3	5.7	9.0	15.4
Lipid	ND	1–2	1–2	ND	ND	17.0
Protein	ND	8.9	9.7	5.1	1.3	7.2
Ash	ND	4.0	5.0	29.5	0.5	4.9

[1] Hamilton and Knight (1962).
[2] Applegarth (1967).
[3] Applegarth and Bozoian (1968).
[4] Grisaro et al. (1968).
[5] Unger and Hayes (1975).

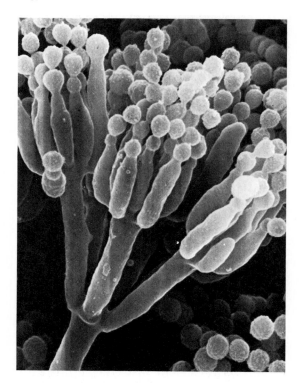

FIGURE 13.1   Conidiophore of *P. citrinum* showing the symmetrically biverticillate penicillus: SEM ×2300. (Reproduced by kind permission of Dr. J. I. Pitt.)

nidia) in a basipetal sequence from a phialide (Fig. 13.1) (Cole and Kendrick, 1969). The morphology of the conidiophore, or penicillus, is a major taxonomic character in these anamorphic forms. Several species produce highly modified macroscopic asexual structures called coremia, for example, *Penicillium claviforme*, and synnemata, for example, *Penicillium duclauxii*. These structures are made up of aggregates of aerial hyphae that grow up together and differentiate into complex spore heads.

A number of species also exhibit sexual development, the teleomorph or perfect state, which involves the formation of cleistothecia (Fig. 13.2) containing spherical asci. In accordance with current nomenclature codes the teleomorphic forms are named as species of the genera *Eupenicillium* and *Talaromyces*. Frequently, the teleomorph has been described later in time than the anamorphic stage, and the convention is to use the same species names for both forms (J. I. Pitt, 1979).

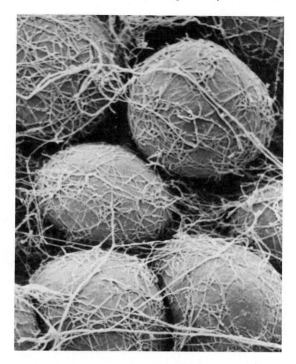

FIGURE 13.2   Cleistothecia of *Eupenicillium shearii*: SEM × 50. (Reproduced by kind permission of Dr. J. I. Pitt.)

III. PHYSIOLOGICAL ASPECTS OF DEVELOPMENT

A. Spore Germination

Studies on spore germination in the penicillia have been done using conidia. These studies have demonstrated that, as in many other fungi, the process involves two phases, the swelling phase and that of germ tube development. Both events are dependent on intense metabolic activity, but certain physiological and biochemical differences exist between them. Studies on conidia of *Penicillium atrovenetum* (Gottlieb and Tripathi, 1968) and *Penicillium notatum* (Martín and Nicolás, 1970) have indicated that the swelling process has a requirement for several exogenous nutrients including glucose, phosphate, nitrogen, and oxygen. Both respiratory and synthetic functions were also very active during this first phase of germination. In *P. atrovenetum* (Gottlieb and Tripathi, 1968), synthesis of DNA, RNA, protein, lipid, and cell wall components was detected.

Changes in ultrastructure and chemical composition of the cell walls of germinating conidia were investigated by Martín et al. (1973a, b). In mature conidia the spore wall is comprised of four layers, and during the swelling phase the outermost layer is released. As the germ tube is formed, its wall arises from the inner two spore wall layers (Martín et al., 1973a). A similar situation was observed during the germination of conidia of *Penicillium griseofulvum* (Fletcher, 1971). Organellar changes were also observed in *P. notatum* (Martín et al., 1973a); large mitochondria are present in the mature spores, and these divide during germ tube development. The conidia lack an endoplasmic reticulum, but this develops and increases substantially during the swelling phase.

The chemical changes observed (Martín et al., 1973b) relate mainly to the amino acid and sugar components of the spore walls. Conidial walls have a higher amino acid content than hyphal walls, although the profiles of the acids are very similar. Significant differences were observed in both the neutral sugar (Table 13.2) and amino sugar (Table 13.3) contents of spore and germ tube walls.

Gross changes in other cell constituents were also found in germinating conidia of *P. notatum* (Martín et al., 1974). During the swelling phase the total N content of conidia increases and attains a maximum before germ tubes develop. As germination proceeds, total N declines to a lower, constant value in the developing hyphae. A similar pattern was observed for protein content, although an increase occurs during the later stage of germ tube formation. Total lipid and carbohydrate contents decrease during the swelling phase, and further accumulation was not found until the later period of hyphal growth. In contrast to these general decreases during swelling and germ tube formation the nucleic acid content of the spores rises sharply, attaining a maximal level just prior to germ tube emergence. A steady decrease then follows during germ tube development and subsequent hyphal elongation.

B. Vegetative Growth

Studies on growth kinetics have focused on *P. chrysogenum* and primarily on the performance of the fungus at low growth rates in a chemostat (Righelato et al., 1968). At specific growth rates of 0–0.11 h^{-1} the hyphae show several signs of differentiation, including increased vacuolation and sporulation, in compar-

TABLE 13.2   Changes in the Neutral Sugar Composition of Conidial Walls of *P. Notatum* during Germination

	Percent Cell Wall Dry Weight			
	Glucose	*Galactose*	*Mannose*	*Rhamnose*
Mature spores	13.3	22.2	8.8	trace
Swollen spores	17.9	7.3	6.1	—
Germinating spores	24.5	8.2	5.7	—
Hyphal walls	22.1	8.6	6.2	—

After Martín et al. (1973b).

TABLE 13.3   Changes in the Amino Sugar Composition of Conidial Walls of P. notatum during Germination

	Percent Cell Wall Dry Weight	
	Glucosamine	Galactosamine
Mature spores	0.9	—
Swollen spores	8.7	—
Germinating spores	28.2	2.9
Hyphal wall	35.1	7.2

After Martín et al. (1973b).

ison with hyphae from cultures grown at higher rates. The balance of macromolecular contents is also affected, the mycelium from slow growing cultures having a reduced protein content and a higher carbohydrate content than that from high growth rate cultures.

The penicillia are not particularly fastidious with respect to nutrient utilization. Most of the common carbohydrates are assimilated, and only certain polymers, for example, cellulose, chitin, and lignin, are generally unavailable. Some species grow on C_5 sugars, a feature that is not common in fungi. Similarly, a broad range of nitrogen sources are readily utilized, including ammonium ions, nitrite, nitrate, urea, and amino acids. A few species do not have these catholic nutritional profiles and are unable to assimilate nitrate or have specific growth factor requirements.

The genus is a group of aerobic species, but no detailed investigation of the oxygen requirement for growth has been made. Growth of these fungi occurs over a fairly broad pH range (pH 3–8), but again there are no detailed data.

Temperature relationships have been collated, to some extent, by Mislivec and Tuite (1970). Species that grow at 5°C and at 37°C were listed. Using as a criterion for growth at 5°C the attainment of a colony 3 mm diameter in 7 days on Czapek yeast autolysate agar (CYA), some 29 species fall into this low temperature category (J. I. Pitt, 1979). At 37°C the criterion was a colony size of 30 mm on CYA in 7 days, and 34 species were identified in this group (J. I. Pitt, 1979). No species was found in both groups. The majority of species grow over a temperature range intermediate between these extremes with optima of 20–25°C.

Some species are xerophilic, growing satisfactorily at minimum water activities of less than 0.85 (Hocking and Pitt, 1979). This property is of obvious ecological significance, since these fungi are able to grow on dry substrates and, in so doing, are frequently the cause of biodeterioration problems.

C. Conidiogenesis

Conidiation in *P. chrysogenum*, as described above, is a function of limited nutrient supply and low growth rate (Righelato et al., 1968). In contrast, experiments with *P. notatum* (D. Pitt and Poole, 1981) indicated that under certain conditions in submerged culture, for example, enhanced levels of Ca^{2+} ions in the medium, extensive conidiation could occur prior to nutrient limitation and during the phase of vegetative growth. The timing of the addition of Ca^{2+} ions appears to be critical; when it was done later in the growth cycle, sporulation was reduced. Conidiation in these cultures is also sensitive to oxygen tension, reflecting the high oxygen consumption by mycelia involved in conidium development. In later studies on *P. cyclopium* (Ugalde and Pitt, 1983) a similar effect of Ca^{2+} ions was observed. A similar influence of Ca^{2+} ions on sporulation in *P. notatum* was also indicated in studies carried out much earlier by Hadley and Harrold (1958a, b).

Biochemical changes during conidiogenesis in *P. notatum* were investigated by D. Pitt (1969a, b, 1970). Using cytochemical techniques, SH groups were detected in the walls of phialides but not at the apices of vegetative hyphae (D. Pitt, 1969a). It was suggested that the presence of the SH groups is significant with respect to the mechanism of conidium formation, which shows similarities to the budding of yeast cells. The differential effects of selenite and telurite on vegetative growth and conidiogenesis was related to the presence of SH groups in the phialide walls. At concentrations of these inhibitors that caused a retardation of colony radial growth, sporulation was profuse (D. Pitt, 1970). In further studies a major compositional change was found in the cytoplasmic ribonucleoprotein content of the phialides during the period of rapid mitosis that is associated with conidium formation (D. Pitt, 1969b). Using fast green stain, only nuclear-DNA-bound histone was detected when conidiogenesis was complete. A possible use for the degraded nucleoproteins is for synthesis of basic proteins for incorporation into the spores.

The initiation and development of coremia are undoubtedly more complex. External nutrients are required for initiation of coremial primordia, but further differentiation into the spore heads is independent of nutrient supply (Watkinson, 1975). Initiation is influenced by nutrient concentration, the number of primordia formed being proportional to the logarithm of nutrient concentration. As the primordia develop, they become photosensitive, and reserves are accumulated to support later conidial development. In a subsequent study (Watkinson, 1977) it was indicated that primordia development, both number per unit area of mycelium and rate of development, is stimulated by casein hydrolysate or several amino acids. This suggested that sites for primordium initiation might arise as a consequence of local variations in nitrogen metabolism. Coremium development in *P. claviforme* shows an endogenous periodicity being formed in concentric rings (Faraj-Salman, 1970). This pattern of development could be a consequence of the changes in nitrogen metabolism previously referred to and brought about by the advance of vegetative hyphae from behind the growing zone, through the agar, to form a new actively growing zone leaving the original growth zone as the site of the next region of coremia formation.

Other environmental factors affect the formation of coremia. Light has a role in induction; for example, in *Penicillium isariiforme*, blue light stimulates formation (Piskorz, 1967), but the mechanism involved and the role of potential photoreceptors have not been explained. A requirement for CO_2 in coremium development in *P. isariiforme* has been suggested by Graffmans (1973).

D. Microcyclic Conidiation

Microcyclic conidiation has been described in several *Penicillium* species; it arises when spores are germinated under specific environmental conditions that prevent normal germination. Zeidler and Margalith (1973) devised a medium in which conidia of *P. digitatum* germinate synchronously but, after 28 h, develop conidiophores from the germ tubes. The process could be accelerated by the addition of glutamic or aspartic acid as nitrogen source. In *Penicillium urticae*, microcyclic conidiation follows a period of spherical growth when the spores are incubated at 37°C (Sekiguchi et al., 1975a). When the conidia are transferred to a medium with reduced nitrogen, synchronous germination is induced, leading to the formation of conidiophores. In a later study it was shown that a temperature shock could also induce the cycle (Sekiguchi et al., 1975b).

E. Sexual Development

Observations on *Penicillium baarnense* indicated that the pathways of asexual and sexual development are alternatives (Bouvier, 1967, 1968). Wild-type isolates produce abundant cleistothecia and few conidiophores; however, both spontaneous and induced mutants have been isolated in which this pattern is modified. These fall into three classes: (1) mutants that are totally sterile, (2) mutants that produce abundant conidia, and (3) defective sexual mutants. Strains carrying two of these mutations display the existence of a hierarchy in reproductive development; sterile mycelium preempts conidiation, which in turn preempts sexual development. The genetic basis of these mutations is unresolved; the induced mutations behave as nuclear genes, but those of spontaneous origin have characteristics of cytoplasmically inherited genes.

IV. METABOLISM

A. Primary Metabolism

The penicillia have not been used extensively for studies on primary metabolism in fungi; consequently, our knowledge of this aspect of their biology is fragmentary. In the metabolism of carbohydrates these organisms utilize both the glycolytic and hexose monophosphate pathways. In some species, for example, *P. notatum*, oxidation of carbohydrates may be incomplete, and glucose is converted to gluconic acid (Bodmann and Walter, 1965). The glucose oxidase is also known as notatin.

As referred to earlier, some species of *Penicillium* are examples of the few fungi that can metabolize xylose and other C_5 sugars. The reaction involves three steps catalyzed by polyol dehydrogenases and a kinase (Chiang and Knight, 1959, 1960a, b).

$$\text{D-xylose} + \text{NADPH} + \text{H}^+ \rightarrow \text{xylitol} + \text{NADP}^+$$

$$\text{xylitol} + \text{NAD}^+ \rightarrow \text{D-xylulose} + \text{NADH} + \text{H}^+$$

$$\text{D-xylulose} + \text{ATP} \rightarrow \text{D-xylulose-5-phosphate} + \text{ADP}$$

In *Penicillium corylophilum*, xylose is converted to xylonic acid (Ikeda and Yamada, 1963). L-Arabinose is also converted to xylulose-5-phosphate by *P. chrysogenum* (Chiang and Knight, 1961).

Some species of *Penicillium* produce polysaccharases. The most detailed study has been made on the β-1,3-glucanases of *P. italicum* by Nombela and co-workers. When this fungus is grown on a defined medium, a cell-bound β-1,3-glucanase and an extracellular β-1,6-glucanase are produced (Santos et al., 1977). When the glucose concentration is limiting or less readily metabolized sugars are supplied, growth is reduced, and enhanced production of both enzymes is found. In the presence of excess readily metabolized sugars, active growth occurs, and production of the enzymes is reduced. Substrates for the β-1,3-glucanase cause no inductive effect, a result suggesting that the enzyme is constitutive.

In a subsequent study (Santos et al., 1978), evidence of the mechanism of the glucose effect on synthesis of the β-1,3-glucanase was presented. Mycelia from a glucose-limited medium that were actively producing the enzyme continued to do so when transferred to a medium containing excess glucose. However, these derepressed mycelia cease to produce the enzyme in the presence of cycloheximide and trichodermin. When 8-hydroxyquinoline is added to the mycelia, synthesis continues for some time and then stops. These results led to the conclusion that the glucose effect is exerted at the pretranslational level. More recently (Santos et al., 1979), a morphogenetic role for the β-1,3-glucanase in mobilizing wall glucans was elucidated. Walls from derepressed mycelia had a higher autolytic activity than walls from actively growing mycelia. Incubation of wall material from the derepressed mycelia caused a reduction of the wall β-1,3-glucan content of more than 30%. Purification of the enzyme by gel filtration revealed three fractions, and these were subsequently shown to have different distribution patterns in the mycelia. Almost half the total enzyme was present in the periplasmic space. This component was comprised only of fractions II and III, which are both capable of releasing glucose from isolated wall material in vitro. Fraction I was retained in the cytoplasm and possibly requires posttranslational modification before it is secreted.

Two other polysaccharases described in *Penicillium* species are a chitosanase in *Penicillium islandicum* (Fenton and Eveleigh, 1981) and an endo-β-1,4-

glucanase in *Penicillium janthinellum* (Rapp et al., 1982). The chitosanase has been purified and shown to cleave chitosan but not chitin. Maximum activity is found against polymers with 30–60% acetylation. Synthesis of the β-1,4-glucanase by *P. janthinellum* is repressed by glucose, sophorose, and glycerol. Purification on DEAE Sephadex A-50 resulted in two components, both of which are competitively inhibited by glucose and cellobiose.

Lipolytic enzymes of *P. cyclopium* were investigated by Okumura et al. (1980). The fungus produces two enzymes that could be fractionated on DEAE Sephadex A-50. The two components differ with respect to specificity toward triglycerides and partial glycerides. The partial glyceride hydrolase shows greatest activity toward glycerides of oleic acid. Triolein is only 20% hydrolyzed to give oleic acid and the diglycerides 1,2 and 2,3 diolein. Dioleins and monoolein are rapidly hydrolyzed.

Various aspects of nitrogen metabolism in the penicillia have been investigated. Ammonia transport was studied in *P. chrysogenum* by using ^{14}C-labeled methylammonium as an analog to characterize the system (Hackette et al., 1970). In nitrogen-sufficient cells, transport of methylammonium is low but increases 800-fold in nitrogen-starved mycelium as a result of derepression and/or reversed inhibition.

Amino acid transport systems in *P. chrysogenum* were examined by Segel and co-workers. Kinetic evidence suggested that the fungus has several highly specific transport systems for particular amino acids, including L-methionine (Benko et al., 1967), L-cysteine and L-cystine (Skye and Segel, 1970), L-lysine, L-arginine, and L-proline (Hunter and Segel, 1971). Group specific systems for acidic amino acids and basic amino acids were also found (Hunter and Segel, 1971). Nonspecific general amino acid transport systems were observed in older carbon- and/or nitrogen-starved cultures of *P. chrysogenum* (Hunter and Segel, 1971) and in *P. griseofulvum* (Whitaker and Morton, 1971).

In the few species investigated, it is apparent that the pathways of amino acid biosynthesis are identical to those in other molds, which with a few exceptions—for example, lysine—are similar to bacteria. A pool of free amino acids that responds rapidly to changes in environmental conditions was found in *P. chrysogenum*. The pool is dominated by L-glutamate, and the levels of L-alanine, L-aspartate, and L-glutamine are also high (Hunter and Segel, 1971). Growth on a medium containing a specific amino acid enriches the pool for that particular compound.

Information on amino acid catabolism is equally limited. *P. chrysogenum* and *P. janthinellum* were shown to assimilate L-threonine as both carbon and nitrogen source (Willetts, 1972).

$$\text{L-threonine} + \text{NAD} \xrightarrow{\substack{\text{NAD threonine} \\ \text{dehydrogenase}}} \text{2-amino-3-oxobutyrate} + \text{NADH}^+$$

$$\text{2-amino-3-oxobutyrate} \xrightarrow{\substack{\text{2-amino-3-} \\ \text{oxobutyrate ligase}}} \text{L-glycine} + \text{acetyl-CoA}$$

L-Glycine is initially metabolized via deamination by glycine-pyruvate amino transferase to yield glyoxylate, which is further metabolized by the glycerate pathway.

In view of the importance of certain species in cheese ripening, studies on proteolytic activities are of interest. *P. roqueforti* was shown to produce a complex of extracellular proteases that includes two endopeptidases (Zevaco et al., 1973; Modler et al., 1974; Gripon and Hermier, 1974) and three exopeptidases (Gripon and Debest, 1976; Gripon, 1977a, b). One of the endopeptidases is an acid protease (aspartyl protease) and is typical of mold enzymes of this type. The second is a metalloprotease. Production of the two enzymes is affected by the pH of the medium; in acid media the acid protease is produced in greatest quantity at pH 4.0, and the metalloprotease at pH 6.0.

B. Secondary Metabolism

Penicillia, in common with many fungi, are known to produce a diverse range of secondary metabolites. These products are generally accumulated when active growth and proliferation cease. They are anabolic products, and their synthesis taps the energy and reducing power resources generated during primary metabolism and active growth. For a detailed review on the general aspects of secondary metabolism the reader should consult a recent treatise on the subject (e.g., Rose, 1979). The majority of the penicillia probably produce secondary metabolites under the relevant environmental conditions. Turner (1971) listed 72 species that produce secondary metabolites, and many more have been described in the intervening period. The metabolites identified to date are representative of all the major chemical groups of secondary products and include polyketides, alkaloids, terpenes, sterols, β-lactams, derivatives of shikimic acid, and derivatives of fatty acids.

The importance of many *Penicillium* species is a consequence of the secondary metabolites they produce. Only two, penicillin and griseofulvin, are of current industrial interest as antibiotics, but others could have potential as new applications are developed. In contrast, many other metabolites have achieved importance or potential importance because they are mycotoxins or have been shown to have toxic properties.

Antibiotics. *Penicillins*: The consequences of the discovery of the penicillins, both in their value to humanity and in the development of the fermentation industry, are well known (Florey et al., 1949; Demain, 1966). Despite the discovery of many other antibiotics, some with more advantageous properties, the penicillins, together with other β-lactams are still the major products. More information on the penicillins may be obtained from recent reviews (Queener and Swartz, 1979; Demain, 1981; O'Sullivan and Ball, 1983).

Griseofulvin: This metabolite, a heptaketide, is important as an antifungal antibiotic. Griseofulvin, in common with many other *Penicillium* metabolites, was first isolated in Raistrick's laboratory (Oxford et al., 1939). The product was developed for use by oral therapy to control dermal fungal infections. Details of the biosynthesis and industrial production of griseofulvin may be found in the review by Rhodes (1963).

Several other metabolites with antimicrobial properties have been described, for example, radicicol from *Penicillium luteoaurantium*, meleagrin from *Penicillium meleagrinum*, orsenellic acid from *Penicillium madriti*, and cyanein from *Penicillium cyaneum*, but these have not been developed commercially for various reasons. Two other metabolites, citrinin and patulin, also have antimicrobial activities; however, they are of greater significance as mycotoxins.

Mycotoxins.   More than 60 species of *Penicillium* have been described as mycotoxin producers (P. M. Scott, 1977). Many of these reports arise from laboratory toxicity trials; only a few mycotoxins have been actually isolated from field samples. However, the potential hazards and dangers of these products in a wide range of food materials cannot be minimized in view of the many reports of the incidence of these fungi in food grains and similar substrates. Six species are commonly encountered in food materials such as cereal grains, beans, soybeans, and animal feeds; these and the toxins they produce will be described below. Detailed reviews on the whole field of *Penicillium* mycotoxicoses may be found in several review articles (Ciegler et al., 1971; P. M. Scott, 1977).

Citrinin: This pentaketide is one of the few mycotoxins that has been isolated from food materials. The compound was first described by Hetherington and Raistrick (1931) in a culture of *Penicillium citrinum*. It was subsequently described as a metabolite of other penicillia as well as of *Aspergillus* species. Its significance as a mycotoxin was realized in 1951, when it was detected in rice exported from Thailand to Japan (Saito et al., 1971a, b). *P. citrinum* was later found in rice samples from many other countries. Citrinin is a nephrotoxin affecting renal functions and causing swelling and dilatation of the kidney tubules (Sakai, 1955).

Patulin: This mycotoxin has also been isolated from food products (P. M. Scott and Kennedy, 1973; Ware et al., 1974). It was first discovered as a metabolite of *P. claviforme* and was later shown to be produced by several other species including *P. expansum* and *Penicillium patulum* (Ciegler, 1977). After its discovery it was used as an antibiotic; however, it was quickly dropped because of toxic side effects.

Patulin is a lactone and has several effects when administered to laboratory animals (P. M. Scott, 1977). At the cellular level, patulin affects nuclear activities, for example, mitostatic effects on chromosomes (Keilova-Rodova, 1949; Steinegger and Leupi, 1956) and effects on spindle structure (Rondanelli et al., 1957; Sentein, 1955); however, controversy exists over some of these observations.

Cyclopiazonic acid: *P. cyclopium*, the producer of this metabolite, is a common contaminant of stored grain and cereal products (D. B. Scott, 1965). It was first implicated by Albright et al. (1964) as a mycotoxin causing problems in cattle. The symptoms induced following oral dosage in chickens and ducks and intraperitoneal dosage in rats indicate that cyclopiazonic acid is a nephrotoxin (Purchase, 1969), but oral administration to rats induced several other effects that may have resulted from the slow absorption following this method of application.

Cyclochlorotine and islanditoxin: These toxins, isolated from culture filtrates from *P. islandicum* (Marumo, et al., 1955; Tatsuno et al., 1955), are chlorine-containing peptides. They are both hepatotoxins and thought to affect glycogen catabolism in liver. Mice given subcutaneous injections develop cirrhosis and hepatic lesions (P. M. Scott, 1977). Cyclochlorotine is the more toxic, causing the same lesions when administered orally.

Luteoskyrin: This mycotoxin was obtained from strains of *P. islandicum* isolated from rice (Tatsuno et al., 1955). An anthraquinone, it is also an hepatotoxin, but in vitro studies suggested a wider effect as an inhibitor of oxygen uptake in a range of rat tissues including heart muscle, kidney and liver. Enzymes associated with oxidative metabolism, for example, cytochrome C oxidase and succinate dehydrogenase, are also inhibited by luteoskyrin. Rat liver mitochondria show reduced oxidative phosphorylation activity when treated with the metabolite (Ueno et al., 1964), and mitochondria from livers of rats fed with the toxin are swollen, and their cristae are severely damaged (Saito, 1959).

Penicillic acid: This was one of the first *Penicillium* metabolites to be described (Alsberg and Black, 1913). It occurs in two tautomeric forms as a γ-keto acid or as a γ-hydroxylactone. It is produced by *P. cyclopium* and several other species as well as by some aspergilli. Subcutaneous injection of the metabolite into rats induces local malignant tumors (Dickens and Jones, 1965), and oral administration to mice causes lesion formation in liver, kidney, and thyroid (Kobayashi et al., 1971). Moldy corn, supporting growth of a known producer strain of *P. cyclopium*, caused death when fed to rats (Kurtzman and Ciegler, 1970).

Ochratoxin: This dihydroisocoumarin metabolite, produced by *Penicillium viridicatum*, is more commonly associated with *Aspergillus ochraceous* and other aspergilli. However, animal feeds contaminated with the *Penicillium* have been implicated in mycotoxicoses of swine in Denmark (Krogh et al., 1973). In these animals the toxin causes damage to the kidneys, but a variety of other effects have also been described.

Rubratoxin: Circumstantial evidence exists for this metabolite from *Penicillium rubrum* having toxic activity (Burnside et al., 1957). Rubratoxin exists in two forms, A and B, and is an anhydride. Histopathological studies on the organs of several animals have demonstrated a number of effects, although generally the liver and kidneys are the most seriously affected. Hemorrhaging is also quite common.

Biological Significance of Secondary Metabolites.   Further comments on secondary metabolites in *Penicillium* species will be confined to their biological role. For discussions on the biosynthesis of these metabolites and the regulatory aspects of synthesis the reader is directed to the several recent reviews on these topics (McCorkindale, 1976; Towers, 1976; Turner, 1976; Wright and Vining, 1976; Demain, 1981; Calam, 1982; Demain et al., 1983; O'Sullivan and Ball, 1983).

Several functions have been proposed for secondary metabolism in microorganisms (see Campbell et al., 1982, for a recent summary). In the penicillia an involvement of metabolites in the asexual sporulation process has been proposed and investigated in several laboratories. The concept was first proposed for bacteria (Bernlohr and Novetelli, 1963) and subsequently considered for the penicillia by Nover and Luckner (1974).

During alkaloid production by *P. cyclopium* a phase-dependent synthesis of the metabolites was observed (Fig. 13.3) (Luckner, 1980) and was concurrent with conidiation. Correlation between sporulation and the production of verrulculogen by *Penicillium simplicissimum* grown in fermenter culture was described by Mantle and Wertheim (1982). As described previously for other species, Ca^{2+} ions are also required for conidiation in this species and in this case for metabolite production also.

Campbell and co-workers have adopted various surface culture techniques in an attempt to correlate secondary metabolism with morphogenetic development in several *Penicillium* species. Thus 6-methyl salicylic acid synthesis by *P. patulum* and mycophenolic acid, brevianimide A and B, and ergosterol production by *Penicillium brevicompactum* begins only after the development of aerial mycelia (Bartman et al., 1981; Bird et al., 1981; Peace et al., 1981). When the formation of aerial mycelia is inhibited by growing the mycelia between two layers of cellophane, metabolite production is blocked. In *P. brevicompactum* a chronology in production was detected with the synthesis of all the metabolites (except the brevianamides) prior to conidia development, and mutants blocked in conidiogenesis were found to be defective in the synthesis of the brevianamides (Campbell et al., 1982).

Other workers have presented contrary observations. In one case the production of a metabolite already mentioned, verruculogen, by a different species (*Penicillium raistrickii*) occurs actively in the absence of sporulation (Mantle and Wertheim, 1982). Sekiguchi and Gaucher (1977) reported similar findings for *P. urticae* and the production of patulin and griseofulvin. Conditions favoring sporulation in this species, for example, high concentrations of Ca^{2+} ions, inhibit metabolite synthesis, and mutants blocked in patulin biosynthesis produce conidia normally.

A variable feature of the observations recounted is clearly the system of culture adopted; however, this cannot account for the significant and totally opposite observations made. Campbell et al. (1982) argue the merits of surface culture for studies on the significance of secondary metabolism in fungal biology. However, for this approach to continue, more careful studies on the morphogenetic aspects, which require more reliable methods for obtaining synchrony of development, are required. These approaches, linked with the

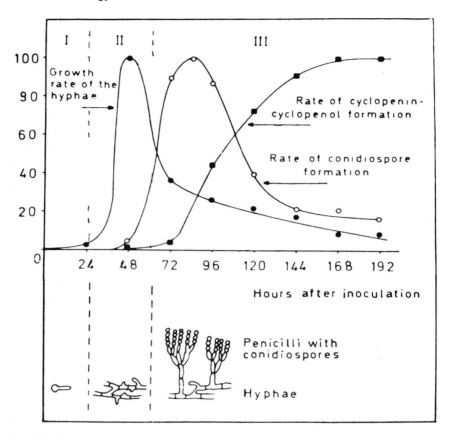

FIGURE 13.3   Alkaloid biosynthesis and morphogenesis in *P. cyclopium*.
● hyphal growth rate, ■ rate of conidiospore production, ▲ rate of alkaloid production. (Redrawn from Luckner, 1980.)

use of mutants blocked in conidiogenesis and secondary metabolite production, could provide some meaningful answers.

V. GENETICS

Because the penicillia are for the most part anamorphic species, the genetic studies that have been made have involved the parasexual cycle. Not surprisingly, most of the work has concerned *P. chrysogenum* and has, to a large extent, related to penicillin production. However, the parasexual cycle has also been described in *Penicillium janchewskii* (Alikhanian et al., 1960), *P. expansum* (Barron, 1962; Beraha and Garber, 1966), and *P. italicum* (Beraha and Garber, 1980). Studies with the latter two species were directed toward the genetics of fungicide resistance.

The parasexual cycle in *P. chrysogenum* was first described by Pontecorvo and Sermonti (1954). The early attempts to use the technique for breeding and strain improvement have been reviewed in depth by several authors (Macdonald and Holt, 1976; Macdonald, 1983). A noticeable feature of many of the crosses carried out in these "breeding" programs was the occurrence of parental genome segregation, which was assumed to arise from the lack of or limited homology between the genomes of the strains used in the cross.

Because of the specific goals of most of this early work, our understanding of genome organization in *P. chrysogenum* is still limited. Recent work has tentatively established the existence of six linkage groups in a wild strain of this species (I. D. Normansell, 1978; Blake, 1980). In earlier work (Ball, 1971), linkage relationships were demonstrated in an industrial strain; mitotic crossing over was also detected. Birkett and Rowlands (1981) isolated chlorate-resistant mutants in *P. chrysogenum* that were similar in their biochemical properties to those originally obtained in *Aspergillus nidulans* (Cove, 1976a, b). The mutations were not mapped, but their potential use in genetic analysis and possibly breeding was recognized.

An important group of mutants of *P. chrysogenum*, of interest for investigations into the regulation of penicillin biosynthesis, are those with impaired antibiotic production, the *npe* mutants (P. J. Normansell et al., 1979). Work on 12 of these indicated an organization of five complementation groups, which have been designated V, W, X, Y, and Z. Seven of the mutants belonged to the Y group. Parasexual analysis showed these mutations to be distributed on several linkage groups, and W, Y, and Z were found to be linked (P. J. Normansell et al., 1979).

P. chrysogenum was one of several fungi used by Anné and Peberdy (1975) for their initial studies on protoplast fusion in these organisms. This technique has been used subsequently for investigations into the possibilities for hybridization between species of the genus. The methods and underlying principles have been discussed previously (Peberdy, 1980).

Hybrids between *P. chrysogenum* and *Penicillium cyaneo-fulvum* were prepared by using nutritionally complementing strains (Peberdy et al., 1977). The fusion products were heterokaryons; and on prolonged culture, more vigorous, hybrid colonies developed. Despite the close relationship between the species (members of the same species group according to Raper and Thom (1949) and the same species according to Samson et al. (1977)), the hybrids bore little if any resemblance to the somatic diploid of either parent. Their behavior was compatible with that of an allodiploid, although this could not be confirmed, since the genetic marking of the parental strains was limited. When grown in the presence of haploidization agents, the hybrids segregated a spectrum of recombinant types.

Anné and co-workers have investigated the prospects for hybridization in more divergent crosses. Fusion products obtained between *P. chrysogenum* and *P. roqueforti* proved to be a mixture of heterokaryons and hybrid types (Anné et al., 1976). The hybrids were formed at a very low frequency, about 0.1% of the fusion products, and produced prototrophic spores. Later examination of the hybrid fusion products indicated that a range of colony phenotypes were formed (Anné and Peberdy, 1981). Some of these were unstable, giving sectors of *P. roqueforti* sporulation; others behaved similarly when selection was relaxed. However, a significant number were quite stable in the presence of haploidization agents. To assess the extent of the contribution of the two parental genomes to the formation of these phenotypes, a study of isoenzyme profiles was made. The patterns obtained were identical to one or the other parent or a mixture of both, and in a few cases, novel bands were observed, for example, for catalase and glucose-6-phosphate dehydrogenase (Anné and Peberdy, 1981).

Several other hybrids have also been produced, for example, *P. chrysogenum* + *P. citrinum* (Anné and Eyssen, 1978), *P. chrysogenum* + *Penicillium stoloniferum* (Anné, 1982a), *P. chrysogenum* + *P. patulum*, and *P. chrysogenum* + *Penicillium verrucosum* var. *cyclopium* (Anné, 1982b). The pattern of behavior of the progeny from these crosses followed one of the two types mentioned already. The first cross gave a hybrid form that segregated parental and nonparental types, that is, similar to the *P. chrysogenum* + *P. cyaneofulvum* hybrid. The other crosses showed patterns of hybridization similar to that of *P. chrysogenum* + *P. roqueforti*. Crosses have also been carried out between *P. chrysogenum* and *P. baarnense*, two very divergent species in taxonomic terms, and a range of hybrid phenotypes has also been obtained in this case (Fig. 13.4) (F. Mellon and J. F. Peberdy, unpublished data).

Anné (1982b) investigated the pattern of penicillin production in interspecies hybrids. *P. chrysogenum* + *P. cyaneo-fulvum* and *P. chrysogenum* + *P. citrinum* hybrids gave penicillin titers similar to the parental strains (all three species produce the antibiotic). Segregants derived from the hybrids showed a range of titers. The hybrid of *P. chrysogenum* + *P. patulum* (the latter is also a penicillin producer) produced reduced levels of the antibiotic as well as a significant change in the amounts of the different forms. Where the hybrid involved a nonproducer parent, for example, *P. chrysogenum* + *P. roqueforti*, *P. chrysogenum* + *P. stoloniferum*, and *P. chrysogenum* + *P. verrucosum* var. *cyclopium*, penicillin production by the hybrid was less than that of the parent. Changes in the proportions of benzyl-, heptyl-, and pentyl-penicillins were also found.

The potential of protoplast fusion in strain improvement programs has been discussed by Elander and Chang (1979) and Elander (1982). Chang et al. (1982) described the recovery of stable haploid recombinants as products of protoplast fusion crosses between auxotrophs of production strains of *P. chrysogenum*. It was concluded that the application of this technique resulted in a modification of the parasexual process involving transient heterokaryotic and diploid phases, a situation similar to that described for *Cephalosporium acremonium* (Hamlyn and Ball, 1979; Hamlyn, 1982). However, since no other classes of recombinants were isolated in these crosses, further studies are necessary to substantiate the view that protoplast fusion causes changes in the timing of parasexual events in *P. chrysogenum*.

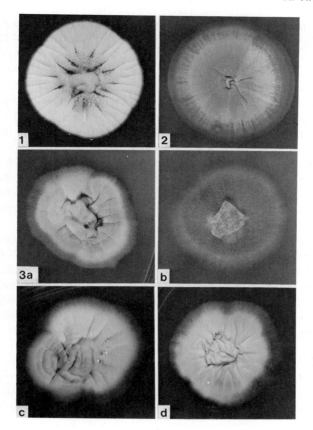

FIGURE 13.4   Interspecies hybridization in penicillia. (1) *P. chrysogenum*; (2) *P. baarnense*; (3a–d) hybrid types following protoplast fusion between these two species. (F. M. Mellon and J. F. Peberdy, unpublished.)

Since the first reports on protoplast fusion in fungi there has been an interest in the possibilities for intergeneric hybridization particularly between *P. chrysogenum* and *C. acremonium* to bring together the hydrophobic and hydrophilic branches of the β-lactam pathways of the two organisms. Recent reports have claimed success with this cross (Chang et al., 1982; Patent Hoechst AG EP 65-257). However, since these are from industrial laboratories, the experimental data presented are limited, and it is therefore difficult to comment meaningfully on them.

VI. VIRUSES

Several species have been found to have viruses, including *P. brevicompactum* (Wood et al., 1971), *P. chrysogenum* (Wood and Bozarth, 1972; Nash et al.,

1973), *P. cyaneo-fulvum* (Banks et al., 1969), and *P. stoloniferum* (Bozarth et al., 1971; Van Frank et al., 1971; Buck and Kempson-Jones, 1973). In each case the virus capsid is polyhedral in morphology, is 30–40 nm in diameter, and contains dsRNA. In electron micrographs the virus particles appear primarily in older hyphae, where they occur as crystalline aggregates enclosed within vesicles (Border et al., 1972). In several, a protein component that shows as a minor component on electrophoresis gels is thought to be a replicase (Morgan and Chater, 1974). RNA replicase activity has been demonstrated in a *P. chrysogenum* virus (Nash et al., 1973).

One isolated report claimed the presence of phagelike dsDNA viruses in several species (Tikchonenko et al., 1974). Mycelial extracts from the fungi added to bacteria-induced plaques that contained the phagelike viruses. This report is unconfirmed, and the possibility of contamination by phage-infected bacteria cannot be ruled out.

The biological significance of viruses in the penicillia is unresolved. In many species they are latent. Studies on submerged cultures of *P. stoloniferum* (Detroy et al., 1973) showed that the presence of virus had no effect on growth rate; similarly, penicillin production is not affected by the presence of virus (Volkoff et al., 1972; Lemke et al., 1973). In contrast, in *P. brevicompactum* and *P. stoloniferum*, mycophenolic acid occurs only in virus-free strains (Detroy et al., 1973).

VII. CONCLUDING REMARKS

In common with those of many other industrially important microorganisms, our knowledge and understanding of the biology of the penicillia are fragmentary. This is perhaps even more surprising for these fungi in view of the fact that many species are important for reasons anything but beneficial. It is clear, however, that the area of most extensive information, namely, on secondary metabolites, does reflect the situation. More fundamental studies in all aspects of Penicillium biology are required, but possibly most of all in the genetics of these fungi. Only with this expertise will our complete understanding of the metabolism of this organism come about. The newer techniques in genetics involving protoplast systems and recombinant DNA are as relevant to the penicillia as any other organism, and their application will surely further our appreciation of these very important fungi.

REFERENCES

Albright, J. L., Aust, S. D., Byers, J. M., Fritz, T. E., Brodie, B. O., Olsen, R. E., Link, R., Simon, J., Rhoades, H. E., and Brewer, R. L. (1964) *J. Amer. Vet. Med. Assoc. 144*, 1013–1019.
Alikhanian, S. I., Kameneva, S. V., and Krylov, V. N. (1960) *Microbiologiya* (Moscow) *29*, 820–825.

Alsberg, C. L., and Black, O. F. (1913) *U.S. Dept. Agr. Bull. Bur. Plant Ind.* No. 270.

Anné, J. (1982a) *FEMS Microbiol. Lett. 14*, 191–196.

Anné, J. (1982b) *Eur. J. Appl. Microbiol. Biotechnol. 15*, 41–46.

Anné, J., and Eyssen, H. (1978) *FEMS Microbiol. Lett. 4*, 87–90.

Anné, J., Eyssen, H., and De Somer, P. (1976) *Nature* (London) *262*, 719–721.

Anné, J., and Peberdy, J. F. (1975) *J. Gen. Microbiol. 22*, 413–417.

Anné, J., and Peberdy, J. F. (1981) *Trans. Brit. Mycol. Soc. 77*, 401–408.

Applegarth, D. A. (1967) *Arch. Biochem. Biophys. 120*, 471–478.

Applegarth, D. A., and Bozoian, G. (1968) *Can. J. Microbiol. 14*, 489–490.

Ball, C. (1971) *J. Gen. Microbiol. 66*, 63–69.

Banks, G. T., Buck, K. W., Chain, E. B., Derbyshire, J. E., and Himmelweit, F. (1969) *Nature* (London) *223*, 155–158.

Barron, G. L. (1962) *Can. J. Bot. 40*, 1603–1613.

Bartman, C. D., Doefler, D. L., Bird, B. A., Remaley, A. T., Peace, J. N., and Campbell, I. N. (1981) *Appl. Environ. Microbiol. 41*, 729–736.

Benko, P. V., Wood, T. C., and Segel, I. H. (1967) *Arch. Biochem. Biophys. 129*, 498–508.

Beraha, L., and Garber, E. D. (1966) *Phytopathology 56*, 870–871.

Beraha, L., and Garber, E. D. (1980) *Bot. Gaz. 141*, 204–209.

Bernlohr, R. W., and Novetelli, G. D. (1963) *Arch. Biochem. Biophys. 103*, 94–104.

Bird, B. A., Remaley, A. T., and Campbell, I. M. (1981) *Appl. Environ. Microbiol. 42*, 521–525.

Birkett, J. A., and Rowlands, R. T. (1981) *J. Gen. Microbiol. 123*, 281–285.

Blake, C. E. M. (1980) "Recombination in *Penicillium chrysogenum*," 230 pp, Ph.D. thesis, Polytechnic of Central London, London.

Bobbit, T. F., and Nordin, J. H. (1978) *Mycologia 70*, 1201–1211.

Bodmann, O., and Walter, M. (1965) *Biochim. Biophys. Acta 110*, 496–506.

Border, D. J., Buck, K. W., Chain, E. B., Kempson-Jones, G. F., Lhoas, P., and Ratti, G. (1972) *Biochem. J. 127*, 4P–6P.

Bouvier, J. (1967) *C. R. Acad. Sci., Ser. D 265*, 1305–1308.

Bouvier, J. (1968) *C. R. Acad. Sci., Ser. D 266*, 220–223.

Bozarth, R. F., Wood, H. A., and Mandelbrot, A. (1971) *Virology 45*, 516–523.

Brunswick, H. (1924) in *Botanische Abhandlugen* Vol. 5 (Goebel, K., ed.), Gustav Fishes, Jena.

Buck, K. W., and Kempson-Jones, G. F. (1973) *J. Gen. Virol. 18*, 223–235.

Burnside, J. E., Sippel, W. L., Forgacs, J., Carll, W. T., Atwood, M. B., and Doll, E. R. (1957) *Amer. J. Vet. Res. 18*, 817–824.

Calăm, C. T. (1982) in *Overproduction of Microbial Products* (Krumphanzl, V., Sikyta, B., and Vaněk, Z., eds.), pp. 89–96, Academic Press, New York.

Campbell, I. M. Doefler, D. L., Bird, B. A., Remaley, A. T., Rosato, L. M., and Davis, B. N. (1982) in *Overproduction of Microbial Metabolites* (Krumphanzl, V., Sikyta, B., and Vaněk, Z., eds.), pp. 141–151, Academic Press, New York.

Chang, L. T., Terasaka, D. T., and Elander, R. P. (1982) *Dev. Ind. Microbiol. 23*, 21–29.

Chiang, C., and Knight, S. G. (1959) *Biochim. Biophys. Acta 35*, 454–463.

Chiang, C., and Knight, S. G. (1960a) *Biochem. Biophys. Res. Comm. 3*, 554–559.

Chiang, C., and Knight, S. G. (1960b) *Nature* (London) *188*, 79–81.

Chiang, C., and Knight, S. G. (1961) *Biochim. Biophys. Acta 46*, 271–278.

Ciegler, A. (1977) in *Mycotoxins in Human and Animal Health* (Rodricks, J. V., Hesseltine, C. W., and Mehlman, M., eds.), pp. 609–624, Pathotox Publishers, Park Forest, Ill.

Ciegler, A., Kadis, S., and Ajl, S. J. (eds.) (1971) *Microbial Toxins,* Vol. VI, *Fungal Toxins,* 563 pp., Academic Press, New York.

Cole, G. T., and Kendrick, B. (1969) *Can. J. Bot. 47,* 779–789.

Cove, D. J. (1976a) *Heredity 36,* 191–203.

Cove, D. J. (1976b) *Mol. Gen. Genet. 146,* 147–159.

Demain, A. L. (1966) in *Biosynthesis of Antibiotics,* Vol. I (Snell, J. F., ed.), pp. 30–94, Academic Press, New York.

Demain, A. L. (1981) in β-lactam Antibiotics: Mode of Action, New Developments and Future Prospects (Salton, M. R. J., and Shockman, G. D., eds.), pp. 567–583, Academic Press, New York.

Demain, A. L., Aharonowitz, Y., and Martin, J. F. (1983) in *Biochemistry and Genetic Regulation of Commercially Important Antibiotics* (Vining, L. C., ed.), pp. 73–94, Addison-Wesley, Reading, Mass.

Detroy, R. W., Freer, S. W., and Fennell, D. I. (1973) *Can. J. Microbiol. 19,* 1459–1462.

Dickens, F., and Jones, H. E. H. (1965) *Brit. J. Can. 19,* 392–403.

Elander, R. P. (1982) in *Overproduction of Microbial Metabolites* (Krumphanzl, V., Sikyta, B., and Vaněk, Z., eds.), pp. 353–369, Academic Press, New York.

Elander R. P., and Chang, L. T. (1979) in *Microbial Technology,* Vol. II, 2nd ed. (Peppler, H., and Perlman, D., eds.), pp. 243–302, Academic Press, New York.

Eschete, M. L., King, J. W., West, B. C., and Oberle, A. (1981) *Mycopathology 74,* 125–127.

Faraj-Salman, A.-G. (1970) *Kulturpflanze 18,* 89–97.

Fenton, D. M., and Eveleigh, D. E. (1981) *J. Gen. Microbiol. 126,* 151–165.

Fletcher, J. (1971) *Ann. Bot. 35,* 441–449.

Florey, H. W., Chain, E. B., Heatley, N. G., Jennings, M. A., Sanders, A. G., Abraham, E. P., and Florey, M. E. (1949) in *Antibiotics,* Vol. II, pp. 631–671, Oxford University Press, London.

Girbardt, M. (1955) *Flora 142,* 540–543.

Girbardt, M. (1957) *Planta* (Berlin) *50,* 47–59.

Gottlieb, D., and Tripathi, R. K. (1968) *Mycologia 6,* 207–211.

Graffmans, W. D. J. (1973) *Arch. Mikrobiol. 91,* 67–76.

Gripon, J. C. (1977a) *Ann. Biol. Anim. Biochim. Biophys. 17,* 283–298.

Gripon, J. C. (1977b) *Biochimie 59,* 679–686.

Gripon, J. C., and Hermier, J. (1974) *Biochimie 56,* 1323–1332.

Gripon, J. C., and Debest, B. (1976) *Lait 56,* 423–438.

Grisaro, V., Sharon, N., and Barkai-Golan, R. (1968) *J. Gen. Microbiol. 51,* 145–150.

Hackette, S. L., Skye, G. E., Burton, C. E., and Segel, H. (1970) *J. Biol. Chem. 245,* 4241–4250.

Hadley, G., and Harrold, C. E. (1958a) *J. Exp. Bot. 9,* 408–417.

Hadley, G., and Harrold, C. E. (1958b) *J. Exp. Bot. 9,* 418–425.

Hamilton, P. B., and Knight, S. G. (1962) *Arch. Biochem. Biophys. 99,* 283–287.

Hamlyn, P. F. (1982) "Protoplast Fusion and Genetic Analysis in *Cephalosporium acremonium,*" 164 pp., Ph.D. thesis, University of Nottingham, Nottingham, England.

Hamlyn, P. F., and Ball, C. (1979) in *Third International Symposium on the Genetics of Industrial Microorganisms* (Sebek, O. K., and Laskin, A. I., eds.), pp. 185–191, American Society for Microbiology, Washington, D.C.

Hennebert, G. L., and Weresub, L. K. (1977) *Mycotaxon 6*, 207–211.

Hetherington, A. C., and Raistrick, H. (1931) *Phil. Trans. Roy. Soc. London B220*, 269–295.

Hocking, A. D., and Pitt, J. I. (1979) *Trans. Brit. Mycol. Soc. 73*, 141–145.

Hunter, D. R., and Segel, I. H. (1971) *Arch. Biochem. Biophys. 144*, 168–183.

Ikeda, S., and Yamada, K. (1963) *Nippon Nogei Kagaku Kaishi 37*, 514–517.

Keilova-Rodova, H. (1949) *Experientia 5*, 242.

Kobayashi, H., Tsunoda, H., and Tatsuno, T. (1971) *Chem. Pharm. Bull. 19*, 839–842.

Krogh, P., Hald, B., and Pederson, E. (1973) *Acta Pathol. Microbiol. Scand. 81*, 689–695.

Kulik, M. M. (1968) *Agricultural Handbook No. 351*, U.S. Government Printing Office, Washington, D.C.

Kurtzman, C. P., and Ciegler, A. (1970) *Appl. Microbiol. 20*, 204–207.

Lemke, P. A., Nash, C. H., and Peper, S. W. (1973) *J. Gen. Microbiol. 76*, 265–275.

Luckner, M. (1980) *J. Nat. Prod. 43*, 21–40.

Macdonald, K. D. (1983) in *Biochemistry and Genetic Regulation of Commercially Important Antibiotics* (Vining, L. C., ed.), pp. 25–47, Addison-Wesley, Reading, Mass.

Macdonald, K. D., and Holt, G. (1976) *Sci. Prog. Oxf. 63*, 547–573.

Mantle, P. G., and Wertheim, J. S. (1982) *Trans. Brit. Mycol. Soc. 79*, 345–350.

Martín, J. F., and Nicolás, G. (1970) *Trans. Brit. Mycol. Soc. 55*, 141–148.

Martín, J. F., Uruburu, F., and Villanueva, J. R. (1973a) *Can. J. Microbiol. 19*, 797–801.

Martín, J. F., Nicolás, G., and Villanueva, J. R. (1973b) *Can. J. Microbiol. 19*, 789–796.

Martín, J. F., Liras, P., and Villanueva, J. R. (1974) *Arch. Microbiol. 97*, 39–50.

Marumo, S., Miyao, K., and Matsuyama, A. (1955) *Bull. Agr. Chem. Soc. 19*, 262–266.

McCorkindale, N. J. (1976) in *The Filamentous Fungi*, Vol. II (Smith, J. E., and Berry, D. R., eds.), pp. 369–422, Edward Arnold, London.

Mislivec, P. B., and Tuite, J. (1970) *Mycologia 62*, 75–88.

Modler, H. W., Brunner, J. R., and Stine, C. M. (1974) *J. Dairy Sci. 57*, 528–534.

Morgan, D. H., and Chater, K. F. (1974) *Heredity 33*, 133.

Nash, C. H., Douthart, R. J., Ellis, R. F., Van Frank, R. M., Burnett, J. P., and Lemke, P. A. (1973) *Can. J. Microbiol. 19*, 97–103.

Normansell, I. D. (1978) "The Genetics of *Penicillium chysogenum* with Particular Reference to Penicillin Synthesis," 235 pp., Ph.D. thesis, Polytechnic of Central London, London.

Normansell, P. J., Normansell, I. D., and Holt, G. (1979) *J. Gen. Microbiol. 112*, 113–126.

Nover, L., and Luckner, M. (1974) *Biochem. Physiol. Pflanzen 166*, 293–305.

Okumura, S., Iwai, M., and Tsujisaka, Y. (1980) *J. Biochem. 87*, 205–211.

O'Sullivan, J., and Ball, C. (1983) in *Biochemistry and Genetic Regulation of Commercially Important Antibiotics* (Vining, L. C., ed.), pp. 73–94, Addison-Wesley, Reading, Mass.

Oxford, A. E., Raistrick, H., and Simonart, P. (1939) *Biochem. J. 33*, 240–248.

Patent HoechstAG EP 65-257. Production of hybrid fungi for antibiotic production — by interspecific of intergeneric protoplast fusion.

Peace, J. N., Bartman, C. D., Doerfler, D. L., and Campbell, I. M. (1981) *Appl. Environ. Microbiol. 41*, 1407–1412.

Peberdy, J. F. (1980) *Enzyme Microb. Technol. 2*, 23–29.

Peberdy, J. F., Eyssen, H., and Anné, J. (1977) *Mol. Gen. Genet. 157*, 281–284.

Piskorz, B. (1967) *Acta Soc. Bot. Polon. 36*, 123–131.

Pitt, D. (1969a) *J. Gen. Microbiol. 59*, 257–262.

Pitt, D. (1969b) *Trans. Brit. Mycol. Soc. 53*, 304–307.

Pitt, D. (1970) *Trans. Brit. Mycol. Soc. 55*, 325–327.

Pitt, D., and Poole, P. C. (1981) *Trans. Brit. Mycol. Soc. 76*, 219–230.

Pitt, J. I. (1979) *The Genus Penicillium and Its Teleomorphic States, Eupenicillium and Talaromyces*, 634 pp., Academic Press, New York.

Pontecorvo, G., and Sermonti, G. (1954) *J. Gen. Microbiol. 11*, 94–104.

Purchase, I. F. H. (1969) cited by C. W. Holzapfel in *Microbial Toxins*, Vol. VI, *Fungal Toxins* (Ciegler, A., Kadis, S., and Ajl, S. J., eds.), pp. 435–457, Academic Press, New York.

Queener, S. W., and Swartz, R. (1979) in *Economic Microbiology* Vol. 3, *Secondary Products of Metabolism* (Rose, A. H., ed.), pp. 35–122, Academic Press, New York.

Raper, K. B., and Thom, C. (1949) *A Manual of the Penicillia*, 875 pp., Williams and Wilkins, Baltimore.

Rapp, P., Knobloch, U., and Wagner, F. (1982) *J. Bacteriol. 149*, 783–786.

Rees, H., and Jinks, J. L. (1952) *Proc. Roy. Soc. Med. Ser. B 140*, 100–106.

Rhodes, A. (1963) *Prog. Ind. Microbiol. 4*, 165–187.

Righelato, R. C., Trinci, A. P. J., Pirt, S. J., and Peat, A. (1968) *J. Gen. Microbiol. 50*, 399–412.

Rondanelli, E. G., Gorini, P., Strosselti, E., and Picorari, D. (1957) *Haematologica 42*, 1427–1440.

Rose, A. H. (1979) in *Economic Microbiology* Vol. 3, *Secondary Products of Metabolism* 595 pp., Academic Press, New York.

Saito, M. (1959) *Acta Pathol. Japan 9*, 785–790.

Saito, M., Enomoto, M., and Tatsuno, T. (1971a) in *Microbial Toxins* Vol. VI, *Fungal Toxins* (Ciegler, A., Kadis, S., and Ajl, S. J., eds.), pp. 299–380, Academic Press, New York.

Saito, M., Ohtsubo, K., Umeda, M., Enomoto, M., Kurata, H., Udagawa, S., Sakabe, F., and Ichinoe, M. (1971b) *Japan J. Exp. Med. 41.*

Sakai, F. (1955) *Nippon Yakurigaku Zasshi 51*, 431–442.

Samson, R. A., Hadlok, R., and Stolk, A. C. (1977) Antonie van Leeuwenhoek. *J. Microbiol. Serol. 43*, 169–175.

Santos, T., Villanueva, J. R., and Nombela, C. (1977) *J. Bacteriol. 129*, 52–58.

Santos, T., Villanueva, J. R., and Nombela, C. (1978) *J. Bacteriol. 133*, 542–548.

Santos, T., Sanchez, M., Villanueva, J. R., and Nombela, C. (1979) *J. Bacteriol. 137*, 6–12.

Scott, D. B. (1965) *Mycopath. Mycol. Appl. 25*, 213–222.

Scott, P. M. (1977) in *Mycotoxic Fungi, Mycotoxins and Mycotoxicoses* (Wyllie, T., and Morehouse, L., eds.), pp. 283–356, Marcel Dekker, New York.

Scott, P. M., and Kennedy, B. (1973) *J. Assoc. Off. Anal. Chem. 57*, 861–865.

Segretain, G. (1959) *Mycopath. Mycol. Appl. 11*, 327–353.

Sekiguchi, J., and Gaucher, G. M. (1977) *Appl. Environ. Microbiol. 33*, 147–158.

Sekiguchi, J., Gaucher, G. M., and Costerton, J. W. (1975a) *Can. J. Microbiol. 21*, 2059–2068.

Sekiguchi, J., Gaucher, G. M., and Costerton, J. W. (1975b) *Can. J. Microbiol. 21*, 2069–2083.

Sentein, P. (1955) *C. R. Soc. Biol. 149*, 1621–1622.

Skye, G. E., and Segel, I. H. (1970) *Arch. Biochem. Biophys. 138*, 306–318.

Steinegger, E., and Leupi, H. (1956) *Pharm. Acta Helv. 31*, 45–51.

Tatsuno, T., Tsujiioka, M., Sakai, Y., Suzuki, Y., and Asami, Y. (1955) *Pharm. Bull.* (Tokyo) *3*, 476–477.

Tikchonenko, T. I., Velikodvorskaya, G. A., Bobkova, A. F., Bartoshevich, Yu. E., Lebed, E. P., Chaplygina, N. M., and Maksimova, T. S. (1974) *Nature* (London) *249*, 454–456.

Towers, G. H. N. (1976) in *The Filamentous Fungi*, Vol. II (Smith, J. E., and Berry, D. R., eds.), pp. 460–474, Edward Arnold, London.

Turner, W. B. (1971) *Fungal Metabolites*, 446 pp., Academic Press, New York.

Turner, W. B. (1976) in *The Filamentous Fungi*, Vol. II (Smith, J. E., and Berry, D. R., eds.), pp. 445–459, Edward Arnold, London.

Ueno, Y., Ueno, I., Tatsuno, T., and Uraguchi, K. (1964) *Japan J. Exp. Med. 34*, 197–209.

Ugalde, U., and Pitt, D. (1983) *Trans. Brit. Mycol. Soc. 80*, 319–325.

Unger, P. D., and Hayes, A. W. (1975) *J. Gen. Microbiol. 91*, 201–206.

Van Frank, R. M., Ellis, L. F., and Kleinschmidt, W. J. (1971) *J. Gen. Virol. 12*, 33–42.

Volkoff, O., Walters, T., and Dejardin, R. A. (1972) *Can. J. Microbiol. 18*, 1352–1353.

Ware, G., Thorpe, C., and Pohland, A. (1974) *J. Assoc. Off. Anal. Chem. 57*, 861–865.

Watkinson, S. C. (1975) *J. Gen. Microbiol. 87*, 292–300.

Watkinson, S. C. (1977) *J. Gen. Microbiol. 101*, 269–275.

Whitaker, A., and Morton, A. G. (1971) *Trans. Brit. Mycol. Soc. 56*, 353–369.

Willetts, A. J. (1972) *J. Gen. Microbiol. 73*, 71–83.

Wood, H. A., and Bozarth, R. F. (1972) *Virology 47*, 604–609.

Wood, H. A., Bozarth, R. F., and Misilvec, P. B. (1971) *Virology 44*, 592–598.

Wright, J. L. C., and Vining, L. C. (1976) in *The Filamentous Fungi*, Vol. II (Smith, J. E., and Berry, D. R., eds.), pp. 475–502, Edward Arnold, London.

Zeidler, G., and Margalith, P. (1973) *Can. J. Biochem. 19*, 481–483.

Zevaco, C., Hermier, J., and Gripon, J. C. (1973) *Biochimie 55*, 1352–1360.

Cephalosporium acremonium: A β-Lactam Antibiotic- Producing Microbe

Claude H. Nash, III
Rajanikant J. Mehta
Christopher Ball

I. INTRODUCTION

The fungus described in this chapter was isolated by Professor Giuseppe Brotzu in 1945 while examining the microbial flora of seawater near a sewage outlet in the Gulf of Cagliari, Sardinia (Brotzu, 1948). Cultures of this Cephalosporium-like hyphomycete produced substances with broad antibacterial activity. These metabolites and the microbe that produces them have occupied the energies of many scientists for the past three and a half decades. The major biologically active metabolites, cephalosporin P_{1-5}, penicillin N and cephalosporin C, were isolated by Professor E. P. Abraham and his co-workers at Oxford from 1948 to 1956 (Burton and Abraham, 1951; Crawford et al., 1952; Abraham et al., 1953; Newton and Abraham, 1956; Abraham and Newton, 1961). Extensive chemical modification of cephalosporin C has led to at least 18 commercially useful derivatives (Hoover and Nash, 1982).

Biological studies with *Cephalosporium acremonium*, the primary organism used for manufacture of cephalosporin C, have closely paralleled the chemical studies. Biological research with this organism has been directed toward resolving the biosynthetic pathway for the β-lactam antibiotics and improving the antibiotic-producing potential of the microbe. The chemistry and fermentation of cephalosporin C and β-lactam biosynthesis have been extensively reviewed (Lemke and Brannon, 1972; Hoover and Dunn, 1979; Demain et al., 1982; Hoover and Nash, 1982; Queener and Neuss, 1982; Elander and Aoki, 1982).

II. TAXONOMY

The Cephalosporium-like fungus, originally designated as *Cephalosporium acremonium* corda sensu Brotzu CMI 49137, is a member of fungi imperfecti. It produces two forms of asexual spores, conidia and arthrospores. The typical conidia are oblong structures about 15 μm long and 8 μm wide with a mean spore volume of 5.06 μm³ (Pisano, 1963). The conidia are abstricted successively and held together in a slime head. Arthrospores are formed by rounding off the contents of one or more of the hyphal segments, followed by thickening of the cell wall. *C. acremonium* is homothallic and lacks a sexual cycle. Hyphae are organized as uninucleate segments (Treichler et al., 1972; Hamlyn, 1982). A photomicrograph of a typical hyphal filament stained with chromomycin (Hamlyn, 1982) is shown in Fig. 14.1.

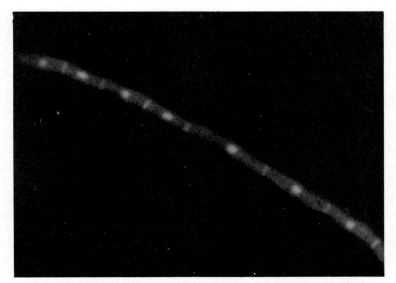

FIGURE 14.1   Photomicrograph of a hyphal filament showing individual nuclear regions and septa. (Photo courtesy of Dr. Paul Hamlyn.)

The genus *Cephalosporium* is a rather heterogeneous group that has been the subject of considerable taxonomic dispute (Gams, 1971; Chesson et al., 1978; Raituair and Tarasov, 1975). Morphologically distinct antibiotic-producing strains of *Cephalosporium chrysogenum* express similar physiological and biochemical properties. This precludes the use of these parameters for distinguishing strains (Pisano, 1970). In his 1971 monograph on Cephalosporium-like hyphomycetes, Gams recommended that the β-lactam-producing fungus designated *Cephalosporium acremonium* should more accurately be named *Acremonium chrysogenum* Brotzu. This recommendation was based on the characterization of the type culture of the antibiotic-producing microbe, which bears only limited resemblance to the phytopathogenic *Cephalosporium acremonium* strains (Sabet et al., 1970; Lah and Saksena, 1977). The new name, *Acremonium chrysogenum*, has received sporadic acceptance by scientists working with the cephalosporin C producer. For historic reasons we will continue to use the original designation *Cephalosporium acremonium* in this chapter.

III. ECOLOGICAL DISTRIBUTION

The *Cephalosporium* species associated with plant, animal, and human disease are widely distributed in the environment. Representatives of this genus have been isolated from such diverse habitats as the mid-gut of fiddler crabs (Pitts and Cowley, 1974), leaves of plants (Lindsey and Pugh, 1976), wood chips (Greaves, 1975), and, of course, soil (Pisano, 1963). *C. acremonium* strains have been implicated as the etiological agent in onychomycosis, an infection of the nails in humans, and in stripe disease of corn. These strains are morphologically distinctive from the antibiotic-producing fungus (Gams, 1971). The ability of certain species of *Cephalosporium* to survive in a variety of environments may explain their wide distribution in diverse soils (acid, basic, sandy, etc.). Although antibiotic-producing strains of Cephalosporium-like hyphomycetes have routinely been isolated by various pharmaceutical companies, the taxonomic identity of these strains has not been adequately defined.

IV. MORPHOLOGY AND SUBCELLULAR STRUCTURE

The hyphae and spores of *C. acremonium* have a typical eukaryotic fine structure. On solid media the type culture *C. acremonium* corda Brotzu; ATCC 11550 grows rapidly and forms a large number of conidia. Typical conidia and conidiophores are shown in Fig. 14.2.

In addition, this strain produces a soluble yellow pigment that diffuses into solid medium. Submerged cultures of *C. acremonium* contain four morphological cell types (hyphae, arthrospore, conidia, and germlings), which represent stages in the growth cycle (Nash and Huber, 1971).

Fawcett et al. (1973) described the fine structure of hyphae grown on defined media. The cell wall is 170 nm thick and is composed of five layers, including a diffuse outer layer and an inner layer associated with the cytoplasmic membrane. The cytoplasm contains packets or ribosomes, amorphous areas devoid of ribosomes, vacuoles, electron-dense inclusion bodies, oval-shaped mitochondria, and membrane-bound nuclei. Tubular invaginations were observed at the cytoplasmic membrane in some hyphae.

The differentiation of hyphae into multicellular and ultimately unicellular arthrospores is more evident in the high antibiotic-producing mutants (Queener and Ellis, 1975; Matsumura et al., 1980; Drew et al., 1976). Except for the thickened cell wall, these swollen hyphal fragments contain typical subcellular organelles. Typical hyphae and swollen arthrospores are shown in Fig. 14.3. The unicellular arthrospores continue to propagate by generation of germ tubes, which swell and fragment prior to the formation of a hyphal filament. Germinating arthrospores have thick cell walls, rough outer surfaces, and indentations on the outer wall. The internal structure of arthrospores resembles that of swollen hyphae and typical hyphal filaments. Filaments contain fewer lipid-filled vacuoles than either the swollen hyphae or the arthrospores.

Unicellular arthrospores, formed when a high antibiotic-producing mutant of *C. acremonium* is grown on glycerol, contain large numbers of fatty bodies (Mehta et al., 1981). The detailed fatty acid composition of *C. acremonium* grown in complex media is similar to other fungi (Huber and

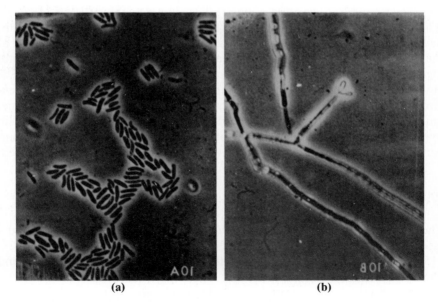

(a)   (b)

FIGURE 14.2   Typical (a) conidia and (b) conidiophore of *C. acremonium*. (Photos courtesy of Dr. Stephen Queener, Eli Lilly & Co., Indianapolis, IN.)

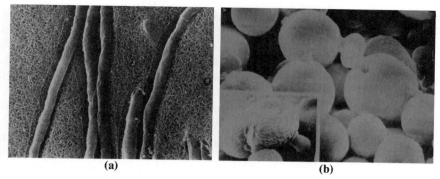

(a)

(b)

FIGURE 14.3 Scanning electron micrographs of (a) hyphae and (b) unicellular arthrospores. (Photos courtesy of Drs. Stephen Queener and Lee Ellis, Eli Lilly & Co., Indianapolis, IN.)

Redstone, 1967). Four lipid classes (sterol esters, triacylglycerides, free fatty acids, and sterols) were reported in *Cephalosporium falciforme* and *Cephalosporium riliense*, the causative agents of mauduromycosis. Both qualitative and quantitative alterations in these lipids occur as a function of temperature (Sawicki and Pisano, 1977). The overall lipid composition is similar to other fungi.

The conidia of *C. acremonium*, like other fungi, progress through three morphological states: dormant spores, swollen spores, and germlings (Nash and Pieper, 1974). An exogenous supply of carbon, nitrogen, magnesium, and phosphate is essential for germination. The conidia of *C. acremonium* germinate over a broad temperature range of 27–32°C but have a sharp pH optimum of 8.0. Spores incubated at a low pH (4.0–5.5) swell but do not form germ tubes. When the pH is adjusted to 7.5, the swollen spores proceed to form germ tubes at about twice the rate as dormant spores.

Sporulation is highly dependent on the specific strain being studied. The high antibiotic-producing mutants that were ultimately derived from strain M8650 appear defective in sporulation or at least spore maturation.

Viruslike particles have been observed in strains of *C. acremonium*. These particles are similar to other fungal viruses, the few particles observed being about 50 nm in diameter and spherical in shape (Day and Ellis, 1971). Crystalline bodies have also been observed by electron microscopy in the organism (Mason and Crosse, 1975).

V. NUTRITION AND PHYSIOLOGY

C. acremonium is heterotrophic and aerobic. The organism grows on diverse carbon sources including glucose, galactose, fructose, sucrose, lactose, ribose, glycerol, mannitol, soluble starch, maltose, arabinose, and methyl oleate.

Both inorganic and organic nitrogen can be utilized by *C. acremonium.* Nitrate, ammonium sulfate, ammonium chloride, peptones, yeast extract, and various complex proteins have been utilized as nitrogen sources for growth and antibiotic synthesis (Demain and Newkirk, 1962; Mehta and Nash, 1979). Sporulation and mycelial growth on solid media are dramatically influenced by available light (Propst-Ricciuti and Kenney, 1975).

    C. acremonium grows over a wide temperature (20–35°C) and pH range (4.0–9.0). Significant sporulation occurs within a pH range of 6.0–7.0. Czapek-Dox agar with 3.5% lactose supports good sporulation. The most productive strains sporulate better on Czapek-Dox media with added ribose and 0.2% nitrate. Several defined media have been described for production of cephalosporin C. These media typically contain glucose, ammonium sulfate, and various salts (Demain and Newkirk, 1962; Nash, 1979; Mehta et al., 1979).

    The commercial production of cephalosporin C involves utilization of a complex medium. Glucose, sucrose, beet molasses, and methyl oleate are used as carbon sources in various commercial processes. The primary nitrogen sources include peanut meal, fish meal, and soy flour. Sulfur is typically supplied as $(NH_4)_2SO_4$, calcium sulfate, or methionine (Pan et al., 1982; Niss and Nash, 1973; Komatsu et al., 1975; Drew and Demain, 1975b; Matsumura et al., 1981, Matsumura and Taguchi, 1980). The media and fermentation conditions used for manufacture of cephalosporin C are a function of the specific strain used. High cephalosporin C production requires efficient agitation and aeration of the complex media (Feren and Squires, 1969).

VI. METABOLISM AND REGULATION

Most research on *C. acremonium* intermediary metabolism has focused on the primary and secondary pathways relating to antibiotic biosynthesis. *C. acremonium* strains produce two classes of antibiotics: β-lactams and steroids (Lemke and Brannon, 1972). The structures for these compounds are shown in Fig. 14.4. Other β-lactam-containing compounds, such as D-5-amino-5-carboxyvaleramide-(5-formyl-4-carboxy-2H, 3H, 6H tetrahydro-1, 3 thiazinyl), glycerine, 7 beta-(5-D-aminoadipamido)-3 beta-hydroxy-3 alpha-methylcepha-M-4-alpha-carboxylic acid, dimers α-(L-α-aminoadipyl-L-cysteinyl-D-valine and a S-methylthio derivative 7 α-(4-carboxybutanamide)-cephalosporin, and 7-α-(D-5-amino-5-carboxy-N-valeramide)-3-(1,1 dimethyl-2-amino-2 carboxy-ethyl) thiomethyl-3-cephem-4-carboxylic acid are reported to accumulate in certain mutants (Kitano et al., 1976; Shirafugi et al., 1979; Fujisawa and Kanzaki, 1975a; Spry et al., 1981). In addition, some *C. acremonium* strains catalyze the conversion of progesterone to androstenedione (Sardinas and Pisano, 1967).

    The biosynthetic pathway for the major α-aminoadipic acid containing β-lactams, cephalosporin C, and penicillin N has been elucidated by incor-

Cephalosporin P_{1-5}

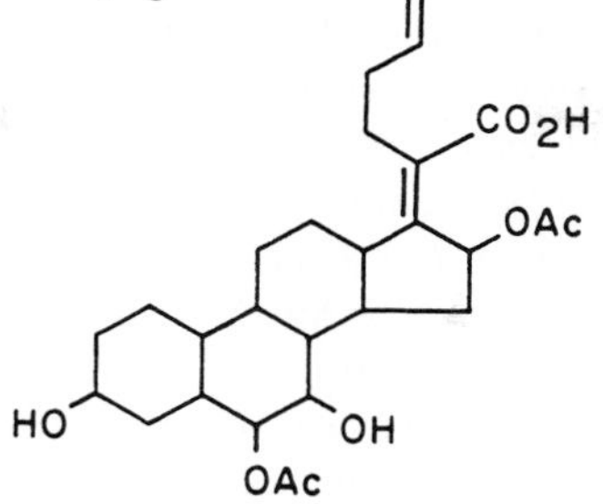

Cephalosporin C

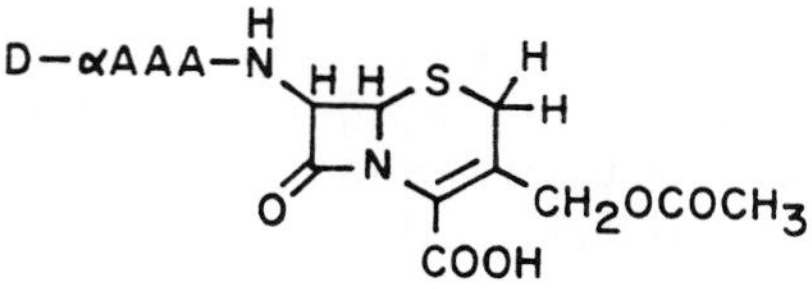

Deacetyl Cephalosporin C

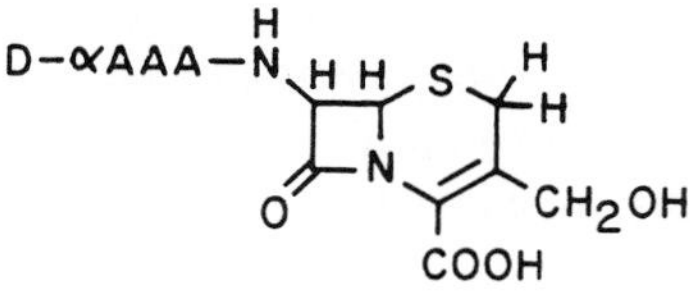

Deacetoxy Cephalosporin C

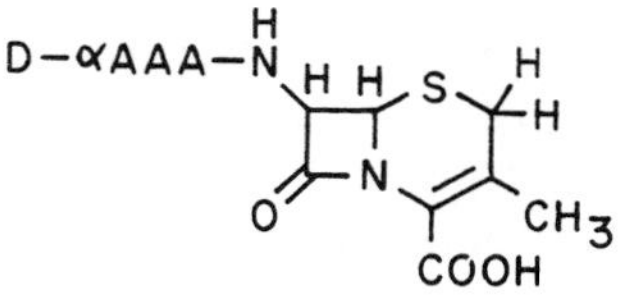

Penicillin N

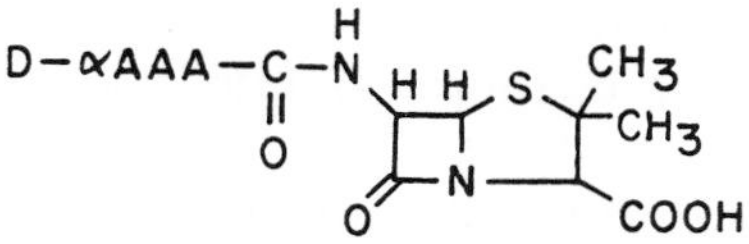

FIGURE 14.4 Secondary metabolites produced by *Cephalosporium acremonium*.

poration of C^{13}- and C^{14}-labeled precursors, analysis of blocked mutants, and biosynthesis with cell-free systems (Shirafugi et al., 1979; Bahadur et al., 1981; Jayatilake et al., 1981; Neuss et al., 1971, 1973; Fujisawa et al., 1977; Bost and Demain, 1977; Fawcett et al., 1976; Kohsaka and Demain, 1976; Yoshida et al., 1978; Felix et al., 1981; Baldwin et al., 1981). These structurally bicyclic tripeptides are derived from the primary metabolites α-aminoadipic acid, cysteine, and valine. The tripeptide intermediate α-(L-aminoadipyl)-L-cysteinyl-D-valine appears to be the precursor of penicillin N. The cephalosporin ring structure is formed by the conversion of penicillin N to deacetoxycephalosporin C, which is then converted to deacetylcephalosporin C and finally by acetylation to cephalosporin C.

The accumulation of the β-lactam antibiotic is markedly influenced by the biosynthesis of the primary metabolites α-aminoadipic acid, cysteine, and valine (Drew and Demain 1975a, b).

C. acremonium can obtain sulfur for cysteine either from sulfate by the sulfate reduction pathway or from methionine by the reverse transulfuration pathway. Thus sulfate and methionine are alternative sulfur sources for cysteine and, in turn, cephalosporin C (Matsumura et al., 1980; Drew and Demain, 1975b, c; Suzuki et al., 1980; Komatsu and Kodaira, 1977; Niss and Nash, 1973). Recent studies indicate that activated cystathionine, and not cysteine, is the immediate precursor for sulfur in the formation of β-lactam antibiotics (Doebeli and Neusch, 1980; Gygax et al., 1980). Any mutation that enhances the accumulation of cystathionine would, in turn, facilitate antibiotic biosynthesis. An alternative pathway from H_2S to cystathionine via homocysteine appears to be the optimal biosynthetic route when sulfate is the main sulfur source. *C. acremonium* also utilizes O-acetyl-L-serine sulfhydrylase in the biosynthesis of cysteine from H_2S.

Lysine is synthesized via the homocitric acid pathway in *C. acremonium*. The biosynthetic pathway was elucidated by using blocked mutants, radiolabeled substrates, and isolated enzyme preparation. Exogenous lysine, at high concentrations, appears to inhibit the first enzyme in the pathway, homocitrate synthetase; at low concentrations it inhibits or represses the conversion of α-aminoadipic acid to lysine (Nash et al., 1974; Mehta et al., 1981; Lemke and Nash, 1972; Abraham and Newton, 1961).

The third amino acid involved in biosynthesis of cephalosporin C, valine, is believed to be synthesized by a pathway similar to the one found in bacteria. This pathway also appears to be subject to feedback regulation (Nash et al., 1974).

Several enzymes have been isolated from *C. acremonium* and purified to various degrees (Hinnen and Neusch, 1976; Fujisawa and Kanzaki, 1975b; Benz et al., 1971; Dennen at al., 1971; Fujisawa et al., 1977; Oleniacz and Pisano, 1968; Queener et al., 1975). These enzymes are listed in Table 14.1. Enzymatic activity has also been observed against casein, gelatin, milk, hemoglobin, human plasma, starch, and N-acetyl-β-D-glucose (Pisano et al., 1963).

TABLE 14.1   Enzymes Described from *Cephalosporium acremonium*

Enzyme	Cellular Location
Acetyl hydrolase	Extracellular
Acetyl transferase	Intracellular
D-aminoacid oxidase	Intracellular
Arylamidase	Extracellular
Acylase	Extracellular
Sulfatase	Intracellular
DOCPC hydroxylase	Intracellular
Glutamate dehydrogenase	Intracellular
Protease	Extracellular

The fibrinolysis and esterase activities were more extensively characterized (Oleniacz and Pisano, 1968a; Pisano and Capone, 1967). A polyol dehydrogenase from *C. chrysogenum* was similar to the homologous enzyme from other microbes (Birken and Pisano, 1976). Except for the protease, the enzyme activities have been described with crude cellular preparations.

The regulation of metabolic pathways in *C. acremonium* appears to be similar to that in other microorganisms. Biosynthetic pathways for primary metabolites are subject to end product inhibition or repression. Furthermore, the production of antibiotics is in some way tied to cellular differentiation.

Regulation of cysteine biosynthesis is complicated by the presence of multiple biosynthetic pathways. Cysteine pool sizes available for antibiotic synthesis vary according to the sulfur source used in the medium (Drew and Demain, 1975c; Niss and Nash, 1973). Potentiation of cephalosporin C synthesis by methionine appears to be multifaceted. Part of the activity can be attributed to the rapid conversion of methionine to cystathionine. In addition, methionine appears to act directly (or indirectly) on the β-lactam antibiotic biosynthetic enzymes (Drew and Demain 1975b, c).

The α-aminoadipic acid available for cephalosporin C synthesis is regulated by lysine concentration. The molecular nature (inhibition or repression) of this regulatory mechanism remains unresolved (Mehta et al., 1979).

Antibiotic synthesis can be dramatically affected by carbon, nitrogen, phosphate, and sulfur sources. For example, glycerol totally inhibits the synthesis of cephalosporin C. The mechanism for this nutrient action has not been elucidated (Mehta et al., 1981). Cephalosporin C synthesis is also affected by glucose and phosphate (Kuenzi, 1980; Jaramillo and Kuenzi, 1980; Kennel and Demain, 1978; Demain and Kennel, 1978; Martín et al., 1982). Carbon catabolite regulation of cephalosporin C synthesis seems to be exerted by blocking the conversion of penicillin N to cephalosporin C. Time of addition studies suggest suppression of enzyme synthesis rather than inhibition of enzyme activity. The presence of restrictive levels of inorganic phosphate in the medium decreases both penicillin N and cephalosporin C, a result that suggests an effect on the overall flux of precursors (Martín et al., 1982).

VII. GENETICS

Although *C. acremonium* has been the primary focus for many industrial strain improvement programs, only limited information is reported on its fundamental genetics. Neusch et al. (1973) described the conditions for mutagenesis of conidia with N-methyl-N'-nitro-N-nitrosoguanidine (NTG). After NTG treatment these investigators isolated 164 single-auxotrophic mutants representing a wide spectrum of vitamin, amino acid, and nucleic acid requirements. Certain specific auxotrophs, however, occurred with a frequency over ten times greater than others. Hamlyn and Ball (1979) and Hamlyn (1982) described a number of UV-induced auxotrophic mutants that also contained a disproportionate number of certain mutant classes. Susceptibility of *C. acremonium* to 100 antimetabolites that may be useful in the isolation of deregulated mutants was described by Lemke (1969).

Elander et al. (1979) described the kinetics of UV induction of morphological mutants. The optimal frequency of mutants was observed at 1% survival. The frequency of mutants in this study was similar to results of Neusch et al. (1973) and Hamlyn (1982). The optimal conditions for induction of qualitative mutation (e.g., morphological or auxotrophic) were different from those for quantitative mutation, specifically, enhanced cephalosporin C productivity (Elander et al., 1979). Ultraviolet light was more effective for the isolation of superior antibiotic-producing strains than ten chemical mutagens, including β-propiolactone, ethyl methane sulphonate, and NTG. The optimum dose of UV for recovery of superior cephalosporin C producing mutants, at least 20% increase in titer, was at a 20–30% survival of spores. The selection of morphological mutants on agar did not enhance the recovery of superior antibiotic producers. Mutations increasing antibiotic synthesis appear to be extremely diverse, reflecting the complex interrelationship of a multigenic family. The interplay of these mutants was succinctly described by Rowlands (1982) as follows:

> Consider the set of titer-increasing mutations to be a carefully balanced house of playing cards. To improve the titer still further, one has to randomly balance another card from the gene pool onto it. This is easy while the house is small, but the bigger it gets the more difficult it becomes to add another card without causing partial collapse of the existing structure.

As an alternative to random screening, Chang and Elander (1979) describe the isolation of a number of specific *C. acremonium* mutants with superior cephalosporin C–producing potential (Chang et al., 1982). Representative amino acid analog–resistant mutants, mutants sensitive to antimetabolites and growth inhibitors, mutants resistant to metallic ions, specific morphological mutants, and revertants of auxotrophs were screened for antibiotic-producing potential. Superior antibiotic-producing strains were recovered more frequently after direct selection of specific mutants than by random screening. In another study involving selenolysine and selenomethionine, resistant mutants were reported to be superior antibiotic-producing strains.

There is substantial evidence supporting the existence of a parasexual cycle in *C. acremonium*. The classical parasexual cycle can be summarized as the following sequence of events: (1) cell fusion, (2) nuclear fusion, (3) segregation. The cell fusion events are generally rare and therefore can be detected only when the parental strains carry suitable markers that permit the selection of heterokaryon or recombinants. The product of cell fusion is a heterokaryon (i.e., a cell that contains more than one nuclear species in a common cytoplasm). The second event, nuclear fusion, is also detected by selection. For this purpose, spores from the heterokaryon are plated onto selective medium, and the rare stable prototrophs that arise as putative diploids are recovered. The third stage, segregation, involves induction of the diploid state to form segregants that can be haploid, diploid, or aneuploid. These segregants can then be classified by using any or all of several criteria such as marker classification, spore size, or DNA content.

The work of Neusch et al. (1973) was very important in emphasizing the difficulty in establishing a true heterokaryotic state with *C. acremonium*. This fungus typically accommodates only one nucleus per cell, a feature that selects against the formation of a heterokaryon. Protoplast fusion circumvents the

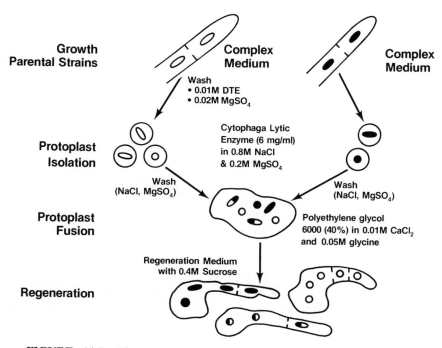

FIGURE 14.5 Diagrammatic procedure for protoplast fusion with *Cephalosporium acremonium*.

normal incompatibilities of *C. acremonium* (Anné and Peberdy, 1976), a feature that makes this technique almost obligatory in studying the genetics of *C. acremonium*. The three stages of the classical parasexual cycle can be cut short in *C. acremonium*, and segregants can be isolated directly following protoplast fusion (Hamlyn and Ball, 1979). The conditions for protoplast fusion are diagrammatically depicted in Fig. 14.5. Protoplasts are prepared by using cytophaga lytic enzymes in 0.8 M of NaCl and 0.2 M of MgSO₄. The protoplasts are fused by using polyethylene glycol, and recombinants are recovered by selection on a suitable regeneration medium. Phase contrast micrographs of protoplast and a germinating protoplast (solid media) are shown in Fig. 14.6. The protoplasts have a typical spherical shape in liquid media but take on an irregular appearance during germination on solid media. Ball et al. (1979) described a number of detailed conditions for achieving efficient protoplast fusion in *C. acremonium*, and Hamlyn (1982) has made a very thorough study of optimal conditions for protoplast preparation, regeneration and fusion in this organism.

Hamlyn and Ball (1979) studied 40 conventional and eight protoplast fusion crosses; no stable diploids were recovered from any of these crosses. Stable diploid of *C. acremonium* was described by Neusch et al. (1973); but in later studies, Hamlyn (1982) found no evidence for stable diploids in nine protoplast fusion crosses. These results, together with indications from other studies (Elander et al., 1979), suggest that stable diploids, if they do exist in *C. acremonium*, are very rare and that the most frequent nuclear fusion products in this organism are spontaneously unstable.

Further studies with one of the unstable diploids reported by Hamlyn and Ball (1979) gave rise to recombinants with industrially relevant properties — namely, growth rate, sporulation, cephalosporin C titer, and production of cephalosporin C from sulfate as the sole sulfur source. Other unstable diploids have also been studied by Hamlyn et al. (1983). Their data, together with complementary data based on direct segregation and recovery by plating onto a

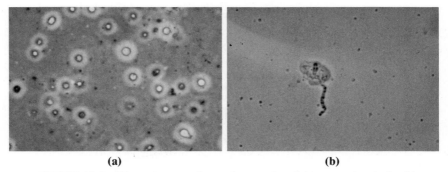

(a)   **(b)**

FIGURE 14.6   Phase contrast photomicrographs of (a) protoplast in liquid media and (b) a germinating fused protoplast on solid media.

variety of selective media, has enabled the detection of eight linkage groups in *C. acremonium*, including four cases of linkage.

Although protoplast fusion has proven to be useful with *C. acremonium*, the utility of interspecific protoplast fusion still requires critical evaluation with the organism. Additional research on the basic genetics of *C. acremonium* should facilitate the utility of genetic recombination and rational strain selections with this organism.

VIII. CONCLUSIONS

The biology of the industrially important strains of *Cephalosporium acremonium* has been elucidated in concert with development of optimal fermentation conditions, resolving the biosynthetic pathways of antibiotics and construction of superior antibiotic-producing strains. Although the cephalosporin C titers have been increased from about 10 μg/ml to near 20,000 μg/ml (Pan et al., 1982), basic knowledge of the biology of *C. acremonium* remains in its infancy. The taxonomic classification of this industrially important fungus is the subject of considerable controversy, and the ecological distribution is defined only by the assumed association with other *Cephalosporium* species. The nutritional requirements for optimal antibiotic synthesis and the biosynthetic pathways leading to the β-lactam antibiotics have been extensively studied.

This organism proved to be a difficult system for genetic manipulation. The strains used in industry today have resulted from extensive research programs using mutation followed by random screening for superior antibiotic production. Attempts to utilize microbial breeding via protoplast fusion have overcome the strict homothallic life style of this fungus and recent efforts have suggested that both recombination and direct selection will be useful methods for construction of future strains.

REFERENCES

Abraham, E. P., and Newton, G. G. F. (1961) *Biochem. J. 79*, 377–378.

Abraham, E. P., Newton, G. G. F., Crawford, K., Burton, H. S., and Hale, C. W. (1953) *Nature 171*, 343–353.

Anné, J., and Peberdy, J. F. (1976) *J. Gen. Microbiol. 92*, 413–417.

Bahadur, G. A., Baldwin, J. E., Usher, J. J., Abraham, E. P., Jayatilake, G. S., and White, R. L. (1981) *J. Amer. Chem. Soc. 103*, 7650–7651.

Baldwin, J. E., Keeping, J. W., Singh, P. D., and Vallejo, C. A. (1981) *Biochem. J. 194*, 649–651.

Ball, C., Collins, J. S., and Hamlyn, P. F. (1979) U.K. Patent Application CB 2 021 640 A.

Benz, F., Liersch, M., Nuesch, J., and Treichler, A. (1971) *Eur. J. Biochem. 20*, 81–83.

Birken, S., and Pisano, M. A. (1976) *J. Bacteriol. 125*, 225–232.

Bost, P. E., and Demain, A. L. (1977) *Biochem. J. 162*, 681–687.

Brotzu, G. (1948) *Richerche sudi un nuovo antibiotico*, pp. 1–11, Labori Dell' Instituto D'Igiene du Cagliari.

Burton, H. S., Abraham, E. P. (1951) *Biochem. J. 50*, 168–173.

Chang, L. T., and Elander, R. P. (1979) *Dev. Ind. Microbiol. 20*, 367–379.

Chang, L. T., Terasaka, D. T., and Elander, R. P. (1982) *Dev. Ind. Microbiol. 23*, 21–29.

Chesson, A., Morgan, J. J., and Codner, R. C. (1978) *Trans. Brit. Mycol. Soc. 70*, 345–361.

Cole, M. (1966) *Appl. Microbiol. 14*, 98–104.

Crawford, K., Heatley, N. G., Boyd, P. F., Hale, C. W., Kelly, B. K., Miller, G. A., and Smith, N. (1952) *J. Gen. Microbiol. 6*, 47–52.

Day, L. E., and Ellis, L. F. (1971) *Appl. Microbiol. 22*, 919–920.

Demain, A. L., and Kennel, Y. M. (1978) *J. Ferment. Technol. 56*, 323–328.

Demain, A. L., and Newkirk, J. (1962) *Appl. Microbiol. 10*, 321–325.

Demain, A. L., Kupka, J., Shen, Y.-Q., and Wolfe, S. (1982) in *Trends in Antibiotic Research* (Umezawa, H., Demain, A., Hata, T., and Hutchinson, C. R., eds.), pp. 233–247, Japan Antibiotics Research Association, Tokyo.

Dennen, D., Allen, C., and Carver, D. (1971) *Appl. Microbiol. 21*, 907–915.

Doebeli, H., and Nuesch, J. (1980) *Antimicrob. Agents Chemother. 18*, 111–117.

Drew, S., and Demain, A. L. (1975a) *Antimicrob. Agents Chemother. 8*, 5–10.

Drew, S. W., and Demain, A. L. (1975b) *Eur. J. Appl. Microbiol. 1*, 121–128.

Drew, S., and Demain, A. L. (1975c) *J. Antibiot. 28*, 889–895.

Drew, S., Winstanley, D. J., and Demain A. L. (1976) *Appl. Environ. Microbiol. 31*, 143–145.

Elander, R. P., and Aoki, H. (1982) in *Chemistry and Biology of β-lactam Antibiotics*, Vol. III (Morin, R. B., and Gorman, M., eds.), pp. 83–146, Academic Press, New York.

Elander, R. P., Corum, C. J., DeValeria, H., and Wilgus, R. M. (1979) in *Second International Symposium on the Genetics of Industrial Microorganisms* (McDonald, K. D., ed.), pp 253–271, Academic Press, New York.

Fawcett, P. A., Loder, P. B., Danoan, M. J., Beesley, T. J., and Abraham, E. P. (1973) *J. Gen. Microbiol. 79*, 293–309.

Fawcett, P. A., Usher, J. J., Huddleston, J. A., Bleaney, R. C., Nisbet, J. J., and Abraham, E. P. (1976) *Biochem. J. 157*, 651–660.

Felix, H. R., Peter, H. H., and Treichler, H. J. (1981) *J. Antibiot. 34*, 657–675.

Feren, C. J., and Squires, R. W. (1969) *Biotech. Bioeng. 11*, 583–590.

Fujisawa, Y., and Kanzaki, T. (1975a) *J. Antibiot. 27*, 372–378.

Fujisawa, Y., and Kanzaki, T. (1975b) *Agr. Biol. Chem. 39*, 2043–2048.

Fujisawa, Y., Kihachi, M., and Kanzaki, T. (1977) *J. Antibiot. 30*, 775–777.

Gams, W. (1971) *Cephalosporium-artige schimmelpilze (Hyphomycetes)* Gustav Fischer, Stuttgart, West Germany.

Greaves, H. (1975) *Aust. J. Biol. Sci. 28*, 315–322.

Gygax, D., Doebeli, H., and Nuesch, J. (1980) *Experientia 36*, 487.

Hamlyn, P. F. (1982) "Protoplast Fusion and Genetic Analysis in *Cephalosporium acremonium*," Ph.D. thesis, University of Nottingham, Nottingham, England.

Hamlyn, P. F., and Ball, C. (1979) in *Genetics of Industrial Microorganisms* (Sebek, O. K., and Laskin, A. I., eds.), pp. 185–191, American Society for Microbiology. Washington, D.C.

Hamlyn, P. F., Birkett, J. A., and Peberdy, J. F. (1984) *J. Gen. Microbiol.* (in press)

Hinnen, A., and Nuesch, J. (1976) *Antimicrob. Agents Chemother. 9*, 824–830.

Hoover, J. R. E., and Dunn, G. L. (1979) in *Med. Chem.* 4th ed., Part II, (Wolff, M. E., ed.), Chapter 15, John Wiley, New York.

Hoover, J. R. E., and Nash, C. H. (1982) in *Antibiotics, Chemotherapeutics and Antibacterial Agents for Disease Control* (Bryson, M., ed.), John Wiley, New York.

Huber, F. M., and Redstone, M. O. (1967) *Can. J. Microbiol. 13*, 332–334.

Jaramillo, A., and Kuenzi, M. T. (1980) *Experientia 36*, 1455–1460.

Jayatilake, G. S., Huddleston, J. A., and Abraham, E. P. (1981) *Biochem. J. 194*, 645–647.

Kennel, Y. M., and Demain, A. L. (1978) *Exp. Mycol. 2*, 234–238.

Kitano, K., Fujisawa, Y., Katamoto, K., Nara, K., and Nakuo, Y. (1976) *J. Ferment. Technol. 54*, 712–719.

Kohsaka, M., and Demain, A. L. (1976) *Biochem. Biophys. Res. Comm. 70*, 465–473.

Komatsu, K., and Kodaira, R. (1977) *J. Antibiot. 30*, 226–233.

Komatsu, K., Mizuno, M., and Rodaira, R. (1975) *J. Antibiot. 28*, 881–888.

Kuenzi, M. T. (1980) *Arch. Microbiol. 128*, 78–83.

Lah, S., and Saksena, H. K. (1977) *Seed Res.* (New Delhi) *5*, 52–60.

Lemke, P. A. (1969) *Mycopathol. Mycol. Appl. 38*, 49–59.

Lemke, P. A., and Brannon, D. R. (1972) in *Cephalosporins and Penicillins, Chemistry and Biology* (Flynn, E. H., ed.), pp. 370–437, Academic Press, New York.

Lemke, P. A., and Nash, C. H. (1972) *Can. J. Microbiol. 18*, 255–259.

Lindsey, B. I., and Pugh, G. J. F. (1976) *Trans. Brit. Mycol. Soc. 67*, 427–434.

Martín, J. F., Revilla, G., Zanca, D. M., and Lopez-Nieto, M. J. (1982) in *Proceedings of the Trends in Antibiotic Research* (Umezuwa, H., Demain, A., Hata, T., and Hutchinson, R., eds.), Japan Antibiotic Research Association, Tokyo.

Mason, P. J., and Crosse, R. (1975) *Trans. Brit. Mycol. Soc. 65*, 129–134.

Matsumura, M., Imanaka, T., Yoshida, T., and Taguchi, H. (1980) *J. Ferment. Technol. 58*, 197–204.

Matsumura, M., and Taguchi, H. (1980) *J. Ferment. Technol. 58*, 205–214.

Matsumura, M., Imanaka, T., Yoshida, T., and Taguchi, H. (1981) *J. Ferment. Technol. 59*, 115–123.

Mehta, R. J., and Nash, C. H. (1979) *Can. J. Microbiol. 25*, 818–821.

Mehta, R. J., Speth, J. L., and Nash, C. H. (1979) *Eur. J. Appl. Microbiol. 8*, 117–182.

Mehta, R. J., Speth, J. L., Oh, Y. K., and Nash, C. H. (1981) *Dev. Ind. Microbiol. 21*, 227–232.

Nash, C. H. (1979) *Ann. Rev. Ferment. Proc. 3*, 197–214.

Nash, C. H., and Huber, F. M. (1971) *Appl. Microbiol. 22*, 6–10.

Nash, C. H., and Pieper, R. L. (1974) *Mycopath. Mycol. Appl. 54*, 369–376.

Nash, C. H., De la Higuera, N., Neuss, N., and Lemke, P. A. (1974) *Dev. Ind. Microbiol. 15*, 114–123.

Neusch, J., Treichler, H. J., and Liersch, M. (1973) in *Genetics of Industrial Microorganisms*, Vol. II, (Vanek, Z., Hostalek, Z., and Cudlin, J., eds.), pp. 309–334, Elsevier, Amsterdam.

Neuss, N., Nash, C. H., Lemke, P. A., and Grutzner, J. B. (1971) *J. Amer. Chem. Soc. 93*, 2337–2339.

Neuss, N., Nash, C. H., Baldwin, J. F., Lemke, P. A., and Grutzner, J. B. (1973) *J. Amer. Chem. Soc. 95*, 3797–3798.

Newton, G. G. F., and Abraham, E. P. (1956) *Biochem. J. 62*, 651–660.

Niss, H. F., and Nash, C. H. (1973) *Antimicrob. Agents Chemotherapy 4*, 474–478.

Oleniacz, W. S., and Pisano, M. A. (1968a) *Appl. Microbiol. 16*, 90–96.

Oleniacz, W. S., and Pisano, M. A. (1968b) *Lloydia 31*, 197–201.

Pan, C. H., Vost-Speth, S., McKemp, E., and Nash, C. H. (1982) *Dev. Ind. Microbiol. 23*, 315–323.

Pisano, M. A. (1963) *Trans. N.Y. Acad. Sci. 25*, 716–730.

Pisano, M. A. (1970) *Antonie van Leeuwenhoek 36*, 455–457.

Pisano, M. A., and Capone, J. J. (1967) *Dev. Ind. Microbiol. 8*, 417–423.

Pisano, M. A., Oleniacz, W. S., Mason, R. T., Fleischman, A. I., Vaccaro, S. E., and Catalano, G. R. (1963) *Appl. Microbiol. 11*, 1111–1115.

Pitts, G. Y., and Cowley, G. T. (1974) *Mycologia 66*, 669–675.

Propst-Ricciuti, C., and Kenney, C. M. (1975) *Dev. Ind. Microbiol. 17*, 233–240.

Queener, S. W., and Ellis, L. F. (1975) *Can. J. Microbiol. 21*, 1981–1996.

Queener, S. W., and Neuss, N. (1982) in *Chemistry and Biology of β-lactam Antibiotics*, Vol. III. (Morin, R. B., and Gorman, M., eds.), pp. 1–81, Academic Press, New York.

Queener, S. W., McDermott, J., and Radue, A. B. (1975) *Antimicrob. Agents Chemotherapy 5*, 646–651.

Raituair, A. G., and Tarasov, K. L. (1975) *Mikol. Fitopatol. 9*, 203–207.

Rowlands, R. T. (1982) in *Fungal Technology*, Vol. IV, Filamentous Fungi (Smith, J. E., Berry, D. R., and Kristiansen, B., eds.), pp. 346–372, Edward Arnold, London.

Sabet, K. A., Zaher, A. M., Samra, A. S., and Monsour, I. M. (1970) *Ann. Appl. Biol. 66*, 257–263.

Sardinas, J. L., and Pisano, M. A. (1967) *Appl. Microbiol. 15*, 277–284.

Sawicki, E. H., and Pisano, M. A. (1977) *Lipids 12*, 125–127.

Shirafugi, H., Fujisawa, Y., Kida, M., Kanzaki, T., and Yoneda, M. (1979) *Agr. Biol. Chem. 43*, 155–160.

Spry, D. O., Miller, R. D., Huckstep, L. L., Neuss, N., and Kokolja, S. (1981) *J. Antibiot. 43*, 1078–1079.

Suzuki, M., Fujisawa, Y., and Veheda, M. (1980) *Agr. Biol. Chem. 44*, 1995–1997.

Treichler, H. J., Neusch, J., and Lieisch, M. (1972) *Pathol. Microbiol. 38*, 19–21.

Yoshida, M., Konomi, T., Kohsaka, M., Baldwin, J. E., Herchen, S. S., Singh, P., Hunt, N. A., and Demain, A. L. (1978) *Proc. Nat. Acad. Sci. U.S.A. 75*, 6253–6257.

Biology of *Claviceps*

Willard A. Taber

I. INTRODUCTION

Ergot alkaloids are indole-containing alkaloids that have four rings arranged as shown in Fig. 15.1. Exceptions are the less common chanoclavines in which the top or fourth ring is not closed by formation of a covalent bond between the adjacent carbon and the methylated nitrogen. Ergot alkaloids that possess a carboxyl group at the indicated position are known as lysergic acid alkaloids if the indicated hydrogens are *trans* and as isolysergic acid ergot alkaloids if they are *cis* (Fig. 15.1). The names of the former isomers end in -ine (e.g., ergokryptine) and the latter in -inine (e.g., ergokryptinine); the former are of commercial interest. Ergot alkaloids possessing a more reduced carbon at the site of the carboxyl group are referred to as clavine ergot alkaloids; they have much less physiological activity. Certain of the ergot alkaloids are used to reduce postpartum hemorrhaging, to induce labor, and to relieve the symptoms of migraine headache. Although ergot alkaloids in the form of the purple spurs (Fig. 15.2) were used in some locations by midwives dating back to at least the 14th century, ergot was not entered into the U.S. Pharmacopoeia until 1820 and into the British Pharmacopoeia until 1836. Alkaloids that are amide derivatives of lysergic acid are: ergine, lysergic acid *alpha*-hydroxyethylamide,

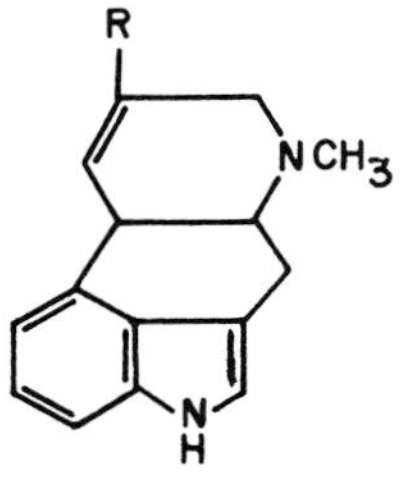

CLAVINE
ERGOT ALKALOID

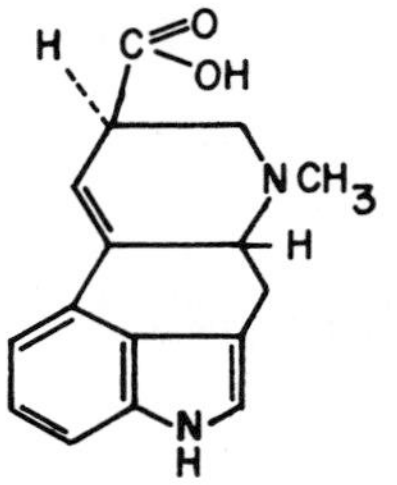

LYSERGIC ACID
ERGOT ALKALOID

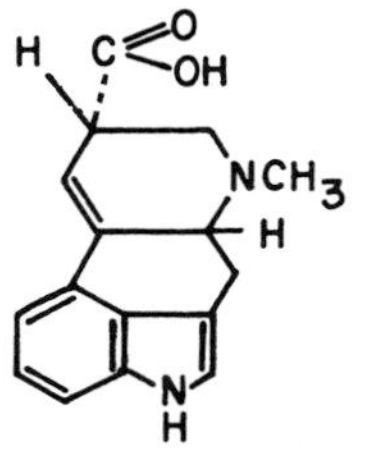

ISOLYSERGIC ACID
ERGOT ALKALOID

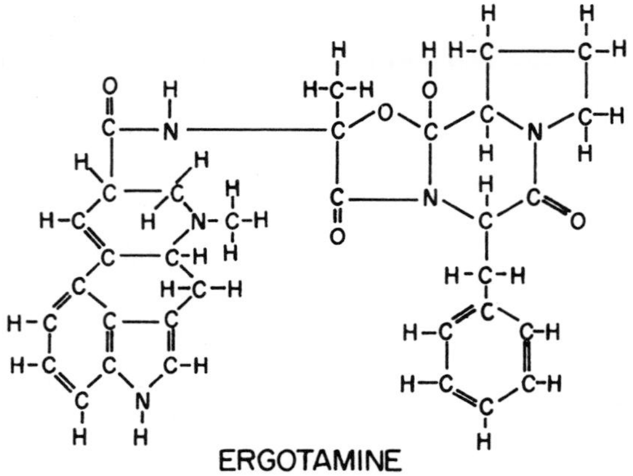

ERGOTAMINE
PEPTIDE LYSERGIC ACID ALKALOID

FIGURE 15.1   Structures of the clavine-type ergot alkaloids that do not possess a carboxyl group, the two isomers of lysergic acid, and an example of a peptide type lysergic acid ergot alkaloid.

ergometrine, ergovaline, ergotamine, ergosine, ergocristine, ergocornine, ergokryptine, *beta*-ergokryptine, and ergostine. While some cause contraction of the smooth muscles (hence their action on capillaries), the ergotoxine group has an inhibitory effect on the sympathetic nervous system. Dihydroderivatives of certain of the peptide lysergic acid alkaloids are used to treat depression and similar symptoms in the elderly. Certain of the ergot alkaloids and derivatives cause marked regression in experimental mammary tumors and bromoergocryptine is used to treat acromegaly and pituitary tumors. Some of the amide derivatives cause euphoria, and the semisynthetic lysergic acid diethylamide (LSD-25) is probably the most potent. Such compounds eventually may find use in human chemotherapy. Many thousands of people have died and many

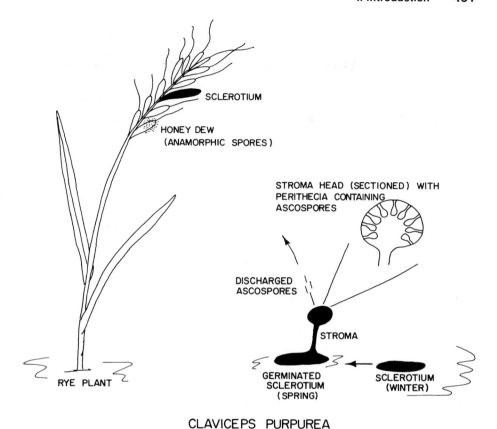

SCLEROTIUM

HONEY DEW
(ANAMORPHIC SPORES)

STROMA HEAD (SECTIONED) WITH
PERITHECIA CONTAINING
ASCOSPORES

DISCHARGED
ASCOSPORES

STROMA

RYE PLANT

GERMINATED
SCLEROTIUM
(SPRING)

SCLEROTIUM
(WINTER)

CLAVICEPS PURPUREA

FIGURE 15.2   Life cycle of *Claviceps purpurea* parasitizing rye.

more have been crippled over the centuries from eating bread made from ergotized rye flour when rye was used unintentionally or intentionally along with wheat and then in place of wheat in those parts of Europe where a frequently moist and cool climate favored growth of the rye plant over that of wheat (Christensen, 1975). Ergotism can assume the form of convulsions or gangrene, the latter because of the ability of some ergot alkaloids to cause contraction of smooth muscles. Convulsions are often accompanied by hallucinations, and it has been suggested that it is the result of vitamin D deficiency. Ergotism was thought by some to be a punishment by God, and it became known variously as "St. Anthony's fire" or "holy fire." The Tulasne brothers finally proved the purple body present in some rye grain to be a fungus. For historical accounts, see Barger (1931), Ramsbottom (1953), Rolfe and Rolfe (1966), and Bove (1970). Renewed interest has developed recently following reexamination of the circumstances of the Salem witchcraft trials of 1692. This reexamination prompted the theory that the hysteria that caused some girls to accuse some of the citizenry of being witches was brought about by the unintentional consumption of ergotized rye (Matossian, 1982, Caporael, 1976). For a countercharge, see Spanos and Gottlieb (1976). Other explanations for this hysteria have also been advanced (Demos, 1982). Commercial ergot alkaloids are extracted either from ergot spurs collected from fields that have been intentionally infected with *Claviceps purpurea* (Fr.) Tul. or from mycelium grown in large fermentors. In both cases, special laboratory-developed strains are employed. Fermentor yields are probably in excess of 4 g/l of culture. Ergot alkaloids are also produced by a few other organisms, namely, species and strains of *Penicillium* (Vining, 1973; Vining et al., 1982; Taber and Vining, 1960; Scott et al., 1976), *Aspergillus* and *Rhizopus* (El-Rafai et al., 1970; Spilsbury and Wilkinson, 1961), *Balansia* (Bacon et al., 1981), certain wild morning glories (Hofmann and Tscherter, 1960), and two varieties of commercial morning glories (Taber et al., 1963). New ergoline alkaloids continue to be discovered (Vining et al., 1982; Bianchi et al., 1982; Scott et al., 1976; Abe et al., 1970; Ohmomo et al., 1975; Tscherter and Hauth, 1974; Atwell and Mantle, 1981).

II. TAXONOMY

Claviceps is a pyrenomycetous ascomycete fungus placed in the order Clavicipitales and family Clavicipitaceae (Rogerson, 1970). *Claviceps* is the type genus of this family, which embraces 23 genera. *Claviceps* produces filiform ascospores (sexual spores or teleomorphic spores) that are discharged violently from the ascus (Fig. 15.2). The asci are borne in a fleshy perithecium that is embedded in a perithecial stroma. The stroma develops in the spring from the germinating sclerotium, which does not survive in soil more than a few months (Cunfer and Seckinger, 1977). Asexual or anamorphic spores are also produced, and they are formed on the conidial stroma that develops in the freshly infected rye floret just before the sclerotium that contains the ergot alkaloids is produced. The ostiolate perithecium possesses periphyses in the perithecial neck, and the asci arise from a distinct layer of hyphae situated on the floor of the flask-shaped perithecium. *Claviceps* is a homothallic fungus; that is, all stages of the life cycle can develop from one germinated ascospore. Typically, the ascogonium or female structure receives a nucleus from the male antheridium as evidenced by the sudden doubling of nuclei in ascogenous hyphae that arise from the ascogonium (Oddo and Tonolo, 1967). In an important species, *Claviceps paspali*, no such rapid increase in numbers of nuclei takes place, and it is believed that fertilization occurs by autogamy; antheridia are not produced.

TABLE 15.1 Species of *Claviceps*

C. annulata	C. nigricans
C. cinerea	C. orthocladae
C. cynodontis	C. paspali
C. cyperi	C. phalaridis
C. diadema	C. platytricha
C. digitariae	C. purpurea
C. euphorbi oides	syn. C. microcephala,* C. sesleriae,
C. flavella	C. setulosa, C. litoralis
C. fusiformis	C. pusilla
C. gigantea	C. queenslandica
C. glabra	C. ranuncoloides
C. grohii	C. sulcata
C. hirtella	C. tripsaci
C. inconspicua	C. uliana
C. lutea	C. viridis
C. maritima	C. yanagawensis
C. maximensis	C. zizaniae

* May be synonymous with *C. fusiformis* (Siddiqui and Khan, 1973).

Depending upon the authority, there are approximately 30–36 species of *Claviceps* (Table 15.1). According to Bove (1970), *Claviceps* parasitizes approximately 600 species of plants with about 100 additional plants parasitized by a sphacelial (asexual) phase that is assumed to be an anamorphic state of *Claviceps*. These plants are the sources of the *Claviceps* strains that are screened for alkaloid production in the laboratory. These host plants are monocots in the families Gramineae (600 species), Juncaceae (4 species), and Cyperaceae (17 species). The most important commercial species, *C. purpurea*, parasitizes ~400 species of the Gramineae, most of which are members of the tribes Festuceae, Hordeae, Aveneae, and Agrostideae. Except for *C. purpurea*, most species — as currently visualized — are tribe-specific (Loveless, 1971). The type(s) of alkaloid produced by a given *Claviceps* is characteristic of that culture; hence knowledge of the host is required to locate a given *Claviceps*. If one desires isolates of *C. paspali* (which are high producers of simple amides), one must go to paspalum grass. Physiological races are said to exist, but some doubt exists as to whether the best criteria are presently employed to define and delimit species and races. *C. paspali* differs from all other species in possessing a yellowish tan cauliflower-shaped sclerotium rather than the purplish banana-shaped sclerotium, and it has been suggested that this species be transferred to the genus *Mothesia* (Oddo and Tonolo, 1967). The size of anamorphic spores of *C. purpurea* grown on various species of grasses differs significantly from one another (Loveless, 1971), a feature suggesting that dimensions of spores in honey dew appearing on infected plants would not suffice to identify a species.

For a discussion of species, see Langdon (1954). *C. purpurea* isolated from wild grass (*Arthraxon lancifolius*) produced different shaped sclerotia and also slightly different peptide alkaloids when inoculated onto rye (Janardhanan et al., 1982).

III. LIFE CYCLE

A. Infection of Host

Infection of the plant floret occurs in the spring when the glumes of the florets nave opened and the ascospores from "synchronized" *Claviceps* are wafted onto the stigma of the style of the female plant ovary. A comprehensive account of infection has been detailed by Luttrell (1980), some aspects of which follow (Fig. 15.3). The germ tube of the ascospore grows intercellularly down to the ovary. Infection is quite rapid; the stigma is penetrated within four hours, and within two days intercellular hyphae appear in the inner wall of the ovary. By day 3, hyphae have extended from the ovary wall to the ovule, which they grow around. Noticeable disintegration of plant cells commences from this day. Glucanases and pectic enzymes apparently destroy the cells (Shaw and Mantle, 1980). Hyphae also extend downward into the central vascular bundle to the level of the origin of the lodicules. By the fourth day, hyphae have replaced ovary wall cells, although the ovule has not yet been invaded. Fungal tissue breaks through the ovary cuticle, and these hyphae produce the sphacelial phialides that bear the conidia of the honey dew; this tissue *is* the co-nidial stroma, and, as will be seen, it differs metabolically from the sclerotium, which next develops and is the site of ergot alkaloid synthesis. By the fifth or sixth day, plectenchymatous fungal tissue has replaced plant tissue of the ovary, and the fluid honey dew is noticeable. The ovule, though apparently in-tact, is embedded in fungal conidial tissue. Xylem cells in the tip of the rachilla are destroyed, and sclerotium tissue begins to develop at the foot of the tissue and begins pushing the ovule, ovary cap, and conidial stroma outward (Fig. 15.3). This foot takes up the nutrients even after the sclerotium has formed (Bacon and Luttrell, 1982). Conidia in honey dew ooze from the floret by days 6–10. By the twelfth day the sclerotium protrudes from the floret, bearing the withered remnants of the conidial stroma at the tip. The sclerotium possesses the general shape of the fruit, qualifying the disease as a replacement tissue disease (Luttrell, 1980); the size can vary, however (Frey and Brack, 1968). The host can exert some influence on the kind of alkaloid produced (Kybal and Kleinerova, 1973). Throughout these 28 or so days, hyphae do not penetrate downward beyond the rachilla; this may be due to formation of polyphenols that inactivate pectolytic enzymes via polyphenol oxidase activity (Shaw and Mantle, 1980) and are inhibitory to some fungi (Barz et al., 1976). The disease can spread throughout a field of rye or other grain by the honey dew conidia,

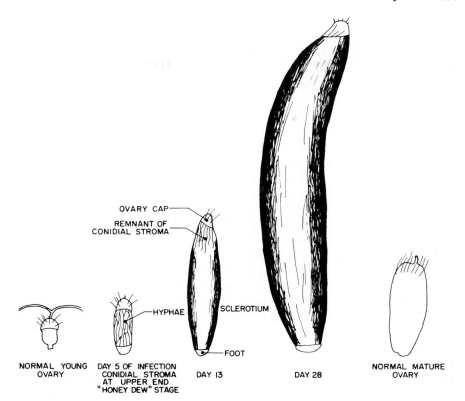

CLAVICEPS   PURPUREA

FIGURE 15.3   Development of *Claviceps purpurea* in the ovary of the rye floret. Ergot alkaloids are produced in the sclerotium. The conidial stroma forms first, and then below it the sclerotium begins to develop, pushing the older conidial stroma and the remnants of the ovary to the tip.

which are carried by flies attracted by the sweet fluid. In commercial production in the field, machines carrying needles dipped in laboratory-produced conidial suspensions are used to inoculate florets of the appropriate age. With the exception of *C. paspali* (Luttrell, 1977) the sclerotium becomes purple in the outer rind, the inner cortex remaining white or tinted with pink. Ergot alkaloids are formed; these are degraded by light, but the purple pigment may protect them, a feature suggesting that the alkaloids offer survival value.

B. Honey Dew
Honey dew consists of fungal conidia suspended in a sugary fluid having an osmotic pressure 25 times that of neighboring plant tissue (Mower and Hancock,

1973). This low water content may prevent the conidia from germinating before they have been transported to fresh host tissue. The fluid contains some 19 amino acids (Gröger, 1958), and the sugars described below are produced by the fungus from sucrose; pure cultures grown in vitro also produce them. Sucrose is the principal carbon source supplied by the plant (Bassett et al., 1972). The honey dew sugars consist mainly of fructose and fructose oligosaccharides. Present are fructose oligosaccharides containing arabinitol or mannitol, trifructosyl mannitol, 6-0-*beta*-D-fructofuranosyl-D-glucose, trisaccharides and tetrasaccharides containing glucose and fructose, free sucrose, glucose, fructose, D-arabinitol, D-mannitol, and others (Mower and Hancock 1973; Mower and Hancock, 1975; Arcamone et al., 1970a). The high sucrose gradient in the conidial stroma causes translocation by diffusion from the host to the fungus (Mower and Hancock, 1975). When sucrose is used by the fungus, glucose is released as a consequence of transferring fructose to other fructose molecules by a *beta*-fructofuranosidase, which is not a normal invertase but rather is cell-bound and inducible (Bassett et al., 1972; Dickerson, 1973). Free fructose is used when glucose is no longer present. For a contrast between the fungal tissue that has replaced the ovary and mature grain, see Kent-Jones (1967) and Bland (1971). Honey dew and mature sclerotia of *C. paspali* are often coral colored because of growth of *Fusarium heterosporum*, which produces the trichothecene mycotoxins (Cole et al., 1981). Germination of diluted honey dew spores is promoted by germinating pollen (Williams and Colotelo, 1975) and starch, glycerol, sucrose, glucose, and lactose (Hiruma et al., 1961). Under other conditions, such nutrients do not favor germination (Garay, 1956). The optimum pH for germination is 4.8, and germination is best at a pressure of 3.8 atm (Garay, 1956). These conidia are at first mononucleate, but in time some become multinucleate (Jung and Rochelmeyer, 1960). Conidia also can be produced in vitro (Pažoutová et al., 1978). Honey dew spore dimensions are of some use in identifying the species of Claviceps (Loveless, 1964).

C. Conidial Stroma and Sclerotium

The fungal tissue succeeds in diverting plant nutrients into it (Corbett et al., 1974; Mower and Hancock, 1975; Dickerson, 1976). The last group demonstrated this in experiments in which plants bearing infected and uninfected florets were exposed to $^{14}CO_2$; most of the radioactivity was found in the water-soluble fractions of the fungus, most activity being in mannitol and lesser amounts in trehalose. Seeds adjacent to infected florets were smaller, again indicating the ability of the fungus to divert much of the translocatable nutrients of the plant to it. This phenomenon has been observed in other seed diseases (see Dickerson et al., 1976). Mantle's group (Nisbet et al., 1977; Corbett et al., 1974) compared the sphacelial stage that is borne on the conidial stroma with the subsequently developing sclerotium. The conidial stroma is the fungal tissue that is present for the first 6–10 days of infection. As described above, the sclerotium develops below it from the hyphal foot lying over the vascular system and is recognizable by approximately 14 days. The shape may be determined in part by phosphate starvation (Kybal, 1964). Respiration of the two tissues (stroma and sclerotium) was examined using ^{14}C-labeled substrates and excised tissues of various ages in the Warburg respirometer. Stromal tissue (that up to 14 days) possessed a low endogenous respiration but also possessed a high RQ (respiratory quotient, μl $CO_2/\mu l$ O_2) of 1.0, which indicates sugar respiration.

$$C_6H_{12}O_6 + 6\ O_2 \longrightarrow 6\ CO_2 + 6\ H_2O; \quad RQ = 6/6 = 1.0$$
Glucose

$$C_8H_{16}O_2 + 11\ O_2 \longrightarrow 8\ CO_2 + 8\ H_2O; \quad RQ = 8/11 = 0.72$$
Fatty acid

The transition from sphacelial stroma to sclerotia was characterized also by a fall in RNA, malate, polyols, and phosphate concentrations (Corbett et al., 1974). The RQ decreased with time to 14 days, this decrease indicating that the new tissue (sclerotium) was respiring lipids. By 50 days the RQ had decreased to 0.5–0.67. This mature sclerotium was 30–50% lipid. This change in substrate utilization with the change from predominance of conidia-bearing stroma to alkaloid producing sclerotium was attributed at least in part to the fact that the sclerotium became progressively isolated from the source of sucrose—the host. As will be seen later, the information obtained from these host–parasite relationships is translatable to metabolism of in vitro–grown alkaloid-producing cultures (Řeháček, 1974; Řeháček and Kazova, 1975). Sclerotial development, then, includes synthesis of massive amounts of triglycerides, including the unique ricinoleic acid, from sucrose as well as their utilization in respiration. This respiration as measured by O_2 uptake is inhibited by exogenous fumarate (Garay, 1958), and the inhibition is reversed by added succinate. It was later found that during this period the amount of both malate (Corbett et al., 1974) and malate dehydrogenase decreased drastically (Kleinerova and Kybal, 1980). The loss of malate dehydrogenase may be equivalent to adding exogenous fumarate, suggesting regulation by accumulation of certain acids of the citric acid cycle; backup of the carbon train may favor ergot alkaloid synthesis by causing accumulation of sufficient amounts of one of the acids to block the condensing enzyme (Fig. 15.4), thus shunting acetyl-CoA from the cycle to mevalonate and lipds, the former of which is an intermediate in alkaloid biosynthesis (Řeháček et al., 1975; Kybal et al., 1981).

Although the sclerotial tissue respires largely at the expense of lipids, excised pieces possess a greater *capacity* for sugar respiration than does the conidial stroma. This was determined (Nisbet et al., 1977) by measuring gas exchange and respiration of ^{14}C-sugars by excised tissues of various ages. The exogenous

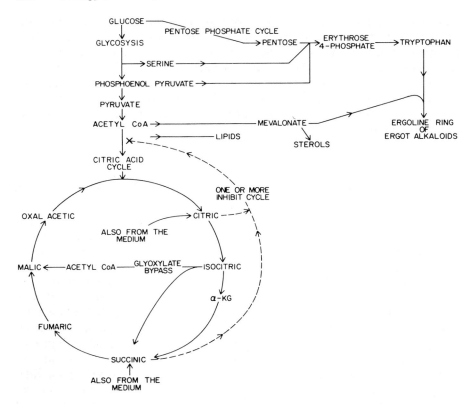

FIGURE 15.4   Scheme representing draw-off of intermediates for synthesis of ergot alkaloids and possible explanation for stimulation of production by adding citric acid intermediates to the production medium.

plus endogenous RQ of maturing sclerotial tissue increased to 2.0 or higher. This was attributed to fermentation of part of the added sugar, since comparison of the yield of $^{14}CO_2$ with the actual gas exchange revealed that O_2 uptake was not sufficient to account for total oxidation to CO_2, assuming mole for mole exchange of O_2 and CO_2. Unlike animal and plant tissues, rapidly growing fungi generate a RQ of 2.0 or better; this is probably due to the fact that fungi cannot take up oxygen at a rate sufficient to keep pace with respiration.* As a result, some of the NADH is oxidized by passing the protons and electrons to a sugar, thus generating polyols, which are of almost universal occurrence in fungi (Lewis and Smith, 1967; Pfyffer and Rast, 1980). The vigorous respiration of exogenous sugars eventually declines after completion of sclerotium development. On a weight basis the conidial stroma possesses the highest

---

* Working with laboratory shake flask cultures, Schultz (1964) demonstrated that even a cotton plug retarded oxygen uptake.

endogenous respiration, but cells in the transition stage and young sclerotial tissue are capable of higher respiration of exogenous sugars. Mannitol, which is employed in fermentation media, where it is slowly used for growth, is not rapidly respired. Access to host sucrose and accumulation of it may explain the low response of excised conidial stroma to exogenous glucose or sucrose. These studies identify some of the metabolism that precedes and is commensurate with alkaloid production in infected tissue; this will be considered in relation to alkaloid synthesis later.

D. Mature Sclerotium

The banana-shaped sclerotium or spur (Figs. 15.2 and 15.3) is dark purple, although albino forms have been observed (Tonolo et al., 1961). The mature sclerotium contains ergot alkaloids (up to 0.5% of the air-dried weight of the sclerotium), and the alkaloid does not disappear during germination. Almost all field sclerotia contain ergot alkaloids, yet few isolates obtained from them produce them in vitro (Corbett et al., 1974). They also contain the unusual hydroxylated fatty acid, ricinoleic acid, whose presence in grains is a certain indication of contamination by *Claviceps* (Mantle, 1977). Approximately 30–50% of the sclerotium is composed of triglyceride oil, and some 30–40% of the fatty acids of this oil is ricinoleic acid. By contrast, uninfected rye grain is 1.7% oil or fat (Kent-Jones, 1967). The ergot alkaloids present in the sclerotium may differ from those produced by pure cultures in vitro (Breuel et al., 1982; Janardhanan et al., 1982). *C. purpurea* DM 595 produced ergoline, ergotamine, and ergocristine when grown on rye but ergocornine and ergokryptine in vitro. With respect to in vitro production, Gröger (1972) recognizes three principal chemical races of *Claviceps*: the *C. purpurea* type, which produces peptide lysergic acid ergot alkaloids and clavines; the *C. paspali* type, which produces clavines and simple amides and is hence of commercial importance; and the *Pennisetum-Agropyron* type, which produces clavines and cannot convert elymoclavine to lysergic acid (i.e., convert the reduced carbon to carboxylic acid). Abe and Yamatodani (1964) recognize five strains. In naturally infected plants, total alkaloid content often is variable even between sclerotia on the same plant head; in one study, the percentages ranged from 0.01% to 0.45% (Young, 1981a, b). This difference may be accounted for by the fact that florets are not all infected by honey dew conidia from the same plant that was infected earlier in the spring by ascospores. The kinds of alkaloid present in sclerotia collected in western Canada differed from those of eastern Canada. Such variation may or may not justify recognition of *Claviceps* races (Darlington et al., 1977). Mutants or races can develop in the field even when the plants have been inoculated with spores from a pure culture (Mantle, 1967). Even when artificially infecting fields, yields can vary from year to year (Bojar et al., 1967); environmental conditions could be expected to play a role here. The mature sclerotium contains many compounds in addition to alkaloids and oils. Among these are mannitol, arabitol, inositol, trehalose, ergothioneine (Heath and Wildy, 1956), most of the natural amino acids, and pigments. In general, the carbohydrate content is low. Pure cultures of *Claviceps* can be isolated from sclerotia, if not over a year old, by teasing out the white to pink internal tissue after first treating the sclerotium as follows. The whole sclerotium is immersed in 70% ethanol for a few seconds to wet the surface, immersed in 10% Clorox for ~1–2 min, and washed in sterile water; the free water is removed by rolling the sclerotium in a dry paper towel that has been previously autoclaved; and the sclerotium is broken open by use of a forcep and dissecting needle. The fragment of hyphae removed from the sclerotium then can be transferred to an agar medium such as potato dextrose agar. "Identification" of the resulting colony is usually made by noting the absence of traits typical of contaminants (e.g., absence of sporulating structures typical of other kinds of fungi) and noting the colony type that is most prevalent. Young colonies are usually white and cottony; and if they do sporulate, the spores will be colorless and of irregular length. They can also be tested for authenticity by determining whether or not they can infect the appropriate grass. Sclerotia in the laboratory or in soil do not remain viable for more than a few months (Cunfer and Marshall, 1977; Cunfer and Seckinger, 1977).

E. Perithecial Stroma

While respiring most of the time, the sclerotium appears quiescent over late summer, fall, and winter, but in the spring the sclerotium that had fallen earlier to the ground or remained trapped in stubble germinates. The term "germination" refers to the upward growth of the perithecial stroma (not to be confused with the conidial stroma from the previous year) from the sclerotium (Fig. 15.2). This stroma consists of a stalk or stipe and a head or capitulum, which consists of plectenchymatous tissue bearing the sunken perithecium, which in turn bears the sexual or teleomorphic ascospores in the asci. The young capitulum contains free female ascogonia and male antheridia (except for *C. paspali*). As the fertilized ascogonium, ascogenous hyphae, and, ultimately, asci develop, surrounding hyphae grow up and around them, forming a flask-shaped enclosure, the perithecium, which remains embedded in the capitulum. The completed perithecium possesses an opening, the ostiole, which allows the violently discharged ascospores to escape from the perithecium into the air, where they can be carried to exposed florets of the appropriate host. This perithecial stroma resembles the conidial stroma in respiring sugars but, in addition, contains high concentrations of polyols that are osmotically active (Yancey et al., 1982) and that may create the hydrostatic system required to discharge the ascospores (Cooke and Mitchell, 1970).

The sclerotium, in early spring, cannot germinate until after it has taken up water and then has been chilled at 1–10° C for 6–18 weeks (Mitchell and Cooke, 1968a, b). While the plant hormone auxin reduces chilling time (Cooke and Mitchell, 1970), nothing happens that leads to germination unless both treatments have taken place in the proper order. In practice, laboratory germination is achieved by soaking sclerotia with or without surface sterilization by Clorox or $HgCl_2$ overnight in water, then placing them on moist sand or filter paper and incubating a few weeks in the refrigerator. Chilling activates lipid metabolism and growth, ultimately leading to formation of the stroma, which begins under the rind. Considerable water is taken up during the soaking, and some sugars and polyols are lost by leaching. Increase in water content at the end of the chilling period markedly increases on about the fifteenth day (Mitchell and Cooke 1968a, b), and respiration (as represented by O_2 uptake and CO_2 evolution) increases rapidly on about the twentieth day after the chilling treatment is over.

The rate of lipid loss as measured by weight loss of the sclerotium increases sharply in about 10 days. The first evidence of perithecial stroma pushing through the outer rind of the sclerotium appears on about day 20–23. The RQ of activated sclerotia is ~0.7–0.8, indicating use of lipids in respiration (Garay, 1958; Mitchell and Cooke, 1968a, b). One theory holds that the lipid content of the activated sclerotium can decrease by 60%, which is equivalent to 25% of the dry weight. Another holds that unless a large number of stromata are being produced, the loss is less as one stroma draws upon nutrients only from the region immediately below it (Corbett et al., 1975; Oddo and Tonolo, 1967); the alkaloid content does not decrease, however (Corbett et al., 1975). *C. paspali* sclerotia do not require low temperature, but germination is much higher following two months storage at 5–10°C followed by incubation at 20° for four days (Cunfer and Seckinger, 1977). Development of the perithecial stroma after the capitulum has broken through the rind is favored by a slightly higher temperature than for chilling, that is, 10–25° (Mitchell and Cooke 1968a, b). At this time the amino acid content of the sclerotium increases, especially alanine, threonine, serine, and glutamate (Corbett et al., 1975). Drastic changes occur within the perithecial stroma as it grows and expands above the supporting sclerotium. The RQ is approximately 1.0 (Garay, 1958; Mitchell and Cooke, 1968a, b), indicating that this stroma, as does the conidial stroma, operates via sugar metabolism. Polyols are synthesized and accumulate, but their utilization is slow. Alanine accumulates to a concentration seven times that of the sclerotium, which is already quite high (Corbett et al., 1975). While conidial stroma, sclerotium, and perithecial stroma appear to be quite similar structurally, profound differences exist between sclerotium and stroma: (a) the stromata operate via sugar respiration, while sclerotia use lipids; (b) stromata, wherever they occur, produce spores of some kind, while sclerotia do not; and (c) ergot alkaloids are produced and stored in the heavily pigmented sclerotia.

IV. GENETICS

Claviceps is homothallic, and this might appear to preclude the possibility of obtaining recombinants. However, genetically different nuclei can be present at karyogamy as a result of mutation and perpetuation of a nucleus bearing the mutation and, more importantly, as a result of anastomosing or hyphal fusion between genetically different strains of the same species. The resulting heterokaryon possesses different nuclei just as if sexuality had taken place. Hyphal fusion readily takes place in *Claviceps* (Amici et al., 1967a). The ascogonium and antheridium originating from common hyphae could then contain genetically different nuclei. The sudden doubling of nuclei in ascogenous hyphae observed by Oddo and Tonolo (1967) was taken as evidence that nuclei were introduced into the female ascogonium by the antheridium.

Obviously for laboratory studies, it would be more convenient to use heterothallic fungi that are self-sterile (i.e., have mating types with one or more loci involved), since the two nuclei would then have to be different. Recombination has been demonstrated by Esser's group (Esser and Tudzynski, 1978; Tudzynski et al., 1982). Genetically different strains of one parent culture were obtained by using mutations to introduce genetic markers. Ultraviolet light at an exposure to give 99% kill was used to produce mutants, of which some were auxotrophic for lysine and others were resistant to fungicides. Conidia from the two lines were introduced into rye florets. The marker genes segregated in the ascospores in Mendelian pattern, as would be expected for a two-factor cross, that is, 50% parental and 50% meiotic recombinants. The markers were unlinked. To explain these results, anastomosing of hyphae (hyphal fusion) in the fungal tissue growing in the floret is required. The hyphae of both the sclerotium that developed in the floret and the perithecial stroma that developed from it were recombinants. Some of the recombinant lines produced higher yields of ergot alkaloids than either parent strain, and all produced more than the lowest-producing parent. No correlation between colony morphology and alkaloid yields was detected. All sclerotia of a single ascospore line had an identical alkaloid spectrum. Heterokaryosis appeared not to be required either for completion of the life cycle or for alkaloid production.

Some earlier studies that did not involve the sexual cycle nevertheless did involve the heterokaryotic state and/or development of mutants. Amici et al. (1967a) noticed that giant colonies developing on agar plates from a laboratory pure culture developed late morphological sectors, which appeared as wedges originating from close to the periphery of the colony rather than from the center inoculation point. Isolations from three sectors were made: V, C, and W, referring to pigmentation. Each strain produced little or no alkaloid; but when V and C were cultured together, allowing anastomosing, the resulting culture produced as much alkaloid as the parent culture from which the sectors had developed. It seemed possible that the parent was a natural heterokaryon that had segregated on the agar. Many hyphal bridges were seen in the crossing, establishing that anastomosing had occurred. Mixing the two cultures together in the production medium, which would allow little time for formation of a heterokaryon, did not result in alkaloid production equivalent to the parent. Both the parent and the partially reconstructed parent averaged five nuclei per hyphal compartment (e.g., protoplasm incompletely isolated by perforated cross walls or septa). Of course uninucleate conidia would be mono-karyotic; but in nature, heterokaryons could develop as described above. In the laboratory, heterokaryons can be preserved by using a strain that does not produce conidia, by using culture media that do not allow conidiation, or by using large amounts of the culture in transferring it. Other studies (Strnadova and Kybal, 1974; Robbers et al., 1974) report that hyphal compartments are monokaryotic and suggest that sectoring could also result from mutations in a monokaryotic culture. Such sectoring was observed in colonies of *C. purpurea* that had developed from uninucleate anamorphic conidia. Robbers et al. (1974) reported that aged hyphae or slow-growing hyphae of *C. paspali* had uninucleate cells. It should be recalled that ascomycete hyphae are not con-structed of real "cells" but rather are coenocytic, that is, they are divided into compartments by perforated septa that can allow migration of nuclei, mito-chondria, and nutrients. Nuclei can migrate even through highly organized basidiomycete hyphae (Davis, 1966). Heterokaryotic hyphae that can express genes of both nuclei and thus act as "recombinants" can also set the stage for the parasexual cycle (Roper, 1966). The parasexual cycle exists when nuclei in ordinary hyphae fuse; they can remain diploid for a time or undergo haploidi-zation. Crossing over can produce recombinants (Sermonti, 1969). These pro-cesses together with true sexuality (and probably also insertion of plasmids such as the 2-μm plasmid of yeast) provide means of developing new strains. Spalla and Marnati (1982) produced recombinants between species using pro-toplast fusion. Genetically fixed strains can also be produced simply by induc-ing mutations; in fact this method has been used to produce commercially important alkaloid-producing strains.

Mutants that produce higher alkaloid yields than the parent and that pro-duce different alkaloids have been developed (Kobel and Sanglier, 1973). The following mutagens have been employed: ultraviolet light, sodium nitrite, nitrosoguanidine, ethyl methanesulfonate, and methyl-bis-(chloroethyl)amine. (See the subsection titled "Strains of *Claviceps*" in Section V.)

Frequent serial transfer of producing strains can lead to degeneration in which substrains arise lacking the capacity to produce alkaloids in vitro and to infect the host (Kobel, 1969). Passing strains through the host appears to pre-serve certain traits, and cultures can be preserved for years by storing under mineral oil or liquid nitrogen or by freeze-drying (Hatt, 1980; Hwang, 1966; Little and Gordon, 1967; Mizrahi and Miller, 1968). However, some strains preserved under oil die within one year when refrigerated but last several years when stored at about 25°C.

V. LABORATORY CULTURES

A. Strains of *Claviceps*

Despite the fact that virtually all of the sclerotia from fields contain ergot alka-loids, most of the isolates from them do not. Consequently, production studies must have to start by screening such isolates or by obtaining strains from others. *C. purpurea* produces the pharmacologically active peptide derivatives or amide, and some *C. paspali* strains can produce simple amides from which derivatives can be made, as well as clavines (Gröger, 1972). Since this information was not available at the time, early screens examined various species from wild grasses. These screens (Abe, 1951), which were the foundations for all subsequent screens by others, detected the inactive clavines. Testing of isolates, most of which were *C. purpurea*, led to isolation of lysergic acid alkaloid producers (Taber and Vining, 1957b, 1958, 1960). Yields were quite low, but others succeeded in isolating high-producing strains of lysergic acid alkaloids (Arcamone et al., 1961; Amici et al., 1966, 1967b). Screens generally employ cultures grown in liquid media that are incubated either stationary or with shaking. Alkaloids can be in the mycelium, culture filtrate, or both (usually 90% mycelial). Consequently, the whole culture is extracted either directly (Amici et al., 1967a) or after freeze-drying (Vining and Taber, 1959). A wet solvent such as diethylether is made alkaline by addition of ammonium hydroxide or other alkali to ionize tryptophan so that it will not partition into the solvent, and after back-extracting into H_2SO_4 the alkaloids are reacted with the *p*-dimethylaminobenzaldehyde reagent, producing a bright blue color within seconds. A standard curve of a pure ergot alkaloid is prepared, and results are expressed as equivalents of that alkaloid. Peroxide in diethyl ether is removed by a preliminary wash with ferrous ammonium sulfate, and all solutions are protected from light. Paper chromatography employs the same reagent except that HCl is used in place of H_2SO_4, which can produce a nonspecific blue color. Screens first employed the natural isolates, then laboratory-induced mutants (above references; Mary et al., 1965; Pacifici et al., 1963; Gröger and Tyler, 1963, Strnadova, 1967, 1976; Kobel and Sanglier, 1973; Tudzynski et al., 1982; Pažoutová et al., 1978). One screen (Chi et al., 1964) detected production in 181 of 800 isolates, and another (Abe et al., 1967) in 19 out of 1000 isolates. Some laboratory strains produce several peptide ergot alkaloids (Vining and Taber, 1963b). Use of mutagens can yield mutants producing either a different alkaloid from the parent or higher yields (Kobel and Sanglier, 1973).

Mutation rates after treatment of conidia with UV light, nitrosoguanidine, and methyl-bis-(chloroethyl)amine ranged between 0.1% and 1%, and the mutants tended to be stable and to retain their virulence toward the host plant. Wherever possible, young (hence uninucleate) conidia should be used; most strains will produce these on the appropriate medium. The medium composition has to be determined by trial and error. Some strains sporulate in test tube slants consisting of a mannitol-soybean medium. Many papers referred to in this section and Luttrell (1980) describe media used in production of conidia on agar slants or in shake flask cultures. If spores are not produced, one can briefly homogenize mycelia obtained from shake flask cultures in a Waring blender and mutagenize these hyphal fragments (Taber and Vining, 1959a). Numbers of nuclei per compartment or hyphal tip can be determined by staining (Hareven and Koltin, 1970; Robbers et al., 1974; Jung and Rochelmeyer,

1960; Pažoutová et al., 1978; Robinow and Bakerspigel, 1965; Olive, 1965), and by electron microscopy (Vorisek et al., 1974a, b; Shaw and Mantle, 1980; Luttrell, 1980).

B. Fine Structure

Claviceps is an ascomycete; and although it possesses a rather unique life cycle, vegetative hyphae appear to be quite conventional. For general information on ascomycete structure, see Ghosh (1977) and Taber and Taber (1977). Photographs of fine structure can be seen in the work of the following: Luttrell (1980), Vorisek et al. (1974a, b), and Shaw and Mantle (1980). *Claviceps* hyphae possess ascomycetous perforated septa and probably also Woronin bodies, which are always associated with ascomycetous septa and probably act as a valve or stopper for the simple septum. Vacuoles are present in older hyphae and are usually present in alkaloid-producing hyphae. Vacuoles can contain polyphosphates and amino acids (see below) and enzymes (Vorisek and Sajdl, 1981).

C. Composition and Intermediary Metabolism

Constituents of mycelium can be divided into various groups according to their formation (Table 15.2). Knowledge of composition of cells is important, as is knowledge of changes that take place with time or a particular experimental variable, because such information can indicate which pathways are in

TABLE 15.2   Groupings of Constituents and Pathways of Fungi

Constituents and Pathways	Examples
Cellular polymers: structural and catalytic	cell wall, RNA, DNA, membranes, dehydrogenases, other proteins
Intermediary metabolism	
a. balanced	pyruvate, amino acids, vitamins
b. unbalanced, overflow metabolism	excess citrate, vitamins, amino acids
Shunt* metabolism (diverted intermediates)	
a. Primary shunt metabolism (universally produced, stable production)	polyols, trehalose, polyphosphates, neutral glycerides, polysaccharides
b. Secondary shunt metabolism (synthesized from intermediates, production rare and unstable)**	alkaloids, antibiotics, pigments, mycotoxins
Degradation products and products of NADH oxidation	amines, acids, ethanol, lactate, perhaps polyols
Epicellular and extracellular enzymes	cellulase, amylase, lipase

* *In sensu* Foster (1949).
** Possibly all fungi produce some secondary metabolites; the difference may lie in the kinds synthesized and in the relative amounts (Bu'Lock, 1980).

operation and possibly which ones have changed in magnitude or direction. Correlation, or lack of it, between some pathways and ergot alkaloid production can reveal how the culture might be manipulated to increase yields. It has been shown over and over that distinct changes in composition occur with changes in nutrition such as kind and concentration of carbon or nitrogen source and amount of phosphate present (Borrow et al., 1961; Taber and Vining, 1963; Taber, 1964; Řeháček, 1974). As Bu'Lock (1975a) has stated, "the medium is the message." A prime example of changes that can occur in mycelial tissues such as sclerotium and stroma that have the same genes has already been considered. The following paragraphs describe various aspects of hyphal composition.

Cell Wall.   Unlike animals, fungi have cell walls that function in part at least as corsets and allow a cell to grow in osmotically insulting media such as are required for natural product formation. Gröger's group (Schmauder and Gröger, 1978; Schmauder et al., 1978) has compared the composition of cell walls of strains producing alkaloid and one not producing alkaloid. Approximately 10–15% of the purified wall was protein, and the producing strain had the highest concentration of proline, valine, and leucine (which incidentally are constituents of some of the ergot alkaloids). Serine, glutamate, threonine, and aspartate were the amino acids present in highest concentration of the 15 identified in the walls. Neutral sugars present were glucose, galactose, and mannose, glucose constituting more than half. Glucosamine was present, no doubt as chitin, which is present in virtually all ascomycetous fungi. Uronic acid made up ~5–8% of the wall, and ash made up 1.5–2%. Fatty acids were represented in the main by lauric, myristic, and palmitic acids. The composition of media was found to affect the composition of the walls, reflecting the dynamic state of microbial composition. Cell walls of *Claviceps* can be removed with snail gut enzymes (Stahl et al., 1977; Keller et al., 1980), thus generating protoplasts that could be used in genetic engineering. The protoplasts survive in suspending agents such as 0.8 M of sucrose containing 10 mM each of $CaCl_2$ and $MgCl_2$. Vacuolate protoplasts were observed to produce some alkaloid. Spalla and Marnati (1982) produced protoplasts that were capable of fusion.

Amino Acid Pool.   *Claviceps* possesses an amino acid pool that is not readily leached from the mycelium (Kirsten et al., 1966; Hartmann, 1965a, b; Berman and Younken, 1958; Steiner and Hartmann, 1964). A new amino acid, *alpha*-aminoheptanoic acid, occurs in at least some species. Regulation of biosynthesis of some amino acids has been observed. Three 2-keto-3-deoxy-D-arabinoheptulosonate-7-phosphate synthetases have been separated, each sensitive to phenylalanine, tyrosine, or tryptophan, respectively (Lingens et al., 1967). Negative feedback–like activity may be uniform, while repressionlike activity may differ with the strain. For reviews on regulation, see Demain (1968), Floss et al.,

(1974), and Vining and Taber (1979). Tryptophan can be degraded to indole-acetic acid, 5-hydroxyindole acetic acid, indole-isopropionic acid, nicotinic acid, 2,3-dihydroxybenzoic acid, and 3-hydroxyanthranilic acid (Yamano et al., 1962; Teuscher, 1965; Erge et al., 1962; Tyler et al., 1964; Gröger and Erge, 1964; Teuscher and Teuscher, 1965). High-producing strains do not divert much tryptophan to these pathways, however (Vining and Taber, 1963b); and high-producing strains take up tryptophan more rapidly than nonproducers (Teuscher, 1964).

Although amino acids of the pool are mainly fed into protein synthesis or shunted into secondary shunt metabolism, other activities can also take place. The amounts and kinds of amino acids added to an in vitro culture can determine the proportions of peptide ergot alkaloids formed (Baumert et al., 1977, 1982). Transaminase activity occurs, including a specific one transferring the tryptophan amino group to *alpha*-ketoglutarate (Teuscher, 1970). Decarboxylation of amino acids also can occur and isoamylamine, *n*-hexylamine, and traces of other amines have been detected (Hartmann, 1964, 1965a, b; Kirsten et al., 1966; Molnar and Tyihak, 1966). Amides also can be produced at least by some strains (ApSimon et al., 1965a).

Pigments.   Purplish pigments are present in all strains but *C. paspali*, and some are produced in aged stationary and shake flask cultures. Many pigments have been characterized (ApSimon et al., 1965b; Aberhart et al., 1965; Baross, 1966; Franck, 1980; Franck and Zemer, 1965; Franck et al., 1965, 1967; Kornhauser and Logar 1965; Gröger et al., 1965; Perenyl et al., 1966). At least 26 papers on ergot pigments (by Franck) have been listed in *Chemical Abstracts* since 1959. Some strains produce the colorless 2,3-dihydroxybenzoic acid, which becomes red when chelated with iron (Kelleher et al., 1971). Unlike other iron chelators (e.g., siderophores made from hydroxamic acid), the amount of chelator *increases* with increasing iron in the medium.

Carbohydrates.   Polysaccharides that accumulate extracellularly can be produced on glucose media (Perlin and Taber, 1963). The disaccharide trehalose probably is produced by all higher fungi and is present in significant amounts in *Claviceps* (Taber, 1964; Taber and Vining, 1963; Vining and Taber, 1964); concentrations as high as 14% have been observed.

Polyphosphates.   Polyphosphates (Kulaev, 1979; Harold, 1966) are also present in all higher fungi and are formed by *Claviceps* (Kulaev, 1979; Kulaev and Uryson, 1965; Taber, 1964; Taber and Vining, 1963). Kulaev detected over 25 phosphate compounds in addition to RNA and DNA. Phosphate may be transferred between polyphosphate and ADP/ATP; turnover can be low, however (Lusby and McLaughlin, 1980). Studies on other fungi indicate that polyphosphates are mainly located in vacuoles along with basic amino acids, which may bring about electrical neutrality (Duerr et al., 1979). The amount of polyphosphate present in fungal cells varies with the organism, medium, age, and other factors but usually averages about 0.07–2% of the total dry weight as phosphorus.

Polyols.   Polyols formed by reduction of sugars (Hult et al., 1980) are present in all fungi except the oomycetes (Pfyffer and Rast, 1980). Some mycologists hypothesize that this group of fungi evolved rather directly from certain algae and therefore may be expected to differ in some respects from other fungi. Mannitol is the most common (Lewis and Smith, 1967; Pfyffer and Rast, 1980). Polyols may function as food reserves (although they are often used slowly when given as exogenous carbon source), oxidants of NADH in absence of sufficient O_2 (i.e., hydrogen sink) (Taber and Siepmann, 1966), or osmotic determinants (Yancey et al., 1982; Slokoska et al., 1981; Puc and Socic, 1977). Claviceps strains produce several polyols, and the kinds vary with the medium composition and the strain (Vining and Taber, 1964). Their production is associated at least in part with rapid growth on rich media, but *Claviceps* can produce some even when growing very slowly on succinic acid (Taber and Siepmann, 1966). *Claviceps* mycelia are approximately 10–17% polyol by dry weight, whereas values for many fungi vary around the range of 1–10% (Pfyffer and Rast, 1980).

Lipids.   Neutral lipids and fatty acids usually accumulate in patterns similar to those for trehalose and polyols and would seem to fit the category of primary shunt metabolites (Table 15.2). Fungi are rather unique in accumulating free fatty acids and yet are often inhibited by a variety of exogenous fatty acids ranging from acetate and propionate up to much larger ones. Onset of alkaloid formation is associated with synthesis or accumulation of lipids. (Řeháček, 1974; Taber and Vining, 1963). Accumulation of large amounts of triglycerides usually is associated with growth on media rich in carbon source and poor in nitrogen source, at which times they may constitute 50% of the dry weight of mycelium or yeast cells (Borrow, et al., 1961). Membranes are largely composed of phospholipids; these and the neutral lipids are often in a dynamic state. When washed mycelium of *Claviceps sp.* was exposed for seven minutes to ^{14}C-succinic acid (Raju and Taber, unpublished), radioactivity appeared in various lipid fractions as follows (percent radioactivity in each fraction): free fatty acids, 4; steroids, 7; diglycerides, 9; phospholipids, 16; and triglycerides, 65. Although the above incorporations are very small, they may indicate that mycelium exposed to succinic acid switches from sugar metabolism (cells were grown on glucose before washing) to production of all constituents from succinic acid. Arcamone et al. (1961) noted a striking difference in lipid content between *Claviceps* and other filamentous fungi. The phospholipid content of

C. purpurea mitochondria, expressed as percent of total phospholipid, was found by Anderson et al. (1964) to be as follows: phosphatidylinositol 28; phosphatidylcholine 27; phosphatidylserine 13; phosphatidylethanolamine 18; polyglycerol 7; and residue 7. Ergosterol was also present.

Nucleic Acids.   The G + C content of *Claviceps* DNA is 52.6%, which places it in the ascomycete range (Normore, 1973). Filamentous fungi contain about 0.5% DNA and 5% RNA (Gottlieb and van Etten, 1966). RNA values for young (9-day-old) stationary cultures of *Claviceps* were 2.6% of dry weight when the medium contained 750 mg KH_2PO_4/l and 3.2% when it contained 250 mg (high phosphate probably favored synthesis of nonphosphate containing compounds as polyols, lipids, and trehalose, thus lowering the percentage of nucleic acids (Taber and Vining, 1963). The values for 30-day-old cultures were 2.6 and 1.5, respectively.

Intermediary Metabolism.   *C. purpurea* produces all of the enzymes required to oxidize glucose by way of glycolysis, the pentose phosphate cycle, and the citric acid cycle (McDonald et al., 1960, 1963; Glund et al., 1981). When growing on a medium containing sucrose, succinic acid, yeast extract, and salts, *C. purpurea* employs the glycolysis–citric acid cycle for about 90% of glucose catabolism and the pentose phosphate cycle for 10%. Isolated mitochondria were more delicate than those of beef heart, and neither isocitrate dehydrogenase nor *alpha*-ketoglutarate dehydrogenase activities were detected (King et al., 1960; McDonald et al., 1963; Anderson et al., 1964). Since acetate is toxic to *Claviceps* (Taber and Vining, 1981), the glyoxylate path may be absent or induced only under certain conditions. The respiratory particles were 26% lipid, 58% protein, and 16% nucleic acids. The succinic acid dehydrogenase was truly soluble and unusually stable (McDonald et al., 1963) in comparison with more conventional ones (Hederstedt and Rutberg, 1981).

D. Nutrition and Growth

Nutrition.   For general discussions, see Lilly and Barnett (1951), Cochrane (1958), Burnett (1968), and Taber and Taber (1977). *Claviceps* will grow in vitro in chemically defined media containing only glucose and biotin as organic constituents (some strains may not even require exogenous biotin). A representative medium is described (Table 15.3). Carbon dioxide is required to initiate growth (Taber and Vining, 1957a), although atmospheric CO_2 has to be removed before this requirement can be demonstrated.

1. *Vitamins.* Biotin can be replaced by biotin-L-sulfoxide and partially replaced by biocytin, desthiobiotin, and aspartate (Taber and Vining,

TABLE 15.3 Representative Chemically Defined Medium for Growth

Medium	Amount (grams per liter)
Glucose (autoclaved separately or filter sterilized	15
NH_4NO_3	1.1
KH_2PO_4	5.20 (much less can be used)
$MgSO_4 \cdot 7H_2O$	0.3
$FeSO_4 \cdot 7H_2O$	0.005*
$ZnSO_4 \cdot 7H_2O$	0.0044
$MnSO_4 \cdot H_2O$	0.003
$CaCl_2 \cdot 2H_2O$	0.006
$(NH_4)_6Mo_7O_{24} \cdot 4H_2O$	0.0018
$CuCl_2 \cdot 2H_2O$	0.00003
$CoCl_2 \cdot 2H_2O$	0.00002
NaCl	0.0025**
H_3BO_4	0.002
Biotin	0.00001†
Thiamine	0.0001†

Distilled water to 1 liter; adjust to desired pH with KOH before autoclaving

* Iron salt can be added directly, but small ppt will form; so can be autoclaved separately or chelated with 0.5 g/1 sodium citrate. Also, phosphate and magnesium salts can be autoclaved separately. Citrate chelation is low at pH values exceeding 6 (Cline et al., 1982).
** Sodium ion appears not to be required by fungi.
† Filter sterilized through a UF fritted glass filter or 0.22-μm membrane filter.
Note: Ammonium succinate or ammonium citrate makes an ideal nitrogen source as the anion often can be metabolized, thus reducing the effect of the nitrogen source on pH. The acid is added to the medium, and then ammonium hydroxide is added to the desired pH.

1957a). Homobiotin, norbiotin, pimelate, oleate, and Tween 80 have little or no biotin-sparing activity. A biotin requirement is difficult to demonstrate, especially when using hyphal fragments as the inoculum, because only a few micrograms of biotin per liter medium are required. Biotin may be unintentionally added to the medium as a contaminant of the water, glassware, nutrients, pipets, or as a constituent of the inoculum or medium the inoculum source was cultured in. It is therefore preferable to use a very small inoculum (Taber, 1957) and to autoclave the basal medium in individual culture flasks covered only with aluminum foil while autoclaving the gauze-wrapped cotton plugs separately in a basket. When the autoclave is opened, the foil is replaced with the plugs (cotton can contain biotin, which may drop into the medium as a result of moisture condensation on the cotton plugs).

2. *Nitrogen source.* Ammonium salts, urea, glutamate, and aspartate can be used as sole nitrogen sources for growth, but nitrate is only poorly used (Taber and Vining, 1957a). A nitrogen source can affect growth by its relative availability and by its ability to cause a change in pH as it is metabolized. For example, use of ammonium sulfate as a nitrogen source

leads to acid pH because less sulfur than nitrogen is used, leaving SO_4^{2-} in the medium. Electrical neutrality is attained from protons excreted into the medium, thus generating H_2SO_4. Similarly, use of KNO_3 results in an eventual alkaline pH after much of the nitrate ion is taken up as residual K^+ is balanced by excreted HCO_3^- and possibly OH , although membranes probably are not very permeable to hydroxyl ions.

3. *Carbon source.* Three strains that were examined (Taber and Vining, 1957a, b, 1958) grew on glucose, fructose, mannose, mannitol, ribose, sucrose, and cellobiose. Growth was very limited on xylose, and there was virtually none on maltose, starch, cellulose, carboxymethylcellulose, or methylcellulose. Galactose was slowly used if a small amount of glucose was present, no doubt allowing sufficient protein synthesis for production of whichever limiting enzyme was inducible. *C. paspali* grows on mannitol, sorbitol, glucose, sucrose, maltose, galactose, lactose, or dextrin, but poorly on starch or glycerin (Arcamone et al., 1961).

4. *Temperature. C. paspali* grows well within the range of 21–30°C. Commercial-scale fermentors are generally operated at 24° (Arcamone et al., 1970b).

5. *pH. C. paspali* grows well in a narrow range around 5.5 to 5.7 (Arcamone et al., 1961) in a medium containing ammonium ion as the nitrogen source and in a range of 5.2 to 6.7 when nitrate is used.

6. *Inorganic ions and elements.* Growth requires C, N, P, H, O, S, Fe, Zn, Mg, and Mn and probably also Ca, Cu, and Mo if nitrate is the nitrogen source. Cells may also require Ni, Co, and perhaps even As, Se, and chromium. Many of these are present in water and glassware.

Growth. Pure cultures of *Claviceps* strains cultured with shaking grow as wefts, pellets, sclerotia, or chlamydospore-like bodies. On the basis of studies of other ascomycetes (Borrow et al., 1961) the composition of mycelium in the exponential growth phase is constant; but as various nutrients of the medium are used up, the composition changes, and the nature of the change depends upon which nutrient is first used up. As mycelium gets older, an increased percentage of the biomass dry weight is made up of lipids, polyols, etc.; and since the amount changes with age and nutrient conditions, these substances should not be considered an integral part of growth. Perhaps it would be more meaningful to consider such substances as being noncellular (Taber and Siepmann, 1965); if so, growth would be represented by the residue remaining after the mycelium had been extracted with hot water and ether.* In the least, total dry weight should be referred to as total biomass rather than as growth. Rate

---

* A person at age 65 who suddenly puts on weight would not be perceived as one who has suddenly started growing again.

of biomass formation is exponential and can be represented by either of the following expressions:

$$k = \frac{\log_{10} B_1 - \log_{10} B_0}{0.434t}$$

$$M = \frac{\sqrt[3]{B_1} - \sqrt[3]{B_0}}{t_1 - t_0}$$

where B_0 is the dry weight of the inoculum or early growth sample, and B_1 is the dry weight after passage of time, t. The former no doubt represents exponential growth best, although the latter focuses attention on the three-dimensional nature of branching mycelial growth. Pellet growth, which is not common for *Claviceps* but otherwise is perhaps the most common growth form for shake flask cultures, may be best represented by the cube root expression because of the unique aspects of the pellet, that is, internal tissue contributes to dry weight but after a time can become oxygen starved as a result of diffusion limitations. If only the growing periphery is considered, the first formula probably is most appropriate. Borrow et al. (1961, 1964a, b) and Pirt (1966, 1973) discuss these problems, as do Koch (1975), Trinci (1974), Righelato (1975), and Bazin (1982). In general, Monod's growth rate expression,

$$u = u_{\max} \frac{S}{K_s + S}$$

where u is growth rate, $u_{\max}$ the maximal growth rate, S the nutrient concentration, and K_s a constant representing the nutrient concentration at which growth rate is half maximal, represents the effect of nutrient concentration on growth rate; it appears not to take into consideration the effect of diffusion through the cell wall pores on rate of uptake by a permease in the cytoplasmic membrane (Koch and Wang, 1982) of bacteria. The problem could exist also in filamentous growth, since ascomycete hyphal walls also have pores (Trevithick et al., 1966; Scherrer et al., 1974).

E. Ergot Alkaloid Production

The biology of *Claviceps* when producing ergot alkaloids is considered separately because it is demonstrably different from that of *Claviceps* when not producing ergot alkaloids. For reviews on production, see Gröger (1972), Taber (1967), and Vining and Taber (1963a, 1979). Secondary shunt metabolism leading to ergot alkaloid production often commences after total biomass formation is over or almost over. High-producing cultures, however, often synthesize and accumulate ergot alkaloids along with biomass formation (Amici et al., 1967a, b, 1969; Arcamone et al., 1961, 1970b; Kobel et al., 1964). Two patterns, then, have emerged: In the first, alkaloid production occurs after biomass formation is over, and in the second, alkaloid production occurs along with restricted growth (Arcamone et al., 1970b; Brar et al., 1968). By appropriate manipulation of the production medium, one strain can be made to produce by both patterns (Brar et al., 1968). Growth can be terminated early by use of low phosphate concentration (Taber and Vining, 1958), and reduced growth rate can be effected by use of slowly utilized carbon source, such as sucrose, mannitol, or a mixture of galactose and glucose (Taber and Vining, 1958).

The sigmoid growth curve characteristic of all forms of life can be divided not only into lag, exponential, and maximum stationary phrases but also into additional ones depending upon which nutrient first becomes used up (Borrow et al., 1961). Composition of the mycelium is constant during the exponential phase. This biomass formation and growth are often simplified by considering two phases to exist as regards secondary shunt metabolism: trophophase and idiophase (Bu'Lock, 1975a, b, 1980). The first represents biomass formation, during which mycelial composition is constant, and the latter represents later activities, including secondary shunt metabolism when compositional changes begin to occur because of nutritional or environmental imbalance. Bu'Lock considers these phases to be part of a continuum of events.

Inoculum. Both the size of the inoculum used to seed the production medium and the history of the inoculum source can affect subsequent alkaloid yields (Arcamone et al., 1961; Gröger and Tyler, 1963; Taber, 1957; Taber and Vining, 1958; 1959a). A system of trial and error is used to determine the optimal inoculum size, the fluid it should be suspended in, the age, and the medium on which it should be cultured. As with other experimental variables, factorial design and analyses can be used to search for desirable interactions between two or more variables (Brar et al., 1968; Taber, 1957; Taber and Vining, 1958). Laboratory studies usually employ either washed conidia produced on slants or in shake flask cultures or blended and washed vegetative mycelium produced in shake flask cultures (Gröger and Tyler, 1963; Taber and Vining, 1958, 1959a). If the inoculum source is not washed, critical nutrients may be carried into the production medium.

Carbon Source. Both mannitol and sucrose are used as carbon sources for production, although mixtures such as galactose plus glucose can be beneficial for production by some strains (Abe, 1951; Arcamone et al., 1961; Brar et al., 1968; Taber and Vining, 1958; Vining and Nair, 1966). The concentration of carbon source as well as the species is important. Since reducing sugars tend to caramelize in the presence of heat, ammonium ion, and phosphate, the carbon source for use in laboratory studies is usually either autoclaved separately or sterilized by filtration using an UF (ultra-fine) fritted glass filter (Morton bacteriological filter) or a 0.22-μm pore membrane filter. Carbon sources are usually added in the range of 50–300 g/l medium (Amici et al., 1967b, 1969; Arcamone et al., 1961, 1970b; Řeháček et al., 1977; Vining and Nair, 1966).

After sucrose hydrolysis, glucose is preferentially used, fructose is poly-merized and then used after glucose is consumed (see discussion under Honey Dew). Secondary shunt metabolism in fungi often is favored by use of a carbon source at a slow rate, which in effect starves the culture and prevents metabolism from becoming committed to biomass formation (i.e., reduces competition between biomass formation and shunt metabolism). This can be achieved by using a poorly available carbon source or an oligosaccharide that is slowly hydrolyzed or by drip feed where small amounts are fed to the fermentor by a carefully monitored process. Certain starchy plant extracts also can be used (Samburthy and Rao, 1971).

Nitrogen Source.   The nitrogen source is usually an ammonium salt of suc-cinic, citric, or fumaric acid or an amino acid closely related to the citric acid cycle (Arcamone et al., 1961, 1970b; Řeháček et al., 1977; Taber and Vining, 1958; Vining and Nair, 1966). Alkaloid production has been linked with nitrogen metabolism but also with citrate utilization (Arcamone et al., 1970b; Rothe and Fritsche, 1967). Metabolism of citric acid cycle acids is important to production because they are usually the anion accompaning the ammonium ion. The organic acid could be used as a carbon source for growth (Taber and Siepmann, 1966; Taber and Vining, 1981), as a regulator or inhibitor, or as a buffer. Various organic acids can be used for production and strain differences exist. The pH of the medium and the pK of the acids will determine the percen-tages of the various species of the acid that are present (Table 15.1) (Taber, 1973). The importance of this can be seen in the facts that alkaloid production is best within a certain pH range and that growth readily occurs on un-dissociated succinic acid but not on the dissociated form even after one month of incubation (Taber, 1976; Taber and Vining, 1981). Succinate is not toxic, since addition of glucose after a few weeks results in rapid growth.* Succinic acid is taken up from a high external concentration (41 mM) by diffusion, and the un-ionized and monoionized forms are taken up most rapidly (Taber, 1973, 1976; Taber and Vining, 1981). Uptake from high external concentrations is not sensitive to temperature, the internal and external concentrations roughly equilibrate, and the rate of uptake is linear over a long range of concentra-tions. Uptake from a low external concentration (0.83 mM) is probably by ac-tive transport, since succinic acid appears to be concentrated and uptake is temperature sensitive. Further, uptake is reduced by the addition of fumaric acid. Alternatively, uptake from low concentrations is significantly favored by metabolic draw (Taber and Vining, 1981).

The acids also serve, no doubt, as buffers, and some can also be respired (Taber, 1968) and converted to amino acids (Taber, 1973); hence gluconeogenic

---

* The culture also fails to grow on acetate, but it is toxic; that is, growth does not follow the addi-tion of glucose.

reactions connect exogenous citric acid cycle acids to intermediary metabolism. Growth can occur by using succinic acid as the sole carbon source (Taber and Siepmann, 1966).

There appears to be a correlation between the amount of alkaloid produced and the amount of organic acid taken up and used (Amici et al., 1967b). The concentrations of organic acids added to media range from 3 to 36 grams per liter. Asparagine usually is added at a concentration of 2.12 g/l (Řeháček et al., 1977), but some strains will not produce on it.

Phosphate Concentration.   For most strains, low phosphate concentrations (0.25 g/l) stimulate production (Taber and Vining, 1958) and restrict growth (Arcamone et al., 1961; Vining and Nair, 1966); exceptions exist, however (Brar et al., 1968). High phosphate represses induction of tryptophan synthetase (Robbers et al. 1972).

Minerals.   In the case of both secondary shunt metabolism and overflow metabolism the optimal concentrations required for production may differ from those required for growth or biomass formation (Foster, 1949; Vining and Nair, 1966; Voigt and Reipert, 1965; Rosazza et al., 1967; Weinberg, 1977). The actual amount of ions that must be added depends upon the extent of carryover from the inoculum source, water, nutrients, and glassware and also upon the concentration of carbon and nitrogen in the medium. In the case of iron in particular, it may also depend upon whether or not chelators such as citrate, certain amino acids, or hydroxamic derivatives are present. It would appear that all microorganisms require a specific iron transporter, and these normally are either siderophores (constructed from hydroxamic acid) or phenolics (Neilands, 1974; Weinberg, 1977). A small precipitate often forms in defined media, and this can be avoided by adding ~ 0.5 g of sodium citrate per liter or by autoclaving iron, magnesium, and phosphate salts separately. The optimal pH for various chelating agents is reported by Cline et al. (1982). Ergot alkaloid production is demonstrably sensitive to the amounts of sulfate, magnesium, potassium, iron, and zinc (Rosazza et al., 1967; Vining and Nair, 1966; Voigt and Reipert, 1965). More magnesium, iron, and zinc are required for production than for growth. A measurable amount of copper was required for production; but under the conditions used, a requirement for growth was not demonstrable. Extra requirements for copper, or other ions, may result from the need for activation of dimethylallyl-4-tryptophan synthase (Lee et al., 1976), the first enzyme of the secondary shunt pathway. The addition of B, Cu, Co, Mo, and Mn to a medium to which only iron and zinc were added favored production of clavines (Voigt and Reipert, 1965).

Temperature.   One study revealed that the strain employed produced highest yield at 21–24°C and the greatest amount of growth at 30° (Arcamone et al., 1961).

pH.   The optimum pH range lies close to 5–6 (Arcamone et al.; 1961; Vining and Nair, 1966).

Stimulators.   Addition of certain chemicals either increases titers or decreases the time required for production. Examples are Tween 80 (Řeháček and Basappa, 1971), ethylene glycol, dimethyl sulfoxide, arsenate, a mixture of glycine and acetamide, auxin, and 1,2-propanediol (Kelleher et al., 1969; Kim et al., 1973; Mizrahi and Miller, 1969; Pacifici et al., 1963; Teuscher, 1965; Tyler, 1965).

Some of these may act as permeabilizing agents that favor production because they favor uptake of a rate-limiting nutrient or increase influx of some intermediate or efflux of alkaloid itself, which might otherwise accumulate and inhibit synthesis as a regulatory effector. High osmolality may favor production; if so, the carbon source may serve this function in addition to its nutritional role. In this regard, NaCl may spare some of the sugar requirement (Puc and Socic, 1977). Addition of tryptophan markedly increases production. Analogs of proline added to the medium may lead to the formation of new peptide alkaloids (Baumert et al., 1982) if they gain entry into the mycelium (Baumert et al., 1977).

VI. BIOSYNTHESIS

The topic of biosynthesis will be only touched upon, since it has been dealt with in great detail elsewhere (Floss and Anderson, 1980; Floss et al., 1974; Vining, 1973; Vining and Taber, 1979). Lysergic acid is assembled from tryptophan, mevalonate (derived from Acetyl CoA), and a methyl group that is donated by methionine, although formic acid also can act as methyl donor (Baxter et al., 1961). The first product unique to this path is dimethylallyl-4-tryptophan (Plieninger et al., 1967), and the enzyme catalyzing the condensation has been isolated (Lee et al., 1976).

Competition between growth (or biomass formation) and alkaloid production exists, and Bu'Lock et al. (1974) see this as growth-linked suppression; that is, during growth an effector is produced that can act on several binding sites representing several different nonvegetative products and with varying affinities. The effector blocks the genes for these various products or morphological entities. When growth is over, they are no longer produced or are produced in decreasing amounts, and the genes are expressed. Addition of tryptophan (at ~ pH 6.8) markedly increases yields if added shortly after inoculation (Floss and Mothes, 1964; Taber and Vining, 1958; Vining, 1970). Tryptophan is a precursor of the ergoline ring (Arcamone et al., 1962; Mothes et al., 1958; Taber and Vining, 1959a, b), but it also induces ergot alkaloid synthesis (Bu'Lock and Barr, 1968; Floss and Mothes, 1964) and synthesis of dimethyl-4-tryptophan synthase (Floss and Mothes, 1964; Krupinski et al., 1976).

Except for the carboxyl group, all carbon, nitrogen and hydrogen atoms of tryptophan are incorporated into the alkaloid (Floss et al., 1964). The amino acid pool may not be depleted of tryptophan once growth is over (Rothe and Fritsche, 1967). High phosphate concentration appears to repress synthesis of tryptophan; when phosphate is depleted, synthesis of tryptophan increases.

VII. PHYSIOLOGY OF PRODUCTION

Many isolates do not produce detectable amounts of ergot alkaloids, and those that do will not produce in many culture media, particularly those favoring abundant biomass formation. Many studies have therefore been undertaken to identify the metabolic events that set the stage for, if not actually trigger, ergot alkaloid biosynthesis.

Certain correlations have been detected between metabolism or enzyme levels and production, and the correlations are not always the same for low- and high-producing strains. Many activities could be rate limiting; consequently, correlations might differ even between strains producing equivalent titers. Bu'Lock's (1975a, b; 1980) metabolic grid expresses this situation. Critical enzymes might be constitutive in one case and inducible and/or repressible in others. Rates of flow of carbon skeletons down various pathways could certainly be expected to vary with the strain, and competition between various pathways could be determining factors in the amounts of intermediates that are funneled into secondary metabolism. Some groups have carried out extensive studies on cellular events under circumstances in which alkaloid is or is not produced (Řeháček, 1974, 1982; Řeháček and Kozova, 1975; Kybal et al., 1981). Řeháček sees a requirement for a specific cytodifferentiation to precede the onset of alkaloid synthesis, and some of the correlations he has recorded are presence of high osmotic pressure in medium, unbalanced growth, high protein turnover (see also Bu'Lock, 1975a, b), decreased activities of citric acid cycle and glyoxylate cycle, increased funneling of acetyl-CoA to both alkaloid and lipids, and an inverse relationship between acid pools and alkaloid synthesis. Alternatively, high pools of citric acid intermediates could trigger closing down of the cycle and thus favor shunting of acetyl-CoA to mevalonate and lipids (Fig. 15.4). After the cycle is closed down, the level of endogenous citric acid cycle intermediates would then be low. Kybal et al. (1981) stress uncoupling of glycolysis from citric acid cycle and a break in the cycle at *alpha*-ketoglutarate, which would cause accumulation of citrate. Addition of exogenous citrate or succinate to the production medium (as is routinely done) would contribute both to turning off the citric acid cycle by raising the level of cycle inhibitors and to supplying various carbon skeletons that otherwise would stem from the cycle. Further, the exogenous acids would allow a low rate of respiration that would be required for maintenance activities (Pirt, 1966). Citrate might also play a mundane role in chelating metal ions under certain circumstances, depending upon the medium, which might account for some of the differences between reports, if not among strains. For example, divalent ions activate the first enzyme of ergot alkaloid biosynthesis (Lee et al., 1976). The role of tryptophan in induction (Krupinski et al., 1976) and the role of phosphate have already been considered. Oversimplified, the cell needs conditions favoring the formation and accumulation of tryptophan and mevalonate, protein turnover, and appropriate induction including synthesis of the enzymes peculiar to this particular example of secondary metabolism. At the same time it must maintain the appropriate amount of reducing power and carry out sufficient respiration for maintenance. Figure 15.4 portrays competition between secondary shunt metabolism and operation of pathways for conventional metabolism. Citrate and succinate are shown as possible inhibitors of enzymes at the condensing enzyme site. Production could be limited by a shortage of precursors or a shortage of specific enzymes; the latter is thought to be the most likely possibility (Bu'Lock, 1975a, b; Vining, 1973).

An unresolved aspect of physiology is whether or not any survival value is conferred on the organism by either the presence of ergot alkaloids or the process of synthesizing these alkaloids. The matter is cogently discussed by Bu'Lock (1980), and Vining (1973). It has been argued that alkaloids have no survival value and that they are products of biochemical lesions. It has also been argued that they are the products of detoxification. In neither case would the regulatory mechanisms seem to be required. They could be vestigial molecules that exist today as "fossils." They could play a role in energy balance (Řeháček, 1982; Řeháček et al., 1973). The metabolic activity that leads to their synthesis might have survival value in keeping pathways operating under conditions not favorable for growth at the time; in this case the products themselves would not need to have any value (Bu'Lock, 1980). Though not likely, ergot alkaloids and other secondary shunt metabolites might optimize interactions between the producer and other organisms; they might even be pesticides (Robinson, 1974). Perhaps the answer lies in understanding "why" (i.e., what is the survival value?) the alkaloids are synthesized in the phase of the life cycle "put on hold" (the idiophase?) in which many other compounds are stored and the whole package is wrapped in light protective pigments, maintained in a dormant state by regulation, and isolated from an otherwise favorable environment for growth and metabolism. In vitro production, then, would be an artifact of the laboratory resulting from institution of metabolism that resembles somewhat that of the sclerotium. A likely possibility is that alkaloid production has a function in the survival of *Claviceps* in nature but that we have not yet discovered that function.

GLOSSARY

1. *Antheridium:* Specialized hyphal tip that produces male haploid conidia capable of introducing nuclei into the female ascogonium of the perithecium.

2. *Ascogonium:* Specialized swollen hypha that functions as a female organelle, receiving the male nucleus, but not mitochondria, etc., from either an antheridium spore or an entwining hypha, which functions as a female element. Some homothallic fungi may not require external introduction of the nucleus functioning as a male element.

3. *Ascogenous hyphae:* Vertically oriented hyphae arising from an ascogonium and which will eventually develop into asci. Each compartment is N + N, one nucleus typically coming from a "male" cell.

4. *Ascomycete:* Typically, a filamentous fungus characterized by the production of haploid sexual spores, the teleomorph spore, called an ascospore in an ablong sac, the ascus, and by production of simple cross-walls or septa, which divide the hypha into compartments, and woronin bodies. Woronin bodies are spherical bodies that may serve to plug the pore of the septum when nutrients, mitochondria, etc. are not passing from one compartment to another.

5. *Ascospore:* The haploid sexual spore that is produced in the ascus after karyogamy and meiosis. Most asci possess eight ascospores, mitosis producing eight from the four meiotic products.

6. *Ascus:* The sac-shaped hyphal tip, which is the site of karyogamy, meoisis, and the mature sexual or teleomorphic spore. It arises directly from the ascogenous hypha.

7. *Asexual (anamorphic) spore:* An asexual haploid spore or conidium that is produced either directly on assimilative hyphae or on conidiophores arising from assimilative or vegetative hyphae.

8. *Heterokaryon:* Assimilative or vegetative hyphae containing genetically different nuclei as a result of either anastomosing (i.e., hyphal fusion) of genetically different hyphae of the same species or perpetuation of a resident nucleus that has undergone mutation.

9. *Heterothallic:* Genetic state in which formation of ascospores requires karyogamy between two nuclei, differing from one another by alleles of one or more mating type loci (i.e., A × a is fertile).

10. *Homothallic:* Genetic state in which one uninucleate ascospore can germinate and complete the life cycle. Secondary homothallism exists if one ascospore contains two nuclei, one of each mating type.

11. *Hypha (pl. = hyphae):* Filamentous assimilative or vegetative cells that in the case of all fungi except phycomycetes possess either simple or complex perforated septa. A segment cut off by septa at both ends is not a cell but rather is one compartment of a long cell. The term is more or less synonymous with mycelium (pl. = mycelia).

12. *Ostiole:* A hole at the top of a perithecium that allows escape of ascospores that have been released from their asci. Usually placed at the tip of a neck (as with a chimney).

13. *Perithecium:* Flask-shaped vessel, the ascocarp, that houses the

ascogonium, ascogenous hyphae, and asci. Usually ~0.5–1 mm high. May exist free or embedded in a stroma, depending upon the taxon (see text).

14. *Pyrenomycetes:* Those ascomycetes that bear oblong asci at various levels in a perithecium.

REFERENCES

Abe, M. (1951) *Ann. Rep. Takeda Res. Lab. 10*; 73–234.

Abe, M., and Yamatodani, S. (1964) *Prog. Ind. Microbiol. 5*, 205–229.

Abe, M., Yamatodani, S., Yamano, T., Kozu, Y., and Yamada, S. (1967) *J. Agr. Chem. Soc. Japan 41*, 68–71.

Abe, M., Fukuhara, T., Ohmomo, S., Hori, M., and Tabuchi, T. (1970) *J. Agr. Chem. Soc. Japan 44*, 573–579.

Aberhart, D. J., Chen, Y. S., Demajo, P., and Stothers, J. B. (1965) *Tetrahedron 21*, 1417–1432.

Amici, A. M., Minghetti, A., Scotti, T., Spalla C., and Tognoli, L., (1966) *Experientia 22*, 415–418.

Amici, A. M., Minghetti, A., Scotti, T., Spalla, C., and Tognoli, L. (1967a) *Appl. Microbiol. 15*, 597–602.

Amici, A., Scotti, T., Spalla, C., and Tonolo, L. (1967b) *Appl. Microbiol. 15*, 611–615.

Amici, A., Minghetti, A., Scotti, T., Spalla C., and Tonolo, L. (1969) *Appl. Microbiol. 18*, 464–468.

Anderson, J. A., Sun, F. K., McDonald, J. K., and Cheldelin, V. H. (1964) *Arch. Biochem. Biophys. 107*, 37–50.

ApSimon, J. W., Corran, J. A., Creasy, N. G., Sim, K. Y., and Whaley, W. B. (1965a) *J. Chem. Soc.* (July), 4130–4133.

ApSimon, J. W., Hannaforda, A. J., and Whaley, W. B. (1965b) *J. Chem. Soc.* (July), 4164–4168.

Arcamone F., Chain, E. B., Ferretti, A., Minghetti, A., Pennella, P., Tonolo, A., and Vero, L. (1961) *Proc. Royal Soc. B. 155*, 26–54.

Arcamone, F., Chain, E. B., Ferretti, A., Minghetti, A., Pennella, P., and Tonolo, A. (1962) *Biochim. Biophys. Acta 57*, 174–176.

Arcamone, F., Barbieri, W., Cassonelli, G., and Pol, C. (1970a) *Carbohyd. Res. 14*, 65–71.

Arcamone, F., Cassinelli, G., Gerni, G., Penco, S., Pennella, P., and Pol, C. (1970b) *Can. J. Microbiol. 16*, 923–931.

Atwell, S. M., and Mantle, P. G. (1981) *Experientia 37*; 1257–1258.

Bacon, C. W., and Luttrell, E. S. (1982) *Phytopathology 72*, 1332–1336.

Bacon, C. W., Porter, J. K., and Robbins, J. D. (1981) *Can J. Bot. 59*, 2534–2538.

Barger, G. (1931) *Ergot and Ergotism*, Gurney and Jackson, London.

Baross, L. (1966) *Planta Med. 14*, 232–238.

Barz, W., Schlepphorst, R., and Laimer, J. (1976) *Phytochemistry 15*, 87–90.

Bassett, R. A., Chain, E. B., Corbett, K., Dickerson, A. G. F., and Mantle, P. G. (1972) *Biochem. J. 127*, 3P.

Baumert, A., Gröger, D., and Maier, W. (1977) *Experientia 33*, 881–882.

Baumert, A., Erge, D., and Gröger, D. (1982) *Planta Med. 44*, 122–123.

Baxter, R. M., Kandel, S. I., and Okany, A. (1961) *Chem. Ind. 1961*, 1453–1454.

Bazin, M. J. (ed). (1982) *CRC Series in Mathematical Models in Microbiology: Microbial Population Dynamics*, CRC Press, Boca Raton, Fla.

Berman, M. L., and Younken, H. W., Jr. (1958) *J. Amer. Pharm. Assoc. 47*, 888–893.

Bianchi, M. L., Perellina, N. C., Gioia, B., and Minghetti, A. (1982) *J. Nat. Prod. 45*, 191–196.

Bland, B. F. (1971) *Crop Production: Cereals and Legumes*, Academic Press, New York.

Bojar, O., Dragomirescu-Manuchian, M., Platon, F., Gruia, S., and Pavel, M. (1967) *Farmacia* (Bucharest). *15*, 715–722.

Borrow, A., Jefferys, E. G., Kessel, R. H. J., Lloyd, E. C., Lloyd, P. B., and Nixon, I. S. (1961) *Can. J. Microbiol. 7*, 227–276.

Borrow, A., Brown, S., Jeffreys, E. G., Kessell, R. H. J., Lloyd, E. C., Lloyd, P. B., Rothwell, A., Rothwell, B., and Swait, J. C. (1964) *Can. J. Microbiol 10*, 407–444.

Borrow A., Brown, S., Jefferys, E. G., Kessell, R. H. J., Lloyd, E. C., Lloyd, P. B., Rothwell, A., Rothwell, B., and Swait, J. C. (1964) *Can J. Microbiol 10*, 445–466.

Bove, F. J. (1970) *The Story of Ergot*, S. Karger, Basel, Switzerland.

Brar, S. S., Giam, C. S., and Taber, W. A. (1968) *Mycologia 50*, 806–826.

Breuel, K., Braun, K., Dauth, C., and Gröger, D. (1982) *Planta Med. 44*, 121–122.

Bu'Lock, J. D. (1975a) *Dev. Ind. Microbiol. 16*, 11–19.

Bu'Lock, J. D. (1975b) in *The Filamentous Fungi*, Vol. I, *Industrial Fungi* (Smith, J. E., and Berry, D. R., eds.), pp. 33–58, John Wiley and Sons, New York.

Bu'Lock, J. D. (1980) in *The Biosynthesis of Mycotoxins: A Study of Secondary Metabolism.* (Steyn, P. S., ed.) pp. 1–16, Academic Press, New York.

Bu'Lock, J. D., and Barr, J. G. (1968) *Lloydia 31*, 324–354.

Bu'Lock, J. D., Detroy, R. W., Hostalek, Z., and Munim-al-shakarchi, A. (1974) *Trans. Brit. Mycol. Soc. 62*, 377–389.

Burnett, J. H. (1968) *Fundamentals of Mycology*, St. Martin's Press, New York.

Caporael, L. R. (1976) *Science 192*, 21–26.

Chi, L., Cheng, F., Ispulz, I. I., and Rastit, L. (1964) *Resursov SSSR SA*, 290–293 (*Chem. Abst. 63*, 2345, 1965).

Christensen, C. W. (1975) *Molds, Mushrooms, and Mycotoxins*, University of Minnesota Press, Minneapolis.

Cline, G. R., Powell, P. E., Szaniszlo, P. J., and Reid, C. P. P. (1982) *Soil Sci. Soc. Amer. J. 46*, 1158–1164.

Cochrane, V. W. (1958) *Physiology of Fungi*, John Wiley and Sons, New York.

Cole, R. J., Dorner, J. W., Cox, R. H., Cunfer, B. M., Cutler, H. G., and Stuart, B. R. (1981) *J. Nat. Prod. 44*, 324–330.

Cooke, R. C., and Mitchell, D. T. (1970) *Trans. Brit. Mycol. Soc. 54*, 93–99.

Corbett, K., Dickerson, A. G., and Mantle, P. G. (1974) *J. Gen. Microbiol. 84*, 39–58.

Corbett, K., Dickerson, A. G., and Mantle, P. G. (1975) *J. Gen. Microbiol. 90*, 55–68.

Cunfer, B. M., and Marshall, D. (1977) *Mycologia 69*, 1137–1141.

Cunfer, B. M., and Seckinger, A. (1977) *Mycologia 69*, 1142–1148.

Darlington, L. C., Mathre, D. E., and Johnston, R. H. (1977) *Can. J. Plant Sci. 57*, 729–734.

Davis, R. H. (1966) in *The Fungi*, Vol. II (Ainsworth, G. C., and Sussman, A. S. eds.), pp. 567–588, Academic Press, New York.

Demain, A. L. (1968) *Lloydia 31*, 395–418.

Demos, J. P. (1982) *Entertaining Satan*, Oxford Press, New York.

Dickerson, A. G. (1973) *2nd Int. Congr. Plant Path. U. Minnesota*, Abst. #0103

Dickerson, A. G., Mantle, P. G., and Nisbet, L. J. (1976) *J. Gen. Microbiol.* 97, 267–276.

Duerr, M., Urech, K., Boller, T., Wiemken, A., Schwencke, J., and Nagy, M. (1979) *Arch. Microbiol.* 121, 169–176.

El-Rafai, A. H., Sallam, L. A. R., and Niam, N. I. (1970) *Jap. J. Microbiol.* 14, 91–97.

Erge, P., Gröger, D., and Mothes, K. (1962) *Arch. Pharm.* 295, 474–481.

Esser, K., and Tudzynski, P. (1978) *Theor. Appl. Gen.* 53, 145–149.

Floss, H. G., and Anderson, J. A. (1980) in *The Biosynthesis of Mycotoxins: A Study in Secondary Metabolism* (Steyn, P. S., ed.), pp. 17–67, Academic Press, New York.

Floss, H. G., and Mothes, U. (1964) *Arch. Microbiol.* 48, 213–221.

Floss, H. G., Mothes, U., and Guenther, H. (1964) *Z. Naturforschung 19b*, 784–788.

Floss, H. G., Robbers, J. E., and Heinstein, P. F. (1974) *Recent Adv. Phytochem.* 8, 141–178.

Foster, J. W. (1949) *Chemical Activities of Fungi*, Academic Press, New York.

Franck, B. (1980) in *The Biosynthesis of Mycotoxins: A Study in Secondary Metabolism* (Steyn, P. S., ed.), pp. 157–191, Academic Press, New York.

Franck, B., and Zemer, I. (1965) *Chem. Ber.* 98, 1514–1521.

Franck, B., Baumann, G., and Ohnsarge, V. (1965) *Tetrahedron Lett. 1965*, 2031–2037.

Franck, B., Radtke, V., and Zerdler, U. (1967) *Angew. Chem. Int. Ed. Eng.* 6, 952–953.

Frey, H. P., and Brack, A. (1968) *Herba Hungarica 7*, 149–154.

Garay, A. S. (1956) *Physiol. Plantarum 9*, 350–355.

Garay, A. S. (1958) *Physiol. Plantarum 11*, 48–55.

Ghosh, B. K. (1977) in *CRC Handbook of Microbiology* Vol. II, 2nd ed. (Laskin, A. I., and Lechevalier, H., eds.), pp. 11–69, CRC Press, Boca Raton, Fla.

Glund, K., Kirston, C., and Schlee, D. (1981) *Folia Microbiol.* 26, 398–402.

Gottlieb, D., and van Etten, J. L. (1966) *J. Bacteriol.* 91, 161–168.

Gröger, D. (1958) *Arch. Pharm.* 291, 106–109.

Gröger, D. (1972) in *Microbial Toxins,* Vol. B (Kadis, S., Ciegler, A., and Ajl, S. J., eds.), pp. 321–373, Academic Press, New York.

Gröger, D. and Erge, D. (1964) *Pharmazie 19*, 775–781.

Gröger, D., and Tyler, V. E., Jr. (1963) *Lloydia 26*, 174–191.

Gröger, D., Erge, D., and Floss, H. G. (1965) *Z. Naturforschung 20b*, 856–858.

Harold, F. M. (1966) *Bacteriol. Rev. 30,* 772–794.

Hareven, D., and Koltin, Y. (1970) *Appl. Microbiol. 19*, 1005–1006.

Hartmann, T. (1964) *Planta Med. 12*, 340–346.

Hartmann, T. (1965a) *Planta 66*, 27–43.

Hartmann, T. (1965b) *Planta 66*, 191–206.

Hatt, H. (ed.) (1980) *American Type Culture Collection Methods*, Vol. I, ATCC, Rockville, Md.

Heath, H., and Wildy, J. (1956) *Biochem. J. 64*, 612–620.

Hederstedt, L., and Rutberg, L. (1981) *Microbiol. Rev. 45*, 542–555.

Hiruma, M., Tanda, S., and Matsunami, Y. (1961) *Tokyo Nogo Daigrku Shuho 7*, 1–5.

Hofmann, A., and Tscherter, A. (1960) *Experientia 16*, 414.

Hult, K., Veide, A., and Gatenback, S. (1980) *Arch. Microbiol. 128*, 253–255.

Hwang, S. (1966) *Appl. Microbiol. 14*, 784–788.

Janardhanan, K. K., Gupta, M. L., and Husain, A. K. (1982) *Planta Med. 44*, 166–167.

Jung, M., and Rochelmeyer, H. (1960) *Beitr. Biol. Pflanz. 35*, 343–378.

Kelleher, W. J., Kim, B. K., and Schwarting, A. E. (1969) *Lloydia 32*, 327–333.

Kelleher, W. J., Krueger, R. J., and Rosazza, J. P. (1971) *Lloydia 34*, 188–194.

Keller, V., Zucher, R., and Kleinkauf, H. (1980) *J. Gen. Microbiol. 118*, 485–494.

Kent-Jones, A. (1967) *Modern Cereal Chemistry*, Food Trade Press, London.

Kim, B. K., Auck, S., and Choi, C. (1973) *Korean Biochem. J. 6*, 167–175.

King, T. E., Ryan, C. A., Cheldelin, V. H., and McDonald, J. K. (1960) *Biochim. Biophys. Acta 45*, 398–400.

Kirsten, G., Hartmann, T., and Steiner, N. (1966) *Planta Med. 14*, 241–246.

Kleinerova, E., and Kybal, J. (1980) *Cesk. Farm. 29*, 153–154.

Kobel, H. (1969) *Pathol. Microbiol. 34*, 249–251.

Kobel, H., and Sanglier, J. J. (1973) in *Genetics of Industrial Microorganisms, Actinomycetes and Fungi.* (Vaném, Z., Hostalek, Z., and Cudlin, J., eds.), pp. 421–425, Academia, Prague.

Kobel, H., Schreier, E., and Rutschmann, J. (1964) *Helv. Chim. Acta 47*, 1052–1064.

Koch, A. L. (1975) *J. Gen. Microbiol.* 89:209–216.

Koch, A. L., and Wang, C. H. (1982) *Arch. Microbiol. 131*, 36–42.

Kornhauser, A., and Logar, S. (1965) *Pharmazie 20*, 447–449.

Krupinski, V. M., Robbers, J. E., and Floss, H. G. (1976) *J. Bacteriol. 125*, 158–165.

Kulaev, I. S. (1979) *The Biochemistry of Inorganic Polyphosphates*, John Wiley and Sons, New York.

Kulaev, I. S., and Uryson, S. O. (1965) *Biokhimiya 30*, 282–291.

Kybal, J. (1964) *Phytopathology 54*, 244–245.

Kybal, J., and Kleinerova, E. (1973) *Folia Microbiol. 18*, 348–349.

Kybal, J., Svoboda, E., Strnadova, K., and Kejzlar, M. (1981) *Folia Microbiol. 26*, 112–119.

Langdon, R. F. N. (1954) *Univ. Queensland Papers Dept. Botany 3*, 61–68.

Lee, S.-L., Floss, H. G., and Heistein, P. (1976) *Arch. Biochem. Biophys. 177*, 84–94.

Lewis, D. H., and Smith, D. C. (1967) *New Phytol. 66*, 143–184.

Lilly, V. G., and Barnett, H. L. (1951) *Physiology of Fungi*, McGraw-Hill, New York.

Lingens, F., Goebel, W., and Vessler, H. (1967) *Eur. J. Biochem. 2*, 442–447.

Little, G. N., and Gordon, M. A. (1967) *Mycologia 59*, 733–736.

Loveless, A. R. (1964) *Trans. Brit. Mycol. Soc. 47*, 205–213.

Loveless, A. R. (1971) *Trans. Brit. Mycol. Soc. 56*, 419–434.

Lusby, E. W., and McLaughlin, C. S. (1980) *Mol. Gen. Genet. 178*, 69–76.

Luttrell, E. S. (1977) *Phytopathology 67*, 1461–1468.

Luttrell, E. S. (1980) *Can. J. Bot. 58*, 942–958.

Mantle, P. G. (1967) *Ann. Appl. Biol. 60*, 353–356.

Mantle, P. G. (1977) in *Mycotoxic Fungi, Mycotoxins, and Mycotoxicosis* (Wyllie, T. D., and Morehouse, L. G., eds.), pp. 83–89, Marcel Dekker, New York.

Mary, N. Y., Kelleher, W. J., and Schwarting, A. E. (1965) *Lloydia 28*, 218–229.

Matossian, M. K. (1982) *Amer. Sci. 70*; 355–357.

McDonald, J. K., Cheldelin, V. H., and King, T. E. (1960) *J. Bacteriol. 80*, 51–71.

McDonald, J. K., Anderson, J. A., Cheldelin, V. H., and King, T. E. (1963) *Biochim. Biophys. Acta 73*, 533–549.

Mitchell, D. T., and Cooke, R. C. (1968a) *Trans. Brit. Mycol. Soc. 51*, 721–729.

Mitchell, D. T., and Cooke, R. C. (1968b) *Trans. Brit. Mycol. Soc. 51*, 731–736.

Mizrahi, A., and Miller, G. (1968) *Appl. Microbiol. 16*; 1100–1101.

Mizrahi, A., and Miller, G. (1969) *J. Bacteriol. 97*, 1155-1159.

Molnar, G., and Tyihak, E. (1966) *Herba Hungarica 5*, 268-274.

Mothes, K., Weygand, F., Gröger, D., and Grisebach, H. (1958) *Z. Naturforschung 13b*, 41-44.

Mower, R. L., and Hancock, J. G. (1973) *2nd Int. Congr. Plant Pathol. U. Minnesota* Abst. #0366.

Mower, R. L., and Hancock, J. G. (1975) *Can. J. Bot. 53*, 2826-2834.

Neilands, J. B. (1974) *Microbial Iron Metabolism*, Academic Press, New York.

Nisbet, L. J., Dickerson, A. G., and Mantle, P. G. (1977) *Trans. Brit. Mycol. Soc. 69*, 11-14.

Normore, W. M. (1973) *CRC Handbook of Microbiology*. Vol. II, 1st ed. (Laskin, A. I., and Lechevalier, H., eds.) pp. 585-740, CRC Press, Boca Raton, Fla.

Oddo, N., and Tonolo, A. (1967) *Ist Super. Sanita Ann. 3*, 16-26.

Ohmomo, S., Sato, T., Utawaga, T., and Abe, M. (1975) *J. Agr. Chem. Soc. Japan 49*; 615-622.

Olive, L. S. (1965) in *The Fungi* Vol. I (Ainsworth, G. C., and Sussman, A. S., eds.) pp. 143-161, Academic Press, New York.

Pacifici, L. R., Kelleher, W. J., and Schwarting, A. E. (1963) *Lloydia 26*, 161-173.

Pažoutová, S., Řeháček, Z., and Pokorny, V. (1978) *Folia Microbiol. 23*, 376-378.

Perenyl, T., Upuardy, E. M., Wick, K., and Baross, L. (1966) *Planta Med. 14*, 42-48.

Perlin, A. S., and Taber, W. A. (1963) *Can. J. Chem. 41*, 2278-2282.

Pfyffer, G. E., and Rast, D. M. (1980) *Exp. Mycol. 4*, 160-170.

Pirt, S. J. (1966) *Proc. Roy. Soc. London Ser. B 106*, 369-373.

Pirt, S. J. (1973) *J. Gen. Microbiol. 75*, 245-247.

Plieninger, H., Immel, H., and Vokl, A. (1967) *Liebigs Ann. der Chem. 706*, 223-229.

Puc, A., and Socic, H. (1977) *Eur. J. Appl. Microbiol. 4*, 283-287.

Ramsbottom, J. (1953) *Mushrooms and Toadstools*, Collins, London.

Řeháček, Z. (1974) *Z. Bakt. Abt. II. 129*, 20-49.

Řeháček, Z. (1982) in *Overproduction of Microbial Products* (Krumphanzl, V., Sikyta, B., Vanék, Z., eds.) pp. 221-234. Academic Press, New York.

Řeháček, Z., and Basappa, S. C. (1971) *Folia Microbiol. 16*, 110-113.

Řeháček, Z., and Kozova, J. (1975) *Folia Microbiol. 20*, 112-117.

Řeháček, Z., Sajdl, P., and Kremen, A. (1973) *Biotechnol. Bioeng. 15*; 207-211.

Řeháček, Z., Desai, J. D., Sajdl, P., and Pažoutova, S. (1977) *Can. J. Microbiol. 23*, 596-600.

Righelato, R. C. (1975) in *The Filamentous Fungi*, Vol. I, *Industrial Mycology*, (Smith, J. E., and Berry, D. R., eds.), pp. 79-103, John Wiley and Sons, New York.

Robbers, J. E., Jones, L. G., and Krupinski, V. M. (1974) *Lloydia 37*, 108-111.

Robbers, J. E., Robertson, L. W., Hornenann, K. M., Jindra, A., and Floss, H. G. (1972) *J. Bacteriol. 112*, 791-796.

Robinow, C. F., and Bakerspigel, A. (1965) in *The Fungi*, Vol. I (Ainsworth, G. C., and Sussman, A. S., eds.), pp. 119-142, Academic Press, New York.

Robinson, T. (1974) *Science 184*, 430-435.

Rogerson, C. T. (1970) *Mycologia 70*, 865-910.

Rolfe, R. T., and Rolfe, F. W. (1966) *The Romance of the Fungi*, Johnson and Johnson Reprint Co., New York.

Roper, J. A. (1966) in *The Fungi*, Vol. II (Ainsworth, G. C., and Sussman, A. S., eds.), pp. 589-617, Academic Press, New York.

Rosazza, J. P., Kelleher, W. J., and Schwarting, A. E. (1967) *Appl. Microbiol. 15*, 1270-1283.

Rothe, K., and Fritsche, W. (1967) *Arch. Mikrobiol. 58*, 77–91.

Samburthy, K., and Rao, L. N. (1971) *Biotech. Bioeng. 13*, 331–334.

Scherrer, R., Louden, L., and Gerhardt, P. (1974) *J. Bacteriol. 118*, 534–540.

Schmauder, H. P., and Gröger, D. (1978) *Biochem. Physiol. Pflanz. 173*, 139–145.

Schmauder, H. P., Gröger, D., Losecke, W., Rudolph, A., and Grimmecke, H. D. (1978) *Biochem. Physiol. Pflanz. 173*, 295–310.

Schultz, J. S. (1964) *Appl. Microbiol. 12*, 305–310.

Scott, P. M., Merrien, M. A., and Polonsky, J. (1976) *Experientia 32*, 140–142.

Sermonti, G. (1969) *Genetics of Antibiotic Producing Microorganisms*, Wiley Interscience, New York.

Shaw, B., and Mantle, P. G. (1980) *Trans. Brit. Mycol. Soc. 75*, 77–90.

Siddiqui, M. R., and Khan, I. D. (1973) *Trans. Brit. Mycol. Soc. 60*, 195–198.

Spalla, C. and Marnati, M. P. (1982) in *Overproduction of Microbial Products*: (Krumphanzl, V., Sikyta, B., and Vanék, Z., eds.) pp. 563–568. Academic Press, NY.

Slokoska, L., Grigorov, I., Angelova, T., Angelova, M., Kozlovskii, A., Arinbasarov, M., and Solov'eva, T. (1981) *Acta Microbiol. Bulg. 9*, 22–28.

Spanos, N. P., and Gottlieb, J. (1976) *Science 194*, 1390–1394.

Spilsbury, S. F., and Wilkinson, S. (1961) *J. Chem. Soc. 1961*, 2085–2091.

Stahl, C. H., Neumann, D., Schmauder, H. P., and Gröger, D. (1977) *Biochem. Physiol. Pflanz. 171*, 363–368.

Steiner, M., and Hartmann, T. (1964) *Biochem. Z. 340*, 436–440.

Strnadova, K. (1967) *Flora Abt. A 157*, 517–523.

Strnadova, K. (1976) *Folia Microbiol. 21*, 455–458.

Strnadova, K., and Kybal, J. (1974) *Folia Microbiol. 19*, 272–280.

Taber, W. A. (1957) *Can. J. Microbiol. 3*, 803–812.

Taber, W. A. (1964) *Appl. Microbiol. 12*, 321–326.

Taber, W. A. (1967) *Lloydia 30*, 39–66.

Taber, W. A. (1968) *Mycologia 50*, 345–355.

Taber, W. A. (1973) *Mycologia 65*, 558–579.

Taber, W. A. (1976) *Can. J. Microbiol. 22*, 245–253.

Taber, W. A., and Siepmann, R. (1965) *Appl. Microbiol. 13*, 827–829.

Taber, W. A., and Siepmann, R. (1966) *Appl. Microbiol. 14*, 320–327.

Taber, W. A., and Taber, R. A. (1977) in *CRC Handbook of Microbiology*, Vol. II, 2nd ed. (Laskin, A. I., and Lechevalier, H. eds.), pp. 97–205, CRC Press, Boca Raton, Fla.

Taber, W. A., and Vining, L. C. (1957a) *Can. J. Microbiol. 3*, 1–12.

Taber, W. A., and Vining, L. C. (1957b) *Can. J. Microbiol. 3*, 55–60.

Taber, W. A., and Vining, L. C. (1958) *Can. J. Microbiol. 4*, 611–626.

Taber, W. A., and Vining, L. C. (1959a) *Can. J. Microbiol. 5*, 418–421.

Taber, W. A., and Vining, L. C. (1959b) *Chem. Ind.* 1218–1219.

Taber, W. A., and Vining, L. C. (1960) *Can. J. Microbiol. 6*, 355–365.

Taber, W. A., and Vining, L. C. (1963) *Can. J. Microbiol. 9*, 1–14.

Taber, W. A., and Vining, L. C. (1981) *Proc. Nova Scotia Inst. Sci. 31*, 237–249.

Taber, W. A., Vining, L. C., and Heacock, H. A. (1963) *Phytochemistry 2*, 65–70.

Teuscher, E. (1964) *Pharmazie 20*, 778–784.

Teuscher, E. (1965) *Phytochemistry 4*, 341–343.

Teuscher, E. (1970) *Allg. Mikrobiol. 10*, 137–146.

Teuscher, G., and Teuscher, E. (1965) *Phytochemistry 3*, 47–51.

Tonolo, A., Scotti, T., and Barcellona, V. (1961) *Sci. Rep. Ist. Super Sanita 1*, 404–422.

Trevithick, J. R., Metzenberg, R. C., and Costello, D. E. (1966) *J. Bacteriol. 92*, 1016–1020.

Trinci, A. P. F. (1974) *J. Gen. Microbiol. 81*, 225–236.

Tscherter, H., and Hauth, H. (1974) *Helv. Chim. Acta 57*, 113–121.

Tudzynski, P., Esser, K., and Groeschel, H. (1982) *Theor. Appl. Genet. 61*, 97–100.

Tyler, V. E., Jr. (1965) U. S. Patent 3,224,943.

Tyler, V. E., Jr., Mothes, K., Gröger, D., and Floss, H. G. (1964) *Tetrahedron Lett. 1964*, 593.

Vining, L. C. (1970) *Can. J. Microbiol. 16*, 473–480.

Vining, L. C. (1973) in *Genetics of Industrial Microorganisms*, Vol II, *Actinomycetes and Fungi* (Vanék, Z., Hostalek, S., and Cudlin, J., eds.), pp. 405–419, Elsevier, New York.

Vining, L. C., and Nair, P. M. (1966) *Can. J. Microbiol. 12*, 915–931.

Vining, L. C., and Taber, W. A. (1959) *Can. J. Microbiol. 5*, 441–451.

Vining, L. C., and Taber, W. A. (1963a) in *Biochemistry of Industrial Microorganisms* (Rainbow, C., and Rose, A. H., eds.), pp. 341–378, Academic Press, New York.

Vining, L. C., and Taber, W. A. (1963b) *Can. J. Microbiol. 9*, 291–302.

Vining, L. C., and Taber, W. A. (1964) *Can. J. Microbiol. 10*, 647–657.

Vining, L. C., and Taber, W. A. (1979) in *Economic Microbiology*, Vol. III (Rose, A. H., ed.), pp. 389–420, Academic Press, New York.

Vining, L. C., McInnes, A. G., Smith, D. G., Wright, J. L. C., and Taber, W. A. (1982) in *Overproduction of Microbial Products* (Krumphanzl, V., Sikyta, B., and Vanék, Z., eds.), pp. 243–251, Academic Press, New York.

Voigt, R., and Reipert, S. (1965) *Pharmazie 20*, 785–788.

Vorisek, J., and Sajdl, P. (1981) *J. Ultrastruct. Res. 75*, 269–275.

Vorisek, J., Judvik, J., and Řeháček, Z. (1974a) *J. Bacteriol. 118*, 285–294.

Vorisek, J. Judvik, J., and Řeháček, Z. (1974b) *J. Bacteriol. 120*, 1401–1408.

Weinberg, E. D. (ed.) (1977) *Microorganisms and Minerals*, Marcell Dekker, New York.

Williams, J. R., and Colotelo, N. (1975) *Can. J. Bot. 53*, 83–86.

Yamano, T., Kishino, K., Yamatodami, S., and Abe, M. (1962) *Ann. Rep. Takeda Res. Lab. 21*, 83–87.

Yancey, P. H., Clark, M. E., Hand, S. C., Bowlus, R. D., and Somerow, G. N. (1982) *Science, 217*, 1214–1222.

Young, J. C. (1981a) *J. Env. Sci. Health Part B 16*, 83–111.

Young, J. C. (1981b) *J. Env. Sci. Health Part B 16*, 381–394.

Trichoderma

Douglas E. Eveleigh

Trichoderma species are soil fungi that occur ubiquitously. They have major potential industrial attributes: production of extracellular enzymes (e.g., cellulase), the ability to transform a wide variety of complex materials both of natural and xenobiotic origin, a potential role as biocontrol agents of plant fungal diseases, production of a scintillating array of secondary metabolites, minimal nutritional requirements, rapid growth and ability to conidiate profusely. This chapter addresses aspects of their taxonomy, nutrition and physiology, ecology, secondary metabolite production, and genetics.

I. TAXONOMY

The taxonomy of *Trichoderma* is as follows (Ainsworth et al., 1973):

Trichoderma	*Hypocrea*
anamorph[1] (asexual)	teleomorph[1] (sexual)
Deuteromycotina	Ascomycotina
Hyphomycetes (syn. Moniliales)	Pyrenomycetes
	Hypocreales
T. hamatus[2]	*H. semiorbis*
T. viride[2]	*H. rufa*
T. harzianum[2]	*H. vinosa*

[1]Hennebert and Weresub, 1977.
[2]J. Dingley, pers. comm.

Trichoderma was first described by Persoon in 1794 as a conidiating (asexually reproducing) fungus. The relationship between the anamorph of *Trichoderma viride* Pers. ex S. F. Gray to the teleomorph *Hypocrea rufa* (Pers. ex Fr.) Fr. was noted by the Tulasne brothers in 1865 (Tulasne and Tulasne, 1865). Since then, there has been mayhem regarding the taxonomic status of *Trichoderma*. Three points must be noted in considering their current taxonomic status.

1. Ascomycetous fungi often exhibit sexual (teleomorph) and asexual (anamorph) reproductive modes. Yet *different teleomorphic genera* can exhibit morphologically indistinguishable anamorphs. Thus sexually distinct genera such as *Hypocrea* and *Podostroma* can each possess indistinguishable anamorphs of *Trichoderma*.

2. A single teleomorph can exhibit different anamorphs. Thus *Hypocrea* species have predominantly *Trichoderma* anamorphs, but some species are genotypically different and possess *Gliocladium* and *Verticillium* anamorphs.

3. The morphologic differences between *Trichoderma* species even when grown with standardized media and environmental conditions, are not apparent to the average microbiologist; it requires both an expert eye and experience to delineate each species.

This taxonomic complexity has sometimes led to any green-spored *Trichoderma*-like molds being indiscriminantly accorded the name *T. viride*. The resultant confusion is illustrated by the classical misnaming of *Gliocladium* cultures as strains of *T. viride* (Weindling, 1934), and the incorrect assertions that *T. viride* produces the antibiotics *gliotoxin* (Brian, 1944) and *viridin* (Brian et al., 1946). This misinformation continues to be perpetuated (e.g., *American Type Culture Collection Catalog*, 1983; Bu'Lock and Leigh, 1975). However, Webster and Lomas (1964) clearly showed that 40 strains of *T. viride* do not produce these antibiotics, and they also illustrated that the misnamed *Gliocladium* (*sic* "*Trichoderma*") species yielded gliotoxin and viridin. Their conclusion was reconfirmed by Dennis and Webster (1971a) and is supported by the findings that antibiotic production by *Trichoderma* species (Brewer et al., 1982; Okuda et al., 1982) does not include the formation of gliotoxin or viridin. There is one recent exception (Hussain et al., 1975), but it requires reexamination in light of the present taxonomic discussion. In a similar taxonomic vein, it should be noted that most commercial cellulase preparations are "claimed" to be prepared from *T. viride*. The classic cellulase producing *Trichoderma* strain from the U.S. Army Natick Laboratories is unique, *T. reesei* (Simmons, 1977). Simmons authenticated cultures received from T. M. Wood and G. Halliwell as *T. koningii* and from K.-E. Eriksson as *T. viride* but considered the strains of *T. viride* from E. Galun, E. Morozova, K. Nisizawa, and N. Toyama to require further taxonomic study.

The classification of *Trichoderma* species has evolved slowly due especially to the more recent efforts of J. Webster, H. Doi and M. A. Rifai. Webster (1964) initially showed the distinction between *H. rufa* and *H. gelatinosa* by indicating their respective anamorphs to be *Trichoderma* and *Gliocladium*, while over the last 20 years, Doi has devoted his studies to clarifying further distinctions between such teleomorphs and anamorphs (for review, see Doi and Doi, 1979). In noting the complexities of morphologic speciation, Rifai (1969) presented a classification guide to the genus *Trichoderma* based on nine "species aggregates." This has been widely accepted as a practical guide, with incorporations of further species by Domsch et al., 1980; and Doi and Doi, 1979, while retaining the "species aggregate" concept. Further revisions are in progress (Bissett, 1984). For general discussion of the teleomorph, anamorph, and holomorph (combined sexual and asexual states) concepts the reader is referred to G. T. Cole and Kendrick (1981), G. T. Cole and Samson (1979), Hennebert and Weresub (1977), Kendrick (1977), and Subramanian (1983) and, with respect to *Trichoderma*, to Domsch et al. (1980) and Tribe (1983). In view of the complexity of the classification of *Trichoderma* species and of different genotypes from such teleomorphs as *Hypocrea*, *Podostroma*, and *Thuemenella* it is advisable to gain confirmation of the taxonomic status of strains before embarking on major research enterprises.

II. MORPHOLOGY

In view of the morphologic variation that arises through nutritional and environmental influence, it is essential to standardize cultural conditions in order to be able to gain meaningful comparison, e.g., the use of oatmeal agar and culture in daylight (Domsch et al., 1980). *Trichoderma* colonies grow rapidly, initially forming a smooth white surface and becoming compact following conidiation. Colonies may produce yellow pigments, but mature colonies are green from the color of the conidial masses. Conidia form on branched aerial conidiophores (e.g., *T. viride*, Fig. 16.1), speciation being based in part on conidiophore morphology (Rifai, 1969). Conidia (phialospores) are produced inside flask-shaped phialides and are released at the tip (Fig. 16.2) to form clusters, that is, basipetal succession via production of enteroblastic conidia. The initial phialide cell wall is ruptured on release of the first conidium; and as successive conidia are released, a ridged collarette comprised of rings of residual cell wall material builds up inside the rim of the phialide tip (Hammill, 1974a). The conidia lack distinctive characteristics apart from those of *T. viride*, which have rough walls (Figure 16.1).

In *Hypocrea* species the mycelium aggregates to form a stroma, which is often bright colored—white, yellow, or red. Many *Hypocrea* species are heterothalllic, but *H. psychrophila* is homothallic and *H. pulvinata* readily forms perithecial stroma in culture (W. Gams, pers. comm.). Perithecia, in which the sexual organs develop, are borne on the stroma. In such fruiting

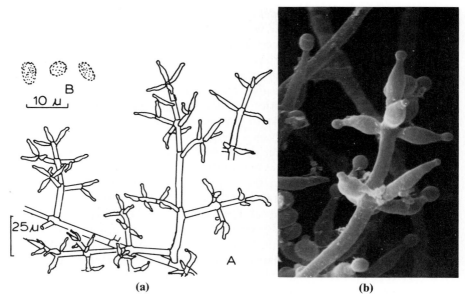

FIGURE 16.1 (a) *Trichoderma viride* (A) phialides and (B) philalospores. (From Rifai (1969), with permission.) (b) Scanning electron micrograph of *Trichoderma* sp. to illustrate conidia emerging from phialides. Magnification ×2000. (Photo courtesy Erica Rowe. From Griffin (1981), copyright 1981 by John Wiley and Sons, Inc. Used with permission.)

cultures, the stromata are formed from mycelial tufts on the culture surface; coiled oogonia form within these tufts with resultant stimulation of perithecial wall formation. Ascogenous hyphae are formed at the base of the cavity of the perithecia. This is typical of the *Nectria* type of development (Webster, 1980). Perithecia develop well in cultures grown at 16–20°C but should be maintained up to six weeks. The ascospores develop in long thin-walled asci and are generally two-celled. In comparison to the anamorph, the teleomorph is noted to occur relatively rarely in nature (Webster, 1964, 1980), though this could be due to a sparsity of total observations.

III. NUTRITION AND PHYSIOLOGY

Trichoderma species are metabolically versatile and can utilize a diverse range of substrates. Carbon sources include many sugars and polysaccharides such as cellulose, chitin, laminaran, pectin, starch, and xylan (Danielson and Davey, 1973c; Domsch et al., 1980). Considerable variation among species occurs in the utilization of inulin, melezitose, raffinose, sucrose, and tannic and gallic acids (Danielson and Davey, 1973c). Rhamnose, *i*-inositol and α-methyl-

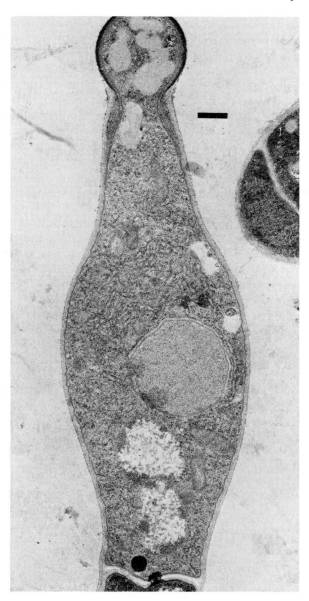

FIGURE 16.2   Phialide of *Trichoderma saturnisporum* with young conidial initial. Note the electron-opaque outer wall layer and the accumulation
of lipid globules of the conidial initial, the collarette and lower thick-walled
neck region of the phialide, the single nucleus lower in the phialide, and the
septal pore with pore plug at the base of the phialide. Bar equals 0.5 μm.
(From Hammill (1974a). Used with permission.)

D-glucoside are exceptions and are poorly used. It is somewhat enigmatic that in spite of the marked cellulolytic nature of most *Trichoderma* species, their ability to degrade wood (lignocellulose) is relatively weak. Strains can colonize wood by using nonstructural carbohydrates. They attack loblolly pine logs but, in doing so, destroy only the ray parenchymatous cells and the half-bordered pits (Hulme and Stranks, 1970). These actions can be put to good use to promote enhanced penetration and thus improved application of timber preservatives (Johnson and Gjovik, 1970). However, the attack is species specific, and no action is shown toward Douglas fir. In pure culture studies, several *Trichoderma* species were relatively ineffective in degrading dogwood leaves and loblolly pine needles (Danielson and Davey, 1973c), beechwood (Butcher, 1968), and birch and pine blocks (Bergman and Nilsson, 1971).

Besides being able to metabolize a variety of polymeric materials, *Trichoderma* species are one of the few groups of organisms that can metabolize C_1 compounds. Thus *T. lignorum* was considered as a source of single-cell protein based on its ability to metabolize methanol, but it was abandoned because it grows slowly (0.029 generation/h), grows optimally only at a low (0.16%) methanol concentration and gives low cell yields (Tye and Willets, 1977). *Trichoderma* strains can degrade hydrocarbons (Davies and Westlake, 1979), and they are major components of populations from soils polluted with oil (Gudin and Chater, 1977; Llanos and Kjoller, 1976; Pinholt et al., 1979). Many nitrogen sources (ammonium compounds, L-alanine, L-aspartate, L-glutamic acid, and proteins) are all readily utilized, although nitrate assimilation is often poor and is species dependent (Danielson and Davey, 1973c). From the above nutritional patterns, Okuda et al. (1982) suggest that the utilization of sucrose, raffinose, melezitose, and nitrate, and their reactions toward tannic and gallic acids, can be used as an aid in classification.

IV. PHYSIOLOGY

Trichoderma species grow readily on solid or in liquid media, in pure (Pitt and Bull, 1982a) or mixed continuous cultures (Senior et al., 1976). Cell composition is typical of ascomycetes and the cell composition is modified routinely in response to carbon or nitrogen limitation (Pitt and Bull, 1982a, b). The cell walls are comprised of β-glucans and chitin (Benitez et al., 1975b). Hammill (1974b) showed that in *T. saturnisporum* the cross-walls contain septal pores, which became blocked by septal pore plugs both in the hypha and in the conidiophore (Fig. 16.2; Hammill, 1974a). This occurs commonly in *Trichoderma* species. The composition and function of these plugs are unknown. Chitin synthesis occurs at the hyphal tip. This was determined by the inhibition of growth of hyphal tips and of germinating conidia by application of wheat germ agglutinin, a lectin specific for β-(1,4)-linked N-acetylglucosamine residues (Mirelman et al., 1975). Cell wall synthesis by protoplasts can result in aberrant wall structures composed of β-1,3- and β-1,6-glucans without chitin

(Benitez et al., 1975a). Benitez et al. (1976) also reported that the major components of *T. viride* conidia are β-1,6-glucans, β-1,3-glucans, and melanins (35%, 10% and 21%, respectively), and emphasized the absence of chitin. *Trichoderma* has been considered as a source of single-cell protein (Chahal and Gray, 1969). Protein synthesis appears to be maximal at the hyphal tip, and ribosomes are not transported along the hypha as the tip advances (Stavy et al., 1970). Since limited amounts of endoplasmic reticulum (ER) are found in fungi (Hawker, 1965), it is significant that relatively large amounts of ER occur in the conidia of *T. saturnisporum* (Hammill, 1974a) and of other species. Enhanced amounts of ER occur in the mycelium of hypersecreting mutants of *T. reesei* (Ghosh et al., 1982).

Trichoderma species are excellent producers of enzymes (Table 16.1). Cellulase has received particular attention in the utilization of biomass as a source of feedstock chemicals (Bungay, 1981). *Trichoderma* cellulase is noteworthy in that it attacks crystalline cellulose through the synergistic action of three major types of enzyme: cellobiohydrolase, endoglucanase, and cellobiase. Cellulase is inducible by cellulose and also by lactose and sophorose, this latter feature not being commonly found in fungi. Through the use of hypercellulolytic mutants and controlled fermentation conditions, extraordinarily high yields of cellulase (2% extracellular protein, 70% of which is cellulase) have been obtained. Scale-up to 150 liters has also produced good titers (Watson and Anziska, 1983). As was noted above, the high productivity is probably related to the enhanced amounts of endoplasmic reticulum, at least in the mutant *T. reesei* RUT-C30. The cell walls also influence the release of certain extracellular enzymes. Strains of *T. pseudokoningii* and *T. viride*, which produce higher amounts of β-1,3-glucanase (an enzyme associated with

TABLE 16.1   *Trichoderma Enzymes*[1]

Enzyme	Species	Reference
Cellulase	*T. reesei*	Petterson et al. (1981)
		Tangnu et al. (1981)
		Warzywoda et al. (1983)
	T. koningii	Halliwell and Vincent (1981)
		Wood (1981)
Xylanase	*T. reesei*	Nhlapo (1982)
	T. pseudokoningii	Baker (1977)
α-(1,3)-glucanase	*T. viride*	Hasegawa and Nordin (1969)
β-(1,3)-glucanase	*T. reesei*	Bamforth (1980)
	T. viride	Bielecki and Galas (1977)
Fungal cell wall lytic complexes	*T. harzianum*[2]	Wessels and Sietsma (1979)
L-Lysine-α-oxidase	*T. viride*	Midorikawa (1979)

[1]Representative examples are given. Most *Trichoderma* species produce cellulase, xylanase, and chitinase.
[2]Originally cited as *T. viride*, but later changed.

cell wall mobilization), release greater amounts of cellobiase into the culture medium (Kubicek, 1983). Treatment of *T. reesei* cell walls with laminaranase or chitinase also releases cellobiase, while trypsin and α-mannanase release endo-β-1,4-glucanase (Kubicek, 1981). Several recent reviews of *Trichoderma* cellulase production are available (Bailey et al., 1975; Becker et al., 1981; Brown and Jurasek, 1979; Cuskey et al., 1983; Ghose, 1981; Goksøyr and Eriksen, 1980; Lee and Fan, 1980; Mandels, 1982; Montenecourt, 1983; RAPAD, 1983; Ryu and Mandels, 1980; Shoemaker et al., 1981; Wood, 1981; Woodward and Wiseman, 1982; Zhu et al., 1982).

Cell wall lytic enzymes from *Trichoderma* spp. are used in the preparation of protoplasts from fungi (Wessels and Sietsma, 1979) and plants (Evans and Bravo, 1983).

Conidiation of *Trichoderma* is an excellent model for analysis of differentiation and has received considerable study. A short burst of light results in phialospore formation at the periphery of the colony, with initial changes being visible under the microscope by three hours and full conidiation occurring within 24 hours. Sporulation in *T. viride* is induced by light (350 to 510-nm wavelength) with maximal response at 380–390 and 440–450 nm (Fig. 16.3) (Gressel and Hartman, 1968; Jenssen, 1970; Kumagai and Oda, 1969) and with a minor response at 320 nm (Kumagai and Oda, 1969).

Minuscule amounts of light are required for conidial induction, half maximal response being achieved with 6.6×10^{-10} Einsteins/cm^2 at 447 nm (a few seconds of daylight) (Gressel et al., 1971). 5-Fluorouracil (Galun and Gressel, 1966) and other inhibitors of RNA synthesis (Betina and Zajakova, 1978) inhibit conidiation, the implication being that de novo RNA synthesis is required for the induction process, although it has not been possible to detect changes in RNA species following photoinduction (Stavy et al., 1972). Gressel et al. (1971) noted that acetylcholine in the presence of a choline esterase inhibitor, eserine, could mimic photoinduction of conidiation. Thus this fungal response is perhaps analogous to the blue light receptor systems of higher plants, but such a relationship remains speculative at present. When certain strains of *T. viride* are grown with limiting amounts of nitrogen, induction of sporulation requires a contact stimulus as well as light. Additional nitrogenous substrates abolish this contact requirement (Ellison et al., 1981). Studies of sporulation have practical import if large-scale production of *Trichoderma* is envisaged, both for the production of inocula or in biocontrol applications. Light slightly retards the growth rate of *Trichoderma* on solid media (Jenssen, 1970; Wishard, 1957). It was also noted that increased hyphal branching occurs in light-grown cultures. These effects on growth rate were not observed in liquid culture. Other physiological aspects of *Trichoderma* biology are less well documented. In one of the rare reports of energy charge in fungi, Pitt and Bull (1982b) showed that with *T. aureoviride* in continuous culture the adenylate energy charge is independent of specific growth rate, although the ATP content falls at dilution rates of less than 0.07/h. Other physiological aspects include the observation that isocitric lyase functions anapleurotically (Ohmori

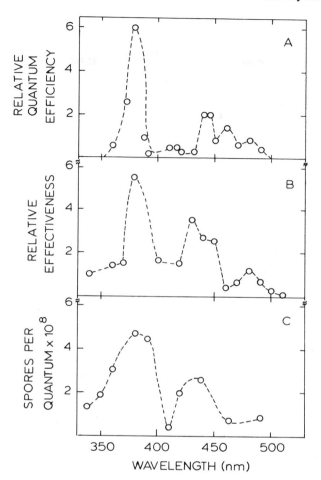

FIGURE 16.3 Action spectra for photo-induced sporulation of *Trichoderma viride* (From Jenssen (1970). Used with permission.)

and Gottlieb, 1965) and that nitrate reductase is inhibited by chloride at high ionic strength (Balderston et al., 1977).

The biochemical versatility of *Trichoderma* has been well documented, for example, the hydroxylation of the aromatic ring of L-tyrosine derivatives to produce intermediates for L-dopa (L-3,4-dihydroxyphenylalanine) synthesis (Sih et al., 1969), and the ability to transform a variety of chemically diverse pesticides such as allyl alcohol (Woodcock, 1971), Arachlor (Chahal et al., 1976), DDT, Aldrin, and Dieldrin (Matsumura and Boush, 1971), Malathion (Matsumura and Boush, 1966), Dalapon (Senior et al., 1976), and also *Aspergillus* aflatoxin (Mann and Rehm, 1976). Other reports include the hydroxylation of indole alkaloids, production of 6-aminopenicillanic acid, cyano-group cleavage of 2,3-dichlorobenzonitrile, and transformation of mycophenolic acid, vanillin, and fenclozic (Kieslich, 1976).

With the ability to utilize such a variety of substrates and to effect such diverse transformations, in combination with being able to survive under relatively adverse conditions, it is not too surprising to note that *Trichoderma* spp. are seen to be general spoilage organisms. They have been found in microcosms associated with the deterioration of paintings, masonry, rubber, plasticizers, polyethylenes (Pitt, 1981; Rose, 1981) and jet fuel (Sheridan, 1974). On paints they may simply be growing on surface detritus. Conversely, from their degradative attributes they have been recommended as control fungi in the evaluation of fungicides (*Trichoderma* sp. ATCC 12668).

V. ECOLOGY

A. Soil Inhabitants

Trichoderma spp. comprise a group of fast-growing Hyphomycetes that are extremely common in agricultural, prairie, forest, salt marsh, and desert soils in all climatic zones (Danielson and Davey, 1973a; Domsch et al., 1980). They are particularly prevalent in the litter of humid, mixed hardwood forests, comprising a minor component of the microbiota in the initial colonization but subsequently becoming more dominant in the H and F horizons. Most records of their occurrence are obtained by use of soil dilution plate counts (a method that emphasizes their conidial state), with reports that they constitute up to 3% of the total fungal propagules in a wide range of forest soils (Danielson and Davey, 1973a) and 1.5% of the fungi in pasture soils (Brewer et al., 1971). Their active hyphal state in soil has been recognized by use of the direct soil inoculum plating protocol (Waksman, 1916) and use of buried inert substrates (Parkinson et al., 1971).

There is some correlation between species distribution and environmental conditions. *Trichoderma polysporum* and *T. viride* occur in cool temperature regions, while *T. harzianum* is characteristic of warm climates. This correlates with optimal temperature requirements for each species (Danielson and Davey, 1973b). *T. viride* from shiitake mushroom logs grows optimally at 30°C (Hashioka et al., 1961). In general, *Trichoderma* species appear to be more prevalent in acidic soils, and Gochenaur (1970) was able to correlate the occurrence of *T. viride* with acid soils from cooler regions in Peru. *Trichoderma hamatum* and *T. pseudokoningii* are more tolerant of excessive moisture conditions than are other species. However, as a group, *Trichoderma* spp. are relatively intolerant of low moisture levels, and this has been suggested as one factor that contributes to the relatively low numbers of *Trichoderma* in the drier forest litter layers (Danielson and Davey, 1973a). In summary, the distribution of species is dependent on a variety of factors, for example, though *T. harzianum* may be characteristic of warm climates, it occurs widely in New Zealand and, in contrast, *T. viride* is relatively uncommon (J. Dingley, pers. comm.).

The widespread occurrence and effective colonization potential of *Trichoderma* species can be associated with several factors, including their metabolic versatility, their resistance to microbial inhibitors, and their antagonism to other microbes. Their ability to degrade a diverse variety of organic compounds has already been addressed. *Trichoderma* species are also tolerant of growth inhibitors. Strains grow on "staled agar media" (media on which other microorganisms have already been grown and which contain general growth inhibitors) (Dwivedi and Garrett, 1968). The germination of *Trichoderma* spores can be relatively insensitive to fungistasis (Emmatty and Green, 1966). In contrast, in reclaimed Dutch polder soils, the *Trichoderma* population has been noted to decline after a few years. This is thought to be in response to antagonism by the bacterial population (Pugh and van Emden, 1969). *Trichoderma* species are relatively resistant to synthetic chemicals such as carbon disulfide (Bliss, 1951; Webster, 1964), captan and chloropicrin (Anderson, 1962–1964; Smith, 1939), formalin (Warcup, 1951), dichloropropane-dichloropropylene (D-D) (Altson, 1950), and allyl alcohol, methyl bromide, and Semesan (2-chloro-4-hydroxy-mercuriphenol) (Woodcock, 1971). The dominance of *Trichoderma* species in soil following fumigation is well known. It is probably due to their inherent resistance to fumigants plus their enhanced ability to colonize in the absence of competitive microorganisms. This postfumigation dominance is of special significance in the use of *Trichoderma* as a biological control agent (Bliss, 1951; see below). The inherent general resistance has resulted in the incorporation of biologically active chemicals in media designed for the selective isolation of *Trichoderma* spp., for example, the use of sodium 4-dimethylamino-benzenesulfonate (Lesan), pentachloronitrobenzene (Terraclor), tetrachlorotetraiodofluorescein (rose bengal), and N-(trichloro-methyl-thio)-cyclohex-4-ene 1,2-dicarboximide (Captan) (Elad and Chet, 1983). Presumably, this remarkably broad-based tolerance of Trichodermas to growth inhibitors of both microbial and xenobiotic origin facilitates their effective colonization of soil.

B. Fungal Antagonism and Potential as Biocontrol Agents

Trichoderma has been noted to be antagonistic to several fungi. Bliss (1951) found that following fumigation of soil with carbon disulfiide there was widespread killing of *Armillaria mellea*. He proposed that the cause of death of the residual *A. mellea* population was destruction by *T. viride* that became dominant as a result of resistance to fumigation and reduction in the population of competitive soil microbes. Altson (1950) made a similar proposal concerning the destruction of *Rigidoporus microporus (Fomes lignosus), Phellinus noxius (F. noxius)*, and *Ganoderma pseudoferreum* by *T. viride* following soil treatment with D-D. In an analogous manner, *T. viride* is an effective antagonist to the growth of *Heterobasidion (Fomes) annosum* on pine root surfaces in acid soils (Rishbeth, 1951). On the basis of these results, trials for evaluating *Trichoderma* species as biological control agents for root disease fungi have been made. Mughogho (1968) noted that large populations of *Trichoderma*, which arose following the fumigation of soils, were ineffective in controlling *Armillaria mellea* or *Rhizoctonia solani*. The most effective species was *T. viride*, but it was concluded that there was considerable variation in the antagonistic properties of the individual strains. Kelley (1976) found that *T. harzianum* was ineffective in controlling *Phytophthora cinnamomi*. However, *Trichoderma* has been successfully applied as a biological control agent for a range of plant pathogens (Table 16.2). These results involve only small-scale demonstrations, but, larger field trials are in progress. Ricard (1977) demonstrated the effectiveness of *T. polysporum* (ATCC 20,475) against dry bubble disease (*Verticillium fungicola*) of mushrooms (*Agaricus* sp.) and gained registration by the French Government for this strain and also for *T. viride* ATCC 20,476 as fungicides (Bailly and Dubois, 1979). It should be noted that *T. viride* can also be a nuisance by inhibiting the growth of the shiitake mushroom (*Lentinus edodes*) (Komatsu and Inada, 1969). In fact, benomyl has been used to selectively inhibit *Trichoderma* in order to protect the shiitake crop (Tabata and Kondo, 1977). Nevertheless, there continues to be an active interest in developing the use of *Trichoderma* in biocontrol measures for plant pathogens (Nelson and Hoitink, 1983; Papavizas and Lewis, 1983). Indeed, the Binab Corporation (Sigtuna, Sweden) has marketed Scytalidium and Trichoderma biocontrol agents for a decade (Ricard, 1981).

The antagonistic nature of *Trichoderma* is not defined and can fall into any or all of the following categories: (1) competition for nutrients, (2) hyphal interaction/lysis, and (3) inhibition by antibiotic products.

1. *Competition for nutrients*: Hulme and Stranks (1970) showed that after invasion by *T. viride*, beechwood resisted colonization by Basidiomycetes. They suggested that the inhibition of the secondary colonizers was a result of the prior rapid removal of nonstructural carbohydrates (sugars and starch) by *Trichoderma*. A similar mechanism has been proposed to explain why *Trichoderma* becomes the dominant fungus on Southern pine logs following their treatment with sodium fluoride. Dominance in this instance was regarded as due to the resistance of *Trichoderma* to the preservative and the rapid utilization of nonstructural carbohydrates (Lindgren and Harvey, 1952). The invasion results in little effect on the mechanical properties or physical appearance of the wood. Both groups of workers proposed that the treatment of logs with *Trichoderma* can be used prophylactically to inhibit attack by wood decay and blue stain fungi.

2. *Hyphal interaction and lysis*: The action of *Trichoderma* as a mycoparasite is well known and was perhaps first recorded on *Sclerotinia* (Ezekiel, 1927). Subsequent reports include attack of *Armillaria mellea* (Aytoun, 1953) and *Mycena citricolor* (Salas, 1970). The production of lytic enzymes by *Trichoderma* that are active toward other fungi is well known (Table 16.1). Penetration and lysis of *Phycomyces* and *Rhizoctonia solani* (Durrell, 1966) and of *Pellicularia sasaki* (Hashioka, 1973) by *T. viride* and *T. longibrachiatum* respectively, could be considered common. How-

TABLE 16.2 Attempts at Application of *Trichoderma* spp. for Biological Control

Fungus	Species	Comment	Reference
Botrytis cinerea	*T. harzianum*	Antagonism demonstrated under laboratory conditions	Wells et al. (1972)
Botrytis cinerea	*T. pseudokoningii*	Infection of apples and strawberries suppressed	Tronsmo and Dennis (1978)
Ceratocystis ulmi	*T. viride*	Field observations of Dutch elm disease	Ricard (1983)
Hymenomycetes	*T. harzianum*	Hymenomycete wound invasion inhibited up to 21 months following inoculation of injured *Acer rubrum* by *T. harzianum*	Pottle et al. (1977)
Hymenomycetes (*Peniophora*) and blue stain fungi	*T. viride*	Protection of logs obtained in the field following use of a preservative (sodium fluoride) pretreatment	Lindgren and Harvey (1952)
Lentinus lepideus	*T. polysporum*	Reduction in infection of creosote treated poles by wood-rotting fungi	Ricard (1976)
Rhizoctonia solani	*T. harzianum*	Control of "damping-off" of radish, combined with the additive effect of pentachloronitrobenzene	Henis, et al. (1978)
Rhizoctonia solani	*T. harzianum*	Control of stem rot of carnations	Elad et al. (1981)
Sclerotinia trifoliorum	*T. harzianum*	Antagonism demonstrated under laboratory conditions	Wells et al. (1972)
Sclerotium rolfsii	*T. harzianum*	Control in the field of the disease of blue lupins, tomatoes, and peanuts	Wells et al. (1972)
Stereum purpureum	*T. viride*	Control of silver leaf disease following pruning obtained by application of a spore inoculum	Grosclaude et al. (1973)
Stereum purpureum	*T. harzianum* ATCC-20476 *T. polysporum* ATCC-20475	A 7 year study on which the British Pesticides Safety Precautions Clearance 12776 was based.	Corke (1979)
Verticillium fungicola	*T. polysporum*	Effective mycocide used with mushrooms against the dry bubble disease agent	Ricard (1977) deTrogoff and Ricard (1976)

ever, the interaction between *Trichoderma* and other fungi is not always lytic in nature. Dennis and Webster (1971c) noted that *Trichoderma* rarely penetrated *Heterbasidion annosus* cells but that *Trichoderma* hyphae coiled around the basidiomycete's hyphae with resultant marked inhibition of growth of the latter. "Coiling" is a classical action of *Trichoderma* toward the hyphae of other fungi, but the mechanism of inhibition has yet to be defined.

3. *Production of antibiotics*: *Trichoderma* produces a variety of antibiotics, several of which have antifungal properties. These antibiotics may or may not be volatile and can affect growth and germination of spores. It is assumed that these compounds are produced to some extent under natural conditions and contribute to the colonizing potential of *Trichoderma*. The antibiotics are discussed further below.

It is unclear which of these antagonistic traits are the most critical in aiding the dominance of *Trichoderma*. The reader is referred to a detailed review by Hawksworth (1981), which considers fungal–fungal interactions in considerable depth. One well-defined instance of pathogenicity toward humans is recorded (Loeppky et al., 1983).

VI. MYCOTOXINS AND OTHER SECONDARY METABOLITES

Trichoderma species produce a range of mycotoxins, which can be broadly considered in three main classes: trichothecenes, cyclic peptides, and isocyanide-containing metabolites.

A representative trichothecene is trichodermin, a 12,13-epoxytrichothecene (Figure 16.4; Abrahamsson and Nilsson, 1966; Godtfredsen and Vangedal, 1965). The trichothecenes exhibit activity toward eukaryotes (Carrasco et al.,

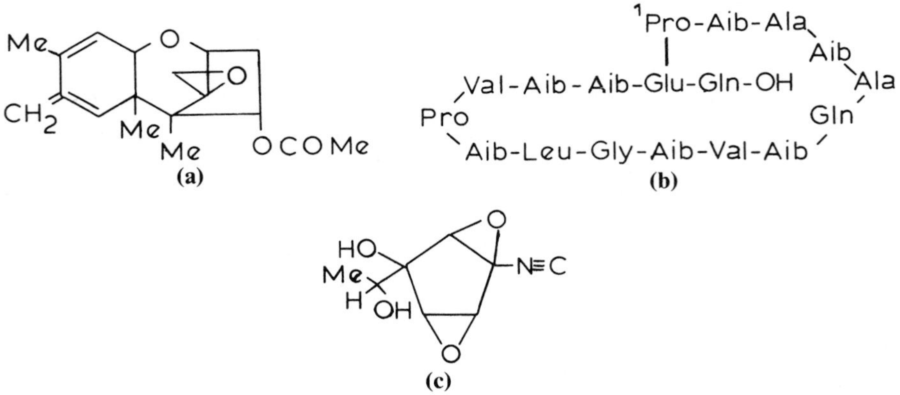

FIGURE 16.4 *Trichoderma* mycotoxins. (a) Trichodermin: 4, β-acetoxy-12,13-epoxy-Δ'-trichothecene. (b) Alamethicin (antibiotic U-22324) cyclic peptide. (c) Trichoviridin.

1973; Cutler and LeFiles, 1978) through inhibition of peptidyl transferase action. Mutants resistant to trichodermin can provide a genetic marker for a 60S ribosomal function (Grant et al., 1976). Impairment of growth of higher plants by trichodermin implies that its production in soil could reduce crop yields (Cutler and LeFiles, 1978). The trichothecenes can also provide a template for the production of medicinals and plant growth regulators (Cutler and LeFiles, 1978). There is a single report of the production of T-2 toxin by *T. lignorum* (R. C. Cole and R. H. Cox, 1981), but this could possibly be due to misnaming of the culture.

Cyclic peptide mycotoxins include alamethicin (antibiotic U-22324) (Fig. 16.4; Jung and Dubischar, 1975; Meyer and Reusser, 1967), suzukacillin (Ooka and Takeda, 1972), trichotoxins (Hou et al., 1972), trichopolyns (Fuji et al., 1978), and trichorzianine (Bachet et al., 1983). These lipophilic peptides are inhibitory to both prokaryotic and eukaryotic cells. They intercalate within the phospholipid membranes to modify ionic permeability and also promote lysis.

The third group of toxic metabolites are isocyanides, such as trichoviridin (Fig. 16.4; Nobuhara et al., 1976). Isocyanides have been thought to occur for quite some time; but owing to their instability, they have only recently been isolated (Brewer et al., 1982). Trichoviridin is an exception and is fairly stable. These toxins occur widely in strains of *T. hamatum*, which is a major component of the microbial soil population of certain sheep pastures. As these isocyanide metabolites markedly inhibit cellulolytic rumen microbes, they have been implicated as factors promoting ill-thrift of sheep (Brewer et al., 1971).

Trichoderma strains produce a variety of secondary metabolites. Anthroquinone pigments are major products, for example, *pachybasin* (1-hydroxy-3-methyl-9,10-anthraquinone), *chrysophanol* (1,8-dihydroxy-3-methyl-9,10-anthroquinone) and *emodin* (1,6,8-trihydroxy-3-methyl-9,10-anthroquinone) (Fig. 16.5) (Slater et al., 1967; Jenssen, 1970). Light (even a ten minute exposure on the third day of culture) reduced the production of these anthraquinones twofold to fivefold (Jenssen, 1970). No function has yet been clearly ascribed to the anthroquinones.

Other major secondary metabolites include the volatile 6-pentyl-α-pyrone (Fig. 16.5), which in part gives strains of *T. viride* (Collins and Halim, 1972; Kikuchi et al., 1974; Moss et al., 1975), and some *T. hamatum* (Okuda et al., 1982) strains a characteristic coconut odor. *Trichoderma* strains also produce a range of volatile inhibitory compounds that may aid in their colonization of soil (Dennis and Webster, 1971b). Other volatiles promote self-fertilization of *Phytophthora infestans* (Brasier, 1971). When fed to chickens, an unknown metabolite from the culture filtrate of *T. viride* modifies their cholesterol metabolism and results in larger eggs with 20% less cholesterol. Pullets began egg laying 3–6 weeks earlier than control birds (Qureshi et al., 1984). Production of antiviral (*T. todica*, ATCC 36936), antitumor (*T. viride* ATCC 20538 and 36387) and anti-weed (Charudattan and Lin, 1974) agents have been reported. Other metabolites are listed in Table 16.3.

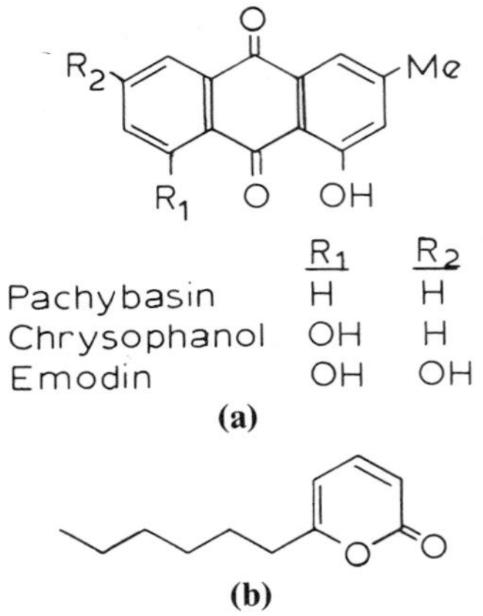

(a)

	R₁	R₂

	R_1	R_2
Pachybasin	H	H
Chrysophanol	OH	H
Emodin	OH	OH

(b)

FIGURE 16.5 *Trichoderma* secondary metabolites. (a) anthraquinone pigments. (b) 6-pentyl-α-pyrone (coconutlike aroma).

It is obvious that any project utilizing *Trichoderma* species should take into account the capability of these fungi to synthesize a variety of potent secondary metabolites. Yet it should be noted that cellulase preparations from *T. viride loc. sit.* have been used as a digestive aid in the preparation of geriatric foods in Japan for several years. Furthermore, *T. polysporum* preparations are registered as fungicides for the prevention of mushroom diseases. These strains have a negligible toxicity to rats: an LD_{50} in excess of 4 g/kg (Ricard, 1976).

VII. GENETICS

An interest in the genetics of *Trichoderma* has recently arisen, in relation to attempts to clone the cellulase genes; earlier studies are sparse. The emphasis on

TABLE 16.3 *Trichoderma* **Metabolites**[1]

2-(2-Amino-4-chlorophenoxy)-benzyl alcohol
Benzoquinones (thermophyllin)
Cardinanes (avocettin)
Citric acid
Dihydrocoumarins
Fatty acid derivatives (methyl-2,4,6-triene-1-carboxylate)
Fungal sex factor[2]
Gentisic acid
Itaconic acid
Polyacetylene-trichodermene[3]
Sterols (pyrocalciferol)

[1]Compiled from the *American Type Culture Collection Catalog* (1983), Turner, 1971, and Turner and Aldridge (1983)
[2]An unknown metabolite can induce self-fertilization in strains of *Phytophthora infestans*
[3]This is apparently the first record of a branched-chain polyacetylene

sexual reproduction mainly deals with the relationships between teleomorphs and anamorphs (Doi and Doi, 1979; Webster, 1964), besides showing that certain strains (e.g., *Hypocrea rufa*) were heterothallic. Strains of the opposite mating type are essential for most studies of classical sexual genetics. This is impossible, for *T. reesei* since only one strain (QM 6a) was isolated and maintained at the U.S. Army Natick Fungal Collection. Presumably, the opposite mating strain, if heterothallic, is still in the soil in the Solomon Islands, the original site of isolation. Genetic studies have therefore been limited to mutational approaches, although protoplast fusion and recombinant DNA methodologies have recently come to the fore.

Trichoderma is amenable to mutational analysis, since it has uninucleate conidia (see the electron micrographs of Rose et al. (1974) and Hammill (1974a, b) and it has no major nutritional limitations. Auxotrophs are readily isolated following mutagenesis, though in our experience leaky auxotrophs are common and double auxotrophs are thus preferred (Picataggio et al., 1983). Growth inhibitors are routinely used to retard the spreading of the colonies. Oxgall (2-deoxycholate) is our routine choice. Phosfon D was used initially but is not recommended, since Phosfon D–resistant mutants are recovered that have abnormal physiological characteristics. A wide variety of techniques for selection of hypercellulolytic strains have been used worldwide (for review, see Cuskey et al., 1983). From these studies, certain mutants were shown to yield coordinate increases in cellobiohydrolase, endoglucanase, and cellobiase, thus indicating a general mechanism for control of synthesis of the cellulase complex. However, cellulase negative/cellobiase positive strains have been isolated and also mutants with changes in the proportions of the three glycosidases (Shoemaker et al., 1981). This argues for additional discrete regulatory controls for synthesis of each enzyme. Heterokaryon formation between nonconidiating (con⁻) mutants resulted in complementation, but the conidia always produced colonies of the parental type (Weinman-Greenshpan and Galun, 1969). Protoplast fusion, which permits efficient heterokaryon formation, has produced more promising results. Protoplasts from *Trichoderma* can be readily prepared by using "home-made lytic enzyme cocktails" (Benitez et al., 1975a; Toyama et al., 1983) or via commercially available enzymes (Picataggio et al., 1983). Toyama et al. (1984) have developed this technique for protoplast fusion of *T. reesei* and gained high numbers of new phenotypes from conidia of the fusants. Prototrophs were obtained from auxotrophic parents. One recombinant strain produced twice the amount of cellulase than the cellulase-producing parent strain (the other parent was cellulase negative). The approach thus has industrial potential. Furthermore, *T. reesei* can be fused with *T. viride* and/or *T. pseudokoningii*, illustrating that the genetic differences between these species are minor (Toyama et al., 1984).

Genetic studies have also focused on recombinant DNA approaches. By "shotgun cloning" *T. reesei* DNA into a yeast cosmid system, complementation of yeast auxotrophs (ura-3, trp-1, and met-14 loci) was obtained (Picataggio et al., 1982). However, the large pieces of cloned *Trichoderma* DNA (30 Kb) in the cosmid were unstable in yeast. With the knowledge of the amino acid sequence of cellobiohydrolase I (Fägerstam et al., 1984; Shoemaker et al., 1983) it is possible to synthesize radioactive DNA probes to permit identification of cellulase-containing clones of *Trichoderma* gene banks. With this type of impetus, two groups have successfully cloned the cellobiohydrolase I gene in *E. coli* (Shoemaker et al., 1983, Teeri et al., 1983). Shoemaker et al. (1983) also showed the presence of two introns, which from their sequence data, are amenable to removal via the yeast-splicing system. These recent genetic advances via protoplast fusion and recombinant DNA approaches will rapidly result in new insights into the concepts of the holomorphic *Hypocrea/Trichoderma* system.

Acknowledgments

This article of the New Jersey Agriculture Experiment Station, Publication No. F-01111-02-84, was supported by State Funds, the U.S. Hatch Act and the Department of Energy Contracts DE-AS05-80ER10702 and XX-3-03045-01. I thank J. H. Dingley, W. Gams, and J. Ricard for their critical review of the manuscript.

REFERENCES

Abrahamsson, S., and Nilsson, B. (1966) *Acta Chem. Scand. 20*, 1044–1052.

Ainsworth, G. C., Sparrow, F. K., and Sussman, A. S., (1973) *The Fungi: An Advanced Treatise*, Vol. IV A, 621 pp., Academic Press, New York.

Altson, R. A. (1950) *Diseases of the Root System*, pp. 96–190, Rept. Rubber Research Institute, Malaya, 1945–1948.

Anderson, E. J. (1962–1964) in *Proceedings of the Annual Conference on Control of Soil Fungi, San Francisco and San Diego, Cal. 9*, 17: *10*, 13–14. (Cited in Baker, K. F., and Cook, R. J., in *Biological Control of Plant Pathogens*, pp. 124–293, W. H. Freeman and Sons, San Francisco.)

Aytoun, R. S. C. (1953) *Trans. Bot. Soc. Edinb. 36*, 99.

Bachet, B., Brassy, C., Morize, I., Surcouf, E., Mornon, J. P., Bodo, B., and Rebuffat, S. (1983) *J. Molec. Biol. 170*, 795–796.

Bailey, M., Enari, T.-M., and Linko, M. (eds.) (1975) *Symposium on Enzymatic Hydrolysis of Cellulose*, The Finnish National Fund for Research and Development (SITRA), Aulanko, Finland.

Baitty, R. and Dubois, G. (1979) *Index phytosanitaire*. Assoc. Coordin. Technique Agricole, 149 Rue de Bercy, F-75579, Paris Cedex 12, France.

Baker, C. J. (1977) *Phytopathology 67*, 1250–1258.

Balderston, W. L., Rowe, J. J., and Payne, W. J. (1977) *J. Bacteriol. 129*, 1657–1658.

Bamforth, C. W. (1980) *Biochem. J. 191*, 863–866.

Becker, D. K., Blotkamp, P. J., and Emert, G. H., (1981) in: *Fuels from Biomass and Wastes* (Klass, D. L. and Emert, G. H., eds.), pp. 375–391, Ann Arbor Science, Ann Arbor, Mich.

Benitez, T., Ramos, S., and Garcia Acha, I. (1975a) *Arch. Microbiol. 103*, 199–203.

Benitez, T., Villa, T. G., and Garcia Acha, I. (1975b) *Arch. Microbiol. 105*, 277–282.

Benitez, T., Villa, T. G., and Garcia Acha, I. (1976) *Can. J. Microbiol. 22*, 318–321.

Bergman, O., and Nilsson, T. (1971) *Dept. For. Prod. Roy. Coll. For. Stockholm. Res. Notes 71*, 54.

Betina, A., and Zajakova, J. (1978) *Folia Microbiol. 23*, 460–464.

Bielecki, S., and Galas, E. (1977) in *Proceedings of the Bioconversion Symposium* Vol. II (Ghose, T. K., ed.), pp. 203–225, Academic Press, New York.

Bissett, J. (1984) *Can. J. Bot. 62*, 924–931.

Bliss, D. E. (1951) *Phytopathology 41*, 665–683.

Brasier, C. M. (1971) *Nature (New Biology) 231*, 283.

Brewer, D., Calder, F. W., MacIntyre, T. M., and Taylor, A. (1971) *J. Agr. Sci. 71*, 465–477.

Brewer, D., Feicht, A., Taylor, A., Keeping, J. W., Taha, A. A., and Thaller, V. (1982) *Can. J. Microbiol. 28*, 1252–1260.

Brian, P. W. (1944) *Nature 155*, 667–668.

Brian, P. W., Curtis, P. J., Hemming, H. G., and McGowan, J. C. (1946) *Ann. Appl. Biol. 33*, 190–200.

Brown, R. D., Jr., and Jurasek, L. (eds.) (1979) *Hydrolysis of Cellulose, Advances in Chemistry Series 181*, American Chemical Society, Washington, D.C.

Bu'Lock, J. D. , and Leigh, C. (1975) *J. Chem. Soc. Comm. 75*, 628–629.

Bungay, H. R. (1981) *Energy — The Biomass Options*, 347 pp., John Wiley and Sons, New York.

Butcher, J. A. (1968) *Can. J. Bot. 46*, 1557–1589.

Carrasco, L., Barbacid, M., and Vasquez, D. (1973) *Biochim. Biophys. Acta 312*, 368–376.

Chahal, D. S., and Gray, W. D. (1969) in *Biodeterioration of Materials* (Walters, A. H., and Elphick, J. J., eds.), pp. 584–594.

Chahal, D. S., Bans, I. S., and Chopra, S. L. (1976) *Plant Soil 45*, 689–692.

Charudattan, R., and Lin, C. (1974) *Hyacinth Control J. 12*, 70–73.

Cole, G. T., and Kendrick, B. (1981) *Biology of Conidial Fungi*, Vol. I, 486 pp., Academic Press, New York: Vol. II, 660 pp., Academic Press, New York.

Cole, G. T., and Samson, R. A. (1979) *Patterns of Development in Conidial Fungi*, 190 pp., Pittman, London.

Cole, R. C., and Cox, R. H. (1981) *Handbook of Toxic Fungal Metabolites*, 937 pp., Academic Press, New York.

Collins, R. P., and Halim, A. F. (1972) *J. Agr. Food Chem. 20*, 437–438.

Corke, A. T. K. (1979) *Long Ashton Research Station Annual Report*, pp. 190–198, University of Bristol, U.K.

Cuskey, S. M., Montenecourt, B. S., and Eveleigh, D. E. (1983) in *Liquid Fuel Developments* (Wise, D. L., ed.), pp. 31–48, CRC Press, Boca Raton, Fla.

Cutler, H. G., and LeFiles, J. H. (1978) *Plant and Cell Physiol. 19*, 177–182.

Danielson, R. M., and Davey, C. B. (1973a) *Soil Biol. Biochem. 5*, 485–494.

Danielson, R. M., and Davey, C. B. (1973b) *Soil Biol. Biochem. 5*, 495–504.

Danielson, R. M., and Davey, C. B. (1973c) *Soil Biol. Biochem. 5*, 505–515.

Davies, J. S., and Westlake, D. W. S. (1979) *Can. J. Microbiol. 25*, 146–156.

Dennis, C., and Webster, J. (1971a) *Trans. Brit. Mycol. Soc. 57*, 25–29.

Dennis, C., and Webster, J. (1971b) *Trans. Brit. Mycol. Soc. 57*, 41–48.

Dennis, C., and Webster, J. (1971c) *Trans. Brit. Mycol. Soc. 57*, 363–369.

deTrogoff, H., and Ricard, J. L. (1976) *Plant Dis. Rep. 60*, 677–680.

Doi, N., and Doi, Y. (1979) *Bull. Nat. Sci. Mus. Tokyo Ser. B. (Bot.) 5*, 117–123.

Domsch, K. H., Gams, W., and Anderson, T.-H. (1980) *Compendium of Soil Fungi*, Vol. I., Academic Press, New York.

Durrell, L. W. (1966) *Mycopath. Mycol. Appl. 35*, 138–144.

Dwivedi, R. S., and Garrett, S. D. (1968) *Trans. Brit. Mycol. Soc. 51*, 95–101.

Elad, Y., and Chet, I. (1983) *Phytoparasitica 11*, 55–58.

Elad, Y., Hadar, Y., Hadar, E., Chet, I., and Henis, Y. (1981) *Plant Disease 65*, 675–677.

Ellison, P. J., Harrower, K. M., Chilvers, G. A., and Owens, J. D. (1981) *Trans. Brit. Mycol. Soc. 76*, 441–445.

Emmatty, D. A., and Green, R. J., Jr. (1966) *Can. J. Microbiol. 13*, 635–642.

Evans, D. A., and Bravo, J. E. (1983) in *International Review of Cytology*, Suppl. 16, *Plant Protoplasts* (Giles, K. L., ed.), Academic Press, New York.

Ezekiel, W. N. (1927) *Phytopathology 17*, 791–792.

Fägerstam, L. G., Pettersson, L. G., and Engström, J. A. (1983) *FEBS Letters 167*, 309–315.

Fuji, K., Fujita, E., Takaishi, Y., Fujita, T., Arita, I., Komatsu, M., and Hiratsuka, N. (1978) *Experientia 34*, 237–239.

Galun, E., and Gressel, J. (1966) *Science 151*, 696–698.

Ghosh, A., Al-Rabiai, S., Ghosh, B. K., Trimino-Vasquez, H., Eveleigh, D. E., and Montenecourt, B. S. (1982) *Enzyme Microbial Technol. 4*, 110–113.

Ghose, T. K. (ed.) (1981) *Bioconversion and Biochemical Engineering Symposium 2. BERC IIT Delhi, India, March 3–5, 1980*, Raj Bandhu Industrial Co., New Delhi, India.

Gochenaur, S. E. (1970) *Mycopath. Myco. Appl. 42*, 259–272.

Godtfredsen, W. O., and Vangedal, S. (1965) *Acta Chem. Scand. 19*, 1088–1103.

Goksøyr, J., and Eriksen, J. (1980) in *Economic Microbiology*, Vol. VI (ed. A. H. Rose), pp. 283–330, Academic Press, New York.

Grant, P., Schindler, G. D., and Davies, J. E. (1976) *Genetics 83*, 667–673.

Gressel, J. B., and Hartmann, K. M. (1968) *Planta 79*, 271–274.

Gressel, J. B., Strausbauch, L., and Galun, E. (1971) *Nature 232*, 648.

Griffin, D. H. (1981) *Fungal Physiology*, John Wiley and Sons, New York.

Grosclaude C., Ricard, J., and Dubos, B. (1973) *Plant Disease Rep. 57,* 25–28.

Gudin, C., and Chater, K. W. A. (1977) *Environ. Pollut. 144*, 1–4.

Halliwell, G., and Vincent, R. (1981) *Biochem. J. 199*, 409–417.

Hammill, T. M. (1974a) *Amer. J. Bot. 61*, 15–24.

Hammill, T. M. (1974b) *Amer. J. Bot. 61*, 767–771.

Hasegawa, S., and Nordin, J. H. (1969) *J. Biol. Chem. 244*, 5460–5470.

Hashioka, Y. (1973) *Forsch. Geb. Planzenkrakn. 8*, 179–190.

Hashioka, Y., Komatsu, M., and Arita, I. (1961) *Rep. Tottori Mycol. Inst. 1*, 1–8.

Hawker, L. E. (1965) *Biol. Rev. 40*, 52–92.

Hawksworth, D. L. (1981) in *Biology of Conidial Fungi* (Cole, G. T., and Kendrick, B., eds.) Vol. II, pp. 171–244, Academic Press, New York.

Henis, Y., Ghaffar, A., and Baker, R. (1978) *Phytopathology 68*, 900–907.

Hennebert, G. L., and Weresub, L. K. (1977) *Mycotaxon 6*, 207–211.

Hou, C. T., Ciegler, A., and Hesseltine, C. W. (1972) *Appl. Microbiol. 23*, 183–185.

Hulme, M. A., and Stranks, D. W. (1970) *Nature 226*, 469–470.

Hussain, S. A., Noorani, R., and Qureshi, I. H. (1975) *Pak. J. Sci. Ind. Res. 18*, 221. (Cited in Turner and Aldridge, 1983.)

Jenssen, W. D. (1970) "Morphogenesis in *Trichoderma*: Pigmentation and Sporulation," 138 pp., Ph.D. thesis, Rutgers University, New Brunswick, N. J.

Johnson, B. R., and Gjovik, L. R. (1970) *Amer. Wood-Preserver Assn. 66*, 233–242.

Jung, G., and Dubischar, N. (1975) *Eur. J. Biochem. 54*, 395–409.

Kelly, W. D. (1976) *Phytopathology 66*, 1023–1027.

Kendrick, B. (ed.) (1977) *The Whole Fungus: The Sexual-Asexual Synthesis*, Vol. I, 410 pp., National Museum of Nature Sciences, National Museums of Canada, and the Kananaskis Foundation. Ottawa, Canada.

Kieslich, K. (1976) *Microbial Transformations of Non-Steroid Cyclic Compounds*, 1261 pp., G. Thieme, Stuttgart, West Germany.

Kikuchi, T., Mimura, T., Harimaya, K., Yano, H., Arimoto, T., Masada, Y., and Inoue, T. (1974) *Chem. Pharm. Bull. 22*, 1946–1948.

Komatsu, M., and Inada, S. (1969) *Rep. Tottori Mycol. Inst. 7*, 19–26.

Kubicek, C. P. (1981) *J. Appl. Microbiol. Biotechnol. 13*, 226–231.

Kubicek, C. P. (1983) *Can. J. Microbiol. 29*, 163–169.

Kumagai, T., and Oda, Y. (1969) *Plant Cell Physiol. 10*, 387–392.

Lee, Y.-H., and Fan, L. T. (1980) *Adv. Biochem. Eng. 17*, 101–129.

Lindgren, R. M., and Harvey, G. M. (1952) *J. For. Prod. Res. Soc. 2*, 250–256.

Llanos, C., and Kjoller, A. (1976) *Oikos 27*, 377–382.

Loeppky, C. B., Sprouse, R. F., Carlson, J. V., and Evrett, E. D. (1983) *Southern Med. J. 76*, 798–799.

Mandels, M. (1982) in *Annual Report on Fermentation Processes* (Tsao, G. T., ed.), Vol. V, pp. 35–78. Academic Press, New York.

Mann, R., and Rehm, H. J. (1976) *Europ. J. Appl. Microbiol. 2*, 297–306.

Matsumura, F., and Boush, G. M. (1966) *Science 153*, 1278.

Matsumura, F., and Boush, G. M. (1971) in *Soil Biochemistry*, Vol 2 (McLaren, A. D., and Skujins, J., eds.), 527 pp., Academic Press, New York.

Meyer, P., and Reusser, F. (1967) *Experientia 23*, 85–86.

Midorikawa, Y. (1979) *Agr. Biol. Chem. 43*, 1749–1752.

Mirelman, D., Galun, E., Sharon, N., and Lotan, R. (1975) *Nature 256*, 414–416.

Montenecourt, B. S. (1983) *Trends Biotechnol. 1*, 156–160.

Moss, M. O., Jackson, J. M., and Rogers, D. (1975) *Phytochemistry 14*, 2706–2708.

Mughogho, L. K. (1968) *Trans. Brit. Mycol. Soc. 51*, 441–459.

Nelson, E. B., and Hoitink, H. A. J. (1983) *Phytopathology 73*, 274–278.

Nhlapo, Sipho David. (1982) "Fungal Degradation of Cellulose," Ph.D. thesis, Rutgers University, New Brunswick, N. J.

Nobuhara, M., Tazima, H., Shudo, K., Itai, A., Okamoto, T., and Iitaka, Y. (1976) *Chem. Pharm. Bull. 24*, 832–834.

Ohmori, K., and Gottlieb, D. (1965) *Phytopathology 55*, 1328–1336.

Okuda, T., Fujiwara, A., and Fujiwara, M. (1982) *Agr. Biol. Chem. 46*, 1811–1822.

Ooka, T., and Takeda, I. (1972) *Agr. Biol. Chem. 36*, 112–119.

Papavizas, G. C. and Lewis, J. A. (1983) *Phytopathology 73*, 407–411.

Parkinson, D., Gray, T. R., and Williams, S. T. (1971) *Methods for Studying the Ecology of Micro-organisms*. Blackwell Scientific Publications, Oxford, U. K.

Pettersson, G., Fägerstam, L., Bhikhabhai,R., and Leandoer, K. (1981) in *Ekman-Days International Symposium on Wood and Pulping Chemistry*, Vol. III, pp. 39–42, Stockholm, Sweden.

Picataggio, S. K., Morris, D., Noti, J., and Szalay, A. (1982) *Proc. Amer. Soc. Microbiol.* H85.

Picataggio, S. K., Schamhart, D. H. J., Montenecourt, B. S., and Eveleigh, D. E. (1983) *Eur. J. Appl. Microbiol. Biotechnol. 17*, 121–128.

Pinholt, Y., Struwe, S., and Kjoller, A. (1979) *Holarctic Ecol. 2*, 195–200.

Pitt, D. E., and Bull, A. T. (1982a) *Exp. Mycol. 6*, 41–51.

Pitt, D. E., and Bull, A. T. (1982b) *J. Gen. Microbiol. 128*, 1517–1527.

Pitt, J. I. (1981) in *Biology of Conidial Fungi* (Cole, G. T., and Kendrick, B. eds.), Vol. II, pp. 111–142, Academic Press, New York.

Pottle, H. W., Shigo, A. L., and Blanchard, R. O. (1977) *Plant Dis. Rep. 61*, 687–690.

Pugh, G., and van Emden, J. (1969) *Neth. J. Pl. Pathol. 75*, 287–295.

Qureshi, A. A., Prentice, N., Din, Z. Z., Burger, W. C., Elson, C. E. and Sunde, M. L. (1984) *Lipids 19*, 250–257.

RAPAD (1983) *Research and Development on Synfuels*, Annual Technical Report, 87 pp., Research Association for Petroleum Alternatives Development, Tokyo.

Ricard, J. L. (1970) *Stud. For. Suec. 84.*

Ricard, J. (1976) *Inst. Wood Sci. J.* (U.K.) *7* (No. 4), 6–9.

Ricard, J. (1977) *Neth. J. Plant Path. 83*, (Suppl. 1), 443–448.

Ricard, J. (1979) *Amer. Soc. Microbiol. News 45*, 122–124.

Ricard, J. (1981) *Biocontrol News and Information 2*, 95–98.

Rifai, M. A. (1969) *Mycol. Pap. 116*, 1–56.

Rishbeth, J. (1951) *Ann. Bot. Lond. 15*, 221–146.

Rose, A. H. (ed.) (1981) *Economic Microbiology*, Vol. VI, 516 pp., Academic Press, New York.

Rosen, D., Edelman, M., Galun, E., and Danon, D. (1974) *J. Gen. Microbiol. 83*, 31–49.

Ryu, D. D. Y., and Mandels, M. (1980) *Enzyme Microbial Technol. 2*, 91–102.

Saksena, S. B. (1960) *Trans. Brit. Mycol. Soc. 43*, 111–116.

Salas, J. A. (1970) Unpublished observation cited in Baker and Cook (1974), p. 40. (See Anderson reference).

Senior, E., Bull, A. T., and Slater, J. H. (1976) *Nature 263*, 470–479.

Sheridan, J. E. (1974) *Int. Biodet Bull. 10*, 105–107.

Shoemaker, S. P., Raymond, J. C., and Bruner, R. (1981) in *Trends in the Biology of Fermentations for Fuels and Chemicals* (Hollaender, A., ed.), pp. 89–109, Plenum Press, New York.

Shoemaker, S., Schweickart, V., Ladner, M., Gelfand, D., Kwok, S., Myambo, K., and Innis, M. (1983) *Bio/technology 1*, 691–696.

Sih, C., Foss, P., Rosazza, J., and Lemberger, M. (1969) *J. Amer. Chem. Soc. 91*, 6204.

Simmons, E. G. (1977) in *Abstracts of the 2nd International Mycology Congress, Tampa, Fla*, p. 618.

Slater, G. P., Haskins, R. H., Hogge, L. R., and Nesbitt, L. R., (1967) *Can. J. Chem. 45*, 92–96.

Smith, N. R. (1939) *Proc. Soil Amer. 3*, 188.

Stavy, R., Stavy, L., and Galun, E. (1970) *Biochim. Biophys. Acta 217*, 468–476.

Stavy, R., Galun, R., and Gressel, J. (1972) *Biochim. Biophys. Acta 259*, 321–329.

Subramanian, C. V. (1983) *Hyphomycetes: Taxonomy and Biology*, 502 pp., Academic Press, New York.

Tabata, T., and Kondo, T. (1977) *Mokuzai Gakkaishi 10*, 504–508 (cited in *Chem. Abst. 90*:088050330042). (See also *Chem. Abst.* 07913074923F and 082010010410).

Tamara, A., Kontani, H., and Naruto, S. (1975) *J. Antibiot. 28*, 161–162.

Tangnu, S. K., Blanch, H. W., and Wilke, C. R. (1981) *Biotech.Bioeng. 23*, 1837–1849.

Teeri, T., Salovvori, I., and Knowles, J. (1983) *Bio/technology 1*, 696–699.

Toyama, H., Shinmyo, A., and H. Okada, (1983) *J. Ferment. Technol. 61*, 409–411.

Toyama, H., Yamaguchi, K., Shinmyo, A., and Okada, H. (1984) *App. Environ. Microbiol. 47*, 363–368.

Tribe, H. T. (1983) *Bull. Brit. Mycol. Soc. 17*, 94–103.

Tronsmo, A., and Dennis, C. (1978) *Trans. Brit. Mycol. Soc. 71*, 469–474.

Tulasne, L. R., and Tulasne, C. (1865) *Selecta fungorum carpologia* (Eng. transl. by W. B. Grove) *3*, 27–35.

Turner, W. B., and Aldridge, D. C. (1983) *Fungal Metabolites*, Vol. II, 631 pp., Academic Press, New York.

Tye, R., and Willetts, A. (1977) *Appl. Environ. Microbiol. 33*, 758–761.

Waksman, S. A. (1916) *Soil Sci. 2*, 103–155.

Warcup, J. H. (1951) *Trans. Brit. Mycol. Soc. 34*, 519–532.

Warzywoda, M., Ferre, V., and Pourquie, J. (1983) *Biotechnol. Bioeng. 25*, 3005–3012.

Watson, T. G., and Anziska, K. G., (1982) *South Afr. Food Rev. 9*, (Suppl.), 102–104.

Webster, J. (1964) *Trans. Brit. Mycol. Soc. 47*, 75–96.

Webster, J. (1980) *Introduction to Fungi*, 2nd ed., 669 pp., Cambridge University Press, Cambridge, England

Webster, J., and Lomas, N. (1964) *Trans. Brit. Mycol. Soc. 47*, 535–540.

Weindling, R. (1934) *Phytopathology 24L*, 53–79.

Weinman-Greenshpan, D., and Galun, E. (1969) *J. Bacteriol. 99*, 802–806.

Wells, H. D., Bell, D. K., and Jaworski, C. A. (1972) *Phytopathology 62*, 442–447.

Wessels, J. G. H., and Sietsma J. H. (1979) in *Fungal Walls and Hyphal Growth* (Burnett, J. H., and Trinci, A. P. J., eds.), pp. 27–48, Cambridge University Press, Cambridge, England.

Wishard, R. H. (1957) "Effect of Visible Light on the Growth of Certain Fungi," 42 pp., M. S. thesis, George Washington University, Washington, D. C.

Wood, T. M. (1981) in *Proceedings of the International Symposium on Wood and Pulping Chemistry*, Vol. III, pp. 31–38, The Ekman Days, Stockholm, Sweden.

Wood, T. M., and McCrae, S. I. (1975) in *Symposium on Enzymatic Hydrolysis of Cellulose* (Bailey, M., Enari, T.-M., and Linko, M., eds.) pp. 231–254, Aulanko, Finland.

Woodcock, D. (1971) in *Soil Biochemistry*, Vol. II (McLaren, A. D., and Skujins, J., eds.), 527 pp., Academic Press, New York.

Woodward, J., and Wiseman, A. (1982) *Enzyme Microbial Technol. 4*, 73–79.

Zhu, Y. W., Wu, Y. Q., Chen, W., Tan, C., Gao, J. H., Fei, J. X., and Shih, C. N. (1982) *Enzyme Microbial Technol. 4*, 3–12.

The Biology of *Saccharomyces*

G. G. Stewart
I. Russell

I. INTRODUCTION

Yeasts are, without a doubt, both quantitatively and economically the most important group of microorganisms commercially exploited by humans. The total amount of yeast produced annually, including that formed during brewing and in distilling practices, is on the order of a million tons. This group of microorganisms is by traditional agreement limited to fungi in which the unicellular form is predominant and vegetative reproduction is usually, but not always, achieved by budding (Fig. 17.1). This group does not constitute a taxonomic entity, although it comprises subdivisions of narrowly related species. Compared to other major groups of microorganisms — algae, bacteria, and protozoa — the yeasts are represented by comparatively few genera and species. It is thought that there are only about 400 species grouped into 39 genera; however, the accepted text on yeast taxonomy is 1400 pages long (Lodder, 1970).

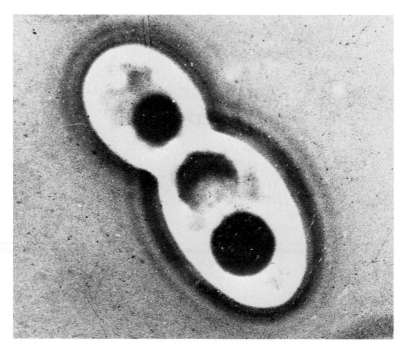

FIGURE 17.1   Phase contrast micrograph of a budding cell of *Saccharomyces cerevisiae*. (Photograph courtesy of C. F. Robinow, University of Western Ontario, London, Ontario.)

Phaff et al. (1978) have reviewed the etymology of "yeast" and equivalent terms in other languages and point out the consistent relationship to fermentation and ability to metabolize a wide range of sugars. Some biochemists and geneticists incorrectly use the term yeast as synonymous with *Saccharomyces cerevisiae*. Although most of the biochemical and genetic studies have been conducted on this species, there is a rich variety of yeast genera and species that await biological investigation; indeed, the most exotic species may offer advantages for experimental study. In this chapter the authors have drawn heavily on data obtained with *S. cerevisiae* and related species, but whenever possible reference has been made to other species of the genus *Saccharomyces*.

No other group of microorganisms has been more intimately associated with the progress and well-being of the human race than *S. cerevisiae* and its closely related species. Their contribution to human progress has been based very largely on the capacity of these yeasts to effect a rapid and efficient conversion of sugars to ethanol and carbon dioxide, thus conducting an ethanolic fermentation of sugary liquids such as grape juice, grain extracts, and milk.

Some biotechnologists feel that in the future the industrial importance of this group of microorganisms will diminish because they produce few secondary metabolites of commercial interest (e.g., they do not produce any biologically active antibiotics). However, many species of yeast (particularly of the genus *Saccharomyces*) produce the very important primary metabolite ethanol used both as a beverage and for industrial/fuel purposes. A second major contribution that yeasts have made to human progress has been their use in the elucidation of the basic biochemical, metabolic and genetic processes of living cells. Rose (1977) stated the following:

> Biochemistry was born out of the yeast *Saccharomyces cerevisiae.* Attention was then turned to the bacterium *Escherichia coli;* however, the cycle is now coming full circle, for increasing numbers of biologists are returning to *S. cerevisiae* which is seen as a most valuable microbe.

S. cerevisiae is currently favored as an experimental organism because it is an eukaryote that can be grown and analyzed genetically with similar ease as bacteria. The power of the yeast genetic system is a critical and important parameter. Cells grow stably either as haploids or as diploids (Fig. 17.2). It is easier to obtain mutants in the haploid phase than in diploid or polyploid cells of higher plants or animals. It is possible in the diploid phase to determine whether mutants are dominant or recessive, to classify them into complementation groups, and to map the mutation in the genome. During the past 30 years, genetic studies have yielded a great deal of information on yeast genes — well over 300 loci have been mapped to 17 linkage groups (Mortimer and Schild, 1981). In addition, considerable effort has been made to manipulate and modify the yeast genome using classical techniques, mainly mutation and hybridization. This knowledge has provided the essential background for gene cloning and host–vector development in *S. cerevisiae.*

Many investigators emphasize the fact that, although yeasts can be manipulated like prokaryotes (bacteria and viruses), they are essentially similar to higher eukaryotes (plants and animals) in cellular structure and behavior. For instance, it is often pointed out that yeasts have a nucleus bounded by a nuclear membrane, mitochondria, vacuoles, and Golgi bodies (Fig. 17.3). In addition, macromolecular synthesis, chromosome replication, and segregation in yeasts are similar to these processes in higher organisms. However, Carter (1978) has pointed out that there are features of yeast structure and behavior that are rather different from those of higher eukaryotes. For example, during mitosis in yeast, spindle formation is quite different from that observed in higher eukaryotes, and the nuclear membrane does not break down. In addition, in yeast, unlike other eukaryotes, mitosis and cell segregation are temporarily distinct. Yeast is unusual in that mitosis occurs in both haploid and diploid cells. Cell division by budding and nuclear migration into the isthmus of mother cell and bud prior to nuclear division is not a common feature of

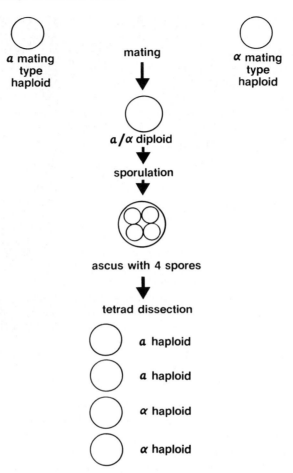

FIGURE 17.2   Haploid/diploid life cycle of *Saccharomyces* sp.

eukaryotic cells. Nevertheless, studies on yeast have and will continue to provide valuable information that is pertinent to all eukaryotes.

Antonie van Leeuwenhoek is credited with being the first to have observed yeasts microscopically. In 1680 he sent descriptions and drawings of yeast cells from the Netherlands to the Royal Society in London. Leeuwenhoek was not a scientist but a draper by profession, and it was left to the botanists of the day to lay the firm foundations of yeast microbiology. However, it was not until the first half of the 19th century that significant progress was made toward understanding the biology of yeasts; through this developed an appreciation of yeast physiology and biochemistry.

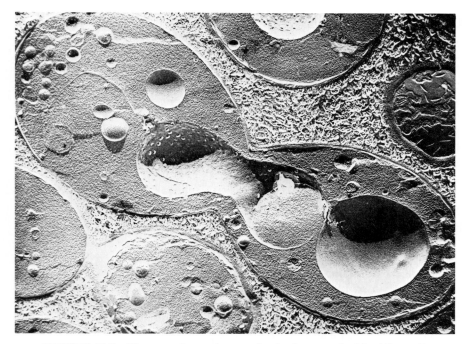

FIGURE 17.3   Electron photomicrograph of a freeze-etched budding cell of *Saccharomyces cerevisiae*. Bar represents 1 μm. (Photograph courtesy of C. F. Robinow, University of Western Ontario, London, Ontario.)

Cagniard-Latour in 1837 demonstrated that beer contains spherical bodies that are able to multiply and that belong to the vegetable kingdom. In this he received support from Schwann, who termed yeast "Zuckerpilz" or "sugar fungus" from which the name *Saccharomyces* originates. However, this cellular or vitalistic theory of fermentation was vehemently, and occasionally mockingly, attacked by a trio of chemists — Liebig, Wöhler, and Berzelius. The euphoria that stemmed from the success of the newly recognized branch of organic chemistry in explaining hitherto complex and mysterious organic processes convinced this trio that chemical reactions, rather than the activities of living cells, could perfectly well explain the alcoholic fermentation of sugars.

It finally fell to Louis Pasteur to prove that fermentation is due to living cells — thereby winning the argument for a vitalistic theory of fermentation. Pasteur continued to make masterly contributions to yeast microbiology and to understanding fermentation in general, particularly in his "Etudes sur le Vin" (Pasteur, 1866) and "Etudes sur la Biere" (Pasteur, 1876), in which he clarified much concerning the effect of oxygen on alcoholic fermentation by yeast.

As previously described, the association of yeast with the science of biochemistry is an old and well-established one, and in fact, biochemistry was born out of yeast technology. The Buchner brothers, working in Germany in 1897, were interested in preparing extracts of yeast for medicinal purposes. To do this, they ground brewer's yeast with sand and then squeezed out the juice. In an effort to preserve their extract, they tried adding large amounts of sucrose, since it was known that solutions containing high concentrations of sugar are less prone to microbial infection. To their surprise, they discovered that the sugar was rapidly fermented by the yeast juice. This chance encounter set off a whole series of studies in an effort to determine the nature of the steps in the fermentation of glucose to ethanol and carbon dioxide. This work spanned a number of years and is associated with the names of many outstanding biochemists.

The industrial uses of yeast can be classified into three groups according to the relationship of the product to the biochemistry of the organism:

1. Cell constituents
    a. food and fodder yeasts (single-cell protein)
    b. macromolecular constituents—lipids, proteins and peptides, enzymes, coenzymes, nucleic acids, and vitamins
2. Excretion products: beer, wine, spirits, cider, industrial ethanol, glycerol, and carbon dioxide (for carbonated beverages and in the leavening of bread and cakes with bakers yeast)
3. Enzyme–substrate interaction
    a. whey (lactose) utilization—*Kluyveromyces fragilis* or *lactis*
    b. starch/dextrin utilization—*Saccharomyces diastaticus*

The two main yeast genera of commercial interest are *Saccharomyces* and *Candida*, especially *S. cerevisiae* and closely related species. So many strains of the same species have been selected and, in these days of genetic manipulation, are being bred for a variety of purposes, that taxonomy in the classical sense becomes almost meaningless. Strain differentiation is in many cases more important than species classification and, in the future, will become a matter of extremely accurate quantitative measurement with the advent of patents protecting new hybrids. However, van der Walt (1970) has defined the genus *Saccharomyces* as a group of microorganisms in which vegetative reproduction is solely by multilateral budding; the cells are spheroidal, ellipsoidal, cylindrical, or elongate and may form pseudomycelia, but mycelia are absent. The ascospores are spheroidal, and the asci do not rupture on reaching maturity. Usually, one to four ascospores are formed per ascus, but on rare occasions more will form. As was previously mentioned, the genus is fermentative, and lactose, higher paraffins, and nitrate are not utilized. This rather general definition of the genus has a number of shortcomings due to the inadequacy of the currently available diagnostic criteria; therefore it is impossible to formulate separate genera on the basis of mutually exclusive lines, completely eliminating the misclassification of more advanced elements of these lines caused by possible convergent evolution. On the other hand, this is by no means a new feature of *Saccharomyces*, since this taxon has been phylogenetically heterogeneous almost since the inception of yeast systematics. The present demarcation of the genus is thus utilitarian and aims merely at providing a method of reference and communication.

Although at present only *S. cerevisiae* and related species are of major industrial importance, some of the many yeast species could certainly assume a considerable role in industry, as a source of protein, for producing ethanol, and for carrying out other chemical transformations such as those involved in synthesizing precursors of important natural products. The following research would further such ends.

1. The carbohydrate utilization capabilities of the more versatile yeasts should be characterized genetically and biochemically.
2. Artificial means of genetically manipulating yeasts should be developed further to make a wider range of yeasts available. It may no longer be essential to choose yeasts having an amenable natural system of genetic recombination; instead, the useful physiological characteristics found in various strains that have been little studied can be transferred, by gene cloning or fusion, to a commonly used but less versatile strain.
3. To assist this endeavor, more must be understood about the genetic and physiological differences between yeast genera and species.

II. CARBOHYDRATE METABOLISM

S. cerevisiae has the ability to take up and ferment a wide range of sugars — for example, sucrose, glucose, fructose, galactose, maltose, and maltotriose. In addition, the closely related species *Saccharomyces diastaticus* and *Saccharomyces uvarum* (*carlsbergensis*) are able to utilize dextrins and melibiose, respectively. The first step in the utilization of any sugar (or carbohydrate) by yeast is usually either

1. its passage intact across the cell membrane, or
2. initial hydrolysis outside the membrane followed by entry into the cell by some or all of the hydrolysis products.

Unlike some bacteria, which ferment a wide range of organic compounds, yeasts (with a few notable exceptions) ferment only those metabolized via the Embden-Meyerhof and Entner-Doudoroff pathways. Furthermore, cytochrome-deficient mutants apart, a yeast that uses sugar anaerobically also uses it aerobically (Barnett, 1976, 1981).

S. cerevisiae is constitutive for the ability to metabolize glucose, mannose, and fructose. With the exception of certain mutant strains (Lobo and Maitra,

1977a, b), all yeasts are able to utilize these sugars. In *S. cerevisiae* a number of transport systems for sugars have been described: a constitutive system common for glucose, fructose, and mannose and inducible systems for galactose, α-methyl-D-glucoside, maltose, and maltotriose. The transport of hexoses by the constitutive and inducible galactose system is equilibriative and does not require energy. On the other hand, the α-methyl-D-glucoside, maltose, and maltotriose systems have been characterized as active transport. For example, maltose transport, although independent of the intracellular ATP level, is coupled to the electrochemical gradient of protons (Serrano, 1977).

Sucrose is hydrolyzed outside the cell membrane by the extracellular enzyme invertase (β-D-fructofuranoside fructohydrolase) (E.C. 3.2.1.26) to glucose and fructose. Six genes, *SUC 1, 2, 3, 4*, and *5* and *SUC80*, code for the production of invertase, and a number of them have been assigned chromosome location (Mortimer and Schild, 1981).

The disaccharide maltose and the trisaccharide maltotriose possess independent uptake mechanisms (maltose permease and maltotriose permease), which transport the two sugars across the cell membrane into the cell. Once inside the cell, both sugars are hydrolyzed to glucose units by the α-glucosidase system (Hautera and Lovgren, 1975; Stewart et al., 1979). The transport, hydrolysis, and fermentation of maltose are particularly important in brewing, distilling, and baker's yeast strains, since maltose is the major sugar component of brewing wort, spirit mash, and wheat dough. There are (at least) five unlinked polymeric genes that control the ability of yeast to produce α-glucoside permease in response to maltose: *MAL1, MAL2, MAL3, MAL4*, and *MAL6*. Strains carrying an active allele at any one of these loci are inducible; strains carrying inactive alleles at all loci are uninducible but produce basal levels of α-glucosidase (Khan et al., 1973). The role of the *MAL* genes is still not fully understood. It has been suggested that the *MAL* loci are either (1) structural genes for α-glucosidase, (2) regulatory genes controlling both α-glucosidase and α-glucoside permease, or (3) complex loci containing both regulatory and structural elements.

It has been proposed that the *MAL* loci include the structural genes for α-glucosidase and maltose permease. Naumov (1971, 1976) has analyzed the progeny of crosses between various naturally occurring maltose-negative and maltose-positive isolates of *Saccharomyces* and concluded that maltose fermentation requires the presence of two genes termed *MALp* and *MALg* and that each *MAL* locus is composed of these two linked genes. A maltose-negative strain could arise if it lacked either a functional *MALp* or a functional *MALg* gene. A strain carrying *MALg* and *MALp*, at either linked or unlinked loci, allows the cell to utilize maltose as a carbon source. Federoff et al. (1982) have constructed a clone bank from a donor yeast strain that carries the *MALpMALg* locus. The cloned yeast DNA was then amplified in *E. coli* K-12 strain RR101. The recombinant plasmid DNA isolated from pools of 200 *E. coli* (*p*YT) transformants was used to transform a *Mal⁻* yeast strain. *Mal⁺* colonies were identified as transformants. Most of the *Mal⁺* colonies were found to be highly unstable during mitotic growth on nonselective media as they segregated mostly Mal^- clones. DNA prepared from several Mal^+ yeast transformants grown under selective conditions were transferred to *E. coli*, and a new plasmid, *pMAL26*, was identified by this approach. This plasmid was shown by positive hybridization translation, as well as Southern and Northern blot experiments, to carry a *MAL6* structural gene for α-glucosidase.

D-Galactose utilization by extracts of galactose-adapted yeast can be described in the following reactions (Cardini and Leloir, 1953). The first three are often referred to as the Leloir pathway.

$$\text{Galactose} + \text{ATP} \longrightarrow \text{Gal-1-P} + \text{ADP} + \text{H}^+$$
$$\text{galactokinase (E.C. 2.7.1.6)}$$

$$\text{Gal-1-P} + \text{UDPG} \longrightarrow \text{Gluc-1-P} + \text{UDP-galactose}$$
$$\text{galactose-1-P uridylyltransferase}$$
$$\text{(E.C. 2.7.7.10)}$$

$$\text{UDP-galactose} \longrightarrow \text{UDP-glucose}$$
$$\text{UDP-D-galactose 4-epimerase}$$
$$\text{(E.C. 5.1.3.2)}$$

$$\text{Gluc-1-P} \longrightarrow \text{Gluc-6-P}$$
$$\text{phosphoglucomutase (E.C. 2.7.5.1)}$$

The genetic system that encodes the enzymes responsible for the conversion of exogenous galactose to endogenous glucose 1-phosphate in *Saccharomyces* has been well studied. After transport across the cell membrane by the *GAL2*-encoded permease (Douglas and Hawthorne, 1966), galactose is converted to glucose 1-phosphate by reactions catalyzed by the products of the *GAL1*, *GAL7*, and *GAL10* genes, that is, galactokinase, galactose-1-P uridylyltransferase, and UDP-D-galactose 4-epimerase, respectively. These three genes are located in a tightly linked cluster, six map units from the centromere of chromosome II (Douglas, 1961). The genetic order of these genes has been determined by Bassel and Douglas (1968) as centromere$-GAL7$-$GAL10$-$GAL1$.

The enzymes encoded by the galactose gene cluster are coordinately induced by the addition of galactose to appropriately adapted cultures (Hartwell, 1970). Broach (1979) has shown that the galactose enzymes are translated as separate polypeptides. These data are consistent with those of Hopper et al. (1978), which showed that the *GAL1* and *GAL7* genes specify distinct and separate mRNAs and are the result of de novo synthesis after induction.

Pentoses (D-xylose and L-arabinose) comprise nearly one third of the reducing sugars obtained from lignocellulose hydrolysis (Sciamanna et al., 1977). Several researchers (Barnett, 1976; Suomalainen and Oura, 1971) have reported that yeasts are able to assimilate aldopentoses oxidatively but are unable to ferment them to ethanol. Species of the genus *Saccharomyces* are unable to utilize D-xylose. However, when strains of *S. cerevisiae* were tested on D-xylulose (D-threopentulose), an intermediate of D-xylose metabolism, surprisingly good growth was obtained. Owing to the inability of this species to metabolize D-xylose, methods have been sought to convert D-xylose to D-xylulose in a fermenter. To this end, a method has been developed (Schneider et al., 1981) that involves the inclusion of glucose isomerase in the medium, an enzyme that can isomerase D-xylose to D-xylulose. Chiang et al. (1981) have employed the technique of cell immobilization for ethanol production from pentoses. The results indicate a slow fermentation rate when pentose was used as substrate. As expected, *S. cerevisiae* had a superior fermentation rate on D-xylulose, and this system could be employed to produce ethanol, at high concentrations, from D-xylose by employing immobilized glucose isomerase and yeast cells in separate columns.

It has already been stated that the first step in the utilization of many sugars and carbohydrates by yeast is their initial hydrolysis outside the membrane followed by entry into the cell by some or all of the hydrolytic products. This initial hydrolysis is facilitated by the presence of extracellular enzymes. These enzymes have been discussed in detail by Arnold (1981) and fall into two classes: those which cleave substrates that do not permeate the plasma membrane and those enzymes involved with the turnover of envelope polymers during growth. Extracellular enzymes that fall into the first category that are involved in the hydrolysis of sugars and carbohydrates in *Saccharomyces* species include invertase (β-fructofuranosidase E.C. 3.2.1.26), glucoamylase (E.C. 3.2.1.3), and melibiase (E.C. 3.2.1.22); α-amylase (E.C. 3.2.1.1) is produced by some yeasts but not by *Saccharomyces* species (Sills and Stewart, 1982). Extracellular enzymes involved in the hydrolysis of noncarbohydrate material in *Saccharomyces* would include acid phosphatase (E.C. 3.1.3.2), catalase (E.C. 1.11.1.6), and phospholipase (E.C. 3.1.1.4).

Brewer's wort contains, in addition to fermentable sugars, unfermentable carbohydrates collectively termed dextrins. As the wort is fermented by the brewing yeasts *S. cerevisiae* and *S. uvarum* (*carlsbergensis*), the wort sugars are utilized, and as a result, the specific gravity falls. The attenuation limit, which is a fixed characteristic of the wort and the yeast strain, is the lowest specific gravity that can be attained. Occasionally, a beer with a specific gravity much lower than the usual attenuation limit is obtained, and this phenomenon has been called superattenuation. This phenomenon is often due to the presence in the fermentation of the yeast *S. diastaticus*, which can metabolize or assimilate glucose, fructose, galactose, maltose, maltotriose, sucrose, one third of the raffinose molecule, dextrin, or starch. The amylolytic enzyme system of this yeast species has been identified as a glucoamylase (syn. amyloglucosidase, α-1,4-glucan glucohydrolase, E.C. 3.2.1.3). To date, six genes associated with glucoamylase activity in *S. diastaticus* have been identified: *DEX1*, *DEX2*, and *DEX3* by Erratt and Stewart (1981a) and *STA1*, *STA2*, and *STA3* by Tamaki (1978). Studies have been conducted to determine whether allelism exists between any of these genes. Both *DEX3* and *STA3* were isolated from *S. diastaticus* strains that were brewing contaminants, and tetrad analysis has revealed that *DEX3* and *STA3* are allelic.

Glucoamylase hydrolyzes successive glucose units from the nonreducing ends of starch molecules and as a consequence glucose is the sole hydrolysis product. Glucoamylase from *S. diastaticus* is an extracellular enzyme. The enzyme located in the culture medium has been purified from a strain containing a single *DEX1* gene. Two mannoproteins have been isolated with nearly identical molecular weights ($\sim 312,000$ daltons); however, when a number of substrates were compared (Erratt and Stewart, 1981b), their ratios of carbohydrate to protein differed, as did their rates of carbohydrate hydrolysis. Neither enzyme possessed significant activity toward the hydrolysis of the α-1,6 bond of isomaltose or pullulan. The temperature optimum was 50°C, and the pH optimum at 40°C was 5.2. When maltose was used as the substrate the K_m was 1.0×10^{-2} M.

III. AMINO ACID UPTAKE AND METABOLISM

Amino acids, along with peptides, not only provide a source of nitrogen for yeast growth, but they or their metabolic products also affect the flavor of beers, wines, and spirits. As brewer's wort contains 19 amino acids, a significant amount of research on the uptake and subsequent metabolism of amino acids by *Saccharomyces spp.* has been conducted in the brewing context. The current understanding of the manner in which amino acids are taken up from wort was formulated in the mid to late 1960's. Jones and Pierce (1964) exploited the then novel automatic column-chromatographic technique for amino acid analysis to follow the time course of removal of each wort amino acid during the course of a brewery fermentation. It was found that, irrespective of the fermentation conditions employed, amino acids were removed from wort at different rates. They were able to classify wort amino acids into four classes depending on the rate of removal (Table 17.1). Under semi-anaerobic conditions (those encountered in brewing situations), proline, the most plentiful amino acid present in wort, is scarcely assimilated. While 95% of the other amino acids have disappeared by the end of the fermentation, there is a considerable amount of proline in the finished product (~ 200–300 μg/ml). Under

TABLE 17.1   The Composition of the Jones and Pierce Groups for Removal of Amino Acids from Brewer's Wort

Group A	Group B	Group C	Group D
Arginine	Histidine	Alanine	Proline
Asparagine	Isoleucine	Glycine	
Aspartic acid	Leucine	Phenylalanine	
Glutamic acid	Methionine	Tryptophan	
Glutamine	Valine	Tyrosine	
Lysine			
Serine			
Threonine			

aerobic laboratory conditions, however, proline is assimilated after exhaustion of the other amino acids.

The inability of *S. cerevisiae* and related species to assimilate proline under semi-anaerobic (brewery) conditions is the result of several phenomena (Bourgeois and Thouvenot, 1972). As long as other amino acids or ammonium ions are present in the medium, the activity of proline permease (the enzyme that catalyzes the transport of proline across the cell membrane) is repressed; as a consequence of this permease repression, proline absorption is slight. The first catabolic reaction of proline, once it is inside the cell, involves proline oxidase, which requires the participation of cytochrome *c* and molecular oxygen. By the time all of the other amino acids have been assimilated, however, thus removing the repression of the proline permease system, conditions are strongly anaerobic; as a consequence, the activity of proline oxidase is inhibited, and proline uptake is once again inhibited.

Recent research on amino acid uptake mechanisms into strains of *S. cerevisiae* has employed yeast grown in defined media (Eddy, 1982; Rose and Keenan, 1981). It is now established that *S. cerevisiae* has two classes of mechanism for transporting amino acids across the plasma membrane. There is a general amino acid permease (GAP), which can transport all basic and neutral amino acids except proline. In addition, *S. cerevisiae* can synthesize a range of at least 11 transport systems that are specific for just one or a small number of amino acids.

These data were not available when attempts were made to explain the Jones and Pierce (1964) sequence of amino acid removal from wort. However, current data would suggest that in the first 20–24 hours of a wort fermentation with *S. cerevisiae*, the GAP is not synthesized. Growth with ammonium ions as the sole source of nitrogen represses synthesis of GAP, and in most wort fermentations there is an appreciable concentration of free ammonia present for this period. Even if this ammonia was not taken up in sufficient quantity to act as a repressor of GAP, deamination of specifically transported amino acids probably would repress GAP. Specific transport systems for Group A amino acids such as asparagine and glutamine in two strains of *S. cerevisiae* have been established. Furthermore, competition studies have indicated that preferential uptake of Group A amino acids over Group C amino acids would not be expected if the GAP were operating (Woodward and Cirillo, 1977).

IV. ETHANOL TOLERANCE

It has long been appreciated that strains of *S. cerevisiae* differ in their ability to remain viable in the presence of ethanol (Brachvogel, 1907). In general, strains used in brewing have only moderate tolerance (Day et al., 1975a), while those employed in distilleries, predictably, have a greater tolerance (Harrison and Graham, 1970). Unfortunately, there is no recognized method for measuring the ethanol tolerance of yeast strains, and considerable confusion pervades this area of investigation (Rose, 1982). The simplest method is to assess the effect of ethanol, incorporated into the medium, on batch growth of *S. cerevisiae*. The first data on ethanol tolerance of *S. cerevisiae* were obtained in this manner. Other workers have elected to assess the ability of *S. cerevisiae* to tolerate ethanol by determining the effect of the ethanol on the fermentative activity of cells. For example, Nojivo and Ouchi (1962) found that the tolerance of yeast strains examined fell in the range of 20–30% (w/v); surprisingly, differences in the tolerance of saké and distilling yeasts, as compared with brewer's and baker's yeasts, were difficult to find.

Research on the physiological basis of ethanol tolerance in *S. cerevisiae* and related species was minimal until the upsurge of interest in high-gravity brewing and the manufacture of industrial ethanol by fermentation. Two areas of cell physiology are likely to be influenced by the presence of high concentrations of this narcotic. The first of these are intracellular macromolecules, and in particular glycolytic enzymes. Until recently, hardly anything was known of the effect of ethanol on the activity of the dozen enzymes that catalyze reactions leading to the conversion of glucose into ethanol and carbon dioxide in *S. cerevisiae*. Nagadawithana et al. (1977) reported noncompetitive inhibition of the activity of yeast hexokinase by ethanol; they also found that phosphofructokinase was not affected under the conditions of their study. A more extensive study of the effect of ethanol on glycolytic enzymes of *S. cerevisiae* has recently been published by Miller et al. (1982). Fructose 1,6-diphosphate aldolase, glyceraldehyde phosphate dehydrogenase, and pyruvate decarboxylase were most sensitive to the presence of ethanol, while triose phosphate isomerase and, somewhat predictably, ethanol dehydrogenase were least so.

The second area of cell–ethanol interaction concerns the extent and manner in which the narcotic becomes associated with cell membranes, and in particular the plasma membrane. Ethanol is an amphipathic molecule, but, as is indicated by its membrane-buffer partition coefficient, it is much more hydrophilic than hydrophobic. Nevertheless, there are data showing that ethanol can become concentrated at all domains in the depth of biological membranes, although the ethanol concentration is greatest in the more aqueous regions of the membrane—that is, at and just below the membrane surfaces—and lowest in the middle of the bilayer (Chin and Goldstein, 1981).

In recent years a number of researchers have turned to a study of the effect of ethanol on transport processes in the plasma membrane. As part of a study of the relationship between lipid composition and function in the plasma membrane of a strain of *S. cerevisiae*, Rose (1980, 1982) has reported that cells in which the plasma membrane was enriched in linoleyl ($C_{18:2}$) residues were more resistant to the lethal effects of one molar ethanol than cells enriched in oleyl ($C_{18:1}$) residues. Further, exponential growth of linoleyl residue–enriched cells was inhibited to a smaller extent when cultures were supplemented with ethanol than were cells in cultures enriched with oleyl residues.

Another factor that affects the ethanol tolerance of *S. cerevisiae* is growth temperature. It has been known for some years that *S. cerevisiae* tends to be more sensitive to the inhibitory effects of ethanol as the growth temperature is raised. However, these observations were essentially qualitative. In an attempt to gain a deeper understanding of this phenomenon, van Uden and de Cruz Duarte (1981) have employed a respiratory-deficient mutant of *S. cerevisiae* and found that ethanol concentrations above 3% (w/v) decreased the maximum temperature for growth of the mutant in batch culture. At a concentration of 6% (w/v), ethanol depressed the optimal temperature for growth from 37°C to 35°C, the final maximum temperature for growth from 40°C to 33°C, and the initial maximum temperature for growth from 44°C to 36°C. Thus during a batch fermentation these cardinal temperatures are variables that decrease as ethanol accumulates in the culture.

V. THE CELL ENVELOPE

The cell envelope consists of the plasma membrane, the periplasmic space, the cell wall, and (for some species — not *Saccharomyces* species) a slime layer (Arnold, 1981). Yeast cells can accommodate substantial changes in their environment; this is due in large part to physical properties of their cell envelopes. Accordingly, the live yeast cell can withstand an extended range of pH, ionic strength, or osmolality in the bathing medium. On the converse side, the industrial or experimental extraction of internal yeast components is often hampered by cell envelope barriers; the cell wall is particularly resistant to disruption.

The plasma membrane of the yeast cell, as in any plant and animal cell, protects the cell from the environment, assisted in some measure by the overlying cell wall. The plasma membrane is vulnerable to stresses imposed by environmental factors, and studies on the manner in which various properties of this membrane are affected by the lipid and protein composition of the structure are proving to be an invaluable avenue in the quest for an understanding of the way in which cells in higher eukaryotic forms respond to environmental stimuli.

All regions of the yeast cell envelope contain polysaccharides — as more or less fluid components of the plasma membrane, as soluble glycoproteins in the periplasmic space, and as the fabric of the cell wall. The cell wall of yeasts, like those of most other plants, is composed almost entirely of polysaccharides. In place of cellulose, which is universally present in the cell walls of higher plants, they contain other β-glucans, in which (1-3) and (1-6) linkages predominate, and occasionally α-glucans; both these types are also found in the cell walls of filamentous fungi. Equally important in yeast cell walls are polysaccharides in which mannose is the predominant sugar component, that is, the mannans. Small amounts of chitin are present in most species, and amino sugars also occur as links between carbohydrate and polypeptide. The peptide content of yeast cell walls, though small, is intimately associated with the polysaccharides, and so are esterified phosphate groups. The cell walls of species of the genus *Saccharomyces* are virtually devoid of lipid.

The yeast cell wall has three main roles. It acts first as a semirigid envelope, providing both the compressional and tensile strength of the cell. Second, it must expand and accommodate as the cell grows and buds. Third, it must provide a passage for materials for cell growth and locations for enzyme activity.

A further function of the yeast cell wall is control of a particular culture's aggregation or flocculation properties. Flocculation is the phenomenon wherein yeast cells adhere in clumps and sediment rapidly from the medium in which they are suspended. There is little doubt that differences in the flocculation characteristics of various yeast cultures are primarily manifestations of the yeast culture's cell wall structure. Studies in this and other laboratories have failed to reveal any meaningful differences in gross composition between the walls of the two culture types that could be directly correlated to the phenomenon of flocculation (Stewart and Russell, 1981). Consequently, it has been stated on many occasions that only when the yeast cell surface is examined with an electron microscope will any meaningful difference in the microstructure of flocculent and nonflocculent yeast cell surfaces be revealed. When ether-washed cells were shadowed, differences between flocculent and nonflocculent cultures became immediately apparent (Fig. 17.4). Whereas the cells of nonflocculent cultures appeared to possess no extracellular projections, cells from flocculent cultures were covered with an extensive layer of fimbriae or hairlike protuberances (Day et al., 1975b). Collaborative studies between Rose and his co-workers and this laboratory have shown that carboxyl groups in wall proteins are the anionic groups involved in floc formation (Beavan et al., 1979). It is generally agreed that floc formation in *Saccharomyces* involves formation of bridges between calcium ions and anionic polymers on the cell surface.

Genetic studies on yeast flocculation date from the early 1950's (Gilliland, 1951; Thorne, 1952). By using a flocculent haploid yeast strain a single dominant gene for flocculation has been identified (Stewart and Russell, 1977; Russell et al., 1980). Mapping studies have revealed that it is linked to *ade1* and thereby located on chromosome I. This flocculation gene has been designated *FLO1*. *FLO1* has been found to be located 45 cm from the chromosome I centromere and 40 cm from *ade1*.

It would seem from the existing data that either yeast plasma membranes are atypical of eukaryotic cells generally or else artifacts are involved (Arnold, 1981). The protein, carbohydrate, and lipid contents of reasonably pure plasma membranes have been the subject of many analyses. The protein content of these preparations (47–63%) undoubtedly includes certain enzymes and transport proteins and is consistent with other eukaryotic plasma membranes (Guidotte, 1972), whereas the lipid content is somewhat lower, and the carbohydrate fraction is commensurately higher. A repeated but remarkable finding is the high proportion of neutral lipids and the correspondingly lower fraction of phospholipids. The whole lipid extract of *S. cerevisiae* contains about 52% phospholipid.

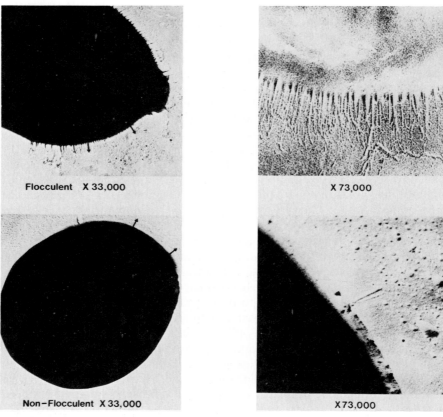

Flocculent   X 33,000

X 73,000

Non–Flocculent X 33,000

X 73,000

FIGURE 17.4   Electron photomicrographs of flocculent and nonflocculent strains of *Saccharomyces cerevisiae*.

There are basically three ways in which the lipid composition of microorganisms can be altered as a prelude to a study of the relationship between lipid composition of microbial membranes and their function. These are by altering environmental conditions, such as growth temperature and the concentration of dissolved oxygen, using mutant strains of a microbe that are auxotrophic for a membrane component such as a sterol and by using drugs that affect specific reactions on pathways leading to synthesis of membrane lipids.

Rose (1977) has employed two of these methods to bring about specific changes in the lipid composition of the plasma membrane of a strain of *S. cerevisiae*. The first of these exploits revealed that, when grown under strictly anaerobic conditions, this yeast, in common with all strains of *S. cerevisiae*, becomes auxotrophic for a sterol and an unsaturated fatty acid, both requirements being fairly nonspecific (Andreasen and Stien, 1954). With this technique, cells were grown under conditions that lead to extensive enrichment of the plasma membrane with residues of oleic, linoleic, or γ-linolenic acid or with any one of several different sterols. Cells were converted into spheroplasts and the resistance of the plasma membrane to stretching was assessed by examining the stability of the spheroplasts in hypotonic solutions of buffered sorbitol (Alterthuss and Rose, 1973). These experiments established that resistance to stretching in the yeast plasma membrane is confined by the presence of monounsaturated rather than polyunsaturated fatty-acyl residues in the phospholipids and by sterols that have an unsaturated (e.g., ergosterol or stigmasterol) as compared with a saturated (e.g., cholesterol or sitosterol) side chain. The first effect has been attributed to the constraints that unsaturation in fatty-acyl residues impose on the packing of the fatty-acyl chains of phospholipids in membranes, while the stabilizing effect of sterols with unsaturated side chains has been explained by a greater interaction between the side chain and fatty-acyl chains in membrane phospholipids.

Cells enriched in γ-linolenic acid residues are more resistant to conversion into spheroplasts using a basidiomycete β-glucanase than cells enriched in linoleic acid residues, which in turn are more difficult to convert into spheroplasts than cells enriched in oleic acid residues. These findings suggest an effect of fatty-acyl unsaturation in the plasma membrane on yeast cell wall composition. Lysates of spheroplasts enriched in the more unsaturated fatty-acyl residues have a diminished capacity to incorporate mannose residues from GDP-mannose into a polymeric form compared with those enriched in oleic acid residues. These results certainly suggest a role for fatty-acyl unsaturation in phospholipids during mannan synthesis.

The second method employed to change the lipid composition of the yeast plasma membrane involves including a low concentration of choline or ethanolamine in a defined medium that causes the yeast to synthesize increased proportions of phosphatidylcholine and phosphatidylethanolamine, respectively (Hossack et al., 1977). Spheroplasts from cells with plasma membranes enriched with phosphatidylcholine are more susceptible to osmotic lysis than those with plasma membranes enriched in phosphatidylethanolamine. The enhanced stability of membranes enriched in phosphatidylethanolamine has been explained by the greater difficulty with which the polar head group of this phospholipid is hydrated.

The periplasmic space is the region of the cell envelope that is bounded by the plasma membrane and the inner aspect of the cell wall. This space includes invaginations in the plasma membrane and outward excursions into the inner aspect of the cell wall. The periplasmic space is the locale for several digestive enzymes that act upon substrates that require hydrolysis prior to assimilation by the cytoplasm (e.g., invertase).

VI. THE INFLUENCE OF OSMOTIC PRESSURE AND SUBSTRATE CONCENTRATION ON *SACCHAROMYCES*

Yeasts capable of growth and fermentation at high solute concentrations have been classified as osmophilic, osmotolerant, or, more correctly, xerotolerant. Indeed, by strict definition a xerotolerant organism is one that grows better at high solute concentrations than at low solute concentrations. The most common and also the most xerotolerant osmophilic yeast is *Saccharomyces rouxii*, which resists high concentrations of both sugar and salt. The only other truly xerotolerant species of *Saccharomyces* is *Saccharomyces bailii*. There are many other species of yeast that are xerotolerant, but they are not classified in the genus *Saccharomyces* (Tilbury, 1980).

The mechanisms of xerotolerance in yeasts has been studied in detail by Brown (1976, 1978). Xerotolerant yeasts (and other eukaryotes) accumulate high concentrations of polyols in response to a water stress; when the stress is extreme, the polyol is usually glycerol. It is also known that polyols, especially glycerol, confer a remarkable degree of protection on enzymes at high solute concentrations; hence they function as the essential compatible solutes. They also function as osmoregulators under water stress. The problem is that nontolerant species of the same eukaryotic genera can also accumulate glycerol in response to water stress.

Xerotolerant yeasts accumulate at least one acyclic polyol, commonly arabitol, when grown in a dilute medium; nontolerant yeasts do not. When grown at high solute concentrations, species such as *S. rouxii* respond by accumulating glycerol while the arabitol content remains constant. *S. cerevisiae*, which is not xerotolerant, also responds to water stress by accumulating glycerol. Over the range of solute concentrations that it can tolerate, *S. cerevisiae* accumulates as much glycerol as does *S. rouxii* under the same conditions.

There is however, a major difference in the mechanisms by which glycerol accumulation is regulated in the two yeasts. *S. rouxii* synthesizes relatively small amounts of additional glycerol in response to high solute concentrations; its major response is to retain within the cell a progressively greater proportion of the glycerol that it synthesizes. On the other hand, *S. cerevisiae* leaks to the growth medium a very high proportion of the glycerol that it produces. It responds to increasing solute concentrations by synthesizing much more glycerol and retaining an approximately constant proportion within the cell.

In this laboratory (Stewart et al., 1982) the effect of elevated glucose concentrations on fermentation rate in a synthetic medium has been studied with strains of both *S. rouxii* and *S. uvarum* (*carlsbergensis*) (Fig. 17.5). The *S. rouxii* strain was able to ferment the sugar even at 56% (w/v), at which concentration fermentation by the *S. uvarum* (*carlsbergensis*) strain was totally inhibited. However, with both yeast cultures, elevated sugar concentrations had an inhibitory effect upon the rate of fermentation.

A major cause of the different tolerance ranges of the two yeasts is probably that *S. cerevisiae* eventually directs an unacceptably high proportion of its biosynthetic capacity to glycerol production. Another important difference between the yeasts is that *S. cerevisiae* seems to be generally more permeable than *S. rouxii* to a variety of solutes, a factor that itself could be important in initiating a major metabolic response to water stress in *S. cerevisiae*.

Although strains of *S. cerevisiae* and related species are susceptible to high solute concentrations, studies have been ongoing in this laboratory to study

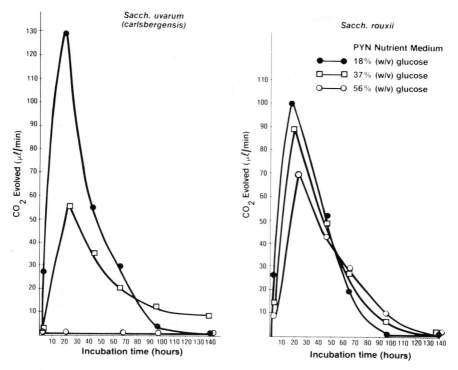

FIGURE 17.5 Effect of elevated glucose concentrations in PYN nutrient medium on the fermentation rate of strains of *Saccharomyces uvarum* (*carlsbergensis*) and *Saccharomyces rouxii* at ●——● 18% (w/v) glucose, □——□ 37% (w/v) glucose, and ○——○ 56% (w/v) glucose.

the influence of substrate concentration on fermentation rate (Panchal and Stewart, 1981). Different concentrations of sucrose have been fermented with a strain of *S. uvarum* (*carlsbergensis*). As the sucrose concentration was increased up to 25% (w/v), increased fermentation rates occurred; but further increases up to 40% (w/v) sucrose resulted in a significantly decreased fermentation rate. In order to study the influence of elevated solute concentrations further, cells of the same strain of *S. uvarum* (*carlsbergensis*) were subjected to fermentation in a growth-promoting medium containing 10% (w/v) sucrose as the substrate but with varying amounts of sorbitol in the medium in order to increase the osmotic pressure (sorbitol is a sugar not metabolized by *Saccharomyces* species). Intracellular ethanol levels were analyzed by performing enzymatic assays of perchloric acid extracts of cells. These values were then compared with those obtained from analysis of ethanol in the cell-free medium (Fig. 17.6). In the early stages of the fermentation, as the osmotic pressure was increased, the proportion of intracellular ethanol increased considerably. Thus with a 30% (w/v) sorbitol-containing medium (approximately 50 atm osmotic

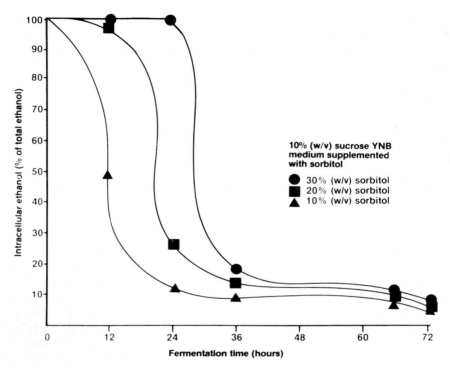

FIGURE 17.6 Retention of intracellular ethanol by cells of *Saccharomyces uvarum* (*carlsbergensis*) cultured in 10% (w/v) sucrose YNB medium supplemented with ●——● 30% (w/v) sorbitol, ■——■ 20% (w/v) sorbitol, and ▲——▲ 10% (w/v) sorbitol.

pressure) at 24 hours into the fermentation, almost all of the ethanol was located intracellularly. During the course of the fermentation, however, this proportion decreased rapidly; after 73 hours of fermentation, no more than 5% of the ethanol was located intracellularly. With a medium of low osmotic pressure, most of the ethanol diffused out of the cell very rapidly.

VII. GENETICS AND MOLECULAR BIOLOGY OF *SACCHAROMYCES*

As was previously discussed, species of the genus *Saccharomyces*, particularly *S. cerevisiae* and related species, are amenable to genetic manipulation and analysis. With the advent of novel genetic manipulation methods and development of a transformation system in *Saccharomyces* (Hinnen et al., 1978) a powerful tool has been presented to researchers. Yeast can now serve as a functional assay system for DNA sequences introduced into cells as part of chimeric plasmids. In this way it has been possible to clone and characterize dozens of yeast genes that complement mutations in the transformed host cells. It has also been possible to examine the behavior of other DNA segments, for example, replication origins and centromeres, which serve quite different functions in the genome. As a consequence of these developments, the yeast transformation system can be used to develop cloning vectors for higher eukaryotes and also to manufacture foreign proteins that may possess medical or industrial potential. The molecular biology of *Saccharomyces* is a complex and much studied field; it is too complex for detailed discussion in this document. However, there is a two-volume treatise edited by Strathern et al. (1981, 1982) that discusses in considerable depth recent developments in this field.

There are a number of methods that can be employed in the genetic research and development of *Saccharomyces* species. These include hybridization, mutation, spheroplast (or protoplast) fusion and transformation, and, associated with it, DNA recombination. The life cycle of *S. cerevisiae* and related species normally alternates between diplophase and haplophase. Both ploidies can exist as stable cultures. In heterothallic strains the haploid cultures are of two mating types, *a* and *α*. Mating of *a* and *α* cells results in *a*/*α* diploids that are unable to mate but can undergo meiosis (Fig. 17.2). The four haploid products resulting from meiosis of a diploid cell are contained within the wall of the mother cell (the ascus). Digestion of the ascus wall and separation of the spores by micromanipulation yield four clones that represent the four haploid meiotic products. Analysis of the segragation patterns of different heterozygous markers among the four spores constitutes tetrad analysis. Linkage between two genes or between a gene and its centromere is revealed by this kind of analysis. In addition, both chiasma and chromatid interference can be assayed. Chromosome aberrations can also be analyzed by examination of the patterns of spore lethality in a sample of tetrads.

Industrial yeast strains are often polyploid or even aneuploid and, as a consequence, do not possess a mating type; they have a low degree of sporulation and poor spore viability, rendering genetic analysis of such strains extremely difficult. It would appear, however, that the widespread use of polyploid yeasts for industrial purposes is no accident. Owing to their multiple gene structure, polyploids are genetically more stable and less susceptible to mutational forces than either haploid or diploid strains, thus enabling such strains to be used routinely with a high degree of confidence.

Tubb et al. (1981) and Young (1981) have studied the rare-mating technique as a possible means to introduce cytoplasmic genetic elements into brewing strains without transfer of nuclear genetic elements from the nonbrewing parent. This process is called cytoduction and requires that the nonbrewing parent carries a *kar1* mutation that impairs nuclear fusion (Conde and Fink, 1976). Cytoduction can be used to achieve three main goals, the first of which is substitution of the mitochondrial genome. This genetic element is involved primarily in respiratory functions, which are not active during fermentation. Second, cytoduction can be used to transfer double-stranded RNA species that encode production of toxins that kill other yeasts and also confer resistance to the effects of such toxins (Wickner, 1979). Therefore double-stranded RNA can be introduced into brewing and other industrial yeast strains in order to increase their resistance to wild-yeast infections. Finally, cytoduction can be used to introduce DNA plasmids into which specific genetic characters have been inserted by the techniques of gene-cloning and transformation.

Other techniques that show great promise and potential as aids in the genetic manipulation of brewer's yeast strains are spheroplast or protoplast fusion and transformation. Neither of these methods is dependent on ploidy or mating type; consequently, they have applicability to industrial yeast strains because of their polyploid/aneuploid nature and absence of mating type characteristic.

The term protoplast has been defined as:

> the resulting form of a plant cell or microorganism when its cell wall has been mechanically or enzymatically removed. This "naked" cell can easily be plasmolyzed by osmotic shock, but is capable of cell wall regeneration and subsequent cell division. A spheroplast has all the properties of a protoplast but contains remnants of the cell wall.

The first step in spheroplast fusion is the removal of the yeast cell wall with lytic enzymes such as extracts of snail gut or enzymes from various microorganisms. Removal of the yeast cell wall leaves only the membrane (with possible remnants of the cell wall) surrounding the cytoplasm, that is, a spheroplast. Such structures are osmotically fragile and remain intact only if maintained in a medium of high osmotic pressure, usually 0.8–1.2 M of sorbitol in buffer. After thorough washing to remove traces of the spheroplasting enzyme the spheroplasts are suspended in the fusing agent, which consists of polyethylene glycol (PEG) and calcium ions in buffer, and then are mixed with spheroplasts from a yeast strain with different genetic characteristics.

After fusion (interchange of DNA material) the fused spheroplasts must be induced to regenerate their cell walls (this is achieved in solid media containing 3% (w/v) agar and sorbitol) and to begin cell division. The action of PEG as the fusing agent is not fully understood. It is believed to act as a polycation that, together with calcium ions, induces the formation of small aggregates of spheroplasts, thus facilitating the interchange of DNA material (Russell and Stewart, 1979).

Studies in this laboratory have resulted in a number of successful fusions (Stewart, 1981); however, it has been concluded that fusion is not specific enough to genetically modify industrial yeast strains in a controllable fashion. The fusion product is nearly always very different from both original fusion partners because the genome of both donors becomes integrated. Consequently, it would be difficult to selectively introduce a single trait into a yeast strain.

Transformation is a technique that offers a means of overcoming the nonspecificity of fusion. It is another nonsexual technique for achieving genetic recombination. Although some success has been achieved with native DNA (Russell and Stewart, 1980), the use of plasmid vector systems together with recombinant DNA techniques shows greatest promise in the genetic manipulation of yeast strains, particularly when it comes to the production of foreign proteins by yeast. There now exist simple and general methods for isolating and amplifying virtually any yeast gene, although these methods generally require an intermediate step in *E. coli*. Powerful and exquisitely sensitive hybridization methods have been developed that allow direct physical analysis of any chromosomal region containing a gene that has been molecularly cloned. Most important, it is now possible to return to yeast, by transformation with DNA, cloned genes by using a variety of selectable marker systems developed for this purpose. These technological advances have combined to make feasible truly molecular as well as classical genetic manipulation and analysis in yeast.

An early success employing rDNA techniques to an industrial context has been described by Henahan and Klausner (1983). Molecular geneticists working at the research department of a brewery in Dublin, Eire, have built the β-glucanase gene from *Bacillus subtilis* into a strain of *S. cerevisiae*. The gene has been found to be stable in the yeast for five months. The transplanted gene has yet to express itself in the yeast; but when it is removed from the yeast and transformed into *E. coli*, the gene produces endo-β-1,4-glucanase.

In order to improve the uptake of DNA into spheroplasted yeast cells, studies in this laboratory have investigated the use of liposomes to deliver the DNA into the yeast spheroplast (Russell et al., 1983). It has been found that whereas less than 0.5% of ^{14}C-labeled DNA without liposomes is taken up into spheroplasts, when the DNA was encapsulated in liposomes composed of phosphatidylserine, cholesterol, and stearylamine, 25% of the nucleic acid was taken up or bound so tightly to the spheroplast that it could not be removed by repeated washes.

VIII. CONCLUSIONS

The fact that *Saccharomyces* species plays two important roles — as an industrial and an experimental microorganism — has hopefully been adequately described in this chapter. The number of strains of *S. cerevisiae* and related species that are available in research laboratories, both academic and industrial, throughout the world is very large. Rose (1981) has made a persuasive case for nominating one strain, or at least a selected number of strains, upon which research should be concentrated. Genetic considerations dictate that the strain or strains selected must be haploid. Over a decade ago, Rose and Harrison (1971) borrowed a concept that Frances Crick had applied to research on *E. coli* and proposed that the long-term goal in research on eukaryotic cells should be the complete solution of the molecular biology of *S. cerevisiae*. This was termed Project Y, and it was determined that this project should concentrate on one or a small number of selected haploid strains. While these authors are in full sympathy with this proposal, the profound differences between haploid/diploid strains and polyploid/aneuploid industrial yeast strains should *not* be ignored.

REFERENCES

Alterthuss, F., and Rose, A. H. (1973) *J. Gen. Microbiol. 77*, 371.

Andreasen, A. A., and Stien, T. J. B. (1954) *J. Cell Comp. Physiol. 43*, 271.

Arnold, W. N. (1981) *Yeast Cell Envelopes: Biochemistry, Biophysics and Ultrastructure*, CRC Press, Boca Raton, Fla.

Barnett, J. A. (1976) *Adv. Carbohyd. Chem. Biochem. 34*, 125.

Barnett, J. A. (1981) *Adv. Carbohyd. Chem. Biochem. 39*, 347.

Bassel, J., and Douglas, H. C. (1968) *J. Bacteriol. 95*, 1103.

Beavan, M. J., Belk, D. M., Stewart G. G., and Rose, A. H. (1979) *Can. J. Microbiol. 25*, 888.

Bourgeois, C. M., and Thouvenot, D. R. (1972) *J. Inst. Brew. 78*, 270.

Brachvogel, J. K. (1907) *Industrial Alcohol*, Crosby Lockwood and Son, New York.

Broach, J. R. (1979) *J. Mol. Biol. 131*, 41.

Brown, A. D. (1976) *Bacteriol. Rev. 40*, 803.

Brown, A. D. (1978) in *Advances in Microbial Physiology*, Vol. XVII (Rose, A. H., and Morris, J. G., eds.), p. 181, Academic Press, New York.

Cardini C. E., and Leloir, L. F. (1953) *Arch. Biochem. Biophys. 45*, 55.

Carter, B. L. A. (1978) in *Advances in Microbial Physiology*, Vol. XVII (Rose, A. H., and Morris, J. G., eds.), p. 244, Academic Press, New York.

Chiang, L. C., Hsiao, H. Y., Ueng, P. R., Chen, L. F., and Tsao, G. T. (1981) *Biotechnol. Bioeng. Symp. 11*, 263.

Chin, J. H., and Goldstein, D. B. (1981) *Mol. Pharmacol. 19*, 425.

Conde, J., and Fink, G. R. (1976) *Proc. Nat. Acad. Sci. U.S.A. 73*, 3651.

Day, A., Anderson, E., and Martin, P. A. (1975a) in *Proceedings of the 15th Congress European Brewery Convention*, p. 377. Nice, France. Elsevier Scientific Publishers, Amsterdam.

Day, A. W., Poon, N. H., and Stewart, G. G. (1975b) *Can. J. Microbiol. 21*, 558.

Douglas, H. C. (1961) *Biochim. Biophys. Acta 52*, 209.

Douglas, H. C., and Hawthorne, D. C. (1966) *Genetics 54*, 911.

Eddy, A. A. (1982) in *Advances in Microbial Physiology*, Vol. XXIV (Rose, A. H. and Morris, J. G., eds.), p. 260, Academic Press, New York.

Erratt, J. A., and Stewart, G. G. (1981a) in *Current Developments in Yeast Research* (Stewart, G. G., and Russell, I. eds.), p. 177, Pergamon Press, Toronto, Canada.

Erratt, J. A., and Stewart, G. G. (1981b) *Dev. Ind. Microbiol. 22*, 577.

Federoff, H. J., Cohen, J. D., Eccleshall, T. R., Needleman, R. B., Buchferer, B. A.,

Giacalone, J., and Marmur, J. (1982) *J. Bacteriol. 149*, 1064.

Gilliland, R. B. (1951) in *Proceedings of the 3rd Congress European Brewery Convention* p. 35. Brighton, Great Britain, Elsevier Scientific Publishers, Amsterdam.

Guidotte, G. (1972) *Ann. Rev. Biochem. 41*, 731.

Harrison, J. S., and Graham, J. C. J. (1970) in *The Yeasts* (Rose, A. H., and Harrison, J. S., eds.), p. 283, Academic Press, New York.

Hartwell, L. H. (1970) *Ann. Rev. Genet. 4*, 373.

Hautera, P., and Lovgren, T. (1975) *J. Inst. Brew. 81*, 309.

Henahan, J., and Klausner, A. (1983) *Bio/technology 1*, 462.

Hinnen, A., Hicks, J. B., and Fink, G. R. (1978) *Proc. Nat. Acad. Sci. U.S.A. 75*, 1929.

Hopper, J. E., Broach, J. R., and Rowe, L. B. (1978) *Proc. Nat. Acad. Sci. U.S.A. 75*, 2878.

Hossack, J. A., Sharpe, V. J., and Rose, A. H. (1977) *J. Bacteriol. 129*, 1144.

Jones, M., and Pierce, J. S. (1964) *J. Inst. Brewing 70*, 307.

Kahn, N. A., Zimmerman, F. K., and Eaton, N. R. (1973) *Mol. Gen. Genet. 124*, 365.

Lobo, Z., and Maitra, P. K. (1977a) *Genetics 86*, 727.

Lobo, Z., and Maitra, P. K. (1977b) *Mol. Gen. Genet. 157*, 297.

Lodder, J. (1970) *The Yeasts, A Taxonomic Study*, North Holland, Amsterdam.

Miller, D. G., Griffiths-Smith, K., Algar, E., and Scopes, R. K. (1982) *Biotech. Lett. 4*, 601.

Mortimer, R. K., and Schild, D. (1981) *Microbiol. Rev. 44*, 519.

Nagadawithana, T. W., Whitt, J. T., and Cutaia, A. J. (1977) *J. Amer. Soc. Brewing Chemists 35*, 179.

Naumov, G. I. (1971) *Genetika 7*, 141.

Naumov, G. I. (1976) *Genetika 12*, 87.

Nojivo, K., and Ouchi, K. (1962) *J. Soc. Brewing* (Japan) *27*, 824.

Panchal, C. J., and Stewart, G. G. (1981) in *Current Developments in Yeast Research* (Stewart, G. G., and Russell, I., eds.), p. 9, Pergamon Press, Toronto, Canada.

Pasteur, L. (1866) *Etudes sur la Vin*, Imprinceiss Imperials, Paris.

Pasteur, L. (1876) *Etudes sur la Biere*, Gauthier-Villars, Paris.

Phaff, H., Miller, M. W., and Mrak, E. M. (1978) *The Life of Yeasts*, 2nd ed., Harvard University Press, Cambridge, Mass.

Rose, A. H. (1977) in *Alcohol Industry and Research* (Forsander, O., Eriksson, K., Oura, E., and Jounela-Eriksson, P., eds.), p. 179, Alkon Keskaslaboratorio, Helsinki, Finland.

Rose, A. H. (1980) in *Biology and Activities of Yeasts* (Skinner, F. A., Passmore, S. M., and Davenport, R. R., eds.), p. 103, Academic Press, New York.

Rose, A. H. (1981) in *Current Developments in Yeast Research* (Stewart, G. G., and Russell, I., eds.), p. 645, Pergamon Press, Toronto, Canada.

Rose, A. H. (1982) in *Ethanol from Biomass* (Duckworth, H. E., and Thompson, E. A., eds), p. 458, Royal Society of Canada, Ottawa, Canada.

Rose, A. H., and Harrison, J. S. (1971) *The Yeasts*, Academic Press, New York.

Rose, A. H., and Keenan, M. H. J. (1981) *Proceedings of the 18th Congress European Brewery Convention*, p. 207. Copenhagen, Denmark, IRL Press Ltd. Oxford, Great Britain.

Russell, I., and Stewart, G. G. (1979) *J. Inst. Brew. 85*, 95.

Russell, I., and Stewart, G. G. (1980) *J. Inst. Brew. 86*, 55.

Russell, I., Stewart, G. G., Reader, H. P., Johnston, J. R., and Martin, P. A. (1980) *J. Inst. Brew. 86*, 120.

Russell, I., Jones, R. M., Weston, B. J., and Stewart, G. G. (1983) *J. Inst. Brew. 89*, 136.

Schneider, H., Wang, P. Y., Johnson, B. F., and Shopsis, C. (1981) in *Current Developments in Yeast Research* (Stewart, G. G., and Russell, I., eds.), p. 81, Pergamon Press, Toronto, Canada.

Sciamanna, A. F., Freitas, R. P., and Wilkie, C. R. (1977) *Composition and Utilization of Cellulose for Chemicals from Agricultural Residues*, Publication 5966, Lawrence Berkeley Laboratories, Berkeley, Calif.

Serrano, R. (1977) *Eur. J. Biochem. 80*, 97.

Sills, A. M., and Stewart, G. G. (1982) *J. Inst. Brew. 88*, 313.

Stewart, G. G. (1981) *Can. J. Microbiol. 27*, 973.

Stewart, G. G., and Russell, I. (1977) *Can. J. Microbiol. 23*, 441.

Stewart, G. G., and Russell, I. (1981) in *Brewing Science*, Vol. II (Pollock, J., ed.), p. 61, Academic Press, New York.

Stewart, G. G., Erratt, J., Garrison, I., Goring, T., and Hancock, I. (1979) *MBAA Tech. Quart. 16*, 1.

Stewart, G. G., Panchal, C. J., Russell, I., and Sills, A. M. (1982) in *Ethanol from Biomass* (Duckworth, H. E., and Thompson, E. A., eds.), p. 4, Royal Society of Canada, Ottawa, Canada.

Strathern, J. N., Jones, E. W., and Broach, J. R. (1981) *The Molecular Biology of the Yeast Saccharomyces—Life Cycle and Inheritance*, Cold Spring Harbor Laboratory, Cold Spring Harbor, New York.

Strathern, J. N., Jones, E. W., and Broach, J. R. (1982) *The Molecular Biology of the Yeast Saccharomyces—Metabolism and Gene Expression*, Cold Spring Harbor Laboratory, Cold Spring Harbor, New York.

Suomalainen, H., and Oura, E. (1971) in *The Yeasts*, Vol. II (Rose, A. H., and Harrison, J. S., eds.), p. 3, Academic Press, New York.

Tamaki, H. (1978) *Mol. Gen. Genet. 164*, 205.

Thorne, R. S. W. (1952) *C. R. Trav. Lab. Carlsberg, 25*, 101.

Tilbury, R. H. (1980) in *Biology and Activities of Yeast* (Skinner, F. A., Passmore, S. M., and Davenport, R. R., eds.), p. 153, Academic Press, New York.

Tubb, R. S., Brown, A. J. P., Searle, B. A., and Goodey, A. R. (1981) in *Current Developments in Yeast Research* (Stewart, G. G., and Russell, I., eds.), p. 75, Pergamon Press, Toronto, Canada.

van der Walt, J. P. (1970) in *The Yeasts, a Taxonomic Study* (Lodder, J., ed.), p. 555, North-Holland, Amsterdam.

van Uden, N., and de Cruz Duarte, H. (1981) *Z. Allg. Microbiol. 21*, 743.

Wickner, R. B. (1979) *Plasmid 2*, 303.

Woodward, J. R., and Cirillo, V. P. (1977) *J. Bacteriol. 130*, 714.

Young, T. W. (1981) *J. Inst. Brew. 87*, 292.

Biology of Yeasts
Other Than *Saccharomyces*

Herman J. Phaff

This chapter is intended to furnish biological information on industrially important yeasts other than species of *Saccharomyces*. The number of yeast species that have been used in industry or show potential in future applications is relatively large. Because of space limitation a choice needed to be made; this chapter covers the biology of four species: *Saccharomycopsis lipolytica*, *Hansenula polymorpha*, *Kluyveromyces fragilis*, and *Candida utilis*. These four organisms reveal different metabolic pathways, depending on the substrate used, different applications, and they represent quite different taxonomic entities. The coverage of each species is divided approximately into taxonomy (including morphological and physiological properties), natural habitat, genetics, biochemistry and physiology, and a brief statement on industrial application.

I. *SACCHAROMYCOPSIS LIPOLYTICA*

A. Taxonomy
This species has had a long and rather complicated history. The original strain representing this species was first isolated in 1921 from spoiled margarine in

the Netherlands by Kneteman. It was designated in 1923 as *Torula lipolytica* by Jacobsen (also in the Netherlands); but since he published no description of this yeast, the name had no standing, a *nomen nudum*. The first report in the literature was made by Harrison (1928), who named the strain *Mycotorula lipolytica*. Harrison noted especially that this species did not ferment any sugars and demonstrated a strong proteolytic activity on gelatin and casein. Diddens and Lodder (1942) restudied the original strain and transferred it to the genus *Candida*; it was then known under the name *Candida lipolytica* (Harrison) Diddens et Lodder. They included in their description its strong hydrolytic action on triglycerides. Diddens and Lodder recognized several later described species as synonyms of *C. lipolytica*. Lodder and Kreger-van Rij (1952) and van Uden and Buckley (1970) retained the name *C. lipolytica* and concluded that *Candida olea*, isolated from olives in Italy, was also a synonym of *C. lipolytica*.

Next, Wickerham et al. (1970) discovered a sexual state of *C. lipolytica*, and they named the ascosporogenous form *Endomycopsis lipolytica*. Some strains were isolated as diploids, which sporulated directly on suitable media, but others were haploids, which sporulated only upon mixing of complementary mating types. The yeast is therefore heterothallic. The ascospore shape was strain-dependent and included morphologies ranging from spheroidal to hemispheroidal and shallow bowl-shaped with a ledge. The number of spores per ascus ranges from one to four, and the spores are usually liberated rapidly by rupture of the ascus. Colonies of the various strains studied by Wickerham and co-workers were quite pleomorphic, and their morphology was strongly influenced both by the growth medium and by the conditions of maintenance prior to growth on a particular medium. Besides budding, the organism produces under appropriate conditions pseudomycelium with sparse formation of blastoconidia as well as true septate mycelium. Kreger-van Rij and Veenhuis (1973) subsequently showed by electron microscopy that the septa contain a single, central connection between the two cells on either side of the cross wall.

van der Walt and Scott (1971) pointed out that according to the International Code of Botanical Nomenclature (which also governs the nomenclature of yeasts and other fungi), the genus name *Endomycopsis* Stelling-Dekker was illegitimate because *Saccharomycopsis* Schiönning had priority. Yarrow (1972) therefore renamed *Endomycopsis lipolytica* Wickerham et al., *Saccharomycopsis lipolytica* (Wickerham et al.) Yarrow. A final name change was made by van der Walt and von Arx (1980), who placed the ascigerous teleomorph of *Candida lipolytica* in a new genus, *Yarrowia*, with the type species *Yarrowia lipolytica* (Wickerham et al.) van der Walt et von Arx. This change was prompted by the unique structure of the ascospores, its coenzyme Q-9 system, and its relatively high nuclear G + C content of 49.5–50.2 (Nakase and Komagata, 1971). In this review I shall continue to use the genus name *Saccharomycopsis* under which it is best known in the biochemical and industrial literature.

B. Physiological Properties

S. lipolytica is unusual among yeasts by its strong ability to hydrolyze gelatin and casein, its lipolytic activity, and its ability to grow on a number of *n*-alkanes. It lacks the ability to ferment sugars. Its ability to assimilate various carbon compounds is shown in Table 18.1. Among the hydrocarbons the following are assimilated strongly: *n*-decane, *n*-dodecane, *n*-tridecane, *n*-tetradecane, *n*-pentadecane, *n*-hexadecane, *n*-heptadecane, *n*-octadecane, *n*-nonadecane, *n*-eicosane, 1-hexadecene, and 1-octadecene (Wickerham et al., 1970). *n*-Docosane, 1-dodecene, and 1-tetradecene are assimilated slowly, and 1-decene is not utilized.

Other properties are as follows: KNO_3 is not assimilated; growth in vitamin-free medium is weak; thiamine (100–200 $\mu g/l$) stimulates growth to normal levels; maximum temperature for growth is 33–37°C; sodium chloride tolerance is 10–14% (w/v); growth in the presence of 100 p.p.m. cycloheximide is strongly positive.

C. Natural Habitat

Strains of *S. lipolytica* are more often isolated from substrates containing lipids, hydrocarbons, or proteins than from those containing sugars or polysaccharides. This is probably the result of its strong extracellular protease production, lipolytic activity, and ability to oxidize hydrocarbons, while its sugar-assimilating ability is very limited (Table 18.1). Most strains have been isolated from spoiled margarine, olive pulp, steep water or other sources in corn-processing plants, meat products with high fat content (e.g., sausage), kerosine, various petroleum products, and crude oil spills.

D. Genetics

Wickerham et al. (1970), who demonstrated heterothallism in *S. lipolytica*, made genetics possible with this yeast. A problem with strains isolated from various natural sources was that spore viability and the percentage of four-spored asci were so low that genetic analysis was virtually impossible (Ogrydziak et al., 1978). Another problem was the very low mating frequencies between haploid cells of opposite mating type (Bassel and Ogrydziak, 1979), which are about 2–4% zygotes for crosses between prototrophs and 10^{-3}–10^{-4} for crosses between auxotrophs. Thus matings must be forced by using complementary nutritional markers and they take 7–8 days to complete. Sporulation itself takes 12–14 days, and dissection is more difficult than spore isolation from asci of *Saccharomyces* (Ogrydziak et al., 1978). However, an extensive program of inbreeding in Mortimer's laboratory has led to greatly increased spore viability (up to 85%) as well as an increased number of four-spored asci. Similar improvements in spore viability have been reported from Heslot's laboratory (Gaillardin et al., 1973). However, incompatibilities between in-

TABLE 18.1   Assimilatory Properties of Four Yeast Species

	S. lipolytica	*C. utilis*	*K. fragilis*	*H. polymorpha*
D-glucose	+	+	+	+
D-galactose	− (rarely +)	−	+	− or latent
L-sorbose	− (sometimes weak)	−	+ or −	−, or weak, latent
maltose	−	+	−	+
surcrose	−	+	+	+
cellobiose	− (rarely +)	+	+	+ (rarely −)
trehalose	−	+ (or weak)	−	latent
lactose	−	−	+	−
melibiose	−	−	−	−
raffinose	−	+	+	−
melezitose	−	+	−	+, weak, latent, or −
inulin	−	+	+	−
soluble starch	−	−	−	−
D-xylose	−	+ (or weak)	+	+, latent, or −
L-arabinose	−	−	+	−
D-arabinose	−	−	−	−
D-ribose	− or +	−	+ or −	latent
L-rhamnose	−	−	−	− or latent
ethanol	+	+	+	+
glycerol	+	+	+ or weak	+
erythritol	+	−	−	+
ribitol	− (rarely +)	−	+ (rarely −)	latent
galactitol	−	−	−	latent
D-mannitol	+ (rarely −)	+ or −	+	+
D-glucitol	+ (rarely −)	− (or weak)	+	+
methyl-α-D-glucoside	−	+ (or weak)	−	weak, latent
salicin	− (rarely +)	+	+	+ (rarely −)
D-gluconate	+ (rarely −)	+	−	−
2-ketogluconate	−	−	−	−
5-ketogluconate	−	−	−	−
D-glucarate	−	−	*	−
lactate	+	+	+	−
succinate	+	+	+	weak or −
citrate	+	+	+ or −	+ or weak
meso-inositol	−	−	−	−
D-glucosamine	−	−	−	−
n-hexadecane	+	−	−	−
methanol	*	−	−	+ or weak

*Data not available.

bred strains from different laboratories still exist (Bassel and Ogrydziak, 1979). Nevertheless, the inbreeding programs mentioned above have made it possible to study the genetics of *S. lipolytica* by tetrad analysis, so a good deal of information has accumulated on this subject. Ogrydziak et al. (1978) detected 22 cases of linkage among 278 gene pairs and developed a genetic map consisting of six linkage fragments. Since no centromere markers have been identified as yet, no exact information on the number of chromosomes is available. However, the fact that the first six markers (for which there are sufficient data) show no centromere linkage suggests that *S. lipolytica* may have just a small number of long chromosomes. More recently, Ogrydziak et al. (1982a) reported further work on genetic map development by tetrad and random spore analysis, and they identified mutations in 23 additional nuclear genes. Eight genes were located on linkage fragment 1, four on fragment 2, two on fragment 5, and three on fragment 6. Linkage fragments 3 and 4 were found to be linked, and this fragment now contains 12 markers. Although markers exhibiting possible centromere linkage were identified, the tentative nature of the evidence does not allow conclusions as to the number of chromosomes in *S. lipolytica*.

Ogrydziak and co-workers also have conducted an extensive study on alkaline extracellular protease (AEP) production by *S. lipolytica*. Mutations (*xpr*) in at least 16 genes result in reduced ability to produce extracellular protease (Ogrydziak and Mortimer, 1977). Later, Simms and Ogrydziak (1981) isolated mutants that were temperature sensitive for AEP production but not for growth. Biochemical studies had indicated that under conditions used for the genetic studies a single alkaline and no neutral extracellular proteases were produced. Thirty-three mutants were temperature sensitive for protease production, and one (*xpr*-32) produced a temperature-sensitive protease. Genetic analysis showed that *xpr*-32 was located in gene *XPR*2, and gene dosage efforts and other genetic evidence revealed that *XPR*2 is the structural gene for the alkaline extracellular protease. The fact that *xpr*2 mutants grew on minimal medium under conditions where no protease was produced indicated that AEP activity is not essential for growth on minimal medium. In addition to those in the structural gene, other *xpr* mutations could be in regulatory genes or in genes involved in the processing and secretion of the enzyme. The finding by Ogrydziak et al. (1982b) that the *xpr* mutants, with the exception of the structural gene mutant discussed above, produced lower levels of the wild-type enzyme suggested that the mutations were in genes affecting regulation or secretion. Several pleiotropic *xpr* mutants also produced lower levels of other extracellular enzymes (such as acid proteases and RNases), and these results suggested that such mutations may affect secretion.

Other genetic studies with *S. lipolytica* include the isolation of mutants defective in lysine catabolism (Gaillardin et al., 1976), alkane utilization (Bassel and Ogrydziak, 1979), and acyl-coenzyme A synthetase (Kamiryo et al., 1977).

E. Biochemistry

S. lipolytica produces higher levels of extracellular proteases than do most other yeast species (Ahearn et al., 1968; Meyers and Ahearn, 1977). Different strains appear to produce various combinations of extracellular proteases; for example, *S. lipolytica* AJ 4555 produces an alkaline protease (Tobe et al., 1976), whereas *S. lipolytica* 37-1 produces acid protease(s) and one neutral protease but no alkaline protease (Abdelal et al., 1977). When grown at neutral pH, *S. lipolytica* CX161-1B (an inbred strain) produces a single alkaline extracellular protease, low levels of acid protease(s), and no neutral protease (Ogrydziak and Scharf, 1982). These authors purified the alkaline protease to homogeneity and reported its physicochemical properties. It is probably a serine protease with alanine as the *N*-terminal amino acid. More recently, Yamada and Ogrydziak (1983) purified three extracellular acid proteases when strain CX161-1B was grown in a medium containing glycerol, proteose peptone (Difco), and mineral salts at pH 3.4. The enzymes were excreted during exponential growth, and their physicochemical properties were determined.

S. lipolytica is also known as a lipase-producing yeast. The lipase is induced by growing the yeast in the presence of olive oil, 3-hydroxymyristic acid, or unsaturated long-chain fatty acids (e.g., oleic acid), and most of the enzyme is cell bound (Ota et al., 1968; Sugiura et al., 1975). Zvyagintseva et al. (1980) demonstrated that the enzyme is mainly present in the cell wall and on the outer surface of the cytoplasmic membrane. Ota et al. (1978) discovered that alkaline extracts of defatted ground soybeans, when boiled and added to the growth medium of *S. lipolytica*, greatly increased the production of extracellular lipase, which was accompanied by a decrease in cell-bound lipase activity. Fractions in such extracts with high exolipase-enhancing activity were rich in protein and carbohydrate. Ruschen and Winkler (1982) then found that potassium hyaluronate (K-HA), when added to an induced cell suspension of *S. lipolytica*, causes an immediate and strong increase in extracellular lipase activity, even though K-HA cannot be used by the cells as a carbon or energy source. Its effect is consistent with the "detachment hypothesis," which states that certain polysaccharides (such as K-HA) promote the detachment (solubilization) of preexisting lipase molecules from the surface of the yeast cells. Ruschen and Winkler also presented evidence that de novo lipase synthesis is not responsible for the rise in extracellular level. Ota et al. (1982) solubilized cell-bound lipase from intact cells by suspending induced, pregrown cells in a medium containing Emulgin 950 (polyoxyethylene nonylphenol ether, average polymerization of ethylene oxide 50, Kao Atlas Co.) for 2 h at 30°C. Two types of lipase were isolated and purified from the extract and their properties determined. The two lipases were similar in enzymological properties, and one form was thought to represent an enzymatically modified form of the other. The purified enzymes require an activator, such as oleic acid, for hydrolysis of triglycerides, but the cell-bound enzyme(s) does not. The highest activity was obtained with tricaprylin as the substrate.

Another important biochemical property of *S. lipolytica* is its ability to utilize *n*-alkanes for growth. It shares this ability with a number of other yeast species (Bos and de Bruyn, 1973). Fukui and collaborators at Kyoto University (see Fukui and Tanaka, 1979a, b, for reviews) observed that yeast cells grown on alkanes contain a large number of specific organelles with a homogeneous matrix and surrounded by a unit membrane. Because of their high content of catalase, these organelles were designated as peroxisomes or microbodies. It should be noted that the function of peroxisomes in alkane-utilizing yeasts is quite different from those in methanol-grown yeasts (which are discussed in the section on *Hansenula polymorpha*). The peroxisomes of yeasts grown on alkanes play an essential role in the degradation of fatty acids derived from alkane substrates. After *n*-alkanes diffuse across the cytoplasmic membrane, the first step in their degradation is a hydroxylation of the terminal methyl group (Ratledge, 1978) by a particulate (microsomal?) fraction in the cytoplasm. This monooxygenase system (ω-hydroxylase) requires O_2 and NADPH and contains cytochrome P-450 and a NADPH-cytochrome P-450 (cytochrome *c*) reductase (Delaissé et al., 1981; Marchal et al., 1982). The oxidation of the fatty alcohol to fatty acid takes place by soluble, nonparticulate alcohol dehydrogenase and aldehyde dehydrogenase in the cytoplasm (Marchal et al., 1982). Fukui and Tanaka (1979a, b) have reviewed and described the role peroxisomes play in the further metabolism of the fatty acids derived from *n*-alkanes. When the fatty acids enter the peroxisomes, they are converted to the corresponding CoA esters by an acyl-CoA synthetase with the expenditure of ATP. The acyl-CoA synthetase in the peroxisomes is different from that in the mitochondria and supplies acyl-CoA only to be degraded by β-oxidation *in the peroxisomes*. H_2O_2 formed at the step of acyl-CoA dehydrogenation by FAD is decomposed by catalase, a process accounting for the high level of this enzyme in peroxisomes. After *trans-α, β* acyl-CoA is converted to β-hydroxy-acyl-CoA, this intermediate is oxidized by NAD to β-keto acyl-CoA, and the resulting NADH appears to be reoxidized by the concerted action of NAD-linked glycerol-3-phosphate dehydrogenase in the peroxisomes in conjunction with FAD-linked glycerol-3-phosphate dehydrogenase in the mitochondria:

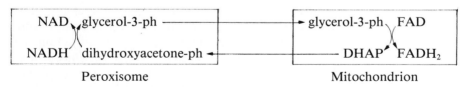

Peroxisome   Mitochondrion

Glycerol-3-ph/dihydroxyacetone phosphate shuttle

A thiolytic cleavage of the β-keto acyl-CoA yields acetyl-CoA and a new molecule of R'acyl-CoA. The acetyl-CoA formed by β-oxidation is then transported to a mitochondrion as O-acylcarnitine; after passing the inner membrane it is converted back to acetyl-CoA in the mitochondrial matrix, where it can enter the citric acid cycle by reacting with oxalacetate.

Another remarkable interaction between peroxisomes and mitochondria is the absence of the two key enzymes of the glyoxylate bypass cycle in the mitochondria of alkane-grown yeast and their presence in the peroxisomes. The mitochondria supply isocitrate to the peroxisomes, where it is cleaved by isocitrate lyase to succinate and glyoxylate; the latter reacts with acetyl-CoA (generated in the peroxisome by β-oxidation) to form malate. Succinate and malate are then thought to return to the mitochondria, where they can contribute to gluconeogenesis. This unique interaction between a microsomal fraction and two cell organelles illustrates the complex pathway from *n*-alkanes to cell material. Osumi et al. (1978) have reported the presence of a small DNA molecule in *Candida tropicalis* peroxisomes; DNA could therefore be present in *S. lipolytica* peroxisomes as well. Its coding potential is as yet unknown.

F. Applications

Reed (1982) has reviewed industrial processes for microbial biomass production. Yeasts of the genus *Candida* and *S. lipolytica* have been grown industrially or in pilot operations on *n*-alkanes and gas oil. Successes in this industry have been modest, mainly because of economic factors and concerns of governmental agencies in certain countries about safety of the products.

A much more important use of *S. lipolytica* is in the production of citric acid. Although molds, in particular *Aspergillus niger*, have been successfully used since 1919 for the microbial production of citric acid, the discovery in the late 1960's that very high yields of citric acid from *n*-alkanes were produced by *S. lipolytica* allowed this yeast to be competitive with *A. niger* (see Kapoor et al. (1982) for literature on this subject). Since aconitase, which converts citrate to isocitrate, requires Fe^{2+} for activity, the iron content of the growth medium is adjusted to very low concentrations to increase the citrate/isocitrate ratio. Because low concentrations of iron are difficult to maintain in complex media, fluoroacetate-sensitive strains with low aconitase activity have been developed; they produce increased amounts of citric acid. Recently, Furukawa et al. (1982) reported that limitation of phosphate increases citric acid yield. In addition, 0.1–0.2 mg/l of $CuSO_4 \cdot 5H_2O$ (an inhibitor of aconitase) and 0.1 mg/l of $Na_2B_4O_7 \cdot 10H_2O$ were found to inhibit formation of isocitrate and increase citrate production. They also obtained mutants with increased citrate synthetase and lower aconitase activity. Some of these mutants accumulated ammonium citrate in high concentration. The use of NH_4OH as a neutralizer has the advantage of forming soluble NH_4-citrate rather than the less soluble Ca-citrate with $CaCO_3$ as the neutralizer. Furukawa and Ogino (1982) described a semicontinuous cell recycle system that prolongs the production phase, minimizes product inhibition, and shortens the lag phase preceding the production phase.

Citric acid can also be produced from glucose under aerobic conditions by *S. lipolytica* and a number of other yeast species, such as *Candida zeylanoides* and *Candida citrica* (Kapoor et al., 1982). Briffaud and Engasser (1979a, b) grew *S. lipolytica* on glucose in a stirred, aerated fermentor. Citric acid production started at the end of the exponential growth phase (N-exhaustion) and continued for 80 h at a constant rate of 0.7 g/l/h. In this process the proportion of isocitrate was low, and the specific oxygen consumption rate was nearly threefold lower than with *n*-alkanes as the substrate. A trickle flow fermentor (in which the growth phase is linear rather than exponential) was about 30% less efficient. A metabolic model for the production of citrate from glucose by *Candida lipolytica* was described by Aiba and Matsuoka (1979).

II. *HANSENULA POLYMORPHA*

A. Taxonomy

This species is included in this chapter because of its good growth on methanol (a potentially useful industrial substrate) as a sole carbon source; several studies on methanol metabolism of this yeast have been conducted. *H. polymorpha* was originally described as *Hansenula angusta* by Teunisson et al. (1960), who provided a Latin description nine years after Wickerham (1951) first gave an English description. However, a paper with a Latin description by de Morais and Maia (1959), describing the same yeast from a different source, had priority, reducing *H. angusta* Teunisson et al. 1960 to a synonym of the valid name *Hansenula polymorpha* de Morais et Maia 1959.

B. Ascospore Formation

H. polymorpha exists in nature predominantly as haploid cells. Sporulation is usually preceded by conjugation between a cell and its bud, seldom between independent cells. The zygotes normally contain four small, hat-shaped spores with distinct downward bent brims. The spores are rapidly released from the asci. Heavily sporulating strains (best on malt extract agar) turn pink because of the presence of pulcherrimin in the spores. The ascospores of this species are unusually heat resistant (Teunisson et al., 1960; Wickerham, 1970). In culture, diploid cells may arise from conjugated haploid cells, and such diploid cells can be directly converted into asci.

C. Morphological Properties

On most media the haploid cells are quite small, measuring approximately 1–3.5 × 2–4.5 μm, and they usually occur singly and in pairs. Diploid cells are somewhat larger and may measure 3.4–5.2 μm. Pellicles on liquid media are very thin. Pseudomycelium is not produced on corn meal agar, either aerobically or anaerobically.

D. Physiological Properties

Glucose is fermented, but galactose, maltose, sucrose, lactose, and raffinose are not. The ability to assimilate various carbon compounds is given in Table 18.1. Other characteristics are as follows.

Assimilation of nitrogen compounds: $(NH_4)_2SO_4$ +; KNO_3 +; $NaNO_2$ −; ethylamine +; lysine +; cadaverine +.

Growth in vitamin-free medium: negative.

Sodium chloride tolerance: 5–8% (w/v) depending on the strain.

Growth in the presence of 50% (w/w) glucose: negative.

Maximum temperature for growth: 45–47°C.

Acid production on chalk agar: weakly positive.

Growth in the presence of 100 p.p.m. of cycloheximide: positive.

Urea hydrolysis: negative.

Gelatin hydrolysis: negative.

Casein hydrolysis: negative.

E. Natural Habitat

The literature on its ecology suggests that *H. polymorpha* is associated with certain plant materials both in temperate climatic zones and in hot tropical or desert areas. The type strain, isolated by de Morais and Maia (1959) from soil irrigated with wastewater from a distillery in Brazil, and the strain that Wickerham (1951) intended to use as the type strain, isolated from concentrated orange juice, do not appear to have come from typical habitats but more likely represent cases of accidental contamination of these substrates. In the following accounts of isolation sources, both the names *H. angusta* (used primarily before 1960) and *H. polymorpha* (used shortly after the validation of the specific epithet *polymorpha*) were used in the literature citations given, but in this chapter I shall use the correct name *H. polymorpha*. Teunisson et al. (1960) stated that this species is widely distributed among certain deciduous trees, for example, in insect frass of *Liquidambar styraciflua* L. (Louisiana), *Nyssa sylvatica* Marsh. (South Carolina), *Quercus nigra* L. (Louisiana), and, according to Shehata et al. (1955), in one slime flux of *Quercus kelloggii* Newberry in the Yosemite area of the California Sierra Nevada. In a subsequent study of slime fluxes of all tree species occurring in the same Yosemite area, Phaff and Knapp (1956) identified no strains of *H. polymorpha* among 136 yeasts isolated, but Phaff et al. (1956) isolated 19 strains from the alimentary canal of five species of *Drosophila* and one species of *Aulacigaster* in the same region. Shehata et al. (1955) isolated six additional strains from *Drosophila pseudoobscura* in a yellow pine area on Mount San Jacinto in southern California and three more strains from *D. pseudoobscura* in central California. Although these wild species of *Drosophila* appear to feed regularly on substrates containing *H. polymorpha*, their identity in these mountain areas remains unknown. More recently, Starmer and Phaff (unpublished data) discovered that *H. polymorpha* is a common species in rotting tissue of *Opuntia* cacti in the southwestern deserts of the United States. Cactus rots are utilized by a number of cactus-specific *Drosophila* species for feeding as well as breeding. More than 30 strains of *H. polymorpha* were isolated from *Opuntia phaeacantha*, and two strains from *Opuntia ficus-indica* in southern Arizona. Seven additional strains came from rotting *Opuntia lindheimeri* in Texas. It is interesting that *H. polymorpha* was not found among 721 yeasts isolated in necrotic tissue of columnar cacti in the North American Sonoran Desert in Mexico or in *Drosophila* species breeding in them (Starmer et al., 1982). The strains from *Opuntia* cacti are all heavily sporulating, giving the colonies a pink appearance. It is possible that the high heat resistance of the ascospores, mentioned earlier, is responsible for selection of spores over vegetative cells in the cacti of the hot Sonoran Desert.

Other sources of *H. polymorpha* listed by Wickerham (1970) include acorns that are host to larvae feeding on the tissue, compost of bagasse (sugar cane residue), and an ambrosia beetle.

F. Genetics

Single ascospores of *H. polymorpha* give rise to sporulating colonies. This, together with the fact that haploid vegetative cells conjugate with their own buds prior to meiosis and sporulation, suggests that this species is homothallic. This property precludes genetic crosses between haploid progenies of single spores derived from strains with different assimilatory properties, followed by tetrad analysis.

The strains from rotting *Opuntia* tissue, discussed above, differ from those isolated from noncactus substrates (represented by the type strain from concentrated orange juice and strain DL-1 isolated from an enrichment culture inoculated with soil and river water samples by Levine and Cooney (1973)) in that the cactus strains assimilate methanol much more slowly and weakly than the noncactus strains. Other differences are the slightly lower nuclear DNA base composition of the noncactus strains (47.3 mol % G + C for both the type strain and strain DL-1 versus 48.1–49.0 mol % G + C for seven strains from *Opuntia* cacti). DNA–DNA reassociation reactions between the cactus strains and the other two strains mentioned above were conducted in our laboratory (H. J. Phaff and Joanne Tredick, unpublished results). We found that seven strains from *Opuntia* cacti (including two from Australian species) exhibited 99–100% DNA complementarity with one of the American strains as the reference labeled with [125]I. The type strain of *H. polymorpha* showed 64% DNA homology with the reference strain, and strain DL-1 was 72% homologous. It appears therefore that the cactus strains with their highly specific habitat have diverged somewhat from the strains from other sources. By current views of DNA complementarity values, all strains clearly belong to the same species.

G. Biochemistry and Physiology

H. polymorpha is one of the yeast species that has attracted attention from biochemists because of its good growth on methanol as the sole carbon source. It shares this ability with approximately 25 other yeast species, among which *Candida boidinii* has also been studied extensively (van Uden and Buckley, 1970; Sahm, 1977). Ogata et al. (1969) published the first report of a methanol-utilizing yeast, which they referred to as *Kloeckera* sp. No. 2201 but which was subsequently reidentified as *C. boidinii* in my laboratory. No species of *Kloeckera* or its perfect state *Hanseniaspora* have been demonstrated to utilize methanol.

Quayle (1980) has reviewed the initial developments in the elucidation of the metabolic reactions involved in the assimilation of methanol. These early studies showed that the bulk of the carbon of methanol is assimilated at the level of formaldehyde via hexose phosphates, and the evidence at that time suggested the operation of a ribulose monophosphate cycle involving fixation of formaldehyde to yield hexulose phosphate, as occurs in some bacteria. Quayle and colleagues confirmed that [^{14}C] methanol was incorporated via formaldehyde into hexose phosphates by *H. polymorpha* and *C. boidinii* but were unable to detect the two key enzymes of the bacterial ribulose monophosphate cycle, that is, hexulose monophosphate synthase and phosphohexulo-isomerase, in cell-free extracts of these two yeasts. The Quayle group then considered the possibility that a pentose phosphate–dependent fixation of formaldehyde in their crude yeast extracts might be catalyzed by transketolase, using xylulose 5-phosphate as ketol donor and formaldehyde as acceptor, which would result in the synthesis of dihydroxyacetone (DHA) as the initial product. Further metabolic steps would require phosphorylation of DHA, its conversion to fructose 1,6-bisphosphate by enolase, and hydrolysis to fructose 6-phosphate by fructose 1,6-bisphosphatase.

van Dijken et al. (1978) then supported this postulated pathway by the demonstration of high levels of DHA kinase and fructose 1,6 bisphosphatase in methanol-grown *H. polymorpha* and *C. boidinii* cells. Moreover, O'Connor and Quayle (1979) found that mutants of these two species lacking DHA kinase activity were unable to grow on methanol. Revertants of the *H. polymorpha* mutant to wild-type growth on methanol regained the ability to synthesize DHA kinase. Similarly, a mutant of *H. polymorpha* with lowered levels of fructose bisphosphatase showed weaker growth on methanol, while revertants to normal growth on methanol regained the wild-type levels of fructose bisphosphatase.

Quayle's group then focused attention on the postulated transketolase reaction involved in the synthesis of DHA. Waites and Quayle (1981) showed that *C. boidinii* grown on glucose, D-xylose, or ethanol synthesized a single transketolase of the classical type, which was heat stable and showed no appreciable activity with formaldehyde as acceptor. However, when *C. boidinii* was grown on methanol, two transketolases were synthesized: one with the classical activity and a new, unstable enzyme that showed dihydroxyacetone synthase activity as well as classical transketolase activity, that is, it could use both aldose phosphates (e.g., ribose 5-phosphate) and formaldehyde as acceptors of the glycolyl group from the donor substrate xylulose 5-phosphate (see also Kato et al. (1979), Waites and Quayle (1980, 1983), and O'Connor and Quayle (1980) for further details on this special type of transketolase). In contrast to the transketolase, the fructose 1,6-bisphosphatases of *H. polymorpha* purified from methanol- and ethanol-grown cells were found to be similar in all properties investigated (Attwood and van Dijken, 1982).

The reactions so far discussed are catalyzed by cytoplasmic enzymes. However, the oxidation of methanol to formaldehyde occurs in special organelles, termed microbodies or peroxisomes. The synthesis of these organelles is induced only upon transfer of cells into methanol-containing media. The reader is referred to the excellent reviews by Sahm (1977) and Fukui and Tanaka (1979a, b) for history of the discovery of peroxisomes in all methanol-grown yeasts so far investigated. Briefly, a variable number of these organelles, surrounded by a single unit membrane, occur in such yeasts, and cytochemical-staining techniques have shown that they contain a flavin adenine dinucleotide (FAD)-dependent alcohol oxidase and catalase. Spheroplast lysates of *C. boidinii* subjected to noncontinuous Ficoll density gradient centrifugation yield a fraction containing the spheroidal peroxisomes, which can then be observed by transmission electron microscopy. These organelles, 0.4–0.6 μm in diameter, contain crystalloid inclusions. Osumi et al. (1982) recently reported on the ultrastructure of these crystalloid inclusions by a cryosectioning technique, together with electron diffraction, optical diffraction, and computer simulation and concluded that the crystalloids are cubic in structure and composed of two types of particles, large and small ones, arranged alternately, the large particles representing alcohol oxidase molecules (an octamer with a molecular weight of 673,000) and the small particles thought to be catalase molecules.

In summary, the following reactions lead to the conversion of the one-carbon compound methanol to glucose 6-phosphate in methylotrophic yeasts:

$$2\ CH_3OH + 2\ FAD \rightarrow 2\ HCHO + 2\ FADH_2 \quad \text{Alcohol oxidase}$$
$$2\ FADH_2 + 2\ O_2 \rightarrow 2\ H_2O_2 + 2\ FAD$$
$$\underline{2\ H_2O_2 \rightarrow 2\ H_2O + O_2} \qquad\qquad\qquad\ \text{Catalase}$$

Sum: $2\ CH_3OH + O_2 \rightarrow 2\ HCHO + 2\ H_2O$ 　　Reactions in
　　　　　　　　　　　　　　　　　　　　　　　　　　　　　peroxisomes

Cytoplasmic reactions (see text for enzymes involved):

Xylulose 5-ph + HCHO $\rightarrow$ dihydroxyacetone + glyceraldehyde 3-ph

dihydroxyacetone + ATP $\rightarrow$ DHA-phosphate + ADP

Glyceraldehyde 3-ph + DHA-ph $\leftrightharpoons$ fructose 1,6-bisphosphate

fructose 1,6-bisphosphate $\rightarrow$ fructose 6-ph + Pi

fructose 6-ph $\leftrightharpoons$ glucose 6-ph
_____

Sum: xylulose 5-ph + HCHO + ATP $\rightarrow$ glucose 6-ph + ADP + Pi

H. Applications

H. polymorpha and approximately 25 additional yeast species (Sahm, 1977) are known for their ability to grow on methanol as the sole carbon source. The increasing demand for proteins with high nutritive value for human food and animal feed has focused attention on microorganisms as a source of such proteins, usually referred to as single-cell protein (SCP). In this connection, yeasts have some advantages over bacteria: they are easier to harvest because of their larger size, they have a lower nucleic acid content, and they are psychologically more acceptable for human consumption than bacteria. Cooney and Levine (1975) pointed out that among yeast substrates for SCP production, methanol has certain advantages, including its miscibility with water, its high purity, its restricted use by microorganisms (lessening contamination problems), and its lower oxygen requirements and lower heat generation per unit weight of cells in comparison with the more reduced hydrocarbons as substrates. Disadvantages of the use of methylotrophic yeasts include a lower protein content and a lower specific growth rate than some methylotrophic bacteria (Sahm, 1977). An additional disadvantage compared to yeasts growing on carbohydrates is the low cell yield on methanol, which is probably caused by the requirement of one mole of ATP to convert methanol to glucose 6-phosphate and the additional requirement of one mole of phosphorylated xylulose.

In addition to the use of methanol as a substrate for SCP, Sahm (1977) has listed some processes for the production of metabolites from methanol by methylotrophic yeasts; these include L-alanine, L-glutamate, L-lysine, L-threonine, L-valine, citric acid, α-ketoglutaric acid, and hypoxanthine. More recently, Denenu and Demain (1981a, b) isolated several classes of mutants of *H. polymorpha* that were deregulated to enhance the flow of aromatic intermediates through the tryptophan biosynthetic branch, which resulted in the accumulation of tryptophan and/or tryptophan metabolites in the extracellular fermentation broth. Later Longin et al. (1982) expanded this work and determined the effect of various environmental and cultural conditions on the production of indole-containing metabolites from methanol by this yeast.

III. *KLUYVEROMYCES FRAGILIS*

A. Taxonomy

This yeast was first isolated by Jörgensen (1909) from kefyr. He named it *Saccharomyces fragilis* on the basis of the "feeble power of resistance of the cellwall" during the formation of the bean- or kidney-shaped ascospores, which are very rapidly liberated from the asci (diploid cells are converted directly into asci). The placement of this yeast in the genus *Saccharomyces* at that time is perhaps not surprising because *S. fragilis* was a fermenting, budding yeast that did not form a pellicle on liquid media; and with these properties it fitted well in the broad diagnosis of *Saccharomyces* accepted at that time. Lodder and Kreger-van Rij (1952) also accepted *S. fragilis* as a member of the genus *Sac-*

charomyces. The first authors to challenge this taxonomic placement were Wickerham (1955) and Wickerham and Burton (1956a, b). They pointed out the heterogeneity of the genus *Saccharomyces* and announced their intention to establish a separate genus (to be named *Dekkeromyces*) to accommodate *S. fragilis, Saccharomyces marxianus, Saccharomyces lactis*, and a few other species that appeared to be interfertile among each other but not with true *Saccharomyces* species. The proposed new genus contained species with crescentiform or bean-shaped ascospores (e.g., *S. fragilis* and *S. marxianus*) as well as species with spheroidal spores (e.g., *S. lactis*); but in contrast to true *Saccharomyces* species, asci ruptured readily at maturity, liberating the ascospores. Perhaps because Kudriavzev (1954) had made a similar earlier proposal to reclassify *S. fragilis* and related species, Wickerham and Burton neither gave a formal description nor designated a type for the envisaged genus, so the name *Dekkeromyces* according to the International Code of Botanical nomenclature constitutes a *nomen nudum*. Kudriavzev (1954), in a Russian edition of his monograph, proposed placing the diploid species (such as *S. fragilis*) in a new genus *Fabospora* and species with haploid vegetative cells (such as *Zygosaccharomyces marxianus*) in another genus *Zygofabospora*. However, six years passed before Kudriavzev (1960) formally diagnosed these two genera, resulting in a valid name, *Fabospora fragilis* (Jörgensen) Kudriavzev 1960. The genus *Kluyveromyces* was originally established by van der Walt (1956a) for a fermentative, budding yeast that formed large multispored asci containing numerous reniform to long oval spores and that was named *Kluyveromyces polysporus*. In the same year, van der Walt (1956b) described a second species, *Kluyveromyces africanus*, which differed in physiological properties and in forming not more than 16 spores per ascus. Later, van der Walt (1965) argued that the difference between four-spored species (*Fabospora, Zygofabospora*) and multispored species (resulting from supernumerary mitotic divisions of haploid nuclei following meiosis) was not sufficiently fundamental to retain these as separate genera, and he united the genera *Kluyveromyces, Fabospora*, and *Zygofabospora* into a single taxon. According to the International Code, the oldest name *Kluyveromyces* must be retained; van der Walt (1965) therefore emended the diagnosis of *Kluyveromyces* by the proviso that "the number of ascospores per ascus ranges from one to numerous." Thus the new name of *Fabospora fragilis* became *Kluyveromyces fragilis* (Jörgensen) van der Walt 1909.

The next nomenclatural development occurred as a result of a study by Bicknell and Douglas (1970), who determined by DNA–DNA reassociation experiments that *K. fragilis* and *K. marxianus* share 93% of their DNA sequences and are therefore synonymous. The two species differed chiefly in that lactose is not, or is only seldom weakly and latently, fermented by *K. marxianus*, whereas *K. fragilis* ferments lactose rapidly. These results were fully confirmed by Phaff et al. (1978), who reported between 90% and 100% DNA complementarity not only between *K. fragilis* and *K. marxianus* but also between *Kluyveromyces bulgaricus, Kluyveromyces cicerisporus*, and *Kluyveromyces wikenii* with *K. marxianus* as the reference species. Because *K. marxianus* is the oldest name, it has priority over the other four species, which all become synonyms of *K. marxianus* (Hansen) van der Walt 1888. Because of its long history and the common usage in industrial literature of the specific epithet *fragilis*, I shall continue to use the name *K. fragilis* in this chapter. The reader, however, should keep in mind that the correct name according to the Botanical Code is *K. marxianus* and that *K. fragilis* strains of this species are good fermenters of lactose in contrast to strains of *K. marxianus*, which are not.

Two imperfect forms are known. van Uden and Buckley (1970) have given a description of *Candida pseudotropicalis* (Castellani) Basgal 1911 as the imperfect state of *K. fragilis* and a description of *Candida macedoniensis* (Castellani et Chalmers) Berkhout 1919 as the imperfect state of *K. marxianus*. Since *K. fragilis* is now a synonym of *K. marxianus*, only one imperfect form remains. The earliest name has priority, that is, *C. pseudotropicalis*. The close relationship between the two pairs of imperfect *Candida* species and their perfect *Kluyveromyces* counterparts was further confirmed by Yamazaki and Komagata (1982) on the basis of electrophoretic mobility of ten metabolic enzymes in polyacrylamide gels and by antigenic cell wall structures (Tsuchiya et al., 1965). Since *K. fragilis* and *K. marxianus* have been shown to be homothallic, the possibility that the asporogenous *Candida* species represent clones of haploid heterothallic mating types can be eliminated. Because van der Walt (1970) in his restudy of the genus *Kluyveromyces* found a number of previously sporogenous strains of *K. fragilis* and *K. marxianus* to be asporogenous, it appears that diploid strains of these species may lose their sporulating ability in culture.

B. Morphological Properties
In malt extract the cells are subglobose, ellipsoidal to cylindrical, approximately 2–5 × 3–13 μm. The cells occur singly, in pairs, or in short chains, and pseudomycelium is variably developed. A pellicle is lacking, but a thin ring may be present. Growth on corn meal agar usually results in abundant formation of pseudomycelium, especially under anaerobic conditions. Formation of ascospores takes place directly in diploid cells or after conjugation of two haploid cells. One to four crescentiform, reniform or bean-shaped, or spheroidal to prolate-ellipsoidal spores are formed and are rapidly released from the asci by ascus wall lysis; the spores tend to agglutinate. Sporulation is usually abundant on most media. Single spores develop into sporulating clones; the yeast is therefore homothallic.

C. Physiological Properties
The fermentation of sugars is as follows: glucose + ; D-galactose + ; sucrose + ; maltose − ; cellobiose − ; lactose + strong, or latent and weak, or − ; trehalose − ; melibiose − ; raffinose + (1/3); melezitose − ; methyl-α-D-

glucoside − ; inulin + or slow; soluble starch − . The ability to assimilate various carbon compounds is given in Table 18.1. Other properties are as follows:

Assimilation of nitrogen compounds: $(NH_4)_2SO_4$ + ; KNO_3 − ; $NaNO_2$ − ; ethylamine + .

Growth in a vitamin-free medium: negative. Niacin is an absolute requirement. Biotin and pantothenate stimulate growth.

Sodium chloride tolerance: 8–10% (w/v).

Growth in the presence of 50% (w/w) of glucose: negative.

Maximum temperature for growth: 45–47°C.

Acid production on chalk agar: negative or weak.

Growth in the presence of 100 p.p.m. of cycloheximide: positive.

Urea hydrolysis: negative.

Gelatin hydrolysis: negative (may be weakly positive in heavily sporulating clones).

Casein hydrolysis: negative (may be weakly positive in heavily sporulating clones).

D. Natural Habitat

The natural habitat of *K. fragilis* and its synonyms discussed above does not appear highly specific on the basis of the origin of the strains of the respective species of *Kluyveromyces* studied by van der Walt (1970) and of *Candida* studied by van Uden and Buckley (1970). The ability to ferment or assimilate lactose of many of the strains may account for their occurrence in various dairy products, such as cheeses, yoghurt, kumiss, and milk from a cow with mastitis. However, many strains have been reported from nonlactose-containing substrates, such as sewage of a sugar factory, a decaying sisal leaf, South African bantu beers, maize meal, compressed yeast, and canned apples. Although *K. fragilis* and its synonyms are not recognized as pathogens, many strains have been isolated from clinical materials or from patients suffering from various infectious diseases, for example, sputum, pus, feces, a case of bovine mastitis, tubercular lung, lesion of tonsils, thoracic cavity, infected nails, and a generalized yeast infection in humans. Possibly, the yeast's ability to grow up to 45–47°C is a contributing factor to its association with disease processes in warm-blooded animals and humans. Some of the infections may have been opportunistic in immuno-incompetent hosts as recently shown by Lutwick et al. (1980). In this case, *K. fragilis* caused a pulmonary infection in an immunosuppressed cardiac transplant patient. Even in this immuno-compromised host the yeast was poorly pathogenic, and the patient recovered without sequelae after treatment with amphotericin B. Other sources of *K. fragilis* (= *K. marxianus*) reported since van der Walt's (1970) review of the genus *Kluyveromyces* include a small number of strains isolated from rotting cactus tissue in the hot southwestern deserts of the United States. Starmer et al. (1982) reported on one strain from rotting giant saguaro tissue. Subsequent work by Starmer and Phaff (unpublished data) yielded two strains from *Opuntia ficus-indica* growing in the city of Tucson, Arizona; one strain from *Opuntia phaeacantha* in the Tucson Mountains; one strain from *Opuntia lindheimeri* in Texas; and one strain each from agria and cochal in the Mexican Sonoran Desert. Shehata et al. (1955) earlier reported the isolation of six strains of *K. fragilis* from *Drosophila pseudoobscura* crops at Pinon Flats, a desert area in southern California. Because rotting cactus tissue is utilized by *Drosophila* spp. as a food source, it is possible that the *K. fragilis* strains in the digestive tract of the flies came from nearby rotting cactus tissue.

E. Biochemistry

Production of an extracellular endo-polygalacturonase ($1 \rightarrow 4$ α-D-polygalacturonide glycanohydrolase) in strains of *K. fragilis* that latently ferment lactose was discovered by Luh and Phaff (1951), who named it "yeast polygalacturonase." The enzyme was produced constitutively during growth in standing cultures with glucose as the energy source (Luh and Phaff, 1954; Phaff, 1966). Pectin methylesterase (usually accompanying fungal polygalacturonases) is not produced by the yeast. The enzyme catalyzes a random hydrolysis of polygalacturonic acid to a mixture of D-galacturonic acid and digalacturonic acid in an approximate molar ratio of 4:3, at which stage 70% of the bonds in the linear polygalacturonic acid are cleaved (Demain and Phaff, 1954a). The purified enzyme catalyzes an initial, linear rate of hydrolysis of polygalacturonic acid, during which approximately 25% of the glycosidic bonds are hydrolyzed, after which the reaction rate slows down. Reaction rate measurements are described by Phaff (1966). Purification of the enzyme by affinity adsorption and ammonium sulfate precipitation was described by Phaff and Demain (1956). Its optimum pH with polygalacturonic acid as the substrate is 4.5, with tetragalacturonic and trigalacturonic acids 3.5, while digalacturonic acid is not hydrolyzed. The following series of reactions take place during hydrolysis of polygalacturonic acid (Demain and Phaff, 1954b):

(a) polygalacturonic acid $\rightarrow$ penta- + tetra- + tri- + digalacturonic acids
(b) pentagalacturonic acid $\rightarrow$ tetra- + galacturonic acid
                     and $\rightarrow$ tri- + digalacturonic acid
(c) tetragalacturonic acid $\rightarrow$ tri- + galacturonic acid
(d) trigalacturonic acid $\rightarrow$ di- + galacturonic acid

Reactions (a) and (b) are rapid (25% of total hydrolysis), while reaction (c) is about 26 times slower than (a) and (b). After about 50% of total hydrolysis, (d) is the main reaction (pH 3.5), and its rate is approximately 1450 times slower than (a) and (b). At 70% of total hydrolysis the final products are the dimer and monomer. Patel and Phaff (1959) showed that reactions (c) and (d) occur exclusively at the glycosidic bond next to the terminal reducing group and that pentagalacturonic acid is the smallest oligogalacturonide that shows a tendency to undergo random cleavage, although mainly at the reducing end of the molecule.

A limited hydrolysis of pectin is possible, depending on its degree of esterification. Luh and Phaff (1954) showed that the extent of hydrolysis of pectins of increasing methoxyl content is an inverse linear function of the methoxyl content between 68% and 0% esterification.

Yeast endopolygalacturonase has been particularly useful for the preparation oligogalacturonides from polygalacturonic acid in high yields. Phaff and Luh (1952) described the preparation of trigalacturonic and digalacturonic acids via their lead and strontium salts, respectively; Demain and Phaff (1954a) described the isolation of tetragalacturonic acid via its cupric salt.

It is curious that *K. fragilis* produces a constitutive, extracellular polygalacturonase (which accounts for 92% of the total protein in the culture medium), but the yeast is unable to assimilate either polygalacturonic acid or any of its enzymatic hydrolysis products.

When grown with aeration in a medium with inulin as the carbon source, *K. fragilis* produces an extracellular and a cell-associated β-fructofuranosidase capable of hydrolyzing sucrose as well as β-2 ← 1 (inulin type) and β-2 ← 6 (levan type) furanosidic linkages (Snyder and Phaff, 1960, 1962). Degradation occurs by stepwise liberation of fructose from nonreducing chain ends, and the rate of hydrolysis falls with increasing chain length. The extracellular and cell-associated enzymes were partially purified by DEAE-cellulose adsorption and elution, followed by acetone precipitation in the cold, and were found to have identical properties. The ratio of hydrolysis rates with sucrose (at the optimum pH of 4.2) and with inulin (at the optimum pH of 5.0) was approximately 25 and approximately 14,000 with invertase from *Saccharomyces cerevisiae*. *K. fragilis* does not appear to contain significant amounts of invertase, and it ferments sucrose, raffinose, and inulin with the aid of a nonspecific fructan hydrolase. In a study of the pattern of action of inulinase, Snyder and Phaff (1962) made use of the fact that the inulin molecule ends in a glucose molecule. By following the rate of glucose liberation in relation to the overall rate of hydrolysis at pH 5.1 and 30°C, they showed that the degradation of inulin occurs largely by the single-chain mechanism in which the enzyme completely hydrolyzes one substrate molecule at a time.

The ability of *K. fragilis* to ferment lactose at various rates has led to a number of studies of the lactase of this yeast. Mahoney et al. (1975) found that 41 strains labeled *K. fragilis* varied 60-fold in ability to produce lactase. One of their best strains (NRRL-Y-1109) gave maximum lactase production at the beginning of the stationary phase in a medium with 15% lactose and an aeration rate of at least 0.2 mmol oxygen/liter/min. Mahoney and Whitaker (1978) described the purification of the enzyme following toluene autolysis of a cell mass and centrifugation. β-Galactosidase in the extract was purified to apparent homogeneity by acetone precipitation, pseudo-affinity chromatography, hydroxylapatite chromatography, and DEAE-Sephadex A-50 chromatography. The purified enzyme was free of carbohydrate, had a molecular weight of 201,000, and had 9–10 subunits based on electron microscopic observations. The amino acid composition was quite different from that of *Escherichia coli* β-galactosidase. The enzyme was rapidly inactivated by mercaptide formation with *p*-chloromercuribenzoate. There appear to be six rapidly reacting sulfhydryl groups per mole of protein. Mahoney and Whitaker (1977) described the following additional properties of the enzyme purified to electrophoretic, chromatographic, and immunochemical homogeneity. Potassium ions were specifically required for stability, and $MnCl_2$ increased the stability. Mn^{2+} at 0.1 mM in potassium phosphate buffer (0.05 M, pH 7.0) also gave the highest activation. Activity was completely inhibited by ethylendia-mine-tetraacetate and partially restored by the addition of $MnCl_2$. Greatest stability was at pH 6.5–7.5 and markedly less below 6.5 and above 8.5. The authors also reported kinetic constants of the enzyme with lactose and *o*-nitrophenyl β-D-galactopyranoside as substrates. D-Galactono 1,4-lactone was a strong competitive inhibitor.

F. Applications

Because of their ability to utilize lactose, *K. fragilis* and its imperfect forms (see taxonomy) are grown commercially on whey for the production of food yeast. Strains of *K. fragilis* that ferment lactose rapidly are used for the production of fuel alcohol from concentrated whey permeate. Details of these processes are discussed by Bernstein et al. (1977) and by Friend and Shahani (1979). Some *K. fragilis* strains are good sources of the enzyme lactase (Mahoney et al., 1975). The purification, enzymatic properties, and stability of this enzyme have been reported by Mahoney and Whitaker (1977, 1978), and some applications of lactase in the food industry have been described by Pomeranz (1964). The ability of certain *K. fragilis* strains to ferment inulin has led to recent studies on the use of Jerusalem artichokes (*Helianthus tuberosus* L.) as an energy crop for fuel alcohol production. The tubers of this plant are rich in inulin, and tuber pulp mashes can be fermented directly by *K. fragilis* or after acid hydrolysis by *S. cerevisiae* (Sachs et al., 1981; Williams and Ziobro, 1982a, b; Ziobro and Williams, 1983).

IV. *CANDIDA UTILIS*

A. Taxonomy

Henneberg (1926) found that this yeast was present as a contaminant in nearly all German yeast factories. He gave a brief description and named it *Torula utilis*. Because Persoon in 1796 had already used the genus name *Torula* for a fungus with true mycelium and dark-colored conidia, *Torula* Persoon has priority and thus cannot be used for budding yeast species. Lodder (1934) therefore changed the name to *Torulopsis utilis* (Henneberg) Lodder. Lodder and Kreger-van Rij (1952) transferred *T. utilis* to the genus *Candida* because of its ability to form a more or less well-developed pseudomycelium, especially under anaerobic conditions (growth on corn meal agar under a coverslip). Strains differ in degree of pseudomycelium formation as well as in blastoconidial differentiation. The full name of the asporogenous form now is *Candida utilis* (Henneberg) Lodder et Kreger-van Rij 1926. Wickerham (1970) pointed out that because of the similarity of fermentation and assimilation reactions, *C. utilis* appeared to be closely related to *Hansenula jadinii*, the difference being that *H. jadinii* Sartory produces very few hat-shaped ascospores and *C. utilis* is asporogenous. Kurtzman et al. (1979) conducted experiments on DNA base sequence homology to resolve the question of relatedness between the two species. The G + C contents of the nuclear DNAs were almost identical: 45.8 mol % for *C. utilis* and 45.1 mol % for *H. jadinii*. DNA–DNA reassociation reactions showed two different strains of *C. utilis* to give 101% reassociation with each other and 85% and 86%, respectively, with the type strain of *H. jadinii*. Reactions with *Hansenula petersonii*, another phenotypically similar species, gave 5% or less reassociation. On the basis of these experiments, Kurtzman et al. (1979) concluded that *C. utilis* represents the imperfect state of *H. jadinii*. Manachini (1979) fully confirmed the findings of Kurtzman et al. by a different technique of DNA–DNA reassociation. He also determined that the proton magnetic resonance spectra of the cell wall mannans of the two species were indistinguishable. Further evidence for their synonymy was given by Yamazaki and Komagata (1982), who showed that the electrophoretic mobilities of nine metabolic enzymes out of ten were identical. A small difference in mobility was noted only for fructose-1,6-bisphosphate aldolase.

B. Ascospore Formation in *H. jadinii*

Absence of conjugation suggests that the vegetative cells that become asci are diploid. Ascospores are hat shaped with a broad and fairly thick brim, one to four per ascus. The asci rupture at maturity, releasing the spores. Kurtzman et al. (1979) estimated only about one spore per 10^4 vegetative cells under the best conditions. They isolated 25 single spores by micromanipulation, of which only six were viable. Four spores gave colonies that sporulated no better than the parent, while the other two produced asporogenous clones. Treatment of a sporulating culture for 10 min at 45°C to selectively kill vegetative cells also did not result in improved sporulation. *H. jadinii* thus appears to be homothallic, but its poor sporulating ability virtually precludes further genetic studies.

C. Morphological Properties

Young cells in glucose–yeast autolysate–peptone are ovoidal to cylindrical, approximately 3.5–4.5 × 7–13 μm. A pellicle is not formed. Pseudomycelium on corn meal agar is well developed in most strains, although there is often little differentiation between blastoconidia and pseudomycelial cells.

D. Physiological Properties

Glucose, sucrose, and raffinose (1/3) are fermented. Inulin may be weakly fermented. Galactose, maltose, cellobiose, trehalose, lactose, melibiose, melezitose, soluble starch, and methyl-α-D-glucoside are not fermented. The assimilation of carbon substrates is shown in Table 18.1. Other properties are as follows:

> KNO_3, KNO_2, ethylamine, and lysine as nitrogen sources in Yeast Carbon Base (Difco) give good growth.
>
> Growth in vitamin-free medium: some strains are vitamin independent; others are stimulated by thiamine.
>
> Sodium chloride tolerance: 6–8% (w/v).
>
> Growth in the presence of 50% (w/w) of glucose: negative.
>
> Maximum temperature for growth: 39–43°C.
>
> Acid production on chalk agar: positive.
>
> Growth in the presence of 100 p.p.m. of cycloheximide: negative; at 10 p.p.m.: positive.
>
> Urea hydrolysis: positive (weak).
>
> Gelatin hydrolysis: negative.

Because the original strain of Henneberg had rather small cells and was difficult to harvest under commercial conditions, Thaysen and Morris (1943) grew this strain on malt agar plates containing 0.3% camphor. Among the colonies that developed were some whose cells had doubled in volume (from 318 μm^3 to 644 μm^3), probably owing to a doubling in ploidy. This change, which was stable, permitted more efficient recovery during centrifugal harvesting. The new strain, named as the variety *major*, did not differ in other respects from the parental strain.

Lodder and Kreger-van Rij (1952) found that *Candida guilliermondii* var. *nitratophila* is a synonym of *C. utilis*. This variety was described by Diddens and Lodder (1942) for a strain labeled *Mycotorula muhira* (Mattlet) Cif. et Red. This strain resembled *C. guilliermondii* in many respects but differed in having cylindrical cells and assimilating nitrate.

E. Natural Habitat of *H. jadinii* and *C. utilis*

The relatively few strains that have been isolated since the original discovery of *C. utilis* in 1926 and *H. jadinii* in 1932 do not give clear indications on their natural habitats. One of the two extant strains of *H. jadinii* was isolated along with a staphylococcus from pus in an abscess of a young woman, and the second strain was isolated from the udder of a cow with mastitis. Strains of the imperfect state, *C. utilis*, were originally isolated by Henneberg (1926) as contaminants in several German yeast factories. In my laboratory we also isolated it in 1945 from contaminated baker's yeast in a plant in California. In this factory, *C. utilis*, starting as a minor contaminant, outgrew baker's yeast during the production process and finally amounted to 10% of the final compressed yeast. Subsequently isolated strains of *C. utilis* were obtained from the following sources: one strain from a yeast deposit in a distillery; two strains from the digestive tract of cows in Portugal; two strains from sputum in the Netherlands; one strain from a vaginal discharge in Germany; one strain from a flower in Denmark (van Uden and Buckley, 1970). The occasional occurrence of either *H. jadinii* or *C. utilis* in diseased tissues of humans or animals should not be regarded with alarm in view of the use of *C. utilis* for the production of feed and food yeast. There is no evidence in the literature for pathogenicity of either species, and its occasional occurrence in clinical materials (as is the case even with *S. cerevisiae*) is probably due to chance contamination, to its ability to grow strongly at 37°C, and possibly to lowered resistance of the host.

F. Applications

Strains of *C. utilis* have been used with an excellent history of safety for more than four decades in the industrial production of "Torula yeast" that is incorporated in animal feed formulas and in human diets (Thaysen, 1957; Kihlberg, 1972; Chen and Peppler, 1978; Litchfield, 1979). *C. utilis* was selected during World War II in Germany for the production of food and feed yeast from wood hydrolysates because it utilizes the xylose in such hydrolysates, it does not require amino acids or B-vitamins for growth, and it has a high growth rate with ammonium salts as the nitrogen source. Presently, sulfite waste liquor (a by-product of the paper industry with 15–22% hexose and pentose sugars on the dry weight basis) is the most commonly used substrate for growing *C. utilis*. Ethanol is another suitable substrate, and at least one plant for the continuous production of *C. utilis* on ethanol has been built in Minnesota (Litchfield, 1979).

REFERENCES

Abdelal, A. T. H., Kennedy, E. H., and Ahearn, D. G. (1977) *J. Bacteriol. 130*, 1125–1129.

Ahearn, D. G., Meyers, S. P., and Nichols, R. A. (1968) *Appl. Microbiol. 16*, 1370–1374.

Aiba, S., and Matsuoka, M. (1979) *Biotechnol. Bioeng. 21*, 1373–1386.

Attwood, M. M., and van Dijken, J. P. (1982) *J. Gen. Microbiol. 128*, 2313–2317.

Bassel, J., and Ogrydziak, D. M. (1979) in *Genetics of Industrial Microorganisms* (Sebek, O. K., and Laskin, A. I., eds.), pp. 160–165, American Society for Microbiology, Washington, D.C.

Bernstein, S., Tzeng, C. H., and Sisson, D. (1977) *Biotechnol. Bioeng. Symp. 7*, 1–9.

Bicknell, J. N., and Douglas, H. C. (1970) *J. Bacteriol. 101*, 505–512.

560 Biology of Yeasts Other Than *Saccharomyces*

Bos, P., and de Bruyn, J. C. (1973) *Antonie van Leeuwenhoek 39*, 99–107.
Briffaud, J., and Engasser, J.-M. (1979a) *Biotechnol. Bioeng. 21*, 2083–2902.
Briffaud, J., and Engasser, J.-M. (1979b) *Biotechnol. Bioeng. 21*, 2093–2111.
Chen, S. L., and Peppler, H. J. (1978) *Dev. Ind. Microbiol. 19*, 79–94.
Cooney, C. L., and Levine, D. W. (1975) in *Single-Cell Protein*, Vol. II (Tannenbaum, S. R., and Wang, D. I. C., eds.), pp. 402–423, The MIT Press, Cambridge, Mass.
Delaissé, J.-M., Martin, P., Verheyen-Bouvy, M.-F., and Nyns, E.-J. (1981) *Biochim. Biophys. Acta 676*, 77–90.
Demain, A. L., and Phaff, H. J. (1954a) *Arch. Biochem. Biophys. 51*, 114–121.
Demain, A. L., and Phaff, H. J. (1954b) *J. Biol. Chem. 210*, 381–393.
de Morais, J. D. F., and Maia, D. (1959) *Anais da Escola Superior de Quimica da Universidade Recifi 1*, 15–20.
Denenu, E. O., and Demain, A. L. (1981a) *Appl. Environ. Microbiol. 41*, 1088–1096.
Denenu, E. O., and Demain, A. L. (1981b) *Appl. Environ. Microbiol. 42*, 497–501.
Diddens, H. A., and Lodder, J. (1942) *Die anaskosporogenen Hefen, Zweite Hälfte*, North-Holland, Amsterdam.
Friend, B. A., and Shahani, K. M. (1979) *N. Z. J. Dairy Sci. Technol. 14*, 143–155.
Fukui, S., and Tanaka, A. (1979a) *J. Appl. Biochem. 1*, 171–201.
Fukui, S., and Tanaka, A. (1979b) *Trends Biochem. Sci. 4*, 246–249.
Furukawa, T., and Ogino, T. (1982) *J. Ferment. Technol. 60*, 377–380.
Furukawa, T., Ogino, T., and Matsuyoshi, T. (1982) *J. Ferment. Technol. 60*, 218–286.
Gaillardin, C. M., Charoy, V., and Heslot, H. (1973) *Arch. Microbiol. 92*, 69–83.
Gaillardin, C. M., Fournier, P., Sylvestre, G., and Heslot, H. (1976) *J. Bacteriol. 125*, 48–57.
Harrison, F. C. (1928) *Trans. Roy. Soc. Canada, Ser. III, 22*, (Section V), 187–225.
Henneberg, W. (1926) *Handbuch der Gärungsbakteriologie*, Vol. II, 2nd ed., p. 56, Berlin.
Jörgensen, A. (1909) *Die Mikroorganismen der Gärungsindustrie*, 5th ed., p. 377, Berlin [English transl. by S. H. Davies (1911) pp. 371–372, Charles Griffin and Co., Ltd., London].
Kamiryo, T., Mishina, M., Tashiro, S.-I., and Numa, S. (1977) *Proc. Nat. Acad. Sci. U.S.A. 74*, 4947–4950.
Kapoor, K. K., Chaudhary, K., and Tauro, P. (1982) in *Prescott & Dunn's Industrial Microbiology*, 4th ed. (Reed, G., ed.), pp. 709–747, AVI, Westport, Conn.
Kato, N., Nishizawa, T., Sakazawa, C., Tani, Y., and Yamada, H. (1979) *Agr. Biol. Chem. 43*, 2013–2015.
Kihlberg, R. (1972) *Ann. Rev. Microbiol. 26*, 427–466.
Kreger-van Rij, N. J. W., and Veenhuis, M. (1973) *Antonie van Leeuwenhoek 39*, 481–490.
Kudriavzev, W. I. (1954) *Sistematica Drozzei*, Akademiya Nauk, Moscow (in Russian).
Kudriavzev, W. I. (1960) *Die Systematik der Hefen*, Academie Verlag, Berlin.
Kurtzman, C. P., Johnson, C. J., and Smiley, M. J. (1979) *Mycologia 71*, 844–847.
Levine, D. W., and Cooney, C. L. (1973) *Appl. Microbiol. 26*, 982–990.
Litchfield, J. H. (1979) in *Microbial Technology*, 2nd ed. (Peppler, H. J., and Perlman, D., eds.) pp. 93–155, Academic Press, New York.
Lodder, J. (1934) *Verh. Koninkl. Akad. Wetensch., Amsterdam, Afd. Natuurk XXXII* (2nd Sect.), 1–256.
Lodder, J., and Kreger-van Rij, N. J. W. (1952) *The Yeasts—A Taxonomic Study*, North-Holland, Amsterdam.

Longin, R., Cooney, C. L., and Demain, A. L. (1982) *Appl. Biochem. Biotechnol. 7*, 281–293.

Luh, B. S., and Phaff, H. J. (1951) *Archiv. Biochem. Biophys. 33*, 212–227.

Luh, B. S., and Phaff, H. J. (1954) *Archiv. Biochem. Biophys. 48*, 23–37.

Lutwick, L. I., Phaff, H. J., and Stevens, D. A. (1980) *Sabouraudia 18*, 69–73.

Mahoney, R. R., and Whitaker, J. R. (1977) *J. Food Biochem. 1*, 327–350.

Mahoney, R. R., and Whitaker, J. R. (1978) *J. Food Sci. 43*, 584–591.

Mahoney, R. R., Nickerson, T. A., and Whitaker, J. R. (1975) *J. Dairy Sci. 58*, 1620–1629.

Manachini, P. L. (1979) *Antonie van Leewenhoek 45*, 451–463.

Marchal, R., Metche, M., and Vandecasteele, J.-P. (1982) *J. Gen. Microbiol. 128*, 1125–1134.

Meyers, S. P., and Ahearn, D. G. (1977) *Mycologia 69*, 646–651.

Nakase, T., and Komagata, K. (1971) *J. Gen. Appl. Microbiol. 17*, 259–279.

O'Connor, M. L., and Quayle, J. R. (1979) *J. Gen. Microbiol. 113*, 203–208.

O'Connor, M. L., and Quayle, J. R. (1980) *J. Gen. Microbiol. 120*, 219–225.

Ogata, K., Nishikawa, H., and Ohsugi, M. (1969) *Agr. Biol. Chem. 33*, 1519–1520.

Ogrydziak, D. M., and Mortimer, R. (1977) *Genetics 87*, 621–632.

Ogrydziak, D. M., and Scharf, S. J. (1982) *J. Gen. Microbiol. 128*, 1225–1234.

Ogrydziak, D. M., Bassel, J., Contopoulou, R., and Mortimer, R. (1978) *Mol. Gen. Genet. 163*, 229–239.

Ogrydziak, D. M., Bassel, J., and Mortimer, R. (1982a) *Mol. Gen. Genet. 188*, 179–183.

Ogrydziak, D. M., Cheng, S.-C., and Scharf, S. J. (1982b) *J. Gen. Microbiol. 128*, 2271–2280.

Osumi, M., Kazama, H., and Sato, S. (1978) *FEBS Lett. 90*, 309–312.

Osumi, M., Nagano, M., Yamada, N., Hosoi, J., and Yanagida, M. (1982) *J. Bacteriol. 151*, 376–383.

Ota, Y., Suzuki, M., and Yamada, K. (1968) *Agr. Biol. Chem. 32*, 390–391.

Ota, Y., Morimoto, Y., Sugiura, T., and Minoda, Y. (1978) *Agr. Biol. Chem. 42*, 1937–1938.

Ota, Y., Gomi, K., Kato, S., Sugiura, T., and Minoda, Y. (1982) *Agr. Biol. Chem. 46*, 2885–2893.

Patel, D. S., and Phaff, H. J. (1959) *J. Biol. Chem. 234*, 237–241.

Phaff, H. J. (1966) *Meth. Enzymol.-Complex Carbohydr. 8*, 636–641.

Phaff, H. J., and Demain, A. L. (1956) *J. Biol. Chem. 218*, 875–884.

Phaff, H. J., and Knapp, E. P. (1956) *Antonie van Leeuwenhoek 22*, 117–130.

Phaff, H. J., and Luh, B. S. (1952) *Arch. Biochem. Biophys. 36*, 231–232.

Phaff, H. J., Miller, M. W., Recca, J. A., Shifrine, M., and Mrak, E. M. (1956) *Ecology 37*, 533–538.

Phaff, H. J., Lachance, M. A., and Presley, H. L. (1978) in *Abstracts of the XIIth International Congress of Microbiology, Munich, Sept. 3–8*, S27-3.

Pomeranz, Y. (1964) *Food Technol. 18*, 690–697.

Quayle, J. R. (1980) *Trans. Biochem. Soc. 8*, 1–10.

Ratledge, C. (1978) in *Developments in Biodegradation of Hydrocarbons* (Wiseman, A., ed.), pp. 1–46, Applied Science Publishers, Ltd., London.

Reed, G. (1982) in *Prescott & Dunn's Industrial Microbiology*, 4th ed. (Reed, G., ed.), pp. 541–592, AVI, Westport, Conn.

Ruschen, S., and Winkler, U. K. (1982) *FEMS Microbiol. Lett. 14*, 117–121.

Sachs, R. M., Low, C. B., Vasavada, A., Sully, M. J., Williams, L. A., and Ziobro, G. C. (1981) *Calif. Agr. 35*, 4–6.

Sahm, H. (1977) *Adv. Biochem. Eng. 6*, 77–103.

Shehata, A. M. El-Tabey, Mrak, E. M., and Phaff, H. J. (1955) *Mycologia 47*, 799–811.

Simms, P. C., and Ogrydziak, D. M. (1981) *J. Bacteriol. 145*, 404–409.

Snyder, H. E., and Phaff, H. J. (1960) *Antonie van Leeuwenhoek 26*, 433–452.

Snyder, H. E., and Phaff, H. J. (1962) *J. Biol. Chem. 237*, 2438–2441.

Starmer, W. T., Phaff, H. J., Miranda, M., Miller, M. W., and Heed, W. B. (1982) *Evol. Biol. 14*, 269–295.

Sugiura, T., Ota, Y., and Minoda, Y. (1975) *Agr. Biol. Chem. 39*, 1689–1694.

Teunisson, D. J., Hall, H. H., and Wickerham, L. J. (1960) *Mycologia 52*, 184–188.

Thaysen, A. C. (1957) in *Yeasts* (Roman, W., ed.), pp. 155–210, Academic Press, New York.

Thaysen, A. C., and Morris, M. (1943) *Nature* (London) *152*, 526–528.

Tobe, S., Takami, T., Ikeda, S., and Mitsugi, K. (1976) *Agr. Biol. Chem. 40*, 1087–1092.

Tsuchiya, T., Fukazawa, Y., and Kawakita, S. (1965) *Mycopathol. Mycol. Appl. 26*, 1–15.

van der Walt, J. P. (1956a) *Antonie van Leeuwenhoek 22*, 265–272.

van der Walt, J. P. (1956b) *Antonie van Leeuwenhoek 22*, 321–326.

van der Walt, J. P., (1965) *Antonie van Leeuwenhoek 31*, 341–348.

van der Walt, J. P. (1970) in *The Yeasts — A Taxonomic Study*, 2nd ed. (Lodder, J., ed.), pp. 316–378, North-Holland, Amsterdam.

van der Walt, J. P., and Scott, D. B. (1971) *Mycopathol. Mycol. Appl. 43*, 279–288.

van der Walt, J. P., and von Arx, J. A. (1980) *Antonie van Leeuwenhoek 46*, 517–521.

van Dijken, J. P., Harder, W., Beardsmore, A. J., and Quayle, J. R. (1978) *FEMS Microbiol. Lett. 4*, 97–102.

van Uden, N., and Buckley, H. (1970) in *The Yeasts — A Taxonomic Study*, 2nd ed. (Lodder, J., ed.), pp. 893–1087, North-Holland, Amsterdam.

Waites, M. J., and Quayle, J. R. (1980) *J. Gen. Microbiol. 118*, 321–327.

Waites, M. J., and Quayle, J. R. (1981) *J. Gen. Microbiol. 124*, 309–316.

Waites, M. J., and Quayle, J. R. (1983) *J. Gen. Microbiol. 129*, 935–944.

Wickerham, L. J. (1951) *Taxonomy of Yeasts*, U.S. Department of Agriculture Tech. Bull. No. 1029, pp. 1–56, U.S. Department of Agriculture, Washington, D.C.

Wickerham, L. J. (1955) *Nature* (London) *176*, 22.

Wickerham, L. J. (1970) in *The Yeasts — A Taxonomic Study*, 2nd ed. (Lodder, J., ed.), pp. 226–315, North-Holland, Amsterdam.

Wickerham, L. J., and Burton, K. A. (1956a) *J. Bacteriol. 71*, 290–295.

Wickerham, L. J., and Burton, K. A. (1956b) *J. Bacteriol. 71*, 296–302.

Wickerham, L. J., Kurtzman, C. P., and Herman, A. I. (1970) in *Recent Trends in Yeast Research* (Ahearn, D. G., ed.), Spectrum, Vol. I, pp. 81–92, School of Arts and Sciences, Georgia State University, Atlanta, Ga.

Williams, L. A., and Ziobro, G. C. (1982a) *Biotechnol. Lett. 4*, 45–50.

Williams, L. A., and Ziobro, G. C. (1982b) in *Proceedings of the 5th International Alcohol Fuel Symposium*, Vol. 1, pp. 55–61. Hohn McIndoe Ltd., Dunedin, NZ.

Yamada, T., and Ogrydziak, D. M. (1983) *J. Bacteriol. 154*, 23–31.

Yamazaki, M., and Komagata, K. (1982) *J. Gen. Appl. Microbiol. 28*, 119–138.

Yarrow, D. (1972) *Antonie van Leeuwenhoek 38*, 357–360.

Ziobro, G. C., and Williams, L. A. (1983) *Dev. Ind. Microbiol. 24*, 313–319.

Zvyagintseva, I. S., Dmitriev, V. V., Ruban, E. L., and Fikhte, B. A. (1980) *Mikrobiologiya 49*, 293–298.